FIELD GUIDE

to the

FISHES

of the

AMAZON, ORINOCO & GUIANAS

FIELD GUIDE
to the
FISHES
of the
AMAZON,
ORINOCO
& GUIANAS

Edited by Peter van der Sleen and James S. Albert
Principal illustrator Peter van der Sleen

Princeton University Press

Princeton and Oxford

Published by Princeton University Press, 41 William Street, Princeton, New Jersey 08540

In the United Kingdom: Princeton University Press, 6 Oxford Street, Woodstock, Oxfordshire OX20 1TR

press.princeton.edu

Cover images (top to bottom): *Arapaima* sp. Courtesy of Donald J. Stewart. *Paracheirodon axelrodi. Sternarchorhynchus goeldii*. Courtesy of Peter van der Sleen. *Ancistrus* sp. Courtesy of Mark Sabaj Pérez. *Pygocentrus cariba*. Courtesy of Ivan Mikolji. *Cichla melaniae*. Courtesy of Mark Sabaj Pérez.

Photograph previous page: Upper Rio Negro in Brazil.

With illustrations by Peter van der Sleen unless otherwise indicated.

ISBN (pbk.) 978-0-691-17074-9

Library of Congress Control Number: 2017942252

British Library Cataloging-in-Publication Data is available

This book has been composed in Adobe Garamond Pro and Avenir.

Printed on acid-free paper. ∞

Designed by D & N Publishing, Baydon, Wiltshire, UK

Printed in China

10 9 8 7 6 5 4 3 2 1

CONTENTS

FOREWORD

MICHAEL J. GOULDING

A predatory fish, the red-bellied piranha, is the Amazon's most famous species of any animal or plant group, and almost singularly because of popular writings and later films became the symbol of the region as a dangerous place teeming with fishes. The nature of Amazonian fishes, however, has historically also been discussed at the highest political levels, beginning with Louis Agassiz in the mid-1860s in his anti-Darwinian letters to the intellectual Dom Pedro II, the last emperor of Brazil, right up to Brazilian biologist Ronaldo Barthem of the Goeldi Museum in Belém being summoned by President Lula to explain how to mitigate the impact of dams in the Madeira River on migratory goliath catfishes. An even more politically sensitive issue has gradually arisen since the 1970s as taxonomists described hundreds of new Amazonian fish species, ecologists elicited their tight links to forests and other wetlands, and conservationists raised concerns of the possible impacts of large-scale infrastructure development leading to river impoundment, headwater deforestation, wetland modification, and pollution from mining operations. Infrastructure development has in fact outpaced the ability of scientists to understand the true diversity and distribution of Amazonian fishes. It is therefore a great pleasure to introduce the much-needed *Field Guide to the Fishes of the Amazon, Orinoco, and Guianas*, organized by Peter van der Sleen and James S. Albert, with contributions from 50 specialists spanning three continents.

The word Amazon or Amazonia has various definitions, and the authors logically opt for a largely biogeographical framework that includes the Amazon and Orinoco basins and the Guianas (thus, the AOG region). This region encompasses approximately 8.5 million km² and is largely covered by upland rainforest through which flow numerous rivers, including the world's largest, the Amazon River, and tens of thousands of streams. There are also huge savannas, vast lowland wetlands such as floodable forests, and high mountainous rivers and streams, some originating in Andean glaciers. The known fish fauna of this vast AOG region has approximately 3,000 species, and considering that dozens of new species are described each year, there may be at least 4,000 species. More than 90% of the fish fauna is made up of catfishes (Siluriformes), characiforms such as piranhas and tetras (Characiformes), cichlids (Perciformes), and killifishes and relatives (Cyprinodontiformes). This *Field Guide* brings to light that catfishes and characiforms each have approximately 1,100 known species in the AOG and together make up two-thirds of the known fish fauna. Although the AOG area is approximately that of the continental United States, the former has a fish species density (species/km²) about 7.1 times greater.

In an area as large, diverse, and unknown as that of the Amazon, nearly all scientists working with biodiversity ultimately depend on taxonomists to identify or verify species. On a personal note, many of those on the impressive list of authors of this *Field Guide* have been essential to my own studies and to the teams I have worked with since the 1970s. The *Field Guide to the Fishes of the Amazon, Orinoco, and Guianas* represents a major step forward in bringing the Amazon fish fauna to a wider audience; with summaries of evolution, diversity (including drawings), ecology, and conservation, it provides under one cover the basic information needed to grasp our planet's most diverse freshwater fish fauna. As students of Neotropical fish we still have a long way to go to catch up with ornithological taxonomy and guides, but we also have a much more difficult task because the animals we are attracted to live submerged lives and are not easily observed and photographed in the wild. As a final note, I see this *Field Guide* as an auspicious baseline that can be used to attract citizen scientists, including professionals, aquarists, and adventurers, to photograph the Amazon fish fauna in all its splendid colors and forms in the wild, help map its distribution, and last but not least, stimulate young students interested in biodiversity and conservation to see their future as much-needed Neotropical fish biologists.

PREFACE

LUIZ R. MALABARBA

The Amazon is a world of superlatives: the longest river of the world with the greatest annual water discharge, draining the largest freshwater basin on Earth with the largest extent of flooded wetlands. The altitudinal variation apportioned by the Andes to the west and by the Guiana and Brazilian shields in the north and south, respectively, has produced a wide range of soil types and sediments from different geological formations, and a variety of ecosystems, resulting in a great assortment of landscapes and environments. As reviewed by Peter van der Sleen and James S. Albert in this *Field Guide*, Greater Amazonia exhibits the highest concentration of species in many groups of plants and animals, including an unrivaled number of freshwater fishes.

The seemingly endless waterways of Greater Amazonia support a tremendous variety of life forms, of which more than 3,000 species and 564 genera of fishes are compiled in this *Field Guide*. Although inspiring scientific investments by generations of fish taxonomists and ecologists, the exceptional diversity of Amazonian fishes has actually inhibited initiatives for the production of regional field guides such this one. The large number of species in many groups, like the small-bodied tetras *Hyphessobrycon* and *Moenkhausia*, has yielded a proliferation of confusing synonyms. Further, the many new species descriptions published every year quickly make identification guides incomplete or even obsolete.

Yet identifying and naming species is the first step toward knowing them, and authenticated species lists offer the first line of defense for conserving them as well. In *One Hundred Years of Solitude*, the celebrated novel by Gabriel García Márquez, Colonel Aureliano Buendía finds solitude after the wars by molding and melting down a seemingly endless stream of tiny golden fishes. Collecting fishes in the Amazon confronts the investigator with a similar experience—a seemingly endless array of small-bodied silvery tetras (characids), armored catfishes (loricariids), and colorful cichlids, all with subtle but noticeable differences in shapes, sizes, counts, and color patterns. Most of the information previously available for identifying Amazon fishes is distributed among numerous specialized reports or published in technical journals in several languages and deals with taxonomic revisions of genera, species, or more rarely families.

What has been lacking until now is a single, comprehensive, and portable reference to aid the patient field worker in attaching names to these seemingly endless forms. A practical guide for identifying Amazonian fishes has long been coveted by scientists, resource managers, aquarists, and curious members of the general public, anyone interested in the natural history and sheer beauty of Amazonian aquatic systems. However, until very recently such an undertaking had not been logistically possible because of a lack of basic taxonomic information and the immense diversity and complexity of the fauna. Accordingly, this *Field Guide* is by no means the last word on this topic, which the authors readily acknowledge, and many opportunities exist for future work. The taxonomic scope of this *Field Guide* is necessarily limited to the genus (not the species) level, and the geographic scope is limited to Greater Amazonia, excluding adjacent areas of tropical America.

The *Field Guide to the Fishes of the Amazon, Orinoco, and Guianas* is the first of its kind, paving the way for future contributions. Building on recent compilations of alpha taxonomy at regional scales and on numerous taxonomic revisions of the many individual groups, this *Field Guide* employs an efficient strategy to deal with the identifications of so many different Amazonian fishes. The editors gathered contributions from 50 taxonomic specialists for 74 chapters representing all the families and subfamilies, with dichotomous keys, illustrations, and color photographs that help the reader to quickly identify fishes to order, family, and genus levels. The reader is further assisted with a compilation of information about the diversity, geographical distribution, body size, diagnostic characters, sexual dimorphism, habitats, feeding ecology, and behavior of each family or subfamily. Keys to genera are furnished separately for large families and subfamilies, and all genera are represented schematically by at least one illustration presenting the general appearance of each taxon. Further information on each genus is given regarding diagnosis, species diversity, literature for species identification, distribution, habitat, and biology.

Although targeting the ichthyofauna of Greater Amazonia, this *Field Guide* also represents an important contribution to the field of Neotropical ichthyology as a whole. Most of the information provided for identifying large groups of Amazonian fishes applies to other regions of South and Central America. The *Field Guide to the Fishes of the Amazon, Orinoco, and Guianas* therefore represents a large step forward for the community of Neotropical ichthyologists, in its multicentury mission to document the greatest continental fish fauna on Earth. More generally, this *Field Guide* is a welcome addition to the library of anyone fascinated with the diversity of life on Earth.

ACKNOWLEDGMENTS

First and foremost, we thank all the authors for their hard work and contributions to this book. Without their incredible knowledge on Amazonian fishes this book would not exist. Many thanks also to several experts who provided excellent reviews, thereby improving the quality of the chapters: Frank Pezold (Eleotridae and Gobiidae), Roberto Reis (Characidae *incertae sedis*; and key to the families), Luiz Malabarba (key to the families), Lilian Casatti (Sciaenidae), and Jansen Zuanon (all chapters).

Peter van der Sleen thanks his wife Tessa de Vries for her immense support, encouragement, and patience during the six years it took to compile this book. Inge Smit and Gerard Engels are thanked for their help with converting distribution data to maps. PvdS received financial support from Treub Maatschappij and the KNAW Fonds Ecologie. PvdS also thanks Roberto Reis for facilitating a visit to the fish collection of the PUCRS, José Birindelli for arranging the financial means to visit the Universidade Estadual de Londrina, Martien van Oijen and Bert Hoeksema of the Natural Biodiversity Center, Leiden, the Netherlands, for their advice, and the Forest Ecology and Management Group of the Wageningen University, the Netherlands, for partially hosting me when working on this book. B. Lehner, at the Conservation Science Program, World Wildlife Fund US, generously provided the elevation map used in figure 4 in chapter 1. Peter thanks James Albert for his encouragement, guidance, and friendship.

James Albert is indebted to George Barlow, William Eschmeyer, William Fink, William Gosline, Robert Miller, and Gerald Smith for an education in the science of ichthyology, and to William Crampton and Roberto Reis for help in learning the Amazonian ichthyofauna. JSA thanks Samuel Albert and Sara Albert for critical support over many years, and the following people for camaraderie in the field and lab, where the species, the "atoms of biodiversity," are learned one at a time: Jon Armbruster, Maxwell Bernt, Paulo Buckup, Ricardo Campos-da-Paz, Tiago Carvalho, Prosanta Chakrabarty, Bibiana Correa, Jack Craig, Tom de Benedetto, Brian Dyer, Jesica Espino, Kory Evans, Flora Fernandes-Matioli, Michael Goulding, Sven Kullander, Steven Ivanyisky, Lesley Kim, Flávio Lima, Hernán López-Fernández, Nathan Lovejoy, Nathan Lujan, John Lundberg, Francisco Mago-Leccia, Luiz Malabarba, Javier Maldonado, Claudio Oliveira, Hernán Ortega, Lawrence Page, Edson Pereira, Paulo Petry, Franco Provenzano, Blanca Rengifo, Fábio Roxo, Ramiro Royero, Mark Sabaj, Donald Taphorn, Roberto Quispe, Robson Ramos, Victor Tagliacollo, Brandon Waltz, and Kirk Winemiller. JSA acknowledges support by US National Science Foundation awards 0614334, 0741450, and 1354511. JSA is grateful to Peter van der Sleen for the wonderfully brazen idea of compiling a comprehensive field guide to all Amazonian fishes, and for the perseverance to bring this idea to completion.

Leo Nico, Michel Jégu, and Marcelo Andrade thank Stephen J. Walsh, Lisa Jelks, and Howard L. Jelks for reviewing drafts of the manuscript and providing constructive remarks and suggestions, and Donald C. Taphorn for sharing his knowledge of serrasalmids. Original photographs used as templates for some of the illustrations were generously provided by Mark Sabaj Pérez, Ivan Mikolji, and Howard L. Jelks. Robert H. Robins kindly allowed LGN use of Florida Museum of Natural History ichthyological collection. Leo Nico's affiliation, the US Geological Survey, states that any use of trade, product, or firm names is for descriptive purposes only and does not imply endorsement by the US Government. Rodrigo Caires thanks Frank Pezold, Sven O. Kullander, and Mark Sabaj-Perez for their advice. Brian Sidlauskas, Michael Burns, and Ben Frable were supported by NSF grant DEB-1257898 to Brian Sidlauskas and thank the late Richard Vari for many years of support, mentorship, and collaboration, without which they would never have been in a position to contribute to this volume. Research activities by Paulo Andreas Buckup are supported by grants from CNPq (307610/2013–6, 564940/2010–0, 476822/2012–2), FAPERJ (E-26/111.404/2012, E 26/200.697/2014), and CAPES. Hernán López-Fernández was financially supported by the Royal Ontario Museum, the Natural Sciences and Engineering Council of Canada, the US National Science Foundation, and National Geographic Society. He thanks the following people for their advice: Donald C. Taphorn, Elford A. Liverpool, Jessica Arbour, Sarah E. Steele, Frances E. Hauser, Nathan K. Lujan, Karen M. Alofs, Carmen G. Montaña, Stuart C. Willis, Nathan R. Lovejoy, Javier Maldonado, Devin Bloom, Erling Holm, Mary Burridge, Margaret Zur, Jonathan Armbruster, David Werneke, Mark Sabaj-Perez, Jansen Zuanon, Lucia Rapp Py-Daniel, Izeni P. Farias, Jan Mol, and Calvin R. Bernard. José Birindelli was supported by CNPq grant 478900/2013–9 and a Fundação Araucária research grant. He thanks Heraldo Britski and Mark Sabaj Pérez for various discussions. Veronica Slobodian was supported by CNPq grant 156515/2010–5 and FAPESP grant 2013/18623–4. Tiago Carvalho was supported by the Conselho Nacional de Desenvolvimento Científico e Tecnológico (CNPq # 229355/2013–7) and PNPD-CAPES. Michael Burns thanks Stan Weitzman for help with the key to Gasteropelecidae genera. Jonathan Armbruster and Nathan K. Lujan were financially supported by NSF grants DEB-0107751, DEB-0315963, and DEB-1023403 to JWA and OISE-1064578 (International Research Fellowship) to NKL, with additional support to NKL from DEB-1257813 (the iXingu Project), National Geographic Committee for Research and Exploration Grant #8721–09, the Coypu Foundation of New Orleans, and the Canada Department of Fisheries and Oceans. Bruno Melo thanks Claudio Oliveira for support and general discussions and was financially supported by FAPESP grants 11/08374–1 and 13/16436–2, and CNPq PDJ 40258/2014–7. Alexandre P. Marceniuk thanks the National Science and Technology Council of the Brazilian Federal Government (CNPq grant no.

152782/2007–9) and the Pará State Research Foundation (FAPESPA grant no. 350790/2012–4). Pedro Amorim and Pedro Bragança thanks Wilson J. E. M. Costa for his advice. Flávio A. Bockmann and Veronica Slobodian were funded by Conselho Nacional de Desenvolvimento Científico e Tecnológico, Federal Government (CNPq, Proc. No. 440621/2015–1, 312067/2013–5, and 562268/2010–3 to FAB, and Proc. No. 156515/2010–5 to VS) and Fundação de Amparo à Pesquisa do Estado de São Paulo (FAPESP, Proc. No. 2013/18623–4 to VS).

The editors also thank Hernán Ortega and Roberto Reis for detailed reviews of the list of common names, and the many people who made photographic material available for this publication, in particular, Tiago Carvalho and Mark Sabaj Pérez. The photos taken by Mark Sabaj were supported in part by iXingu Project, NSF DEB-1257813.

Last but not least, we thank Jansen Zuanon at the Instituto Nacional de Pesquisas da Amazônia for his careful review of the entire manuscript.

CONTRIBUTORS

Alberto Akama   Setor de Ictiologia, Museu Paraense Emílio Goeldi, Belém, Brazil. E-mail: aakama@gmail.com

James S. Albert   Department of Biology, University of Louisiana, Lafayette, USA. E-mail: jalbert@louisiana.edu

Pedro F. Amorim   Laboratório Sistemática e Evolução de Peixes Teleósteos, Depto. de Zoologia, Instituto de Biologia, Universidade Federal do Rio de Janeiro, Brazil. E-mail: pedro_f_a@hotmail.com

Marcelo C. Andrade   Laboratório de Biologia Pesqueira e Manejo dos Recursos Aquáticos, Universidade Federal do Pará, Cidade Universitária Prof. José Silveira Netto, Belém, Brazil. E-mail: andrademarcosta@gmail.com

Jonathan Armbruster   Department of Biological Sciences, Auburn University, Auburn, USA. E-mail: armbrjw@auburn.edu

José L. O. Birindelli   Departamento de Biologia Animal e Vegetal, Universidade Estadual de Londrina, Londrina, Brazil. E-mail: josebirindelli@yahoo.com

Maxwell J. Bernt   Department of Biology, University of Louisiana, Lafayette, USA. E-mail: mjbernt@gmail.com

Flávio A Bockmann   Laboratório de Ictiologia de Ribeirão Preto, Departamento de Biologia, Universidade de São Paulo, Ribeirão Preto, Brazil. E-mail: fabockmann@ffclrp.usp.br

Devin Bloom   Department of Biological Sciences & Environmental and Sustainability Studies Program, Western Michigan University, Kalamazoo, USA. E-mail: devin.bloom@wmich.edu

Pedro H. N. Bragança   Laboratório de Sistemática e Evolução de Peixes Teleósteos, Departamento de Zoologia, Instituto de Biologia, Universidade Federal do Rio de Janeiro, Brazil. E-mail: pedrobra88@gmail.com

Paulo A. Buckup   Departamento de Vertebrados, Museu Nacional, Universidade Federal do Rio de Janeiro, Rio de Janeiro, Brazil. E-mail: buckup@acd.ufrj.br

Michael D. Burns   Department of Fisheries and Wildlife, Oregon State University, Corvallis, USA. E-mail: michael.burns@oregonstate.edu

Rodrigo Caires   Museu de Zoologia da Universidade de São Paulo, São Paulo, Brazil. E-mail: rodricaires@yahoo.com.br

Tiago Carvalho   Academy of Natural Sciences, Drexel University, Philadelphia, USA. E-mail: carvalho.ictio@gmail.com

Raphael Covain   Département d'herpétologie et d'ichtyologie, Muséum d'histoire naturelle, Genève, Switzerland. Mail: Raphael.Covain@ville-ge.ch

Jack M. Craig   Department of Biology, University of Louisiana, Lafayette, USA. E-mail: jack.m.craig@gmail.com

Kory M. Evans   Department of Biology, University of Louisiana, Lafayette, USA. E-mail: kxe9300@gmail.com

Luis Fernández   CONICET-IBN, Instituto Miguel Lillo and Universidad Nacional Catamarca, Argentina. E-mail: luis1813@yahoo.com

Benjamin Frable   Department of Fisheries and Wildlife, Oregon State University, Corvallis, USA. E-mail: bwfrable@gmail.com

John P. Friel   Alabama Museum of Natural History, University of Alabama, Tuscaloosa, USA. E-mail: jpfriel@ua.edu

Aurycéia Guimarães da Costa   Laboratório de Genética e Biologia Molecular, Instituto de Estudos Costeiros, Universidade Federal do Pará, Pará, Brazil. E-mail: auryceia@yahoo.com.br

Michel Jégu   Institut de Recherche Pour le Développement, Biologie des Organismes et Ecosystèmes Aquatiques, and Laboratoire, d'Icthyologie, Muséum National d'Histoire Naturelle, Paris, France. E-mail: michel.jegu@gmail.com

Fernando Jerep   Departamento de Biologia Animal e Vegetal, Universidade Estadual de Londrina, Londrina, Brazil. E-mail: fjerep@gmail.com

Lesley Y. Kim   Department of Biology, University of Louisiana, Lafayette, USA. E-mail: lyk4468@louisiana.edu

Sven O. Kullander   Department of Zoology, Swedish Museum of Natural History, Stockholm, Sweden. E-mail: sven.kullander@nrm.se

Francisco Langeani   Laboratório de Ictiologia, Departamento de Zoologia e Botânica, Instituto de Biociências, Universidade Estadual Paulista, São José do Rio Preto, Brazil. E-mail: langeani@ibilce.unesp.br

Flávio C. T. Lima   Museu de Zoologia da Universidade Estadual de Campinas "Adão Jose Cardoso," Campinas, Brazil. E-mail: fctlima@gmail.com

Hernán López–Fernández   Department of Natural History, Royal Ontario Museum, and Department of Ecology and Evolutionary Biology, University of Toronto, Toronto, Canada. E-mail: h.lopez.fernandez@utoronto.ca

Nathan R. Lovejoy   Department of Biological Sciences, University of Toronto Scarborough, Toronto, Canada. E-mail: lovejoy@utsc.utoronto.ca

Paulo H. F. Lucinda   Laboratório de Ictiologia Sistemática, Universidade Federal do Tocantins, Porto Nacional, Brazil. E-mail: lucinda@mail.uft.edu.br

Nathan K. Lujan   Department of Ecology and Evolutionary Biology, University of Toronto, Toronto, Canada. E-mail: nklujan@gmail.com

Luiz R. Malabarba   Departamento de Zoologia and Programa de Pós-Graduação em Biologia Animal, Universidade Federal do Rio Grande do Sul, Porto Alegre, Brazil. E-mail: malabarb@ufrgs.br

George M. T. Mattox   Departamento de Biologia, Universidade Federal de São Carlos, campus Sorocaba, Brazil. E-mail: gmattox@ufscar.br

Alexandre P. Marceniuk   Museu Paraense Emilio Goeldi, Belém, Brazil. E-mail: a_marceniuk@hotmail.com

Bruno F. Melo   Departamento de Morfologia, Universidade Estadual Paulista, Botucatu, Brazil. E-mail: brunfmelo@hotmail.com

Cristiano R. Moreira   Museu Nacional/Universidade Federal do Rio de Janeiro, Setor de Ictiologia, Departamento de Vertebrados, Rio de Janeiro, Brazil. E-mail: moreira.c.r@gmail.com

Leo G. Nico   US Geological Survey, Gainesville, USA. E-mail: lnico@usgs.gov

André Netto-Ferreira   Laboratório de Ecologia e Conservação, Instituto de Ciências Biológicas, Universidade Federal do Pará, Belém, Brazil. E-mail: alnferreira@gmail.com

Osvaldo Takeshi Oyakawa   Museu de Zoologia da, Universidade de São Paulo, São Paulo, Brazil. E-mail: oyakawa@usp.br

Roberto E. Reis   Faculdade de Biociências, Pontifícia Universidade Católica do Rio Grande do Sul, Porto Alegre, Brazil. E-mail: reis@pucrs.br

Oscar A. Shibatta   Centro de Ciências Biológicas, Universidade Estadual de Londrina, Londrina, Brazil. E-mail: shibatta@uel.br

Brian L. Sidlauskas   Department of Fisheries and Wildlife, Oregon State University, Corvallis, USA. E-mail: brian.sidlauskas@oregonstate.edu

Peter van der Sleen   Marine Science Institute, University of Texas at Austin, Port Aransas, USA, and Forest Ecology and Management Group, University of Wageningen, Wageningen, the Netherlands. E-mail: j.p.vandersleen@gmail.com

Veronica Slobodian   Laboratório de Ictiologia, Museu de Zoologia da Universidade de São Paulo, São Paulo, Brazil, E-mail: verorp@gmail.com

Leandro M. de Sousa   Universidade Federal do Pará, Campus Altamira, Altamira, Brazil. E-mail: l.m.sousa@gmail.com

Victor Tagliacollo   Programa de Pós Graduação em Ciências do Ambiente, Universidade Federal do Tocantins, Palmas, Brazil. E-mail: victor_tagliacollo@yahoo.com.br

Mônica Toledo-Piza   Departamento de Zoologia, Instituto de Biociências, Universidade de São Paulo, Brazil. E-mail: mtpiza@usp.br

Marcelo Salles Rocha   Universidade do Estado do Amazonas, Curso de Ciências Biológicas, Manaus, Brazil. E-mail: marcelo.inpa@gmail.com

Brandon T. Waltz   Department of Biology, University of Louisiana, Lafayette, USA. E-mail: btwaltz09@gmail.com

Angela M. Zanata   Departamento de Zoologia, Instituto de Biologia, Universidade Federal da Bahia, Campus de Ondina, Salvador, Brazil. E-mail: zanata.angela@gmail.com

Amazonia is a vast and complex landscape, with a biodiversity unrivaled on the surface of the Earth. The Amazon River is the largest in the world by any measure, including maximum length from mouth to most distant headwater tributary (6,712 km or 4,195 mi), total catchment area (7.05 million km² or 2.72 million mi²), area of seasonally flooded wetlands (250,000 km² or 96,530 mi²), average annual water discharge (219,000 m³ second⁻¹), and proportion of global river surface area (25–28%) (Goulding et al. 2003). Near its mouth as it approaches the Atlantic Ocean, the Amazon is so wide that one cannot see across it from one bank to the other. Here the Amazon flows inexorably like an inland freshwater sea, discharging a volume of water into the Atlantic so immense that it accounts for about one-sixth to one-fifth of all the Earth's river water, depending on the year.

Many Amazonian headwaters arise as glacier and snow melt high in the Andes (>5,000 m or >16,400 ft), eroding the steep mountain slopes as they fall, and carrying with them a high sediment load. Other headwaters arise from clayey and sandy soils deep in the rainforest, where they are stained red by acidic plant compounds. Yet others originate on the crystalline granites of the Brazilian and Guiana shields, where the waters run clear. Some of these black- and clearwater headwaters of the Amazon basin are connected by permanent rivers (e.g., Casiquiare Canal), seasonally flooded swamps (e.g., Rupununi savanna), or occasional stream capture events, to headwaters of the adjacent Orinoco River and coastal rivers of the Guianas. Altogether these river basins constitute a biodiversity province known as Greater Amazonia (figs. 1, 2).

Greater Amazonia extends over more than 8.4 million square kilometers (3.2 million mi²) of northern South America. This enormous region is drained by hundreds of thousands of kilometers of terra firme (nonfloodplain) streams and small rivers that

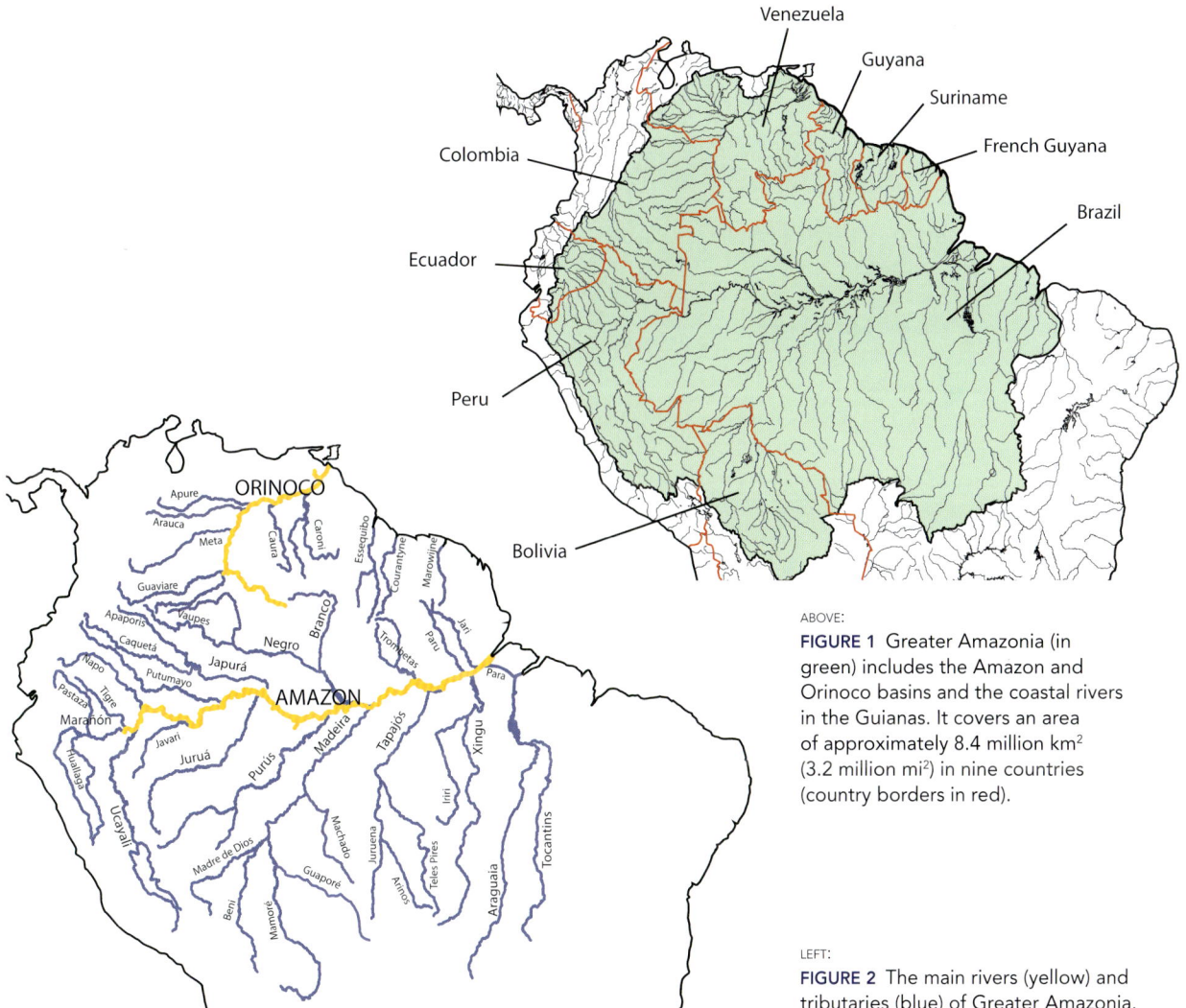

ABOVE:

FIGURE 1 Greater Amazonia (in green) includes the Amazon and Orinoco basins and the coastal rivers in the Guianas. It covers an area of approximately 8.4 million km² (3.2 million mi²) in nine countries (country borders in red).

LEFT:

FIGURE 2 The main rivers (yellow) and tributaries (blue) of Greater Amazonia.

flow under a closed forest canopy, and tens of thousands of kilometers of larger lowland rivers that meander across broad and sunlit floodplains. At the start of the twenty-first century, most of Greater Amazonia remains covered with dense tropical rainforests. The region also includes other distinct ecosystems, such as the cloud forests in the Andean piedmont and the tabletop mountains (*tepuis*) of the Guiana Shield, seasonally flooded wetlands (Llanos) in the central Orinoco basin, seasonally burned tropical savannas (Cerrado) in central Brazil or Lavrado (also called Gran Sabana) in the western Guiana Shield, and coastal estuaries at the mouths of the Amazon (Marajó) and Orinoco (Amacuro) rivers.

The ecosystems of Greater Amazonia are home to the greatest concentration of species on Earth. This region is the global center of highest species richness for many groups of organisms, including flowering plants, insects, birds, mammals, reptiles, and amphibians. Greater Amazonia is also the center of diversity for continental (freshwater) fishes. From the torrential headwaters cascading off the Andes, to the murky waters of the large lowland river channels and floodplains, the fishes of Greater Amazonia thrive in astonishing abundance and diversity. To date more than 3,000 fish species have been described from Greater Amazonia, and dozens of new species are described every year.

This field guide provides descriptions and identification keys for all the known genera of fishes that inhabit Greater Amazonia. It summarizes our current state of knowledge on the taxonomy, species richness, and ecology of these fish groups, and provides references to relevant literature for species-level identifications. It is our sincere hope that the *Field Guide to the Fishes of the Amazon, Orinoco, and Guianas* will be useful to anyone interested in quickly and accurately identifying Amazonian fishes, including aquarists, aquatic biologists, ecotourists, environmental engineers, sport fishers, and fish taxonomists.

Evolutionary History of Amazonian Fishes

Paleogene Origins of Major Groups

Amazonian fishes trace their evolutionary origins to the super-greenhouse world of the Late Cretaceous and Early Cenozoic (120–50 million years ago or mya; Albert and Reis 2011). This was a time of Earth history without polar ice sheets, when tropical climates extended to high latitudes, and tropically adapted organisms like palm trees and crocodilians lived in the lands we call Greenland and Alaska today. Neotropical fishes diversified in concert with the major groups of plants and animals that dominate modern tropical rainforest ecosystems (Lundberg et al. 1998).

Over these immense time periods, different groups of Amazonian fishes diversified under a wildly diverse set of environmental conditions. Some of the most important general influences were the relatively stable, warm, and wet climates that prevailed globally at low latitudes for most of the Paleogene (66–23 mya), regional hydrological and climatic changes associated with rise of the Northern Andes and formation of the modern river basins during the Miocene (22–5 mya), and global cooling and eustatic sea-level changes during the Pliocene (5–2.6 mya) and Pleistocene (2.5–0 mya) (Albert and Reis 2011).

In terms of species richness, total abundances, and fish biomass, the Amazonian fish fauna is dominated by three major groups: Characiformes (including piranhas, tetras, and relatives), Siluriformes (catfishes of diverse sizes, shapes, and natural histories), and Cichlidae (including peacock basses, freshwater angel fishes, oscars, and relatives). Fossils from each of these groups ascribed to modern families and genera (*Tremembichthys, Corydoras, Gymnogeophagus*) have been discovered in Paleogene sediments (López-Fernández and Albert 2011). These fossils are direct evidence that at least some Neotropical fishes had diversified to modern forms more than 40 million years ago.

The formation of Amazonian fish fauna was thus a long, long time in the making. These myriad forms accumulated over the course of tens of millions of years, and across a geographical arena that included the whole continent of South America. In other words, Amazonian fishes did not arise as the result of a recent or rapid adaptive radiation. The great antiquity of Amazonian fish lineages is perhaps surprising, as they are much older than the rivers and drainage basins in which they live. The modern Amazon and Orinoco basins are in fact relatively young features of the South American landscape, having achieved their modern configurations from tectonic and erosional processes only in the past 10 million years or so, in association with the rise of the Northern Andes (Hoorn et al. 2010).

Formation of Megadiverse Fish Species Assemblages

The ecological and evolutionary reasons for the incredible diversity of Amazonian fishes have been debated for more than a century. Among the most influential hypotheses are the great age and size of the drainage system (Lovejoy et al. 2011), habitat succession and niche diversity (Lowe-McConnell 1987), the high proportion of lowlands with stable environmental conditions (capable of supporting a large abundance of fishes; Henderson and Crampton 1997), and a long and diverse history of river capture events (see section below).

As with most species-rich ecosystems worldwide, the Amazon is both a museum and a cradle of biodiversity. From the perspective of biodiversity, a museum is a place where species accumulate through dispersal and persist by resisting extinction, and a cradle a place where species are born through the process of speciation. The principal evolutionary forces that affected the formation of fish species assemblages are speciation and dispersal, which in combination serve to increase the number of species in a region or a basin, and local extinction, which serves to reduce the number of species. These three processes are interrelated: all else being equal, a species with a higher dispersal rate has more gene flow and therefore a lower chance of becoming genetically fragmented. That is to say, dispersal acts to reduce the chances for a population near the periphery of a species' geographic range to become genetically isolated and eventually form a new (daughter) species. By the same token, dispersal and gene flow reduce the chances that a species will become geographically restricted, thereby reducing its risk of extinction. By this reasoning, species with higher dispersal capacities are expected to have lower rates of both speciation and extinction, or in other words, lower species turnover. Contrariwise, species with low dispersal abilities should have higher rates of species turnover.

The effect of dispersal capacity on evolutionary diversification is well illustrated by comparing the catfish families Loricariidae and Pimelodidae. Loricariids attain relatively small adult body sizes (avg. 16.2 cm SL; size data in Albert et al. 2009), and most loricariid species have relatively low dispersal capacities, increasing the chances for geographic isolation and speciation. Indeed, more than 400 loricariid species are known in Greater Amazonia. These numbers stand in strong contrast to the large-bodied migratory catfishes in the family Pimelodidae (59 species with an avg. size of 60.5 cm SL), many of which are widely distributed over much of the Amazon or Orinoco basins. Reduced dispersal capacity has in fact been implicated in the diversification of all three of the most species-rich AOG fish families: Loricariidae, Characidae, and Cichlidae. Species in these families generally have relatively small adult body sizes (<20 cm SL) and restricted geographic distributions (Albert et al. 2011a). Of course many other factors may contribute to differences in the diversity of fish groups. For example, species can differ in their inherent capacity to diversify morphologically. The exceptional species richness of the Loricariidae may also be related to adaptations of the oral jaws, facilitating specializations in feeding ecology (Schaefer and Lauder 1996, Lujan and Armbruster 2012).

River and Stream Capture

How do fish populations get isolated and form new species, in one of the wettest continental regions on Earth? For freshwater organisms, landscapes are divided naturally into discrete drainage basins by watersheds, boundaries that separate different river basins. Watersheds are not, however, constant landscape features but rather change through time. River or stream capture occurs when an upstream portion of one river drainage is diverted into the downstream portion of an adjacent basin, shifting the location of the watershed divide. River capture can occur from multiple geophysical causes, including headward or lateral erosion, differential geophysical uplift or subsidence, or natural damming from landslides or mudflows.

River capture by erosion is a perennial Earth history process that episodically isolates and reunites portions of adjacent

waterways and their resident biotas. By merging portions of adjacent basins, river capture allows species ranges to expand and increases gene flow among populations, thereby lowering extinction risk. By isolating other portions of the same basins, river capture also divides species ranges and reduces gene flow, thereby increasing genetic isolation and the chance for speciation. These twin effects of river capture on biotic diversification are especially pronounced in regions like the Amazon and Orinoco basins, with low (flat) topographic relief and readily eroded sediments, and therefore higher rates of river capture. Over the course of millions of years, river capture likely made a strong contribution to the assembly of basinwide fish faunas, and the accumulation of astonishing fish species diversity.

Amazonian Fish Diversity

Major Classes of Fishes

Continental freshwater fishes worldwide may be classified into three categories based on ecological and physiological criteria (Myers 1966, Matamoros et al. 2015). Primary (obligatory) freshwater fishes have little or no tolerance to salt or brackish water, inhabiting water with less than 0.5 grams total dissolved mineral salts per liter (i.e., <0.5 ppt). As a result, marine water is an important barrier to dispersal in primary freshwater fishes. Primary freshwater fishes of Greater Amazonia include the Ostariophysi and four other families that originated and diversified in freshwater habitats and were isolated in South America on its final separation from the remaining portions of the ancient supercontinent Gondwana. The Ostariophysi comprise approximately 75% of all freshwater fishes on Earth, and in South America this group includes the Characiformes (>1,700 species), Siluriformes (>2,000 species), and Gymnotiformes (>200 species). Primary freshwater fishes of Amazonia also include the paiche or pirarucu *Arapaima gigas* (Arapaimidae), the largest freshwater fish with scales in the world, two species of arowanas *Osteoglossum* (Osteoglossidae), three species of leaf-fishes (Polycentridae), and the South American lungfish *Lepidosiren paradoxa* (Lepidosirenidae).

Secondary freshwater fishes have greater tolerance to brackish waters but normally occur in continental aquatic systems rather than in the sea, and they are capable of occasionally crossing narrow marine barriers. Secondary freshwater fishes of Greater Amazonia include the Cichlinae (>500 species), Rivulidae (>270 species), Cyprinodontidae (>60 species), and livebearers of the families Anablepidae (about 17 species) and Poeciliidae (>250 species).

Peripheral freshwater fishes are members of otherwise marine groups and exhibit high salt tolerance. Peripheral freshwater fishes of Greater Amazonia include representatives of many otherwise marine families that invaded and specialized for life in freshwaters at different times during the history of the continent. Most of these groups include one or a few species. Peripheral freshwater fishes of Greater Amazonia include freshwater anchovies (Engraulidae), drums (Sciaenidae), flatfishes (Achiridae), gobies (Gobiidae), needlefishes (Belonidae), puffers (Tetraodontidae), and stingrays (Potamotrygonidae).

Composition of the Amazonian Fish Fauna

As with most faunas, Neotropical fish species are not distributed equally among higher taxa. More than 90% of the fish species in Greater Amazonia belong to only four taxonomic orders (fig. 3): Siluriformes (catfishes), Characiformes (tetras, piranhas, and relatives), Perciformes (e.g., cichlids), and Cyprinodontiformes (killifishes and relatives). Such an unbalanced pattern of species richness is also observed at other levels of the taxonomic hierarchy (table 1; fig. 4). For example, a single characiform family, Characidae, includes 581 species in the AOG region, or about 53% of all characiform species in the AOG region. Similarly, the siluriform family Loricariidae includes 403 species in the AOG region, or about 36% of all siluriform species in the AOG region. At the other end of the diversity spectrum, 24 fish families in the AOG region are represented by fewer than 10 species each, containing a total of just 99 species in all (3% of the AOG fauna). These species-poor families include eight families known from only one or two species, like Arapaimidae (2 species), Batrachoididae (2 species), and Lepidosirenidae (1 species).

Of course many new species are described each year, and additional species may eventually be described in these species-poor families as well. Yet the larger pattern is inescapable: just a few families in the AOG region are very species rich, while most of the families are not. As in most parts of the Tree of Life on Earth, species-rich higher taxa are rare, and species-poor higher taxa are common.

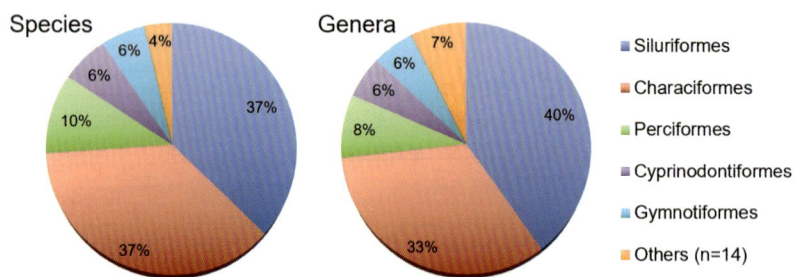

FIGURE 3 Species (left) and genus (right) composition of fish orders in the AOG region. The colored slices represent the relative contribution of different fish orders to the total fish diversity (in %). Data as compiled in this book.

BELOW:

FIGURE 4 Taxa with greatest species richness in the AOG fish fauna. (A) 19 orders representing all AOG fish species. (B) Top 20 families representing 87% of AOG fish species. (C) Top 20 genera representing 32% of AOG fish species.

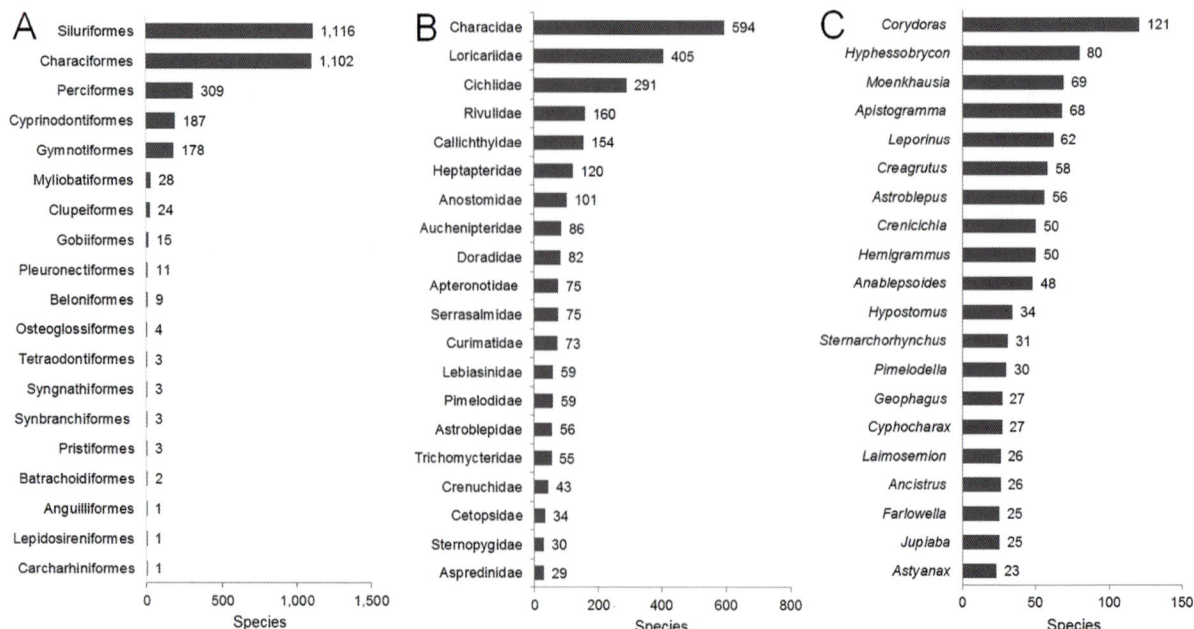

TABLE 1 Classification and Diversity of the AOG Fish Fauna

Class	Order	Species	Genera	Families
Chondrichthyes	Carcharhiniformes	1	1	1
	Pristiformes	3	1	1
	Myliobatiformes	28	4	1
Sarcopterygii	Lepidosireniformes	1	1	1
Actinopterygii	Anguilliformes	1	1	1
	Osteoglossiformes	4	2	2
	Clupeiformes	24	11	3
	Characiformes	1,102	187	19
	Siluriformes	1,116	226	13
	Gymnotiformes	178	32	5
	Cyprinodontiformes	187	31	4
	Perciformes	309	46	3
	Gobiiformes	15	8	2
	Batrachoidiformes	2	2	1
	Beloniformes	9	4	2
	Synbranchiformes	3	1	1
	Syngnathiformes	3	2	1
	Pleuronectiformes	11	3	1
	Tetraodontiformes	3	1	1
3	19	3,000	564	63

Note: Taxa arranged in a conventional phylogenetic sequence.

Comparisons with Other Faunas

The AOG fish fauna listed in this *Field Guide* comprises 3,000 species that inhabit an area of about 8.4 million km², or an area slightly larger than that of the contiguous United States (8.08 million km²), or slightly smaller than that of Europe west of the Ural Mountains (10.18 million km²). These numbers correspond to a fish species density (species/km²) in the AOG region about 7.1 times greater than that of the contiguous United States, or 6.1 times greater than Europe's (fig. 5). This spectacular diversity is reflected in the large number of genera (564) to which these species have been assigned, compared with other biogeographic provinces.

Despite the high diversity of fish species and genera in the AOG region, this diversity represents relatively few distinct evolutionary lineages, as represented by higher taxonomic categories like orders and classes. This overall pattern reflects the geographic isolation of South America throughout most of the Cenozoic, a condition referred to as "splendid isolation" (Simpson 1983).

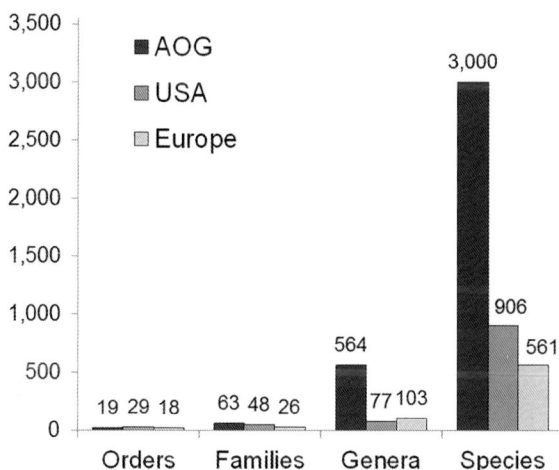

FIGURE 5 Diversity of freshwater fishes by taxonomic group, among three major biogeographic provinces. Data for obligate freshwater, native species only. Data for contiguous USA from Page and Burr (2011); for Europe west of the Ural Mountains from Kottelat and Freyhof (2007); for the Amazon-Orinoco-Guiana (AOG) region as compiled in this book.

Amazonian Fish Ecology

The Amazon and Orinoco basins contain innumerable rivers, lakes, streams, beaches, floodplains, and flooded forests. Nothing is permanent. Seasonal fluctuations of water levels induce the appearance and disappearance of many landscape formations. Yet, many fish species exhibit physiological specializations that confine them to a narrow range of habitats throughout their geographical range.

A few species are ecological generalists that can tolerate a wide variety of local environmental conditions. Here we describe some of the main features of the habitats in which Amazonian fishes live.

Major River Types

The water chemistry of Amazonian rivers is influenced by the dominant vegetation cover and soil types of their headwaters. These water types have important consequences for many fishes and other aquatic organisms. Rivers of Greater Amazonia can be roughly divided into three main types based on the chemical and physical properties of the water:

(1) Whitewater rivers have high sediment and nutrient loads and a neutral pH (e.g., Junk 1997, Goulding et al. 2003). They drain soils formed on sediments from the Andes (fig. 6). The Andes are a relatively young mountain range, and heavy erosion causes these rivers to be loaded with a large amount of suspended silt, giving them a pale muddy color (figs. 7, 8). Major whitewater tributaries include the Marañón, Meta, Madeira, and Napo rivers. The whole Solimões-Amazon and Orinoco river system exhibits whitewater as well as receiving other water types from various tributaries.

(2) Blackwater rivers have a low sediment load, low pH, and tannin-rich tea-colored water (Goulding et al. 1988). They drain sandy soils that originate from an area known as the Guiana Shield, a remnant of an ancient (Precambrian) mountain range (fig. 6). Hundreds of millions of years of erosion have left only the hardest rocks behind. From these rocky outcrops, little minerals erode and soils in these areas are extremely nutrient poor. They are, nonetheless, covered by dense forests. One prominent adaptation of this vegetation is an increased investment in the protection against herbivores, as low nutrient levels in the soil make it costly to replace lost

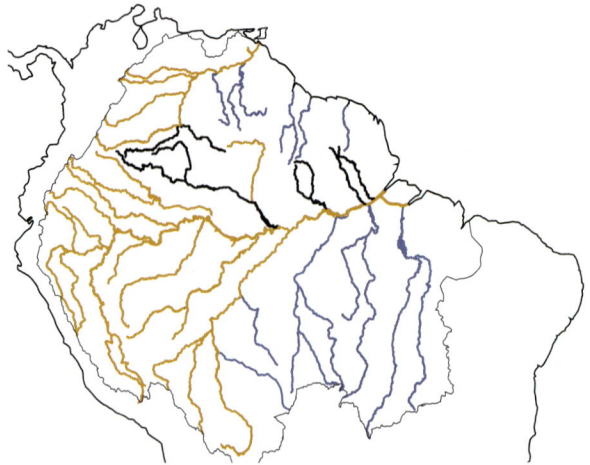

FIGURE 7 Water type of main tributaries: some rivers draining the Guiana Shield are dark-colored; rivers draining the Brazilian Shield have relatively clear water, whereas rivers originating in the Andes are sediment laden and have a brown color.

leaves. Most plants here protect their leaves with a mixture of toxic chemicals and produce stems at a higher density than elsewhere in the Amazon (e.g., ter Steege et al. 2006, Stropp et al. 2014). Toxic leaf compounds in combination with extremely low soil fertility hamper and slow decomposition of plant remains. Many leaf compounds, such as humic and fulvic acids, leach out with rainwater and accumulate in streams and eventually rivers, giving them a dark color (e.g., Atabapo, Negro, and Tefé rivers; figs. 7, 8).

(3) Clearwater rivers have a low sediment load, high transparency, and neutral pH (Goulding et al. 2003). They drain sandy soils in the interior of lowland rainforests and the crystalline granites of the Brazilian Shield (fig. 6). Similar to the Guiana Shield is the Brazilian Shield, formed from the remains of an ancient (Precambrian) mountain range. Soils in the Brazilian Shield are, however, less nutrient poor than those in the Guiana Shield. As such, plants in this region produce fewer protective compounds, the decomposition rates are faster,. and waters have high transparency. The Xingu, Tocantins, Tapajós, and upper Orinoco are major clearwater rivers in Amazonia (fig. 7). Many lowland Amazonian rivers exhibit hybrid properties or change seasonally from one category to another.

Major Aquatic Habitat Types

Amazonian fishes inhabit a broad range of aquatic habitats, ranging from mountain lakes and torrential cascades of the High Andes, to the deep (>40 m) channels of the great lowland rivers, and extensive freshwater marshes of expansive river estuaries. Greater Amazonia also has extensive areas of seasonally flooded

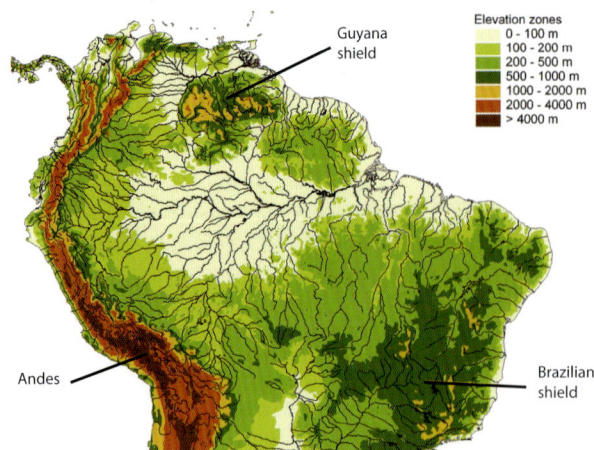

FIGURE 6 Elevation map of northern South America (in meters above sea level). Base map provided by the Conservation Science Program, World Wildlife Fund US.

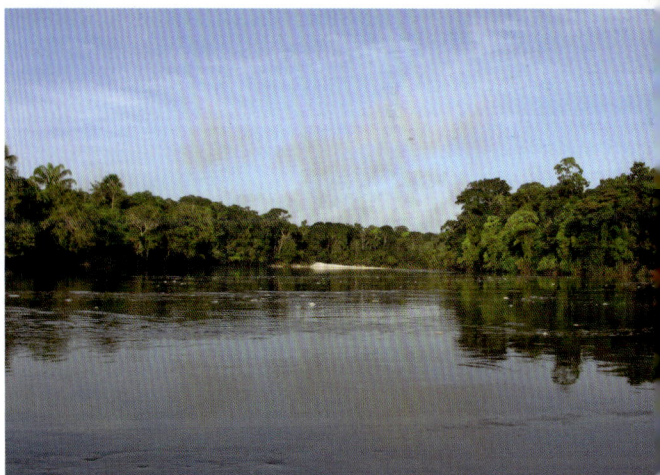

FIGURE 8 From headwater to lowland river. Top left: headwaters in the Andean mountains (Bolivia). Top right and middle: upper reaches of the Piray River, Bolivia. Lower left: whitewater lowland river (Mamore River in the Bolivian Amazon). Lower right: blackwater lowland river (Içana River in the Brazilian Amazon).

wetlands, including the Llanos of the central Orinoco basin (300,000 km²; 116,000 mi²), the Pantanal Setentrional of the central Negro basin (82,000 km²; 31,700 mi²), the Mamirauá wetlands in the central Amazon (57,000 km²; 22,000 mi²), and the Pacaya-Samiria wetlands in the Ucamara basin of the western Amazon (20,000 km²; 7,700 mi²). Based on structural properties and on the presence of recognizable groups of fish species, lowland aquatic systems have been divided into four main habitat types (Crampton 2011): upland streams, lowland terra firme (nonfloodplain) forest streams, floodplains, and river channels.

— Upland Streams

Upland streams are swiftly flowing and well oxygenated, have rocky or gravel substrates, and either are rich in dissolved minerals (Andean streams; fig. 8) or carry little suspended sediment (Shield streams). They are generally found above the fall line at 200–250 m above sea level, and below 1,000 m. In the shield areas, such streams can be located below the 200 meter contour in areas where streams with riffles, falls, and rocky substrates overlie granite.

— Lowland Terra Firme (Nonfloodplain) Forest Streams

Terra firme streams lie above the upper limit of the seasonal river flooding and compose most of the water surface area in Greater Amazon (Goulding et al. 2003; fig. 9). Electrolytes are scarce in terra firme streams because of a long history of weathering and leaching into the soils/subsoils, and because rainforest trees typically sequester nutrients via root-mycorrhizal complexes (Crampton 2011). Consequently, electrical conductivity is low (c. 5–20 μSc·m^{-1}). Incomplete decomposition of organic matter in the soil horizon on land as well as of leaf litter in many of these streams infuses the water with high concentrations of humic substances, resulting in low pH (3–5) and the characteristic tea-like blackwater coloration. Flow rates are typically low (<0.2 m·s^{-1}), temperatures are low under forest canopy (c. 24–26 °C), and dissolved oxygen is typically in the range 2–5 mg·L^{-1} but can drop lower (Crampton 2011).

Amazonian streams typically have several distinct microhabitats, including sandy riffles, deep pools, leaf-litter banks, and curtains of tree roots growing out of the banks. Aquatic macrophytes are often rare because of canopy shading and low nutrient levels, and algal periphyton and detritus are usually scoured away by flash floods. Terra firme stream fishes prey primarily on a combination of autochthonous (aquatic) and allochthonous (nonaquatic) invertebrates and plant material (Knöppel 1970, Ibañez et al. 2007). Many fishes in terra firme streams are miniatures, reaching a maximum standard length of less than 2.5 cm (Weitzman and Vari 1988). Miniaturized body sizes presumably evolved to allow access to the tiny interstices of underwater structures and because of the abundance of small insect larvae; many species that live in the marginal vegetation of rivers or floodplain floating meadows are also diminutive. But another hypothesis is that small size has evolved to cope with the low food availability in the terra firme streams.

— Floodplains

The Amazon and Orinoco rivers are flanked by low-relief alluvial floodplains. In the Amazon this floodplain forms an almost uninterrupted corridor more than 4,500 km (2,800 mi) long and 20–75 km (12–47 mi) wide called the várzea (Goulding et al. 2003). The annual flood pulse varies by year along the length of the river and depending on local floodplain geometry, but it can range up to 15 meters (50 feet) between maximum low and high. The nutrient-rich floodplains in whitewaters are the largest and most productive floodplain system in the world (Junk 1997). About one-third of the várzea comprises flooded forests, and another third comprises floating meadows of aquatic macrophytes along the margins of lakes and channels (fig. 9). Floodplains of nutrient-poor clearwater and blackwater rivers (called igapós) are much less extensive and productive than those of whitewater rivers. However, because low productivity means low insect abundance, almost all the large cities of the Amazon basin were built on clearwater or blackwater rivers, near the confluences with whitewater rivers to support the fish demand of local human populations; e.g., Belém, Santarem, Manaus, Tefé, Iquitos.

The annual flood pulse exposes fish to extreme fluctuations in the availability of food and shelter, the density of predators and parasites, and the physicochemical properties of water (Lowe-McConnell 1975, Junk 1997). During the dry season, fish are confined to shrinking pools, lakes, and channels, where they are exposed to high levels of predation. During the high-water period, floodplains support enormous autochthonous productivity in the form of phytoplankton, periphyton, and macrophyte growth (much of which decomposes to fine organic detritus), and also allochthonous productivity in the forest canopy and aerial portions of floating meadows (invertebrates, seeds, fruits, and other plant material).

This high productivity explains the high standing crop and turnover of fish; however, access to this seasonal bonanza is usually limited by extreme and persistent hypoxia caused by the decomposition of forest litter and other plant material during the flood season, at least in central and upper Amazonia. Lower (eastern) Amazonian floodplains experience less extreme hypoxia as a result of vertical mixing induced by trade winds. All várzea residents must possess a combination of morphological, physiological, or behavioral adaptations for hypoxia, and these have been intensively studied (Val and de Almeida-Val 1995). Such adaptations include aestivation and air-breathing using lungs (Lepidosiren), air-breathing using a swim-bladder (Arapaima, many catfishes; Gymnotus), vascularized mouth organs (Electrophorus, Synbranchus), aquatic surface-respiration (Osteoglossum, Colossoma, many characids), and physiological alterations to the way hemoglobin binds oxygen (Hoplias). Many floodplain fishes also exhibit physiological specializations for high temperatures, particularly those living in the floating meadows of open lakes, where temperatures routinely exceed 35 °C (de Almeida-val et al. 2005).

Some fishes are permanent floodplain residents, but most species are seasonal or migratory visitors taking advantage of the high productivity at high water for foraging and or breeding. Many floodplain fishes feed on seeds and fruits, notably Brycon, Colossoma, Piaractus, Mylossoma, and some catfishes (e.g., Goulding 1980). Some of these species act as seed dispersers (e.g., Kubitzki and Ziburski 1994, Anderson et al. 2009, Correa et al. 2015b).

FIGURE 9 Other aquatic habitats. Top left: várzea forest (Amazon River near Manaus, Brazil). Top right: floating meadow with some giant waterlilies, *Victoria amazonica* (Amazon River near Manaus, Brazil). Lower photo: blackwater forest stream (upper Negro basin, Brazil).

Omnivory and diet switching are common among floodplain fishes, with increases in the proportion of allochthonous food derived from the rainforest canopy at high water.

— River Channels

Channels of large Amazonian rivers range from 10 to more than 80 m (33–262 ft) deep. Large rivers are defined as those experiencing a seasonal flood cycle and exceeding 3–5 m (10–16 ft) at midchannel. River channels typically host many more species than terra firme streams (Crampton 2011). Large river channels vary in substrate composition, current speed, and depth, which influence species composition. However, in terms of species diversity and biomass, the bottom layer of large Amazonian rivers is dominated by species in two families of gymnotiform electric fishes (Apteronotidae and Sternopygidae) and two catfish families (Doradidae and Pimelodidae). A common characteristic of these groups, possibly explaining their success in this environment, is that they don't depend on light to move around. Catfishes have well developed barbels that allow them to recognized chemical and tactile stimuli, and most catfishes also have taste buds and electroreceptor organs distributed over the entire surface of their head, body, and even fins. Electric fishes also have electroreceptor organs distributed over the head and body, which they use to recognize environmental characteristics and the presence of other fishes and prey items by distortions of the electric fields they generate.

Conservation of Amazonian Fishes

Amazonia is home to the highest concentration of biological diversity in the world. Yet the survival of many of its fish species in the near future is uncertain. This is especially the case for species occurring in the headwaters of Amazonia, many of which have limited distributional ranges. Habitat conditions in headwaters have been severely altered in many regions, particularly the southeastern part of the Amazon basin, by deforestation for large-scale agricultural purposes, timber exploitation, and cattle ranching (Castello and Macedo 2016). With the destruction of the riparian vegetation, soil erosion is often accelerated, and excessive sediment deposition in streams can destroy aquatic food webs. Deforestation of floodplains has a particularly strong effect on fish abundance and diversity, as many Amazonian fish species also depend on floodplain forests for feeding and reproduction (e.g., Goulding 1980, Goulding et al. 1988). Mining poses an additional threat to fish diversity in several regions. In areas where gold, cassiterite (tin), and aluminum are mined, the deposition of inorganic sediments has heavily impacted streams, rivers, and wetlands. Gold mining can be especially damaging for aquatic species, as it often involves the use of hazardous chemicals such as mercury (Akagi et al. 1995, Bastos et al. 2015). Hundreds of thousands of gold diggers have devastated even remote areas. The effects of these activities on fish diversity have been poorly investigated thus far. However, approximately 50% of all AOG species live in headwaters, and there is consequently a danger of broadscale species extinctions (Junk et al. 2007).

To a certain extent, fish species that inhabit large rivers are better protected than those living in small rivers and streams, because most species occur over large river stretches and are thus more resilient to local habitat deterioration. However, the construction of large reservoirs for hydroelectric power plants can have far-reaching effects on fish biodiversity (Castello and Macedo 2016, Winemiller et al. 2016). Large hydropower dams have been constructed in the Amazon (e.g., the Tucuruí Dam near Belém, Balbina Dam near Manaus, and Belo Monte Dam on the Xingu River), Orinoco (the Guri Dam on the Caroni River in Venezuela) and Guianas (e.g., the Afobaka Dam in the Brokopondo district in Suriname), and hundreds of additional dams are planned for construction in the near future. Hydropower dams affect fish by modifying the natural flood regime downriver of the dam, retaining sediments and nutrients in the reservoirs, blocking fish migration routes, and modifying habitat conditions in large areas both up- and downriver of the dam.

The incredible fish diversity in Amazonia has been a very long time in the making, and once damaged or destroyed this vital ecosystem will never be replaced. As with all unique forms of life on Earth, extinction is forever.

Further Reading

- Albert, J. S., and W. G. Crampton. 2010. The geography and ecology of diversification in Neotropical freshwaters. Nature Education Knowledge **1**, 13.
- Albert, J. S., and R. E. Reis, editors. 2011. Historical Biogeography of Neotropical Freshwater Fishes. University of California Press.
- Goulding, M. 1980. The Fishes and the Forest: Exploration in Amazonian Natural History. University of California Press.
- Goulding, M., M. Leal Carvalho, and E. Ferreira. 1988. Rio Negro: Rich Life in Poor Water. Amazonian diversity and foodchain ecology as seen through fish communities. SPB Academic Publishing.
- Goulding, M., R. Barthem, and E. Ferreira. 2003. The Smithsonian Atlas of the Amazon. Smithsonian Books.
- Hoorn, C., and F. Wesselingh, editors. 2010. Amazonia, Landscape and Species Evolution: A Look into the Past. Wiley-Blackwell.
- Junk, W. J. 1997. The Central Amazon Floodplain: Ecology of a Pulsing System. Ecological Studies **126**. Springer.
- Lowe-McConnell, R. H. 1987. Ecological Studies in Tropical Fish Communities. Cambridge University Press.
- Reis, R. E., S. O. Kullander, and C. J. Ferraris Jr., editors. 2003. Check List of the Freshwater Fishes of South and Central America. EDIPUCRS, Porto Alegre, Brazil.
- Val, A. L., and V. M. F. de Almeida-Val. 1995. Fishes of the Amazon and Their Environment. Springer-Verlag.

Concept

The goal of this book is to provide an overview of our current state of knowledge on the taxonomy of fishes in the Amazon and Orinoco basins and the coastal rivers in the Guianas (the biodiversity province known as Greater Amazonia). Because of the very high number of fish species (around 3,000), the large taxonomic uncertainties in many fish groups, and the large number of still undescribed species in the region, this book is not a field guide to fish species, which at the moment is still impossible to compile. We focus here on fish genus and species groups, the next higher levels in the Linnaean taxonomic hierarchy. Some of the species described here as a "species complex" may include one or more "cryptic species" that differ genetically but are difficult to identify using morphological criteria. We provide references to the relevant literature for species-level identification when available. This book is thus both a guide to fish genera and to the taxonomic literature on Amazonian fishes, which is vast, written in more than six languages, and spans more than a century.

Structure

After this chapter, the book provides a taxonomic key to identify the larger groups and families of fishes (which include related fish genera) in the region. After working through this key, readers will be directed to a relevant family chapter, which starts with a short introduction to some of the general features of fishes in the family. This introduction is followed by a key to the genera, and then individual genus accounts that summarize current knowledge regarding diagnostic traits, numbers of species, geographic range distributions, habitat, and ecology. The taxonomy used in this *Field Guide* is based on our current and imperfect understanding of evolutionary (phylogenetic) relationships. Some of the genera presented here are "artificial," meaning the group as recognized by morphological features does not reflect phylogenetic relationships as determined from genetic analyses.

Readers unfamiliar with the fauna are recommended to start with the general key to the fish families in chapter 3. Identifying a species to the right fish family can be aided by consulting the photo plates. These images show some characteristic species in each family and exhibit some key diagnostic features. After identifying the family, the reader will find a key to the genera in the specific family chapter. After a candidate genus is identified from the key, the genus account can be consulted to confirm the identification using additional information, including body size and color pattern, geographic location, and habitat. If species-level identification is required, the references provided in the account (e.g., to published keys to the species) can be looked up. An illustrated glossary of some of the technical terminology is provided at the end of the book.

Using Taxonomic Keys

Taxonomic keys are designed to allow taking a specimen of unknown identity and subjecting it to successive "tests" for the presence or absence of relevant diagnostic characters, and by following these tests eventually deducing the correct identification. The keys in this book are *dichotomous*, allowing the user to simply choose between two alternatives at each step. We recommend always reading both alternatives. Each numbered step in the key, consisting of two alternative statements, is termed a "couplet." By following the couplets, one will eventually arrive at a certain genus as the probable correct identification.

A fish with a body without stripes, with fewer than 40 scales along the lateral line, and with more than 15 anal-fin rays will come out as *Genus D* in the example key below.

Example key

1a.	Dark stripes on body	**Genus A**
1b.	Body without dark stripes	**2**
	[if this is the case, then go to couplet 2]	

2a.	Lateral line scales 45 or more	**Genus B**
2b.	Lateral line scales fewer than 40	**3**
	[if this is the case, then go to couplet 3]	

3a.	Anal fin with 12–15 rays	**Genus C**
3b.	Anal fin with >15 rays	**Genus D**

The Genus Accounts

Following the Latin genus name, the adult size range of the species within that genus is given. Juveniles can of course be much smaller, so note that this information entails maximum adult lengths. The size range is usually expressed as the standard length (SL) which is the horizontal distance from the tip of the snout to the base of the caudal fin (fig. 10).

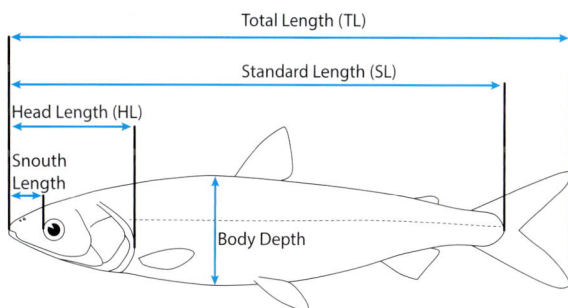

FIGURE 10 Standard measurements used to indicate size (and shape) of fishes.

In each genus account we summarize information in five categories:

DIAGNOSIS The main characters that separate a certain genus from other (closely related) genera. We focus on external characters and avoid internal and osteological features, which have little value for field identification. However, for some genera, internal characteristics might be the best way, or the only way, to distinguish from other genera. Here we also include information on sexual dimorphism if present and known.

SPECIES The total number of species in a genus, and of species in that genus present in Greater Amazonia, a biogeographic province here referred to as the Amazon-Orinoco-Guiana (AOG) region. Relevant literature on species-level identification is provided here if available.

COMMON NAMES Informal, vernacular names for the genus (or for specific species in the genus) in any of the countries in the AOG, as well as popular English names. The names provided are the most widely used but an incomplete overview of common names used in the region.

DISTRIBUTION AND HABITAT For most genera we provide the total geographic distribution of the genus and subsequently focus on its specific distribution in the Amazon-Orinoco-Guiana region. Information in this section overlaps with the distribution maps, although more detailed information is frequently included. We also list the habitats in which species of a certain genus are typically found, when such generalizations can be made.

BIOLOGY Summarizes the knowledge on natural history of species in a certain genus, for example, concerning diet, reproduction, behavior, and migration.

Distribution Maps

The distribution of a genus in the AOG is indicated by its presence or absence within 20 freshwater ecoregions of Greater Amazonia (fig. 11 and table 2). The ecoregion limits are delineated primarily by watershed boundaries (hydrological basins). It is important to note that the distribution maps give a rough estimation of actual distribution in most cases, as a genus might be found in only a specific part of an ecoregion. Furthermore, many fish genera are also found outside the AOG, but this distribution is not included in the maps (regions outside the AOG are shaded in gray). For a few genera with very restricted distributions, we have included a more customized distribution map, since depicting an entire ecoregion would greatly overestimate the actual distribution.

Drawings

A rather simple line drawing is provided for each genus. These drawings give a general impression of overall body shape and the configuration of some external features (e.g., fins, barbels, and body coloration) of species in the genus. For many genera with a high species number, a single line drawing cannot summarize the variability of body shapes and pigmentation patterns. Identifying species to genus using the line drawings alone is therefore not recommended.

Photographic Plates

The photographic plates include a selection of fish species provided as an aid to family-level identification, and to help the reader get a feel for the major groups of Amazonian fishes. Again, the extremely high diversity of fish species in the AOG cannot be captured with a limited selection of photos. These plates are therefore not intended as a guide to identify fish genera.

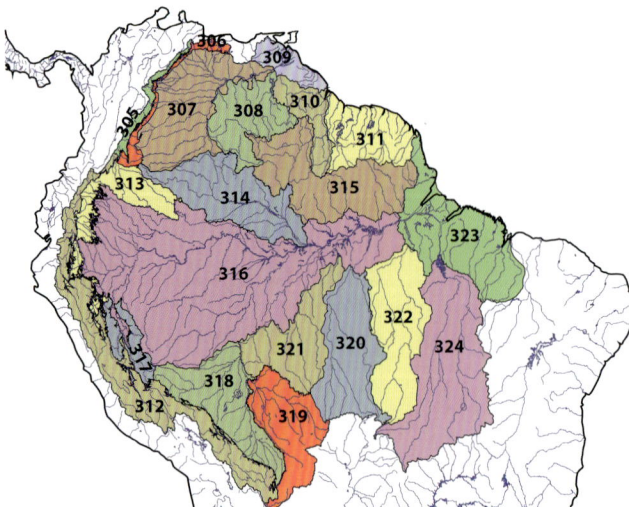

FIGURE 11 Freshwater ecoregions of Greater Amazonia, including the Amazon and Orinoco basins and coastal rivers in the Guianas. Ecoregions after Abell et al. (2008). Ecoregion names and geographical data shown in table 2.

Limitations of This Book

Scientific knowledge of the diversity and taxonomy of Amazonian fishes is still highly imperfect, and so therefore is necessarily the information in this *Field Guide*. About 10% of all currently known Neotropical freshwater fishes were described in just the past 12 years, and the rate of new species descriptions is still very high, being limited mainly by the number of active fish taxonomists and the time it takes to publish new species descriptions. The actual total number of fish species in Greater Amazonia is therefore not known; current estimates place that number at more than 4,000 species, meaning that perhaps 25% of all the fish species in the AOG region are as yet undiscovered and unnamed.

Furthermore, the geographical distribution of most Amazonian fish genera is poorly known, making it hard to provide accurate distribution maps. We have chosen to approach this problem by using absence/presence data for a genus in each of the 20 ecoregions in Amazonia (fig. 11, table 2); however, this methodology likely overestimates the actual distributions in many cases. We therefore stress that the distribution maps provided in this book will give only a rough estimation of actual distributions for most genera.

Certainly another limitation of this book is the geographic scope, which targets only the AOG region. As immense as it is, the Amazonian ichthyofauna is only a portion of a more inclusive Neotropical ichthyofauna that extends from southern Mexico to northern Argentina and has more than twice the number of fish species as does Greater Amazonia. All the families and many of the genera described in this *Field Guide* also inhabit waterways in other parts of tropical South and Central America. It is our hope that future field guides will be able to provide keys, descriptions, and maps for the fish genera of this much broader region.

TABLE 2 The 20 Freshwater Ecoregions of Greater Amazonia

Code	Ecoregion	Area (km²)	Core or periphery	Lowland or Upland
305	Orinoco High Andes	68,148	Periphery	Andes highland
306	Orinoco piedmont	82,491	AOG core	Lowland
307	Orinoco-Llanos	575,142	AOG core	Lowland
308	Orinoco Guiana Shield	348,090	AOG core	Shield upland
309	Orinoco Delta and adjacent coastal rivers	138,602	AOG core	Lowland
310	Essequibo	182,512	AOG core	Shield upland
311	Eastern Guiana	336,492	AOG core	Shield upland
312	Amazonas High Andes	530,073	Periphery	Andes highland
313	Marañón-Napo-Caquetá	258,909	AOG core	Lowland
314	Rio Negro	496,301	AOG core	Lowland
315	Amazonas Guiana Shield	605,130	AOG core	Shield upland
316	Amazon Lowlands	1,909,012	AOG core	Lowland
317	Ucayali-Urubamba	104,605	AOG core	Lowland
318	Mamoré–Madre de Dios	378,174	AOG core	Lowland
319	Guaporé-Itenez	326,437	Periphery	Lowland
320	Tapajós-Juruena	429,427	Periphery	Shield upland
321	Madeira Brazilian Shield	349,019	Periphery	Shield upland
322	Xingu	463,772	Periphery	Shield upland
323	Amazon estuary and coastal drainages	580,379	Periphery	Lowland
324	Tocantins-Araguaia	717,332	Periphery	Shield upland

Source: Abell et al. 2008.

IDENTIFICATION KEY TO FISH FAMILIES

1a. Five pairs of external gill openings; skeleton cartilaginous (sharks and rays)..**2**
1b. Single pair of external gill openings; skeleton bony...**4**

2a. Body shape fusiform, tapering at both ends; gill openings lateral.....................................**Carcharhinidae** (Bull shark, p. 69)
2b. Body shape depressed, dorsoventrally flattened; gill openings ventral..**3**

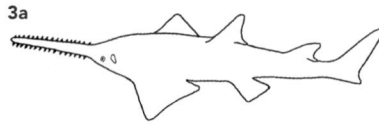

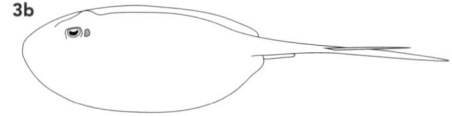

3a. Snout elongate, thin, flat and saw-like, with teeth along lateral margins; body somewhat flattened,
but not disk-shaped ....**Pristidae** (Sawfishes, p. 69)
3b. Snout not elongate, not saw-like; body strongly flattened dorsoventrally and disk-shaped
....**Potamotrygonidae** (Freshwater stingrays, p. 70)

4a. Body very flat (laterally compressed) and asymmetric; mouth asymmetric;
both eyes located on right side of head; usually one side of body darker
and other lighter ....**Achiridae** (American soles, p. 400)
4b. Body and mouth laterally symmetric; eyes situated on each side of head;
body shape variable **5**

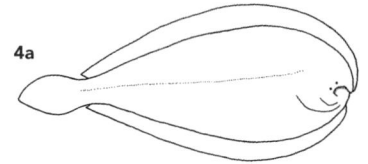

5a. Body long and slender, anguilliform (eel-like), head length <15% body length in
adults; barbels or bony rings absent..**6**
5b. Body generally not anguilliform; if body elongate, then usually with small barbels
(i.e., Trichomycteridae) or body encased in bony rings (i.e., Syngnathidae) **9**

6a. Paired fins absent in adults (pectoral fins present in Synbranchus larvae ≤15 mm) **7**
6b. Paired fins (at least pectoral fins) present **8**

7a. External gill opening as single slit, located in throat; posterior nares located close to eye **Synbranchidae** (Swamp eels, p. 399)
7b. Two external gill openings, located laterally; posterior nares located on upper lip or
inside mouth...**Ophichthidae** (Snake eels, p. 73)

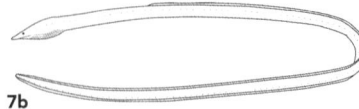

8a. Pectoral and pelvic fins thin and filamentous, without rays; laterosensory canals present as pronounced
grooves on surface of head and body; anal fin restricted to posterior portion of body
....**Lepidosirenidae** (South American lungfishes, p. 72)
8b. Pectoral fins broad, not filamentous; pelvic fins absent; laterosensory canals not as pronounced grooves on
surface of head and body; anal fin extends along most of ventral body margin ...**Gymnotiformes** (Knifefishes) → **see key A**

9a. Body elongate and encased in bony rings; elongate snout; anal fin small to absent; pelvic fins absent;
males with a ventral brood pouch...**Syngnathidae** (Pipefishes, p. 399)
9b. Not combination of characters as above **10**

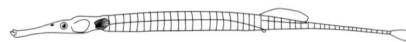

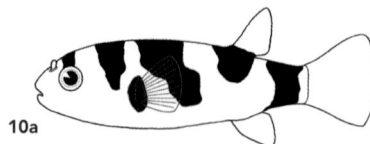

10a. Jaws modified to form a beak of 4 heavy, powerful teeth, 2 above and 2 below; pelvic fins absent;
body rounded (if threatened, capable of inflating body)....................................**Tetraodontidae** (Pufferfishes, p. 402)
10b. Not combination of characters as above **11**

11a. Broad and flat head, often with fleshy projections; eyes located on top of head; two dorsal fins, first consisting of two or three strong spines; second dorsal fin with large number of soft rays; pelvic fins jugular and inserted well in advance of pectoral fins......................**Batrachoididae** (Toadfishes, p. 396)

11b. Not combination of characters as above **12**

12a. Body covered by skin, sometimes with bony plates; barbels usually present; pectoral fin often with strong spine..**Siluriformes** (Catfishes) → **see key B**

12b. Body covered by scales, at least partially, barbels usually absent; pectoral fin usually without spine **13**

13a. Lower jaw elongate; body slender and elongate.. **14**

13b. Lower jaw not elongate; body shape variable **16**

14a. Both jaws elongate into slender, strong toothed beak.. **Belonidae** (*Potamorrhaphis* and *Pseudotylosurus*) (Needlefishes, p. 397)

14b. Lower jaw elongate; upper jaw not elongate **15**

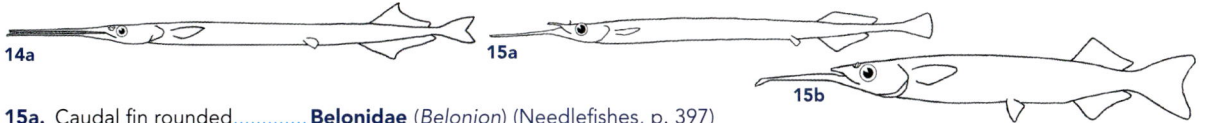

15a. Caudal fin rounded.............**Belonidae** (*Belonion*) (Needlefishes, p. 397)

15b. Caudal fin indented........................**Hemiramphidae** (Halfbeaks, p. 398)

16a. Tongue with well-developed teeth; large adult size, 90–400 cm TL..............................**17** (Osteoglossiformes, Bonytongues)

16b. Tongue without well-developed teeth (few exceptions; however if present, fish lacks a long dorsal fin with many rays); adult size <90 cm, usually much smaller **18**

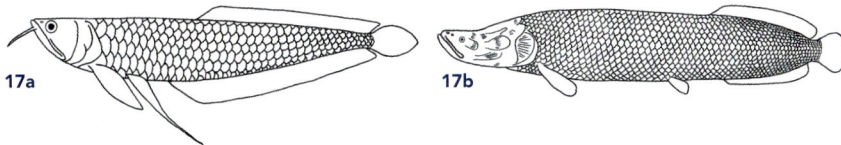

17a. Two barbels at extremity of lower jaw ....**Osteoglossidae** (Arowanas, p. 74)

17b. Barbels absent..**Arapaimidae** (*Arapaima*, p. 73)

18a. Ventral margin of abdomen with sharp keel, usually saw-like; adipose fin absent; lateral line absent; body color silvery or golden, without other colors or marks; caudal fin forked (with lobes of equal size) ....**Clupeiformes** (Herrings, Anchovies, and relatives) → **see key C**

18b. Not combination of characters as above **19**

19a. All dorsal-fin rays soft and flexible **20**

19b. Dorsal fin with spiny and soft portions (as either two separate dorsal fins, or together as one dorsal fin, with spiny anterior portion).. **21**

20a. Caudal fin truncate or rounded; adipose fin absent; protrusible upper jaw bordered by premaxilla; lateral line composed of superficial neuromasts......................**Cyprinodontiformes** (Killifishes and livebearers) → **see key D**

20b. Caudal fin usually forked or emarginate, but rounded in some groups (e.g., Erythrinidae); adipose fin usually present; upper jaw usually not truly protrusible; lateral line often decurved, sometimes incomplete ....**Characiformes** (Characins) → **see key E**

21a. One dorsal fin with spiny anterior and soft posterior portion (deep notch can be present between spiny and soft portions) **22**

21b. Two separate dorsal fins, anterior with spines, posterior with soft rays (note that adipose fin can be differentiated from second dorsal fin by lacking rays)........... **24**

22a. Scales absent on head anterior to eye; only 1 nostril on each side of head; lateral line usually not continuous, with long anterior portion running closer to dorsal region, and short posterior section (but continuous lateral line present in some species) ....**Cichlidae** (Cichlids, p. 359)

22b. Scales present on head anterior to eye; 2 nostrils on each side of head; lateral line absent or present as continuous line along flanks **23**

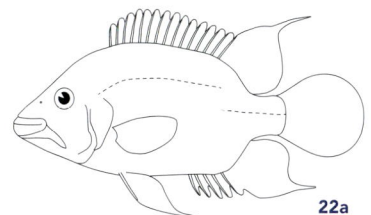

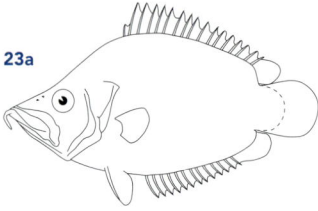
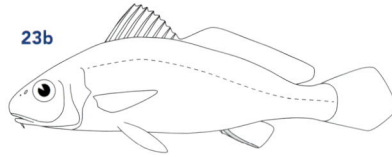

23a. Lateral line absent; anal fin with many spines; soft-ray portion of dorsal fin short, shorter than spiny portion and shorter than anal fin (mimicking dead leaves in body shape, color pattern and behavior)..**Polycentridae** (Leaf-fishes, p. 385)
23b. Lateral line present, continuous along flanks; anal fin with 1 or 2 spines; soft-ray portion of dorsal fin long, longer than spiny portion and longer than anal fin......... **Sciaenidae** (Drums and Croakers, p. 386)

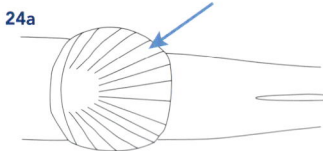
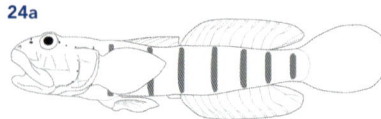
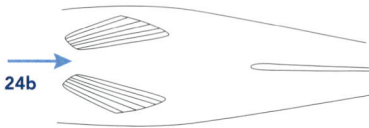

24a. Pelvic fins fused medially to form a cup-like disk.. **Gobiidae** (Gobies, p. 392)
24b. Pelvic fins separate medially ....**Eleotridae** (Sleepers, p. 388)

A KEY TO GYMNOTIFORM FAMILIES

1a. Continuous extension of anal-fin rays along ventral body margin to tip of tail; scales absent; dark brown ground color with orange to red on ventral surface of head and abdomen .... *Electrophorus* (Electric eel, p. 330)
1b. Caudal appendage (with or without caudal-fin rays) posterior to last anal-fin ray; scales always present at least on posterior portion of body; never orange or red on ventral surface of head and abdomen **2**

2a. Distinct caudal fin present; fleshy dorsal organ present (strip of soft tissue on posterior half of middorsum)..**Apteronotidae** (Ghost knifefishes, p. 322)
2b. Caudal fin and fleshy dorsal organ absent...**3**

3a. Lower jaw prognathous, mouth superior; anterior nares in mouth; eye at horizontal with gape...*Gymnotus* (Banded knifefishes, p. 330)
3b. Lower jaw not prognathous; anterior nares on dorsal head surface; eye positioned above horizontal with gape.................**4**

4a. Small villiform oral teeth; eye large, diameter greater than distance between nares; most species without dark pigment bars or saddles (except wide bars in *Japigny kirschbaum* and wide saddles in *Sternopygus astrabes*)..........................**Sternopygidae** (Glass knifefishes and Rattail knifefishes, p. 341)
4b. Oral teeth absent in most species; eye small, diameter less than distance between nares; many species with numerous irregular dark pigment bands, bars, or saddles...**5**

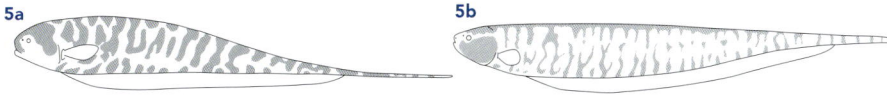

5a. Anal-fin origin at, or anterior to, pectoral-fin base; anal-fin base without
semitransparent tissue fold ....**Rhamphichthyidae** (Painted knifefishes, Sand knifefishes, Trumpet knifefishes, p. 337)

5b. Anal-fin origin ventral, or posterior to, pectoral-fin base; anal-fin base with thick
semitransparent tissue fold ....**Hypopomidae** (Grass knifefishes, p. 334)

B KEY TO SILURIFORM FAMILIES

1a. Body with plates (or scutes) at least partially **2**
1b. Body naked, without plates...**5**

2a. Mouth inferior, lower lip turned back and together with upper lip forming sucker disk (generalized
in figure 2a); body usually completely covered by plates......................**Loricariidae** (Suckermouth armored catfishes, p. 253)
2b. Mouth not modified into sucker disk; plates in one or two rows on side of body **3**

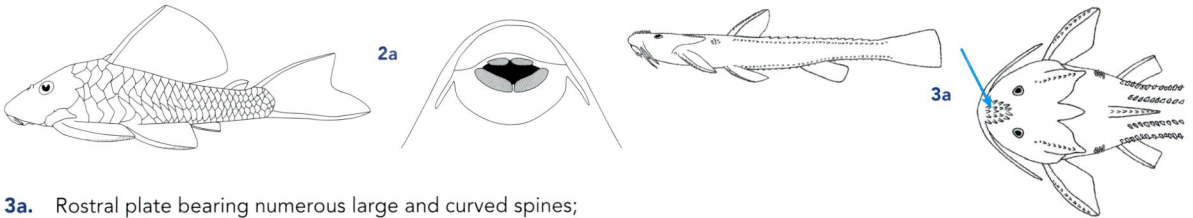

3a. Rostral plate bearing numerous large and curved spines;
miniaturized, with maximum adult size <2 cm.........................**Scoloplacidae** (Spiny dwarf catfishes, p. 310)
3b. Rostral plate without spines; adult size >2 cm (most 4–20 cm) **4**

4a. One row of plates on each side of body, each plate usually with robust, backward-
directed, medial thorn and sometimes with smaller spines and serrations.....................**Doradidae** (Thorny catfishes, p. 222)
4b. Sides of body completely covered by two rows of plates; thorns on
plates absent .... **Callichthyidae** (Callichthyid armored catfishes, p. 216)

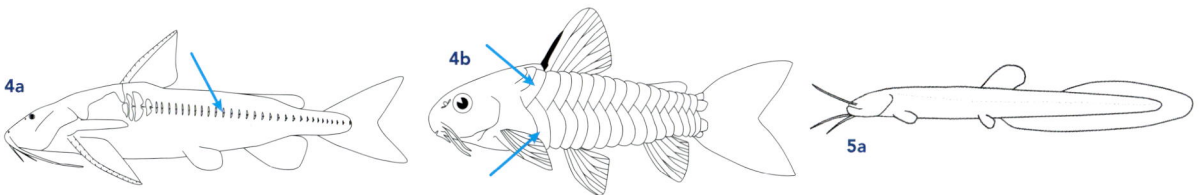

5a. Bright red to pink body coloration; caudal fin ventrally confluent with
anal fin... **genus** *Phreatobius* (family not determined; p. 322)
5b. Body coloration usually not bright red to pink; anal fin separate from caudal fin, or absent...................................**6**

6a. A patch of odontodes (small spines) usually present on opercle and/or interopercle;
lack of pectoral- and dorsal-fin spines (body loach-like)..**Trichomycteridae** (Pencil catfishes, p. 311)
6b. Operculum and interopercle without spines; pectoral and dorsal-fin spines usually present **7**

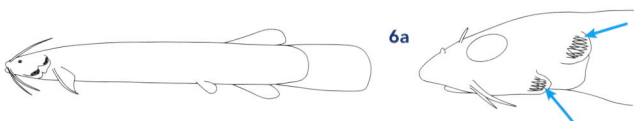

7a. Mouth inferior, lower lip turned back and together with upper lip forming sucker disk (high elevation streams of Andes)...**Astroblepidae** (Naked suckermouth or Climbing catfishes, p. 207)

7b. Mouth not modified into sucker disk **8**

7a

8a

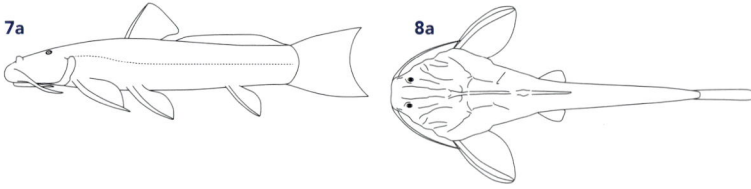

8a. Body broad anteriorly (with depressed head) and elongate posteriorly (with slender caudal peduncle); skin heavily keratinized with rows of large tubercles arranged along dorsal surface and sides of body ....**Aspredinidae** (Banjo catfishes, p. 202)

8b. Body shape variable; skin not strongly keratinized...**9**

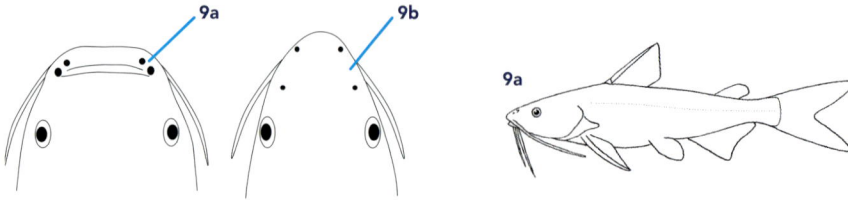

9a   **9b**

9a

9a. Anterior and posterior nasal openings usually very close together; caudal fin deeply forked (estuaries and lower parts of coastal rivers) .... **Ariidae** (Sea catfishes, p. 198)

9b. Anterior and posterior nares openings usually widely separated and distinctly marked off by skin covering head; caudal-fin shape variable, including rounded and forked, sometimes deeply forked..................................**10**

10a. Adipose fin small, its base much smaller than anal-fin base, or adipose fin absent **11**

10b. Adipose fin well developed, its base equal to or larger than base of anal fin..**12**

(Some genera in the following catfish families are difficult to key out by external characters only)

11a. Eyes relatively large, and covered with adipose tissue; dorsal region of body, between head and dorsal-fin origin, covered with bony plates that are sutured together and readily visible beneath thin skin....................................**Auchenipteridae** (Driftwood catfishes, p. 208)

11b. Eyes often very reduced or even absent; no bony plates visible beneath skin in dorsal region of body, between head and dorsal-fin origin ....**Cetopsidae** (Whale catfishes, p. 220)

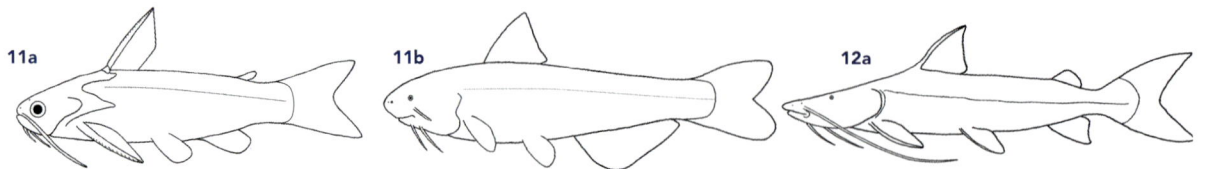

11a   **11b**   **12a**

12a. Dendritic (i.e., heavily branched) arrangement of lateral line tubes in skin of snout, cheek, and nape; barbels long, reaching or surpassing origin of dorsal fin; usually medium to large-sized species (>20 cm SL); lips usually simple, lacking secondary folds.........**Pimelodidae** (Long-whiskered catfishes, p. 299)

12b. Lateral line tubes on head usually simple, with single pores at skin surface; barbels long or short; small adult body size, usually <20 cm SL; upper and lower lips usually subdivided into two (rarely three) fleshy ridges paralleling transverse profiles of snout and chin (ridges may be smooth or variously textured with papillae or plicae, and separated by sulci) **13**

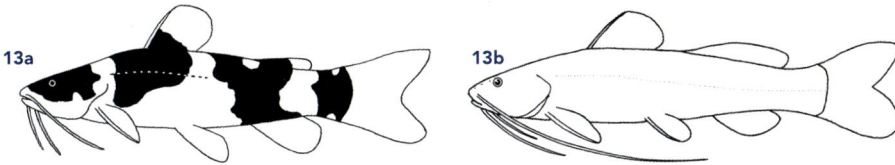

13a. Eyes small, without free orbital margin; pectoral fin with spine serrated anteriorly and posteriorly; barbels short (many species have beautiful patterns and coloration, including characteristic light band running across nape and alternate light and dark blotches over body) ....**Pseudopimelodidae** (Bumblebee catfishes or Dwarf-marbled catfishes, p. 308)

13b. Eyes varying from large to small (sometimes completely absent), with or without free orbital margin; pectoral fin ranging from bearing anterior pungent spine (sometimes serrated posteriorly and anteriorly) to completely flexible and mostly segmented fin ray; barbels long or short...**Heptapteridae** (Three-barbeled catfishes, p. 233)

C KEY TO CLUPEIFORM FAMILIES

ILLUSTRATION AFTER LIDONNICI AND LASTRICO IN WHITEHEAD (1985)

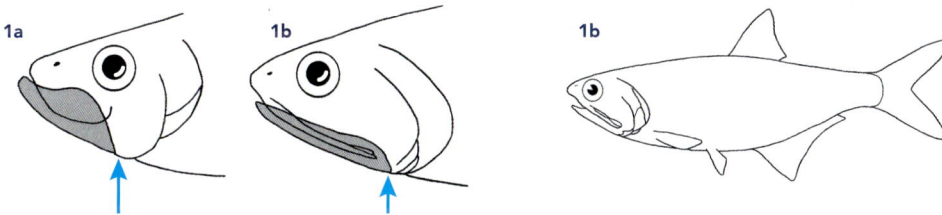

1a. Articulation of lower jaw under or only just behind eye, lower jaw deep; belly scuted **2**
1b. Articulation of lower jaw well behind eye, lower jaw usually slender; belly not scuted ....**Engraulidae** (Anchovies, p. 75)

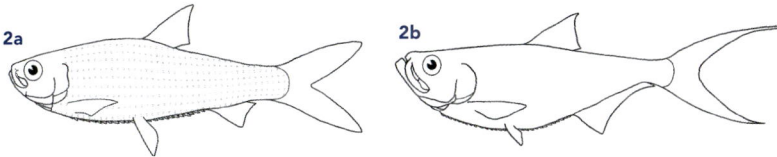

2a. Anal fin moderate, <20 fin rays..**Clupeidae** (Herrings, p. 74)
2b. Anal fin long, ≥30 fin rays.. **Pristigasteridae** (Longfin herrings, p. 79)

D KEY TO CYPRINODONTIFORM FAMILIES

1a. Prominent eyes, elevated above top of head, with pupils horizontally divided (making it appear as if there are two eyes on each side)...**Anablepidae** (Four-eyed fishes, p. 345)
1b. Eyes simple and not prominent...**2**

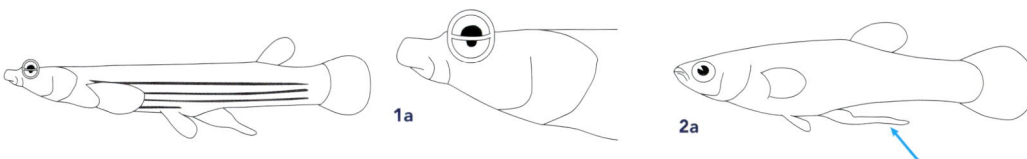

2a. Anal-fin rays 3–5 of adult males modified to form rod-like intromitting organ ....**Poeciliidae** (Livebearers, except *Fluviphylax*, p. 346)
2b. Anal-fin rays of adult males not modified into intromitting organ..**3**

3a. Eyes extremely large (about 50% of head length); miniature adult sizes (<2.0 cm) .... **Poeciliidae** (*Fluviphylax*; p. 348)
3b. Eyes usually large, but <50% of head length; adult sizes from miniature (<2.0 cm) to >15 cm **4**

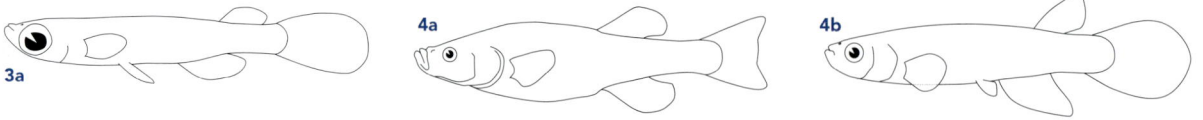

4a. Pelvic fins absent; dorsal median ridge of scales from head to dorsal fin; found only in high-altitude lakes
and tributary streams of Peruvian, Bolivian, and Chilean Andes .... **Cyprinodontidae** (Pupfishes, genus *Orestias*, p. 346)
4b. Pelvic fins usually present; dorsal median ridge of scales absent from head to dorsal fin .. **Rivulidae** (Rivuline killifishes, p. 350)

E   KEY TO CHARACIFORM FAMILIES

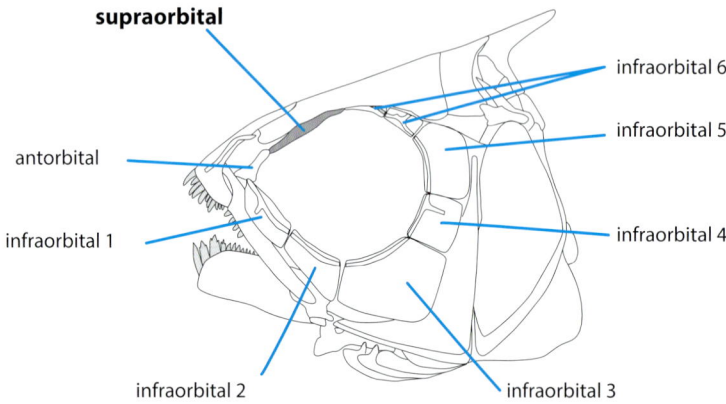

FIGURE 1 **Head bones of a generalized characoid**
The infraorbital bones surround the eye and have a laterosensory canal. Most characiforms have 6 infraorbital bones; the first one is located just ventral to the antorbital and extends to a varying degree along the longitudinal length of this bone. The supraorbital, when present (absent in the Characidae), is situated dorsal or anterodorsal to the orbit. Illustration after Weitzman 1962.

1a. Supraorbital bone absent **2**
1b. Supraorbital bone present (see fig. 1) **4**

2a. Mouth large, often with strong canine teeth; body round in cross section; caudal fin rounded;
medium-sized to large species (SL >10 cm) .... **Erythrinidae** (Wolf-fishes, p. 156)
2b. Mouth small; body round to compressed in cross section; caudal fin usually forked; small species (SL <10 cm) **3**

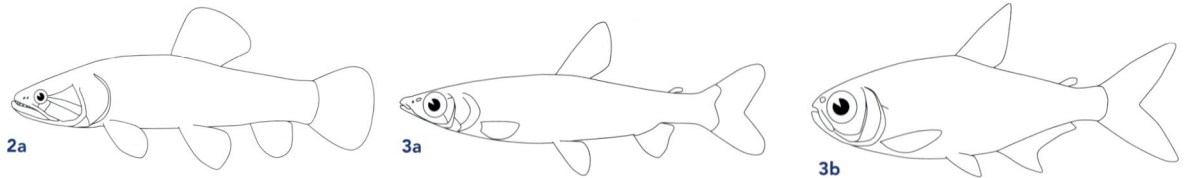

3a. Large and thin scales on surface of posterior portion of head; body cylindrical or fusiform; anal fin
with 9 or 10 branched rays (species adapted to life at water surface or in midwater) .... **Lebiasinidae** (Pencilfishes, p. 165)
3b. Scales lacking on posterior portion of head; body compressed; anal fin usually with
>10 branched rays .... **Characidae** (Tetras and relatives, p. 92)

4a. Scales situated dorsal to lateral line about twice the size of those ventral
to it; relative size of scales along lateral line other than on caudal peduncle
alternatively large and small .... **Chalceidae** (Tucanfish, p. 91)
4b. Dorsal and ventral scales about equal in size; scales along lateral line series
of similar size or gradually decreasing in size posteriorly **5**

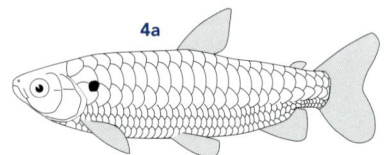

5a. Coracoids expanded to form distinct abdominal keel **6**
5b. Coracoids not expanded in abdominal keel **8**

6a. Body hatchet-shaped; pectoral fin large, usually reaching vertical through
anal-fin origin ....**Gasteropelecidae** (Freshwater hatchetfishes, p. 158)
6b. Body not hatchet-shaped; pectoral fin never reaching vertical through anal-fin origin **7**

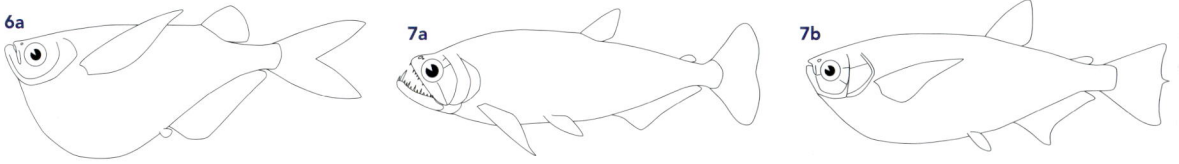

7a. Large fishes; body elongate and strongly compressed; mouth large, very oblique and superior; all teeth
conical or canine, with highly developed pair of dentary canines ....**Cynodontidae** (Dogtooth characins, p. 154)
7b. Medium- to small-sized fishes; body clupeiform (herring-like, but with an adipose fin, which is absent in real
herrings and relatives); mouth small (except *Agoniates*), superior; teeth tri- to multicuspid on premaxilla; teeth
tri- to multicuspid and/or conical on dentary and maxilla .... **Triportheidae** (Elongate hatchetfishes and allies, part, p. 196)

8a. Body very elongate, pike-like; snout long; origin of dorsal fin much nearer to caudal-fin base than to tip of snout **9**
8b. Body not pike-like; snout short; origin of dorsal fin more or less on middle of body **10**

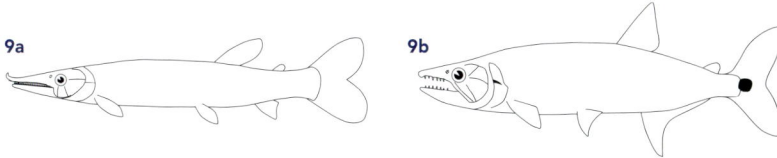

9a. Lower jaw shorter than upper jaw; all teeth identical and barely visible ....**Ctenoluciidae** (Pike characins, p. 148)
9b. Jaws of similar size; conspicuous caniniform teeth of different sizes .... **Acestrorhynchidae** (Freshwater barracudas, p. 81)

10a. Anal fin medium to long with ≥3 unbranched rays and >11 branched rays **11**
10b. Anal fin short with 2 or 3 unbranched rays followed by <11 branched rays **15**

11a. Serrae present (scales forming saw-like margin) on belly, at least in
postpelvic region; predorsal spine present (may be pronounced, or
relatively small and hidden just beneath skin; absent in *Colossoma*,
Piaractus, and *Mylossoma*) ....**Serrasalmidae** (Piranhas and Pacus, p. 172)
11b. Serrae absent on belly; predorsal spine usually missing **12**

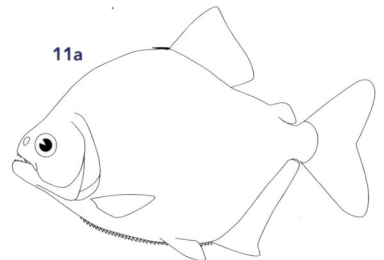

12a. Relatively large size (15–100 cm SL), with trout-shaped body;
teeth of lower jaw in 2 rows, inner row composed of pair of
small conical teeth; premaxillary teeth in upper jaw usually
in 3 rows, with larger teeth in inner row ....**Bryconidae** (Dorados, p. 90)
12b. Relatively small size (usually <15 cm SL); teeth of lower jaw in
one row; premaxillary teeth very rarely in three rows **13**

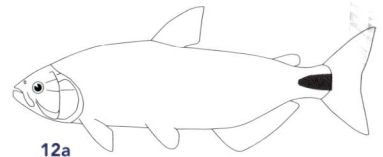

13a. Spiniform projection present on posterior margin of preopercle
(absent in miniature *Priocharax*) ....**Heterocharacinae** (p. 110)
13b. Spiniform projection absent on posterior margin of preopercle **14**

14a

14b

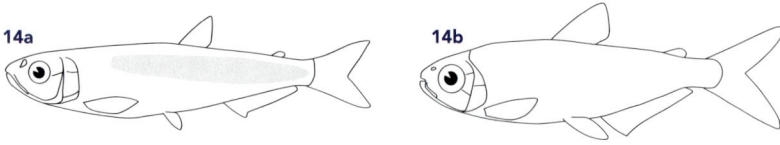

14a. Silvery lateral band present on body ....**Triportheidae** (*Engraulisoma taeniatum*, p. 197)
14b. Silvery lateral band absent on body.. **Iguanodectidae** (Iguanodectid Characiforms, p. 163)

15a. Pectoral fins well developed with 3–5 unbranched rays (except *Crenuchus*, recognized by its oval black
    spot on lower part of caudal peduncle); usually small (<10 cm) bottom-dwelling fishes **16**
15b. Pectoral fins less developed with single unbranched ray; species of medium size **17**

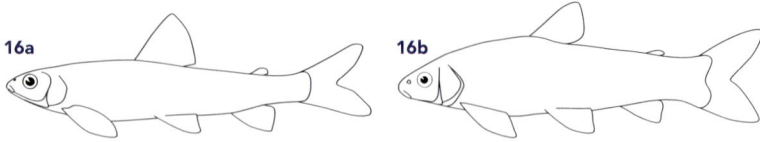

16a

16b

16a. Mouth terminal with numerous conic or tricuspid teeth on both jaws; upper jaw with
    teeth organized in 1 row, teeth in lower jaw usually form 2 rows......................**Crenuchidae** (South American darters, p. 142)
16b. Mouth inferior with 4 (rarely 2) teeth on each premaxilla, with horizontal multicuspid cutting
    border; dentary and maxillary teeth occur in some species, absent in others ....**Parodontidae** (Scrapetooths, p. 169)

17a. Adipose eyelid present ....**Hemiodontidae** (Halftooths, p. 161)
17b. Adipose eyelid absent..**18**

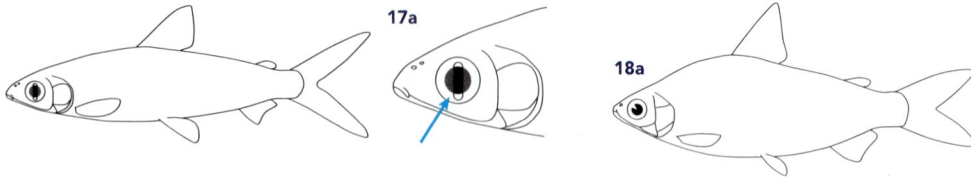

17a

18a

18a. Oral teeth absent ....**Curimatidae** (Toothless characins, p. 148)
18b. Oral teeth present...**19**

19a. Teeth in one row, with 3 or 4 large, firmly fixed teeth on each side of premaxillary
    and dentary ....**Anostomidae** (Toothed headstanders, p. 82)
19b. Teeth small and weakly implanted (movable) on lips...**20**

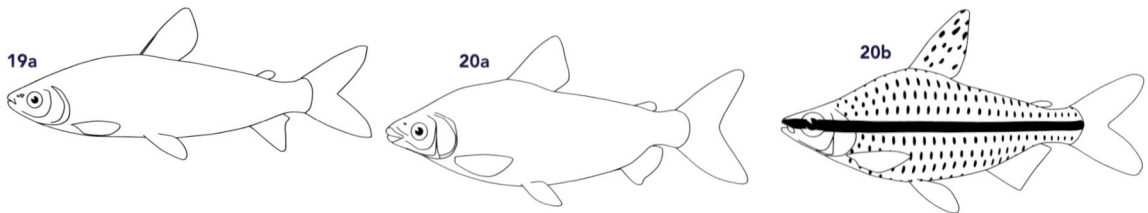

19a

20a

20b

20a. Mouth protrusible, forming suction disk; scales without black spots; bifid spine in front of dorsal
    fin present; numerous, very small teeth set on lips forming comb ....**Prochilodontidae** (Flannel mouth characins, p. 170)
20b. Mouth terminal and not protrusible; black spot on each scale; spine in front of dorsal fin
    absent; teeth less numerous and poorly visible ....**Chilodontidae** (Headstanders, p. 141)

PHOTOGRAPHIC GUIDE TO
FISH FAMILIES

The following plates show a limited selection of fish species and are not a complete overview of the fish genera in the AOG. Fishes on plates are not shown to scale. Families not included: Carcharhinidae (Requiem sharks), Pristidae (Sawfishes), and Hemirhamphidae (Halfbeaks). See also the "Illustrated Key to Fish Families" on pp. 26–34.

POTAMOTRYGONIDAE (RIVER STING RAYS) (p. 70)

Potamotrygon orbignyi
(size not recorded; max. size 38 cm disk width)

LEPIDOSIRENIFORMES

LEPIDOSIRENIDAE (SOUTH AMERICAN LUNGFISH) (p. 72)

Lepidosiren paradoxa
(100 cm TL)

ANGUILLIFORMES

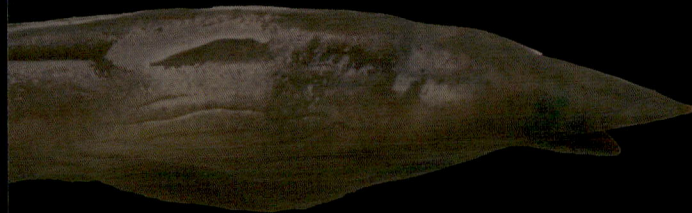

OPHICHTHIDAE (SNAKE EELS) (p. 73)

Stictorhinus potamius
(35 cm TL; preserved specimen)

ARAPAIMIDAE (BONYTONGUES) (p. 73)

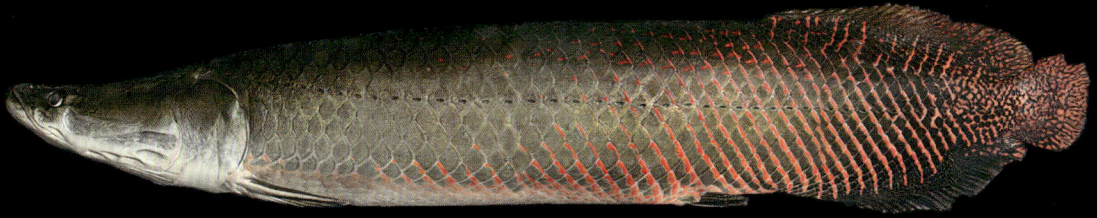

Arapaima sp.
(174 cm SL)

OSTEOGLOSSIDAE (AROWANAS) (p. 74)

Osteoglossum bicirrhosum
(75 cm SL)

CLUPEIFORMES

CLUPEIDAE (HERRINGS) (p. 74)

Rhinosardinia amazonica
(3.4 cm SL; preserved specimen)

ENGRAULIDAE (ANCHOVIES) (p. 75)

Anchovia surinamensis
(size not recorded; max. size 15 cm SL)

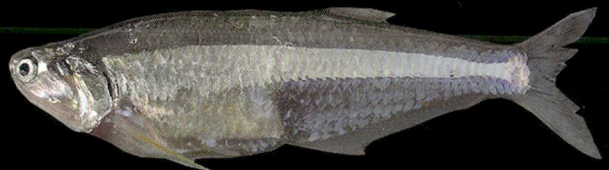

Pterengraulis atherinoides
(size not recorded; max. size 30 cm SL)

PRISTIGASTERIDAE (LONGFIN HERRINGS) (p. 79)

Pellona flavipinnis
(40 cm SL)

Pristigaster cayana
(15 cm SL)

CHARACIFORMES

ACESTRORHYNCHIDAE (FRESHWATER BARRACUDAS) (p. 81)

Acestrorhynchus falcatus
(15 cm SL)

ANOSTOMIDAE (TOOTHED HEADSTANDERS) (p. 82)

Laemolyta taeniata
(size not recorded;
max. size 29 cm SL)

ANOSTOMIDAE (TOOTHED HEADSTANDERS) (p. 82)

Leporellus vittatus
(size not recorded;
max. size 30 cm SL)

Leporinus fasciatus
(size not recorded;
max. size 37 cm SL)

Pseudanos irinae
(size not recorded;
max. size 20 cm SL)

BRYCONIDAE (DORADOS OR JAW CHARACINS) (p. 90)

Brycon amazonicus
(size not recorded;
max. size 46 cm SL)

Salminus sp.
(14 cm SL)

CHALCEIDAE (TUCANFISHES) (p. 91)

Chalceus epakros
(30 cm SL)

CHARACIDAE (TETRAS AND RELATIVES) (p. 92)

Astyanax bimaculatus (11 cm SL)

Aphyocharax sp. (3.0 cm SL)

Brachychalcinus copei (5.8 cm SL)

Paragoniates alburnus (7.7 cm SL)

Creagrutus ungulus (7.8 cm SL)

Brittanichthys myersi (2.2 cm SL)

CHARACIDAE (TETRAS AND RELATIVES) (p. 92)

Galeocharax gulo (14 cm SL)

Moenkhausia oligolepis (4.3 cm SL)

Paracheirodon axelrodi (3.0 cm SL)

Tetragonopterus argenteus (7.8 cm SL)

Roeboides myersi (8.5 cm SL)

Hyphessobrycon jackrobertsi (3.4 cm SL)

CHILODONTIDAE (HEADSTANDERS) (p. 141)

Chilodus punctatus (size not recorded; max. size 8.0 cm SL)

Caenotropus maculosus (7.0 cm SL)

CRENUCHIDAE (SOUTH AMERICAN DARTERS) (p. 142)

Poecilocharax weitzmani (4.0 cm SL)

Characidium zebra (7.0 cm SL)

Leptocharacidium omospilus (4.4 cm SL)

CTENOLUCIIDAE (PIKE-CHARACINS) (p. 148)

Boulengerella cuvieri (50 cm SL)

Boulengerella lateristriga (25 cm SL)

CURIMATIDAE (TOOTHLESS CHARACINS) (p. 148)

Curimatella meyeri
(12 cm SL)

Psectrogaster rutiloides (14 cm SL)

Steindachnerina hypostoma
(10 cm SL)

CYNODONTIDAE (DOGTOOTH CHARACIFORMS) (p. 154)

Cynodon meionactis
(19 cm SL)

Hydrolycus armatus (40 cm SL)

ERYTHRINIDAE (WOLF-FISHES AND YARROWS) (p. 156)

(10 cm SL; preserved specimen)

Hoplias malabaricus (70 cm SL)

GASTEROPELECIDAE (FRESHWATER HATCHETFISHES) (p. 158)

Thoracocharax stellatus (3.0 cm SL)

Carnegiella strigata (3.0 cm SL)

HEMIODONTIDAE (HALFTOOTHS) (p. 161)

Bivibranchia bimaculata (11 cm SL)

Anodus elongatus (size not recorded; max. size 30 cm SL)

HEMIODONTIDAE (HALFTOOTHS) (p. 161)

Hemiodus gracilis (size not recorded;
max. size 16 cm SL)

IGUANODECTIDAE (IGUANODECTID CHARACIFORMS) (p. 163)

Bryconops melanurus (7.6 cm SL)

Piabucus dentatus (12 cm SL)

LEBIASINIDAE (PENCILFISHES) (p. 165)

Copella compta (5.0 cm SL)

Nannostomus nigrotaeniatus (3.3 cm SL)

Copeina guttata
(4.7 cm SL; preserved specimen)

PARODONTIDAE (SCRAPETOOTHS) (p. 169)

Parodon guyanensis (7.0 cm SL)

Apareiodon orinocensis (size not recorded; max. size 13 cm SL)

PROCHILODONTIDAE (FLANNEL MOUTH CHARACIFORMS) (p. 170)

Semaprochilodus kneri (26 cm SL)

Prochilodus nigricans (19 cm SL)

SERRASALMIDAE (PIRANHAS AND PACUS) (p. 172)

Pristobrycon maculipinnis (size not recorded; max. size 25 cm SL)

SERRASALMIDAE (PIRANHAS AND PACUS) (p. 172)

Serrasalmus irritans
(size not recorded;
max. size 14 cm SL)

Pygocentrus cariba
(size not recorded;
max. size 30 cm SL)

Serrasalmus rhombeus
(size not recorded;
max. size 41 cm SL)

Metynnis hypsauchen
(size not recorded;
max. size 15 cm SL)

SERRASALMIDAE (PIRANHAS AND PACUS) (p. 172)

Colossoma macropomum (size not recorded; max. size 100 cm SL)

Myloplus asterias (size not recorded; max. size 25 cm SL)

TRIPORTHEIDAE (ELONGATE HATCHETFISHES AND RELATIVES) (p. 196)

Engraulisoma taeniatum (4.6 cm SL; preserved specimen)

Clupeacharax anchoveoides (6.7 cm SL; preserved specimen)

Triportheus albus (11 cm SL)

Cathorops agassizii (25 cm SL)

Sciades parkeri (45 cm SL)

Bagre marinus (60 cm SL)

ASPREDINIDAE (BANJO CATFISHES) (p. 202)

Bunocephalus knerii (dorsal and lateral views; 13 cm SL)

Aspredo aspredo (dorsal and lateral views; 13 cm SL)

Xyliphius melanopterus (9.4 cm SL)

ASTROBLEPIDAE (ANDEAN HILLSTREAM OR CLIMBING CATFISHES) (p. 207)

Astroblepus sp. (5.3 cm SL)

Astroblepus sp. (6.0 cm SL)

AUCHENIPTERIDAE (DRIFTWOOD CATFISHES) (p. 208)

Ageneiosus ucaylensis (size not recorded; max. size 35 cm SL)

Auchenipterus nuchalis (size not recorded; max. size 27 cm SL)

Tatia intermedia (size not recorded; max. size 12 cm SL)

Trachelyopterus coriaceus (size not recorded; max. size 18 cm SL)

CALLICHTHYIDAE (CALLICHTHYID ARMORED CATFISHES) (p. 216)

Dianema longibarbis (7.3 cm SL)

Corydoras eques (5.0 cm SL)

Callichthys callichthys (6.8 cm SL; preserved specimen)

CETOPSIDAE (WHALE CATFISHES) (p. 220)

Cetopsis oliveirai (3.3 cm SL)

Cetopsis sp. (6.0 cm SL)

Helogenes marmoratus (8.0 cm SL)

Cetopsis coecutiens (6.3 cm SL; preserved specimen)

Acanthodoras cataphractus (size not recorded; max. size 12 cm SL)

Anadoras weddellii (size not recorded; max. size 15 cm SL)

Pterodoras granulosus (size not recorded; max. size 70 cm SL)

Rhynchodoras boehlkei (15 cm SL)

Tenellus ternetzi (13 cm SL)

Megalodoras uranoscopus (50 cm SL)

HEPTAPTERIDAE (THREE-BARBELED CATFISHES) (p. 233)

Leptorhamdia sp. (5.0 cm SL)

Phenacorhamdia anisura (4.6 cm SL)

Pimelodella geryi (11 cm SL)

Rhamdia quelen (14 cm SL; preserved specimen)

Cetopsorhamdia insidiosa (6.0 cm SL)

LORICARIIDAE (SUCKERMOUTH ARMORED CATFISHES) (p. 253)

Hypoptopoma incognitum (7.0 cm SL)

Otocinclus vittatus (2.6 cm SL)

LORICARIIDAE (SUCKERMOUTH ARMORED CATFISHES) (p. 253)

Planiloricaria cryptodon (13 cm SL)

Lamontichthys filamentosus (15 cm SL)

Farlowella nattereri (13 cm SL)

Mouth of *Crossoloricaria bahuaja*

Mouth of an unidentified loricariid

Crossoloricaria bahuaja (15 cm SL)

Ancistrus sp. (14 cm SL)

LORICARIIDAE (SUCKERMOUTH ARMORED CATFISHES) (p. 253)

Leporacanthicus triactis (size not recorded; max. size 25 cm SL)

Panaque schaeferi (60 cm SL)

Pseudolithoxus dumas (dorsal and lateral views; 9.0 cm SL)

Hypostomus emarginatus (21 cm SL)

Pseudorinelepis genibarbis (size not recorded; max. size 35 cm SL)

Brachyplatystoma juruense (size not recorded; max. size 60 cm SL)

Hemisorubim platyrhynchos (size not recorded; max. size 53 cm SL)

Pimelodus blochii (12 cm SL)

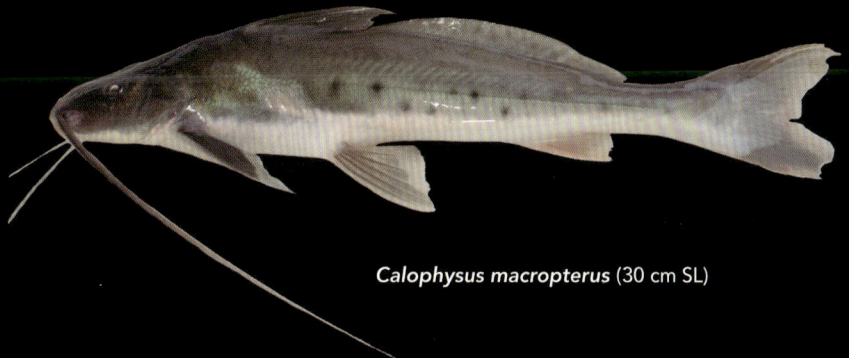

Calophysus macropterus (30 cm SL)

PSEUDOPIMELODIDAE (BUMBLEBEE CATFISHES, DWARF-MARBLED CATFISHES) (p. 308)

Pseudopimelodus bufonius (9.0 cm SL)

Cephalosilurus albomarginatus (8.0 cm SL)

Batrochoglanis vilosus (size not recorded; max. size 15 cm SL)

SCOLOPLACIDAE (SPINY DWARF CATFISHES) (p. 310)

Scoloplax distolothrix (1.0 cm SL; preserved specimen)

TRICHOMYCTERIDAE (PENCIL CATFISHES, TORRENT CATFISHES, AND PARASITIC CATFISHES (CANDIRÚS)) (p. 311)

Pseudostegophilus nemurus (12 cm SL)

TRICHOMYCTERIDAE (PENCIL CATFISHES, TORRENT CATFISHES, AND PARASITIC CATFISHES (CANDIRÚS)) (p. 311)

Henonemus punctatus (7.0 cm SL)

Vandellia sanguinea (6.0 cm SL)

Plectrochilus cf. diabolicus (6.3 cm SL)

Trichomycterus quechuorus (4.8 cm SL)

PHREATOBIUS INCERTAE SEDIS (p. 322)

Phreatobius dracunculus (size not recorded;
max. size 4.0 cm SL)

APTERONOTIDAE (GHOST KNIFEFISHES) (p. 322)

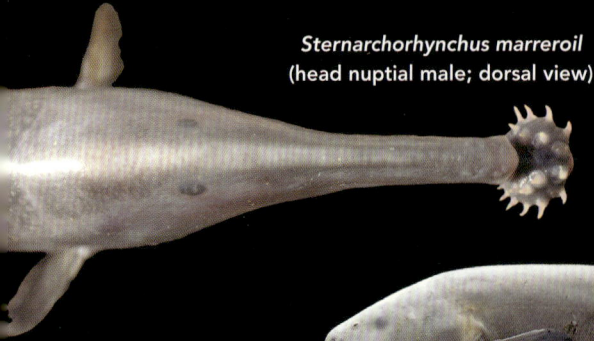

Sternarchorhynchus marreroi
(head nuptial male; dorsal view)

Apteronotus albifrons (20 cm TL;
preserved specimen)

Platyurosternarchus macrostomus
(40 cm TL)

Adontosternarchus balaenops (20 cm TL)

Sternarchorhynchus sp.
(30 cm TL)

Magosternarchus raptor (size not recorded;
max. size 20 cm TL)

Sternarchorhamphus muelleri
(50 cm TL)

GYMNOTIDAE (ELECTRIC EEL AND BANDED KNIFEFISHES) (p. 330)

Electrophorus electricus (size not recorded; max. size 250 cm TL)

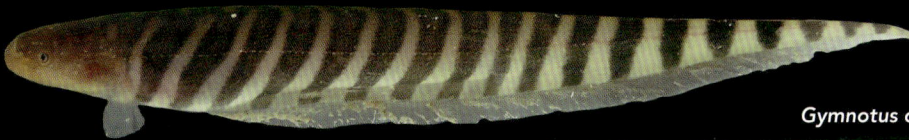

Gymnotus chaviro (8.0 cm TL)

Gymnotus carapo (size not recorded; max. size 60 cm TL)

HYPOPOMIDAE (GRASS KNIFEFISHES) (p. 334)

Brachyhypopomus beebei (5.6 cm TL)

Akawaio penak (22 cm TL; preserved specimen)

RHAMPHICHTHYIDAE (PAINTED KNIFEFISHES, SAND KNIFEFISHES, AND TRUMPET KNIFEFISHES) (p. 337)

Gymnorhamphichthys hypostomus (16 cm TL)

Rhamphichthys heleios (37 cm TL; preserved specimen)

Steatogenys elegans (13 cm TL)

STERNOPYGIDAE (GLASS KNIFEFISHES, RATTAIL KNIFEFISHES) (p. 341)

Eigenmannia virescens (13 cm TL)

Sternopygus macrurus (13 cm TL)

STERNOPYGIDAE (GLASS KNIFEFISHES, RATTAIL KNIFEFISHES) (p. 341)

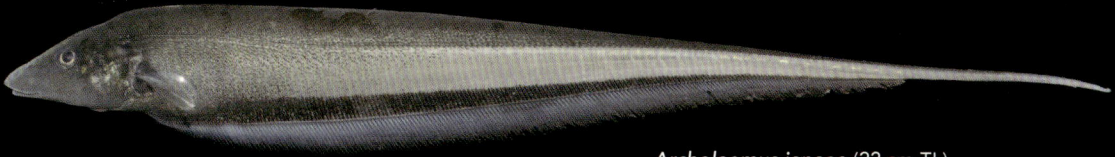

Archolaemus janeae (22 cm TL)

CYPRINODONTIFORMES

ANABLEPIDAE (FOUR-EYED FISHES) (p. 345)

Anableps anableps (14 cm SL; preserved specimen)

CYPRINODONTIDAE (PUPFISHES) (p. 346)

Orestias agassizii (8.5 cm SL)

Orestias agassizii (6.0 cm SL)

POECILIIDAE (LIVEBEARERS) (p. 346)

Fluviphylax sp. (size not recorded; max. size 2.0 cm SL)

Tomeurus gracilis (3.5 cm SL; preserved specimen)

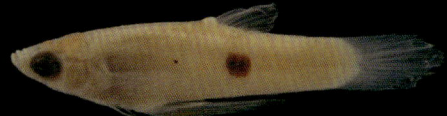

Phalloceros leticiae (1.6 cm SL; preserved specimen)

RIVULIDAE (RIVULINE KILLIFISHES) (p. 350)

Laimosemion geayi (size not recorded; max. size 5.0 cm SL)

Pituna xinguensis (size not recorded; max. size 3.0 cm SL)

Melanorivulus zygonectes (size not recorded; max. size 4.0 cm SL)

Trigonectes rubromarginatus (size not recorded; max. size 12 cm SL)

PERCIFORMES

CICHLIDAE (CICHLIDS) (p. 359)

Acarichthys heckelii (size not recorded; max. size 14 cm SL)

Apistogramma steindachneri (5.0 cm SL)

CICHLIDAE (CICHLIDS) (p. 359)

Crenicichla lugubris (size not recorded; max. size 24 cm SL)

Aequidens michaeli (size not recorded; max. size 20 cm SL)

Crenicichla percna (size not recorded; max. size 22 cm SL)

Cichla melaniae (29 cm SL)

Geophagus argyrostictus (size not recorded; max. size 18 cm SL)

POLYCENTRIDAE (NEW WORLD LEAF-FISHES) (p. 385)

Polycentrus schomburgkii (2.2 cm SL)

Monocirrhus polyacanthus (5.7 cm SL)

SCIAENIDAE (DRUMS OR CROAKERS) (p. 386)

Plagioscion squamosissimus (21 cm SL)

Pachyurus cf. stewarti (7.0 cm SL)

GOBIIFORMES

ELEOTRIDAE (SLEEPERS) (p. 388)

Microphilypnus ternetzi (2.0 cm SL)

Eleotris amblyopsis (5.0 cm SL)

GOBIIDAE (GOBIES) (p. 392)

Awaous flavus (size not recorded; max. size 8.2 cm SL)

Gobioides broussonnetii (size not recorded; max. size 55.3 cm SL)

BATRACHOIDIFORMES

BATRACHOIDIDAE (TOADFISHES) (p. 396)

Thalassophryne amazonica (9.0 cm SL)

BELONIFORMES

BELONIDAE (NEEDLEFISHES) (p. 397)

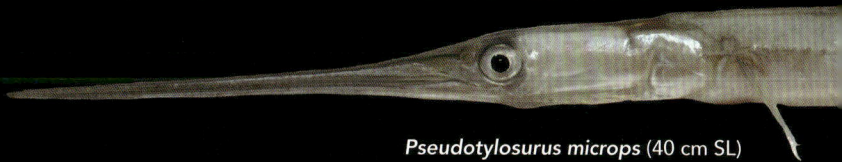

Pseudotylosurus microps (40 cm SL)

Potamorrhaphis guianensis (23 cm SL)

SYNBRANCHIFORMES

SYNBRANCHIDAE (SWAMP EELS) (p. 399)

Synbranchus marmoratus (50 cm TL)

Synbranchus marmoratus (80 cm TL)

SYNGNATHIFORMES

SYNGNATHIDAE (PIPEFISHES) (p. 399)

Microphis brachyurus (14 cm SL)

PLEURONECTIFORMES

ACHIRIDAE (AMERICAN SOLES) (p. 400)

Apionichthys finis (6.6 cm SL)

Hypoclinemus mentalis (18 cm SL)

TETRAODONTIFORMES

TETRAODONTIDAE (PUFFERFISHES) (p. 402)

Colomesus asellus (6.0 cm SL)

FAMILY CARCHARHINIDAE—REQUIEM SHARKS

— *PETER VAN DER SLEEN and JAMES S. ALBERT*

Family includes 59 species in 12 genera worldwide, including several species that enter freshwater. In the Amazon, one species, *Carcharhinus leucas* (the Bull shark), is occasionally encountered.

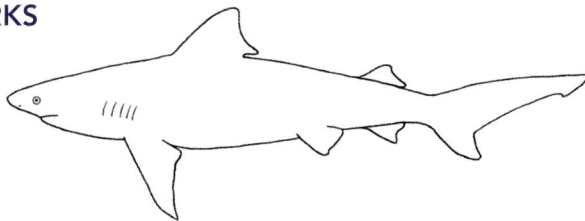

Carcharhinus (420 cm SL)

Distinguished from other sharks by: gray coloration; heavy body; short, rounded snout; large pectoral fins; small round eyes; 5 gill openings, the fifth behind origin of pectoral fin; mouth large and subterminal; upper teeth triangular and serrated; caudal peduncle with a precaudal pit; dorsal fin with lateral undulations along posterior margin. **SPECIES** 35, with only 1 species, *C. leucas* (the Bull shark), in the AOG region. **COMMON NAMES** *Cação fidalgo, Sicuri branco* (Brazil); *Tiburón gris* (Colombia); *Cazón, Tiburón ñato* (Peru). **DISTRIBUTION AND HABITAT** Recorded in Iquitos (Peru), Leticia (Colombia), Manaus, Santarém, and Belém (Brazil); see Thorson (1972) and Soto and Castro-Neto (1998). **BIOLOGY** Bull sharks normally feed on fishes, but are known to attack humans. They are viviparous, producing 1–13 young per litter, generally in estuaries. Length up to 220 cm in the Amazon, but can reach more than 4 m in the ocean.

FAMILY PRISTIDAE—SAWFISHES OR CARPENTER SHARKS

— *PETER VAN DER SLEEN and JAMES S. ALBERT*

Family includes 7 species in 2 genera worldwide: *Pristis* (6 species) and *Anoxypristis* (1 species). In AOG, species of the genus *Pristis* are occasionally encountered.

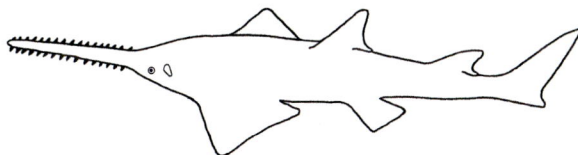

Pristis (250–750 cm TL)

Distinguished by an elongate, blade-like snout, with a single row of large teeth on each side; a large and shark-like body, with anterior margin of pectoral fins attached to head; two large dorsal fins of equal size, widely separated; and the nostrils well anterior, not connected to mouth. **SPECIES** Six, including three in the AOG region: *P. pristis, P. perottetz,* and *P. pectinata.* **COMMON NAMES** *Araguaguá, Peixe serra* (Brazil); *Pez sierra, Pez rastrillo* (Spanish). **DISTRIBUTION AND HABITAT** Mostly tropical and subtropical coastal waters; some enter estuaries and the lower parts of large rivers. *Pristis pristis* has been caught 750 km up the Amazon River; *P. perotteti* in the Amazon River up to Santarém (Thorson 1974, Lovejoy et al. 2006). **BIOLOGY** Feed on fishes and bottom-living animals and are harmless to humans. The saw-like rostrum is covered with sensitive pores that help to detect the movement of buried prey. The rostrum is also used to injure prey items as well as for defense. In the ocean, sawfishes can grow to larger sizes (6–8 m), but individuals caught in the Amazon are usually 1.5–2.0 m. Sawfishes are yolk-sac viviparous. All species are listed as critically endangered (IUCN 2013) as a result of overfishing and habitat degradation (Palmeira et al. 2013).

FAMILY POTAMOTRYGONIDAE—RIVER STINGRAYS

— *PETER VAN DER SLEEN and JAMES S. ALBERT*

DIVERSITY 33 species in 4 genera, with 28 species in the AOG region. River stingrays originated from marine ancestors when shallow seas encroached into the center of the continent, forming large lakes and mega-wetland systems extending across much of the area of the western Amazon basin (Lovejoy et al. 2006). The marine ancestors of river stingrays adapted to freshwater conditions when the Andean uplift eventually caused this seaway to become landlocked and it transformed into the modern Amazon basin.

COMMON NAMES *Raia, Rodeirão* (Brazil); *Raya* (Peru).

GEOGRAPHIC DISTRIBUTION All South American rivers that drain into the Atlantic Ocean or Caribbean Sea. Many river systems in the Amazon have one or two endemic species of stingray and only a few species are widely distributed (e.g., *P. motoro* and *P. orbignyi*).

ADULT SIZES Range in disk diameter from 25 to 110 cm.

DIAGNOSIS OF FAMILY Body strongly flattened dorsoventrally and with a circular (disk-like) form; elongate tail with one or more strong spines. Many species have colorful dorsal patterns and are covered with many denticles, thorns, and tubercles. Family can be distinguished from marine and freshwater stingrays of other families by the well-developed anteromedian projection of their pelvic girdle, called the prepelvic process (Rosa et al. 1987). Other derived characters include blood with low concentrations of urea, and reduction of the rectal gland (Rosa et al. 1987, de Carvalho et al. 2003).

SEXUAL DIMORPHISM Males can be recognized by the presence of claspers on the posterior portion of the pelvic fins.

HABITATS Sandy bottoms in lakes, rivers, and streams, but also found on rocky bottoms and very common in shallow rapids of the Xingu, Trombetas, Tocantins, and Tapajós rivers (J. Zuanon pers. comm.). There might be sexual and ontogenetic segregation in habitat (Rosa et al. 2010). Juvenile rays often stay in shallow waters to avoid predators. Adult rays generally remain in deeper waters. Predominantly female and male rays are caught at specific locations along the reproduction cycle.

FEEDING ECOLOGY Diet includes a wide variety of items: worms, insect larvae, mollusks, crustaceans, and fishes. They find most of their food in the sediment and literally suck it up. Teeth are small rounded molars that form a flat surface designed to grip and crush but not to cut. Review of literature on diet and feeding in Rosa et al. (2010).

BEHAVIOR Potamotrygonids are all ovoviviparous (aplacentally viviparous), and the developing embryos are nourished by uterine milk secreted by trophonemata (Thorson et al. 1983). Number of young produced in each gestation varies among species, but is usually from two to seven (de Carvalho et al. 2003, Charvet-Almeida et al. 2005).

ADDITIONAL NOTES River stingrays are much feared because of their venomous caudal barb or sting. These stings are continuously worn and replaced and up to four stings may be present in one individual. Being stung by a ray can lead to infection and tissue necrosis if left untreated, and can cause weeks of agonizing pain; however, they pose little threat when not stepped on. A common advice of local people is to "shuffle your feet" while walking through shallow water.

KEY TO THE GENERA FROM ROSA ET AL. (2010)

1a. Distance from mouth to anterior of disk relatively long, 2.2–3.3 times in disk width; pelvic fins dorsally covered by disk (fig. 1a) 2

1b. Distance from mouth to anterior of disk relatively short, 3.6–5.6 times in disk width; pelvic fins exposed behind posterior margin of disk (fig. 1b)...3

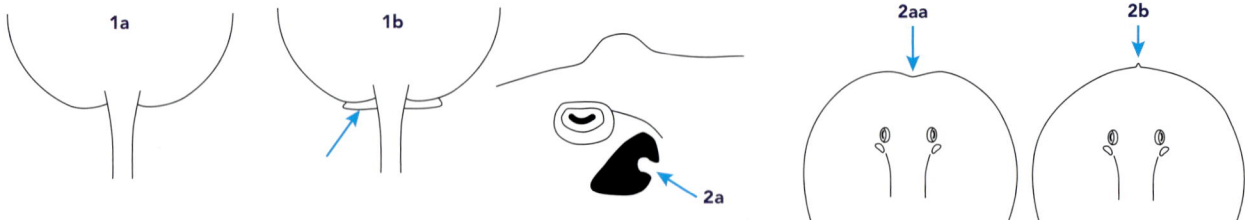

2a. A knob-shaped process on the external margin of the spiracles (fig. 2a); anterior margin of disk concave, lacking anteromedian prominence (fig. 2aa); caudal sting present ....*Paratrygon*

2b. Knob-shaped process on the external margin of the spiracles absent; anterior margin of disk rounded, with a small anteromedian prominence (fig. 2b); caudal sting absent or minute...*Heliotrygon*

3a. Tail relatively short, <2 times disk width; tail with dorsal and ventral fin folds; eyes relatively large and pedunculated, eye diameter usually <4.0 times in interorbital distance..*Potamotrygon*
3b. Tail relatively long and whip-like, >2 times disk width; tail with only a ventral fin fold; eyes minute, nonpedunculate, eye diameter ≥4.4 times in interorbital distance..*Plesiotrygon*

GENUS ACCOUNTS

Heliotrygon (13–80 cm disk width)

Eyes relatively small; dorsal fin fold absent; disk highly circular with a rounded anterior margin; extremely reduced to absent caudal sting; spiracle without a small protrusion. **SPECIES** Two species, *H. gomesi* (13 cm) and *H. rosai* (80 cm), both in the AOG region. Species descriptions in de Carvalho and Lovejoy (2011). **DISTRIBUTION AND HABITAT** Amazon River and lower reaches of its major tributaries. **BIOLOGY** Carnivorous bottom feeders. Females are ovoviviparous.

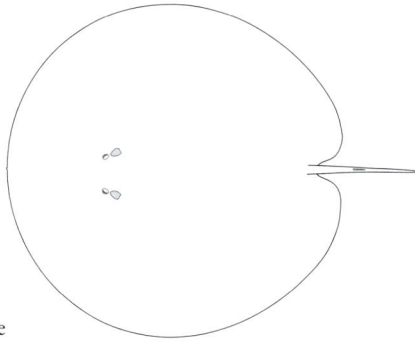

Paratrygon (110 cm disk width)

Eyes relatively small; dorsal fin fold absent; tail very long in juveniles but much reduced in adults; the anterior margin of disk concave; spiracle contains a small protrusion (knob-like structure). **SPECIES** One, *P. aiereba*. **DISTRIBUTION AND HABITAT** Throughout the Amazon (Ucayali, Solimões, Amazon, Negro, Branco, Madeira and its affluents in Bolivia, and Tocantins River) and Orinoco basins. Common in shallow areas and near banks. **BIOLOGY** Carnivorous bottom feeders, prey on insects, crustaceans, and small fishes. Females are ovoviviparous, and in the upper Orinoco bear only two offspring per gestation (Lasso et al. 1997).

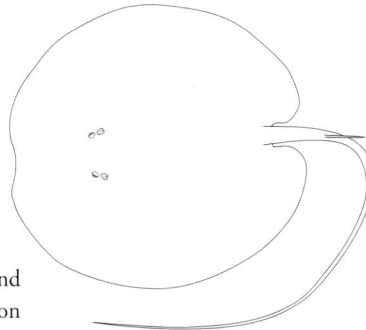

Plesiotrygon (25–58 cm disk width)

Eyes relatively small; dorsal fin fold absent; slender and whip-like tail (much longer than disk length); spiracle without a small protrusion; anterior margin of disk broadly pointed. **SPECIES** Two, *P. iwamae* (58 cm; Rosa et al. 1987) and *P. nana* (25 cm; de Carvalho and Ragno 2011), both in the AOG region. **DISTRIBUTION AND HABITAT** *P. iwamae* from the upper to lower Amazon basin, from Ecuador to Belém; *P. nana* is known from the río Pachitea, a tributary of río Ucayali in Peru, and possibly also found in the western portion of the Brazilian Amazon. **BIOLOGY** Carnivorous bottom feeders that prey on small fishes, insects, crustaceans, and nematodes. Females are ovoviviparous; gestation periods can last up to 8 months (Charvet-Almeida et al. 2005).

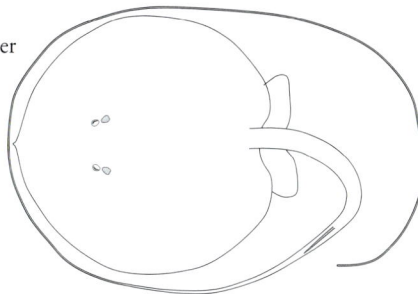

Potamotrygon (30–85 cm disk width)

Eyes large and clearly prominent; tail with both dorsal and ventral fin folds (other genera have only a ventral fin fold); disk usually slightly longer than wide, often with a pointed snout; tail relatively short, usually shorter than disk length. **SPECIES** 28, including 23 species in the AOG region. Photographic overview of species in Ross and Schäfer (2000) and Rosa et al. (2010). **COMMON NAMES** *Raia, Rodeirão* (Brazil); *Raya* (Peru). **DISTRIBUTION AND HABITAT** All South American countries except Chile, in the Atrato, Magdalena, Orinoco, Maracaibo, Amazon (including the Tocantins), Parnaíba, and Paraná-Paraguay basins, as well as coastal rivers in the Guianas. **BIOLOGY** Carnivorous bottom feeders, preying on a variety of animals, from aquatic insects and crustaceans to fishes (including catfishes). Females are ovoviviparous (Thorson et al. 1983).

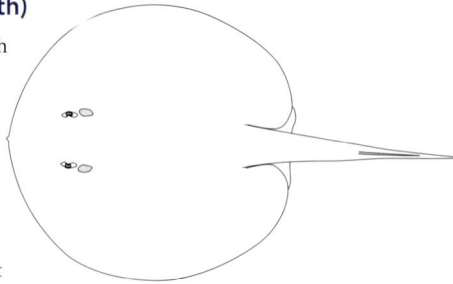

FAMILY LEPIDOSIRENIDAE—SOUTH AMERICAN LUNGFISH

— *PETER VAN DER SLEEN and JAMES S. ALBERT*

Family includes one genus, with a single species, *Lepidosiren paradoxa*.

Lepidosiren (120 cm TL)

Distinguished by: elongate body; thin and filamentous pectoral and pelvic fins, without rays; laterosensory canals present as pronounced grooves on surface of head and body; body color brown to dark gray, often with many darker spots. Juveniles with yellow spots. **SPECIES** One, *L. paradoxa*. **COMMON NAMES** *Pirambóia, Pirarucu bóia* (Brazil); *Anguille tété* (French Guiana); *Pez pulmonado, Paiche machaco* (Peru). **DISTRIBUTION AND HABITAT** Swamps and floodplain lakes in the lowlands of the Amazon, Paraguay, and lower Paraná basins in Argentina, Brazil, Bolivia, Ecuador, Paraguay, and Peru (Arratia, 2003). One record in the Rio Tomo, a headwater tributary of the Orinoco basin in Colombia (Bogota-Gregory and Maldonado-Ocampo 2006). **BIOLOGY** Swim bladder is developed into a highly vascularized lung and lungfishes are obligate air breathers (Johansen and Lenfant 1967, Bemis and Lauder 1986). Lungfishes can survive droughts by making a burrow to a depth of about 30–50 cm, which they seal off with mud and mucus, leaving only some small holes for aeration (Sanchez et al. 2005, Ferreira da Silva et al. 2008). They can survive for months in this chamber by strongly reducing their metabolism. Reproduction takes place at the beginning of the rainy season. Males make a burrow and guard the eggs. The pelvic fins of reproductive males become highly vascularized and feather-like (see illustration above), infusing oxygen into the water where young are developing (Planquette et al. 1996). Young lungfishes resemble amphibian larvae with four external gills that become resorbed after around seven weeks (Boujard et al. 1997). Juveniles are toxic to at least some predators and have bright yellow flecks on the skin, perhaps as a warning coloration to advertise these toxins (W. Crampton, pers. comm.). Juveniles feed on insect larvae and snails. Adults feed on bottom-living crustaceans, mollusks, and small fishes which they capture with a sudden and strong sucking action. They are mainly nocturnal.

FAMILY OPHICHTHIDAE—SNAKE EELS
— *PETER VAN DER SLEEN and JAMES S. ALBERT*

Family includes 314 species in 60 genera worldwide. Snake eels are found in tropical to temperate coastal waters in the Atlantic, Indian, and Pacific oceans. One freshwater species occurs in South America.

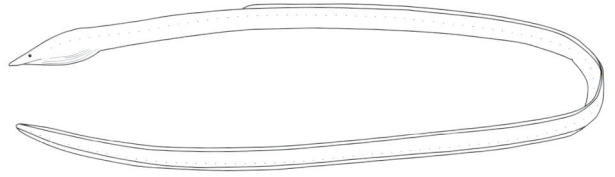

Stictorhinus (35 cm TL)

Distinguished by: dorsal and anal fins present; dorsal-fin origin far behind head, just before midpoint between snout tip and anus; pectoral fins absent; anus slightly in advance of midlength; gill openings straight, placed low, diagonal in lateral aspect, their upper ends posterior, their anterior ends separated by a narrow isthmus; gill membrane without a pouch anterolateral to the gill opening; anterior nostril without a tube, but a hole with lateral fleshy projections on it; posterior nostril wholly inside mouth, its position not marked externally by a groove or notch; a median groove present on undersurface of snout in which there are intermaxillary teeth; teeth pointed (Böhlke and McCosker 1975). **SPECIES** One, *S. potamius*, see description of species in Böhlke and McCosker (1975). **COMMON NAME** *Anguila serpiente* (Peru). **DISTRIBUTION AND HABITAT** Amazon (lower Tocantins) and lower Orinoco basin, and northeastern Brazil. Specimens have been collected in shallow pools during the dry season. **BIOLOGY** No data available. Some other species of Ophichthinae feed mainly on benthic invertebrates, and use their hard and pointed tails to burrow backward.

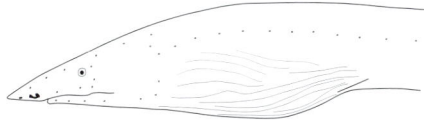

FAMILY ARAPAIMIDAE—BONYTONGUES
— *PETER VAN DER SLEEN and JAMES S. ALBERT*

Family includes three species in two genera: *Arapaima*, with two species in the Amazon and Essequibo basins, and *Heterotis niloticus* from Africa.

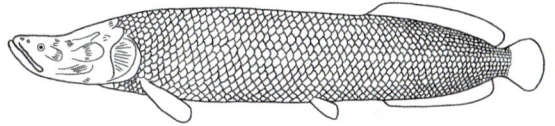

Arapaima (200–450 cm SL)

Among the largest freshwater fishes in the world, with reported lengths of up to 4 meters. Distinguished by: large, thick scales and heavily carved head bones. During the spawning season males can present a dark coloring on the upper part of the head and flanks, and the abdomen and caudal peduncle turn reddish. **SPECIES** At least two species in the Amazon and Essequibo basins: *A. gigas* and *A. leptosoma* (Stewart 2013a, b). **COMMON NAMES** *Pirarucu* (Brazil); *Paiche* (Peru). **DISTRIBUTION AND HABITAT** Amazon and Essequibo basins. Found in rivers and flooded forests, especially seasonally hypoxic oxbow lakes. **BIOLOGY** *Arapaima* is an obligatory air breather, with a modified and enlarged swim bladder composed of lung-like tissue. They routinely surface to engulf air, making a noisy snapping sound that fishermen use to locate individuals. *Arapaima* has been heavily overfished throughout most of its natural range, and most populations are now greatly reduced. Individuals over 2 meters are rare in the wild today, but recent management efforts in the Brazilian Amazon have succeeded in steadily recovering local populations, such as in the Mamirauá Reserve. Several populations have also been introduced into new areas, and *A. gigas* is currently expanding its native range into the upper Madeira basin of southeastern Peru and northwestern Bolivia. *Arapaima* feed mainly on other fishes, crustaceans, and even small mammals and birds if in the water. During the breeding season, *Arapaima* form pairs. Eggs are laid in a small excavated pit on the muddy bottom before the time of high water and hatch during the onset of the flooding season. Both males and females guard the eggs and males mouth-brood the fry after hatching.

FAMILY OSTEOGLOSSIDAE—AROWANAS

— *PETER VAN DER SLEEN and JAMES S. ALBERT*

Family includes two genera: *Osteoglossum*, with two species in South America, and *Scleropages*, with four species in Southeast Asia, Australia, and New Guinea.

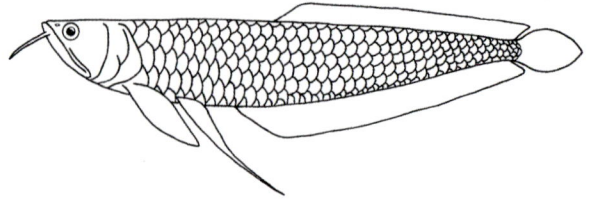

Osteoglossum (90 cm SL)

Distinguished by: elongate, slender bodies with large scales; a large and diagonally positioned mouth, with two barbels at the extremity of the lower jaw and the dorsal and anal fins almost fused with the caudal fin. **SPECIES** Two, *O. bicirrhosum* (Silver arowana) and *O. ferreirai* (Black arowana). **COMMON NAMES** *Arauanã* (Brazil); *Arahuana* (Peru). **DISTRIBUTION AND HABITAT** *O. bicirrhosum* throughout the Amazon basin and Guianas; *O. ferreirai* restricted to the Rio Negro basin. **BIOLOGY** Both species are specialized surface feeders and can jump out of the water to catch invertebrates and even small reptiles and birds from overhanging branches and leaves. They build nests and the male mouth-broods the fry after hatching. Arowanas are obligatory air-breathers and have a modified and enlarged swim bladder.

FAMILY CLUPEIDAE—HERRINGS

— *PETER VAN DER SLEEN and DEVIN BLOOM*

ILLUSTRATION AFTER LIDONNICI AND LASTRICO IN WHITEHEAD (1985)

Family includes 198 species in 54 genera worldwide. Most species are found in coastal marine waters, but at least 57 species known from freshwaters worldwide. Two species from one genus (*Rhinosardinia*) in the AOG region.

Rhinosardinia (5.0–8.0 cm SL)

Distinguished from all other clupeids by: a sharp, backward pointing spine on the upper part of the maxilla, at about eye center; strongly compressed body; sharply keeled belly; hind-border of gill opening evenly rounded; lower jaw articulation located under the eye; deep lower jaw; anal fin short, with about 15–18 fin rays; flanks with or without a distinct silver stripe; two distinct parallel striae on scales; fins without spines; no adipose fin; and no lateral line canal system on the body (Whitehead 1985). **SPECIES** Two, both in the AOG region; *R. amazonica* and *R. bahiensis*. Species information in Whitehead (1985). **COMMON NAMES** *Sardinha da água doce* (Brazil); *Arenque de agua dulce* (Peru); Amazon spinejaw sprat (English). **DISTRIBUTION AND HABITAT** Lower reaches of rivers draining into the Atlantic Ocean from the Orinoco River to northeastern Brazil. Found mostly in large rivers, often caught in seine nets along sandbars; also found in brackish water. **BIOLOGY** Feed on zooplankton.

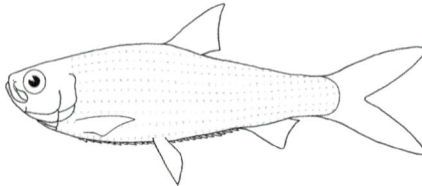

FAMILY ENGRAULIDAE—ANCHOVIES

— *PETER VAN DER SLEEN and DEVIN BLOOM*

ILLUSTRATION AFTER LIDONNICI AND LASTRICO IN WHITEHEAD ET AL. (1988)

DIVERSITY 146 species in 17 genera worldwide, with 14 species in 7 genera in the AOG region: *Amazonsprattus* (1 species), *Anchoa* (1 species), *Anchovia* (1 species), *Anchoviella* (7 species), *Jurengraulis* (1 species), *Lycengraulis* (3 species), and *Pterengraulis* (1 species). There are numerous undescribed freshwater species and the taxonomy of several genera is in need of revision (Bloom and Lovejoy 2012).

COMMON NAMES *Biqueirã, Maiacá, Manjuba, Pititinga* (Brazil); *Anchoa, Anchoveta* (Peru).

GEOGRAPHIC DISTRIBUTION Found along nearly every coastline and ocean except at very high latitudes. Most of the freshwater species occur in South America, with a few species found in freshwaters of southeastern Asia, Australia, and Papua New Guinea. In South America engraulids occur in the Amazon, Orinoco, Essequibo, and São Francisco rivers and their tributaries, as well as coastal rivers of the Guianas.

ADULT SIZES Most engraulids are between 10 and 20 cm SL, with a few species reaching 25–30 cm SL. Engraulids of the AOG region range in size from 2.0 cm SL in the Rio Negro pygmy anchovy *Amazonsprattus scintilla*, to 25 cm SL in the Wingfin anchovy *Pterengraulis atherinoides* from the Amazon River.

DIAGNOSIS: Easily recognized by a prominent snout, with a unique "rostral organ" (Nelson 1984), and a large, inferior mouth. In most species the maxilla reaches well beyond the eye and, in some species, to the posterior end of the head. The body is covered with moderately sized scales; lateral line is absent; fins without spines; dorsal fin is short based and usually situated over the middle of the body; pelvic fin located abdominally; New World species lack abdominal scutes; body often entirely silver, white or translucent with a silver lateral stripe (Kullander and Ferraris Jr. 2003).

SEXUAL DIMORPHISM Not pronounced.

HABITATS Most engraulids are marine and estuarine. A few species are diadromous (Mai and Vieira 2013, Bloom and Lovejoy 2014, Mai et al. 2014) while several species are entirely freshwater and inhabit large and medium-sized rivers.

FEEDING ECOLOGY Many engraulids have high numbers of long gill rakers used for suspension filter feeding on plankton, including appendicularians, cladocerans, fish eggs, copepods, and pteropods. Some engraulids filter-feed via crossflow filtration (Sanderson et al. 2001). A few marine species (e.g., *Cetengraulis*) are obligate suspension feeders, but most are facultative suspension and ram filter feeders (Sanderson and Wassersug 1990, 1993). Some of the larger species including *Anchoa spinifer*, *Lycengraulis* spp., and *Pterengraulis atherinoides* are piscivorous, with the latter also feeding on shrimp (Krumme et al. 2005).

BEHAVIOR Engraulids are primarily schooling fishes. Behavior of Amazonian species is poorly known.

KEY TO THE GENERA BASED ON WHITEHEAD ET AL. (1988)

1a. Adult size larger than 2 cm; pelvic fin with 6 rays **2**
1b. Adult size under 2 cm; pelvic fin with 5 rays .... *Amazonsprattus*

2a. Dorsal-fin origin in front of anal-fin origin or only slightly behind, near the midpoint of body; pectoral fins short, not reaching posteriorly beyond pelvic-fin base..**3**
2b. Origin of dorsal fin well behind anal-fin origin and behind midpoint of the body; pectoral fins long, reaching posteriorly beyond pelvic-fin base .... *Pterengraulis*

3a. Teeth in lower jaw small and evenly spaced or absent **4**
3b. Teeth in lower jaw enlarged, canine-like.. *Lycengraulis*

4a. Lower gill rakers on first arch <45 ....**5**
4b. Lower gill rakers on first arch ≥45 .... *Anchovia*

5a. A few short gill rakers present on hind face of third arch (fig. 5a).................... **6**
5b. No gill rakers on hind face of third arch (very long head and long pointed snout)... *Jurengraulis*

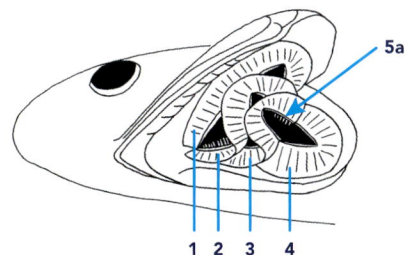

1 2 3 4

6a. Maxilla short, tip blunt, not reaching or just reaching
anterior margin of preopercle (fig. 6a) .... *Anchoviella*
6b. Maxilla long, tip pointed, reaching onto or beyond
preopercle (fig. 6b) .... *Anchoa*

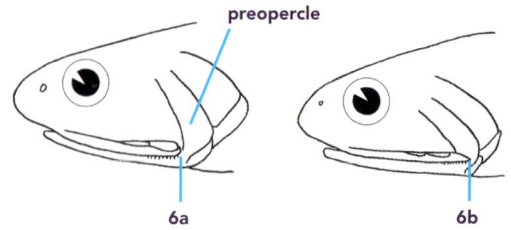

preopercle

6a   6b

GENUS ACCOUNTS

Amazonsprattus (2.0 cm SL)

Characterized by: small adult body size, easily
mistaken for a juvenile clupeoid, but with 5
pelvic-fin rays (juveniles of other anchovies
usually with 6 pelvic-fin rays); color semitransparent to transparent; body without scales;
mouth small, premaxillae absent or minute and toothless; maxilla very short, just reaching to
front border of eye, with two supramaxillae; articulation of lower jaw under hind border of
pupil or just behind; gill rakers elongate, total number 18 or 19; dorsal-fin origin well behind
midpoint of body; anal fin short, with 12–14 fin rays, its origin under first quarter of dorsal-fin
base (Whitehead et al. 1988). **SPECIES** One, *A. scintilla* (Rio Negro pygmy anchovy), species
information in Roberts (1984) and Whitehead et al. (1988). Several undescribed species (D.
Bloom unpublished data). **DISTRIBUTION AND HABITAT** Widespread throughout the Amazon and
Orinoco basins, Essequibo, and rivers of Guianas. **BIOLOGY** Consume dipteran larvae and small
crustaceans, often collected in seines along sandbars.

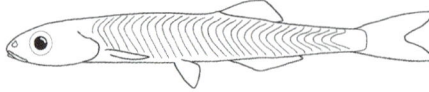

Anchoa (19 cm SL)

Characterized by: compressed body and
pointed snout; most have a long maxilla with
pointed tip (vs. short and blunt in *Anchoviella*),
reaching to or a little beyond edge of gill cover;
fine teeth on lower jaw; gill rakers slender,
lower gill rakers 12–19 (anterior gill rakers mere stumps); gill rakers present on hind face of
third arch; hind border of gill cover with a small triangular projection (on subopercle); pectoral
fins long, reaching beyond pelvic-fin base; anal fin long, with 31–37 fin rays, its origin below
midpoint of dorsal-fin base; smaller individuals with a silver stripe along the flank (Whitehead
et al. 1988). **SPECIES** 35, including one species *A. spinifer* (Spicule anchovy) in freshwaters in the
AOG region (Bloom and Lovejoy 2012, 2014). Further species information in Whitehead et
al. (1988). **DISTRIBUTION AND HABITAT** Most species marine and estuarine along the Atlantic and
Pacific coasts. A few species enter lower reaches of rivers, and one species (*A. spinifer*) inhabits
the lower part of rivers draining into the sea in northern Brazil, the Guianas, and Venezuela.
BIOLOGY *A. spinifer* forms large schools and feeds on small fishes and crustaceans (Whitehead et
al. 1988). Tolerates lower salinity levels (Bloom and Lovejoy 2014).

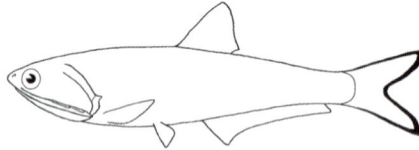

Anchovia (13 cm SL)

Characterized by: laterally compressed and
deep body, with a pointed snout (tip well
above center of eye); a silver stripe along
flank; maxilla short, tip blunt, failing to
reach articulation of lower jaw; subopercle
with angular hind margin (but not forming
a triangular projection as in *Anchoa spinifer*); very fine denticulations on lower jaw; fine and
slender lower gill rakers, 51–62; no gill rakers on hind face of third arch (as in *Jurengraulis*);

dorsal-fin origin at about midpoint of body; anal fin moderate, with 20–25 fin rays, its origin below about middle of dorsal-fin base (Whitehead et al. 1988). **SPECIES** Three, with only *A. surinamensis* (Surinam anchovy) in freshwaters. Species information in Whitehead et al. (1988). **COMMON NAMES** *Arenque-da-noite* (Brazil); *Anchoa, Anchoveta* (Peru). **DISTRIBUTION AND HABITAT** Marine species along western Atlantic and eastern Pacific coastlines; *A. surinamensis* from lower parts of rivers in the Amazon basin (up to Manaus), lower part of the Orinoco and rivers of the Guiana shield. **BIOLOGY** Poorly known. Numerous and long gill rakers suggest *Anchovia* filter-feeds on plankton. Caught in large seines and gill nets.

Anchoviella (4.0–12 cm SL)

Characterized by: body slightly compressed laterally; numerous elongate gill rakers (usually >15 on the lower branch of the first gill arch); gill rakers also present on hind face of third epibranchial; maxilla short (not or only just reaching to front margin of preopercle), tip blunt (vs. long and pointed in *Anchoa*), barely extending beyond tip of second supramaxilla; fine teeth on lower jaw; dorsal-fin origin at about midpoint of body; anal fin short to long with 14–29 rays, its origin usually below dorsal-fin base or behind base of last dorsal-fin ray (Whitehead et al. 1988). **SPECIES** 15, including 8 species in the AOG region (Bloom and Lovejoy 2012, Loeb 2012). Species information in Whitehead et al. (1988), except for *A. juruasanga* (Loeb 2012) and *A. vaillanti* (Loeb and Figueiredo 2014). Key to the species in the Amazon basin in Loeb (2012). **DISTRIBUTION AND HABITAT** Atlantic and Pacific coasts. **BIOLOGY** Some species filter-feed on plankton; those with few gill rakers feed on larger invertebrates and small fishes.

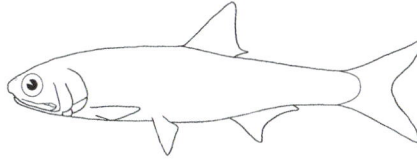

Jurengraulis (16 cm SL)

Characterized by: body long and slender, round in cross section; a very long head; pointed and long snout (rostral organ); maxilla short, tip bluntly rounded, not reaching to front border of preopercle, not projecting beyond tip of second supramaxilla; jaw teeth minute or absent; tip of lower jaw behind nostril; gill rakers fine and numerous (lower gill rakers 53–58), serrae present only along inner edge, a distinct flap present behind gill raker bases on first three arches forming a groove; no gill rakers on hind face of third epibranchial; dorsal fin at about midpoint of body; anal fin short, with 20–22 branched fin rays, its origin under base of last dorsal-fin ray (Whitehead et al. 1988). Distinguished from *Pterengraulis* by the position of the dorsal fin; from *Lycengraulis* by absence of canine-like teeth and from *Anchoviella*, *Lycengraulis*, and *Pterengraulis* by having more gill rakers. Resembles the marine *Cetengraulis* that sometimes enter estuaries (e.g., *Cetengraulis edentulous*) and *Anchovia surinamensis*. Species of *Cetengraulis* are deeper-bodied, have only 8 (long) branchiostegal rays, and have a broad branchiostegal membrane that covers most of the isthmus, whereas species of *Jurengraulis* are tubular in shape, have 9 or more branchiostegal rays, and have a relatively small branchiostegal membrane that does not cover the isthmus. *Anchovia surinamensis* has a much shorter snout, a deeper body, and larger specimens have more gill rakers. In addition, these species are not known to co-occur, as *Cetengraulis* is restricted to estuaries, *A. surinamensis* to lower reaches of rivers, and *Jurengraulis* are known only from the upper Amazon. **SPECIES** One, *Jurengraulis juruensis* (Jurua anchovy). Species information in Whitehead et al. (1988). **DISTRIBUTION AND HABITAT** Middle and upper Amazon: from the Rio Mamoré in Bolivia to Manaus, including the Juruá basin. **BIOLOGY** Feeds on plankton, but little is known.

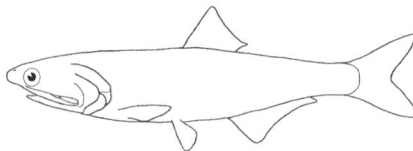

Lycengraulis (15–26 cm SL)

Distinguished from all other Neotropical engraulids by the enlarged and canine teeth, which are well spaced (especially in the lower jaw). Also characterized by: relatively large and compressed body; fewer and shorter gill rakers, with low to moderate counts (12–27 on the first arch), the anterior gill rakers becoming rudimentary in some species; pectoral fins moderate, not or just reaching to pelvic-fin base; dorsal-fin origin only a little behind midpoint of body; anal fin long (21–31 branched fin rays), its origin below about midpoint of dorsal-fin base, which distinguishes it from *Pterengraulis* (Whitehead et al. 1988, Loeb and Alcântara 2013). Sometimes confused with *Anchoa spinifer*, from which it is distinguished by having a small triangular projection on the subopercle. **SPECIES** Five, including three species in the AOG region: *L. grossidens*, *L. batesii*, and *L. figueiredoi*. Species information in Whitehead et al. (1988), except for *L. figueiredoi* (Loeb and Alcântara 2013). **DISTRIBUTION AND HABITAT** Atlantic and Pacific coasts and rivers of South and Central America. Northern populations of *L. grossidens* mostly fully marine, southern populations diadromous, ascending lower parts of coastal rivers in South America, or even fully freshwater (Mai and Vieira 2013, Mai et al. 2014); *L. batesii* is a freshwater species that inhabits the Orinoco, as far up as the Rio Manacacias in Colombia, rivers of the Guianas, and the Amazon at least as far up as Tefé, and possibly even as far as the Ucayali, Huallaga, and Morona rivers of the upper Amazon; *L. figueiredoi* inhabits the Purus, Negro, Trombetas, and Solimões in Brazil. **BIOLOGY** Carnivorous, feeding on small fishes and crustaceans.

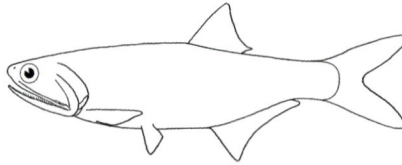

Pterengraulis (25 cm SL)

Characterized by: dorsal fin positioned posteriorly on body, well behind midpoint of body, and well behind the origin of anal fin; and by the long pectoral fins reaching posterior to pelvic-fin base. The teeth in the jaws short and even; lower gill rakers few (12–14), shorter than most engraulids; small tooth patches on upper edge of hyoid bones; anal fin long, with 28–32 branched rays (Whitehead et al. 1988). **SPECIES** One, *P. atherinoides*. Species information in Whitehead et al. (1988). **COMMON NAMES** *Anchoa aletona* (Spanish); Wingfin anchovy (English). **DISTRIBUTION AND HABITAT** Estuaries and lower parts of rivers from the Orinoco delta south to Ceará, Brazil. **BIOLOGY** Consumes shrimp, prawns, and small fish (Krumme et al. 2005).

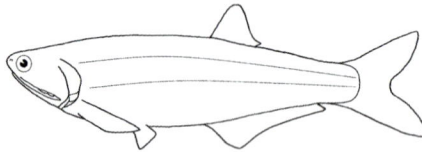

FAMILY PRISTIGASTERIDAE—LONGFIN HERRINGS

— *PETER VAN DER SLEEN and DEVIN BLOOM*

ILLUSTRATION AFTER LIDONNICI AND LASTRICO IN WHITEHEAD (1985)

DIVERSITY 38 species in 9 genera worldwide, mostly marine, with five freshwater species in three genera in the AOG region: *Ilisha amazonica, Pellona castelnaeana, Pellona flavipinnis, Pristigaster cayana,* and *Pristigaster whiteheadi.*

COMMON NAMES *Apapá, Sardinhão* (Brazil); *Arenques de aletas largas* (Peru).

GEOGRAPHIC DISTRIBUTION Throughout the tropics and subtropics in coastal waters. Freshwater species occur in South America and Southeast Asia.

ADULT SIZES Usually about 20–25 cm SL, some *Pellona* species to about 50 cm SL.

DIAGNOSIS OF FAMILY Longfin herrings can be recognized by a lower jaw articulation under or only just behind the eye; deep lower jaw; two supramaxillae; long anal fin, with at least 30 rays; small pelvic fins that are displaced anteriorly (so that the tip of the pectoral fin reaches or surpasses the vertical through the base of the pelvic fin); and a complete series of scutes along the belly (Whitehead 1985).

SEXUAL DIMORPHISM Not pronounced.

HABITATS Marine, estuarine, freshwater.

FEEDING ECOLOGY Omnivorous to piscivorous.

BEHAVIOR Poorly known. Usually found in schools.

KEY TO THE GENERA BASED ON WHITEHEAD (1985)

1a. Hypomaxilla toothed (fig. 1a)... *Pellona*
1b. Hypomaxilla not toothed (fig. 1b) **2**

2a. Body very deep (depth about 50% of standard length); pelvic fins absent, or small pelvic fins sometimes present .... *Pristigaster*
2b. Body not very deep, but moderately elongate; pelvic fins present.. *Ilisha*

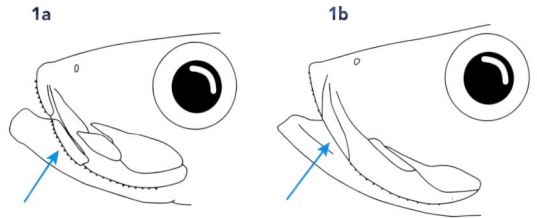

GENUS ACCOUNTS

Ilisha (17 cm SL)

Characterized by: moderately elongate body; belly with 25 or 26 scutes; large eye; lower jaw projecting; 17–22 lower gill rakers; dorsal-fin origin clearly before midpoint of body; pelvic fins present; and a long anal fin, with 47–52 fin rays (Whitehead 1985). Can be distinguished from species of the genus *Pellona* by lack of a toothed hypomaxilla and from *Pristigaster* by its body shape, the latter being much deeper-bodied. **SPECIES** 16, including one species in the AOG region, *Ilisha amazonica*. Species information in Whitehead (1985). **COMMON NAMES** *Sardinhão* (Brazil); *Panshin* (Peru); Amazon ilisha (English). **DISTRIBUTION AND HABITAT** Most species are marine and distributed throughout the world in coastal waters. *Ilisha amazonica* is found from the mouth of the Amazon River up to Iquitos (Peru). It is a pelagic riverine species often found in floodplain lakes. **BIOLOGY** Feeds on small fishes, shrimps, and insects (Zuanon and Ferreira 2008).

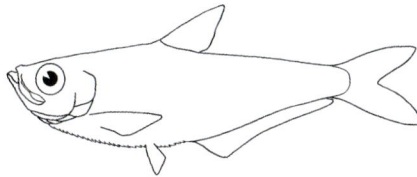

Pellona (47–55 cm SL)

Similar to *Ilisha*, but distinguished by the small toothed hypomaxilla (vs. smooth in *Ilisha*). The toothed hypomaxilla can be detected by rubbing a finger along the lower edge of the jaw. Also characterized by: a moderately deep and compressed body; a sharp keel with 32–37 scutes along belly; large eyes; a projecting lower jaw; mouth directed upward; jaw teeth small or minute, usually with a distinct gap at center of upper jaw; lower gill rakers short and robust, 12–31 (in fishes up to 50 cm; less in larger fishes); anal fin long, its origin under dorsal-fin base and with 34–46 fin rays; dorsal fin located more or less at the midpoint of body; pelvic fins present, relatively small and with 6 or 7 fin rays; scales small, about 60–70 in lateral series (Whitehead 1985). **SPECIES** Six, including two species in the AOG region: *P. castelnaeana* and *P. flavipinnis*. Species information in Whitehead (1985). **COMMON NAMES** *Apapá, Apapá amarela* (Brazil); *Peje chino* (Peru); Amazon pellona (English). **DISTRIBUTION AND HABITAT** Marine and inshore in the Indo-Pacific. Freshwater species in South America include *P. castelnaeana* in the Amazon basin in Bolivia, Brazil, Colombia, Ecuador and Peru, and *P. flavipinnis* with a broad geographic distribution in South America, including the Amazon, Parnaíba, Orinoco, Paraná-Prata basins and rivers in the Guianas. Prefers main river channels and constant flow (Guennec and Loubens 2004). **BIOLOGY** Both species are important components of fisheries in the AOG (Galvis et al. 2006). *Pellona flavipinnis* is a generalist carnivorous species that feeds mainly on fishes and aquatic insects (Pouilly et al. 2004, Moreira-Hara et al. 2009). More intense feeding activity was recorded at night and in the high water period (Moreira-Hara et al. 2009). Spawning occurs year-round, at least in some areas, with a peak during the dry season (Guennec and Loubens 2004, Ikeziri et al. 2008). Both sexes spawn for the first time during their second year and maximum reported age is seven years for both sexes (Guennec and Loubens 2004).

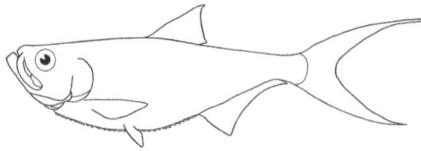

Pristigaster (9.0–15 cm SL)

Relatively small fishes with a very deep body (resembling that of characiform fishes in the family Gasteropelecidae) and a strongly convex belly profile with 29–35 scutes. Distinguished from Gasteropelecidae by: lacking elongate pectoral fins, lacking an adipose fin (adipose fin also absent in *Carnegiella*), and presence of a scuted belly. Mouth directed almost vertically upward; no toothed hypomaxilla; pelvic fins absent in *P. cayana*, but present in *P. whiteheadi*; dorsal fin before midpoint of body; anal fin long, its origin behind dorsal-fin base and with 41–54 fin rays; caudal-fin lobes ending in filaments in larger fishes (Whitehead 1985). **SPECIES** Two, both in the AOG region: *P. cayana* and *P. whiteheadi*. Species information in Whitehead (1985) for *P. cayana* and Menezes and De Pinna (2000) for *P. whiteheadi*. **COMMON NAMES** *Apapá verdadeiro* (Brazil); *Arenque pechito* (Peru); Amazon hatchet herring (English). **DISTRIBUTION AND HABITAT** Amazon basin in Brazil, Colombia, Ecuador, and Peru. **BIOLOGY** Omnivorous. Strongly dependent on insects at low water, switching to plant matter at high water (Galvis et al. 2006). *Pristigaster* species are unlikely able to jump out of the water if threatened (as do the similar-looking species in the characoid family Gasteropelecidae), because of their rather small pectoral muscles and the vertically aligned scute arms (Whitehead 1985).

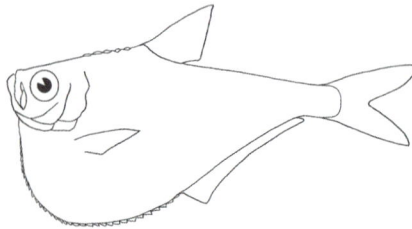

FAMILY ACESTRORHYNCHIDAE—FRESHWATER BARRACUDAS

— *MÔNICA TOLEDO-PIZA*

Family includes 14 species in one genus, *Acestrorhynchus*.

TAXONOMIC NOTE Based on morphological characters, Lucena and Menezes (1998) proposed that Acestrorhynchidae is close to Cynodontinae and Roestinae. Oliveira et al. (2011a) proposed that Acestrorhynchidae also includes Heterocharacinae and Roestinae based on an analysis of molecular data. However, morphological characters have not been found supporting a close relationship between *Acestrorhynchus* and Roestinae or Heterocharacinae (Mirande 2010, Mattox and Toledo-Piza 2012). Mirande (2010) placed *Acestrorhynchus* in the subfamily Acestrorhynchinae within the family Characidae. We provisionally place the genus here in a separate family pending future studies.

Acestrorhynchus (10–40 cm SL)

Pike-shaped characiforms characterized by: a slender, elongate body covered with small silvery cycloid scales; elongate snout with large jaws bearing a single row of alternating conical and caniniform teeth in each jaw; premaxilla with one or two holes (foramina) to accommodate large dentary caniniform teeth; posteriorly placed dorsal fin, originating closer to caudal-fin base than to tip of snout; falcate (sickle-shaped) anal fin with long anterior rays that never bear hooks in mature males; and a black spot at base of caudal fin (Menezes 2003). Additional diagnostic osteological characters include: many small conical teeth on ectopterygoid; first infraorbital covers almost completely the maxilla when the mouth is closed and it extends along the entire ventral margin of the second infraorbital; a branch of the laterosensory canal present on the premaxilla; all gill rakers on the first branchial arch shaped as bony plates and covered with denticles. **SPECIES** 14 species, with 12 species in the AOG region. Species groups readily recognized by differences in color pattern; see information in Toledo-Piza (2007) and Pretti et al. (2009). Key to species in Menezes and Géry (1983), Menezes (1992), Toledo-Piza and Menezes (1996), and López-Fernández and Winemiller (2003). **COMMON NAMES** *Peixe cachorro*, *Pirapuco*, *Uéua* (Brazil); *Dientón* (Ecuador); *Chachorro*, *Peje zorro* (Peru); *Picúa* (Venezuela). **DISTRIBUTION AND HABITAT** Distributed throughout tropical South America east of the Andes. Some species, e.g., *A. falcatus*, *A. microlepis*, and *A. falcirostris*, are widely distributed throughout the AOG region. Many species occur in sympatry and are usually collected together. *Acestrorhynchus* live in many habitats but are most often found in lakes, lagoons, and near the shore of rivers. Species with smaller body size are also found in small streams (Menezes 2003). **BIOLOGY** Most species feed primarily on other fishes, but shrimps and insects have also been reported (Galvis et al. 2006). Diurnal and nonmigratory.

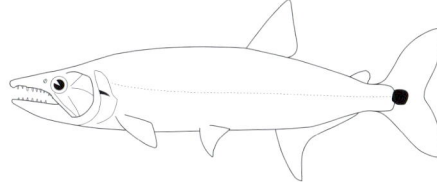

FAMILY ANOSTOMIDAE—TOOTHED HEADSTANDERS

— *BRIAN L. SIDLAUSKAS and JOSÉ L. O. BIRINDELLI*

DIVERSITY Approximately 160 species in 14 genera, with the largest number of species (~100) currently placed in the artificial genus *Leporinus*.

TAXONOMIC NOTE Recent taxonomic revisions include those of *Abramites* (Vari and Williams 1987), *Anostomoides* (Santos and Zuanon 2006), *Laemolyta* (Mautari and Menezes 2006), *Pseudanos* (Sidlauskas and Santos 2005, Birindelli et al. 2012b), and certain components of *Leporinus* (Garavello 2000, Britski and Birindelli 2008, Sidlauskas et al. 2011, Britski et al. 2012, Birindelli and Britski 2013, Birindelli et al. 2013a, Burns et al. 2014). Other taxa requiring substantial taxonomic attention include the widespread species *L. fasciatus* and *L. friderici*, as well as the genera *Schizodon*, *Rhytiodus*, and *Leporellus*.

COMMON NAMES Aracu, Piau, Piava, Piapara (Brazil); Lisa (Peru).

GEOGRAPHIC DISTRIBUTION Inhabit all major drainages of tropical South America, including the Amazon, Orinoco, Essequibo, Paraná-Paraguay, São Francisco, Atrato, and Magdalena basins (Géry 1977, Garavello and Britski 2003). One species of *Leporinus* occurs on the islands of Trinidad and Tobago, and one in trans-Andean Ecuador.

ADULT SIZES 7.7 cm SL in *Leporinus parvulus* to 60 cm SL in *L. macrocephalus*.

DIAGNOSIS: Easily recognized by their dentition, which almost always consists of three or four large and well-developed teeth in a single row on each side of the upper and lower jaws (Garavello and Britski 2003). Tooth forms range from multicuspid to serrate to incisiform (Sidlauskas and Vari 2008) and can frequently diagnose anostomid genera. Anostomids all possess small mouths, triangular heads in lateral view, and relatively short dorsal fins, typically with 10 branched fin rays. Most anostomid species possess relatively large scales (44 or fewer in the lateral line series, with 12 or 16 in the circumpeduncular series), short anal fins with 7 or 8 branched rays, and fusiform bodies, but exceptions to all these generalities exist within the diversity of the family. Additional features of the oral jaws, gill arches, pharyngeal dentition, infraorbitals, opercular series, and other body systems are provided by Vari (1983) and Sidlauskas and Vari (2008).

Family name originates from the Greek words *ano*, meaning up, and *stoma*, meaning mouth, and indeed several of its species possess pronouncedly superior mouths (Myers 1950a), such as the eponymous *Anostomus anostomus*. Other members of the family possess terminal, subterminal, or inferior mouths (Sidlauskas 2007), depending on feeding habits. Anostomids range in body shape from compressiform to elongate fusiform and demonstrate an array of spotted, barred, and striped color patterns (Sidlauskas and Vari 2012). In life, coloration is typically silver, brown, or yellow with dark black markings, with fins ranging from hyaline to red or yellow.

SEXUAL DIMORPHISM Males of some *Leporinus* and *Schizodon* species have thick dorsal portions of the first three ribs and elongate and strongly curved first ribs, possibly providing larger attachment areas for muscles involved in sound production using the gas bladder (Vari and Raredon 1991). This feature has also been reported for Curimatidae and Prochilodontidae, two characiform families closely related to Anostomidae (Castro and Vari 2004). Females of some *Leporinus* and *Schizodon* species grow larger than males in length and weight (Godoy 1975, Santos and Barbieri 1993, Orsi and Shibbata 1999).

HABITATS Anostomids exploit most available habitats in the lowland portions of the AOG region, including open river channels, flooded beaches, lagoons, oxbow lakes, rapids, and backwaters. Many species have a strong affinity for submerged vegetation, rocks and other structures. Larger species (especially of *Leporinus* and *Schizodon*) undertake long spawning migrations to portions of the floodplain (Godoy 1975, Goulding 1981), and juveniles are frequently found in and around mats of floating vegetation (Santos 1980, 1982, Meschiatti et al. 2000). *Pseudanos* can sometimes be found within rotting logs (Sidlauskas and Santos 2005), *Rhytiodus* is very common in floodplain lakes, and *Sartor* occurs in extremely fast rapids (Santos and Jégu 1987), a habitat that also harbors *Petulanos*, *Gnathodolus*, and *Synaptolaemus*.

FEEDING ECOLOGY Anostomids typically occupy herbivorous, omnivorous, or invertivorous niches (Goulding 1980, Santos 1981, 1982, Goulding et al. 1988, Horeau et al. 1998, Pouilly et al. 2003, Melo and Röpke 2004). Some species also feed on filamentous algae, terrestrial insects, small fishes and freshwater sponges (Knöppel 1972, Santos and Rosa 1998), and some captive individuals have even been reported to eat scales from their tank mates (Winterbottom 1980).

BEHAVIOR Anostomids with upturned mouths (as well as members of *Abramites*) adopt a characteristic head-down posture while swimming or resting, particularly when concealing themselves in vegetation. Anostomids with terminal to inferior mouths adopt a horizontal posture while swimming in the middle of the water column. Anostomids with inferior mouths are often found close to the bottom. Species with upturned mouths such as *Pseudanos* and *Petulanos* usually stay hidden inside logs and come out only to feed. Anostomids that inhabit large rivers and floodplain areas are generally schooling fishes; rapids-dwelling species such as *Sartor*, *Synaptolaemus*, and *Gnathodolus* are usually solitary. Some *Leporinus* and *Schizodon* species begin spawning at their first year (Barbieri and Garavello 1981) and can live up to about 10 years (Orsi and Shibbata 1999).

ADDITIONAL NOTES Some large species are harvested as part of artisanal and commercial fisheries (Godoy 1975, Géry 1977).

KEY TO THE GENERA
MODIFIED FROM SIDLAUSKAS AND SANTOS (2005) AND SIDLAUSKAS AND VARI (2012)

1a. Caudal-fin rays largely covered with scales.. *Leporellus*
1b. Caudal-fin rays not covered by scales other than immediately proximate to hypurals **2**

2a. Anal fin with 10–14 branched rays; body deep and diamond-shaped; color pattern of alternating irregular light and dark bands, forming a mottled pattern over the body and median fins .... *Abramites*
2b. Anal fin with 7–9 branched rays; body fusiform; color pattern with horizontal stripes, vertical bars, lateral spots or unadorned, but not as above except in a few species of *Leporinus* **3**

3a. Mouth upturned in individuals of all sizes (fig. 3a)...................................**4**
3b. Mouth terminal (fig. 3b) or downturned in adults (small juveniles <50 mm SL with upturned mouths transitioning downward through ontogeny) **11**

3a upturned mouth   3b terminal mouth

4a. Three or fewer teeth on each dentary **5**
4b. Four teeth on each dentary..**6**

5a. Each dentary with 3 teeth, of which the symphyseal (located where dentaries connect, i.e., teeth in middle of lower jaw) is longest (fig. 5a; after Myers and Carvalho 1959); dentary aligned posterodorsally when mouth is closed.............................. *Sartor*
5b. Each dentary with a single elongate tooth (fig. 5b); dentary aligned vertically when mouth is closed .... *Gnathodolus*

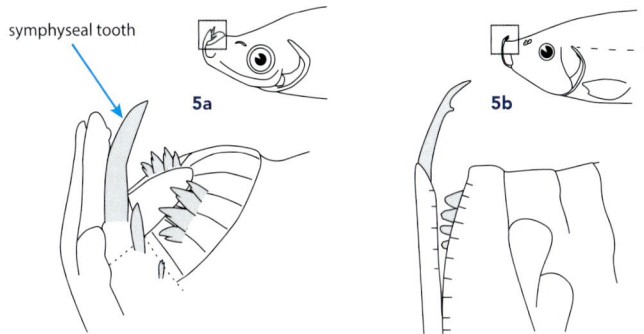

Pseudanos trimaculatus   Anostomoides passionis

Laemolyta garmani   Schizodon fasciatus

symphyseal tooth

5a   **5b**

Anostomoides laticeps   Leporinus lacustris

6a. Color pattern of head and body dark black with thin light vertical bands (light bands red, orange, or yellow in life).. *Synaptolaemus*
6b. Color pattern of head and body generally tan with dark longitudinal stripes, vertical bands, or round blotches. **7**

7a. Symphyseal dentary teeth in individuals >40 mm SL with rounded or incisiform cutting edges; dentary teeth in individuals <40 mm SL multicuspid **8**
7b. Symphyseal dentary teeth with ≥2 cusps of similar size in individuals of all sizes **9**

8a. Four branchiostegal rays; premaxillary teeth with ≥3 distinct cusps of similar size................................... *Laemolyta*
8b. Three branchiostegal rays; premaxillary teeth with round or weakly developed cusps........................ *Anostomoides* (in part)

9a. Body with ≥2 dark longitudinal stripes .... *Anostomus*
9b. Body with a single dark longitudinal stripe, or 1–4 midlateral blotches (faded in some individuals of *Pseudanos varii* >13 cm SL)...**10**

10a. 38–42 scales on lateral line; symphyseal dentary teeth truncate in specimens ≥6.0 cm SL; distance from snout to pelvic-fin origin greater than distance from dorsal-fin origin to caudal-fin origin..................................... *Petulanos*
10b. 42–47 scales on lateral line; symphyseal dentary teeth bicuspid in specimens <15.0 cm SL (teeth truncate in large specimens); distance from snout to pelvic origin equal to or less than distance from dorsal-fin origin to caudal-fin origin ....*Pseudanos*

11a. Teeth of premaxillary and dentary with ≥3 distinct cusps forming serrate or jagged cutting edge; dorsalmost and ventralmost 4–6 principal rays of caudal fin with margins in close contact, thickened and conjoined into rigid plate-like structure in adults **12**
11b. Teeth of premaxillary and dentary with rounded cutting edge or weakly developed cusps; rays of caudal fin not conjoined into rigid plate-like structure in adults **13**

12a. ≥48 scales in lateral line series; body highly elongate, maximum body depth typically 17–23% of SL, up to 26% in *Rhytiodus lauzannei*; central cusp largest on the second and third premaxillary teeth, forming jagged cutting edge .... *Rhytiodus*
12b. <48 scales in lateral line series; body fusiform but not distinctly elongate, maximum body depth typically 25–33% of SL; cusps of premaxillary teeth uniform in size, forming continuous serrate cutting edge .... *Schizodon*

13a. Mouth slightly upturned in adults; body with dark thin longitudinal lines on the posterior half in juveniles ≤ 15 cm SL; 2 dark midlateral blotches, 1 below the dorsal fin and 1 above the anal fin (no dark blotch on the caudal peduncle) in specimens of all sizes .... *Anostomoides* (in part: *Anostomoides passionis*)
13b. Mouth terminal or downturned; body with dark transversal bands, dark longitudinal stripes, or dark blotches of various arrangements (including, but not limited to, examples figured) **14**

14a. Premaxilla with vertical or anterodorsal inclination, such that long axis of premaxillary teeth points ventrally or posteroventrally (fig. 14a; after T. Clark in Sidlauskas and Vari 2008); mouth subterminal to inferior .... *Hypomasticus*
14b. Premaxilla with posterodorsal inclination, such that long axis of premaxillary teeth points anteroventrally (fig. 14b; after T. Clark in Sidlauskas and Vari 2008); mouth terminal to subterminal............................ *Leporinus*

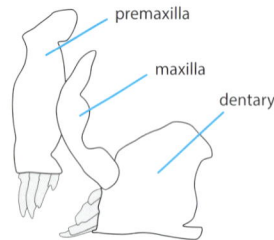
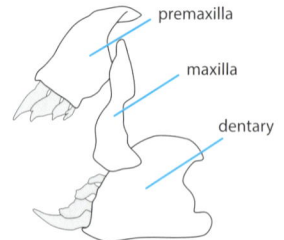

14a premaxilla maxilla dentary
14b premaxilla maxilla dentary

GENUS ACCOUNTS

Abramites (14 cm SL)

Distinguished from other anostomid genera by: deep, laterally compressed body; a postpelvic median keel extending between base of the pelvic fin and the anus; anal fin with 10–12 branched rays; 37–40 pored lateral line scales; 16 scale rows around caudal peduncle; terminal mouth; three teeth on the premaxilla with a bicuspid median tooth; four teeth on the dentary; symphyseal teeth with strong concave distal border; head with two longitudinal dark stripes; body with eight dark bands, the second and third band incomplete and fused in some specimens, the fourth band more conspicuous than others and continuing onto the dorsal and pelvic fins, the fifth band sometimes incomplete (Vari and Williams 1987). **SPECIES** Two, including one (*A. hypselonotus*, Marbled headstander) in the AOG region. **DISTRIBUTION AND HABITAT** Widespread throughout much of cis-Andean tropical South America, in the lowland portions of the Paraguay, lower Paraná, Amazon, and Orinoco basins (Vari and Williams 1987). **BIOLOGY** Slow-swimming headstander, usually found associated with submerged vegetation. *Abramites* are typically herbivorous, but when kept in fish tanks, individuals sometimes become aggressive and eat scales of other fishes (Winterbottom 1980).

Anostomoides (25–31 cm SL)

Distinguished from other anostomids by: mouth terminal to slightly upturned; four teeth on the premaxillary and dentary bones; teeth with straight or mildly tricuspid distal borders; dentary with symphyseal tooth distinctly larger than second tooth; 38–41 lateral line scales; 16 scale rows around the caudal peduncle; body with three vertically elongate dark midlateral blotches (sometimes inconspicuous), and an inconspicuous dark midlateral stripe (sometimes absent) or with two rounded dark midlateral blotches (Eigenmann 1912a, Santos and Zuanon 2006). **SPECIES** Three, all in the AOG region: *A. atrianalis*, *A. laticeps* and *A. passionis*. Key to the species in Santos and Zuanon (2006). **DISTRIBUTION AND HABITAT** *A. atrianalis* endemic to the Orinoco basin in Venezuela; *A. laticeps* widespread in the Amazon and Essequibo basins in Brazil and Guyana; *A. passionis* endemic to the middle Xingu River in Brazil (Santos and Zuanon 2006). **BIOLOGY** *Anostomoides* are usually found close to the river margins and associated with a rocky bottom (Santos et al. 1984). The diet is considered omnivorous and includes insects, fruit, plant material, and sponges (Santos et al. 2004, Santos and Zuanon 2006).

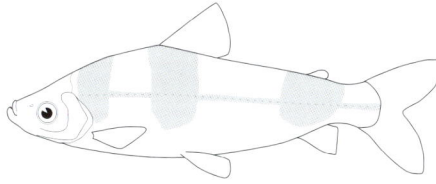

Anostomus (12–16 cm SL)

Distinguished from other anostomids by: mouth distinctly upturned in individuals of all sizes; four teeth with multicuspid borders on each premaxillary and dentary; 38–43 lateral line scales; 16 scale rows around the caudal peduncle; body with multiple light and dark longitudinal bands (Winterbottom 1980). **SPECIES** Three, all in the AOG region: *A. anostomus* (Striped headstander), *A. brevior*, and *A. ternetzi*. The subspecies *A. anostomus longus* Géry 1961 from the western Amazon is sometimes treated as a full species, but its validity needs to be tested in a range-wide revision. Key to the species in Sidlauskas and Vari (2012). **DISTRIBUTION AND HABITAT** *A. anostomus* and *A. ternetzi* are the most widespread species, occurring throughout the AOG region (Winterbottom 1980). Both species may represent species complexes. *Anostomus brevior* is limited to the rivers of eastern Suriname and French Guiana (Sidlauskas and Vari 2012). **BIOLOGY** *Anostomus* are omnivorous and consume a variety of invertebrates, fungi, and plants in their diet (Knöppel 1972, Santos and Rosa 1998). When kept in large aquaria, they exhibit cleaning behavior and can be seen picking parasites from larger fishes. In small aquaria they can become quite aggressive.

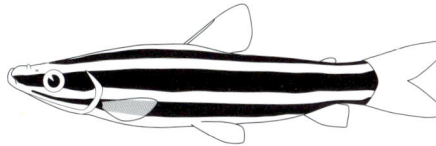

Gnathodolus (12 cm SL)

Distinguished from other anostomids by: mouth superior, with dorsal opening; four teeth with straight distal border on premaxilla; a single elongate sickle-shaped tooth on each dentary; lower lips with large papillae; 38 or 39 pored scales in the lateral line; 16 scale rows around the caudal peduncle; body with two or three dark midlateral blotches and dark fragmented lateral bands (faded in large specimens) (Myers 1950a, Winterbottom 1980, Santos and Jégu 1987, 1996). **SPECIES** One, *G. bidens*. **DISTRIBUTION AND HABITAT** Upper Orinoco basin, including Casiquiare canal in Venezuela and tributaries of Amazon River in Brazil, including the Uatumã, Jamari (a tributary of the Madeira), and Xingu rivers (Santos and Jégu 1987, 1996, Leandro Sousa pers. comm.). The conspecificity of the northern and southern populations has not been tested rigorously. **BIOLOGY** Known from relatively few specimens, all collected in or near areas with rocky bottoms and strong currents (Santos and Jégu 1996, Leandro Sousa pers. comm., pers. obs.).

Hypomasticus (6.4–25 cm SL)

Distinguished from other anostomids by: mouth subterminal or inferior; three or four (rarely five) dentary teeth; three or four premaxillary teeth with a concave or blunt margin, but not serrate; 33–42 lateral line scales; 16 circumpeduncular scales; body with lateral stripes or one or more prominent midlateral blotches (Borodin 1929, Géry 1960a, Sidlauskas and Vari 2008, Birindelli et al. 2013b). **SPECIES** Eight, all in the AOG region. The limits between *Hypomasticus* and *Leporinus* are somewhat unclear when only external morphology is considered (Birindelli et al. 2013b), and osteological and molecular data may eventually reveal that some species of *Leporinus* should be reassigned to *Hypomasticus*. **DISTRIBUTION AND HABITAT** Tributaries of the Amazon and Orinoco rivers, as well the Essequibo and coastal drainages of the Guianas and Brazil. Several species are endemic to small coastal drainages of eastern Brazil. *Hypomasticus* species are usually found in large rivers (J. Zuanon pers. comm.). **BIOLOGY** Omnivorous and focused on the benthos, including insect larvae, sponges, and filamentous algae (Géry 1977, Santos and Jégu 1989).

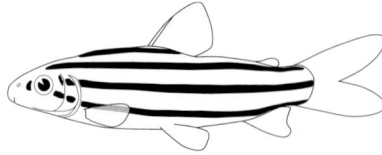

Laemolyta (7.1–29 cm SL)

Distinguished from other anostomids by: mouth slightly upturned; four teeth with four or five cusps on premaxilla; four teeth with straight margin on dentary; dentary with symphyseal tooth slightly larger than second tooth; 40–58 lateral line scales; 16–20 scale rows around the caudal peduncle; body with 3–5 dark vertically elongate blotches and an inconspicuous dark midlateral stripe, or body with a dark midlateral stripe (Mautari and Menezes 2006). **SPECIES** Five, all in the AOG region: *L. fernandezi*, *L. garmani*, *L. orinocensis*, *L. proxima*, and *L. taeniata*. Key to the species in Mautari and Menezes (2006). **DISTRIBUTION AND HABITAT** Widespread in the Amazon, Essequibo, and Orinoco basins, in Brazil, Colombia, Ecuador, Guyana, Peru, and Venezuela. *Laemolyta orinocensis*, *L. taeniata*, and *L. fernandezi* are found in the Orinoco basin, *L. proxima*, *L. garmani*, *L. taeniata*, and *L. fernandezi* in the Amazon basin, and *L. proxima* in the Essequibo basin. Frequently encountered amid submerged vegetation in lakes or near riverbanks. **BIOLOGY** Primarily herbivorous and secondarily invertivorous. Diet includes plant material and periphyton as well as some insect larvae (Santos et al. 1984, Melo and Röpke 2004, Hawlitschek et al. 2013).

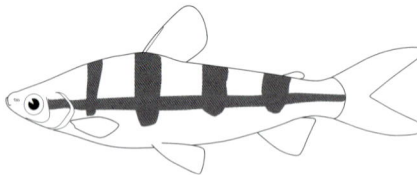

Leporellus (25 cm SL)

Distinguished from other anostomids by: rays of caudal fin heavily scaled; mouth subterminal; four teeth on each premaxilla and dentary; teeth with a blunt or undulating margin; 40–46 lateral line scales; 16 scale rows around caudal peduncle; dorsal fin with a prominent black bar and multiple dark stripes on each caudal-fin lobe; body with multiple dark longitudinal stripes as well as small dark spots centered on each scale (Géry 1977, Géry et al. 1987, Sidlauskas and Vari 2008). **SPECIES** At least one, *L. vittatus* (Valenciennes, 1850), although this is likely a species complex in need of a comprehensive and continent-wide revision. **DISTRIBUTION AND HABITAT** All major drainages of cis-Andean tropical South America, as well as in the Magdalena River of trans-Andean Colombia (Sidlauskas and Vari 2008). Most lots in collections represent only one or a few specimens, suggesting that either population sizes are small or that members of the genus tend to evade standard collection methods. **BIOLOGY** Primarily invertivorous, but include some plant material in their diet (Taphorn 1992, Luz-Agostinho et al. 2006).

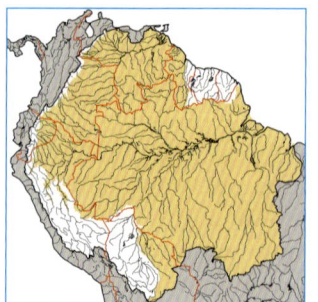

Leporinus (7.7–40 cm SL)

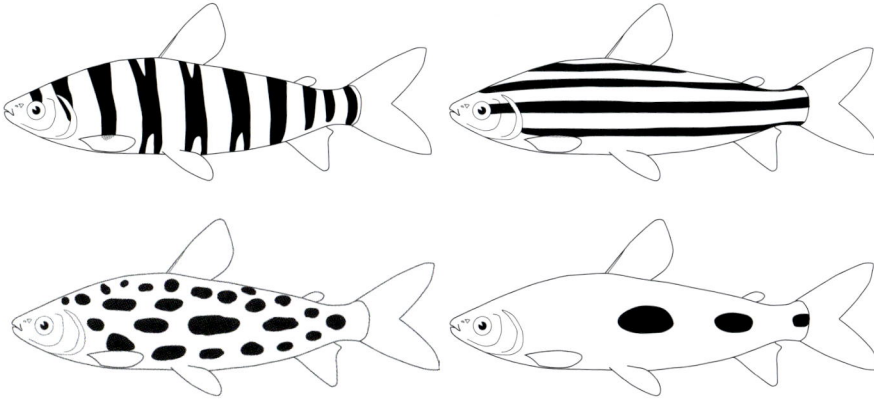

Distinguished from other anostomids by: mouth terminal or subterminal; three or four teeth on each premaxilla and dentary; teeth arranged in a stair-like pattern, with symphyseal teeth distinctly largest; teeth usually incisiform, heavy and thick at base, with blunt distal margin (some teeth of some specimens may have additional small cusps); 33–47 lateral line scales; 12 or 16 scale rows around the caudal peduncle; body with 1 or 4 dark midlateral blotches, which are vertically elongate in some species, or body completely covered in small rounded dark blotches, or body with 1 to 6 dark stripes associated with dark dorsal bands in some species, or body with 7–14 dark lateral bands. **SPECIES** *Leporinus* is the most species-rich anostomid genus, with 62 valid species in the AOG region, including *L. fasciatus* (Banded leporinus) (Birindelli and Britski 2013). **COMMON NAMES** *Piau, Piava, Piapara* (Brazil); *Lisa* (Peru). **DISTRIBUTION AND HABITAT** Widespread throughout the AOG region. **BIOLOGY** Ranging in adult body size from 7.7 cm SL in *L. parvulus* from the Tapajós River in Brazil, to 60 cm SL in *L. macrocephalus* from the Paraguay basin. Several *Leporinus* species in the AOG attain body sizes up to about 40 cm SL. This size diversity encompasses substantial diversity of ecological and reproductive strategies, such as reproductive migration in larger species like *L. trifasciatus* (Carolsfeld et al. 2003). The position of the mouth in adults ranges from terminal to subterminal, and tooth morphology ranges from bluntly rounded to multicuspid or spatulate (Sidlauskas and Vari 2008). The diet also varies, with most species tending toward omnivory, but with the proportion of different items varying considerably among species (Goulding 1980, Santos et al. 1984, Balassa et al. 2004, Melo and Röpke 2004)

Petulanos (9.5–16 cm SL)

Distinguished from other anostomids by: mouth superior in individuals of all sizes; four multicuspid teeth on each premaxilla and four teeth of graduated size on each dentary, symphyseal dentary teeth largest and with

truncate margins; 38–42 pored lateral line scales; 16 circumpeduncular scales; body with multiple dark midlateral blotches plus dark transverse bands on dorsum, transverse band continuing below the lateral line in some species. **SPECIES** Three, all in the AOG region (Sidlauskas and Vari 2008): *P. intermedius*, *P. plicatus*, and *P. spiloclistron*. Key to the species in Sidlauskas and Vari (2012). **DISTRIBUTION AND HABITAT** *P. intermedius* occurs in the southern tributaries of the Amazon, including the Xingu and Madeira basins, while the other two species are restricted to the Essequibo, Corantijn, and Nickerie systems of the Guianas (Winterbottom 1980, Sidlauskas and Vari 2012). **BIOLOGY** The diet is omnivorous and includes plants, fungi, detritus, and insects (Knöppel 1972). *Petulanos* is strongly associated with submerged woody debris.

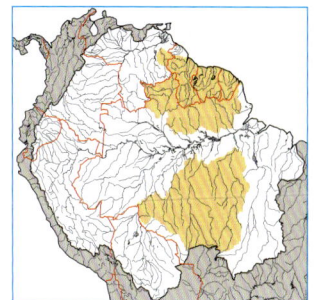

Pseudanos (15–21 cm SL)

Distinguished from other anostomids by: mouth superior in individuals of all sizes; four bicuspid or tricuspid teeth on premaxilla; four rounded, bicuspid or tricuspid teeth on dentary (with symphyseal usually rounded and cusps more developed in lateral-most teeth in larger specimens); upper and lower lips with small papillae; 42–47 pored scales in the lateral line; 16 scale rows around caudal peduncle; body with 3 or 4 dark midlateral blotches or a single dark midlateral stripe, scales of body sometimes with dark centers, dark thin transversal bands in some specimens (Winterbottom 1980, Sidlauskas and Santos 2005, Birindelli et al. 2012b). **SPECIES** Four, all in the AOG region: *P. gracilis*, *P. trimaculatus* (the Threespot headstander), *P. winterbottomi*, and *P. varii*. Key to the species in Birindelli et al. (2012b). **DISTRIBUTION AND HABITAT** Widely distributed in the Amazon, Essequibo, and Orinoco basins; *P. gracilis* occurs in the Amazon basin, *P. trimaculatus* in the Amazon and Essequibo basins, *P. winterbottomi* in the Orinoco and Tapajós basins, and *P. varii* in the upper Orinoco basin and Negro basin (Birindelli et al. 2012b). Some species are often collected in association with submerged plants and roots near the river margins, or inside submerged logs. *Pseudanos* shows a preference for habitats with boulders, leaves, sticks, and logs. **BIOLOGY** The diet consists of plant remains, fungi, algae, detritus, and insects (Knöppel 1972, Sidlauskas and Santos 2005).

Rhytiodus (30–40 cm SL)

Distinguished from other anostomids by: mouth terminal in adults and strongly upturned in juveniles; four multicuspid teeth on each premaxilla and dentary with central cusp largest, forming irregular cutting margin; 48–92 pored scales in lateral line series; 16–28 circumpeduncular scales; body coloration with four transverse blotches with indistinct margins and/or a diffuse dark lateral band; body highly elongate with maximum depth 17–26% of SL (Santos 1980, Géry 1987). **SPECIES** Three, all in the AOG region: *R. argenteofuscus*, *R. lauzannei*, and *R. microlepis*. The taxonomic and geographic boundaries between the latter two species are unclear and merit revision. A fourth species *R. elongatus* was described from a single specimen and sometimes treated as valid. Key to the species in Géry (1987). **DISTRIBUTION AND HABITAT** Widespread throughout and limited to the Amazon basin. Juveniles associate with floating vegetation, and adults adapted to inhabit large and deep river channels. *Rhytiodus* is the only anostomid genus known to be collected from trawls in the deep channels of the Amazon (J. Lundberg pers. comm.). **BIOLOGY** Primarily herbivorous (Santos 1980, Mérona and Rankin-de-Mérona 2004).

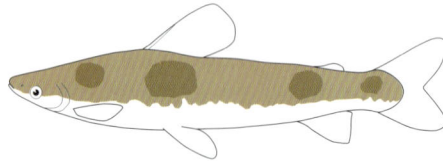

Sartor (8–11 cm SL)

Distinguished from other anostomids by: mouth superior and opening posterodorsally; four slightly tricuspid teeth on premaxilla; three teeth with rounded margin on dentary, of which the symphyseal is substantially larger than adjacent teeth; lower lips with large papillae; 37–42 pored scales in the lateral line; 12 or 16 scale rows around caudal peduncle; body with three dark midlateral blotches and 7–10 dark transversal bands (blotches and bands faded and fragmented in some specimens) (Myers and Carvalho 1959, Winterbottom 1980, Santos and Jégu 1987). **SPECIES** Three, all in the AOG region, but all known from only a few specimens: *S. respectus*, *S. elongatus*, and *S. tucuruiensis*.

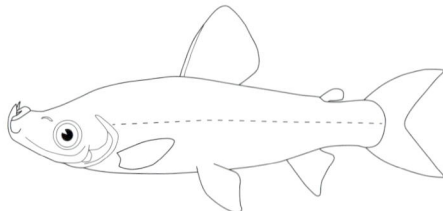

The limits of these species are poorly known. Key to the species in Sidlauskas and Santos (2005).
DISTRIBUTION AND HABITAT Tributaries of the Amazon River in Brazil, including *S. respectus* from the middle and upper Xingu (Myers and Carvalho 1959, Winterbottom 1980, Santos and Jégu 1987), *S. elongatus* from the Trombetas, and *S. tucuruiensis* from the lower and middle Tocantins basins (Santos and Jégu 1987, Lucinda et al. 2007). Found exclusively among large rocks in areas of strong current. **BIOLOGY** Feed on small invertebrates including sponges and bryozoa (Knöppel 1972, Machado et al. 2008). *Sartor tucuruiensis* is considered threatened (Machado et al. 2008, Teixeira 2014).

Schizodon (23–40 cm SL)

Distinguished from other anostomids by: mouth terminal in adults of most species, subterminal or inferior in few; mouth upturned in small juveniles of all species; four strongly multicuspid teeth on each premaxilla and dentary, forming a jagged cutting margin; 40–46 pored scales in the lateral line series; 16–20 circumpeduncular scales; body with multiple dark transverse blotches, a dark midlateral stripe, and/or a distinct dark mark at the base of the caudal fin (Santos 1980, Bergmann 1988, Garavello and Britski 1990, Vari and Raredon 1991, Garavello 1994, Sidlauskas et al. 2007). **SPECIES** 14, including three in the AOG region: *S. fasciatus* in the Amazon basin and Guianas, *S. scotorhabdotus* in the Orinoco basin, and *S. vittatus* in the Amazon, Araguaia, and Essequibo basins. The geographic and morphological limits between *S. fasciatus* and *S. vittatus* are not well described, and these named species may in fact represent color morphs of a single species or complexes of similar species. **COMMON NAMES** *Piau vara*, *Voga* (Brazil); *Lisa* (Peru); *Cotí* (Venezuela). **DISTRIBUTION AND HABITAT** All major drainages of tropical cis-Andean South America, with the majority of species occurring in the Paraná-Paraguay and other southeastern river systems (Garavello and Britski 2003). Juveniles associate with floating meadows and other vegetation in relatively still water, while larger adults can be found in both lentic and lotic environments (Sidlauskas et al. 2007). **BIOLOGY** Primarily herbivorous upon macrophytes (Goulding 1980, Santos 1980, Mérona and Rankin-de-Mérona 2004, Sidlauskas et al. 2007). Adults undergo spawning migrations (Godoy 1975, Goulding 1981).

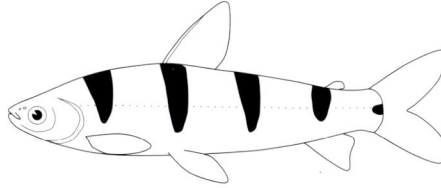

Synaptolaemus (11 cm SL)

Distinguished from other anostomids by: mouth superior, with dorsal opening; four tricuspid premaxillary teeth and four rounded, bicuspid, or tricuspid teeth on dentary (number of cusps varies by specimen size and tooth position); upper and lower lips with small papillae; 37–39 pored scales in the lateral line; 12 circumpeduncular scale rows; head dark with 3 or 4 slender light bands; body dark with 7 slender light bands, bands yellow, red, or orange in life (Myers 1950a, Winterbottom 1980, Santos and Jégu 1987, Britski et al. 2011). **SPECIES** One, *S. latofasciatus*. **DISTRIBUTION AND HABITAT** Orinoco and Amazon basins in Brazil, Colombia, and Venezuela (Britski et al. 2011). Usually found associated with rocky bottoms in areas of strong current. **BIOLOGY** Feeds on insect larvae, sponges, algae, and detritus (Santos and Jégu 1987).

FAMILY BRYCONIDAE—DORADOS OR JAW CHARACINS
— *PETER VAN DER SLEEN, FLÁVIO C. T. LIMA, and JAMES S. ALBERT*

DIVERSITY 49 species in four genera and two subfamilies (Lima 2003a, Oliveira et al. 2011a, Abe et al. 2014). The subfamily Bryconinae includes the genera *Brycon* (43 species), *Chilobrycon* (1 species: *C. deuterodon*) and *Henochilus* (1 species: *H. wheatlandii*). The subfamily Salmininae has only one genus, *Salminus* (4 species).

GEOGRAPHIC DISTRIBUTION *Brycon* has a very wide distribution from southern Mexico to northern Argentina. *Salminus* is mostly confined to the northwestern and southeastern part of tropical South America, not found in most of the central Amazon basin, and absent from the Guiana Shield. *Chilobrycon deuterodon* inhabits the Tumbes River on the Pacific slope of northern Peru. *Henochilus wheatlandii* is known only from the Doce basin in southeastern Brazil.

ADULT SIZES Medium to large sizes. *Brycon* species range from 15 to 70 cm SL; *Salminus* species have relatively larger sizes, ranging from 50 to 100 cm SL.

DIAGNOSIS: Bryconidae is well supported by molecular data (Calcagnotto et al. 2005, Javonillo et al. 2010, Oliveira et al. 2011a, Abe et al. 2014), but the family is poorly diagnosed by morphological characters. Most species can be differentiated from other characiforms by: relatively large adult body sizes, with a fusiform (trout-shaped) body; premaxillary teeth in 2 or 3 (rarely 4) rows; mandibular teeth usually biserial; absence of pterygoid teeth; and breast and abdomen not forming a keel, as such differing from *Triportheus* (Géry 1977, Lima 2003a).

KEY TO THE GENERA BASED ON GÉRY (1977)

1a. Teeth conical or very faintly tricuspid, numerous, in two rows on both jays; maxilla partly covered by the antorbital bone; scales relatively small; lateral line usually at midbody; large adult size (50–100 cm)..........................*Salminus*

1b. Teeth multicuspid, in 3 rows at least on upper jaw, in 2 rows on mandible with usually a pair of conical teeth behind main mandibular series; maxillary less covered by antorbital bone; scales relatively large; lateral line low; medium to large adult size (15–70 cm)..*Brycon*

GENUS ACCOUNTS

Brycon (15–70 cm SL)

Recognized by: three to four rows of teeth on the premaxilla; an inner symphyseal tooth on the dentary; relatively long anal fin, with 18–32 branched rays; and absence of an expanded coracoids (and so differing from the family Triportheidae) (Lima 2004). Many species with a dark blotch on the caudal peduncle, sometimes extending as a stripe on the caudal fin. **SPECIES** 43, with at least 10 species in the AOG region. **COMMON NAMES** *Jatuarana, Matrinxã, Piabanha, Piracanjuba, Pira-putanga, Piratininga* (Brazil); *Sábalo, Sabalito* (Ecuador, Peru); *Bocón, Palambra* (Venezuela). Key to species in the Tocantins basin in Lima (2004). **DISTRIBUTION AND HABITAT** From southern Mexico to Panama, across trans-Andean South American basins from northern Peru to the Maracaibo system in Venezuela, in all major river drainages in cis-Andean South America, and in most Atlantic and Caribbean coastal basins (Lima 2003a). Most species in trans-Andean river basins of Colombia, Ecuador, and Panama. Inhabits streams and rivers of moderate to high current velocity. Many species heavily exploit flooded forest during the rainy season (e.g., Goulding 1980, Horn 1997, Lima and Castro 2000, Azevedo et al. 2011).

BIOLOGY Omnivorous; relying strongly on allochthonous food items, such as fallen fruits, seeds, and insects (e.g., Menezes 1969, Sabino and Sazima 1999, Lima and Castro 2000, Azevedo et al. 2011). Some species are effective dispersers of riparian tree seeds (Gottsberger 1978, Goulding 1980, Horn 1997). Several species make long reproductive migrations (see overview in Carolsfeld et al. 2003). Often found in schools. Sensitive to anthropogenic disturbances and some species are now considered endangered in eastern Brazil (Lima and Castro 2000).

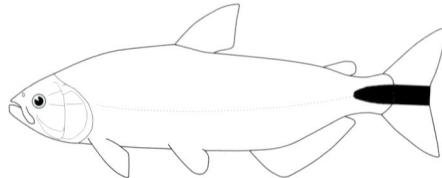

Salminus (50–100 cm SL)

Large predatory fishes recognized by: numerous conical or very feebly tricuspid teeth, in two rows on both jays; maxilla partly covered by the antorbital bone; scales relatively small (like those in *Acestrorhynchus*); and the lateral line usually at midbody (Géry 1977). All species with a dark spot on the caudal peduncle, which continues as a horizontal stripe on the caudal fin; rest of the caudal fin red-yellow. This color pattern is similar to some *Brycon* species. **SPECIES** Four, including two species in the Amazon and Orinoco basins. Key to the species in Gery and Lauzanne (1990). **COMMON NAMES** *Dorado, Dourado, Tabarana* (Brazil), *Picuda, Rubio* (Colombia); *Dama* (Ecuador); *Sábalo macho* (Peru); *Dorada, Saltadora* (Venezuela). **DISTRIBUTION AND HABITAT** *S. brasiliensis* from the La Plata, Jacuí, and upper Madeira basins, and an unnamed species (incorrectly identified in the literature as *S. affinis* or *S. hilarii*) in the Araguaia-Tocantins, upper Amazon, and Orinoco basins (Lima 2006). Inhabits streams and rivers of moderate to high current velocity. **BIOLOGY** Mostly piscivorous as adults, but juveniles have a more varied diet (Rodríguez-Olarte and Taphorn 2006, Lima and Britski 2007). Small individuals of *Salminus* form mixed schools with *Brycon* species and look very similar to these. By mimicking *Brycon*, they increase the chances of surprising prey, which might be less wary of *Brycon* (Bessa et al. 2011). Larger *Salminus* usually school only with others of the same species (Rodríguez-Olarte and Taphorn 2006, Bessa et al. 2011). Mature individuals make annual migrations for reproduction during the rainy season (Lima and Britski 2007).

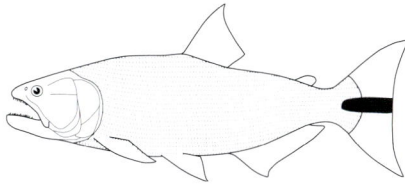

FAMILY CHALCEIDAE—TUCANFISHES
— *PETER VAN DER SLEEN and JAMES S. ALBERT*

Family includes eight species in one genus.

TAXONOMIC NOTE Based on morphological characters, the genus *Chalceus* was placed in the African family Alestidae (Zanata and Vari 2005, Mirande 2010). Molecular data however suggest that *Chalceus* is related to a large assemblage of Neotropical characids (Oliveira et al. 2011a), and the genus has since been placed in its own family, the Chalceidae.

Chalceus (14–25 cm SL)

Readily identified by a bright silvery body, red fins, short anal fin, and large shiny scales. Also recognized by: presence of a supramaxilla; three series of teeth on the premaxilla; internal series of dentary teeth formed by a large symphyseal conical tooth followed by a gap and a series of smaller conical teeth; scales situated dorsal to lateral line much larger than those ventral to it; alternated large and small scales along lateral line other than on caudal peduncle (Zanata and Toledo-Piza 2004). **SPECIES** Eight, all in the AOG region, including *C. erythrurus* (Tucanfish), and *C. macrolepidotus* (Pinktail chalceus). Revision of genus and key to the species in Zanata and Toledo-Piza (2004). **COMMON NAMES** *Araripirá* (Brazil); *Huacamayo challua, San Pedrito* (Peru). **DISTRIBUTION AND HABITAT** Amazon and Orinoco basins, and coastal drainages in the Guianas (Zanata and Toledo-Piza 2004). Usually occur in the middle of the water column and are found in a variety of habitats, including flooded forests, main river channels, and forest streams. **BIOLOGY** Feed on invertebrates, small fishes and vegetable matter (Goulding et al. 1988, Planquette et al. 1996).

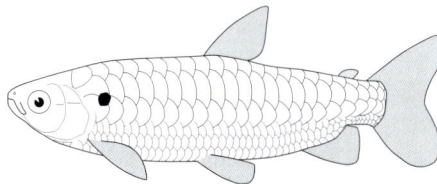

FAMILY CHARACIDAE—TETRAS AND RELATIVES

— PETER VAN DER SLEEN, JAMES S. ALBERT, FLÁVIO C. T. LIMA, ANDRÉ L. NETTO-FERREIRA, GEORGE M. T. MATTOX, and MÔNICA TOLEDO-PIZA.

DIVERSITY More than 1,100 species in about 144 genera, including at least 594 species in 98 genera in the AOG region.

COMMON NAMES *Lambari*, *Piaba* (Brazil); *Mojara*, *Mojarita* (Ecuador, Peru); *Bobita* (Venezuela).

TAXONOMIC NOTE Because of high species richness and morphological diversity, the composition, diagnoses, and phylogenetic relationships of characid genera are as yet incompletely resolved. Characidae has traditionally been composed of characiform species lacking the supraorbital bone (Malabarba and Weitzman 2003, Oliveira et al. 2011a, but see Mirande 2010). Here we restrict Characidae to characiforms in which the supraorbital bone is absent or highly reduced, provisionally including the genera *Roestes*, *Heterocharax*, and *Lonchogenys* (subfamily Heterocharacinae) where the supraorbital is present but very small. Assignment of species to subfamilies is based mainly on Mirande (2010), with modifications for Aphyocharacinae (Tagliacollo et al. 2012), Characinae (Oliveira et al. 2011a, Mattox and Toledo-Piza 2012), Cheirodontinae (Jerep and Malabarba 2011), Heterocharacinae (Mattox and Toledo-Piza 2012), and Stevardiinae (e.g., Thomaz et al. 2015). Because of lack of consensus in the literature, many genera are grouped as Characidae *incertae sedis*, including the subfamilies Stethaprioninae and Tetragonopterinae.

GEOGRAPHIC DISTRIBUTION Southwestern Texas, Mexico, Central America, and South America (to Patagonia, Argentina). Most species are found in the Amazon and Orinoco basins, as well as in the rivers of the Guianas.

ADULT SIZE: Most species have small body sizes (<10 cm SL). Many species are miniaturized, e.g., in the genera *Xenurobrycon* and *Priocharax* with adult sizes <2.0 cm.

DIAGNOSIS OF FAMILY Most species lack a supraorbital bone (see fig. 1 in key to the characiform families on page 32). Other characters that might aid in the recognition of characids: usually small body size (<10 cm SL); long anal fin (with few exceptions), the maxillary bone not very small and usually toothed; dorsal fin with 10–13 rays; caudal fin usually forked; adipose present in most species.

SEXUAL DIMORPHISM Often pronounced, but usually not conspicuous to the naked eye. Sexual dimorphism includes a caudal-fin gland (several genera traditionally grouped into the subfamily Glandulocaudinae; now part of Stevardiinae), gill-derived glands (some species in Stevardiinae, Aphyocharacinae, and Cheirodontinae), and presence of hooks on the anal- and pelvic-fin rays (often also on the caudal fin, and rarely on the dorsal or pelvic fins) in sexually mature males.

HABITATS Highly variable, including large lowland rivers, lakes, flooded forests and savannas, small forest creeks, and fast-flowing mountain streams.

FEEDING ECOLOGY Many species are omnivorous, with diets including plant matter, seeds, insects, crustaceans, and other fishes. The highly diverse teeth morphology and variable arrangement of teeth on the jaws suggest that most species have specialized in feeding on certain food types. However, there is still little evidence for such specializations, as analyses of stomach contents generally show broad and overlapping diets between species. Some species are predominantly carnivorous (e.g., species in the subfamilies Characinae and Heterocharacinae). Species in the genera *Roeboides*, *Bryconexodon*, *Exodon*, and *Roeboexodon* feed on the scales of other fishes and have specialized teeth outside the mouth for ripping of scales.

BEHAVIOR Highly diverse, although poorly known for the majority of species. Most species are externally fertilizing. Some species in the subfamilies Cheirodontinae and Stevardiinae are inseminating (these species usually have a caudal gland). Many species form schools; some make seasonal migrations.

ADDITIONAL NOTES Many characid species are popular ornamental fishes; species of *Paracheirodon* and *Hyphessobrycon* are among the most popular aquarium fishes in the world with tens of millions of wild-caught individuals annually exported from the Amazon and Orinoco basins.

KEY TO THE SUBFAMILIES

1a. Presence of mammiliform (nipple-shaped) teeth outside the mouth, pointing forward (fig. 1a)... **2**
1b. Absence of mammiliform teeth outside the mouth... **3**

2a. Presence of superficial neuromasts (pit-lines) on the head (fig. 2a).....................................**subfamily Characinae** (part: genus *Roeboides*; p. 106)
2b. Absence of superficial neuromasts on the head...**Characidae** *incertae sedis* (genera *Exodon*, *Roeboexodon*, *Bryconexodon*, and *Serrabrycon*; p. 128)

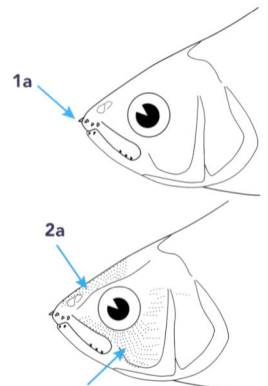

3a. Predatory fishes, armed with conical or canine-like teeth (except in *Phenacogaster*); maxillary elongate and with many teeth...**4**

3b. Omnivorous fishes with small (conical to multicuspid) teeth; maxillary rarely elongate and usually without teeth or with few teeth...**5**

4a. Absence of superficial neuromasts on the head; presence of spiniform projection on posterior margin of preopercle (fig. 4a; absent in the miniature and paedomorphic genus *Priocharax*)...**subfamily Heterocharacinae** (p. 110)

4b. Presence of superficial neuromasts on the head (see fig. 2a); spiniform projection on posterior margin of preopercle usually absent (present in *Acanthocharax*)..**subfamily Characinae** (p. 102)

4a

5a. Infraorbital-3 very large (resembling examples in fig. 5a; after Tagliacollo et al. 2012); pelvic fin with ≤6 branched rays (except *Aphyocharax dentatus* and *Paragoniates alburnus*)..**subfamily Aphyocharacinae** (p. 93)

5b. Infraorbital-3 never very developed; pelvic fin with ≥7 branched rays.....................................**6**

5a

VI V IV III II I

6a

6a. Pseudotympanum usually present (which looks like a bright humeral spot, shaped more or less like an inverse triangle; fig. 6a); premaxillary with a single tooth row...**7**

6b. Pseudotympanum usually absent; premaxilla usually with two or three rows..**8**

7a. Small, uniformly sized, relatively numerous slender teeth on the dentary and premaxillary that are either conic or multicuspid; dark humeral spot often present; typically clear-colored or even translucent in life, with scattered dark pigmentation mostly concentrated at the anal-fin and caudal-fin bases..**subfamily Aphyoditeinae** (p. 97)

7b. Pedunculated teeth that are greatly expanded and compressed distally; premaxillary teeth usually perfectly aligned and similar in shape and cusp number; dark humeral spot always absent ....**subfamily Cheirodontinae** (p. 106)

8a. Two or three rows of teeth on premaxilla (1 row in *Aulixidens*, *Monotocheirodon*, and *Othonocheirodus*); ≤4 teeth in inner premaxillary row (except in some *Chrysobrycon*, *Gephyrocharax*, *Ptychocharax*, *Scopaeocharax*, *Tyttocharax*, and *Xenurobrycon*); usually ≤8 branched dorsal-fin rays (except in some *Chrysobrycon*, *Gephyrocharax*, *Ptychocharax*, *Scopaeocharax*, *Tyttocharax*, and *Xenurobrycon*) ....**subfamily Stevardiinae** (p. 113)

8b. Two rows of teeth on premaxilla (1 row in *Pristella* and *Petitella*); ≥5 teeth in inner premaxillary row; usually ≥9 branched dorsal-fin rays..**Characidae** *incertae sedis* (including subfamilies Tetragonopterinae and Stethaprioninae; p. 128)

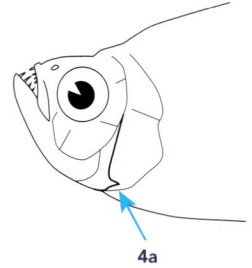

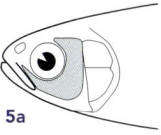

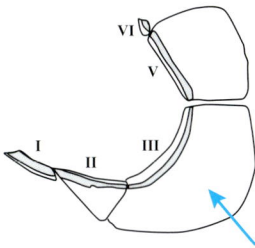

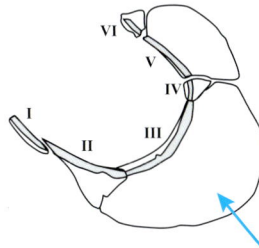

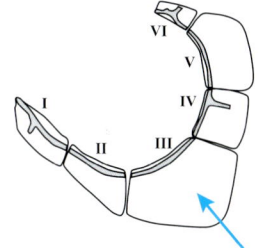

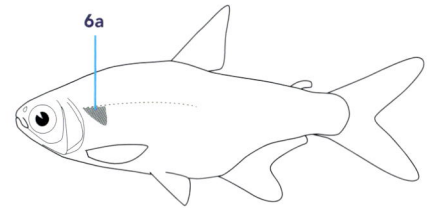

SUBFAMILY APHYOCHARACINAE—BLOODFIN TETRAS

— *PETER VAN DER SLEEN, VICTOR A. TAGLIACOLLO, and JAMES S. ALBERT*

DIVERSITY 22 species in eight genera; *Aphyocharax, Inpaichthys, Leptagoniates, Paragoniates, Phenagoniates, Prionobrama, Rachoviscus,* and *Xenagoniates* (Mirande 2010), including 13 species in six genera in the AOG region.

COMMON NAMES *Lambari cauda vermelha* (Brazil); *Sardinita coliroja* (Peru).

TAXONOMIC NOTE A recent molecular study concluded that *Inpaichthys* and *Rachoviscus* are not closely related to other members of Aphyocharacinae, but that the genus *Aphyocharacidium* should likely be included in the subfamily (Tagliacollo et al. 2012). Pending further studies, we provisionally include here the genera *Inpaichthys* and *Rachoviscus* (latter not in AOG), and exclude *Aphyocharacidium* (included in the subfamily Aphyoditeinae).

GEOGRAPHIC DISTRIBUTION Amazon, Orinoco, Araguaia-Tocantins, and Paraná basins, as well as coastal basins of the Guianas. *Phenagoniates macrolepis* (only species in genus) inhabits the Chucunaque and Atrato rivers and Lake Maracaibo basin, Colombia. The two species in the genus *Rachoviscus* are found in coastal basins of southern Bahia, Paraná, and northern Santa Catarina states in Brazil.

ADULT SIZES Relatively small adult body sizes, ranging from 2.7 cm SL in *Aphyocharax rathbuni* from the Paraguay basin, to 7.6 cm SL in *Leptagoniates steindachneri* from the Amazon basin.

DIAGNOSIS OF THE SUBFAMILY: Aphyocharacine genera excluding *Aphyocharacidium* (i.e., *Aphyocharax*, *Leptagoniates*, *Prionobrama*, *Paragoniates*, *Phenagoniates*, and *Xenagoniates*) can be diagnosed from other characids by several osteological traits that are mostly associated with aspects of neurocranium evolution (Mirande 2010, Tagliacollo et al. 2012). These six genera also share with *Inpaichthys* and *Rachoviscus* the presence of a synchondral articulation between the lateral ethmoid and anterodorsal border of the orbitosphenoid (Mirande 2010). Phylogenetically, however, this hypothesis of sister relationship between the *Inpaichthys* and *Rachoviscus* and all other aphyocharacines has not been corroborated by molecular data (Tagliacollo et al. 2012). All aphyocharacines, except *Aphyocharax dentatus* and *Paragoniates alburnus* possess six or fewer branched pelvic-fin rays. The genera *Leptagoniates*, *Phenagoniates*, and *Xenagoniates* have an elongate body shape and a complete lateral line. *Paragoniates* share with this latter group an elongate anal fin with more than 35 branched rays and usually 8 supraneurals. In *Aphyocharax* and *Prionobrama*, the fourth infraorbital is absent or highly reduced and bordered posteriorly by third and fifth infraorbitals. *Aphyocharax* and *Prionobrama* are popularly known as Bloodfin Tetras because of their bright red dorsal, pelvic, anal, and caudal fins in life.

KEY TO THE GENERA

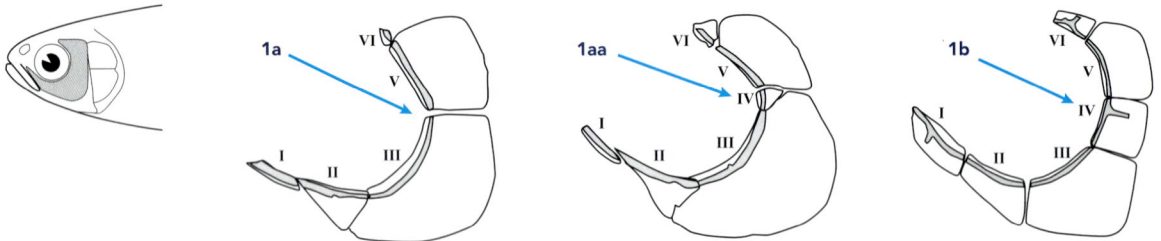

1a. Fourth infraorbital absent (fig. 1a) or much reduced and bordered posteriorly by third and fifth infraorbitals (fig. 1aa); anterior maxillary teeth with a single large cusp (except *Inpaichthys*)..**2**
1b. Fourth infraorbital present and well developed (fig. 1b; after Tagliacollo et al. 2012); presence of 3 or more cusps on anterior maxillary teeth **4**

2a. Lateral line interrupted; single perforated scale on posterior region of caudal peduncle......................................**3**
2b. Lateral line complete (perforated scales on posterior region of caudal peduncle)...*Inpaichthys*

3a. Body more or less elongate and compressed, clupeoid; long anal fin with 29–37 rays; dorsal fin usually behind anal fin origin..*Prionobrama*
3b. Body never elongate and compressed at the same time; relatively short anal fin with 15–22 rays; dorsal fin never behind anal fin origin..*Aphyocharax*

4a. Body relatively short; lateral line incomplete; 43–48 anal-fin rays (about 20 maxillary teeth; 39–46 lateral line scales) ....*Paragoniates*
4b. Body elongate; lateral line complete; 63–68 anal-fin rays **5**

5a. No teeth on palate; 60–72 anal-fin rays; 46–50 lateral line scales...*Leptagoniates*
5b. A few teeth on the ectopterygoid bone; 63–66 anal-fin rays; 50–51 lateral line scales ....*Xenagoniates*

GENUS ACCOUNTS

Aphyocharax (2.7–7.6 cm SL)

Characterized by the bright red dorsal, pelvic, anal, and caudal fins in life. Resembles Prionobrama by having an incomplete lateral line, with only the last scale of the series perforated; a highly reduced or absent fourth infraorbital bone; and a single large cusp on anterior maxillary teeth (Tagliacollo et al. 2012). Differs from *Prionobrama* by a shorter anal fin (12–22 rays) and a dorsal fin near the middle of the body (Géry 1977, Lima 2003c). Also recognized by a terminal to slightly upturned mouth, and mature males with gill glands and hooks on all or on the first pelvic- and anal-fin rays (Lima 2003c, Gonçalves et al. 2005). **SPECIES** 12, with 7 species in the AOG region, including *A. alburnus* (Goldencrown tetra) and *A. anisitsi* (Bloodfin tetra). Key to some of the species is provided in Géry (1977), Souza-Lima (2003), and Willink et al. (2003). **DISTRIBUTION AND HABITAT** Amazon basin (4 species); Orinoco basin (2 species), Essequibo basin (1 species), and Paraná basin (6 species). **BIOLOGY** Likely carnivores. Diet includes worms, small insects, crustaceans, and small fishes (Galvis et al. 2006, Corrêa et al. 2009). Under hypoxic conditions, *A. anisitsi* has been reported to develop dermal extensions on the lower lip that improve the flow of water into the mouth during aquatic surface respiration (Scarabotti et al. 2011).

Inpaichthys (2.8 cm SL)

Readily recognized by its characteristic coloration: bluish body color with a thick, dark blue, lateral stripe running from the mouth to the base of the caudal fin, and reddish dorsal, anal, and caudal fins. *Inpaichthys* shares with *Aphyocharax* and *Prionobrama* a strongly reduced or absent fourth infraorbital bone, bordered posteriorly by third and fifth infraorbitals. However, a combined analysis of molecular and morphological data suggest that *Inpaichthys* is not a member of the Aphyocharacinae, but of a group containing the characid genera *Nematobrycon*, *Carlana*, and *Astyanacinus*, all of which share the following traits: ventral extent of the third infraorbital bone reaching the horizontal arm of the preopercle; ascending process of the premaxilla reaching just anterior of the nasal; four or more maxillary teeth; 10 or fewer gill rakers on the first hypobranchial and ceratobranchial (Tagliacollo et al. 2012). **SPECIES** One, *I. kerri* (the Royal tetra). **DISTRIBUTION AND HABITAT** Aripuanã River, upper Madeira basin, Brazil. **BIOLOGY** No data available.

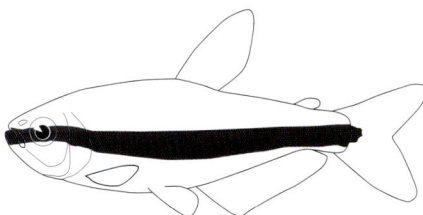

Leptagoniates (3.1–7.6 cm SL)

Includes two species that may not be closely related; *L steindachneri* (top illustration) likely belongs to Aphyocharacinae, and *L. pi* (bottom illustration) to Cheirodontinae (Tagliacollo et al. 2012). Both species share a high number of anal-fin rays (usually >45); a complete lateral line (vs. incomplete in *Paragoniates*); and a toothless ectopterygoid (vs. a few teeth on the ectopterygoid in *Xenagoniates*) (Vari 1978). *Leptagoniates steindachneri* has an elongate and slender body, like that of *Xenagoniates*, whereas the body shape of *L. pi* is much deeper. *Leptagoniates pi* can be readily recognized by

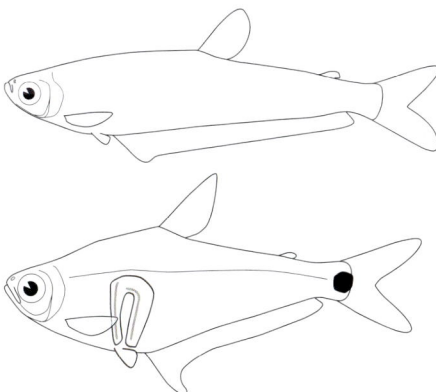

its transparent body in life (milky white in preserved specimens), with large pi-shaped (π) swim bladder. **SPECIES** Two, both in the Amazon basin. Key to the species and species information in Vari (1978). **COMMON NAMES** *Peixe de vidro* (Brazil); *Pez vidrio* (Peru). **DISTRIBUTION AND HABITAT** *L. steindachneri* from the Marañón and Napo basins in Ecuador and the Ucayali basin in Peru; *L. pi* throughout the Solimões-Amazonas system, associated with floating meadows in areas with some current and not in standing waters (J. Zuanon pers. comm.). **BIOLOGY** No data available.

Paragoniates (6.0 cm SL)

Characterized by: body shape relatively short and deep; lateral line incomplete; 43–48 anal-fin rays; about 20 maxillary teeth; 39–46 longitudinal scales. *Paragoniates* is the only species in the aphyocharacines in which the coracoid lamellae are much expanded ventrally and medially, contacting each other and forming a keel. In addition, in *Paragoniates* several rows of scales cover the basal third of the anal-fin base, whereas in most other aphyocharacines only one or two rows of scales cover the anal-fin base (Mirande 2010, Tagliacollo et al. 2012). In life, the fish displays a bluish sheen and red dorsal, anal, and caudal fins. **SPECIES** One, *P. alburnus*. **DISTRIBUTION AND HABITAT** Middle and upper Amazon and Orinoco basins. **BIOLOGY** Omnivorous.

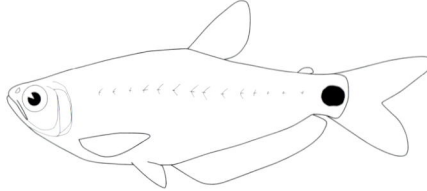

Prionobrama (5.0–6.0 cm SL)

Shares with *Aphyocharax* an incomplete lateral line, with only the last scale of the series perforated; a much reduced or absent fourth infraorbital bone; and a single large cusp on anterior maxillary teeth (Tagliacollo et al. 2012). Differs from *Aphyocharax* by having a more elongate and laterally compressed body, with a longer anal fin containing 29–37 rays (Géry 1977). First ray of anal fin white and elongate filament in males. The only species in the Amazon basin (*P. filigera*) has a translucent to white body in life and a red caudal fin; small scales, 35–41 longitudinally; and 10–16 premaxillary teeth (Géry 1977). **SPECIES** Two, including one in the Amazon basin: *P. filigera* (Glass bloodfin tetra). **DISTRIBUTION AND HABITAT** *P. filigera* in the western Amazon basin; *P. paraguayensis* in the Paraguay and lower Paraná basins. Frequently found in association with *Aphyocharax* species in shallow marginal areas along whitewater rivers (J. Zuanon pers. comm.). **BIOLOGY** *P. filigera* is a terrestrial insectivore, its diet and condition have been studied in relation to deforestation (Bojsen 2005).

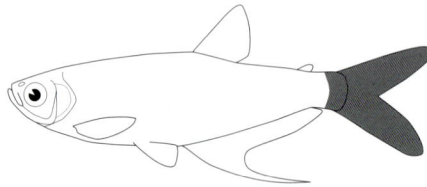

Xenagoniates (6.0 cm SL)

Resembles *Leptagoniates steindachneri* with an elongate body, a high number (>60) of anal-fin rays, and a complete lateral line; however, in *Xenagoniates* an ectopterygoid tooth row is present (vs. absent in *Leptagoniates*). Body is translucent in life. **SPECIES** One, *X. bondi*. **DISTRIBUTION AND HABITAT** Orinoco basin and coastal drainages of Venezuela. **BIOLOGY** No data available.

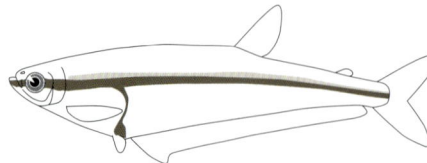

SUBFAMILY APHYODITEINAE—APHYODITEINE TETRAS

— *FLÁVIO C. T. LIMA, PETER VAN DER SLEEN, and CRISTIANO MOREIRA*

DIVERSITY The subfamily Aphyoditeinae, as herein circumscribed, comprises 11 genera and 26 species, including the *incertae sedis* "*Macropsobrycon*" *xinguensis*. Of the 11 genera, 6 are monotypic.

TAXONOMIC NOTES: "Aphyoditeina" of Géry (1973) included the following eight genera: *Aphyodite, Microschemobrycon, Aphyocharacidium, Macropsobrycon, Axelrodia, Brittanichthys, Thrissobrycon,* and *Tyttobrycon*. Géry (1977) added the following seven genera: *Oligobrycon, Atopomesus, Paracheirodon, Parecbasis, Prodontocharax, Oxybrycon,* and *Leptobrycon*. *Paracheirodon* was excluded from Aphyoditeinae by Weitzman and Fink (1983), and Malabarba (1998) transferred *Prodontocharax* and *Macropsobrycon uruguayanae* to Cheirodontinae. Mirande (2010) restricted Aphyoditeinae to the following eight genera: *Aphyocharacidium, Aphyodite, Axelrodia, Leptobrycon, Microschemobrycon, Oxybrycon, Parecbasis,* and *Tyttobrycon*, all of which share the following five characters: (1) dorsal margin of lateral ethmoids aligned, (2) canal of lateral line on caudal fin membrane absent, (3) ≥8 teeth in inner premaxillary row, (4) ≥8 small and slender teeth at front of dentary, and (5) ≤4 supraneurals (except in *A. grammica*, which has ≥5 supraneurals).

GEOGRAPHIC DISTRIBUTION Aphyoditeinae are restricted to the Amazon basin, Orinoco basin and the Essequibo and Demerara rivers in Guyana; a single genus (*Aphyocharacidium*) is reported from Suriname and French Guiana.

ADULT SIZES Most species in the Aphyoditeinae are small to very small (1.5–3.0 cm SL); only *Parecbasis cyclolepis* reaches at least 6.8 cm SL (Lima et al. 2013).

DIAGNOSIS OF SUBFAMILY Representatives of Aphyoditeinae are typically clear-colored or even translucent in life, with scattered dark pigmentation mostly concentrated at the anal- and caudal-fin bases; a conspicuous pseudotympanum is present in most genera. Compared with other characids, they possess small, uniformly sized, relatively numerous and slender teeth on the dentary and premaxillary that may be either conical or multicuspid. The premaxilla has a single tooth row. Some genera (i.e., *Brittanichthys, Leptobrycon, Oxybrycon, Thrissobrycon,* and "*Macropsobrycon*" *xinguensis*) possess a small premaxilla, a well-developed dentary, and a distally enlarged maxilla, resulting in a slightly upturned mouth which resembles that of clupeoid fishes.

SEXUAL DIMORPHISM Most species have little or no apparent sexual dimorphism. Sexual dimorphism is pronounced in *Brittanichthys*, where mature males present a highly modified caudal-fin ray and a richer color pattern; in *Axelrodia lindeae*, where males have a developed anal-fin lobe with thick rays that contain hooks; and in some *Tyttobrycon* species, where males develop hooks on the dorsal-, pelvic-, and/or anal-fin rays and can have developed procurrent caudal-fin rays.

HABITATS Highly variable, with most species having distinct ecological preferences. Habitats include: sandy beaches of large rivers, terra firme streams with abundant submerged vegetation, flooded forests, floodplain lakes, and seasonally flooded savannas.

FEEDING ECOLOGY Poorly known, Goulding et al. (1988) recorded autochthonous and allochthonous invertebrates in stomach contents of *Microschemobrycon casiquiare* and *M. callops*, and autochthonous invertebrates in stomach contents of *Brittanichthys axelrodi*.

BEHAVIOR Also poorly known. *Microschemobrycon* species apparently live in small schools. *Tyttobrycon marajoara* was also observed forming small schools (Marinho et al., 2013) and the same is true for *T. xeruini* (F.C.T. Lima pers. obs.).

KEY TO THE GENERA BASED ON GÉRY (1977)

NOTE Teeth morphology is an important character in the systematics of Aphyoditeinae but might be difficult to examine without clearing and staining, since the teeth are very small in almost all members of the subfamily. As a consequence, characters related to teeth morphology and presence are avoided as much as possible in the key. The key does not include the monotypic genus *Leptobrycon* (with *L. jatuaranae*) since it is very poorly known and few specimens have ever been collected; the species would key with the genera, presenting a distinct upturned mouth and well-developed maxillary.

1a. Dark midlateral stripe absent; black markings restricted to anal- and caudal-fin bases, dorsal fin, humeral spot or small spots along midline; pseudotympanum large, triangular, very conspicuous **2**

1b. Conspicuous stripe along midline present (along the entire trunk or restricted either to the anterior or posterior portion of the body) and/or a conspicuous caudal peduncle blotch present; pseudotympanum ranging from small to large (absent in some genera) **4**

2a. Procurrent dorsal and ventral caudal-fin rays well developed in mature males; two tooth rows on the dentary ....*Aphyocharacidium*

2b. Dorsal and ventral caudal-fin procurrent rays moderately (*Aphyodite*) or poorly (*Microschemobrycon*) developed in mature males; a single tooth row on the dentary **3**

3a. Color pattern consisting of a narrow stripe along anal fin basis and a faint midlateral stripe, but no other distinctive pigment features; lateral line incomplete, with 5–8 perforated scales; caudal fin with small scales on fin lobes (very deciduous and often missing); mouth slightly upturned .... *Aphyodite grammica*

3b. Color pattern consisting of a narrow stripe along anal fin basis, and either a black symmetrical marking at the end of caudal peduncle, a pair of comma-shaped black markings close to the base of each caudal lobe, a humeral blotch, small dark markings along midline, or a dark blotch on dorsal fin; lateral line either incomplete or complete; caudal fin lacking scales; mouth always terminal................................*Microschemobrycon*

4a. Mouth slightly to pronouncedly upturned, with a well-developed maxilla, clupeoid-like..**5**
4b. Mouth terminal or subterminal, maxilla normally developed **9**

5a. Mature males with modified, S-shaped, 12th caudal-fin ray with hooks at its tips (fig. 5a; after Géry 1965a); color pattern in life translucent, with large patches of red pigmentation .... *Brittanichthys*
5b. Caudal-fin rays straight; color pattern distinct from the one described above **6**

6a. A large dark blotch on caudal-fin base **7**
6b. Large blotch on caudal fin absent; dark pigmentation on caudal fin, if present, restricted to two symmetrical dark blotches on lobes..**8**

7a. Small humeral blotch, shaped as a diacritical mark; large dark blotch on caudal fin; relatively large-sized, up to at least 68 mm SL...*Parecbasis cyclolepis*
7b. Humeral blotch absent; a large, approximately square, dark blotch present on caudal fin, preceded by a clear area; small-sized, up to about 30 mm SL... "*Macropsobrycon*" *xinguensis*

8a. Body lacking prominent pigment markings; caudal fin with two symmetrical markings on fin lobes .... *Thrissobrycon pectinifer*
8b. Body with a faint, narrow, midlateral stripe ending at caudal peduncle as a small rounded blotch; no dark markings on caudal fin .... *Oxybrycon parvulus*

9a. Mouth subterminal; dentary and premaxillary teeth well developed; the symphyseal teeth more developed, remaining teeth gradually decreasing in size, teeth arrangement stair-like (resembling *Leporinus*); lateral line complete..*Atopomesus pachyodus*
9b. Mouth terminal; teeth small, conical (tri to pentacuspid in *Tyttobrycon marajoara*), similarly sized; lateral line incomplete.. **10**

10a. Conspicuous color markings, with either a narrow stripe extending from snout to midbody or an asymmetrically positioned dark blotch on caudal peduncle; maxillary with 4–9 teeth .... *Axelrodia*
10b. Conspicuous color markings, with either a dark dorsal fin blotch, a narrow dark stripe along midline, or a symmetrically positioned dark caudal peduncle blotch; maxillary without teeth or with only 1–2 teeth................*Tyttobrycon*

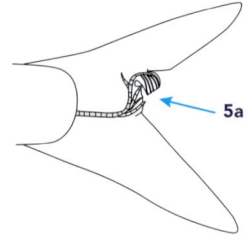

GENUS ACCOUNTS

Aphyocharacidium (2.7 cm SL)

Characterized by: small adult body size; moderately deep body; mouth terminal; lateral line complete or almost complete, with 32–35 scales; pseudotympanum large, very conspicuous; anal fin with 20–21 branched rays; teeth tricuspid, 10–12 on the premaxillary, 16 teeth on dentary; a second, inner row of 16–18 smaller tricuspid teeth on dentary; maxillary with 4–7 teeth, anterior ones tricuspid, posterior ones unicuspid; both dorsal and ventral procurrent caudal-fin rays well developed and forming a keel. Overall color pattern clear, with a small humeral blotch in *A. bolivianum*; a black line across anal-fin basis and symmetrical small black marking at caudal-fin base (Géry 1960b, 1973). **SPECIES** Two, both in the AOG region. **DISTRIBUTION AND HABITAT** Rivers in Guianas (*A. melandetum*) and upper Madeira basin in Bolivia (*A. bolivianum*). Poorly known; apparently scarce. **BIOLOGY** No data available.

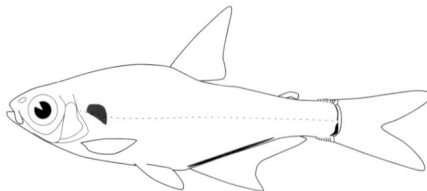

Aphyodite (3.2 cm SL)

Characterized by: small adult body size; body laterally compressed and moderately slender; mouth slightly upturned; maxilla with one or two minute teeth; about 10 premaxillary teeth; a single dentary tooth row; lateral line incomplete, with 5–8 pored scales; anal fin with 16–20 branched rays; overall color pattern clear, translucent in life, with small scattered chromatophores along midline and at bases of anal and caudal fins; pseudotympanum present and conspicuous (Géry 1973, Lima et al. 2013). **SPECIES** One, *A. grammica*. **DISTRIBUTION AND HABITAT** Central Amazon, Madeira, and Essequibo basins. Locally quite common in floodplain lakes of whitewater rivers. **BIOLOGY** No data available.

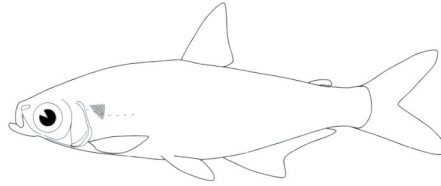

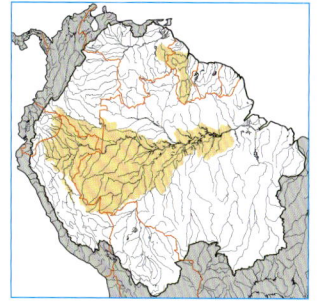

Atopomesus (2.6 cm SL)

Characterized by: small adult body size; body shape laterally compressed and moderately elongate; mouth subterminal; lateral line complete, with 31–33 scales; anal fin with 18–20 branched rays; overall color pattern clear, without conspicuous dark markings, but with a conspicuous pseudotympanum (overall body morphology and color pattern resembling a *Phenacogaster*); dentary and premaxillary teeth relatively robust, symphyseal teeth more developed, remaining teeth gradually decreasing in size, teeth arrangement stair-like (resembling *Leporinus* and as such unique among Characidae). **SPECIES** One, *A. pachyodus*. **DISTRIBUTION AND HABITAT** Known only from middle and lower Negro basin, Brazil. Moderately common in sandy beaches. **BIOLOGY** No data available.

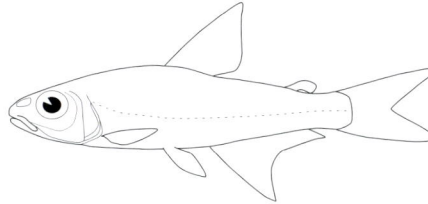

Axelrodia (2.2 cm SL)

Characterized by: small adult body size; uniserial conical teeth; lateral line incomplete; 13–27 branched anal-fin rays (13–16 in *A. stigmatias* and *A. riesei*, 23–27 in *A. lindeae*); 7–15 premaxillary teeth; 4–9 maxillary teeth (Géry 1977, Lima et al. 2013). *A. lindea* sexually dimorphic, males with a well-developed anal-fin lobe and thick anal-fin rays with hooks. **SPECIES** Three, all in the AOG region. Key to the species in Géry (1977). **DISTRIBUTION AND HABITAT** *A. lindeae* from the Central Amazon, including the Curuçamba River (north of Óbidos and near Juruti), middle Madeira, and Branco basin in Brazil (Géry 1973, Ferreira et al. 2007, Lima et al. 2013); *A. riesei* from the upper Meta River, Colombia, and Negro basin, Brazil; *A. stigmatias* from the western Amazon basin in Brazil, Peru, and Colombia and middle Madeira River in Brazil. *A. stigmatias* prefers shaded forest streams with a relatively fast current (Galvis et al. 2006), and *A. lindeae* prefers clearwater terra firme streams with abundant aquatic vegetation. **BIOLOGY** *A. stigmatias* form mixed-species groups of 20 or more individuals (Galvis et al. 2006).

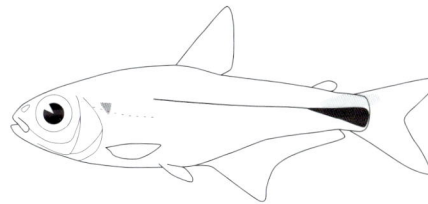

Brittanichthys (3.2 cm SL)

Characterized by: small adult body size; elongate body shape; maxilla enlarged; mouth slightly upturned; dentary very well developed; all teeth conical; no maxillary teeth; lateral line incomplete, with 6–7 pored scales; anal fin relatively elongate, with 18–19 branched

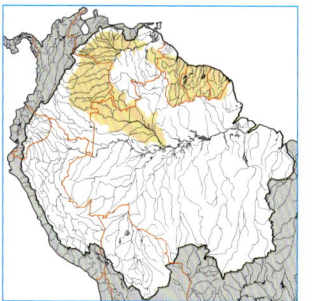

rays; pseudotympanum small. A unique feature of the genus is that mature males possess a highly modified, S-curved caudal-fin ray that contains hooks at the ray's segments (Géry 1965a, Malabarba and Weitzman 1999). Color pattern in life is translucent, with red pigmentation on the dorsal fin basis and on the abdominal region as a broad, curved stripe that extends from behind the dorsal fin to caudal peduncle; caudal peduncle and caudal fin lobes outlined by black pigmentation in *B. axelrodi*; two vertically parallel tiny spots immediately behind opercle in *B. myersi*. **SPECIES** Two, both in the AOG region. **DISTRIBUTION AND HABITAT** Middle and lower Negro basin, upper Orinoco in Colombia and Venezuela, and from Guyana (Demerara River; see Vari et al. 2009). Found in sandy beaches and stream margins. **BIOLOGY** No data available.

Leptobrycon (3.0 cm SL)

Very poorly known species, known only from a few collected individuals. Characterized by: elongate body shape; maxilla enlarged, blade-like, clupeoid, its border thickened, with few or no teeth; mouth slightly upturned; teeth conical and numerous, about 14 on premaxilla; 14 total anal-fin rays; body depth about 3.5 times in standard length; lateral line incomplete, with approximately 6 pored scales. Overall color plain, no blotches or stripes except for a midlateral silvery stripe in the holotype (Eigenmann 1915a). **SPECIES** One, *L. jatuaranae*. **DISTRIBUTION AND HABITAT** First specimens were collected in an uncertain locality ("Jatuarana") in the Brazilian Amazon. Recently, the species has been collected in the Rio Preto do Igapó-Açu at the interfluve Madeira-Purus, Borba municipality, Amazonas state, Brazil (J. Zuanon pers. comm.). **BIOLOGY** Unknown.

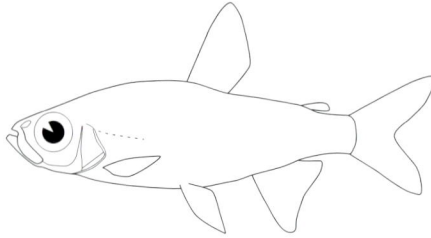

"Macropsobrycon" xinguensis (3.1 cm SL)

Characterized by: small adult body size; moderately elongate body shape; mouth slightly upturned; lateral line incomplete, with 8 perforated scales; anal fin with 17–18 branched rays; pseudotympanum absent; dentary and premaxillary teeth bicuspid to tricuspid, 7–8 on premaxillary, circa 10 teeth on dentary; maxillary toothless. Overall color pattern clear, with a large dark blotch on caudal fin, preceded by a clear area on caudal peduncle. **SPECIES** One. The genus *Macropsobrycon* was transferred to the subfamily Cheirodontinae by Malabarba (1998), however, *"Macropsobrycon" xinguensis* was not considered to belong to the genus and remains within the Aphyoditeinae, although currently not allocated to any of the genera (or a separate genus). **DISTRIBUTION AND HABITAT** Upper Xingu, and Teles Pires (upper Tapajós) and Araguaia (Tocantins) basins. Locally found in small rivers and floodplain lakes, but relatively scarce. **BIOLOGY** No information available.

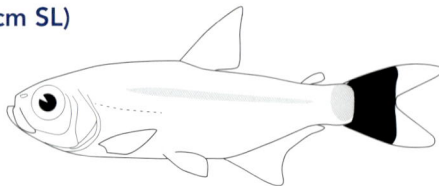

Microschemobrycon (2.0–4.0 cm SL)

Characterized by: small adult body size; usually elongate body shape; color pattern clear, translucent in life, with black markings that are characteristic for each species and consequently useful for species identification: a humeral blotch is present in *M. callops* and *M. geisleri*; a black blotch on dorsal fin is present in *M. callops* and *M. melanotus*; two symmetrical small blotches are present at caudal fin basis in *M. casiquiare*; and a discrete vertical

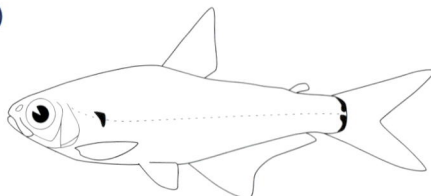

stripe extending from dentary to anterior margin of eye is present in *M. elongatus* (Géry 1973, 1977, Lima et al. 2013). Teeth mostly tricuspid (except in *Microschemobrycon elongatus*, where they are conical and bicuspid); maxilla not thickened, toothed; a conspicuous pseudotympanum present; lateral line complete or incomplete. **SPECIES** Seven, all in the AOG region. Key to the species in Géry (1977), except *M. meyburgi* (Meinken 1975). Genus was reviewed by Cavallaro (2010) in an unpublished PhD thesis. **DISTRIBUTION AND HABITAT** The most widespread and common Aphyoditeinae genus; widely distributed in the Brazilian Amazon, and also recorded for the Orinoco basin (*M. callops* and *M. casiquiare*) and the Essequibo basin (*M. melanotus*). Commonly found in sandbanks in both blackwater and clearwater rivers and streams. **BIOLOGY** No data available.

Oxybrycon (2.1 cm SL)

Characterized by: small adult body size; elongate body shape; large dentary; distinctly upturned mouth; teeth conical, tiny; two tooth rows on the dentary; maxilla toothed; small pseudotympanum present; lateral line lacking; anal fin very short, with 10–13 branched rays; lateral line not complete, with 2–3 pored scales; overall color clear, a very narrow midlateral stripe, broadening into a small caudal peduncle blotch; a narrow stripe along anal fin (Géry 1964a, 1977, Lima et al. 2013). **SPECIES** One, *O. parvulus*. **DISTRIBUTION** Western Amazon near Iquitos and middle Madeira basin (Lima et al. 2013). **BIOLOGY** No data available.

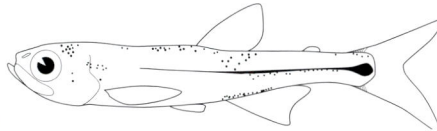

Parecbasis (6.8 cm SL)

Characterized by: relatively large adult body size (the largest in the subfamily); teeth mostly tricuspid; caudal fin scaled; mouth terminal; maxilla not toothed, thickened; a small round humeral blotch and a large caudal fin blotch. **SPECIES** One, *P. cyclolepis*. **DISTRIBUTION** Madeira-Mamoré basin (Bolivia and Brazil) and Amazon River from Peru to near Manaus (Lima and Ribeiro 2011, Lima et al. 2013). Locally common in whitewater rivers, favoring sandy shores (Lima et al. 2013). **BIOLOGY** No data available.

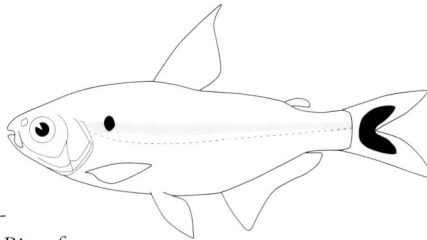

Thrissobrycon (2.8 cm SL)

Characterized by: small adult body size; elongate body shape; mouth slightly upturned; dentary and premaxillary teeth very small; maxillary blade-like and strongly curved, toothless; anal fin short, with 17–18 branched rays; lateral line incomplete, with few perforated scales (perhaps only 2–3 scales; most specimens in collections lack nearly all scales); pseudotympanum absent; gill rakers long and numerous. Overall color clear, without stripes or blotches, except for two dark blotches, symmetrically positioned on the middle portion of each caudal fin lobe. **SPECIES** One, *T. pectinifer*. **DISTRIBUTION** Upper and middle Negro basin, Brazil. Occurs in sandy bank. **BIOLOGY** No data available.

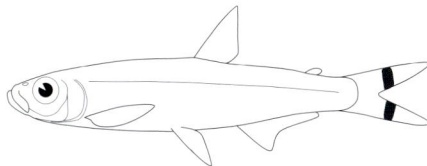

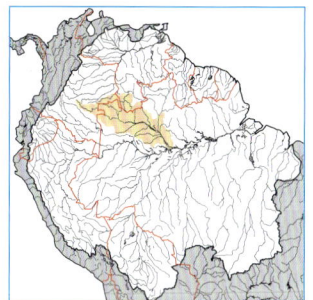

Tyttobrycon (1.5–2.2 cm SL)

Characterized by: very small adult body sizes, among the smallest characiform fishes; elongate body shape (*T. xeruini*) to moderately deep body shape (*T. dorsimaculatus, T. spinosus*); all teeth conical, except in *T. marajoara* (which has tricuspid or pentacuspid teeth); premaxilla bearing fewer than 10 teeth in a single, more or less regular series; a single series of dentary teeth; sides of dentary not raised and with few teeth if any; maxilla not enlarged or thickened and without teeth or with only 1–2 teeth; caudal fin not scaled; infraorbitals reduced; number of fin rays reduced; lateral line very short or lacking; pseudotympanum present, relatively small (Géry 1973, 1977, Lima et al. 2013, Marinho et al. 2013). Two species (*T. hamatus* and *T. spinosus*) lack an adipose fin. Color pattern typically with a conspicuous midlateral stripe, more conspicuous after midbody and extending onto caudal peduncle (*T. xeruini, T. marajoara, T. dorsimaculatus*), but midlateral stripe absent in *T. spinosus*; a conspicuous caudal peduncle blotch; a black blotch on dorsal fin in *T. dorsimaculatus*; and a dark stripe at the anal-fin basis in *T. xeruini* and *T. hamatus*. Sexual dimorphism includes: two large hooks on the anterior anal-fin rays in males of *T. hamatus*; very well-developed procurrent caudal-fin rays in males of *T. spinosus*; small bony hooks on the dorsal-, pelvic-, and anal-fin rays and developed procurrent caudal-fin rays in males of *T. marajoara*; and small bony hooks present at the anal fin of *T. xeruini*. **SPECIES** Five, all in the AOG region. Key to the species in Géry (1977). Recently described species in Marinho et al. (2013). **DISTRIBUTION AND HABITAT** Upper Madeira basin, Bolivia and Peru (*T. dorsimaculatus*); Mamoré basin, Bolivia (*T. spinosus*); upper Amazon, Peru, and Madeira basin, Brazil (*T. hamatus*); Marajó Island (*T. marajoara*); and Negro and Tapajós basins, Brazil (*T. xeruini*). *Tyttobrycon marajoara* has been found in small temporary forest streams (drying completely during the dry season; Marinho et al. 2013). *Tyttobrycon xeruini* inhabits black- and clearwater rivers, typically near submerged vegetation or logs. **BIOLOGY** *T. marajoara* has been found in groups (Marinho et al. 2013), and the same was observed in *T. xeruini* (F.C.T. Lima, pers. obs.). Relatively common locally but often overlooked because of their tiny size.

SUBFAMILY CHARACINAE—CHARACINE TETRAS

— *GEORGE M. T. MATTOX, PETER VAN DER SLEEN, and MÔNICA TOLEDO-PIZA*

DIVERSITY 91 species in 11 genera, including 58 species in 7 genera in the AOG region.

TAXONOMIC NOTE The taxonomic composition of Characinae presented here follows recent studies using morphological (Mattox and Toledo-Piza 2012) and genetic (Oliveira et al. 2011a) data sets. This taxonomy excludes several genera previously included in Characinae, including *Gnathocharax, Heterocharax, Hoplocharax, Lonchogenys*, and *Priocharax* (Lucena and Menezes 2003), which are here assigned to other subfamilies. The taxonomy used here also places three lepidophagous (scale-eating) characids, *Exodon, Bryconexodon*, and *Roeboexodon* (Mirande 2010), as *incertae sedis* within Characidae pending further study. Mirande (2010) tentatively assigned *Priocharax*, a genus of miniature species, to the Characinae, but Mattox and Toledo-Piza (2012) provided evidence that *Priocharax* is more closely related to Heterocharacinae.

GEOGRAPHIC DISTRIBUTION The Characinae inhabit most drainages of cis-Andean South America, with species of *Roeboides* extending into southern Central America.

ADULT SIZES Range in adult body size from 3.6 cm SL in *Phenacogaster ojitatus* from the Xingu basin in Brazil (Lucena and Malabarba 2010) to 40 cm SL in *Cynopotamus magdalenae* from the Magdalena basin in Colombia (Lucena and Menezes

2003), and with *C. tocantinensis* (21 cm SL) as the largest species in the AOG region (Menezes 1987).

DIAGNOSIS OF THE SUBFAMILY: Characinae (*sensu* Mattox and Toledo-Piza 2012) are characterized by one exclusive character: presence of superficial neuromasts on the head, and also by the following combination of characters: absence of an axillary scale on the pelvic fin (present in *Acestrocephalus, Cynopotamus*, and *Galeocharax*); toothed margin of maxilla longer than the edentulous margin; two anteriormost branchiostegal rays slender along their entire length; gill rakers on anterior portion of first branchial arch shaped as bony plates and covered with denticles; and absence of small ossifications associated with first proximal dorsal-fin radial (present in a few species of *Roeboides*).

SEXUAL DIMORPHISM Mature males have bony hooks on the anal fin in all genera except *Acanthocharax*, for which this still needs to be confirmed. Overview of literature on the presence of anal fin hooks for each genus in Mattox and Toledo-Piza (2012).

HABITATS Variable; most species are found in lentic environments, usually associated with vegetation.

FEEDING ECOLOGY Most feed on other fishes and insects (Lucena and Menezes 2003). However, species of *Roeboides* have specialized mammiliform teeth outside the mouth, pointing forward, associated with scale eating (Mattox and Toledo-Piza 2012).

BEHAVIOR *Cynopotamus* may undertake long-distance migrations (Taphorn 1992). *Acestrocephalus* makes short movements but is not considered a migrant.

ADDITIONAL NOTES A few species of *Charax, Phenacogaster,* and *Roeboides* are exploited as ornamental fishes for the aquarium trade.

KEY TO THE GENERA

1a. Cycloid scales 2
1b. Spinose scales 5

2a. Preventral region flat and covered by 2 longitudinal series of elongate scales overlapped in the center and bent at the sides (fig. 2a; based on Galvis et al. 2006); dorsal profile without gibbosity (i.e., no hump) ....*Phenacogaster*
2b. Preventral region round with >2 longitudinal series of scales not overlapped in the center (fig. 2b); dorsal profile with gibbosity (i.e., with a hump; fig. 2b)................................3

gibbosity

2a   2b

3a. Presence of spiniform projection on posterior margin of preopercle (fig. 3a) ....*Acanthocharax*
3b. Absence of spiniform projection on posterior margin of preopercle (fig. 3b)..................4

3a   3b   4a

4a. Presence of mammiliform teeth outside mouth in adults (fig. 4a)............................*Roeboides*
4b. Absence of mammiliform teeth outside mouth in adults...*Charax*

5a. Posteroventral margin of cleithrum notched; inner row on dentary absent or represented by 1–3 small conical teeth......................*Cynopotamus*
5b. Posteroventral margin of cleithrum not notched; inner row on dentary present with 6 or more small conical teeth 6

6a. Presence of laminar expansion lateral to tubular portion of nasal bone (fig. 6a; after Mattox and Toledo-Piza 2012); 77–105 perforated lateral line scales; anal fin with 33–48 branched rays ....*Galeocharax*
6b. Absence of laminar expansion lateral to tubular portion of nasal bone (fig. 6b); 67–79 perforated lateral line scales; anal fin with 25–36 branched rays ....*Acestrocephalus*

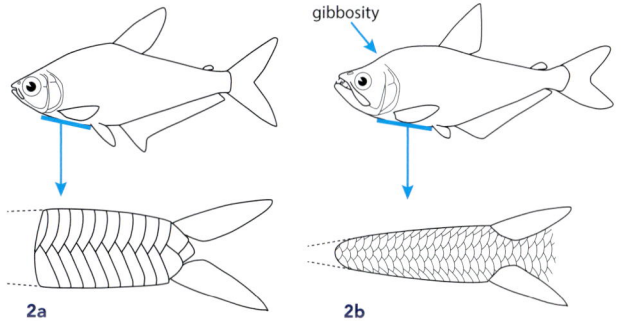

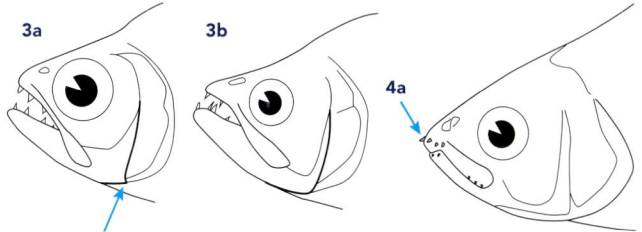

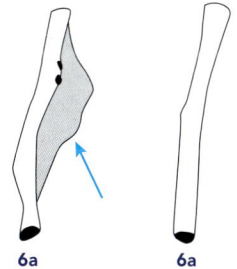

6a   6a

GENUS ACCOUNTS

Acanthocharax (8.5 cm SL)

Characterized by: oblique mouth; very large eye; paddle-shaped maxilla; pectoral fins very large; a spiniform projection on the posteroventral margin of preopercle; and naked predorsal region (i.e., predorsal scales absent). **SPECIES** One, *A. microlepis*. **DISTRIBUTION AND HABITAT** Essequibo and Potaro rivers, Guyana. **BIOLOGY** Carnivorous.

Acestrocephalus (3.0–14 cm SL)

Characterized by (Menezes 2006): body comparatively small; anterior dorsal region not elevated; dorsal profile of body regularly curved from tip of snout to caudal base; lower part of antorbital only in contact with maxilla; first infraorbital relatively short, high on its median part; nasal bone tubular; cleithrum not notched, just with a slight sinuosity along its ventral edge; dentary with two rows of teeth, anterior teeth of external row much larger than posterior ones, first and third anterior teeth canine-like, more developed than the other two; posterior dentary tooth row with 20–40 small conical teeth slightly curved posteriorly, their number tending to increase with fish size; inner row of teeth on dentary formed by 7–14 small conical teeth; scales comparatively large and numerous, perforated lateral line scales 67–79; 10–15 scales above and 9–13 scales below lateral line; anal fin with iv–v + 25–36 rays, its origin situated on vertical line always crossing behind middle of dorsal-fin base length; pectoral fin with i + 11–16 rays. **SPECIES** Eight, including seven in the AOG region. Key to the species and distribution map in Menezes (2006). **COMMON NAMES** *Dientón* (Ecuador); *Dentón* (Peru). **DISTRIBUTION AND HABITAT** Seven species in the Amazon basin, one of which (*A. sardina*) also found in the Orinoco basin. Typically found near the shore of large rivers, usually associated with aquatic vegetation (Taphorn 1992), but also found in lakes (Lima et al. 2005). **BIOLOGY** Carnivorous; Menezes (1976) found fish remains in the stomach of *A. sardina*.

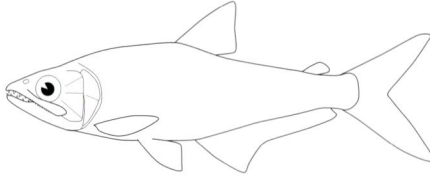

Charax (4.0–14 cm SL)

Distinguished from the similarly appearing *Cynopotamus*, *Galeocharax*, and *Roeboides* by a well-developed notch on the posteroventral margin of the cleithrum, just anterior to the base of the pectoral-fin rays, which forms a long posteriorly directed projection (Mattox and Toledo-Piza 2012). A notch on the posteroventral margin of the cleithrum is also present in *Acanthocharax*, *Cynopotamus*, *Phenacogaster*, and *Roeboides*, but in these genera it is much smaller than in species of *Charax*. The well-developed notch is a specialized character of the genus *Charax* (Lucena 1987, Mattox and Toledo-Piza 2012). Most species bear a conspicuous gibbosity on the dorsal profile of head, which is less pronounced in *C. condei* and *C. rupununi*. **SPECIES** 16, including 15 species in the AOG region. Key to the species and information on species distributions in Menezes and Lucena (2014). **COMMON NAMES** *Giboso* (Brazil); *Dentón*, *Pez giboso* (Peru); *Pez jibao* (Venezuela); Glass headstander (English). **DISTRIBUTION AND HABITAT** Most species occur in the Amazon basin; *C. gibbosus*, *C. hemigrammus*, and *C. rupununi* also occur in coastal drainages of the Guianas and Suriname, and *C. apurensis* and *C. notulatus* are found exclusively in the Orinoco basin. *Charax* inhabit many habitats, but predominantly in lentic environments, usually associated with vegetation (Taphorn 1992, Santos et al. 2004, Lima et al. 2005). **BIOLOGY** Most species are crepuscular or nocturnal, hiding during the day between submerged vegetation and roots. Consume adult and larval insects, shrimp, fish larvae and small fish (Planquette et al. 1996, Galvis et al. 2006).

Cynopotamus (8.5–24 cm SL)

Characterized by: relatively large body sizes; one tooth series on dentary (and sometimes one additional small conical tooth located slightly posteriorly to the canine teeth, near the symphysis); spinous scales; predorsal scales present; a single axillary scale covering only the base of the first pelvic-fin ray; a well-developed infraorbital-4 contributing to two margins of orbital ring; anterior tip of pelvic bone located posterior to vertical through posterior margin of cleithrum. **SPECIES** 12, including 7 species in the AOG region. Key to the species and information on species distributions in Menezes (2007). **DISTRIBUTION AND HABITAT** Three species (*C. amazonum*, *C. juruenae*, and *C. xinguano*) in the Amazon basin, two species (*C. gouldingi* and *C. tocantinensis*) in the Araguaia-Tocantins basin, one species (*C. essequibensis*) in the coastal drainages of the Guianas, and one (*C. bipunctatus*) in the Orinoco basin. Typically found near the shore of large rivers, usually associated with aquatic vegetation (Taphorn 1992). **BIOLOGY** All species are carnivorous. Menezes (1976) found fish in the stomach of most examined specimens of *C. amazonum* and *C. essequibensis*, as well as remains of crustaceans. Possibly capable of making long migrations (Taphorn 1992).

Galeocharax (5.0–25 cm SL)

Distinguished from *Acestrocephalus* and *Cynopotamus* by lateral expansion of tubular portion of nasal bone. In *Galeocharax*, the scales on caudal fin partially cover the middle caudal-fin rays. This is also observed in most *Cynopotamus* species, but *Galeocharax* differ from the former by having two tooth series on dentary (vs. one series in *Cynopotamus*): the first is a more externally (i.e., labially) positioned series of teeth on the anterior portion of the dentary, formed mainly by large canines, and the second tooth series has 6–13 small conical teeth positioned internally. In *Galeocharax* the posteroventral portion of the cleithrum is not notched as in *Cynopotamus*. Other characters that might aid in the identification of the genus are: lamellar projection on anteroventral margin of infraorbital 1, which extends anterior to the anterior margin of the antorbital along the medial surface of the latter ossification; and a well-developed anteriorly directed projection along the anteroventral margin of cleithrum (also present in *Charax*, *Roeboides*, and *Acestrocephalus sardina*) (Mattox and Toledo-Piza 2012). **SPECIES** Four, including two species in the AOG region. **COMMON NAMES** *Dentusca* (Brazil); *Dentón* (Peru). **DISTRIBUTION AND HABITAT** *G. gulo* (25 cm) in the Amazon and Tocantins-Araguaia basins; *G. goeldii* (5 cm) in the Madeira basin. Typically found near the shore of large rivers, usually associated with aquatic vegetation (Taphorn 1992), but also found in streams (Santos et al. 2004). **BIOLOGY** All species are carnivorous and at least *G. knerii*, a species that does not occur in the AOG, is piscivorous (Fugi et al. 2008). Menezes (1976) found mainly fishes in the stomachs of *G. gulo*, but also reported adult and larval insects and crustaceans.

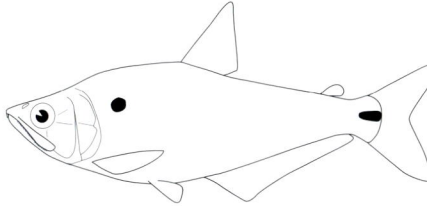

Phenacogaster (3.2–6.0 cm SL)

Distinguished by: dorsal profile of the body not notably humped; preventral region flat and covered by two longitudinal series of elongate and imbricated scales that are different in shape from the remaining body scales and form a zigzag pattern in ventral view. Certain species have a few median interpolated small

scales (Malabarba and Lucena 1995). In addition, in most (but not all) *Phenacogaster* species the external premaxillary tooth row is divided into a medial and a lateral section isolated from each other by a gap, with the medial section having tricuspid teeth and the lateral section conical to tricuspid teeth (Lucena and Malabarba 2010). Some species are translucent in life. **SPECIES** 20, including 15 species in the AOG region. Key to the species and information on species distributions in Lucena and Malabarba (2010). **COMMON NAMES** *Peixes de vidro* (Brazil); *Pez vidrio* (Peru). **DISTRIBUTION AND HABITAT** Streams and margins of small rivers throughout the AOG region. Found mainly in small tributaries of main basins and streams (Taphorn 1992, Santos et al. 2004, Lima et al. 2005). **BIOLOGY** All species are carnivorous. *P. megalostictus* feeds mainly on aquatic invertebrates in the Sinnamary River, French Guiana (Horeau et al. 1998), and *P. pectinatus* feeds on aquatic and terrestrial invertebrates in streams of the Bolivian Amazon (Ibañez et al. 2007).

Roeboides (5.0–19 cm SL)

Distinguished from similar-looking genera (i.e., *Charax*, *Cynopotamus*, *Acestrocephalus*, and *Galeocharax*) by the presence of teeth outside the mouth in adults (also found in *Bryconexodon*, *Exodon*, *Probolodus*, *Roeboexodon*, and *Serrabrycon*, characid genera not included in the Characinae). These teeth are arranged in distinct patterns

on the external surfaces of the premaxilla, maxilla, and dentary. Size and number of these teeth vary among *Roeboides* species. Some species also have mammiliform teeth inside the mouth, along the regular series of conical teeth (Mattox and Toledo-Piza 2012). The external teeth are used to remove scales from other fishes, which are part of their diets. **SPECIES** 21 species in four species groups (Lucena 1998), including 11 species in the AOG region. Key to the species in the *R. affinis* group in Lucena (2007). **COMMON NAMES** *Dentón* (Peru); *Pez jibao* (Venezuela). **DISTRIBUTION AND HABITAT** Widespread in South America and southern Central America; throughout the AOG region. *Roeboides* typically inhabits slow portions of streams and other lentic environments such as ponds, backwaters, and swamps (Taphorn 1992, Santos et al. 2004). **BIOLOGY** All species lepidophagous as adults, feeding on scales of other fish (e.g., Sazima and Machado 1982, Sazima 1983). At least a few species shift diet during ontogeny, with juveniles eating insects and crustaceans when the external mammiliform teeth have not yet developed, and becoming more specialized in eating scales as adults when these teeth develop (Peterson and Winemiller 1997, Peterson and McIntyre 1998, Hahn et al. 2000).

SUBFAMILY CHEIRODONTINAE—CHEIRODONTINE TETRAS
— *FERNANDO C. JEREP, PETER VAN DER SLEEN, and LUIZ R. MALABARBA*

DIVERSITY About 55 species in 17 genera, including 23 species in six genera within the AOG region: *Amazonspinther*, *Cheirodontops*, *Ctenocheirodon*, *Odontostilbe*, *Prodontocharax*, and *Serrapinnus*.

TAXONOMIC NOTE Most cheirodontine species are members of two tribes, Cheirodontini (7 genera: *Amazonspinther*, *Cheirodon*, *Ctenocheirodon*, *Heterocheirodon*, *Nanocheirodon*, *Serrapinnus*, and *Spintherobolus*), and Compsurini (5 genera: *Acinocheirodon*, *Compsura*, *Kolpotocheirodon*, *Macropsobrycon*, and *Saccoderma*), with five genera (*Aphyocheirodon*, *Cheirodontops*, *Odontostilbe*, *Prodontocharax*, and

Pseudocheirodon) treated *incertae sedis* (with uncertain phylogenetic relationships) within the subfamily. Three species of Characidae *incertae sedis* in the Amazon basin are artificially placed in Cheirodontinae: '*Cheirodon*' *luelingi* and '*Cheirodon*' *ortegai* from the Ucayali basin, and '*Macropsobrycon*' *xinguensis* from the Xingu basin. These species do not possess characters of Cheirodontinae, *Cheirodon*, or *Macropsobrycon* (Malabarba 1998, Jerep and Malabarba 2011) and are excluded here.

GEOGRAPHIC DISTRIBUTION Central and South America, from southern Costa Rica to central Argentina and Chile. Most species inhabit Atlantic drainages, with only four species of the genus *Cheirodon* on the Pacific slope of the Andes in southern Chile (Malabarba 2003).

ADULT SIZES Small sizes, from a maximum size of 3.0 cm SL in several *Serrapinnus* species, to 5.6 cm SL in *Odontostilbe pequira* from the Paraguay and lower Paraná basins, and to 5.0 cm SL in females of *O. ecuadorensis* from the western Amazon of Peru and Ecuador, the largest cheirodontine species in the AOG region (Bührnheim and Malabarba 2006).

DIAGNOSIS OF THE SUBFAMILY: Can be recognized by four characters (Malabarba 1998): a large, nearly triangular pseudotympanum formed by a hiatus in the muscles covering the anterior chamber of swim bladder between the first and second pleural ribs; absence of a dark pigmented humeral spot; pedunculate teeth that are expanded and compressed distally; and one series of teeth in the premaxilla with teeth usually perfectly aligned and similar in shape and cusps number. The Cheirodontinae as circumscribed herein (Malabarba 1998) is also supported by genetic analyses (Calcagnotto et al. 2005, Javonillo et al. 2010).

SEXUAL DIMORPHISM Many species are sexually dimorphic, with simple hooks present along the posterior margin of pelvic-fin and anal-fin rays in mature males (e.g., *Odontostilbe*), hypertrophy of dermal tissues and scales in the caudal fin (e.g., Compsurini), expansion and fusion of anal-fin ray segments, or hypertrophy of ventral procurrent caudal-fin rays (e.g., Cheirodontini).

HABITATS Usually inhabit the margins of small lowland rivers, streams, and lakes.

FEEDING ECOLOGY Most species are herbivorous or omnivorous.

BEHAVIOR The Compsurini (*Acinocheirodon*, *Compsura*, *Kolpotocheirodon*, *Macropsobrycon*, and *Saccoderma*) are inseminating fishes, meaning that males transfer sperm from their testis to female ovaries. Inseminating male characids often have hypertrophied dermic tissues at the base of the caudal fin, sometimes associated with hooks and/or modified scales. This caudal "gland" is hypothesized to produce secretions used as chemical signals (pheromones) for communication between males or between the sexes, related to courtship and insemination (Oliveira et al. 2012).

KEY TO THE GENERA
BASED ON MALABARBA (1998), BÜHRNHEIM ET AL. (2008), AND MALABARBA AND JEREP (2012)

1a. Lateral line incompletely pored...**2**
1b. Lateral line completely pored **3**

2a. Pseudotympanum with 2 muscular hiatuses (fig. 2a);
3 dark blotches on the base of the dorsal, caudal, and
anal fins...*Amazonspinther*
2b. Pseudotympanum with a single muscular hiatus (fig. 2b); dark
blotches, if present, on the base of the caudal fin or on the
bases of both dorsal and caudal fins...*Serrapinnus*

3a. 16–19 ventral procurrent caudal-fin rays.. *Ctenocheirodon* (upper Tocantins basin)
3b. 5–11 ventral procurrent caudal-fin rays...**4**

4a. Three anterior dentary teeth large; each dentary tooth with at least 5 cusps, with the three central cusps
larger, compressed into a row forming a sharp cutting edge, cusp tips oriented distally .... *Cheirodontops* (Orinoco basin)
4b. Usually 4–7 anterior large dentary teeth, decreasing in size posteriorly; dentary teeth with tooth cusps,
but usually not arranged into a sharp cutting edge...**5**

5a. Mouth distinctively subterminal; maxilla curved, with tooth-bearing portion continuous to premaxillary
teeth and angled relative to the nontoothed portion of the maxilla; maxilla short, posterior tip reaching
the area of contact between infraorbitals 1 and 2...*Prodontocharax*
5b. Mouth terminal; maxilla with tooth-bearing portion angled relative to the premaxilla; maxilla slightly more
elongate, posterior tip of maxilla reaching one-third of length of infraorbital 2................................. *Odontostilbe*

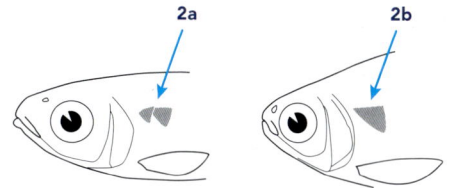

GENUS ACCOUNTS

Amazonspinther (2.0 cm SL)

Distinguished from all characid genera by 3 conspicuous black blotches on the base of the dorsal, anal, and caudal fins (Bührnheim et al. 2008). Distinguished from other cheirodontines by: anteriormost proximal radial of the anal fin has an anteriorly extended lamina entering the abdominal cavity, between the distal portions of the twelfth to fourteenth pleural ribs (vs. short anteriorly extended lamina, not entering the abdominal cavity and not between pleural ribs); and caudal peduncle extremely elongate, corresponding to 27.3–30.2% of SL (vs. 11.0–19.6% of SL in other cheirodontines). **SPECIES** One, *A. dalmata*. **DISTRIBUTION AND HABITAT** Middle Purus and middle and lower Madeira basins. Occupies shallow stretches of both forest and Amazonian savanna streams (J. Zuanon pers. comm.). **BIOLOGY** No data available.

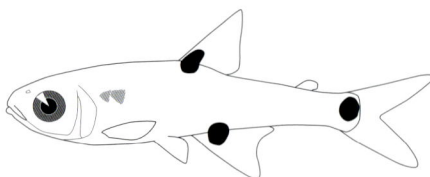

Cheirodontops (4.0 cm SL)

Distinguished from other cheirodontines by: completely pored lateral line; 3 large anterior dentary teeth; each dentary tooth with at least 5 cusps, with the 3 central cusps larger, compressed in a row forming a sharp cutting edge, cusp tips oriented distally (Malabarba 1998). Other characters that might aid identification: maxilla without teeth; 6 premaxillary teeth, each with 3 prominent cups, the middle one longest, with a minute cusp on the side near base of each outer enlarged cusp; caudal fin with scales only at base; 23–25 anal-fin rays; 35–37 lateral line scales, 5 scales above and 4 scales below lateral line (Schultz 1944b, Géry 1977). **SPECIES** One, *G. geayi*. **DISTRIBUTION AND HABITAT** Orinoco basin. **BIOLOGY** External fertilization (Oliveira et al. 2012).

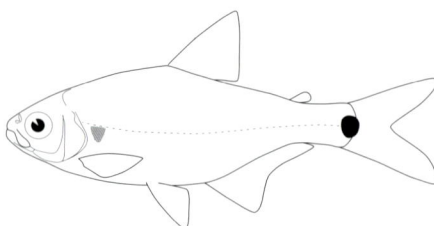

Ctenocheirodon (3.0 cm SL)

Distinguished from other cheirodontines by: a completely pored lateral line; 16–19 ventral procurrent caudal-fin rays; the proximal end of anal-fin rays with lepidotrichia bases extended anteriorly; large ligaments between the enlarged anal-fin rays of males, connecting approximately the midlength of the posterior face of each proximal segment of the lepidotrichia to the proximal anterior face of the lepidotrichia of the subsequent ray, with the ligament diameter nearly equal to the diameter of the expanded anal-fin rays; anterior branched anal-fin rays 1 through 4–8 of males (usually those bearing hooks) slab-shaped and more expanded in the sagittal plane than comparable rays in females; hooks well developed only on the slab-shaped anal-fin rays; anal-fin hooks bilaterally asymmetrical and unpaired with an irregular arrangement, differing in number, size, orientation, and/or position relative to those on the contra-lateral segments of the lepidotrichia; distal tip of ventral procurrent caudal-fin rays conical in adult males (Malabarba and Jerep 2012). **SPECIES** One, *C. pristis*. **DISTRIBUTION AND HABITAT** Upper Tocantins basin, inhabiting clearwater over sand or gravel bottom with slow current and no vegetation (Malabarba and Jerep 2012). **BIOLOGY** No data available.

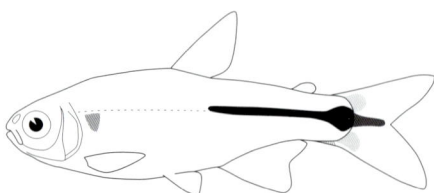

Odontostilbe (3.0–5.6 cm SL)

Distinguished from other cheirodontines by: elongation of the second unbranched dorsal-fin ray in males (except in *O. euspilura*), and elongation of the unbranched pelvic-fin ray in males. The presence of a completely pored lateral line and 5–11 ventral procurrent caudal-fin rays also helps to identify the genus. This combination of characters is also present in *Cheirodontops*, which is differentiated by its tooth morphology, with 3 large anterior dentary teeth versus usually 4–7 large anterior dentary teeth (decreasing in size posteriorly; dentary teeth with tooth cusps not arranged into a sharp cutting edge) in *Odontostilbe* (also in *Prodontocharax*). **SPECIES** 12, including 10 in the AOG region. **DISTRIBUTION AND HABITAT** Six species in the Amazon basin, three species in the Orinoco basin, and *O. pulchra* in the Essequibo basin in Guyana and on Trinidad (Bührnheim and Malabarba 2007). Often found in sandy beaches of large whitewater rivers and lentic sections of rivers, including oxbow lakes and forest creeks (Bührnheim and Malabarba 2006, 2007). **BIOLOGY** Considered herbivorous based on tooth morphology, but most species are omnivorous with seasonal diet shifts (Malabarba 2003, Lima et al. 2012). *Odontostilbe pequira* from the Paraguay basin feeds mainly on plants, but also on animal prey, including the scales of small fishes and possibly the mucus and epidermis of larger fish species (Lima et al. 2012). Reproductive activity for *O. pequira* in the southern Pantanal is associated with the flood pulse (Tondato et al. 2014).

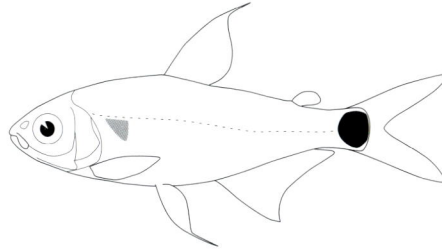

Prodontocharax (4.0–5.0 cm SL)

Distinguished from other cheirodontines by subterminal mouth and forward-pointing dentary teeth, almost horizontal (parallel to the longitudinal body axis). The dentary teeth are somewhat spatulate and shovel-shaped. Other characters useful for identification: maxilla short, with posterior tip reaching the area of contact between the infraorbitals 1 and 2, and maxilla strongly curved, presenting a tooth-bearing portion continuous with the premaxillary teeth and deeply angled relative to the posterior nontoothed portion of the maxilla (Malabarba 1998). **SPECIES** Three, all in the Amazon basin. **DISTRIBUTION AND HABITAT** *P. alleni* in the Amazon and Ucayali basins, *P. howesi* in the Amazon basin, and *P. melanotus* in the Beni, Itenez, and upper Madre de Dios basins. **BIOLOGY** External fertilization (Oliveira et al. 2012).

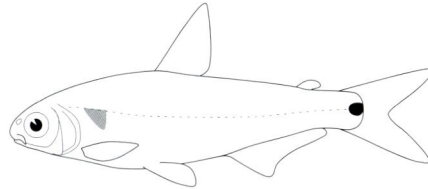

Serrapinnus (2.0–4.0 cm SL)

Distinguished from other cheirodontines by: ray segments of the expanded anal-fin rays progressively fused to one another as males become fully mature; caudal peduncle in adult males deeply arched ventrally, the last vertebrae reaching a 45° angle relative to the first caudal vertebrae; adult males with 2–4 (or sometimes 5) hooks on the posterior border of each anal-fin hook-bearing segment; adult males with spatulate ventral procurrent caudal-fin rays. **SPECIES** 14, including 8 in the AOG region. **DISTRIBUTION AND HABITAT** *S. gracilis* and *S. littoris* in the French Guiana; *S. microdon*

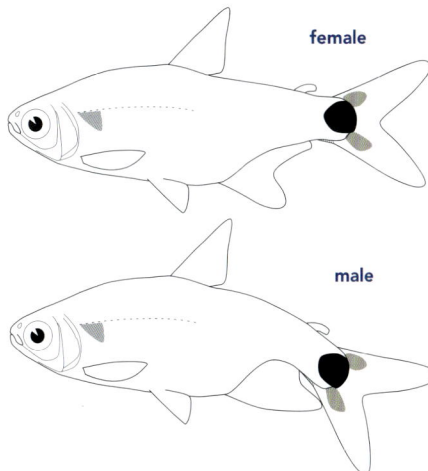

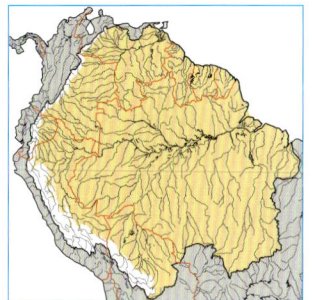

female

male

and *S. micropterus* widespread in most of the Amazon basin; four species in the Tocantins-Araguaia basin: *S. aster*, *S. lucindai*, *S. sterbai*, and *S. tocantinsensis*. Occupy many habitats, including river margins, small forest streams, and oxbow lakes, and *S. gracilis* has been shown to change habitats by season (Planquette et al. 1996). **BIOLOGY** *S. gracilis* was found in groups of 1:1 sex ratio (Planquette et al. 1996). Food items for some species from northeastern Brazil include aquatic insects, microcrustaceans, algae and detritus (Dias and Fialho 2009, Alves et al. 2011).

SUBFAMILY HETEROCHARACINAE—HETEROCHARACINE TETRAS
— *MÔNICA TOLEDO-PIZA and GEORGE M. T. MATTOX*

DIVERSITY 15 species in 7 genera, grouped into two tribes: Roestini (*Roestes* and *Gilbertolus*) and Heterocharacini (*Gnathocharax*, *Heterocharax*, *Hoplocharax*, *Lonchogenys*, and *Priocharax*) (Mattox and Toledo-Piza 2012).

TAXONOMIC NOTE The Heterocharacinae of Mirande (2010) included four genera (*Gnathocharax*, *Heterocharax*, *Hoplocharax*, and *Lonchogenys*), but four other genera (*Roestes*, *Gilbertolus*, *Gnathocharax*, and *Priocharax*) were not included in his analysis. Previous analyses placed *Roestes* and *Gilbertolus* with *Cynodon*, *Hydrolycus*, and *Rhaphiodon* in the Cynodontinae (Menezes and de Lucena 1998), and *Priocharax* in the Characinae (Lucena 1998). The comprehensive morphological study by Mattox and Toledo-Piza (2012) showed that the genera *Lonchogenys*, *Heterocharax*, *Hoplocharax*, and *Gnathocharax* are closely related to *Roestes* and *Gilbertolus*, and that *Priocharax* is more closely related to Heterocharacini than to Characinae.

GEOGRAPHIC DISTRIBUTION Twelve species in six genera inhabit the AOG region. The genus *Gilbertolus*, with three species, is restricted to trans-Andean drainages of Colombia and Venezuela.

ADULT SIZES From 1.5 cm SL in the miniature species of *Priocharax*, to 20 cm SL in species of *Roestes*.

DIAGNOSIS OF SUBFAMILY All heterocharacines share the following salient characters: short snout with relatively large eyes; conical teeth on both jaws of variable size and distribution (useful to help diagnose species); pseudotympanum restricted to the area anterior to the rib of the fifth vertebra; and a notch present on the posterior opercular margin (the latter character absent in the miniature *Priocharax*, and in *Gnathocharax*). Other heterocharacine characters include: spiniform projection on posterior margin of preopercle (absent in the miniature *Priocharax*); dorsal limit of pseudotympanum composed of the *obliquus superioris* muscle; absence of origin of *levator operculi* on hyomandibula due to relative position of hyomandibula and opercle (not examined in the miniature *Priocharax*); posteroventral margin of orbitosphenoid with notch forming posteriorly directed spiniform process; inner arm of *os suspensorium* extending to vertical through second vertebral centrum and aligned in approximately vertical plane (Mattox and Toledo-Piza 2012).

SEXUAL DIMORPHISM Mature males have bony hooks on the anal fin in species of all genera. See Mattox and Toledo-Piza (2012) for full references to hooks on anal fin for each genus. *Gnathocharax steindachneri* also has bony hooks on the pelvic and dorsal fins, and at least one species of *Priocharax* also has bony hooks on the pelvic fins. In addition, the basipterygium (pelvic bone) is more developed in males of the latter genus than in females, reaching further anteriorly and also articulated with the contralateral basipterygium (Mattox et al. 2016).

HABITATS Species of the Heterocharacini usually stay in the water column or close to the shoreline, in or near emergent and marginal vegetation. *Roestes* occurs in places where the substratum is composed predominantly of submerged leaf litter banks.

FEEDING ECOLOGY Feed mostly on aquatic and terrestrial insects, microcrustaceans, and small fishes.

BEHAVIOR At least *Gnathocharax* and *Priocharax* are known to form associations with other species.

ADDITIONAL NOTES Species of *Heterocharax* and *Gnathocharax steindachneri* are exploited as ornamental fishes and are occasionally found in the aquarium trade.

KEY TO THE GENERA

1a. Prepelvic region keeled .... *Gnathocharax*
1b. Prepelvic region not keeled 2

2a. Presence of a larval rayless rounded pectoral fin in adults; miniature species .... *Priocharax*
2b. Pectoral fin with well-developed rays in adults; nonminiature species 3

3a. Presence of two well-developed spines on the posterior margin of the opercle .... *Hoplocharax*
3b. Absence of spines on the posterior margin of the opercle 4

4a. >75 perforated lateral line scales.. *Roestes*

4b. <45 perforated lateral line scales...5

5a. Premaxilla with two series of teeth, 10–12 on the outer series and 8–10 on the inner series; inner row of dentary teeth reaching symphysis...*Lonchogenys*

5b. Premaxilla with one series of teeth or sometimes with only one or two teeth slightly displaced anteriorly from main series; inner row of dentary teeth incipient, not reaching symphysis..............................*Heterocharax*

GENUS ACCOUNTS

Gnathocharax (5.0 cm SL)

Distinguished by: body and mouth shape similar to Roestes, but with a smaller body size and no nuchal hump. Coracoids enlarged, forming a well-developed ventral keel; pectoral fins very large, extending to beyond anal-fin origin; anal fin with <32 branched rays; distal margin of anal fin with a notch on anterior one-third; lateral line incomplete, <36 longitudinal scales; upper jaw with single series of conical teeth; lower jaw with 1 canine medially separated from 3 other canines gradually larger in size laterally, followed by 2 or 3 small conical teeth. **SPECIES** One, *G. steindachneri* (Arowana tetra). **DISTRIBUTION AND HABITAT** Amazonas and Orinoco basins, coastal rivers of the Guianas. *Gnathocharax* is usually found close to river and stream margins, between branches and hanging roots of riparian vegetation (Brejão et al. 2013). **BIOLOGY** *Gnathocharax* is a surface striker, feeding mainly on insects. Occasionally forms mixed schools with species of *Carnegiella* (Brejão et al. 2013).

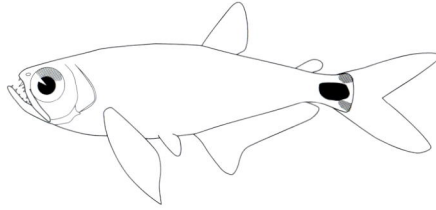

Heterocharax (3.4–4.8 cm SL)

Distinguished by: small adult body sizes; body somewhat diamond-shaped with obtuse angle, more pronounced at anal-fin origin and less so at dorsal-fin origin; mouth terminal to slightly upturned; eyes large, more than 40% of head length; all teeth conical; premaxilla with a single row of 3 or 4 larger anterior teeth followed by a series of up to 9 smaller teeth; lower jaw with 6–8 teeth, medial tooth and two lateral-most teeth largest, bracketing smaller teeth of similar size; a row of small posterior teeth on the dentary of which the anterior ones may be located behind the anterior row, forming an incipient second series of teeth; lateral line complete, with more than 32 perforated scales; more than 28 branched anal-fin rays. **SPECIES** Three, all in the AOG region. Known as Opal tetras. Key to species and information on species distributions in Toledo-Piza (2000b). **DISTRIBUTION AND HABITAT** *H. macrolepis* is widespread throughout the AOG region, while *H. virgulatus* and *H. leptogrammus* are restricted to blackwater streams of the Negro and Orinoco basins. **BIOLOGY** Reported diet for *H. macrolepis* includes detritus and aquatic and terrestrial insects (Roepke et al. 2014). Early developmental stages of *H. macrolepis* studied by Mattox et al. (2014b).

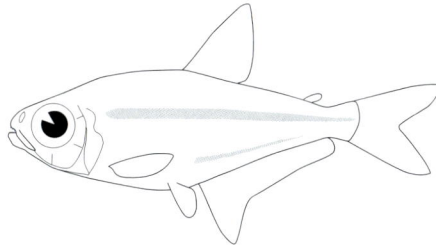

Hoplocharax (3.1 cm SL)

Distinguished by: small adult body size, easily recognized by two well-developed spiniform projections extending posteriorly from the posterior margin of the opercle, with the ventral projection longer and extending beyond the posterior margin of the cleithrum. The first unbranched pectoral-fin ray, the anterior unbranched anal-fin rays, and the dorsal and ventral procurrent caudal-fin rays are transformed into spines; dorsal fin with 7 branched rays; mouth terminal to slightly upturned; eyes large, more than 40% of head length; all teeth conical, arranged in a single row in the upper jaw; anterior row of dentary teeth with 5–6 teeth, of which the medial and lateral-most are the largest; a row of small posterior teeth on the dentary of which the anterior ones are located behind the anterior row, forming an incipient second series of teeth; predorsal scales absent; lateral line incomplete, with more than 45 longitudinal scales; fewer than 28 branched anal-fin rays. **SPECIES** One, *H. goethei*. Species information in Géry (1966). **DISTRIBUTION AND HABITAT** Negro, Madeira, and upper Orinoco basins. Undescribed species in the Purus River (J. Zuanon pers. comm.). Typically found in shallow black- and clearwater streams. **BIOLOGY** Lives in loose mixed schools of similar-sized characins (J. Zuanon pers. comm.).

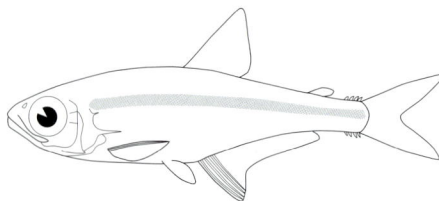

Lonchogenys (7.5 cm SL)

The largest species of Heterocharacini, recognized by two series of conical teeth on both jaws; premaxilla with an outer row of 10–12 and inner row of 8–10 teeth, of which 2 located medially are larger; outer row of dentary teeth with numerous small conical teeth of which the lateral 3 or 4 are larger, followed by a series of smaller teeth; inner row with a medial tooth near the symphysis followed by a series of smaller teeth; lateral line complete, with up to 36 perforated scales; branched anal-fin rays 31–37. **SPECIES** One, *L. ilisha*. Species information in Géry (1966). **DISTRIBUTION AND HABITAT** Blackwater streams in the Negro and upper Orinoco basins. **BIOLOGY** Specimens of *L. ilisha* are usually present in nocturnal samples (Arrington and Winemiller 2003; pers. obs.) and possibly approach shallow margins of rivers after sunset.

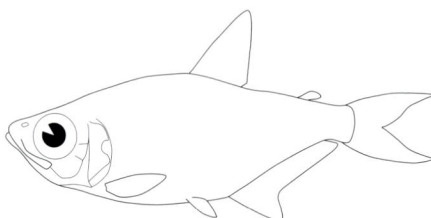

Priocharax (1.5–1.7 cm SL)

Miniature fishes recognized by retention of a rayless larval pectoral fin fold in the adult; jaws with a high number of very small conical teeth arranged in a single series; and 5–6 branched pelvic-fin rays. Body mostly transparent in life with red-brown chromatophores scattered on head and body. **SPECIES** Three, all in the AOG region. Species information in Weitzman and Vari (1987) and Toledo-Piza et al. (2014). **DISTRIBUTION AND HABITAT** Amazon and Orinoco basins. Specimens usually collected in still waters of shaded areas, close to the shoreline in or near emergent and marginal vegetation and leaf litter banks. **BIOLOGY** *P. ariel* is reported to hover in the water column at daytime and lunge at prey on the surface, in midwater, and near the bottom. At night it hovers almost stationary above the leaf litter, usually near an object. It forms associations with translucent-bodied *Microphilypnus* and palaemonid shrimps, suggested to be a protective association similar to numerical or social mimicry (Carvalho et al. 2006).

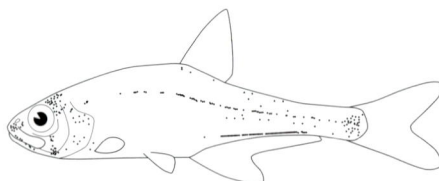

Roestes (14–20 cm SL)

Characterized by: a sharp angle between the lower jaw surface and the prepectoral outline, and its characteristic adult body shape with a pronounced nuchal hump located immediately behind the head. A nuchal hump is also present in some other characins, including *Charax*, *Roeboides*, and *Cynopotamus*. In *Roestes* the top of the head is approximately horizontal and the mouth is superior (upturned). Upper jaw with a single series of conical teeth; lower jaw with 3 caniniform teeth followed by a series of smaller conical teeth; 1 or 2 conical teeth near symphysis posterior to anterior row; eyes very large, always larger than the snout; laterosensory canal of second infraorbital bone with an accessory branch; region of abdomen anterior to pelvic fin rounded, not keeled; pectoral fin large, reaching to anal-fin origin; distal margin of anal fin more or less straight (without a deep notch on anterior one-third); anal fin long, with >38 branched rays; lateral line complete, with >77 perforated scales (Lucena and Menezes 1998). **SPECIES** Three, all in the Amazon and Essequibo basins. Key to species and species information in Menezes and de Lucena (1998). **DISTRIBUTION AND HABITAT** *R. itupiranga* in the Tocantins basin, Brazil; *R. molossus* in the upper Madeira basin, Brazil and Bolivia; *R. ogilviei* in the upper and middle Amazon basin and Essequibo, Branco, and Negro basins. Specimens are usually captured in places where the substratum is composed predominantly of submerged leaf litter banks (Torrente-Vilara et al. 2008). **BIOLOGY** Feed on other fishes, insects and shrimps, mainly at dawn and at night. Spawning season in the Madeira coincides with the beginning of the rising water (Torrente-Vilara et al. 2008).

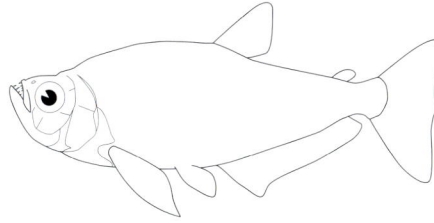

SUBFAMILY STEVARDIINAE—STEVARDIINE TETRAS
— *PETER VAN DER SLEEN, ANDRÉ L. NETTO-FERREIRA, and LUIZ R. MALABARBA*

DIVERSITY Stevardiinae is one of the most diverse subfamilies of Characidae, with nearly 330 valid species in 44 genera (Thomaz et al. 2015), including about 157 species in 29 genera within the AOG region. Most of the species-level diversity in Stevardiinae is included in four genera (*Creagrutus*, *Bryconamericus*, *Hemibrycon*, and *Knodus*) with a combined total of 200 species described in the subfamily (Thomaz et al. 2015).

COMMON NAMES *Piaba*, *Lambari* (Brazil); *Mojarita* (Peru).

TAXONOMIC NOTE The subfamily Stevardiinae comprises the genera in 'Clade A' of Malabarba and Weitzman (2003), including the subfamily Glandulocaudinae *sensu* Weitzman and Menezes (1998) with 19 genera, plus *Cyanocharax* and 18 genera of uncertain relationships previously listed as Cheirodontinae or Tetragonopterinae (Géry 1977) and classified as *incertae sedis* in Characidae by Lima et al. (2003): *Attonitus*, *Boehlkea*, *Bryconacidnus*, *Bryconamericus*, *Caiapobrycon*, *Ceratobranchia*, *Creagrutus*, *Hemibrycon*, *Hypobrycon*, *Knodus*, *Microgenys*, *Monotocheirodon*, *Odontostoechus*, *Othonocheirodus*, *Piabarchus*, *Piabina*, *Rhinobrycon*, and *Rhinopetitia*. Later studies added ten more genera to 'Clade A'. The group has been supported by both morphological (Mirande 2010, Baicere-Silva et al. 2011b, Ferreira et al. 2011, Mirande et al. 2013) and molecular data (Calcagnotto et al. 2005, Javonillo et al. 2010,

Oliveira et al. 2011a, Thomaz et al. 2015). The molecular study by Thomaz et al. (2015) also provided a detailed hypothesis of the internal relations in the subfamily, with the recognition of 7 tribes and 44 genera as valid.

GEOGRAPHIC DISTRIBUTION Throughout southern Central and South America, from Costa Rica to Argentina.

ADULT SIZES Small-sized, ranging from 1.4 cm SL in *Xenurobrycon pteropus* in the central Amazon and *X. polyancistrus* from upper Madeira basin, to 11 cm SL in *Acrobrycon ipanquianus* from the Ucayali and upper Madeira basins in Peru and Bolivia.

DIAGNOSIS OF THE SUBFAMILY: Most stevardiine species have two rows of teeth on the premaxilla, with 4 teeth in the inner premaxillary tooth row; a dorsal fin with 2 unbranched and 8 branched rays (9 branched rays in *Markiana*; 7 branched rays in *Trochilocharax ornatus*), and lack of the epiphyseal branch of the supraorbital canal (present in *Creagrutus* and *Microgenys*).

SEXUAL DIMORPHISM Sexually mature males in the genera *Argopleura*, *Chrysobrycon*, *Corynopoma*, *Gephyrocharax*, *Glandulocauda*, *Hysteronotus*, *Iotabrycon*, *Landonia*, *Lophiobrycon*, *Mimagoniates*, *Phenacobrycon*, *Pseudocorynopoma*, *Pterobrycon*, *Ptychocharax*, *Scopaeocharax*, *Tyttocharax*, and *Xenurobrycon*, have a caudal-fin organ, consisting of glandular tissue contained (in various degrees) in a cavity formed by skin and covered laterally by scales. This glandular tissue produces pheromones (Oliveira et al. 2012). All species with a caudal organ (including the species of *Acrobrycon*, *Diapoma*, *Lepidocharax*, and *Planaltina* in which the caudal organ is observed in both males and females) are inseminating, in that the female retains live sperm in her ovaries introduced by the male. Some other species in the Stevardiinae (e.g., *Attonitus*, *Bryconadenos*, *Monotocheirodon*, *Phallobrycon*, and a few species of *Knodus* and *Creagrutus*) are also inseminating, but lack a caudal organ. Though despite the lack of a caudal-fin organ, hypertrophied tissue can be present on the caudal-fin base in mature males (Weitzman et al. 2005), as well as the anal-fin base. Adult male specimens of *Bryconadenos* and *Phallobrycon* have a distinct anal-fin organ on the anterior lobe (Menezes et al. 2009a). Part of the remaining stevardiines are externally fertilizing, but so far the reproductive mode has been investigated in fewer than a third of the existing species (Thomaz et al. 2015). Other sexual dimorphism includes bony hooks on the pelvic and anal fins in sexually mature males.

HABITATS Typically found in lotic environments, ranging from small forest streams to sandy beaches in large rivers.

FEEDING ECOLOGY Feeding ecology is poorly known, ranging from possibly herbivores in *Bryconacidnus*, *Ceratobranchia*, *Monotocheirodon*, and *Othonocheirodus*, to durophagous in *Creagrutus* species; other species are most generalists tending to insectivory.

BEHAVIOR Several species in Stevardiinae have been shown to be inseminating (including all species with a caudal organ), in which the male transfers sperm directly into the female reproductive tract (see overview in Thomaz et al. 2015).

KEY TO THE GENERA

1a. Caudal fin without modified scales associated with glandular tissues...**2**
1b. Caudal fin with modified scale(s), forming a "pouch," associated with glandular tissues in sexually mature males **22**

2a. Hypertrophied soft tissue present on the anterior part of the anal fin in sexually mature males; pelvic-fin hooks absent in sexually mature males...**3**
2b. Hypertrophied soft tissue absent on the anterior part of the anal fin in sexually mature males; pelvic-fin hooks usually present in sexually mature males...**4**

3a. Urogenital papilla not modified into a copulatory organ in mature males; humeral blotch rounded or absent; broad, dark, longitudinal stripe extending from humeral blotch to caudal-fin base; lateral line canal bordered by a ring of dark chromatophores or at least associated with dark chromatophores; anal-fin organ cup-shaped, not associated with bony hooks or spines in sexually mature males.................................*Bryconadenos*
3b. Urogenital papilla modified into a copulatory organ in mature males; humeral blotch vertically elongate; narrow longitudinal stripe not reaching humeral blotch anteriorly, becoming gradually well marked posteriorly; lateral line canal not associated with dark chromatophores; anal-fin organ not cup-shaped, but with at least 3 spines distinctly larger than anal fin bony hooks.................................*Phallobrycon adenacanthus*

4a. A single conical tooth in the outer premaxillary series; i + 5 pelvic-fin rays (also in *Tyttocharax metae*) ....*Cyanogaster noctivaga*
4b. Outer premaxillary series absent or present, if present with more than one cuspidate tooth; pelvic fin with i + 6–7 rays (except i + 5 in *Tyttocharax metae*) **5**

5a. Single series of teeth on the premaxillary..**6**
5b. Two or three rows of teeth on the premaxillary..**8**

6a. Maxilla lacking teeth; upper and lower lips covering teeth on both jaws; small scales on one-third of length of caudal-fin lobes (upper Rio Negro, Brazil)...*Aulixidens eugeniae*
6b. Maxilla with spatulate teeth forming a single cutting edge with premaxillary teeth; lips reduced, exposing teeth of upper and lower jaws; at least one large, elongate scale on each caudal-fin lobe **7**

7a. Maxilla with 5–9 teeth; adipose fin absent; anal fin with 8–12 branched rays............................. *Monotocheirodon*
7b. Maxilla with 2–3 teeth; adipose fin present; anal fin with 14 branched rays (western Amazon basin, Peru and Ecuador) ....*Othonocheirodus eigenmanni*

8a. Three rows of premaxillary teeth ....*Creagrutus* (in part)
8b. Two rows of premaxillary teeth **9**

9a. 6–20 maxillary teeth in fully grown specimens...**10**
9b. Fewer than 6 maxillary teeth, even in fully grown individuals **12**

10a. Branched anal-fin rays 14 or less..*Creagrutus* (in part)
10b. Branched anal-fin rays 15 or more **11**

11a. Lateral line irregular; scales large, fewer than 4 scale rows between lateral line and anal-fin base; base of caudal-fin lobes scaled.. *Boehlkea*
11b. Lateral line complete; scales small, more than 4 scale rows between lateral line and anal-fin base; base of caudal-fin lobes not scaled (often a red spot on the ventral margin of the caudal peduncle)............................... *Hemibrycon*

12a. Lateral line incomplete... *Bryconacidnus*
12b. Lateral line complete..**13**

13a. Premaxillary teeth of outer series at least as large as those of inner series **14**
13b. Premaxillary teeth of outer series distinctly shorter and more delicate than those of inner series...................**19**

14a. Upper lip reduced, exposing teeth of outer premaxillary series; teeth on premaxillary outer series with spatulate crowns **15**
14b. Upper lip completely covering outer premaxillary teeth; teeth on premaxillary outer series cylindrical with rounded crowns...**16**

15a. Vertically elongate humeral blotch distinctly marked; rounded or slightly elongate snout; adipose eyelid absent, orbit completely free; teeth of premaxillary inner series stalked with distinctly narrow base, and similar in shape to those of outer series... *Ceratobranchia*
15b. Round humeral blotch mostly diffuse or absent; distinctly elongate snout; adipose eyelid present variably enclosing eye and circumorbital bones; premaxillary inner teeth bulky with distinctly wide base, highly contrasting in shape of those of outer series.. *Rhinopetitia*

16a. Mouth terminal... *Microgenys*
16b. Mouth subterminal or ventral...**17**

17a. Round humeral blotch overlapped by dark, variably marked longitudinal stripe; stripe extending from pectoral girdle (at least) to caudal fin median rays; anal fin with 8–10 branched rays (posterior rays of the dorsal fin relatively elongate; series of slender, horizontally elongate scales proximate to the base of the dorsal and anal fins; Tocantins basin) .... *Caiapobrycon tucurui*
17b. Humeral blotch usually absent, if present, diffuse and vertically elongate; longitudinal stripe absent or not as above; anal fin with 11–17 branched rays...**18**

18a. Anal-fin distal margin convex; color pattern consisting at least of a longitudinal stripe at midbody becoming gradually well marked posteriorly; dark markings on fin rays and anal-fin base may also be present (western Amazon, Peru) .... *Attonitus*
18b. Anal-fin distal margin concave; color pattern consisting of only sparse dark chromatophores, longitudinal stripe absent; humeral blotch, if present, diffuse and vertically elongate (Negro basin, Brazil) .... *Rhinobrycon negrensis*

19a. Anal-fin origin anterior to vertical through dorsal-fin rear end..**20**
19b. Anal-fin origin usually posterior to the vertical through origin of posteriormost dorsal-fin ray **21**

20a. Seven or fewer total pelvic-fin rays; dorsal fin with 8 branched rays; anal-fin base covered by a single series of scales; longitudinal rows of dark stripes absent... *Piabarchus analis*
20b. Nine total pelvic-fin rays; dorsal fin with 9 branched rays; anal-fin base densely scaled; longitudinal rows of dark stripes present (more conspicuous in *M. nigripinnis*)... *Markiana*

21a. One or two larger, rounded scales located at the base of the caudal lobes, extending not beyond one-third of the length of the caudal-fin rays, and not covering the procurrent caudal-fin rays (in well-preserved specimens)... *Bryconamericus*
21b. Small and sometimes horizontally elongate scales covering at least two-thirds of the length of the caudal-fin rays as well as the procurrent caudal-fin rays.. *Knodus*

22a. Males with elongate filament on opercle; ventral caudal-fin lobe distinctly longer than upper one; adipose fin absent .... *Corynopoma riisei*
22b. Males without elongate filament on opercle; caudal-fin lobes similar in size; adipose fin present (absent in *Scopaeocharax*) **23**

23a. Naked, scaleless body (except for pouch scale in sexually mature males); dorsal fin with 7 branched rays..*Trochilocharax ornatus*

23b. Body scaled; dorsal fin with ≥8 branched rays...**24**

24a. Caudal organ nearly the same size in mature males and females; three or more series of scales immediately ventral to the lateral line series forming the dorsal border of the pouch opening (fig. 24a: after Weitzman and Menezes 1998) ....*Acrobrycon*

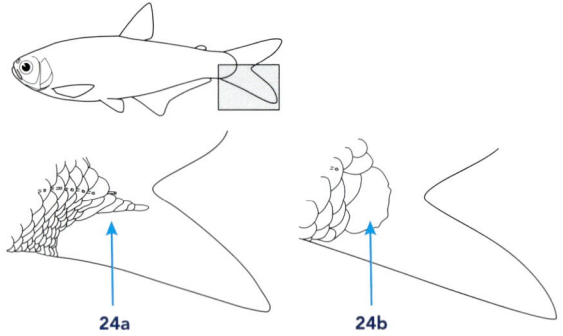

24b. Caudal organ more developed in, or restricted to, mature males; sexually mature males with one very large caudal pouch scale, which is either a lateral line scale or in series with scales of lateral line (e.g., resembling fig. 24b) **25**

24a **24b**

25a. Pouch scale extending from principal caudal-fin ray 12 to the ventral procurrent caudal-fin rays; hypertrophied radii confined to the posterior ventral border of the pouch scale (fig. 25a); caudal-fin ventral procurrent rays modified into a drop-shaped structure (fig. 25a) .....*Gephyrocharax*

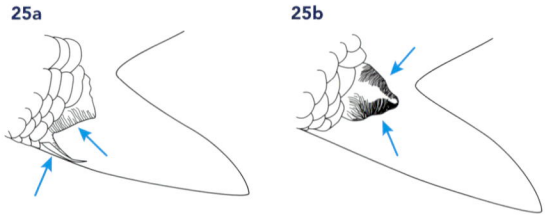

25b. Pouch scale extending from lateral line tube to the middle of the ventral lobe caudal-fin rays; hypertrophied radii spread over the entire posterior border of the pouch scale (fig. 25b); ventral procurrent caudal-fin rays not modified.........................**26**

25a **25b**

26a. Maxillary with 1–4 teeth; more than 18 branched anal-fin rays; anterior jaw teeth tricuspid to multicuspid.........................**27**

26b. Maxillary with 18–30 teeth; 18 or fewer branched anal-fin rays; anterior jaw teeth conical **28**

27a. Posterior portion of the pouch scale with fewer than 35 radii (fig. 27a: after Weitzman and Menezes 1998); pouch scale's mid to distal region and caudal-fin rays bound by relatively loose connective tissue rather than an organized ligament...*Chrysobrycon*

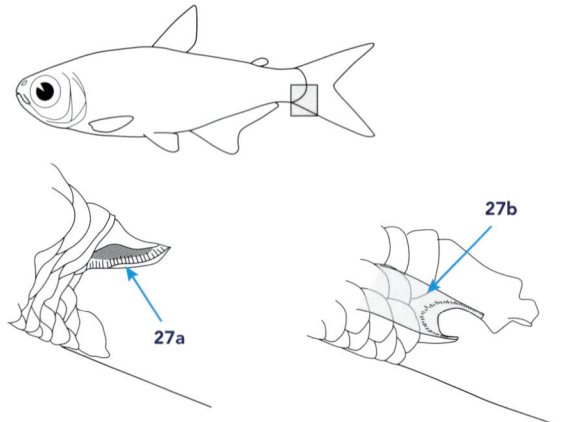

27b. Posterior field of the pouch scale with 35–165 radii; connective tissue of the pouch wall thickened into an extensive ligament (fig. 27b; after Weitzman et al. 1994)...........................*Ptychocharax*

27b

27a

28a. Maxilla of adults with approximately anterior half of its free border toothed; pouch scale of sexually mature males without a prominent anteroventral notch (fig. 28a; after Weitzman and Fink 1985); pelvic fin of sexually mature males approximately one-third of SL, distinctly displaced anteriorly..*Xenurobrycon*

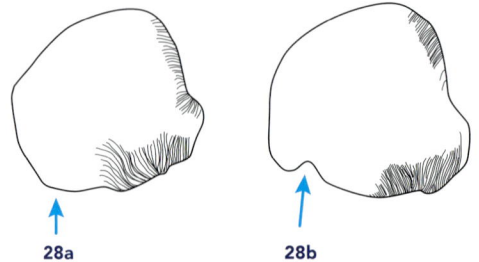

28b. Maxilla of adults three-fourths to fully toothed along its free ventral border; pouch scale of sexually mature males with a prominent anteroventral notch (fig. 28b; after Weitzman and Fink 1985); pelvic fin of sexually mature males less than one-third of SL, at midbody.................................**29**

28a **28b**

29a. Adults with no more than a few exerted teeth on premaxilla and dentary; maxilla long, extending well posterior to anterior border of eye in adults of both sexes; maxilla with a complete or nearly complete row of teeth along its ventral free border; anal fin of sexually mature males with hooks distributed more or less evenly over most of anterior, middle, and sometimes also on nearly all posterior anal-fin rays .... *Scopaeocharax*

29b. Adults with many exerted teeth on premaxilla and dentary; maxilla short, extending to anterior border of eye but not beyond in adults of both sexes; anterior three-fourths of free ventral maxillary border with teeth in sexually mature males; anal-fin hooks of sexually mature males confined to posterior 6–8 rays............... *Tyttocharax*

GENUS ACCOUNTS

Acrobrycon (4.0–11 cm SL)

Differs from other Stevardiinae by: nearly equivalent sizes of the caudal organ in mature males and females (vs. sexually differently sized organs in the remainder of the Stevardiinae); and three or more series of scales immediately ventral to the lateral line series which forms the dorsal border of the pouch opening (versus either lacking such series of scales or having fewer series in the other genera of the Stevardiinae) (Weitzman and Menezes 1998). *Acrobrycon* species from the Amazon also possess the following characters: body elongate, greatest body depth located slightly anterior of dorsal-fin origin; snout rounded; mouth terminal; 22–27 branched anal-fin rays; 9–11 branched pectoral-fin rays; 6–7 branched pelvic-fin rays; lateral line completely pored, with 51–66 scales; premaxillary teeth in two distinct rows, larger teeth in each row with 3 or 5 cusps, smaller teeth with 3 cusps; outer tooth row with 4–6 teeth, inner tooth row with 4, rarely 5 teeth; maxilla with 3–12 teeth; teeth present along more than one-half the length of the dentigerous margin of the maxilla; dentary bearing 4 large anterior teeth with 5 cusps; large teeth followed by 5–10 smaller teeth with 1 or 3 cusps (Arcila et al. 2013). Sexual dimorphism includes hooks on the anal-, pelvic-, and caudal-fin rays in mature males (Arcila et al. 2013). **SPECIES** Three, including two species in the western Amazon basin. Species information, key to the species, and distribution map in Arcila et al. (2013). **DISTRIBUTION AND HABITAT** *A. ipanquianus* in the Ucayali and upper Madeira basins in Peru and Bolivia, and *A. starnesi* in the upper Madeira basin in Bolivia. Additional species (*A. ortii*) occurs in the northwestern region of La Plata basin, Argentina. **BIOLOGY** *A. ipanquianus* is inseminating (Burns et al. 1995, Burns and Weitzman 2005); reproductive mode unknown in remaining species.

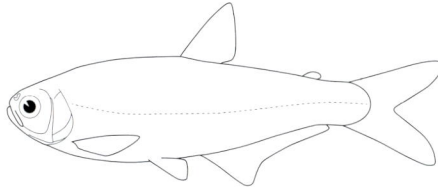

Attonitus (5.0–6.0 cm SL)

Characterized by: a convex expansion of the body wall along the anal fin (anal fin distal margin also convex); inner premaxillary teeth slightly shorter than those on the outer tooth row, with a posterior curvature of the distal portions of the teeth on the inner tooth row and anterior curvature of the distal portions of the teeth in the outer tooth row; an anteroventral curvature of the anterior portion of the dentary with a consequent anterodorsal orientation of the anterior dentary teeth and a distinct concavity of the ventral profile of the anterior portion of the dentary (Vari and Ortega 2000, Weitzman et al. 2005). Furthermore, all *Attonitus* species have club-shaped anal-fin gland cells in sexually mature males, similar in location and structure to those in *Bryconadenos*, but not organized into a glandular organ as in that genus (Weitzman et al. 2005). Other sexual dimorphism includes bony hooks on the anal and pelvic fins in mature males, and differences in body size between males and females. Both males and females have a band

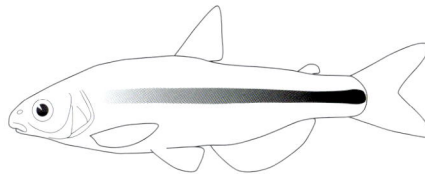

of dark chromatophores on the ventrolateral region of the posterior portion of the body, which is more developed in males (Vari and Ortega 2000). **SPECIES** Three, all in the western Amazon basin. Key to the species and species information in Vari and Ortega (2000). **DISTRIBUTION AND HABITAT** *A. irisae* in the middle portions of the Ucayali basin in the Aguaytia and Pachitea rivers; *A. bounites* in the Madre de Dios basin and upper portions of the Madeira basin in southeastern Peru and northeastern Bolivia; *A. ephimeros* in the southern portions of the Ucayali basin in the Urubamba and Tambo basins. Inhabits clearwater streams bordered by vegetation and with relatively fast current (Vari and Ortega 2000). **BIOLOGY** Feed on aquatic insect larvae (Vari and Ortega 2000). Inseminating (Burns and Weitzman 2005, Weitzman et al. 2005).

Aulixidens (4.3 cm SL)

Characterized by: anterior end of ascending process of maxilla with conspicuous notch; one row of premaxillary teeth; maxillary teeth absent; separation between posterior dentary teeth more than width of these teeth; scales covering caudal-fin lobes one-third of their length. **SPECIES** One, *A. eugeniae*. **DISTRIBUTION** Upper Orinoco basin, Venezuela. Habitats include sandbanks (Jepsen 1997). **BIOLOGY** no data available.

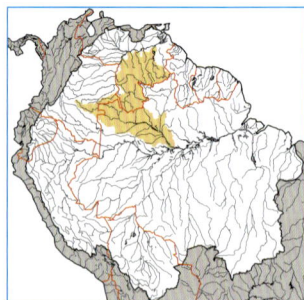

Boehlkea (4.0–5.0 cm SL)

Resembling *Hemibrycon*, but differing by having the base of the caudal fin scaled (Géry 1977). Characterized by: fewer than 4 scales under the lateral line; adipose fin present; premaxilla with two rows of teeth, with 4 inner teeth; maxilla long to rather long, with 6–20 maxillary teeth in full-grown individuals (Géry 1977). One species, *Boehlkea fredcochui*, can be easily recognized by its light metallic blue to purple body color, with the tips of caudal-fin lobes and the adipose fin white. **SPECIES** Two, both in the Amazon basin. **DISTRIBUTION** *B. fredcochui* (*Tetra azul* or Cochu's Blue tetra) inhabits the Amazon basin in Brazil, Colombia, and Peru; *B. orcesi* inhabits the Macuma, Upano, Tayusa, and Pastaza basins in Ecuador. **BIOLOGY** Eggs of *B. fredcochui* are externally fertilized (Thomaz et al. 2015). No data available for *B. orcesi*.

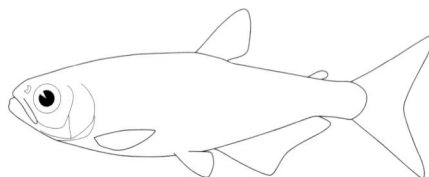

Bryconacidnus (4.0–5.0 cm SL)

Resembles *Microgenys*, but with longer teeth (like *Ceratobranchia*) and an incomplete lateral line, at least lacking a few pores on the caudal peduncle (Géry 1977). Also characterized by: fewer than 6 maxillary teeth; more than 3 dentary teeth; premaxilla with two rows of teeth, with 4 inner teeth; 13 or fewer branched anal-fin rays; snout rounded; mouth more or less terminal (Géry 1977). One species, *B. ellisi*, has the body covered with many small dark dots ("peppered"), at least in certain localities, such as the Beni River in Bolivia (Géry 1977). **SPECIES** Three, all in the western Amazon basin. **DISTRIBUTION** *B. ellisi* inhabits the western Amazon and upper Madeira basins in Bolivia, Ecuador, and Peru; *B. hemigrammus* the upper Madeira basin in Bolivia; and *B. paipayensis*, the Crisnejas River in the upper Marañón River of Peru. *Bryconacidnus ellisi* inhabits sandy bank habitats across its range (Ibarra and Stewart 1989, Albert et al. 2011b). **BIOLOGY** No data available.

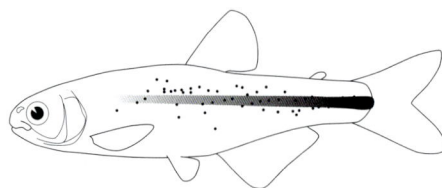

Bryconadenos (4.0 cm SL)

Characterized by: glandular club cells at the surface of the epidermis on the anterior part of the anal fin in sexually mature males, organized into a glandular organ (Weitzman et al. 2005). Other, nonunique characters that might aid identification of the genus (Weitzman et al. 2005): a band of dark chromatophores located on the ventrolateral portion of the body wall above the anal-fin base (more developed in males than females); body wall somewhat convex proximate to anal-fin base; lateral line pores surrounded by a ring of dark chromatophores or at least associated with dark chromatophores (also in *Attonitus*); and sexually mature males without pelvic-fin hooks. Sexual dimorphism includes a cup-shaped anal-fin organ, longer anterior anal-fin lobe, and longer pelvic fins in adult males (Weitzman et al. 2005, Menezes et al. 2009b). **SPECIES** Two, both in the Amazon basin. Thomaz et al. (2015) suggested that *Bryconadenos* are specialized species of *Knodus* and should be included in that genus. Species information in Weitzman et al. (2005) and Menezes et al. (2009b). **DISTRIBUTION AND HABITAT** *B. tanaothoros* in clear and turbid streams with fast-flowing water in the upper Xingu and upper Tapajós basins; *B. weitzmani* lives close to rapids in the Curuá River, a tributary of Iriri River in the Xingu basin. **BIOLOGY** Inseminating (Burns et al. 2000, Weitzman et al. 2005). *Bryconadenos tanaothoros* is solitary and territorial, with a typical way of swimming, being fast, but trembling and quivering in the process (Weitzman et al. 2005); *B. weitzmani* forms schools of 10–15 individuals (Menezes et al. 2009b).

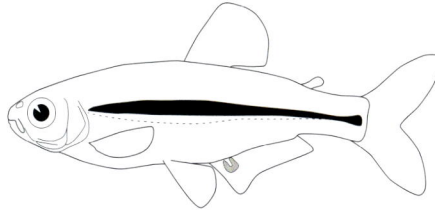

Bryconamericus (3.0–10 cm SL)

The genus *Bryconamericus* is an artificial group (see discussion in Vari and Siebert 1990). The current definition is based on the following characters (Eigenmann 1927, Vari and Siebert 1990): two rows of teeth on the premaxilla, inner tooth row with four teeth larger than the teeth of the outer row; single row of teeth on dentary; few maxillary teeth; gently curved upper jaw; lack of scales on the caudal fin; setiform (bristle-shaped) gill rakers; complete lateral line; and absence of a glandular pouch on the caudal fin in males. None of these characters are unique to *Bryconamericus*, which differs from *Knodus* only by absence of scales covering the base of the lobes of the caudal fin. Sexual dimorphism includes hooks on the anal- and pelvic-fin rays and larger pectoral and pelvic fins in adult males. **SPECIES** Approximately 60 species, with 20 species in the AOG region (Thomaz et al. 2015). Key to the species in Colombia by Román-Valencia et al. (2008b) and Ecuador by Román-Valencia et al. (2013). **COMMON NAMES** *Piaba*, *Lambari piquira* (Brazil); *Mojarita* (Peru). **DISTRIBUTION AND HABITAT** *Bryconamericus sensu stricto* is restricted to the Paraná-Paraguay drainage and rivers draining to the Atlantic in Argentina, Uruguay, South and Southeastern Brazil.

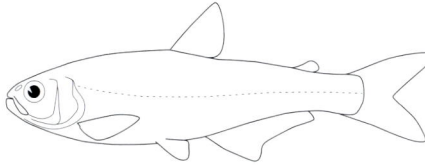

Caiapobrycon (4.5 cm SL)

Characterized by: maxilla expanded vertically immediately posterior to the posterodorsal portion of its attenuate anterodorsal process; ventrally positioned mouth, with the lower jaw shorter than the upper jaw; anterodorsally oriented teeth on the anterior portion of the dentary; and infraorbitals 1 and 2 strongly attached to each other along their common border, and with the longest axis of infraorbital 1 directed anteriorly instead of anterodorsally (Malabarba and Vari 2000). *Caiapobrycon* can also be

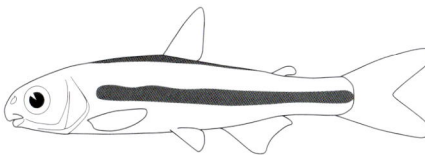

distinguished from most other characid genera by the following characters: a low number of anal-fin rays (ii–iii, 8–10); relatively elongate posterior rays of the dorsal fin; series of slender, horizontally elongate scales proximate to the base of the dorsal and anal fins; and the outer row of premaxillary teeth slightly longer than the inner premaxillary tooth row (Malabarba and Vari 2000). The only external sexual dimorphism observed is bony hooks on pelvic-fin rays of mature males (Malabarba and Vari 2000). **SPECIES** One, *C. tucurui*. **DISTRIBUTION AND HABITAT** Clearwater habitats in the Tocantins basin, with moderate turbidity and with fast water current over a bottom covered with sand and rocks (Malabarba and Vari 2000). **BIOLOGY** No data available.

Ceratobranchia (3.0–6.0 cm SL)

Characterized by: four inner premaxillary teeth with a distinct posterior flexure and cusps directed toward the gullet; inner premaxillary teeth shorter in height than the outer ones, but not necessarily narrower in width, bearing 3–9 cusps; outer premaxillary teeth prominent, with 3–9 cusps, forming a continuous series or biting surface with the maxillary teeth; and 2–5 maxillary teeth, as large as the outer premaxillary teeth, with 3–7 cusps (Chernoff and Machado-Allison 1990). Other characters that might aid identification (Chernoff and Machado-Allison 1990): lateral line complete, even in small individuals; 31–38 lateral line scales, with 4–6 scales above and 3–4 scales below the lateral line; scales present on the base of the caudal fin; snout blunt and rounded; mouth subterminal; two rows of premaxillary teeth (5 on outer, 4 on inner row); dentary with 6–10 teeth, first four prominent, with 3–5 cusps; third infraorbital large, contacting the preopercle at angle; humeral mark well developed, often elongate vertically; branched anal-fin rays 11–19; first gill arch with 4–6 rakers on upper arch, 5–10 rakers on lower arch, and 1–2 rakers at angle. The only sexual dimorphism observed are the tubercles on the surface of the head in breeding males of *C. binghami* (Chernoff and Machado-Allison 1990). **SPECIES** Five, all in the Amazon and Orinoco basins. A review of species and species key in Chernoff and Machado-Allison (1990). **DISTRIBUTION AND HABITAT** Andean piedmont (500–3,150 m) in the western Amazon of Peru and Ecuador (4 species) and Orinoco basin (1 species). **BIOLOGY** Peculiar teeth suggest a specialized diet, possible herbivorous (Géry 1977).

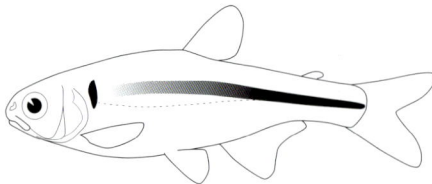

Chrysobrycon (4.3–8.2 cm SL)

Characterized by: a characteristic pouch scale on the ventral lobe of the caudal fin in mature males (see fig. 27a in key). The pouch scale is relatively small, somewhat elongate, curved and horizontally folded so that its lateral face forms a laterally concave, broadly open pocket; the pouch scale is confined to the dorsal region of the pouch opening; an additional curved scale is situated close to, and nearly completely against, the medially placed surface of the pouch scale, while this medially placed scale sometimes has its ventral border laterally curved around the ventral border of the pouch scale (Weitzman and Menezes 1998). *Chrysobrycon* species are also characterized by the presence of gill glands (Bushmann et al. 2002, Vanegas-Ríos et al. 2011). **SPECIES** Four, all in the Western Amazon. Photos and key to the species in Vanegas-Ríos et al. (2011). **DISTRIBUTION** Upper Amazon from about 150 to 600 m above sea level: *Chrysobrycon eliasi* (4.3 cm SL) from the upper Madeira basin in Peru, *C. hesperus* (8.2 cm SL) from the Napo basin in Peru, *C. myersi* (6.4 cm SL) from the Pachitea River, Peru, and *C. yoliae* (5.2 cm SL) from the Ucayali basin, Peru. Possibly also a species in the Leticia region in Colombia (Galvis et al. 2006). See distribution map for species in Vanegas-Ríos et al. (2013b). **BIOLOGY** Inseminating (Burns et al. 1995)

Corynopoma (7.5 cm SL)

Readily distinguished by the strange filamentous extension of the opercle in males. This extension has a paddle-like structure and is enlarged at the end into a shiny flag. Other characters that might aid identification: adipose fin absent; mouth superior, with the lower jaw extending beyond the upper; nostrils near the anterodorsal corner of eye; lateral line initially deflected downward, then continues straight to the base of the caudal fin; dorsal fin placed well back, with its origin over the middle of the anal-fin base; caudal fin deeply forked; body silver to pinkish, with clear fins; and a faint horizontal stripe that is most intense and widest on the caudal peduncle (Taphorn 1992). In addition to the paddle-like extension on the operculum in males, sexual dimorphism includes enlarged anal and dorsal fins and elongation of the lower lobe of the caudal fin in males. **SPECIES** One, *C. riisei*. **DISTRIBUTION AND HABITAT** Foothills of the Venezuelan and Colombian Andes, and Trinidad, in shady, shallow, slow-moving streams. **BIOLOGY** Carnivore, feeding mostly on terrestrial insects, especially ants (Taphorn 1992). During courtship the male displays the flags, one at a time, at a right angle from his body in front of the female. The female often attempts to bite at the flag when it is displayed, suggesting she believes it is a food item. Fertilization is internal in *C. riisei*, and because the female is close to the male when biting at the flag, the ornament has been proposed to facilitate fertilization by positioning the female for successful sperm transfer (Kutaygil 1959, Nelson 1964).

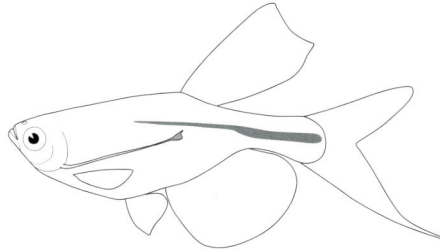

Creagrutus (3.0–11 cm SL)

Characterized by: foreshortened lower jaw; premaxillary teeth arranged in either of the patterns shown in figure to the right (ventral view of left side, anterior at top; with A, triangular cluster of teeth; B, primary tooth row; and C, single lateral tooth; figure after Vari and Harold 2001) and fourth infraorbital whose posterior margin contributes to the posterior margin of the infraorbital series (except *C. cracentis* and *C. maxillaris*) (Vari and Harold 2001). Sexually mature males with bony hooks on the anal- and pelvic-fin rays. **SPECIES** 71, with about 58 species in the AOG region, including *C. beni* (Goldstripe characin). Key to the cis-Andean species in Vari and Harold (2001), with three new AOG species described subsequently: *C. maculosus* (Román-Valencia et al. 2010a), *C. tuyuka* (Vari and Lima 2003), and *C. nigrotaeniatus* (Dagosta and De Lima Pastana 2014). **DISTRIBUTION AND HABITAT** Throughout the humid Neotropics from Panama to Paraguay, with highest diversity in streams and rivers draining the piedmont of the eastern slopes of the central and northern Andes, the Pacific slope of northern Colombia, the Caribbean slopes of Colombia and Venezuela, the coastal ranges of the northern portions of the Orinoco basin in Venezuela, and the uplands of the Brazilian and Guiana shields (Vari and Harold 2001). Distribution of species per subregion of cis-Andean South America in Vari and Harold (2001). Most abundant in moderate to swiftly flowing water bodies, from near sea level to nearly 1,900 m a.s.l., and absent or rare in the low-gradient streams and rivers, most notably in the central portion of the Amazon basin (Vari and Harold 2001). **BIOLOGY** Distinctive jaw and dentition modification allow *Creagrutus* species to exploit specific food items from the bottom, in particular small seeds and aquatic and terrestrial insects (Vari and Harold 2001). In addition, diet can include phytoplankton, mollusks, crustaceans, and, less commonly, fish scales and smaller fishes (Vari and Harold 2001, and references therein).

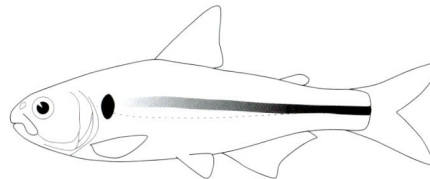

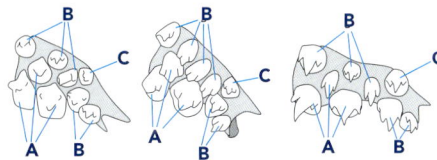

Cyanogaster (1.5 cm SL)

Distinguished from all other members of the Stevardiinae by a reduced number of i + 5 pelvic-fin rays and a single conical tooth in the outer premaxillary tooth series (Mattox et al. 2013). Other characters that could aid identification: lack of maxillary teeth; incomplete lateral line; body transparent and living individuals with a conspicuous blue abdominal region. Mature males have hooks on the pelvic- and anal-fin rays. **SPECIES** One, *C. noctivaga* (Mattox et al. 2013). **DISTRIBUTION AND HABITAT** Negro basin, Brazil. Collected at night, in blackwater with slow current. **BIOLOGY** Analyzed stomachs contained mainly remnants of insect larvae (Mattox et al. 2013).

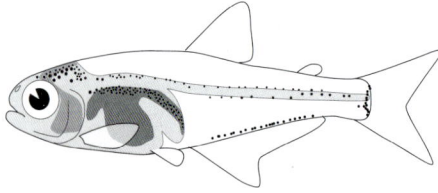

Gephyrocharax (3.0–5.0 cm SL)

Distinguished from all characids by having the second and third ventral procurrent rays of caudal fin hypertrophied, forming a single spur-shaped structure in adult males (vs. rays common or, when modified, forming 2 spur-pointed projections between the second and fourth ventral procurrent rays instead of 1) (Vanegas-Rios 2016). Other characters useful for identification: premaxillary teeth in two distinct series, five teeth in the inner series; second suborbital covering the entire cheek; origin of dorsal fin nearer to caudal fin than the eye, considerably behind the vertical from origin of anal fin; pectoral fins large, overlapping the pelvic fins; males with a modified scale forming a pouch on the lower lobe of caudal fin and with ventral procurrent rays 2 and 3 in form of a claw-shaped structure immediately ventral to ray 9 of lower lobe of the caudal fin (Eigenmann 1912b, Eigenmann and Myers 1929, Vanegas-Ríos et al. 2013a). Sexual dimorphism also includes bony hooks on the rays of the anal, caudal, pectoral, and pelvic fins and a gill gland in mature males (Vanegas-Rios 2016). **SPECIES** 11, including 2 species in the AOG region. Revision of genus and key to the species in Vanegas-Rios (2016). **DISTRIBUTION AND HABITAT** Cis- and trans-Andean basins in southern Central America and the northern half of South America, with *G. major* in the upper Amazon in Bolivia and Peru, and *G. valencia* in the Orinoco basin. Inhabits small rivers and creeks. **BIOLOGY** Inseminating (Burns et al. 1995, Burns and Weitzman 2005).

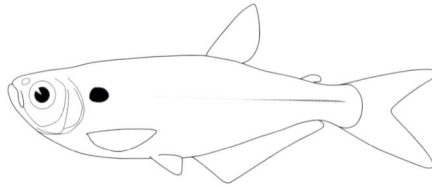

Hemibrycon (4.0–17 cm SL)

Shares several characters with most other genera in the subfamily Stevardiinae: premaxilla with two series of teeth, inner series with four teeth; 8 branched dorsal-fin rays; infraorbital 2 in contact with the lower limb of the preopercle; adipose fin present; anal fin moderate or long; and gill rakers simple. Distinguished from other stevardiines by: more maxillary teeth (6–20), leaving the edentulous portion of the maxilla smaller than toothed portion, and a naked caudal fin (Bertaco and Malabarba 2010). In life there is often a red spot on the ventral margin of the caudal peduncle (Román-Valencia et al. 2010b). Sexual dimorphism includes (but not in all species): bony hooks on dorsal-, pectoral-, anal-, and pelvic-fin rays in mature males, differences in anal-fin shape (slightly convex in males and nearly straight in females), and mature males with gill gland on first gill arch, covering the first branchial filaments (Bertaco and Malabarba 2010). **SPECIES** 42, with about 15 species in the AOG region, including *H. tridens* (Jumping tetra). Key to the cis-Andean species in Bertaco and

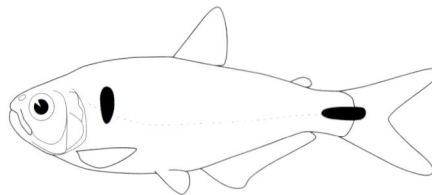

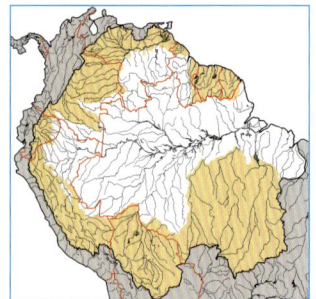

Malabarba (2010). **DISTRIBUTION AND HABITAT** From Panama to Bolivia and Brazil. Cis-Andean species distributed from Caribbean coastal basins of Venezuela to lower Tocantins basin (Bertaco and Malabarba 2010). *Hemibrycon* species have a peculiar distribution in the Amazon and are confined to the periphery of the basin. They inhabit high-gradient streams with clear, fast water and do not occupy slow-running rivers of the lowlands, except *H. surinamensis* (Gery 1962, Bertaco and Malabarba 2010). **BIOLOGY** Feed on aquatic and terrestrial insects, algae, and seeds (Ortaz 1992, Román-Valencia and Botero 2006, Román-Valencia et al. 2008a).

Knodus (3.0–9.0 cm SL)

Differs from *Bryconamericus* only by having a scaled caudal fin. Sexual dimorphism includes bony hooks on the pelvic and anal fins in adult males, but absent in some species (Ferreira and Lima 2006). **SPECIES** 28, including 23 in the AOG region. The validity of the genus *Knodus* has been repeatedly challenged (e.g., Lima et al. 2004) and a recent molecular study indeed shows that the genus as currently conceived is not a natural group (Thomaz et al. 2015). **DISTRIBUTION AND HABITAT** Amazon, Orinoco, Paraná-Paraguay, and São Francisco basins, and the Parnaiba basin of northeastern Brazil. Géry (1977) noted *Knodus* has an Amazonian distribution whereas *Bryconamericus* inhabits southern basins. **BIOLOGY** Some *Knodus* species are inseminating, others are not (Weitzman et al. 2005, Ferreira and Carvajal 2007). Analyzed stomach contents included remains of terrestrial and aquatic insects and vegetal matter (Lima et al. 2004, Ferreira and Lima 2006, Ceneviva-Bastos and Casatti 2007).

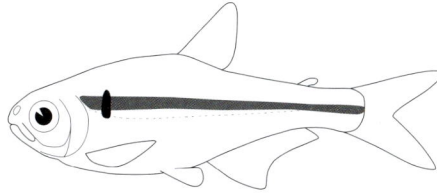

Markiana (10 cm SL)

Characterized by: crenate scales which are very regularly arranged and gradually decreasing in size from the lateral line to the ventral margin of the body; anal fin scaled; a complete lateral line; and longitudinal rows of dark stripes on body (more conspicuous in *M. nigripinnis*) (Eigenmann 1918a, Baicere-Silva et al. 2011a). **SPECIES** Two, both in the AOG region. Key to the species in Eigenmann (1918a). **DISTRIBUTION AND HABITAT** Disjunct distribution with *M. nigripinnis* in the Paraná, Paraguay, and Mamoré (upper Madeira) basins, and *M. geayi* in the Orinoco basin. Inhabit floodplain lagoons in open habitats, such as the Llanos in the Orinoco basin, the Llanos de Mojos in Bolivia, and the Brazilian Pantanal. **BIOLOGY** Possibly territorial; very aggressive toward conspecifics and found as isolated individuals in shallow areas (J. Zuanon pers. comm.).

Microgenys (2.8–7.0 cm SL)

Characterized by: a very short anal fin (10–13 rays) and aquiline snout, resembling young *Creagrutus* but with narrow, tricuspid teeth. The color pattern is quite generalized, with a lateral band, silvery in life and (*M. lativirgata*) with an oval humeral spot (Géry 1977). **SPECIES** Three, including two species in the Amazon basin. Key to the species in Eigenmann (1927). **DISTRIBUTION** *M. lativirgata* (7.0 cm SL) inhabits the upper Marañón basin (Rio Pusoc in the Peruvian Andes), and *M. weyrauchi* (2.8 cm SL) the upper Ucayali basin (Peru, at 1,900 m), but is possibly a junior synonym of *Bryconacidnus ellisi* (Géry 1977). **BIOLOGY** No data available.

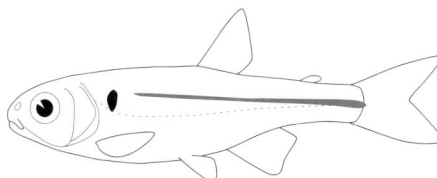

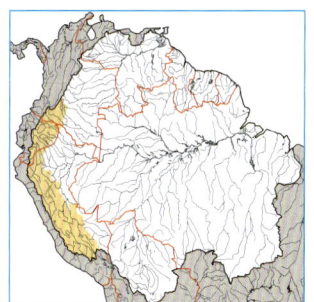

Monotocheirodon (3.0–4.5 cm SL)

Characterized by: an enlarged scale on the basal portion of each caudal-fin lobe; two rows (one external and one internal) of short and slender gill rakers present on each branchial arch; a single row of 4 distally compressed, pedunculate, and multicuspid teeth present on the premaxilla; ascending process of the premaxilla strongly bent ventrally; posterior portion of maxilla strongly bent ventrally; anterior dentary teeth not notably larger than the remaining teeth on the bone, all dentary teeth gradually decreasing in size posteriorly; 2 or 3 longitudinal scale rows from lateral line to pelvic-fin origin; adipose fin absent; anal fin short, with 8–12 branched rays; hooks absent on pelvic and anal fins of males (Menezes et al. 2013). **SPECIES** Three, all in the Amazon basin. Key to the species and species information in Menezes et al. (2013). **DISTRIBUTION AND HABITAT** Andean foothills in Bolivia and Peru: *M. pearsoni* from headwaters of the Bopi River, Beni basin and Iniqui River, Bolivia, at about 5,000 m elevation; *M. drilos* from headwaters of the Tambopata and Madre de Dios Rivers, Madre de Dios basin, Peru; and *M. kontos* in tributaries of the Madre de Dios basin, Peru, between elevations of 350 and 3,200 m (Menezes et al. 2013). **BIOLOGY** All species are inseminating (Burns and Weitzman 2006). Possibly herbivorous based on teeth morphology and dentition pattern.

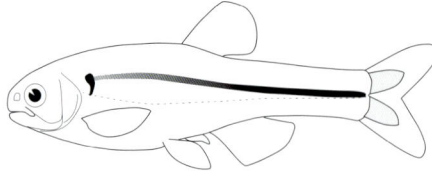

Othonocheirodus (5.0 cm SL)

Characterized by: adipose fin present; lateral line complete; predorsal area scaled; caudal fin naked; suborbital in contact with the preopercle below; caudal lobes equal; teeth 5-pointed, similar in both jaws; horizontal extent of the maxilla small, with 2 teeth, its vertical extent long, extending far below the line of the bases of the dentary teeth, its end large, rounded, and free; premaxilla with 4 similar teeth on each side, in a single, not angulated series, continuous posteriorly with the two maxillary teeth; dentary with 6 similar pentacuspid teeth on each side, grading down slightly in size posteriorly; upper jaw lipless; lower jaw with a thin deep lip that covers not only the dentary teeth but also part, or all, of the premaxillary teeth, when mouth closed; anal fin with 14 rays; dorsal fin with 9 rays; body color brownish, with a conspicuous black humeral spot; a dark lateral band, faint anteriorly, ending at caudal base (Myers 1927). **SPECIES** One, *O. eigenmanni*. **DISTRIBUTION** Western Amazon basin in Peru. **BIOLOGY** Possibly herbivorous based in teeth morphology and dentition pattern.

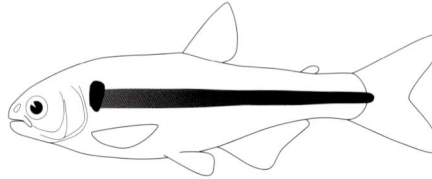

Phallobrycon (4.0 cm SL)

Distinguished from other characids by the presence, in sexually mature males, of two developed spines on the median unbranched portions of the fifth, sixth, and seventh anal-fin rays, associated with swollen glandular tissue on the anterior portion of the anal fin. These spines are larger and separated from the close-set smaller hooks present on distal portions of the anterior anal-fin rays, and different from those found on distal portions of several anal-fin rays of many characids (Menezes et al. 2009a). Other nonunique diagnostic features: urogenital papilla modified into a copulatory organ; absence of pelvic-fin hooks; and glandular tissue on anal fin not organized into an organ (as in *Bryconadenos*). **SPECIES** One, *P. adenacanthus*. **DISTRIBUTION AND HABITAT** Upper Xingu basin, in the main channel of rivers with fast flowing and rather shallow clear waters (Menezes et al. 2009a). **BIOLOGY** No data available.

Piabarchus (3.7 cm SL)

Resembles *Bryconamericus* but with anal-fin insertion just anterior to first dorsal-fin ray (Géry 1977 for *P. analis*). Fewer than 6 maxillary teeth; adipose fin present; more than 3 dentary teeth; snout not pronounced, rounded; mouth more or less terminal; more than 13 branched anal-fin rays; lateral line complete; body color silvery (Géry 1977 for *P. analis*). **SPECIES** Two, including one species is the Amazon basin. **DISTRIBUTION** *P. analis* in the Paraguay and Western Amazon basins, and *P. torrenticola* in the Paraguay basin. **BIOLOGY** No data available.

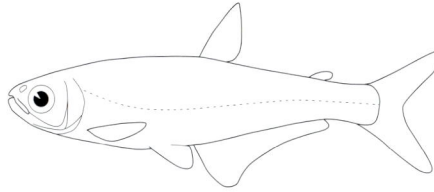

Ptychocharax (6.0 cm SL)

Characterized by: characters of the caudal glandular organ in sexually mature males (females have some of these modifications), see some details in fig. 27b and full description in Weitzman et al. (1994). Other characters that might aid identification: body compressed and moderately elongate; eye large, to about half the length of the head; dorsal fin with ii-9 rays; adipose fin present; anal fin with iv + 21–25 rays, posterior rays split to its base; anal fin with well-developed anterior lobe; principal caudal fin count 10/9; adult females with an enlarged terminal scale of the first horizontal scale row ventral to the lateral line row (= pouch scale in males); lateral line complete, with 32–36 perforated scales; premaxillary teeth in two rows, outer row usually with 3 teeth, inner row with 5 teeth; larger premaxillary teeth with 5–7 cusps, smaller teeth with 5 cusps; maxilla with 2–3 teeth, anterior 1–2 teeth with 5 cusps, posterior teeth with 1–3 cusps; dentary with 5 large teeth with 5–6 cusps and 4–8 smaller posterior teeth with 1–5 cusps (usually tricuspid) (Weitzman et al. 1994). Sexual dimorphism (in addition to a pouch scale in males) includes: bony hooks on anal- and caudal-fin rays, slightly larger body size and brighter coloration in mature males (Weitzman et al. 1994). **SPECIES** One, *P. rhyacophila*. **DISTRIBUTION AND HABITAT** Upper Siapa River in the Casiquiare basin in Venezuela at about 560 m. Inhabit clear streams with sandy or granitic bottoms covered by dense mats of the plant *Apinaga multibranchiata* (Podostemaceae) in areas of fast water (Weitzman et al. 1994). **BIOLOGY** Feeds on aquatic insect larvae (e.g., simuliids, trichopterans), as well as adult ants and spiders (Weitzman et al. 1994). Inseminating.

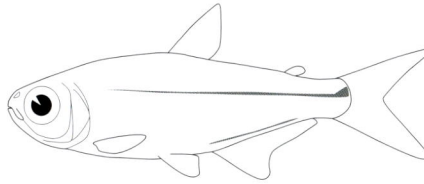

Rhinobrycon (4.0 cm SL)

Characterized by: snout produced and pointed; mouth subinferior; eye oval, somewhat vertically elongate; lateral line complete; upper lip developed, covering the relatively weak outer row of teeth; anterior fontanel moderate; 11–12 branched anal-fin rays (Géry 1977). Another character that might aid identification is that the outer series of premaxillary teeth are as large and long as those of inner series (also in *Bryconacidnus*, *Ceratobranchia*, and *Rhinopetitia*), versus the outer teeth series distinctly smaller and shorter than those of the premaxillary inner series in most characids and Stevardiinae (Netto-Ferreira et al. 2014). **SPECIES** One, *R. negrensis*. **DISTRIBUTION** Negro basin, Brazil. **BIOLOGY** No data available.

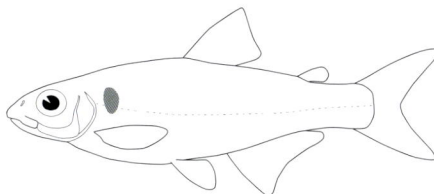

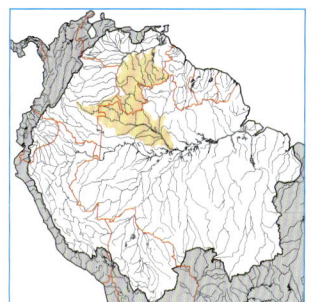

Rhinopetitia (3.0 cm SL)

Characterized by: snout pronounced and pointed with a subinferior mouth; rudimentary upper lips, thereby exposing the outer premaxillary tooth row (also found in *Bryconacidnus*, *Ceratobranchia*, *Monotocheirodon*, *Odontostoechus*, and *Othonocheirodus*); outer premaxillary tooth row with 6 or 7 tricuspid teeth; fewer than 6 maxillary teeth; eyes oval, somewhat vertically elongate; lateral line interrupted on the caudal peduncle; nonsetiform gill rakers; anterior fontanel almost entirely closed; adipose fin present; 15–16 anal-fin rays; adipose eyelid present, variably enclosing eye and circumorbital bones (Géry 1964b, 1977, Netto-Ferreira et al. 2014). Another character that might aid identification is that the outer series of premaxillary teeth are as large and long as those of inner series (also in *Bryconacidnus*, *Ceratobranchia*, and *Rhinobrycon*), versus the outer teeth series distinctly smaller and shorter than those of the premaxillary inner series in most characids and Stevardiinae (Netto-Ferreira et al. 2014). Sexual dimorphism: bony hooks on the anal- and pelvic-fin rays and small gill gland present on the filaments of first gill arch of mature males (Netto-Ferreira et al. 2014). **SPECIES** Two, both in the Amazon basin. Notes on species differences in Netto-Ferreira et al. (2014). **DISTRIBUTION** *R. myersi* in the Araguaia basin, Brazil; *R. potamorhachia* from sandy beaches in the Teles Pires River, a tributary of the Rio Tapajós, Brazil. **BIOLOGY** No data available.

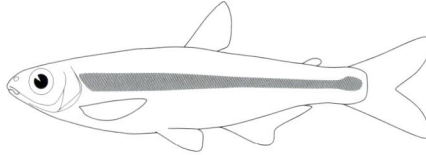

Scopaeocharax (2.0–2.5 cm SL)

Characterized by: very small adult body size; sexually mature males with rugosities along the dorsal margin of principal caudal-fin ray 10 (a distally hypertrophied area located just medial to the posterior process of the pouch scale); sexually mature females with the anterior tip of the pelvic bone located ventral to the distal ends of the anterior pleural ribs; sexually mature males with the medial two or three pelvic-fin rays curved dorsal to the rest of the rays when the fin is relaxed or folded; a short premaxilla, its length from the anteromedial to the posterolateral borders approximately equal to its height, including the teeth; a nearly complete row of teeth present along the ventral border of the maxilla in adults; and adipose fin absent (Weitzman and Fink 1985). **SPECIES** Two, both from the Western Amazon. **DISTRIBUTION AND HABITAT** *S. atopodus* and *S. rhinodus* both in the Huallaga basin, Peru. **BIOLOGY** Inseminating (Burns et al. 1995).

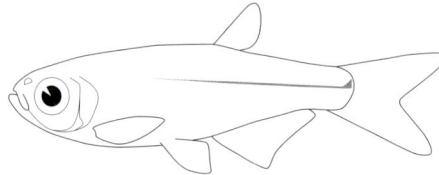

Trochilocharax (1.7 cm SL)

Characterized by: very small adult body size; naked, scaleless body (except for a pouch scale in sexually mature males); dorsal fin with ii + 7 rays; anal fin with iv + 22–24 rays; maxilla long and narrow, with 5–8 conical teeth on the anterior part of the bone; premaxilla and anterior part of lower jaw with irregularly distributed conical teeth, partly protruding out of mouth (Zarke 2010). Males with pouch scale and one relatively large hook on the anterior part of seven (fourth to tenth) anal-fin rays. **SPECIES** One, *T. ornatus* (Zarke 2010). **DISTRIBUTION AND HABITAT** Loreto, Peru. **BIOLOGY** Presumably inseminating.

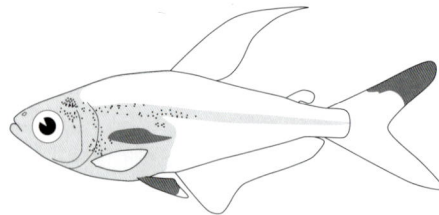

Tyttocharax (1.5–2.0 cm SL)

Characterized by: very small adult body sizes. Also recognized by the following characters of sexually mature males (Weitzman and Fink 1985): bony hooks confined to the posterior 6–8 anal-fin rays, hooks are relatively large and arranged in a vertically elongate clusters; posterior 3–5 proximal anal-fin radials expanded into flattened plates that serve as the origin for the very robust anal-fin erector and depressor muscles. Mature males and females have about 25–40 conical teeth, arranged in 4–7 diagonal rows in each premaxilla; 4–8 rows of dentary teeth, all but the innermost diagonal, with a total of 50–80 or more teeth on each dentary, and with many of the teeth projecting anteriorly or laterally (also in mature females, but usually with fewer teeth). Body color bluish, or with a broad blue lateral stripe. **SPECIES** Four, all in the Amazon basin and Andean foothills. Key to the species in Weitzman and Ortega (1995), except *T. metae* (Román-Valencia et al. 2012). **DISTRIBUTION AND HABITAT** *T. cochui* in swiftly flowing forest streams near Leticia (Colombia), preferring stream margins with submerged vegetation (Galvis et al. 2006); *T. madeirae* in tributaries of the lower and middle Amazon basin; *T. metae* in swiftly flowing blackwater streams in Andean piedmont (264–282 m elevation) in the Orinoco basin of Colombia; and *T. tambopatensis* from the shoreline of small blackwater forest streams with a slow current in the Manu and Tambopata basins of the upper Madeira in Peru (Weitzman and Ortega 1995, Román-Valencia et al. 2012). Species distribution map in Román-Valencia et al. (2012). **BIOLOGY** *T. tambopatensis* was observed near the water surface in schools of about 20–80 individuals (Weitzman and Ortega 1995). Stomach contents of *T. cochui* in the Leticia region in Colombia showed a high percentage (90%) of aquatic insect remains (Galvis et al. 2006). Inseminating (Burns et al. 1995, Burns and Weitzman 2005).

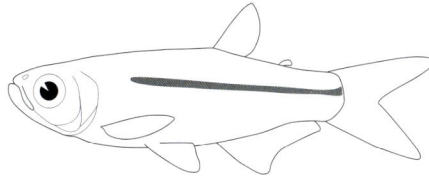

Xenurobrycon (1.4–2.0 cm SL)

Characterized by: very small adult body sizes. Most easily recognized by the pelvic-fin modification in sexually mature males: pelvic fin long, approximately one-third the total length of the fish; interradial membrane extremely expansive in the middle region of the fin; when the fin is extended and spread with forceps it forms an inverted boat-shaped canopy (Weitzman and Fink 1985). Other characters in sexually mature males that can aid identification (Weitzman and Fink 1985): extent of ossification in caudal-fin rays 11, 12, and sometimes 13 interrupted proximally; ventral borders of ventral unbranched principal caudal-fin ray and the adjacent procurrent fin ray are distinctly concave in the region near the terminal few segments of the anteriorly adjacent ray; pouch-scale ligament is relatively weak and loosely organized, its attachment to the pouch scale split primarily into two areas on the apical region of the posterior field of the scale; and pelvic bones are separated posteriorly, so that a space greater than one-third of the length of the pelvic bone extends between the posterior processes (as a result, the bases of ventral fins of males are separated widely and placed rather high on the sides). Sexual dimorphism also includes differences in distance from snout to pelvic-fin origin (shorter in males), depth of caudal peduncle (greater in males), length of anterior anal-fin lobe (greater in males) and bony hooks on the pelvic,- anal,- and caudal-fin rays in males (Weitzman and Fink 1985). Body coloration iridescent blue in *X. coracoralinae* (Moreira 2005). **SPECIES** Five, with four species in the Amazon basin. Key to four species in Weitzman (1987), additional species in Moreira (2005). **DISTRIBUTION AND HABITAT** *X. macropus* from the Paraguay basin; *X. heterodon* from the Caquetá, Napo, Pastaza, and Ucayali basins; *X. pteropus* from the Solimões River; *X. polyancistrus* from the Madeira basin, and *X. coracoralinae* from the Araguaia basin. Species typically found in shallow, relatively still waters (Weitzman 1987, Moreira 2005). **BIOLOGY** *X. coracoralinae* lives in small schools (Moreira 2005).

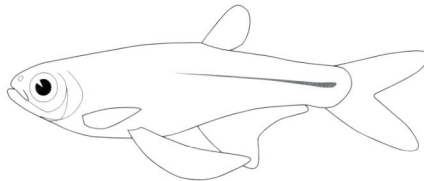

FAMILY CHARACIDAE *INCERTAE SEDIS* INCLUDING SUBFAMILIES TETRAGONOPTERINAE AND STETHAPRIONINAE

— *PETER VAN DER SLEEN and FLÁVIO C. T. LIMA*

DIVERSITY Assembled here are 515 species in 29 genera considered by Mirande (2010) as members of the subfamily Tetragonopterinae and 'Astyanax Clade'. A recent molecular phylogeny of Characidae found evidence for a considerably different classification (Oliveira et al. 2011). Since a consensus classification for these genera is lacking, this chapter addresses a miscellaneous assemblage, including both the subfamilies Stethaprioninae and Tetragonopterinae. The following genera are included here: *Astyanacinus* (4 species), *Astyanax* (144 species), *Bario* (1 species), *Bryconella* (1 species), *Bryconexodon* (2 species), *Ctenobrycon* (5 species), *Dectobrycon* (1 species), *Deuterodon* (9 species), *Exodon* (1 species), *Gymnocorymbus* (4 species), *Hasemania* (8 species), *Hemigrammus* (58 species), *Hyphessobrycon* (138 species), *Jupiaba* (27 species), *Moenkhausia* (76 species), *Paracheirodon* (3 species), *Parapristella* (2 species), *Petitella* (1 species), *Pristella* (1 species), *Roeboexodon* (1 species), *Schultzites* (1 species), *Serrabrycon* (1 species), *Stichonodon* (1 species), *Thayeria* (3 species), and *Tucanoichthys* (1 species), as well as the species included in the subfamily Stethaprioninae, i.e., *Brachychalcinus* (5 species), *Poptella* (4 species), *Stethaprion* (2 species), and Tetragonopterinae, *Tetragonopterus* (9 species).

COMMON NAMES *Piaba, Lambari* (Brazil); *Mojarita* (Peru).

GEOGRAPHIC DISTRIBUTION Distributed in South America mainly from Colombia to Argentina. The Tetragonopterinae are found throughout cis-Andean South America. The remaining genera are widely distributed and found in all major river systems in both cis- and trans-Andean South America, as well as in Central America.

ADULT SIZES 2.0–16 cm SL.

DIAGNOSIS OF FAMILY Grouped together, most species included in this chapter can be recognized from other characids by presenting two tooth rows in the premaxilla (except for the Stevardiinae, most other characids present either a single or, rarely, 3 tooth rows in the premaxilla), and for typically presenting relatively broad, multicuspid teeth (not unicuspid, conical, or compressed). Species of subfamily Stethaprioninae have a deep and compressed (disk-shaped) body, and have an anteriorly directed bony spine preceding the first dorsal-fin ray (Reis 1989). The Tetragonopterinae are recognized by their deep and compressed body and long anal fin, in combination with characters related to dentition and the prepelvic region (see genus description of *Tetragonopterus*).

KEY TO THE GENERA

1a. Specialized mammiliform teeth outside the mouth, pointing forward (fig. 1a) **2**
1b. Specialized mammiliform teeth outside the mouth absent **5**

2a. Complete lateral line; relatively high number of scales in longitudinal series (>31 scales) **3**
2b. Incompletely pored lateral line (pores present on anterior 7–8 scales of laterosensory canal series); relatively small number of scales in longitudinal series (29–31 scales) .... *Serrabrycon*

3a. Prominent snout and mouth completely inferior (head profile reminiscent of a shark); third infraorbital relatively well developed, covering completely the cheek and reaching the horizontal arm of the preopercle..... *Roeboexodon*
3b. Mouth terminal and scarcely oblique; third infraorbital not well developed, not reaching the horizontal arm of the preopercle and leaving a "naked" area between the anterior region of this infraorbital and the preopercle...............**4**

4a. All teeth on premaxillary, maxillary, and dentary conical, caniniform, or mammiliform; dorsal-fin origin anterior to or at vertical through pelvic-fin origin; bony hooks on fin rays absent in adult males; large humeral and caudal spot.. *Exodon*
4b. Premaxillary, maxillary, and dentary with some teeth multicuspid; dorsal-fin origin posterior to vertical through pelvic-fin origin; bony hooks on fin rays (usually in anal and pelvic fins) present in adult males; small humeral and caudal spot.. *Bryconexodon*

5a. Dorsal fin lacking a bony spine at its anterior portion **6**
5b. Dorsal fin presenting a bony spine at its anterior portion (fig. 5b) .... **27** (Stethaprioninae)

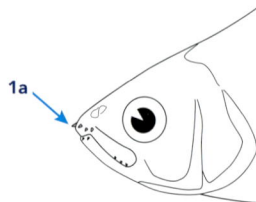

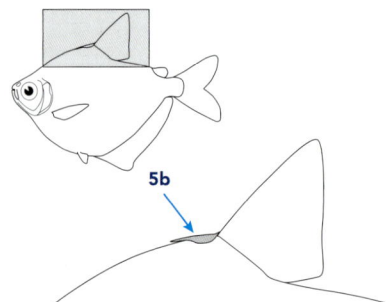

6a. Lateral line complete..**7**
6b. Lateral line pores forming an incomplete line...**16**

7a. Lateral line distinctly curved anteriorly; body high and strongly compressed, with prepelvic region flattened and bordered laterally by distinctly angled scales ....*Tetragonopterus*
7b. Lateral line approximately straight along its entire extension; body ranging from low to high, generally moderately compressed; prepelvic region typically slightly compressed, never flattened or presenting angled scales..........**8**

8a. Caudal fin not scaled, or scales covering only the base of the caudal fin **9**
8b. Scales typically covering one-third the length of the caudal-fin lobes **15**

9a. Anal fin usually typically with <30 branched rays...**10**
9b. Anal fin with >30 branched rays..**13**

10a. Margins of toothed maxilla roughly parallel (fig. 10a) **11**
10b. Margins of toothed maxilla dorsally divergent (fig. 10b); Potaro basin, Guyana..............................*Deuterodon*

11a. Anterior tip of pelvic bone pointed, lacking associated cartilage and frequently projecting outside body wall (fig. 11a)..*Jupiaba*
11b. Anterior tip of the pelvic bone rounded and capped by small cartilages, never projecting outside body wall **12**

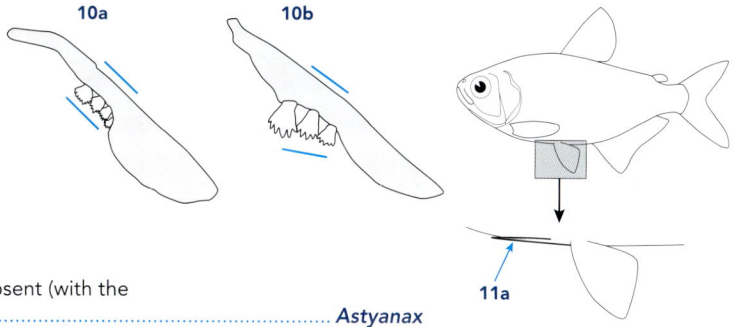
10a 10b 11a

12a. Chevron-like pigmentation along midline absent (with the exception of *A. leopoldi*)..*Astyanax*
12b. Chevron-like pigmentation along midline present ....*Astyanacinus*

13a. Preventral area strongly compressed (fig. 13a) ....*Stichonodon*
13b. Preventral area rounded in cross section, never strongly compressed (fig. 13b).................................**14**

14a. Predorsal line naked or with irregular scale rows (fig. 14a) ....*Gymnocorymbus*
14b. Predorsal series of scales normally developed ....*Ctenobrycon*

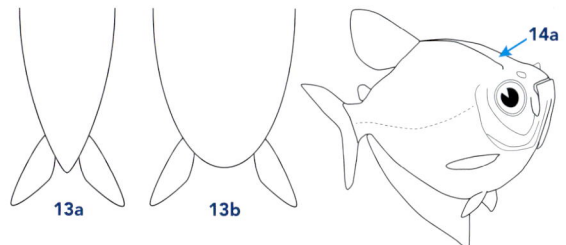
13a 13b 14a

15a. Border of scales crenate, mainly visible in large-sized specimens (fig. 15a).. *Bario steindachneri*
15b. Border of scales not crenate ....*Moenkhausia*

16a. Adipose fin absent .... *Hasemania*
16b. Adipose fin present **17**

17a. Body with brilliant and intense, metallic blue or blue-green lateral body stripe, dense red pigmentation restricted to region ventral to lateral stripe ....*Paracheirodon*
17b. Lateral stripe absent or if present not intense blue or blue-green (but silvery, reddish, pale green or black); red pigmentation in region ventral of lateral stripe absent or if present not very dense **18**

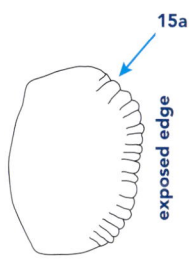
15a exposed edge

18a. Coloration with an intensely red head and the presence of 3 conspicuous black bars in the caudal fin **19**
18b. Coloration without the combination of a red head and 3 conspicuous black bars in the caudal fin....................................**20**

19a. Two rows of premaxillary teeth ....*Hemigrammus bleheri* and *H. rhodostomus*
19b. One row of premaxillary teeth ....*Petitella georgiae*

20a. Lower caudal-fin lobe longer than upper lobe; color of ventral caudal-fin lobe dark brown or black and dorsal caudal-fin lobe hyaline; black color of ventral caudal-fin lobe continuing on caudal peduncle and often as a lateral stripe .... *Thayeria*
20b. Caudal-fin lobes similar in length; caudal-fin lobes similar in color or both hyaline **21**

21a. One series of teeth on premaxilla **22**
21b. Two series of teeth on premaxilla **23**

22a. Very conspicuous black spot on the dorsal fin, underlined by yellow zone and tipped with white (color pattern also present on anal and pelvic fin); body color yellowish, without a broad black lateral band .... *Pristella maxillaris*
22b. No black spot on the dorsal fin; body with a broad black lateral band; red cheek .... *Tucanoichthys tucano*

23a. Caudal fin not scaled .... *Hyphessobrycon*
23b. Caudal fin scaled (at least on its base to a certain extent) **24**

24a. Maxilla long and slender, forming a simple curve with the premaxilla and with 6–12 conical or tricuspid teeth **25**
24b. Maxilla of moderate length, forming an angle with the premaxilla, usually with <6 teeth **26**

25a. Body shape compressed; a broad dark longitudinal band from opercle to the tips of middle caudal-fin rays; anal fin long, with 24–28 branched rays .... *Dectobrycon*
25b. Body shape rather elongate; a humeral sport and a large spot on the base of the middle caudal rays; anal fin short, with 13–16 branched rays .... *Parapristella*

26a. Outer premaxillary tooth row with typically 2–3 teeth, inner premaxillary tooth row with typically 5 (rarely more) teeth .... *Hemigrammus*
26b. Outer premaxillary tooth row with typically 5 teeth, inner premaxillary tooth row with typically 2–3 teeth (in some specimens both rows merging into one) .... *Bryconella*

27a. Scales very small, 59–69 in lateral line; predorsal spine long, spear-shaped (fig. 27a) .... *Stethaprion*
27b. Scales large, 33–38 in lateral line; predorsal spine short (may be concealed under the skin; see fig. 28) **28**

28a. First anal-fin element modified into a strong, forward-directed spine; predorsal spine pointed, forward-directed (fig. 28a) .... *Brachychalcinus*
28b. First anal-fin element simple or slightly laminar, but never as above; predorsal spine resembling a downward-facing shovel, rounded anteriorly (fig. 28b) .... *Poptella*

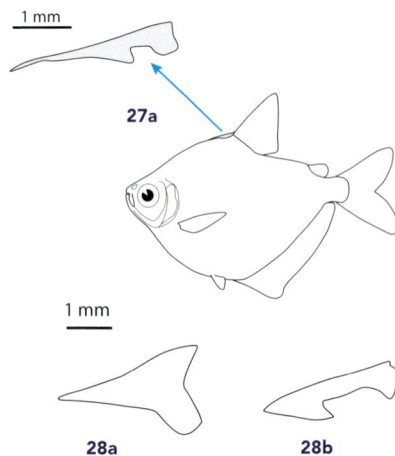

1 mm

27a

1 mm

28a   28b

GENUS ACCOUNTS

Astyanacinus (6.0–14 cm SL)

Characterized by: lower jaw larger than the eye-orbit; 3 or more teeth in maxilla; posterior margin of maxilla posterior to vertical from lateral ethmoid; premaxilla with 2 rows of teeth, the inner one with 5, and the outer one with 4 teeth; dentary with 4 large anterior teeth, followed by many smaller teeth; anteriorly directed chevron-shaped marks on body; the humeral region with 2 diffuse vertical humeral bars and 1 well-defined oval-shaped horizontal mark over the anterior vertical mark; caudal spot horizontally elongate, extending toward distal border of central caudal-fin rays; total number of anal-fin rays 25–36; dorsal-fin rays ii + 9; lack of hooks on fin rays; and pectoral-fin rays i + 11–13 (D'Agosta 2011, Torres-Mejia et al. 2012). **SPECIES** Four, including one species in the AOG region.

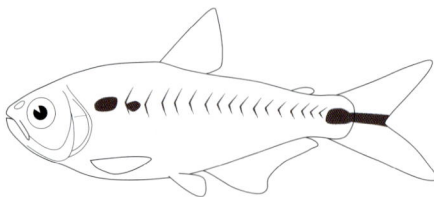

DISTRIBUTION AND HABITAT *A. multidens* from the upper Rio Madeira basin. Three undescribed species from the upper Amazon are mentioned in Torres-Mejia et al. (2012). **BIOLOGY** No data available.

Astyanax (5.0–16 cm SL)

Astyanax is the most species-rich genus in Characidae, but it forms an artificial group (Mirande 2010, Oliveira et al. 2011a). The genus is defined by: 2 rows of premaxillary teeth; 5 teeth in the inner premaxillary series; lateral line complete; adipose fin present; and caudal fin naked (Eigenmann 1921, 1927). Maxillary teeth usually <5. Sexual dimorphism includes (but not for all species) differences in pectoral- and pelvic-fin lengths, differences in the anal-fin shape (being concave in females and nearly straight in males), and bony hooks on the dorsal-, anal-, pectoral-, and pelvic-fin rays in mature males. **SPECIES** 144 valid species, including at least 23 species in the AOG region. *Astyanax* was revised by Eigenmann (1921, 1927), whose accounts still constitute the most extensive review of the genus. However, more than a third of the currently valid species have been described within the last 10 years. **COMMON NAMES** *Piaba*, *Lambari do rabo vermelho* (Brazil); *Chirichigno*, *Colirroja* (Colombia, Ecuador, Peru); *Querepe* (Venezuela). **DISTRIBUTION AND HABITAT** From southern USA to central Argentina. Inhabiting many environments, including mountain areas, lotic and lentic river portions, lakes, and headwater streams. Two closely related *Astyanax* species adapted to subterranean environments and are blind and without pigmentation (*A. jordani* and populations of *A. mexicanus*, both from Mexico). Not particularly diverse in the Amazon basin. **BIOLOGY** Omnivorous. Often found in groups.

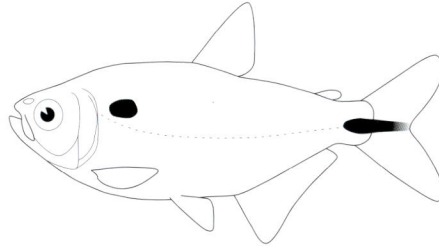

Bario (12 cm SL)

Resembles a *Moenkhausia* species, but differs by having crenate scales (most prominent in fishes larger than 55 mm SL; Géry 1977) versus cycloid in *Moenkhausia*. In addition, in most *Moenkhausia* species the anal-fin base is partially covered by two rows of scales, whereas more than two rows of scales partially cover the anal-fin base in *Bario*. Other characters that may aid identification: body depth about 2.25 times in standard length; complete lateral line, anal fin with 25–26 rays; body color brownish above, silvery on the sides; a series of dark lines between the series of scales; a faint humeral spot and a large conspicuous caudal spot (see e.g., Lima et al. 2013). **SPECIES** One, *B. steindachneri*. **DISTRIBUTION AND HABITAT** Throughout the lowlands of the Amazon basin (Peru, Ecuador, Colombia and Brazil), widespread though not very common. Occurs in small terra firme (nonfloodplain) forest streams. **BIOLOGY** No data available.

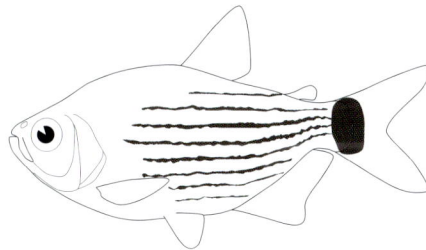

Brachychalcinus (6.9–9.0 cm SL)

Characterized by: body shape rounded in lateral view; first anal-fin element modified into a strong, forward- directed spine; predorsal spine of variable length, roughly triangular and with a sharp, dorsally directed process running along the anterior surface of the first unmodified dorsal-fin ray; scales

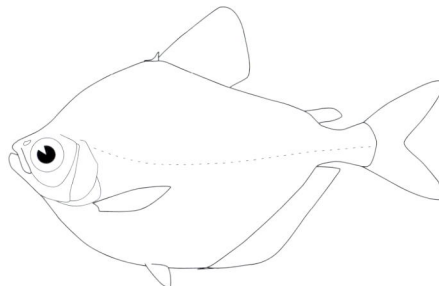

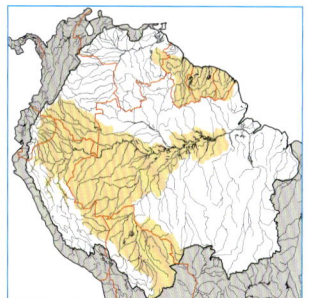

cycloid, large and smooth; 33–38 scales in lateral line; 8–12 scales between lateral line and dorsal-fin origin; 9–12 scales between lateral line and pelvic-fin origin; scales on caudal fin restricted to the base or to the proximal one-third of the rays, but covering about half the dorsal and ventral rays (Reis 1989). Body color clear to silvery in life, with a narrow longitudinal black stripe from the second humeral blotch to the caudal peduncle. A wider but very diffuse dark midlateral band sometimes also present. All fins, except pectorals, with dark chromatophores scattered on interradial membranes. First pelvic-fin rays darker than others. Middorsal scales and the scales above the longitudinal black stripe usually darker (Reis 1989). **SPECIES** Five, including three species in the AOG region. Key to the species, species information, and distribution map in Reis (1989). **DISTRIBUTION AND HABITAT** *B. copei* from the Solimões and Madeira basins in Bolivia, Brazil, and Peru; *B. nummus* from the upper Amazon basin in Colombia, Ecuador, and Peru; and *B. orbicularis* from coastal drainages of Suriname and Guyana. Two other species in Parnaíba and Paraguay basins. **BIOLOGY** Omnivorous, diet includes insects, worms, crustaceans, and plant material (Mol 2012b)

Bryconella (2.5 cm SL)

A *Hemigrammus*-like tetra, distinct from *Hemigrammus* by having very irregular premaxillary teeth. Instead of having 1–3 (or sometimes more) outer teeth and 5 inner teeth (rarely more), *Bryconella* has typically 4 or 5 outer teeth and 3 or 2 inner teeth, which are very close to the anterior row, and in some individuals both rows merge in an irregular row of teeth (Géry 1977). Color pattern consists of a conspicuous brown humeral spot followed by a well-marked longitudinal narrow dark stripe extending to the end of the middle caudal rays (Géry 1977, Lima et al. 2013). **SPECIES** One, *B. pallidifrons*. **DISTRIBUTION AND HABITAT** Widespread in the western portions of the Amazon basin in Brazil, Peru, and Colombia; occurs in clearwater terra firme forest streams. **BIOLOGY** No data available.

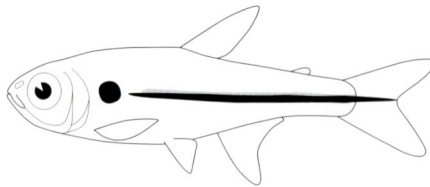

Bryconexodon (11–13 cm SL)

Similar to *Exodon*, but with some multicuspid teeth on the dentary and maxilla, a more posteriorly positioned dorsal fin, and bony hooks on the fin rays in adult males. In addition, *Bryconexodon* species are more dull-colored, and the humeral spot and the spot on caudal peduncle are much smaller than in *Exodon*. **SPECIES** Two, both in the Amazon basin. Species information in Jégu et al. (1991a) and Gery (1980). **DISTRIBUTION AND HABITAT** *B. juruenae* from the upper Tapajós basin, and *B. trombetasi* from the Trombetas basin. *Bryconexodon trombetasi* has been found close to rapids and in fast-flowing waters (Jégu et al. 1991a). **BIOLOGY** Both species feed on the scales of other fishes.

Ctenobrycon (5.0–9.0 cm SL)

Characterized by: deep and compressed body (2.0–2.5 times in SL); long anal fin (39–47 total rays) with its margin nearly straight and its origin behind or below the origin of the dorsal; mouth very small, the maxillary not reaching the eye; spinous scales in the preventral area; cycloid scales on the sides in young, becoming spinous in adults; lateral line complete (extremely

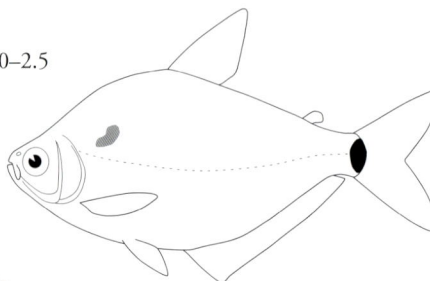

variable number of pored scales, from 36 to 49); caudal fin without scales; a series of tricuspid teeth in the premaxillary and an inner series of pentacuspid teeth, whose cusps are arranged in a U-shaped curve (Eigenmann 1927, Benine et al. 2010). **SPECIES** Four, including three in the AOG region. **DISTRIBUTION AND HABITAT** *C. multiradiatus* in the Amazon basin in Brazil, *C. oliverai* in the Orinoco basin, and *C. spilurus* in the Amazon and Orinoco basin and coastal rivers in the Guianas. An additional species (*C. alleni*) in the Paraguay basin. Abundant in floodplain lakes and other types of lentic environments. **BIOLOGY** Omnivorous, diet includes zooplankton, worms, insects, crustaceans, and plant matter (Mills and Vevers 1989, Planquette et al. 1996).

Dectobrycon (7.0 cm SL)

Characterized by: 5–6 teeth in the inner row of the premaxilla; premaxillary teeth tri- or pentacuspid; 6–11 tricuspid or conical maxillary teeth; dentary teeth pentacuspid and gradually decreasing in size; lateral line incomplete; many small scales on the caudal basis (similar to species of the genus *Markiana*); scales with smooth border; anal fin with iv–v + 23–27 rays; scale formula: 7–8/(15–23) 36–38/6–7; and a dark longitudinal band from gill cover to the tips of middle caudal-fin rays (Zarke and Géry 2006). **SPECIES** One, *D. armeniacus* (Zarke and Géry 2006). **DISTRIBUTION AND HABITAT** Unknown locality in Peru. **BIOLOGY** No data available

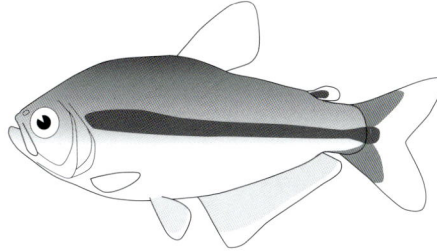

Deuterodon (6.0 cm SL)

Resembling *Astyanax* and *Jupiaba*, but can be recognized by the maxilla. In most characids the margins of the maxilla run roughly parallel in the toothed region. In *Deuterodon*, the margins of toothed region of maxilla are dorsally divergent (Mirande 2010). *Deuterodon* is predominantly vegetarian and has flat, multicuspid and incisiform teeth that are very compressed anteroposteriorly and form a more or less continuous cutting edge (Géry 1977). Dentary teeth gradually decrease in size in *Deuterodon*, while *Astyanax* possess a more common dentition pattern in which the anteriormost 4–5 teeth are much larger than the remaining teeth. Other characters that might aid identification: preventral area rounded or compressed; 24–25 anal-fin rays; 3–5 maxillary teeth, the broadest with 5–7 cusps; 37–40 lateral line scales; a comma-shaped humeral spot; series of dots on scales forming a longitudinal stripe, the one corresponding to the lateral line broader; caudal spot extended onto lower caudal margin (Géry 1977). Presence of bony hooks in the anal fin of males during the reproductive period has been reported for species from southern Brazil (Dala-Corte and Fialho 2013). **SPECIES** Nine, including one species in the AOG region: *D. potaroensis*. **DISTRIBUTION AND HABITAT** Most species inhabit southeastern Brazil, and *D. potaroensis* in the Potaro basin of Guyana. **BIOLOGY** Predominantly vegetarian (Géry 1977).

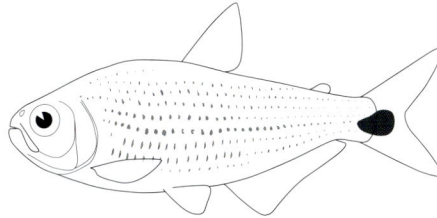

Exodon (7.5 cm SL)

Readily identified by the mammiliform teeth outside the mouth (pointing forward) and its bright coloration, with a yellow caudal fin and body, other fins reddish, and a large humeral and caudal spot. Similar in appearance to *Bryconexodon*, but differs in snout shape and by having only conical, caniniform, or mammiliform teeth on the premaxillary, maxillary, and dentary (vs. some teeth multicuspid in *Bryconexodon*); a more anteriorly positioned dorsal fin; and absence of bony hooks on fin rays in adult males (Géry 1977). **SPECIES** One, *E. paradoxus*.

COMMON NAMES *Lambari do rabo vermelho* (Brazil); *Colirroja* (Colombia, Ecuador, Peru). **DISTRIBUTION AND HABITAT** Amazon basin (Tocantins drainage) and Guyana (Rio Branco drainage). Inhabits rivers with sandy bottoms. **BIOLOGY** Fish scales filled 88% of the stomach contents in *E. paradoxus* collected in the wild (Pereira et al. 2007a).

Gymnocorymbus (5.0–7.5 cm SL)

Characterized by: body rounded in lateral view; predorsal spine absent (as opposed to the genera in the subfamily Stethaprioninae); maxillary teeth absent or only one; predorsal line naked. Body color in wild usually silver or yellowish, and much less spectacular than aquarium developed strains. A second humeral spot is present as a vertical bar (very conspicuous in *G. ternetzi*, but rather inconspicuous in *G. bondi* and *G. thayeri*). **SPECIES** Four, with three in the AOG region, including *G. thayeri* (False-black tetra or Blackskirt tetra). Revision of genus and key for the species in Benine et al. (2015). **DISTRIBUTION AND HABITAT** *G. bondi* in the Orinoco basin, *G. thayeri* in the upper Amazon basin, Guyana, and northeastern Brazil, and *G. flaviolimai* in the Rio Madeira basin (Benine et al. 2015). *Gymnocorymbus thayeri* was found exclusively in forest streams around Leticia, Colombia (Galvis et al. 2006). **BIOLOGY** Diet includes terrestrial insects (Galvis et al. 2006). Popular aquarium fishes, including *G. ternetzi* (Black tetra) from the Paraguay basin.

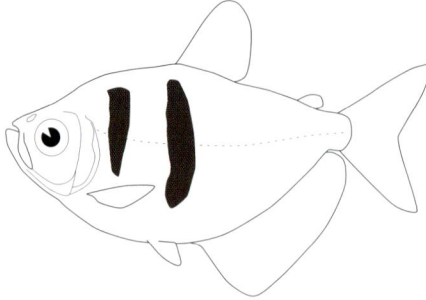

Hasemania (2.5–7.0 cm SL)

Similar to *Hyphessobrycon*, but without an adipose fin. Other characters that could aid identification: two series of teeth in the premaxilla; maxilla with few or no teeth along its exposed anterior margin; lateral line incomplete; caudal fin typically lacking scales, but present in *H. marginata* (Ellis 1911, Bertaco and Carvalho 2010, Zanata and Serra 2010). Sexual dimorphism includes small bony hooks on the anal- and pelvic-fin rays and longer pectoral and pelvic-fin lengths in males. In addition, the anal-fin shape is nearly straight in males and concave in females (*H. kalunga*; Bertaco and Carvalho 2010). **SPECIES** Eight, including two species in the Amazon basin: *H. nambiquara* and *H. kalunga*. **DISTRIBUTION AND HABITAT** *H. nambiquara* in the upper Tapajós basin and *H. kalunga* in the upper Tocantins basin (Bertaco and Malabarba 2007, Bertaco and Carvalho 2010). Species are found in semi-lentic stretches of shallow, clearwater rivers with sand and leaves on the bottom (Bertaco and Carvalho 2010, Zanata and Serra 2010). **BIOLOGY** Omnivorous, consuming insects, arthropods, algae, and organic debris (Bertaco and Carvalho 2010, Zanata and Serra 2010).

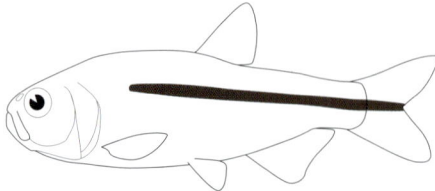

Hemigrammus (2.5–6.0 cm SL)

A species-rich but artificial genus. The type species (*H. unilineatus*) might be a member of the "*Hyphessobrycon* Rosy Tetra Clade" (Weitzman and Palmer 1997). Genus is characterized by two rows of teeth in the premaxilla, incomplete lateral line, and caudal fin partially covered with small scales (Eigenmann 1918a). Most species are 3–5 cm SL. Some species have iridescent spots (mainly on the caudal peduncle, eye, and caudal-fin base), which might serve as recognition signals (Géry 1977). Sexual dimorphism occurs in some species and includes differences in color, pectoral- and pelvic-fin size, anal-fin shape (approximately straight in males and slightly concave in the anterior portion in females), and bony hooks on the fin rays in mature males (Lima and Sousa 2009). **SPECIES** 58, with about 50 species in the AOG region, including many popular aquarium species such as *H. erythrozonus* (Glowlight tetra), *H. falsus* (Head-and-tail-light fish), *H. hyanuary* (January tetra), *H. pulcher* (Garnet tetra), *H. rhodostomus* (Rummy-nose tetra), and *H. unilineatus* (Featherfin tetra). Outdated key to the species in Géry (1977). **DISTRIBUTION AND HABITAT** Widespread throughout South America, occurring from the Orinoco to the Paraguay-Paraná basins. Highest diversity in the Amazon basin and Guianas. Found in many habitats, but most common in slow-flowing small rivers, streams, floodplain lakes, and swamps. **BIOLOGY** Often found in groups. Analyzed stomach contents included plant matter, algae, insects, and other invertebrates (e.g., Planquette et al. 1996, Britski and Lima 2008, Carvalho et al. 2010). Similar to many other characids, *H. erythrozonus* displays a conspicuous "fin flicking" behavior upon detecting a predation threat, which serves as a visual alarm signal to conspecifics and as a deterring signal to potential predators by signaling alertness (Brown et al. 1999).

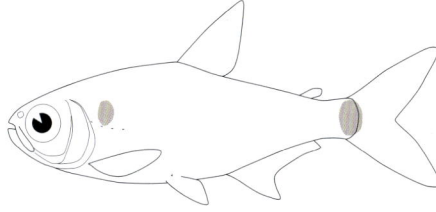

Hyphessobrycon (2.0–5.0 cm SL)

A species-rich but artificial genus (Mirande 2010). Genus is characterized by a small adult body size, resembling *Hemigrammus* but without scales on the caudal fin, and resembling *Astyanax* but with an interrupted (or incomplete) lateral line. Also characterized by: premaxillary teeth in two series, the inner series with 5 teeth; third infraorbital not in contact with the preopercle ventrally; few maxillary teeth; and adipose fin present. Sexual dimorphism varies between species and can include differences in color patterns, the anal-fin border (males with anal-fin border slightly convex to straight, females with distinct anterior lobe), the elongation of the dorsal fin (becoming filamentous), and bony hooks on the fin rays in mature males. **SPECIES** 138 valid species, with at least 80 species in the AOG region, including many species popular in the aquarium trade such as *H. bentosi* (Ornate tetra), *H. eques* (Jewel tetra), *H. erythrostigma* (Bleeding-heart tetra), *H. heterorhabdus* (Flag tetra), *H. loretoensis* (Loreto tetra), *H. peruvianus* (Peruvian tetra), *H. pulchripinnis* (Lemon tetra), *H. rosaceus* (Rosy tetra), and *H. socolofi* (Spotfin tetra). Species have been grouped based on similarities in color patterns (Géry 1977), but these similarities could be mostly artificial (Weitzman and Palmer 1997). The Rosy Tetra Clade includes some 30 species that share a "flag signal" in the dorsal fin, consisting of a black spot, tipped with white and underlined with white or yellow. Eigenmann (1918a, 1921) provided what are still the most extensive reviews of the genus. An outdated key to the species appears in Géry (1977), but more than 60 species have subsequently been included in the genus. **DISTRIBUTION AND HABITAT** Southern Mexico to the Rio de la Plata in Argentina. Highest diversity inhabits the Amazon and Orinoco basins. Found in many habitats including small forest streams, main river channels, and stagnant pools (e.g., Planquette et al. 1996, Galvis

et al. 2006, Mol 2012b), but occurring mostly in streams, small rivers, and floodplain lakes. **BIOLOGY** Consume plant material and small aquatic arthropods and other benthic animals (e.g., Carvalho and Langeani 2013, Teixeira et al. 2013). Found in schools or solitary (Lima and Moreira 2003, Carvalho and Langeani 2013).

Jupiaba (5.0–12 cm SL)

Similar to *Astyanax*, differing in the presence of a spinous pelvic bone that frequently (but not always) projects outside the body wall (Zanata 1997). When not exposed, the spinous pelvic bone is barely covered by thin muscle or skin, and will rip the skin and project out the body wall if pressed. Color pattern very distinctive in some species, with conspicuous humeral blotch(es), midlateral stripes, and/or caudal peduncle blotches. **SPECIES** 27, including 25 species in the AOG region. Key to the species in Netto-Ferreira et al. (2009). **DISTRIBUTION AND HABITAT** Northern cis-Andean South America (*J. acanthogaster* and *J. polylepis* also occur outside this area). Most species in rivers draining the shield regions and the Guianas. Generally found in flowing waters. **BIOLOGY** Omnivorous with diets including insects, detritus, algae, and plant material (e.g., Zanata and Lima 2005, Birindelli et al. 2009, Netto-Ferreira et al. 2009).

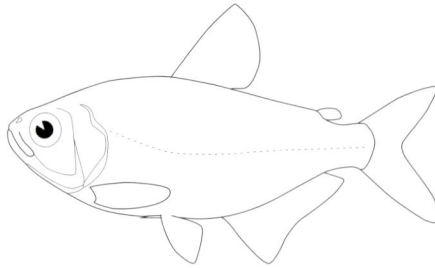

Moenkhausia (4.0–10 cm SL)

A species-rich but artificial genus (Mirande 2010, Mariguela et al. 2013b). Most (but not all) species characterized by: premaxillary teeth in two rows, the inner with five teeth; complete lateral line; and partially scaled caudal fin. *Moenkhausia* species vary greatly in body shapes, including both deep- and shallow-bodied forms, and exhibit many color patterns, including an overall dark reticulated pattern, a dark blotch on the upper caudal-fin lobe, dark zigzag lines on body, and a dark broad stripe on body (Benine et al. 2009, Marinho 2010). Some species have a deep-red eye in life. Observations of sexual dimorphism include fin-ray hook on the anal fin and a longer pelvic fin in adult males (e.g., Benine et al. 2004, Benine et al. 2009, Mol 2012b). Males of *Moenkhausia comma* have breeding tubercles on head and anterior portion of body. **SPECIES** 76, with 69 species in the AOG region, including *M. dichroura* (Bandtail tetra), *M. oligolepis* (Glass tetra), *M. pittieri* (Diamond tetra), *M. sanctaefilomenae* (Redeye tetra). Outdated key in Géry (1977), with many species described afterward. Two nominal monotypic genera, *Gymnotichthys* and *Schultzites*, are very similar and closely related to species currently included within the genus *Moenkhausia*. **DISTRIBUTION AND HABITAT** Widely distributed throughout South America, primarily the Amazon basin and Guianas. Found in a variety of habitats, including slow-flowing sections of rivers, small forest streams, and close to rapids. **BIOLOGY** Omnivorous, stomach contents of examined species included remains of (terrestrial) insects, fishes, algae, fruits, and other plant materials (Planquette et al. 1996, Galvis et al. 2006). Found predominantly at midwater or near the surface. Solitary or in schools. Diurnal. Some species can jump out of the water to avoid predator attacks.

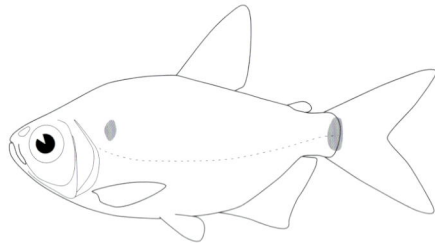

Paracheirodon (3 cm SL)

Readily recognized by a distinct coloration in life, with an intense blue or blue-green lateral body stripe and a dense red pigmentation ventral to the lateral stripe. A stripe of dark brown or black chromatophores underlies and extends dorsal to the blue lateral stripe (Weitzman and Fink 1983). **SPECIES** Three, all in the AOG region. Key to the species and species information in Weitzman and Fink (1983). **DISTRIBUTION AND HABITAT** *P. innesi* (Neon tetra) from black- and clearwater streams in the western Amazon basin, *P. axelrodi* (Cardinal tetra) from the upper Orinoco and Negro basins, and *P. simulans* (Green neon tetra) from the central Amazon and lower Madeira basins. **BIOLOGY** Found mainly in shoals and the middle water layer. Diet includes insects, crustaceans, worms, zooplankton, and plant matter. Among the most popular aquarium fishes in the world and are exported by the millions from their native home range, providing income for hundreds of ornamental fishers (e.g., Chao and Prada-Pedreros 1995, Moreau and Coomes 2007).

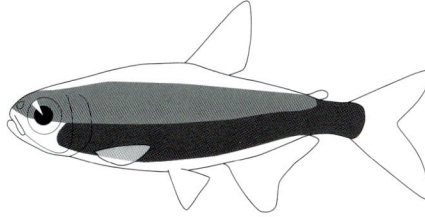

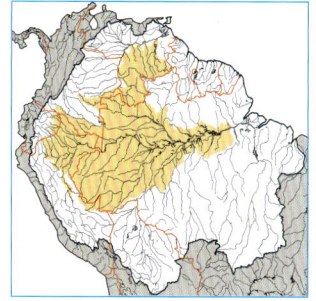

Parapristella (4.5 cm SL)

Characterized by: moderately elongate body shape with a humeral sport; large spot on the base of the middle caudal rays; short anal fin; and caudal-fin lobes in life with red-orange base (Géry 1977). Distinguished from *Pristella* by two rows of premaxillary teeth (vs. one in *Pristella*), and absence of darks spots on the dorsal, anal, and pelvic fins. Resemble *Hemigrammus* species, but with a higher number of maxillary teeth: 6–12 in *Parapristella* vs. usually <6 in *Hemigrammus* (Géry 1977). **SPECIES** Two, both in the AOG region. **DISTRIBUTION AND HABITAT** *P. aubynei* in coastal drainages of Guyana and the Branco River in Brazil, *P. georgiae* in Meta and Aguaro rivers of Orinoco (Colombia, Venezuela) and Negro (Brazil) basins. Occurs in clear- or blackwater streams and flooded forest. **BIOLOGY** No data available.

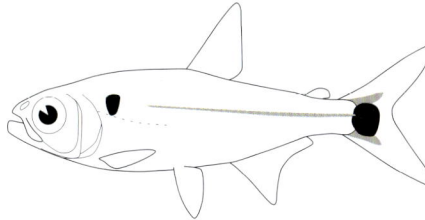

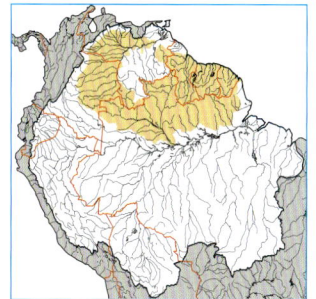

Petitella (4.0 cm SL)

Characterized by: a conspicuous color pattern consisting of an intensely red head and 3 conspicuous black bars on the caudal fin. Differs from *Hemigrammus* by having one row of premaxillary teeth (vs. two rows in *Hemigrammus*). *Hemigrammus bleheri* (Negro and Meta basins, Brazil and Colombia) and *Hemigrammus rhodostomus* (lower Amazon basin in Pará state and Orinoco basin, Brazil and Venezuela) exhibit a similar color pattern and may belong to this genus (Mirande 2010). **SPECIES** One, *P. georgiae* (False Rummy-nose tetra; Géry and Boutiere 1964). **DISTRIBUTION AND HABITAT** Purus, Negro, Solimões, and Madeira basins (Peru, Brazil), mostly in floodplain lakes. **BIOLOGY** Diet includes worms, small crustaceans, and plant matter (Mills and Vevers 1989).

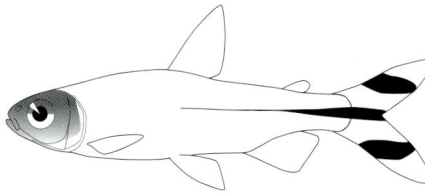

Poptella (7.0–8.0 cm SL)

Characterized by: predorsal spine, of variable length, with a rounded tip and bearing two pointed posteroventrally directed projections (overall form saddle-shaped; Reis 1989); body shape somewhat rounded in lateral view; first anal-fin ray simple and not expanded; scales cycloid, large and smooth, 34–37 in lateral line; 7–10 scales between lateral line and dorsal-fin origin; 8–10 scales between lateral line and pelvic-fin origin; scales restricted to the base of the anal-fin ray; scales on caudal fin covering about two-thirds of the outer rays (Reis 1989). Body color yellowish; darker dorsally, with a narrow longitudinal black stripe from the second humeral blotch to the caudal peduncle; a wider but very diffuse dark band is sometimes present above that stripe; unpaired fins slightly darkened (Reis 1989). Sexual dimorphism includes bony hooks on the anal and pelvic-fin rays (Reis 1989). **SPECIES** Four, including three in the AOG region. Key to the species, species information and distribution map in Reis (1989). **DISTRIBUTION AND HABITAT** *P. brevispina* inhabits the Trombetas, upper Branco, and lower Tocantins basins and the coastal drainages of Guyana, Suriname and Pará, Brazil; *P. compressa* in the Orinoco and Amazon basins, coastal drainages of Venezuela, Guyana, and northeastern Brazil; *P. longipinnis* in the Orinoco basin and coastal drainages of Suriname, and lower Tocantins River. In Guyana, *P. brevispina* has been found in clear water creeks with sandy bottoms (Planquette et al. 1996), and around Leticia (Colombia) *P. compressa* inhabits forest streams (Galvis et al. 2006). **BIOLOGY** Diet includes mainly invertebrates (Wantzen et al. 2002).

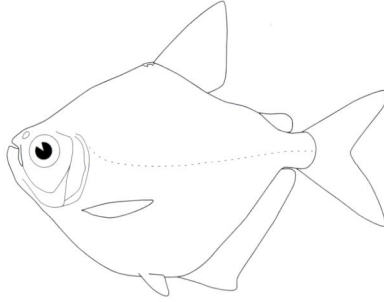

Pristella (3.0 cm SL)

Shares with species of the "*Hyphessobrycon* Rosy Tetra Clade" a conspicuous black spot on the dorsal fin, underlined by a yellow zone and tipped with white. This color pattern is also found on the anterior part of the anal fin and, to a lesser degree, on the pelvic fins. Body color yellowish and with a small round humeral spot. Differentiated from *Hyphessobrycon* by a single series of teeth on the premaxilla (vs. two premaxillary teeth series, the inner series with 5 teeth, in *Hyphessobrycon*). **SPECIES** One, *P. maxillaris*. **COMMON NAMES** *Peixe raio X* (Brazil); *Yaya* (French Guiana); X-ray tetra (English). **DISTRIBUTION AND HABITAT** Lower portion of Amazon and Orinoco rivers, and coastal river drainages of the Guianas and northern Brazil. **BIOLOGY** Omnivore, often found in swampy areas with abundant aquatic vegetation (Planquette et al. 1996).

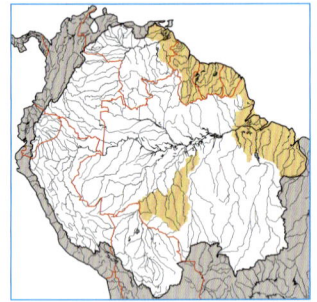

Roeboexodon (12–15 cm SL)

Characterized by: an inferior mouth and conical teeth on the extremity of the prominent shark-like snout. **SPECIES** One, *R. guyanensis; R. geryi* is a junior synonym (Moreira and Lima 2011). **DISTRIBUTION AND HABITAT** Tocantins, Xingu, and Tapajós basins, and rivers of French Guiana and Suriname (Lucena and Lucinda 2004, Moreira and Lima 2011). Inhabits creeks with rapidly flowing water over a sandy substrate (Planquette et al. 1996). **BIOLOGY** Feeds on scales of other fishes.

Serrabrycon (3.0 cm SL)

Characterized by: mammiliform teeth in both jaws; teeth of outermost tooth row of premaxilla directed anteriorly or anteroventrally rather than ventrally, forming irregular series; second and fifth dentary teeth rotated anterodorsally; lateral line pores present only on the anterior 7–8 scales of laterosensory canal series; 29–31 scales in longitudinal series from supracleithrum to hypural joint (Vari 1986). **SPECIES** One, *S. magoi* (Vari 1986); has been collected in small streams. **DISTRIBUTION AND HABITAT** Upper Negro, Casiquiare, and Orinoco basins, Venezuela and Brazil. **BIOLOGY** Stomach completely and exclusively filled with fish scales of different sizes (Vari 1986).

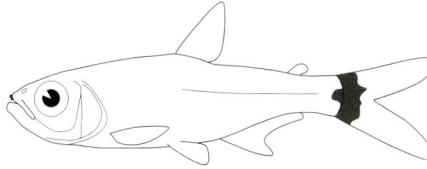

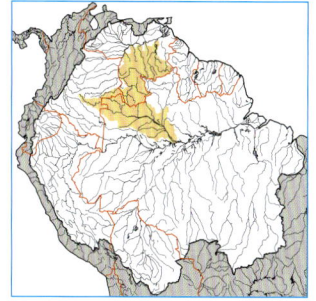

Stethaprion (8.5–9.0 cm SL)

Characterized by: a lanceolate predorsal spine, relatively long (about 12–13 scales flanking it laterally) and with a small, laminar bony process on the posterodorsal portion, making an articulation between the predorsal spine and the first dorsal-fin ray; body shape rounded in lateral view; body with very small scales (18–22 horizontal rows between lateral line and dorsal-fin origin); 59–69 scales in lateral line; 15–20 scales between lateral line and pelvic-fin origin; scales on anal fin covering at least the proximal two-thirds of the rays; scales on caudal fin covering about four-fifths of the rays; first anal-fin rays with a long, anteriorly directed process, giving a triangular shape to the ray (Reis 1989). Body color silvery, with a narrow longitudinal black stripe from the second humeral blotch to the caudal peduncle. Facial bones and sometimes area below longitudinal black stripe densely pigmented with guanine, giving a strong silvery or golden color (Reis 1989). **SPECIES** Two, both in the AOG region. Key to the species, species information, and distribution map in Reis (1989). **DISTRIBUTION AND HABITAT** *S. crenatum* in upper Purus, Madeira, and lower Amazon basins; *S. erythrops* in upper Amazon basin. **BIOLOGY** No data available.

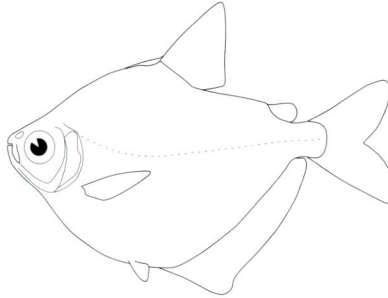

Stichonodon (14 cm SL)

Characterized by: a rounded deep body in lateral view with a strong ventral keel; absence of a predorsal spine; absence of maxillary teeth; scaled caudal and anal fins; long anal fin, with 36 rays; complete lateral line; median predorsal scales leaving a naked area anterior to dorsal fin (but predorsal line not naked and regularly scaled); body color silver with a large black humeral spot above the lateral line extending dorsally as a diffuse stripe; all fins transparent but with a dark border (Géry 1977, Galvis et al. 2006, Mirande 2010). **SPECIES** One, *S. insignis*. **DISTRIBUTION AND HABITAT** Amazon basin. Occurs in floodplain lakes; relatively scarce. **BIOLOGY** Diet includes cladocerans and other small invertebrates (Galvis et al. 2006).

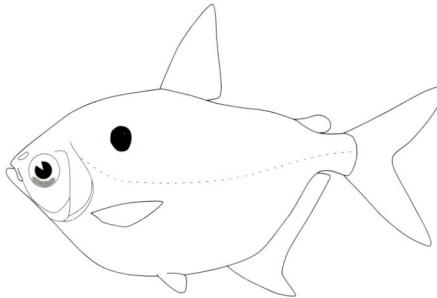

Tetragonopterus (5.0–12 cm SL)

Characterized by: a deep and compressed body; two rows of premaxillary teeth with the inner row generally consisting of 5 or more teeth; transversely flattened prepelvic region that is bordered laterally (particularly proximate to the pelvic-fin insertion) by distinctly angled scales; anal fin with a long base; 7 or more pelvic-fin rays; and a complete lateral line with an anterior portion that is strongly bent downward (Melo et al. 2011, Silva et al. 2013). In addition, Melo et al. (2011) discussed the presence of 3 supraneurals and a branched laterosensory canal on the sixth infraorbital bone as traits of *Tetragonopterus*. Sexual dimorphism includes bony hooks on anal-fin rays in adult males (Silva et al. 2013). **SPECIES** 9, including 8 species in the AOG region. Keys to the species in Silva et al. (2013). **DISTRIBUTION AND HABITAT** Widely distributed throughout cis-Andean South America, including the Amazon and Orinoco basins as well as the coastal rivers in the Guianas. Highest species diversity in the Araguaia basin (Melo et al. 2016). Prefers moderate to fast-flowing waters (mostly in relatively large water bodies), but also found in lakes and slow-flowing water (Galvis et al. 2006, Mol 2012b). **BIOLOGY** Omnivorous, with diet including insects, plants seeds and fruits. Some species feed predominantly on plant materials (*T. argenteus*; Galvis et al. 2006). Diurnal, found in groups (Galvis et al. 2006).

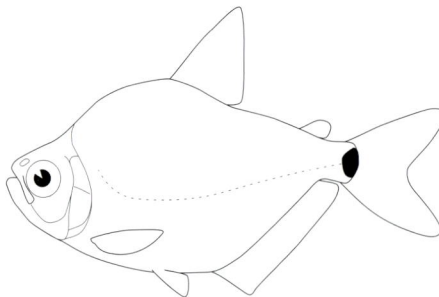

Thayeria (4.0–5.5 cm SL)

Characterized by: an elongate lower caudal-fin lobe, about 10–25% longer than the upper one; and a dark lower caudal-fin lobe that continues (in varying degree) as a dark lateral band on the body (Géry 1977). **SPECIES** Three, all in the AOG region. **COMMON NAME** Penguin tetra. **DISTRIBUTION AND HABITAT** *T. boehlkei* inhabits the upper Tapajós, Xingu, and Araguaia-Tocantins basins; *T. ifati* , the Maroni and Approuague basins, French Guiana, and Suriname, and *T. obliqua*, the middle and upper Amazon and upper Orinoco basins. Species favor flooded forests and shallow areas of large streams and small rivers. **BIOLOGY** All species swim head-up at an angle of about 30° relative to the horizontal. This position, together with the conspicuous black band on the body and lower caudal-fin lobe, disrupts the contour of the animal, making it difficult for predators to recognize it as a fish (Géry 1977). They live in groups. No data on natural diet available.

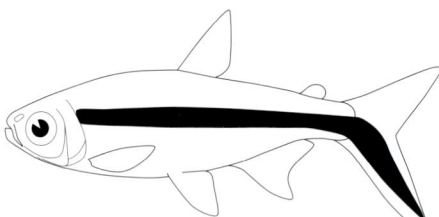

Tucanoichthys (1.7 cm SL)

Characterized by: very small adult body size; jaws relatively large, narrow, set with a single row of conical teeth; premaxilla with a double curve, forming an S; one row of 8 conical premaxillary teeth (resembling some members of Aphyoditeinae or Characinae); maxilla completely toothed (as in the Characinae); predorsal line naked; dorsal fin with 10–11 soft rays; anal fin with 19–21 soft rays; lateral line almost nonexistent; predorsal line broadly naked; and postorbitals lacking (Géry and Römer 1997). Coloration includes a broad black lateral band dorsally delineated by a clear margin, and a red cheek. **SPECIES** One, *T. tucano* (Géry and Römer 1997). **DISTRIBUTION AND HABITAT** Uaupés River in upper Negro basin. **BIOLOGY** No data available.

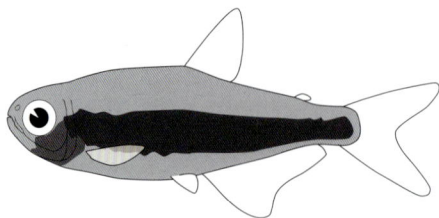

FAMILY CHILODONTIDAE—HEADSTANDERS

— *BRUNO F. MELO and BRIAN L. SIDLAUSKAS*

DIVERSITY Eight species in two genera, *Caenotropus* and *Chilodus*, with four species each (Vari et al. 1995). However, a recent molecular phylogeny suggests the presence of cryptic species within the two most widely distributed species (*Ca. labyrinthicus* and *Ch. punctatus*) and identifies two species (*Ch. zunevei* and *Ch. fritillus*) that may not be valid (Melo et al. 2014).

GEOGRAPHIC DISTRIBUTION Chilodontids live in the Orinoco and Amazon basins, the rivers of the Guianas, and the Parnaíba basin of northeastern Brazil (Vari et al. 1995, Vari and Raredon 2003).

ADULT SIZES 6.4 cm SL in *Chilodus gracilis* to 16.3 cm SL in *Caenotropus mestomorgmatos* (Isbrücker and Nijssen 1988, Vari et al. 1995).

DIAGNOSIS OF FAMILY Chilodontids are small to moderate-sized fishes that possess a striking dark spot of pigmentation at the distal margin of each scale (Isbrücker and Nijssen 1988, Vari et al. 1995). Aside from their distinctive coloration, members of Chilodontidae can also be readily distinguished from other characiform families by having a single row of relatively small teeth movably attached to the fleshy lips that cover their jaws, and the sixth scale along the lateral line distinctly smaller than the other scales of the series (Vari 1983, Vari et al. 1995). Chilodontidae is supported by more than 30 features of the jaws, gill arches, pectoral girdle, ribs, epibranchial organ, and other internal characters (Vari 1983, Vari et al. 1995).

KEY TO THE GENERA ADAPTED FROM VARI ET AL. (1995)

1a. Branched anal-fin rays 6–8; mouth subterminal; dorsal fin with dark pigmentation across distal portions of anterior rays but lacking dark spots on remaining portions of fin... *Caenotropus*

1b. Branched anal-fin rays typically 10 or 11, rarely 9; mouth terminal or slightly superior; dorsal fin with series of dark spots on posterior rays .... *Chilodus*

GENUS ACCOUNTS

Caenotropus (8.0–17 cm SL)

Distinguished from *Chilodus* by fewer branched anal-fin rays (6–8); a subterminal mouth with a moderately developed lower lip; dark pigmentation present only on distal portions of the dorsal fin with the remaining portions hyaline; and posterior margin of scales somewhat serrate (Vari et al. 1995).

SPECIES Four, all in the AOG region. Key to the species and detailed species information in Vari et al. (1995). **DISTRIBUTION AND HABITAT** *Ca. labyrinthicus* (15 cm SL) is relatively abundant and widespread throughout the AOG; *Ca. maculosus* (11 cm SL) is restricted to the Guianas region; *Ca. mestomorgmatos* (16 cm SL) inhabits the Orinoco basin, western Amazon basin, upper Negro basin, and Xingu basin, and *Ca. schizodon* (9.1 cm SL) inhabits the Tapajós, Madeira, and Xingu basins. **BIOLOGY** Adopt a head-down orientation while swimming and feeding. Species possess a prominent epibranchial organ composed mainly of cartilage and connective tissues (Vari et al. 1995) that functions to detect and extract food particles from the engulfed water and substrate (Bertmar et al. 1969). Diet includes small invertebrates, freshwater sponges, and detritus.

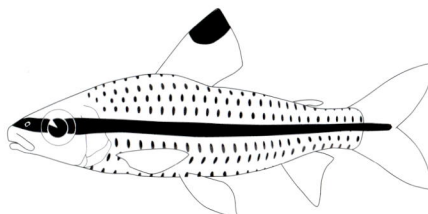

Chilodus (6.0–8.0 cm SL)

Distinguished from *Caenotropus* by higher number (9–11) of branched anal-fin rays; terminal or slightly superior mouth; extensively pigmented dorsal fin; and smooth posterior margin of scales (Vari et al. 1995). **SPECIES** Four, all in the AOG region. The genus was reviewed by Isbrücker and Nijssen (1988). Vari and Ortega (1997) provided a more recent discussion of the genus and described a new species (*Ch. fritillus*). **DISTRIBUTION AND HABITAT** *Ch. punctatus* is the most widely distributed species and well known in the aquarium hobby. It occurs throughout the

Amazon, Orinoco, Essequibo, and Corantijn basins. The other three species are regional endemics: *Ch. gracilis* is apparently restricted to the Negro basin; *Ch. fritillus* to the Madre de Diós basin in the upper Madeira basin; and *Ch. zunevei* to coastal rivers in eastern Suriname and French Guiana. **BIOLOGY** Adopt a head-down orientation while swimming and feeding. Species possess a prominent epibranchial organ composed mainly of cartilage and connective tissues (Vari et al. 1995) that functions to detect and extract food particles from the engulfed water and substrate (Bertmar et al. 1969). Diet includes sponges, detritus, and small invertebrates (Goulding et al. 1988).

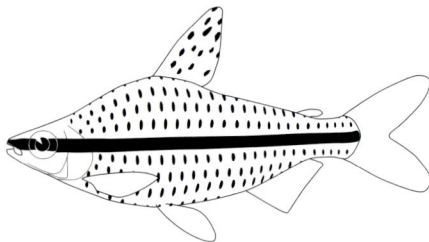

FAMILY CRENUCHIDAE—SOUTH AMERICAN DARTERS

— *PAULO A. BUCKUP and PETER VAN DER SLEEN*

DIVERSITY 93 species in 12 genera and two subfamilies, including 43 species representing all genera in the AOG region. In addition, there are several species that have not yet been described. Most undescribed species belong to the genus *Characidium*. Most species are allocated to the subfamily Characidiinae, which includes the genera *Ammocryptocharax* (4 species), *Characidium* (65 species), *Elachocharax* (4 species), *Geryichthys* (1 species), *Klausewitzia* (1 species), *Leptocharacidium* (1 species), *Melanocharacidium* (9 species), *Microcharacidium* (3 species), *Odontocharacidium* (1 species), and *Skiotocharax* (1 species). The subfamily Crenuchinae includes the genera *Crenuchus* (1 species) and *Poecilocharax* (2 species).

COMMON NAMES Canivete; Mocinha (Brazil); Pez dardo (Peru).

GEOGRAPHIC DISTRIBUTION From eastern Panama to La Plata, Argentina. Most species are found in cis-Andean South America, but a few species of *Characidium* occur in the Pacific slope of Colombia and Panama. Highest species diversity found in the rivers surrounding the Guiana Shield (including the Orinoco basin and the northern tributaries of the Amazon basin), Andean headwaters of the Amazon and coastal streams of southeastern Brazil. Species of *Characidium* occur throughout the geographic range of the family, while the remaining genera are restricted to the watersheds surrounding the Guiana Shield.

ADULT SIZES Small, rarely exceeding 10 cm SL. Many species are miniatures, with adult sizes under 2.5 cm SL.

DIAGNOSIS OF FAMILY Diagnosed by six characters, including one unique to the family (Buckup 1998): paired foramina in the frontal bones located posterodorsally to the orbits, immediately medial to the supraorbital sensory canal, and anterior to the epiphysial branch of that canal (fig. 1). These foramina are associated with a branch of the ophthalmic nerve that traverses the posterior wall of the orbit and reenters the braincase to emerge through the frontal foramen. The foraminae are conspicuous in the Crenuchinae, but are very small in members of the Characidiinae (containing most crenuchids; see fig. 1), and can usually be seen only through careful examination of a cleared and stained specimen. Most characidiines can also be distinguished from other characiform fishes by having more than one unbranched ray in the leading edge of the pectoral fin. Compared with the Characidae and their close relatives, crenuchids have a relatively elongate body and a reduced number of anal-fin rays. Members of Characidiinae also have elongate tricuspid or conical teeth in both jaws, organized as a single row in the upper jaw and two rows in the lower jaw. The inner series in the lower jaw comprises minute teeth that are not implanted in the dentary bone and may be absent in some species (Buckup 2003).

SEXUAL DIMORPHISM Adults may exhibit morphological and color dimorphism associated with sex. Such dimorphism occurs in a few members of the subfamily Characidiinae but is more evident in the subfamily Crenuchinae, including light spots on the anal fin (*Crenuchus spilurus*) or dark sports on the dorsal fin (*Poecilocharax weitzmani*) in sexually mature males. Some *Characidium* species have sexually dimorphic hooks on the branched pelvic- and pectoral-fin rays of males (da Graça et al. 2008), but the number of species that exhibit this trait is not known.

HABITATS Most species of Crenuchidae inhabit fast-flowing small streams, where they hover around pebbles, rock, and submerged

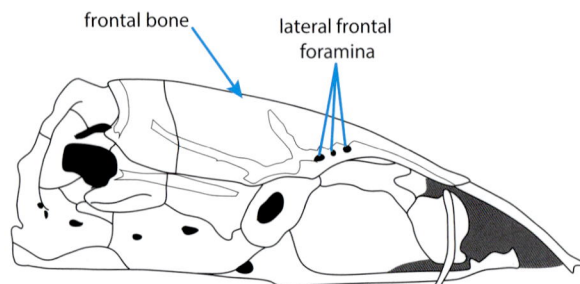

frontal bone

lateral frontal foramina

FIGURE 1 Lateral view of the cranium of *Characidium bolivianum*; after Buckup (1993b).

vegetation (Buckup 2003). Some occur in rapids and small waterfalls, where they are often found adhering to the substrate or clinging to aquatic plants. Some species have been reported to cling to vertical walls (Buckup et al. 2000). Other species, such as *Characidium rachovii*, live in slow-flowing lowland waters.

FEEDING ECOLOGY Most species feed on microinvertebrates; typically close to the bottom.

BEHAVIOR Most crenuchids are translucent or drab-colored bottom dwellers that lie in wait camouflaged against a predominantly sandy or rocky substrate. In the Characidiinae, the camouflage effect is partly achieved by the presence of about nine dark vertical bars on the body. In sand-dwelling species these bars may be fragmented and the body is translucent. Among the exceptions to this general pattern are species of *Ammocryptocharax*, which live on aquatic or submerged plants. *Ammocryptocharax elegans* is able to adopt a bright green color while on submerged plants, which changes to brownish color when posed on submerged twigs and logs. *Ammocryptocharax elegans* and *A. minutus* have curved tips of the pectoral fins allowing it to cling to leaf margins (Zuanon et al. 2006b).

KEY TO THE GENERA MODIFIED FROM BUCKUP (1993c)

1a. Surface of frontal bones evenly convex (subfamily Characidiinae) **2**
1b. Dorsal surface of the head with a ridge across the frontal bones, with soft tissue pad located in front of ridge (fig. 2); tissue pad covering a pair of foramina; each foramen located immediately anterior to the frontal ridge (fig. 2; subfamily Crenuchinae) **11**

2a. Maxillary teeth present.. **3**
2b. Maxillary teeth absent.. **5**

3a. Two unbranched pelvic rays (Amazon, upper Orinoco, and Essequibo basins)... *Ammocryptocharax*
3b. One unbranched pelvic ray **4**

4a. Tricuspid teeth present on the premaxilla and external dentary tooth row; a conspicuous black spot on the medial rays of the pectoral and pelvic fins, near base of fin; a conspicuous black spot on the posterior margin of adipose fin, near its base (upper Amazon basin at border area between Brazil and Peru) .... *Klausewitzia*
4b. All teeth unicuspid; no conspicuous black spot present on the medial rays of pectoral and pelvic fins; no conspicuous black spot on the posterior margin of the adipose fin (Amazon basin, including the Negro basin and the Casiquiare River)... *Odontocharacidium*

5a. Dorsal-fin rays 15 or more.. **6**
5b. Dorsal-fin rays 14 or fewer **7**

6a. Dorsal-fin rays 15–16; discrete bands or spots absent on body, except for a diffuse humeral blotch (Mazaruni and Berbice basins in Guyana) .... *Skiotocharax*
6b. Dorsal-fin rays 17 or more; dorsal fin intensely pigmented with a horizontal dark band near its base associated with a series of darker sports formed by the extension of lateral bands onto the dorsal fin (upper Orinoco and Amazon basins, including the Negro and Madeira rivers) .... *Elachocharax*

7a. Two unbranched rays in leading portion of pelvic fin (Negro and upper Orinoco basins) .... *Leptocharacidium*
7b. One unbranched ray in the leading portion of pelvic fin.. **8**

8a. Pectoral-fin rays 10 or fewer **9**
8b. Pectoral-fin rays 11 or more **10**

9a. Adipose fin absent; suborbital spot absent; number of vertical bars relatively high (17–18); maximum length >3 cm; absence of pterygoid teeth (upper Amazon basin at the region of Iquitos and Ucayali drainage in Peru) .... *Geryichthys*
9b. Adipose fin present; suborbital spot present; number of vertical bars relatively low (around 9); maximum length <3 cm; pterygoid teeth present (upper Orinoco, Amazon, and Guianas)... *Microcharacidium*

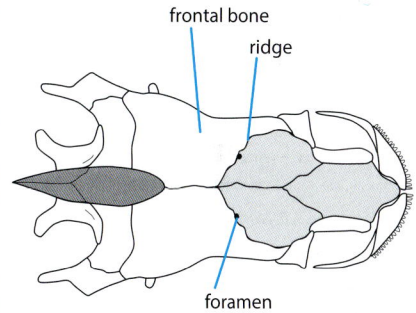

FIGURE 2 Dorsal view of skull of *Crenuchus spilurus*; after Gery (1963).

10a. Supraorbital present (see generalized location of supraorbital bone in figure on p. 32) (Amazon and Orinoco basins, coastal rivers in the Guianas) .... *Characidium*
10b. Supraorbital absent (northern tributaries of the Amazon and the Araguaia River, Orinoco basin, and coastal rivers in the Guianas) .... *Melanocharacidium*

11a. Adipose fin present (Orinoco and Amazon basins, and coastal rivers in the Guianas) .... *Crenuchus*
11b. Adipose fin absent (upper Solimões, Negro, and upper Orinoco basins and the Potaro River) .... *Poecilocharax*

GENUS ACCOUNTS

Ammocryptocharax (2.0–9.0 cm SL)

Recognized by: a subterminal mouth and elongate body shape (more fusiform and elongate than in species of *Characidium*). Additional, but nonunique characters include 2 unbranched pelvic rays; maxillary teeth; and absence of dark patches of chromatophores in prepectoral area (Buckup 1993c). **SPECIES** Four, key to the species in Buckup (1993c). **DISTRIBUTION AND HABITAT** Amazon, Negro, upper Orinoco and Essequibo basins. Species live on aquatic plants or submerged vegetation, often in areas with swift current. **BIOLOGY** Only one species has been studied in detail, *A. elegans*, which is able to adopt a bright green color while on submerged plants and has curved tips of the pectoral fins, allowing it to cling to leaf margins (Zuanon et al. 2006b). This species is also able to twist its head slightly and has remarkably movable eyes, used for foraging on aquatic insects (mainly larvae) in the three-dimensional tangle of submerged vegetation (Zuanon et al. 2006b).

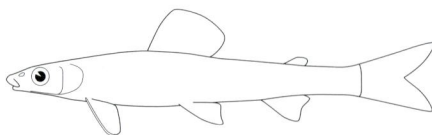

Characidium (2.0–8.0 cm SL)

Recognized by: a black spot near the base of the middle caudal-fin rays, most recognizable in *C. steindachneri, C. zebra, C. fasciatum*, and *C. bahiense*. The spot is formed by a discrete cluster of chromatophores restricted to caudal fin, thus differing from the more diffuse caudal peduncle dark blotches that occur in several other groups of crenuchids. The basicaudal spot may be obscured by superimposed blotches and variegated color patterns or secondarily lost in several species of *Characidium*. Most species have 9–12 vertical bars on the body (but there are several exceptions). Sexual dimorphism has been reported in a few species and includes conspicuous dark anal, pelvic, and pectoral fins in breeding males (Buckup and Reis 1997), as well as sexually dimorphic hooks on the branched pelvic- and pectoral-fin rays of males (da Graça et al. 2008). **SPECIES** 64 valid species, with 26 species in the AOG region. **DISTRIBUTION AND HABITAT** Widely distributed between eastern Panama and northeastern Argentina. Most species occupy the benthic zone in flowing, often swift-water habitats. **BIOLOGY** Insectivorous. Behavioral studies on *Characidium* species (e.g., Sazima 1986, Aranha et al. 2000, Zuanon et al. 2006a) revealed the use of two main foraging tactics: "sit-and-wait predation," in which the fish stays stationary on the bottom and preys on small invertebrates spotted in the bottom or drifted by the current; and "hunting by speculation," in which the fish actively search for prey buried in the upper layer of the substrate (Leitão et al. 2007). Leitão et al. (2007) recorded a variation of the sit-and-wait tactic: a *Characidium* sp. closely following a small armored catfish and catching the particles dislodged by this catfish's grazing activity.

Crenuchus (5.7 cm SL)

Recognized from all crenuchids (except *Poecilocharax*) by a pair of large foramina (holes) in the frontal bone, and two unbranched rays in the leading edge of the pectoral fin (vs. usually 3 or more in the Characidiinae). The paired foramina in the frontal bones are associated with a distinct depression in the frontals, located immediately in front of the foramen (Buckup 1998). *Crenuchus* can be easily distinguished from *Poecilocharax* by the presence of the adipose fin (Buckup 1993a, Planquette et al. 1996). Other characters that can aid identification: body shape is cichlid- or cyprinodont-like; relatively long dorsal fin; color uniformly grayish with a black blotch in the caudal peduncle that is slightly displaced ventrally. Adult males are larger and more brightly colored than females, and have light spots on the unpaired fins (Géry 1977). Courtship-receptive females show a darkened abdominal region (Pires et al. 2016). **SPECIES** One, *C. spilurus*. **DISTRIBUTION AND HABITAT** Orinoco and Amazon basins, and coastal rivers in the Guianas. **BIOLOGY** Life history of *C. spilurus* was studied in detail by Pires et al. (2016). Individuals spend most of the day sheltered under shaded areas among structures such as dead leaves and branches, roots, and plants. Feeding occurs during daylight and includes mainly particulate organic matters that sink slowly through the water column; the fish rarely go to the surface to select food items. Analyzed stomachs showed consumption of flowers, fruits, allochthonous insects, other aquatic invertebrates, and tadpoles. Reproduces by laying eggs on stones or submerged leaves (Boujard et al. 1997). Males exert exclusive parental care of eggs and early larval stages (Planquette et al. 1996, Pires et al. 2016). A peculiar feature of the subfamily Crenuchinae (Crenuchus and *Poecilocharax*) is the presence of an organ on the top of the head, which is housed in the pair of recesses of the frontal bone at eye level (Géry 1977). This organ is formed by rod-shaped cells encircled by a net of capillary blood vessels. Géry (1977) speculated it might function as a receptor of radiation, but up to now its function remains unknown.

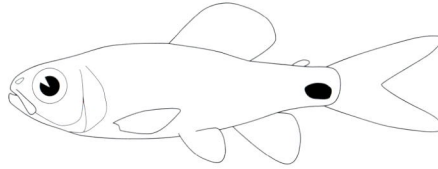

Elachocharax (1.4–2.3 cm SL)

Recognized by: 17 or more dorsal-fin rays; anal-fin origin located below or only slightly posterior to the dorsal-fin base; dorsal fin intensely pigmented with a horizontal dark band near its base associated with a series of darker sports formed by the extension of lateral bands onto the dorsal fin; slap-sided body, characterized by relatively short snout, terminal mouth, deep head with eyes located fairly high, and relatively high caudal peduncle posteriorly prolonged by a more or less emarginate caudal fin (Buckup 1993c). **SPECIES** Four, all in the AOG region. Key to the species in Buckup (1993c). **DISTRIBUTION AND HABITAT** Upper Orinoco and Amazon basins, including the Negro and Madeira rivers. For *Elachocharax pulcher*, Henderson and Walker (1986) note that it dwells among leaf litter and root tangles in sluggish-flowing waters and uses fin undulation to hover. **BIOLOGY** No data available.

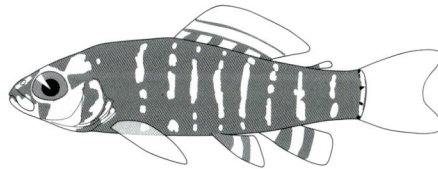

Geryichthys (3.4 cm SL)

Recognized by: maxillary teeth absent; all teeth conical in the premaxilla and dentary; pterygoid teeth absent; adipose fin absent; dorsal-fin rays 14 or fewer; one unbranched ray in the leading portion of pelvic fin; pectoral-fin rays 10 or fewer; 12 scales around the caudal peduncle; 29–32 lateral scales; suborbital spot absent; and 17–18 vertical bands (Zarke 1997). Most of these characters are also found in other crenuchid species, and it is

possible that the single species of *Geryichthys* is actually a member of *Characidium*. **SPECIES** One, *G. sterbai*, see species description in Zarke (1997). **DISTRIBUTION AND HABITAT** Upper Amazon basin in the region around Iquitos and the Ucayali River. Found on the bottom. **BIOLOGY** No data available.

Klausewitzia (2.5 cm SL)

Recognized from other crenuchins (except *Elachocharax*, *Microcharacidium*, *Odontocharacidium*, and *Skiotocharax*) by: fewer than 10 pectoral-fin rays; parietal branch of the supraorbital sensory canal restricted to the frontal; and a dark blotch of chromatophores present at the base of the medial rays of the pectoral fin (Buckup 1993c). Differs from *Elachocharax*, *Microcharacidium*, and *Skiotocharax* by the presence of teeth on the maxilla (all teeth unicuspid), and differs from *Odontocharacidium* by having most head bones normally developed (see list of bone reductions in the diagnosis of *Odontocharacidium*). **SPECIES** One, *K. ritae*, see description in Géry (1965b). **DISTRIBUTION AND HABITAT** Upper Amazon basin at border area between Brazil and Peru. **BIOLOGY** no data available.

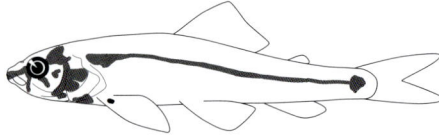

Leptocharacidium (7.0 cm SL)

Shares with *Ammocryptocharax* the presence of 2 unbranched rays in the leading edge of the pelvic fins, and the possession of a wide longitudinal midlateral brown stripe with sharply defined borders. Differs from *Ammocryptocharax* by having a well-defined, narrow dark suborbital stripe extending posteroventrally from the margin of the eye to the angle of the preopercle (Buckup 1993c). **SPECIES** One, *L. omospilus*, see species description in Buckup (1993c). **DISTRIBUTION AND HABITAT** Negro and upper Orinoco basins, Venezuela and Brazil. *Leptocharacidium* dwells on rocky substrates (usually large rock slabs) in fast-flowing streams (J. Zuanon pers. comm.). **BIOLOGY** No data available.

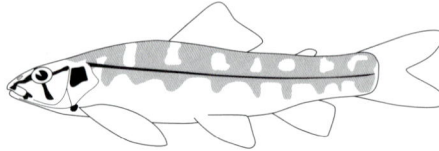

Melanocharacidium (3.0–10 cm SL)

Recognized by: maxillary teeth absent; pectoral-fin rays 11 or more, with 3 or more unbranched rays; dorsal-fin rays 14 or fewer; one unbranched ray in the leading portion of pelvic fin; supraorbital absent; the area near the leading edge and tips of the pectoral and pelvic fins unpigmented, contrasting with a homogeneously dark, barred, or checkered pattern of pigmentation on the remainder of the fin (Buckup 1993c). Differs from *Characidium* (and several other crenuchids) by: absence of a supraorbital bone. Unlike most characiform fishes, the supraorbital bone in *Characidium* is not strongly attached to the skull and is movable along with the supraorbital membrane. **SPECIES** Nine, all in the AOG region. Key to the species in Buckup (1993c), except for *M. auroradiatum* (Costa and Vicente 1994). **DISTRIBUTION AND HABITAT** Orinoco basin, northern tributaries of the Amazon, Araguaia River, and coastal rivers in the Guianas. Species have been found in relatively fast-flowing water over sand, rocks, or plants (Buckup 1993c, Zuanon et al. 2006b). **BIOLOGY** No data available.

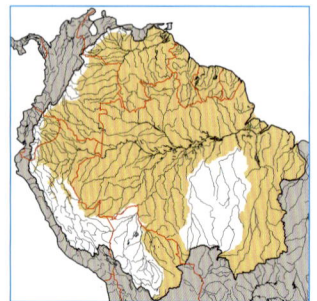

Microcharacidium (1.2–2.4 cm SL)

Recognized by having only 17 principal caudal-fin rays. Some species of *Characidium* have a reduction in the number of caudal-fin rays, but *Microcharacidium* can be easily distinguished by the wide longitudinal midlateral brownish stripe, with sharply defined borders (Buckup 1993c). **SPECIES** Four, all in the AOG region. Key to the species in Buckup (1993c), except *M. geryi* (Zarke 1997). **DISTRIBUTION AND HABITAT** Upper Orinoco, Amazon, and Guianas. Generally found in slow-moving water, but in French Guiana reported in areas where the current is relatively strong (Le Bail et al. 2000). **BIOLOGY** No data available.

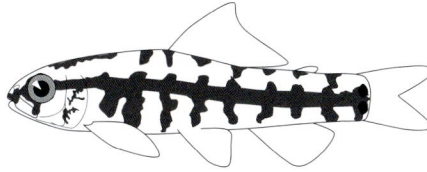

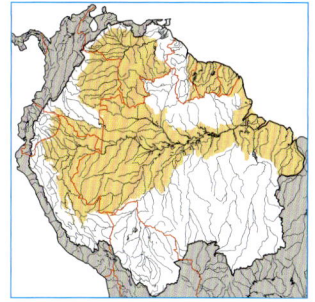

Odontocharacidium (1.6 cm SL)

Recognized from other crenuchins (except *Skiotocharax*) by: absence of posttemporal bone; absence of supratemporal sensory canal; absence of pterotic sensory canal; absence of dermal portions of pterotic, parietal, and sphenotic bones, with the posttemporal fossa limited anteriorly by the sphenotic bone. Distinguished from *Odontocharacidium* by the presence of teeth on the maxilla (Buckup 1993c). Lacks bright colors in life; body musculature rather translucent; back and caudal peduncle olive green in color; head white ventral to eye; a single median horizontal stripe; shoulder spot of variable intensity; and up to 8 transverse, brown dorsal saddle marks across the back (Weitzman and Kanazawa 1977). **SPECIES** One, *O. aphanes*, see description in Weitzman and Kanazawa (1977). **DISTRIBUTION AND HABITAT** Amazon basin, including the Negro basin and the Casiquiare River. **BIOLOGY** No data available.

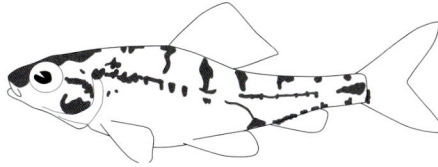

Poecilocharax (4.0–5.0 cm SL)

Recognized from all crenuchids (except *Crenuchus*) by a pair of large foramina in the frontal bone, and 2 unbranched rays in the leading edge of the pectoral fin (vs. usually 3 or more in the Characidiinae). The paired foramina in the frontal bones are associated with a distinct depression in the frontals, located immediately in front of the foramina (Buckup 1998). Distinguished from *Crenuchus* by the absence of an adipose fin (Géry 1977, Buckup 1993a). In males of *Poecilocharax bovallii* the caudal fin is red, decorated with pearl-white spots. Adult males of *P. weitzmani* have a longitudinal black band on which shines the neon-blue center of each scale; superimposed on it is a blood-red band; the large fins are yellow with brown and red, with dark spots on the dorsal and anal fins; the eye is green and red, and the head is marked with a velvet-black suborbital stripe. The teeth, in *P. weitzmani* at least, are tricuspid in females and young and more numerous and acutely conical in adult males (Géry 1977). **SPECIES** Two, *P. bovallii* and *P. weitzmani*, both in the AOG region. **DISTRIBUTION AND HABITAT** *P. bovallii* around Kaieteur Falls on the Potaro River, Guyana; *P. weitzmani* in the upper Solimões, upper Negro, and upper Orinoco basins. **BIOLOGY** Males care for young. The subfamily Crenuchinae (*Crenuchus* and *Poecilocharax*) is characterized by an organ of uncertain function located on top of the head, in the pair of recesses of the frontal bone at eye level, formed by rod-shapes cells encircled by a net of capillary blood vessels (Géry 1977).

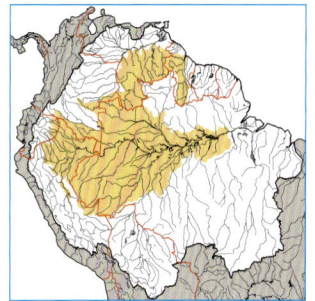

Skiotocharax (3.2 cm SL)

Recognized by: 15 or more dorsal-fin rays; lateral line incomplete; discrete bands or spots absent, except for a diffuse humeral blotch; origin of the anal-fin base located slightly posterior to, or in advance of, a vertical line drawn from the posterior termination of the dorsal fin; all teeth conical; body slab sided, relatively short; supraorbitals absent; a well-defined foramen is present on the posterior region of the medial lamina of the coracoid; circuli present on the posterior or apical field of the lateral body scales (Presswell et al. 2000). **SPECIES** One, *S. meizon*, see species description in Presswell et al. (2000). **DISTRIBUTION AND HABITAT** Mazaruni and Berbice basins in Guyana. **BIOLOGY** no data available.

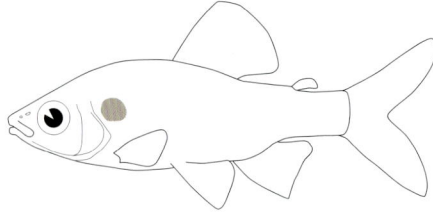

FAMILY CTENOLUCIIDAE—PIKE-CHARACINS

— *PETER VAN DER SLEEN*

Family includes two genera: *Boulengerella* (5 species) and *Ctenolucius* (2 species). The taxonomy and biogeography of the family are reviewed in Vari (1995). Species of *Ctenolucius* occur in the Pacific slope of western Panama, rivers of northwestern and northern Colombia, and Lago Maracaibo basin in Venezuela. *Boulengerella* species are found in the Orinoco, Amazon, and Tocantins basins and the coastal rivers in the Guianas (Vari 1995).

Boulengerella (25–70 cm SL)

Characterized by: an elongate tapering head and body; large gape; numerous relatively small jaw teeth; and posteriorly positioned dorsal fin. *Boulengerella* species also have an elongate fleshy appendage at the anterior tip of the snout (except in very large specimens). Distinguished from somewhat similar looking *Acestrorhynchus* (family Acestrorhynchidae) by: lower jaw shorter than the upper jaw, and presence of many identical teeth that are barely visible (vs. similar sized jaws and conspicuous caniniform teeth of different sizes in *Acestrorhynchus*). *Boulengerella* can be distinguished most easily from the other ctenoluciid genus, *Ctenolucius* (not found in the AOG), by having 87–124 lateral line scales (vs. 45–50 in *Ctenolucius*). **SPECIES** Five, all in the AOG region. Key to the species, species information, and distribution maps in Vari (1995). **COMMON NAMES** *Agulha, Bicuda, Bicuda manchada* (Brazil); *Gazza challua, Picudo* (Peru); *Agujta* (Venezeula). **DISTRIBUTION AND HABITAT** Broadly distributed in the Orinoco, Amazon, and Tocantins basins and the coastal rivers in Guyana, French Guiana, and the states Amapá and Pará in Brazil (Vari 1995). Found in streams, rivers, and lentic water bodies. **BIOLOGY** All are predators, hunting close to the water surface during the day and hiding between submerged vegetation at river margins during the night. They feed primarily on small fishes as adults (Goulding et al. 1988, Montaña et al. 2011). Reproduction of *B. cuveiri* occurs during the period when rivers are rising (Santos et al. 1984, Vazzoler and Menezes 1992).

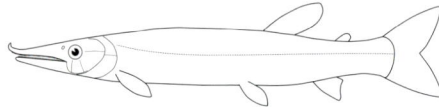

FAMILY CURIMATIDAE—TOOTHLESS CHARACINS

— *BENJAMIN FRABLE*

DIVERSITY 105 species in eight genera, including 73 species in seven genera in the AOG region.

COMMON NAMES *Biru, Saguiru* (Brazil); *Boquiche* (Ecuador); *Chio-chio, Ractafogón, Yahuarachi, Yulilla* (Peru); *Bocachica* (Venezuela).

GEOGRAPHIC DISTRIBUTION Broad geographic distribution throughout tropical lowlands of Central and South America, from southern Costa Rica to south of Buenos Aires, Argentina, and from trans-Andean coastal rivers of Peru throughout the Orinoco, Amazon, Guiana Shield regions, to the Atlantic coast of Brazil (Vari 1989a, 2003).

ADULT SIZES Curimatids span a large size range, with adults of the species *Cyphocharax aninha* reaching 3.9 cm SL (Wosiacki and SilvaMiranda 2013) to *Curimata mivartii* at 32 cm SL. Neither of these species is present in the AOG region.

DIAGNOSIS OF FAMILY Distinguished from other characiforms by absence of teeth in both oral jaws in adults (the hemiodontid genus *Anodus* also lacks oral teeth); absence or reduction of gill rakers; and a suite of internal morphological characters not easily examined in the field (Vari 1989a). Curimatids are generally silver in coloration, sometimes with dark spots, bars, or stripes on the body and fins, and with red or orange pigments in *Curimatopsis*. Species range in body shape from the fusiform *Curimata ocellata* to the laterally compressed and deep-bodied *Cyphocharax abramoides*, with most species possessing a body shape intermediate between these extremes (Vari 1989a).

SEXUAL DIMORPHISM Present in *Curimatopsis* (deeper caudal peduncle in males) and *Curimatella* (females larger than males in at least some species).

HABITATS Most freshwater ecosystems in tropical lowland Central and South America (Vari, 1989a).

FEEDING ECOLOGY Detritivorous, feeding primarily on flocculent debris, filamentous algae, plant matter, small invertebrates, decaying organic matter and fungi (Bowen 1983, Goulding et al. 1988, Vari 1989a, de Mérona et al. 2003, Alvarenga et al. 2006, Sá-Oliveira et al. 2014). As a result, species possess many morphological and physiological adaptations to help process and consume detritus: an elongate gut, loss of oral jaw teeth, modified gill arches, and an enlarged epibranchial organ.

BEHAVIOR Some species, especially the larger *Curimata*, *Psectrogaster*, and *Potamorhina* form massive schools that migrate hundreds of kilometers to reproduce in inundated areas of the floodplain during the wet season (Smith 1981, Goulding et al. 1988, Rodríguez Fernández 1991, Diaz-Sarmineto and Alvarez-León 2003). Some smaller species also migrate shorter distances with the seasons (Godoy 1975, Garavello et al. 2010, Montenegro et al. 2011).

ADDITIONAL NOTES Many curimatids are essential prey for predatory fishes (Hamilton et al., 1992). Additionally, some curimatids, especially migratory species, are harvested for commercial and subsistence fisheries through the Amazon and Orinoco basins (Dahl 1971, Lowe-McConnell 1975, Smith 1981, Cetra and Petrere Jr. 2001, Rodríguez-Olarte et al. 2001, de Jesus and Kohler 2004).

KEY TO THE GENERA ADAPTED FROM VARI (1992a); FIGURES IN KEY BY BENJAMIN FRABLE

1a. Lateral line incomplete in adults; anterior margin of maxilla extending anteriorly to a pronounced degree when lower jaw is depressed; males with pronounced expansion of penultimate principal ray of lower lobe of caudal fin; relative depth of caudal peduncle sexually dimorphic, deeper in males (see figures of species in Vari 1982b); some species with red pigmentation near caudal peduncle .... *Curimatopsis*

1b. Lateral line complete in adults except in a few species of *Cyphocharax*; premaxilla not extending anteriorly to a pronounced degree when lower jaw is depressed; no sexual dimorphism apparent in rays of caudal fin or in relative depth of caudal peduncle **2**

2a. Lateral line scales 85–110; laterosensory canal segment in sixth infraorbital with 4 or 5 branches (fig. 1a); fourth and fifth infraorbitals posteriorly expanded (fig. 2a); 2–3 unbranched and 11–16 branched anal-fin rays..... *Potamorhina*

2b. Lateral line scales between supracleithrum (shoulder region) and hypural joint (base of caudal fin) typically <76, if between 77 and 97, then species is distinctly compressed laterally, with a procumbent spine at base of first dorsal-fin ray and a distinct mid-dorsal keel between rear of dorsal fin and adipose fin; laterosensory canal segment in sixth infraorbital either tripartite or a simple tube (fig. 1b-d); fourth and fifth infraorbitals not posteriorly expanded (fig. 2b,c), 2–3 unbranched and 7–12 branched anal-fin rays **3**

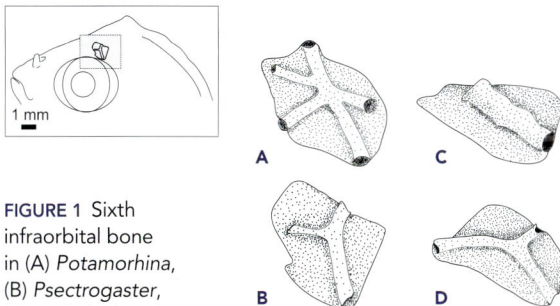

FIGURE 1 Sixth infraorbital bone in (A) *Potamorhina*, (B) *Psectrogaster*, (C) *Steindachnerina*, and (D) *Curimata*. Orientation of sixth infraorbital bone in head of *Psectrogaster* depicted at top left. Modified with permission from Vari (1992a).

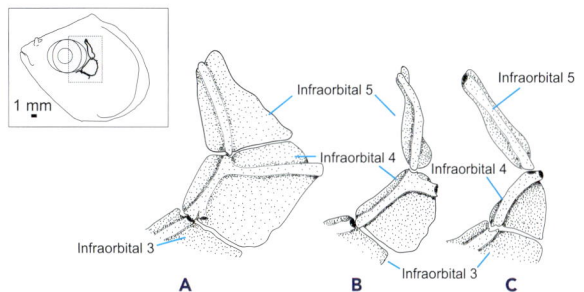

FIGURE 2 Dorsal portion of third, fourth, and fifth infraorbital bones in (A) *Potamorhina*, (B) *Psectrogaster*, and (C) *Curimatella*. Orientation of these bones in head of *Psectrogaster* depicted at top left. Modified with permission from Vari (1992a).

3a. Roof of mouth with 3 prominent fleshy flaps with associated secondary flaps in all but smaller juveniles (fig. 3a) .... *Curimata*

3b. Roof of mouth with 3 simple flaps not paralleled by a series of secondary flaps, or with series of fleshy lobulate bodies over anterior or entire surface in all but smallest individuals (fig. 3b, c) **4**

4a. Laterosensory canal segment in sixth infraorbital tripartite (fig. 1b); laterosensory canal segments in fourth and fifth infraorbitals forming an arch continuous with those in anterior infraorbitals; fourth infraorbital with a distinct posterior branch in the laterosensory canal segment (fig. 2b) .... *Psectrogaster*

FIGURE 3 Ventral view of the roof of the mouth indicating diversity in soft tissue: (A) *Curimata*, (B) *Steindachnerina* and (C) *Cyphocharax*. Reprinted and modified with permission from Sidlauskas and Vari (2012).

4b. Laterosensory canal segment in sixth infraorbital a simple tube (fig. 1c) or canal not present; laterosensory canal segments in fourth and fifth infraorbitals meeting at an acute angle and laterosensory canal on the fourth infraorbital lacking a posterior branch (fig. 2c); or laterosensory canal segments absent or poorly developed in fourth and fifth infraorbitals **5**

5a. Middle rays of upper and lower lobes of caudal fins in mid- to large-size specimens covered with patches of scales smaller than those covering the body; scale patches extending over most of these rays in larger specimens .... *Curimatella*

5b. Middle rays of upper and lower lobes of caudal fins in mid- to large-size specimens without patches of small scales; if scales present on those rays, then they are limited to basal portions of rays and are approximately same size as scales on posterior portion of caudal peduncle **6**

6a. Roof of mouth with series of lobulate processes extending into oral cavity in all but smaller juveniles (fig. 3b), or with multiple thick longitudinal flaps on roof of oral cavity (only in *Steindachnerina seriata*) .... *Steindachnerina* **in part** (all species except *S. argentea*, *S. bimaculata*, *S. binotata*, *S. leucisca* and *S. notograptos*)

6b. Roof of mouth with 3 simple longitudinal flaps without associated fleshy lobulate bodies, longitudinal flaps not thickened (fig. 3c) **7**

7a. A distinct dark spot at base of middle rays of caudal fin and/or one or more longitudinal series of small dark spots on the dorsolateral surface of body with spots irregularly placed over scales; and anus separated from first ray of anal fin by 7–11 scales (except 4–6 in *S. notograptos*) .... *Steindachnerina* **in part** (*S. argentea*, *S. bimaculata*, *S. binotata*, *S. leucisca* and *S. notograptos*)

7b. No distinct dark spot at base of middle rays of caudal fin (but dark spot present on the caudal peduncle in *C. spilurus*, *C. spiluropsis*, *C. gouldingi*) and species lacking longitudinal series of small dark spots irregularly situated on the dorsolateral surface of body (if longitudinal series of dark spots present, spots situated on middle of exposed surface of scale, in 7 or 8 series in *Cyphocharax pantostictos*); anus separated from first ray of anal fin by 2–4 scales .... *Cyphocharax*

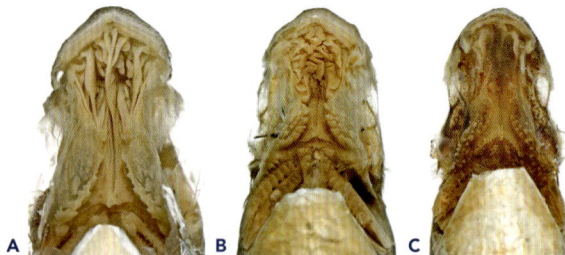

GENUS ACCOUNTS

Curimata (11–32 cm SL)

Distinguished from other curimatids by having 3 fleshy flaps with associated secondary flaps (reduced in small juveniles) on the roof of the mouth (fig. 3a), and a tripartite laterosensory canal on the sixth infraorbital (fig. 1d). The snout in most *Curimata* species is somewhat downturned, with the mouth slightly subterminal; the preventral area is rather flat and covered by a median row of enlarged scales; body is moderately deep, except the fusiform and slender *Curimata ocellata*, which may mimic hemiodontids (Vari, 1989b), and with a silver color, sometimes with dark markings (notably in *C. ocellata* and *C. vittata*). No sexual dimorphism reported. **SPECIES** 13, including 11 in the AOG region. Species key in Vari (1989b),

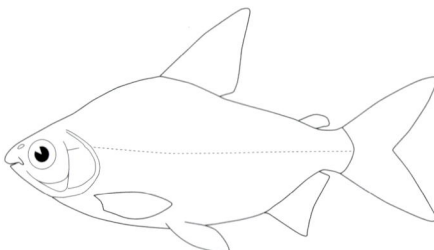

except *C. acutirostris* (see Vari and Reis 1995). **DISTRIBUTION AND HABIT** Present in all AOG ecoregions, in a wide variety of habitats ranging from streams and main channel rivers to lakes and swamps (Goulding et al. 1988, Vari 1989b). **BIOLOGY** All species are detritivorous (Goulding et al. 1988, Jepsen and Winemiller 2002). Some species are seasonal migrants, traveling in large schools over 100–200 km upstream to spawn (Rodríguez Fernández 1991, Diaz-Sarmineto and Alvarez-León 2003). Most species are important in local artisanal and commercial fisheries (Dahl 1971, Goulding et al. 1988, de Jesus and Kohler 2004).

Curimatella (9.3–18 cm SL)

Recognized by: moderately deep body shape; scales covering most of the caudal-fin lobes, scales smaller than those on the caudal peduncle, the extent of scale coverage varies between species and ontogenetically (Vari 1992a); generally silver, only *Curimatella dorsalis* with a small spot on caudal peduncle; other members with some chromatophores on the anterior portions of body scales above lateral line. Larger members of some species of *Cyphocharax* also have scales extending onto the caudal fin (*C. aspilos, C. leucostictus, C. microcephalus,* and *C. magdalenae*); however, these are restricted to the base of the middle caudal-fin rays and do not extend onto the lobes and are similar in size to those on the caudal peduncle. Sexual dimorphism may be present; females tend to be larger and more abundant than males (Alvarenga et al., 2006). **SPECIES** Five, including four in the AOG region (*C. lepidura* excluded). Key to the species of *Curimatella* in Vari (1992a). **DISTRIBUTION AND HABITAT** Present in all ecoregions except the Orinoco piedmont, Suriname, Brazilian Shield Madeira, Tapajós-Juruena, and Xingu. Species sympatric with all four co-occurring in some areas (Pouilly et al. 2004). Genus inhabits diverse habitats such as lagoons, backwater areas, shoreline, and main channels (Vari 1992a). **BIOLOGY** Predominantly detritivorous, known to consume ostracods, small invertebrates, algae, and fungus, especially in dry season (Vari 1992a) (Alvarenga et al. 2006). Migratory behavior between streams and lakes has been reported for *C. meyeri* (Freitas et al. 2012). Some species are important in artisanal fisheries, and some indigenous people recommend consumption of *Curimatella* by the sick (Begossi et al. 1999).

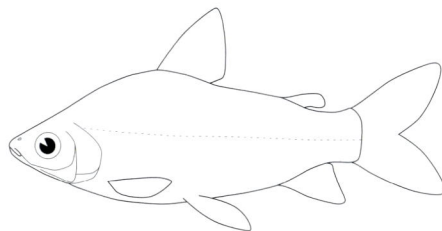

Curimatopsis (3.0–8.9 cm SL)

Recognized by: small adult body sizes; incomplete lateral line; anterior margin of the maxilla distinctly rounded anteriorly and without a distinct notch for articulation with the lateral portion of the premaxilla; anterior margin of the maxilla extending far forward when lower jaw is depressed; caudal peduncle generally deep but sexually dimorphic, deeper in males; significant expansion of the penultimate ray of the caudal fin lower lobe in males. Additionally, a large dark spot is present on the midlateral area of the caudal peduncle in all species (unknown if present in *Curimatopsis microlepis*) extending as a stripe posteriorly on the middle rays of the caudal fin in males. **SPECIES** Five, all occurring in the AOG region. Key to the species in Vari (1992a). **DISTRIBUTION AND HABITAT** Present throughout most of the lowland Amazon and Orinoco region, and absent from upland areas of the Brazilian Shield (e.g., Tocantins-Araguaia, Xingu, and upper Madeira, Mamore-Madre de Dios), as well as the Ucayali-Urubamba, Essequibo, and Orinoco piedmont. Found in a range of habitats from floodplains, backwaters, and swamps to blackwater creeks and river margins (Mol 2012b). **BIOLOGY** Males are

relatively rare, comprising fewer than 20% of individuals present in collections (Vari 1982b). All *Curimatopsis* species are detritivorous (Vari 1982a, b). *Curimatopsis* cf. *crypticus* were observed swimming along the stream bottom scooping fine particulate organic matter from the substrate (Brejão et al. 2013). Brejão et al. (2013) also observed individuals alone or in groups of up to 10 feeding in the day and at dusk.

Cyphocharax (3.8–22 cm SL)

The most species-rich genus of Curimatidae, recognized by: roof of mouth with 3 weak longitudinal fleshy flaps (fig. 3c); body generally ellipsoid, only *C. abramoides* exceptionally deep-bodied; silvery, some species with a small to large dark spot on caudal peduncle; longitudinal stripe or stripes along the length of body; if only one, on the lateral midline; patches of pigmentation on the dorsal, caudal (*C. notatus*), and adipose fins (Vari 1992b). Sexual dimorphism unknown. **SPECIES** 30, including 27 in the AOG region. A partial species key, with subregional keys (excluding *C. aninha, C. biocellatus, C. derhami, C. laticlavis, C. pinnilepis*, and *C. sanctigrabrielis*) provided in Vari (1992b). **DISTRIBUTION AND HABITAT** Present throughout the AOG region except the Ucayali-Urubamba piedmont. Species are found in diverse habitats from inundated backwaters and lakes to streams and open channels of different water types (Vari 1992b). **BIOLOGY** Detritivorous, consuming plant matter, debris, algae, fungi and macroinvertebrates, especially in the dry season (Godoy 1975, Nomura and Hayashi 1980, Sazima 1986, Goulding et al. 1988, Vari 1992b, de Mérona et al. 2003). Some species seasonally migrate short distances to spawn, usually as components of large mixed-species schools (Godoy 1975). Species of *Cyphocharax* are harvested regularly in artisanal and small commercial fisheries (Blanco et al. 2005, Hallwass et al. 2011).

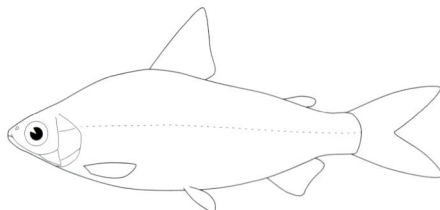

Potamorhina (21–27 cm SL)

Recognized by: largest adult body in Curimatidae; 85–110 lateral line scales; 11–16 unbranched anal-fin rays; laterosensory canal segment in the sixth infraorbital with 4 or 5 branches (fig. 1a); fourth and fifth infraorbitals expanded posteriorly (fig. 2a); moderately deep-bodied; all silvery with darker coloration dorsally. No sexual dimorphism observed. **SPECIES** Five, including four in the AOG region (*P. laticeps* excluded). Key to the species in Vari (1984). *Branquinha cabeça lisa* (Brazil); *Yahuarachi* (Peru). **DISTRIBUTION AND HABITAT** All basins except the Tocantins-Araguaia, Xingu, Tapajós-Juruena, Madeira, Suriname, and Essequibo basins and the Orinoco piedmont. *Potamorhina* inhabits a wide range of habitats such as midchannel, flooded forest, and sandbanks (Goulding et al. 1988). **BIOLOGY** All species are detritivorous, feeding on particulate organic matter on river bottom and sandbanks (Goulding et al. 1988). Species form very large schools and participate in seasonal migration to feed and spawn in inundated floodplains during the flood season (Smith 1981). Individuals produce characteristic vocalizations during mass migrations and can reportedly be heard on land for a distance (Dorn and Schaller 1972; Vari, pers. comm.). Some species are important in commercial and artisanal fisheries in the Amazon basin (Smith 1981).

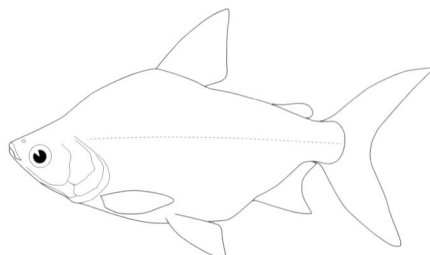

Psectrogaster (12–18 cm SL)

Recognized by: a moderate to very deep body shape in lateral profile; possessing a tripartite laterosensory canal segment in sixth infraorbital (fig. 1b); an arch in the laterosensory canal segments in the fourth and fifth infraorbitals which is contiguous with those in the anterior infraorbitals (fig. 2b); fourth infraorbital with a distinct posterior branch in the laterosensory canal segment (fig. 2b); 3 simple folds on the roof of mouth without secondary flaps (similar to fig. 3b); generally silvery, some species in AOG region (*P. amazonica, P. cuviventris*, and *P. rutiloides*) with pigmentation patches on the caudal fin. No sexual dimorphism is known for this genus. **SPECIES** Eight, including six in the AOG region. Key to the species in Vari (1989c). **COMMON NAMES** *Branquinha cascuda* (Brazil); *Ractacara* (Peru). **DISTRIBUTION AND HABITAT** Present in all AOG ecoregions except the Orinoco piedmont, Suriname, and Xingu. Species can inhabit sandbanks, small streams and main channels, flooded forests, and swamps and occur in white-, black-, and clearwater systems (Vari 1989c). **BIOLOGY** Species are detritivorous (Sá-Oliveira et al. 2014) and schooling. Some species seasonally migrate during the wet season into inundated areas to spawn (Rodríguez-Olarte et al. 2001, Diaz-Sarmineto and Alvarez-León 2003, Correa et al. 2008). Species are of some artisanal and commercial importance (Goulding et al. 1988, Cetra and Petrere Jr. 2001, Rodríguez-Olarte et al. 2001).

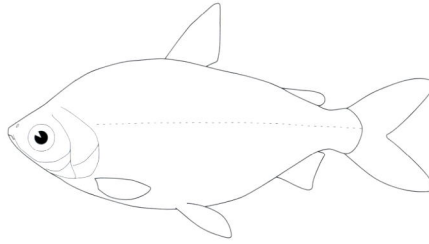

Steindachnerina (5.6–16 cm SL)

The second most species-rich genus within Curimatidae. Difficult to diagnose using characters of external morphology (Vari 1991). All species (except *S. argentea, S. bimaculata, S. binotata, S. leucisca, S. notograptos*, and *S. seriata*) can be distinguished by a lobulate process covering the roof of the oral cavity (fig. 3b). The additional species can be distinguished by having 3 weakly developed, longitudinal folds on the roof of the oral cavity or 3 thick longitudinal folds (only in *S. seriata*); a distinct spot at the base of the middle rays of the caudal fin; and/or one or more longitudinal series of small dark spots on the dorsolateral surface of the body with spots irregularly placed over the scales; anus separated from the first anal-fin ray by 7–11 scales (except 4–6 in *S. notograptos*) (Lucinda and Vari 2009) versus 1–7 scales in the other species of *Steindachnerina*. Members range from fusiform and elongate (*S. hypostoma*) to moderately deep-bodied (*S. guentheri*); generally silvery with some dark pigmentation on the dorsolateral surface, along the lateral midline, and/or on the caudal peduncle. Sexual dimorphism is unknown in this genus (Montenegro et al. 2011). **SPECIES** 24, including 17 in the AOG region. Partial species key provided by Vari (1991). **COMMON NAMES** *Branquinha comum* (Brazil); *Chio-chio* (Peru). **DISTRIBUTION AND HABITAT** Present in all AOG ecoregions. Inhabit a wide range of habitats, but absent or rare in some acidic blackwater systems such as the Rio Negro (Vari 1991). **BIOLOGY** Generally detritivorous with a broad diet ranging from algae and plant matter to scales, zooplankton, and eggs; however, debris and plant matter make up the largest part of the diet (Saul 1975, Nomura and Taveira 1978, Montenegro et al. 2011). Some species migrate seasonally in large mixed schools (Vari 1991, Garavello et al. 2010, Montenegro et al. 2011). Given the generally small size of fishes in this genus, they are of no commercial and of limited artisanal importance; used mainly as bait fish (Brandão et al. 2013). Some indigenous peoples consume *Steindachnerina* to help fight illness (Begossi et al. 1999).

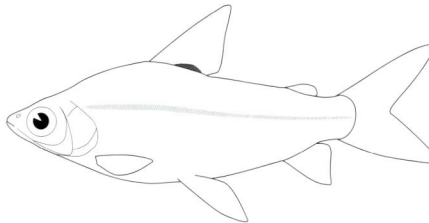

FAMILY CYNODONTIDAE—DOGTOOTH CHARACIFORMS

— *MÔNICA TOLEDO-PIZA*

DIVERSITY Eight species in three genera, all in the AOG region: *Cynodon*, *Hydrolycus*, and *Rhaphiodon* (Toledo-Piza 2000a).

COMMON NAMES Biara, Cachorra, Cachorrinha (Brazil); *Dientón, Chambira, Huapeta, Payara* (Peru); *Payarín* (Venezuela).

TAXONOMIC NOTE The genera *Cynodon*, *Hydrolycus*, and *Rhaphiodon* were until recently included in the family Cynodontidae (as the subfamily Cynodontinae) together with the Roestinae (including *Roestes* and *Gilbertolus*) following hypotheses proposed by Howes (1976) and Lucena and Menezes (1998). However, recent studies based on morphological and molecular characters did not support a close relationship between Cynodontinae and Roestinae (Oliveira et al. 2011a, Mattox and Toledo-Piza 2012). Although Mirande (2010) placed the Cynodontinae in the family Characidae, the molecular study of Oliveira et al. (2011a) placed it alone in the family Cynodontidae. We provisionally follow the later result here.

GEOGRAPHIC DISTRIBUTION All cynodontid species occur in the AOG region. Two species also occur outside the AOG region: *Rhaphiodon vulpinus* in the Paraná-Paraguay and Uruguay basins and *Cynodon gibbus* in the Pindaré River in the state of Maranhão, Brazil.

ADULT SIZES Cynodontids range from 28 cm SL in *Cynodon gibbus* to 66 cm SL in *Hydrolycus armatus*.

DIAGNOSIS OF FAMILY Predatory characiforms with medium to large adult body sizes characterized by a compressed body; an oblique mouth; conical teeth in one series on both jaws, with a highly developed pair of dentary canines that fit into a pair of openings in the upper jaw when mouth is closed; a keeled and relatively expanded pectoral region; well-developed pectoral fins; lateral line scales with many divergent branches of the lateral line canal; gill rakers as flat bony plates with very small spines covering their entire lateral surface. Juveniles possess an extra-oral series of hook-like teeth (Géry and Poivre 1979). Toledo-Piza (2000a) listed 19 osteological characters exclusive to cynodontines among the characiforms.

SEXUAL DIMORPHISM No externally visible characters.

HABITATS Cynodontids inhabit the middle and surface waters of large rivers and streams and are found in the river channel or along the margins (Goulding 1980). Juveniles are typically found in floodplains of large rivers (Taphorn 1992) or within the vegetation along the margins of rivers (Santos et al. 2006). They are associated with areas of clear and transparent waters at least in part of their distribution (de Melo et al. 2009)

FEEDING ECOLOGY Species are primarily piscivorous but sometimes feed on insects and shrimps. They can swallow prey up to 50% of their standard length (Goulding 1980, Santos et al. 2006).

BEHAVIOR Prey are stabbed by the large canine teeth and then swallowed whole, head first (Goulding 1980). Cynodontids usually form groups (Zenaid and Almeida Prado 2012).

ADDITIONAL NOTES Cynodontids are fished commercially in some regions, but they are not much valued because of the numerous intermuscular spines in the flesh and a tendency to spoil rapidly (Taphorn 1992). In Manaus, cynodontids represents less than 1% of total market catch (Santos et al. 2006). Species of *Hydrolycus* and *Rhaphiodon vulpinus* are valued as game fish (Zenaid and Almeida Prado 2012).

KEY TO THE GENERA BASED ON TOLEDO-PIZA (2000A)

1a. Dorsal-fin origin located distinctly anterior to vertical through anal-fin origin .... *Hydrolycus*

1b. Dorsal-fin origin located at, or slightly posterior to, vertical through anal-fin origin 2

2a. Branched anal-fin rays 50 or fewer; dorsal-fin origin located on posterior third of body length (distance from snout to dorsal-fin origin >69% of standard length); body elongate, depth at dorsal-fin origin <20% of SL; 62–68 vertebrae .... *Rhaphiodon*

2b. Branched anal-fin rays 60 or more; dorsal-fin origin located slightly posterior to middle of body length (distance from snout to dorsal-fin origin <59% of standard length); body depth at dorsal-fin origin 20% or more of SL; 51–54 vertebrae...................... *Cynodon*

GENUS ACCOUNTS

Cynodon (28–31 cm SL)

Characterized by: body laterally compressed, distinctly deeper anteriorly, at level of pectoral-fin insertion; anal-fin origin placed approximately at vertical through middle of body; anal fin long, with 60 or more branched rays; 96–109 perforated lateral line scales; pectoral fin with 14–17 branched rays; dorsal fin with ii + 10 rays; pelvic fin with 7 or 8 branched rays, tips of longest rays almost reaching anal-fin origin; total vertebrae 51–55; body silvery with a dark spot just posterior of head. **SPECIES** Three, all in the AOG region. Key to the species and information on species distributions in Toledo-Piza (2000a) and Géry et al. (1999). **COMMON NAMES** *Cachorrinha* (Brazil); *Dientón, Chambira* (Peru); *Payarín* (Venezuela). **DISTRIBUTION AND HABITAT** *C. gibbus* is widely distributed in the AOG region; *C. septenarius* in the central Amazon, upper Orinoco, and Guyana; and *C. meionactis* in the upper Maroni River, French Guiana. **BIOLOGY** Primarily piscivorous, but also eat aquatic insects (Taphorn 1992). Breeding occurs in the beginning of the rainy season.

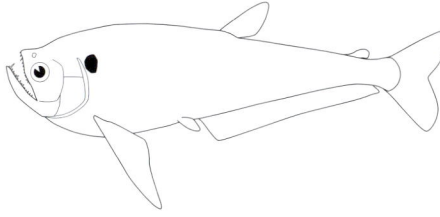

Hydrolycus (34–66 cm SL)

Characterized by: body laterally compressed, deeper anteriorly, at level of pectoral-fin insertion; dorsal fin located distinctly anterior to vertical through anal-fin origin; anal-fin origin located at posterior third of body length; anal fin with 27–47 branched rays; 89–154 perforated lateral line scales; pectoral fin with 14–19 branched rays; dorsal fin with ii + 10 rays; total vertebrae 44–53; elongate blotch of dark pigmentation along posterior margin of opercle. **SPECIES** Four, all only in the AOG region. Key to the species and information on species distributions in Toledo-Piza et al. (1999). **COMMON NAMES** *Icanga, Cachorra, Pirandirá* (Brazil, Venezuela); *Chambira, Payara* (Peru). **DISTRIBUTION AND HABITAT** *H. armatus* and *H. tatauaia* are widely distributed throughout the AOG region; *H. scomberoides* occurs only in the Amazon basin; and *H. wallacei* inhabits the Negro and upper Orinoco basins. Adults are found in, or near, large rivers and creeks. Juveniles are typically found in the floodplain adjacent to larger rivers and sometimes in ponds (Taphorn 1992). Usually live in groups (Lima et al. 2005). *Hydrolycus armatus* and *H. tatauaia* are often found in areas of fast-flowing waters and/or deeper areas in the river associated with trunks and rocks. *Hydrolycus armatus* usually stays in the surface at dawn and dusk and swims to deeper waters during the day (Zenaid and Almeida Prado 2012). **BIOLOGY** Piscivorous. Breeding occurs in the beginning of the rainy season. Spawning occurs in the aquatic vegetation along the shores (Taphorn 1992, Santos et al. 2006). Minimum size of maturation of *H. armatus* is 30 cm. *Hydrolycus armatus* and *H. tatauaia* undertake seasonal migrations (Zenaid and Almeida Prado 2012).

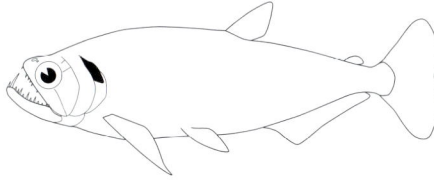

Rhaphiodon (63 cm SL)

Characterized by: body very compressed laterally, elongate and relatively shallow body; body depth approximately the same along entire length; the dorsal fin is placed at posterior third of body length, slightly posterior to the vertical through the anal-fin origin; 121–152 perforated lateral line scales; dorsal fin with ii +10 rays; anal fin with 37–50 branched rays; pectoral fin with 12–17 branched rays; caudal fin with a prominent middle ray; total vertebrae (62–68); body silvery, dorsal portion of pectoral fin covered with dark chromatophores

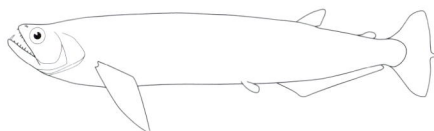

at its base. **SPECIES** One, *R. vulpinus*. **COMMON NAMES** *Biara* (Brazil); *Chambira-challua*, *Huapeta* (Peru). **DISTRIBUTION AND HABITAT** Widely distributed in the AOG region and Paraná-Paraguay and Uruguay basins. Common in rivers, lakes, and flooded forests of all types of water (Goulding 1980). Stay in areas of the river with water coming from different directions and velocities, behind obstacles (trunks and rocks), and often in deeper waters (Zenaid and Almeida Prado 2012). **BIOLOGY** Primarily piscivorous, but also take aquatic insects and shrimps (Almeida et al. 1997). Caught in large numbers during the low-water season, mixed with upstream migrating fishes on which it preys (Goulding 1980). The beginning of its reproductive cycle is associated with the early flooding season (Neuberger et al. 2007). Minimum size of sexual maturation is 24 cm (Zenaid and Almeida Prado 2012).

FAMILY ERYTHRINIDAE—WOLF-FISHES AND YARROWS

— *OSVALDO T. OYAKAWA and GEORGE M. T. MATTOX*

DIVERSITY 12 species in three genera (*Erythrinus*, *Hopleryth-rinus*, and *Hoplias*), including five species in the AOG region (Mattox et al. 2006, Oyakawa and Mattox 2009, Mattox et al. 2014a). Three species, *Hoplias malabaricus*, *Erythrinus erythri-nus*, and *Hoplerythrinus unitaeniatus*, all widely distributed in South America, each of which may represent a species com-plex; i.e., a group of cryptic species (Oyakawa et al. 2013b).

COMMON NAMES *Aimara*, *Lobó*, *Taraíra* (Brazil); *Calabrote*, *Guabina* (Colombia); *Pataka*, *Patakasi* (French Guiana); *Fasaco*, *Huasaco*, *Shuyo* (Peru); *Aimara*, *Guabina* (Venezuela); Wolf-fish (English).

GEOGRAPHIC DISTRIBUTION Inhabit almost all continental waters of Central and South America, from Costa Rica to Argentina.

ADULT SIZES *Erythrinus* and *Hoplerythrinus* are medium-sized fishes, reaching approximately 40 cm standard length (SL). *Hoplias* encompasses medium- to large-sized fishes, ranging from 30–100 cm total length (TL) in *H. lacerdae* and *H. aimara*, being among the largest species of characiforms.

DIAGNOSIS OF FAMILY Subcylindrical body form, rounded cau-dal fin, dorsal fin with 12–16 rays, dorsal-fin origin anterior to a vertical through the anal-fin origin and usually at a vertical through the pelvic-fin origin, anal fin short with 10–11 rays,

no adipose fin, and lateral line with 34–48 scales (Oyakawa 2003, Mattox et al. 2006, Oyakawa and Mattox 2009, Mattox et al. 2014a). Additional useful characters to diagnose repre-sentatives of this family: five branchiostegal rays, a lamellar suprapreopercle, the anterior end of the first infraorbital bifur-cate, and absence of antorbital bone (Oyakawa 2003).

SEXUAL DIMORPHISM Males of *Erythrinus* have longer and more pointed dorsal and anal fins than females. During the breeding season, males of *Hoplerythrinus unitaeniatus* develop an inverted U mark, resembling a "bite mark" above the base of the anal fin, forming an area delimited by a series of scales.

HABITATS Found in a variety of habitats such as lakes, dams, lagoons, and small and large rivers, in both high- and lowland environments.

FEEDING ECOLOGY All species eat mainly other fish as adults, but *E. erythrinus* also consumes terrestrial invertebrates (J. Zua-non pers. comm.).

BEHAVIOR Facultative air breathing in *Hoplerythrinus unitae-niatus* (Graham 1997, Oliveira et al. 2004).

ADDITIONAL NOTES Erythrinids are relatively important as food fishes in many regions of South America. *Erythrinus* is some-times found in the aquarium trade.

KEY TO THE GENERA

1a. Sixth infraorbital as a single plate not divided transversally (fig. 1a); distal end of maxilla with a large dorsal projection (fig. 1aa) ....*Hoplias*

1b. Sixth infraorbital divided transversally into 2 equal parts (fig. 1b); distal end of maxilla with a small dorsal projection (fig. 1bb).. **2**

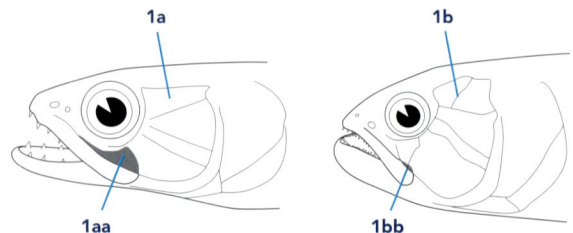

2a. Presence of black round spot on posterodorsal region of opercle; distal end of maxilla always surpassing the vertical through posterior margin of orbit; dark longitudinal midlateral stripe along the body usually present ....*Hoplerythrinus*

2b. Absence of black round spot on posterodorsal region of opercle; distal end of maxilla never surpassing the vertical through posterior margin of orbit; dark longitudinal midlateral stripe along the body rarely present ....*Erythrinus*

GENUS ACCOUNTS

Erythrinus (25 cm TL)

Recognized by a humeral blotch (usually present), and absence of a black, round spot on posterodorsal region of opercle (Oyakawa et al. 2013b). Distinguished from *Hoplerythrinus* by having the distal end of maxilla not surpassing a vertical through posterior margin of orbit, and absence of a dark longitudinal midlateral stripe along body. In some areas, mature males have longer and more pointed dorsal and anal fins than females (e.g., Géry 1977). Mature males may also exhibit more colorful patterns, including several bright orange markings along the flanks of the body or on the dorsal, anal, and caudal fins. **SPECIES** Two, including one (*E. erythrinus*) in the AOG region. **DISTRIBUTION AND HABITAT** *E. erythrinus* is widespread in South America and occurs in most drainages of the AOG region. Another species, *E. kessleri*, inhabits coastal streams of Bahia state, northeast Brazil. **BIOLOGY** Predominantly carnivorous, relying on a variety of items such as insects, shrimps, worms, and small fish, but also occasionally feeds on fruits (e.g., Lima et al. 2005).

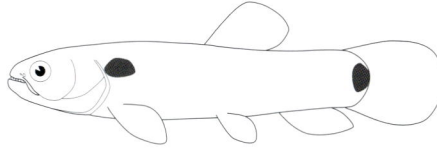

Hoplerythrinus (40 cm TL)

Recognized by a black and round spot on the posterodorsal region of the opercle and by having the distal end of maxilla always surpassing a vertical through the posterior margin of the orbit. A dark longitudinal midlateral stripe along the body is present in most specimens and is useful to further distinguish *Hoplerythrinus* from *Hoplias* and *Erythrinus* (Oyakawa 2003, Oyakawa et al. 2013b). In addition, *Hoplerythrinus* can be recognized by a dentigerous plate of the ectopterygoid that is wider in its anterior portion than its posterior portion and by absence of a row of conical teeth on the lateral margin of the ectopterygoid. At least in part of their distribution, mature males of *H. unitaeniatus* develop an inverted U mark above the base of the anal fin during the breeding season, that resembles a "bite mark" and is delimited by a series of scales (e.g., Eigenmann 1912a, Taphorn 1992). **SPECIES** One, *H. unitaeniatus*. **COMMON NAMES** *Jejú* (Brazil). **DISTRIBUTION AND HABITAT** Occurs in most drainages of South America, including the entire AOG region. Relatively common and found in streams, flooded forests, rivers, lakes, and ponds. **BIOLOGY** Predominantly carnivorous, eating insects, crustaceans and other arthropods, small fish, and worms, but also fruits (e.g., Taphorn 1992, Lima et al. 2005). They are usually active hunters. *Hoplerythrinus unitaeniatus* is capable of breathing air using the swim bladder and has a strong resistance to hypoxia (e.g., Farrell 1978, de Lima Filho et al. 2012).

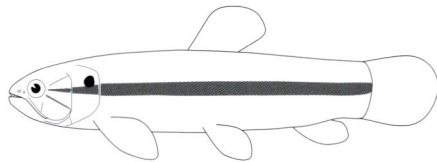

Hoplias (30–100 cm TL)

Recognized among erythrinids by having the sixth infraorbital as a single plate, not divided transversally, and the distal end of maxilla with a large dorsal projection (Oyakawa 2003). In larger specimens, the proximal ends of second and fifth infraorbitals contact each other and exclude the third and fourth infraorbitals from the orbital rim, a condition acquired during ontogeny (Mattox et al. 2006, Oyakawa and Mattox 2009). Most species present vertical stripes along the sides of the body in a chevron-like color pattern, at least in some periods of the day. As other erythrinids, the body is cylindrical and the distal margins of the all fins are round. **SPECIES** Nine, in three species groups: the *H. malabaricus*, *H. lacerdae*, and *H. aimara* groups (Mattox et al. 2006, Oyakawa

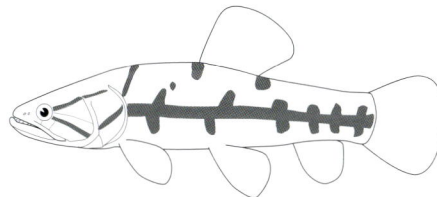

and Mattox 2009, Mattox et al. 2014a). The *H. malabaricus* group is characterized by the form of the medial margins of contralateral dentaries that abruptly converge toward the mandibular symphysis, giving a V-shaped aspect to the dentaries in ventral view; and by tooth-bearing plates on the dorsal surfaces of basihyal and basibranchials, resulting in a rough "tongue." The *H. malabaricus* group includes at least three species *H. malabaricus*, *H. microlepis*, and *H. teres*, of which only *H. malabaricus* and *H. teres* inhabit the AOG region (Mattox et al. 2014a). Cytogenetic studies suggest that *H. malabaricus* represents a species complex (see Mattox et al. 2014a and references therein). The *H. lacerdae* group is characterized by the medial margins of contralateral dentaries gently converging toward each other, resulting in a U-shaped aspect of the dentaries in ventral view. The group includes five species, only one of which, *H. curupira*, occurs in the AOG region (Oyakawa and Mattox 2009). The third species group includes one species, *H. aimara* (*H. macrophthalmus* is a junior synonym), with large eyes and body size. This species is characterized by absence of the accessory ectopterygoid and a vertically elongate dark spot on the opercle membrane. A key to most species in Oyakawa and Mattox (2009). **COMMON NAMES** *Traira*, *Bentón* (Bolivia); *Aimara*, *Traíra*, *Trairão*, *Lobó* (Brazil); *Fasaco*, *Huasaco* (Peru). **DISTRIBUTION AND HABITAT** Four of the nine species occur in the AOG region: *H. malabaricus* is widely distributed along most of the area; *H. aimara* and *H. curupira* have more restricted distributions in coastal rivers of the Guianas and affluents of the middle and lower Amazon basin (i.e., Trombetas, Tapajós, Xingu, Tocantins), with the former also occurring in Rio Madeira (Oyakawa et al. 2013b), and the latter also distributed in the Negro and Orinoco systems; *H. teres* is known only from Lago Maracaibo and adjacent rivers (D. Taphorn pers. comm.). **BIOLOGY** Most species are carnivorous as adults, relying mainly on other fish, and using a lie-and-wait strategy to ambush prey (Taphorn 1992). *Hoplias aimara* is common near waterfalls and rapids (Mattox et al. 2006), being predominantly nocturnal. It is omnivorous as juveniles, becoming piscivorous during ontogeny (Horeau et al. 1998). Although the species is mainly sedentary, some studies report that it can migrate a few kilometers upstream during some periods of the year (e.g., de Morais and Raffray 1999). In affluents of the upper Negro River in Brazil, *H. curupira* is usually found in couples both day and night and prefers large rivers and streams (Lima et al. 2005). At least in part of its distribution, *H. malabaricus* builds nests in shallow water at the bottom of a river that are guarded by the male.

FAMILY GASTEROPELECIDAE—FRESHWATER HATCHETFISHES

— *MICHAEL D. BURNS*

DIVERSITY The family Gasteropelecidae comprises three genera and 11 species, all in the AOG region. Additional undescribed species may be present in *Carnegiella* and *Thoracocharax* (Géry 1977, Schneider et al. 2012, Abe et al. 2013).

COMMON NAMES *Borboleta*, *Borboleta pintada* (Brazil); *Pez pechito* (Peru); *Pechona* (Venezuela).

GEOGRAPHIC DISTRIBUTION Widespread across the AOG region, present in all South American countries except Chile. *Gasteropelecus maculatus* is also found on the Pacific coast of Colombia and in Panama (Weitzman and Palmer 2003).

ADULT SIZES 2.2–9.0 cm SL. *Carnegiella* species are the smallest (2.2–3.5 cm SL) and *Thoracocharax* species the largest (6.7–9.0 cm SL) (Weitzman 1960, Weitzman and Palmer 2003).

DIAGNOSIS OF FAMILY Fishes in the family Gasteropelecidae are easily distinguished from other characiforms by the greatly expanded pectoral girdle. The pectoral girdle has enlarged coracoid bones that expand anteriorly, creating a fan-shaped bone ventral to the head. Each coracoid bone is fused to a single median bone in the pectoral girdle, thereby forming a keel. The frontal bone on the dorsal portion of the head is corrugated, forming a ridge. These characters make the fish very deep-bodied, but relatively thin transversely. Furthermore, species in the family have elongate pectoral fins that project out from the lateral surface of the body, and upturned (superior) mouths for feeding at the surface of the water.

SEXUAL DIMORPHISM None known.

HABITATS Species in the genera *Gasteropelecus* and *Thoracocharax* occur in open habitats of large rivers, streams, and lakes, while smaller-bodied species of *Carnegiella* occupy small streams (Weitzman and Palmer 2003).

FEEDING ECOLOGY All species feed at the water surface. Diet is dominated by allochthonous (terrestrial or aerial) insects (Mérigoux and Ponton 1998, Netto-Ferreira et al. 2007, Roepke et al. 2014).

BEHAVIOR Gasteropelecids are the only fishes known to achieve true powered flight. Many fishes make parabolic jumps out of the water to escape predators, and marine exocoetids (so-called flying fishes) can glide for extended distances over the water surface. Yet at least some gasteropelecids jump out of, and travel above, the water surface for distances of up to several meters (10 feet or more), a distance representing hundreds of body lengths (Myers 1950b). Gasteropelecids achieve true flight by rapidly beating their long pectoral fins at speeds of up to 80 times per second (80 Hz), producing a buzzing sound while in the air (Rayner 1986). This flying behavior is not uncommon to observe in the field, in which the fish's body moves parallel to the water surface, only infrequently or rarely making contact with the water surface while in the air (J. Albert pers. comm.). Dr. Roberto Reis provided the author with the following personal account: "an experience this last May [2014] is still very vivid in my mind... a small school of maybe 25–30 individual *Gasteropelecus* jumped off the surface and flew across a stream for at least 20 feet (6 m). They fly very close to the surface (1–4 inches) and I believe that they may touch the surface with their belly and/or caudal fin during the flight. The buzzing sound is quite perceptible. The trajectory was not at all parabolic, but a straight line from where they first jump off the water to the point where they dive again." Gasteropelecid aerial behavior is used to avoid aquatic predators (Géry 1972b), and possibly also to feed on insects above the water surface (Géry 1977). The diet of *Thoracocharax* is dominated by terrestrial insects, although predation may occur largely at (not above) the water surface (Netto-Ferreira et al. 2007). Curiously, laboratory studies have not yet been able to replicate these field-based observations (Wiest 1995). The fin-beating behavior of *Gasteropelecus* has also been hypothesized to lift the body partly out of the water to reduce drag, rather than to function in true flight (Weitzman and Palmer 1966). It is possible that members in each of the three genera exhibit different behaviors, depending on their absolute body size and relative fin and muscle proportions. For example, *Gasteropelecus* may be too heavy to clear the water, compared with the smaller-bodied *Carnegiella*.

ADDITIONAL NOTES As with all Neotropical fishes species delimitation is very important in gasteropelecids, as several distinct species are frequently collected together and exported as a single variable species in the aquarium trade (Piggott et al. 2011).

KEY TO THE GENERA AFTER (WEITZMAN 1960)

1a. Pelvic rays i + 5 (the 4 forked rays branching twice); dorsal rays ii + 10–15, or iii + 9–14; anal rays ii + 32–42 or iii + 31–41; scales between the posterior lower angle of the coracoid and the anus about 18–28; scales in a longitudinal series 19–22, each with a central circular groove and about 8–10 radial grooves; branchiostegal rays 5; premaxillary with 2 tooth rows, the outer row with 3 teeth .... *Thoracocharax*

1b. Pelvic rays i + 4 (4 forked rays) or v (no forked rays); dorsal rays ii + 8–10 or iii + 7–9; anal rays iii + 19–33; scales between the posterior lower angle of the coracoid and the anus 6–9; scales in longitudinal series 25–37, each without a central circular groove (rarely a small central circular groove is present); there are fewer than 8 radial grooves; branchiostegal rays 4; premaxillary with 1 or 2 tooth rows, the outer row when present with only 1 tooth **2**

2a. Adipose fin present; lateral line developed almost to the origin of anal fin; pelvic rays i + 4; first orbital bone present; sensory canal embedded in the pterotic bone..*Gasteropelecus*

2b. Adipose fin absent; lateral line of 0 to about 3 scales; pelvic rays v; first orbital bone absent; no bony sensory canal in pterotic..*Carnegiella*

GENUS ACCOUNTS

Carnegiella (2.2–3.5 cm SL)

Differs from other gasteropelecids in the following external characteristics: absence of adipose fin; absence of sensory canal in upper region of preopercle; no outer tooth row in premaxillary and 8–11 teeth in inner row; dorsal fin with ii + 6–8 or iii + 5–7 rays; and anal fin with iii + 19–33 rays (Weitzman 1960). **SPECIES** Four, all in the AOG region, including *C. marthae* (Blackwing hatchetfish), *C. myersi* (Pygmy hatchetfish), *C. schereri* (Dwarf hatchetfish), and *C. strigata* (Marbled hatchetfish). See Weitzman (1960) for differences

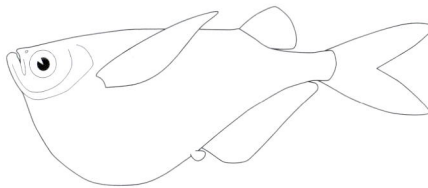

between species. **DISTRIBUTION AND HABITAT** Throughout the AOG region, in small creeks and streams (Weitzman and Palmer 2003). **BIOLOGY** *Carnegiella marthae* feeds primarily on aquatic insects, with a small portion of the gut contents also containing terrestrial insects and aquatic macrophytes (Roepke et al. 2014). *Carnegiella marthae* and *C. strigata* exhibit fright reaction responses to the release of conspecific damage-released alarm signals (Pfeiffer 1977). Two species, *C. myersi* and *C. scherei*, have been hypothesized to have retained paedomorphic features (Weitzman and Vari 1988) with Toledo-Piza et al. (2014) labeling each species as miniatures.

Gasteropelecus (3.5–6.4 cm SL)

Similar to *Carnegiella* but differs in: presence of an adipose fin, 3 radials of the pectoral girdle, dorsal fin with ii + 8–10 or iii + 7–9 rays; anal fin iii + 23–33 rays; a short lateral line segment on the caudal fin, and 28–37 longitudinal scales (Weitzman 1960). **SPECIES** Three, all in the AOG region: *G. levis*, *G. maculatus*, and *G. sternicla* (River hatchetfish). See Weitzman (1960) for differences between species. **DISTRIBUTION AND HABITAT** All species occur throughout the AOG region. **BIOLOGY** *G. sternicla* spawns during the rainy season in an intermittent stream in Trinidad, with ovum morphology possibly indicating synchronized spawning (Alkins-Koo 2000). Feeds primarily on terrestrial insects during both early life history and juvenile stages, with no ontogenetic diet shifts as they mature (Mérigoux and Ponton 1998).

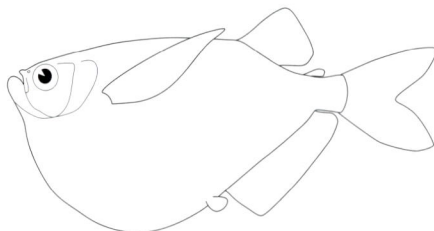

Thoracocharax (6.7–9.0 cm SL)

Largest body size of all members in the family, and can be distinguished from other gasteropelecids by: a sensory canal in the upper portion of the preopercle; lateral line curved downward toward the anal fin; absence of a short lateral line segment on caudal fin; presence of adipose fin; 3 teeth in the outer row of the premaxilla and 7 teeth in the inner row; dorsal fin ii + 10–15 or iii + 9–14 rays; and anal fin with ii + 32–43 or iii + 31–41 rays (Weitzman 1960). **SPECIES** Two, both in the AOG region, including *T. securis* (Giant hatchetfish) and *T. stellatus* (Spotfin hatchetfish). *Thoracocharax stellatus* is composed of distinct genetic lineages in the Orinoco, Amazon, Araguaia, and Paraguay basins (Abe et al. 2013), suggesting this group represents a species complex. See Weitzman (1960) for differences between the two described species. **DISTRIBUTION AND HABITAT** *T. securis* is restricted to the lowland Amazon basin, whereas *T. stellatus* exhibits a very broad geographic distribution across the Amazon and Orinoco basins, the Guianas, and into the La Plata basin as far south as Argentina. **BIOLOGY** *T. stellatus* is abundant across much of its range and one of the most common species along river margins and floodplain lakes in the Amazon lowlands. This species is primarily a diurnal forager with a diet dominated by insects (ants, beetles, and mayflies) mostly of terrestrial origin (Netto-Ferreira et al. 2007). This species exhibits fleshy dermal lip protuberances on the lower jaw during periods of low water and high hypoxia in the Venezuelan Llanos, hypothesized to aid in aquatic surface respiration (Winemiller 1989a). *Thoracocharax securis* grows to a larger body size (9.0 cm SL) and is usually much less abundant than *T. stellatus*, but otherwise the two species exhibit similar ecological and other biological characteristics (Chernoff et al. 2000, Granado-Lorencio et al. 2007).

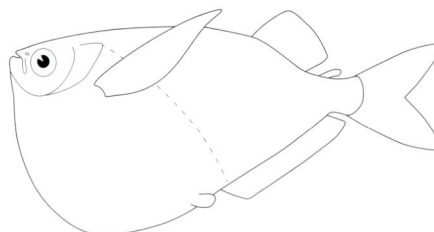

FAMILY HEMIODONTIDAE—HALFTOOTHS

— FRANCISCO LANGEANI

DIVERSITY 31 species in five genera, including 24 species in the AOG region. The subfamily Anodontinae includes *Anodus* (2 species), and *Micromischodus* (1 species), and the subfamily Hemiodontinae includes *Hemiodus* (21 species), *Argonectes* (2 species) and *Bivibranchia* (5 species) (Langeani 2003, Beltrão and Zuanon 2012, Langeani and Moreira 2013).

COMMON NAMES Cruzeiro-do-Sul, Jatuarana, Orana, Piau-banana (Brazil); Yulilla (Peru).

GEOGRAPHIC DISTRIBUTION Most rivers and basins of northern South America, such as the Amazon, Orinoco, and Tocantins basins, coastal rivers in the Guianas, some smaller independent drainages (Amapá, Araguari, Itapecuru, Mearim, and Parnaíba rivers), and the Paraná-Paraguay basin.

ADULT SIZES 7.0–31 cm SL.

DIAGNOSIS OF FAMILY Readily distinguished from other characiforms by their fusiform and streamlined body shape; presence of an adipose fin; well-developed eyelid covering most of the eye, except for a vertically elongate or small circular opening over the pupil; opercle deeply concave dorsally; a suprapectoral sulcus; 9–11 branched pelvic-fin rays; usually a round midlateral pigment spot at a vertical through the posterior portion of the dorsal-fin base and a longitudinal stripe on the lower lobe of the caudal fin (Langeani 1998, 2009, 2013).

SEXUAL DIMORPHISM Males of *Hemiodus atranalis* develop thread-like prolongations in dorsal-fin rays.

HABITATS Most species inhabit open waters in lakes and large rivers (Roberts 1972, 1974, Britski et al. 1999), while some live in smaller rivers and forest streams (Myers 1927, Roberts 1971, Langeani 1999a).

FEEDING ECOLOGY All hemiodontids are pelagic or benthopelagic, feeding on detritus, mud, filamentous algae, vascular plants, and insect larvae (e.g., chironomids, corixids, and ephemeropterans). *Argonectes*, *Bivibranchia*, *Hemiodus*, and *Micromischodus* also consume fish droppings (Menezes and Oliveira e Silva 1949, Roberts 1971, Knöppel 1972), and *Anodus* filters plankton from the water with the aid of numerous elongate gill rakers (Roberts 1972). *Argonectes* and *Bivibranchia* are the only characiform fishes with protractile upper jaws (especially pronounced in *Bivibranchia*), and *Bivibranchia* also has various anatomical specialization on the roof of the mouth, gill arches and gill chamber, used to sort food particles taken with the substrate (Eigenmann 1912a, Roberts 1971, Vari 1985).

BEHAVIOR Hemiodontids are usually social, aggregating into small or large shoals or schools (Roberts 1971, 1974, Vari 1985). Reproduction occurs during rainy season and spawning is total or partial (Santos et al. 2004).

ADDITIONAL NOTES Some *Hemiodus* species are appreciated by aquarists, and some *Anodus*, *Argonectes*, and *Hemiodus* species that attain larger sizes are used as food (Santos et al. 1984).

KEY TO THE GENERA

1a. Teeth absent in both upper and lower jaws; branchial opening very ample, extending to vertical passing through anterior portion of eye; gill rakers very long and numerous (generally >150)...*Anodus*

1b. Teeth present in both upper and lower jaws, or in upper jaw only; branchial opening reduced, extending to vertical, passing through posterior portion of eye; gill rakers short and few (generally <100)..**2**

2a. Teeth unicuspid and pedicellate, barely visible, disposed in a single row in the premaxillary and in 2 rows in the dentary ....*Micromischodus*

2b. Teeth multicuspid or tricuspid in the premaxillary and maxillary; dentary without teeth.......................................**3**

3a. Upper jaw nonprotractile; teeth multicuspid with large crown ....*Hemiodus*

3b. Upper jaw protractile; teeth tricuspid..**4**

4a. Black collar-like mark behind the opercular opening; snout with small, dorsal and transverse folds from its anterior portion to the nares; upper jaw slightly protractile; adipose eyelid covering entire eye except for a minute orifice above pupil; branchial arches normal, their medial portion not extending beyond distal end of gill rakers...*Argonectes*

4b. Absence of a black collar-like mark behind the opercular opening; snout smooth, without dorsal and transverse folds, with a transverse sulcus marking point where upper jaw fits when retracted; upper jaw highly protractile; adipose eyelid with a vertically elongate opening over the pupil; branchial arches very enlarged, their medial portion extending much beyond distal end of gill rakers ....*Bivibranchia*

GENUS ACCOUNTS

Anodus (27–31 cm SL)

Characterized by: absence of teeth in both upper and lower jaws; branchial opening very ample, extending to vertical, passing through anterior portion of eye; gill rakers very long and numerous (generally >150) and with minute ctenii or tooth-like projections (Langeani 1998). **SPECIES** Two, *A. elongatus* and *A. orinocensis*; see Langeani (2013) for a key to the species. **COMMON NAMES** *Charuto*, *Ubarana* (Brazil); *Yulilla sin dientes* (Peru). **DISTRIBUTION AND HABITAT** Main river channels and lakes of the Amazon and Orinoco basins. **BIOLOGY** Planktivorous, consuming algae and microinvertebrates (Santos et al. 2004).

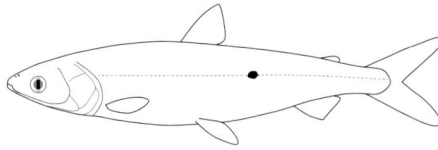

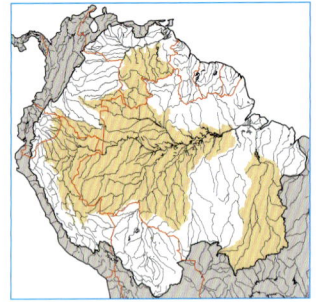

Argonectes (24–29 cm SL)

Characterized by: black collar-like mark behind the opercular opening; upper jaw slightly protractile; tricuspid teeth; snout with small dorsal and transverse folds from its anterior portion to the nares; adipose eyelid covering entire eye except for minute orifice above pupil; branchial arches normal, their medial portion not extending beyond distal end of gill rakers (Langeani 1998). **SPECIES** Two, *A. longiceps* and *A. robertsi*; see Langeani (1999a) for species identification. **COMMON NAMES** *Jatuarana* (Brazil); *Yulilla* (Peru). **DISTRIBUTION AND HABITAT** Main river channels and lakes of the Amazon, Guyana, Suriname, and Orinoco basins. **BIOLOGY** Omnivorous, consuming vegetable matter and invertebrates (Santos et al. 2004).

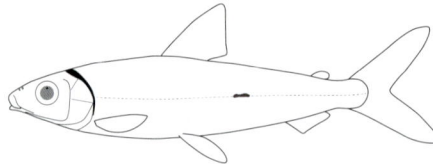

Bivibranchia (7.0–16 cm SL)

Characterized by: upper jaw highly protractile; snout smooth, without dorsal and transverse folds, with a transverse sulcus marking point where upper jaw fits when retracted; adipose eyelid with vertically elongate opening over pupil; branchial arches very enlarged, their medial portion extending much beyond distal end of gill rakers (Langeani 1998). **SPECIES** Five, see Vari and Goulding (1985) and Géry et al. (1991) for species identification. **DISTRIBUTION AND HABITAT** Margins of large rivers, pools, and creeks in the Amazon and Orinoco basins and coastal rivers in the Guianas. Most species are common in areas of sandy beaches and pools (Vari 1985, Vari and Goulding 1985). **BIOLOGY** Specimens were "observed travelling in large schools over sandy beaches, evidently feeding on food items which individuals separate out of the substrate by manipulation of mouthfuls of sand" (Vari 1985), some also form mixed schools with congeneric species (Vari and Goulding 1985). Omnivorous, consuming algae, detritus and aquatic invertebrates (Santos et al. 2004).

Hemiodus (7.0–25 cm SL)

Among hemiodontids, *Hemiodus* is differentiated only by one unique, externally visible trait: the presence of multicuspid teeth (Langeani and Moreira 2013). **SPECIES** 21 species, outdated key to the species in Langeani-Neto (1996). Recently described species in Langeani (1999b), Langeani (2004), Beltrão and Zuanon (2012), and Langeani and Moreira (2013). **COMMON NAMES** *Voador*, *Flexeiro* (Brazil); *Pez volador* (Peru). **DISTRIBUTION AND HABITAT** Main river channels, pools, creeks, and reservoirs of the Amazon, French Guiana, Guyana, Paraguay, Suriname, and

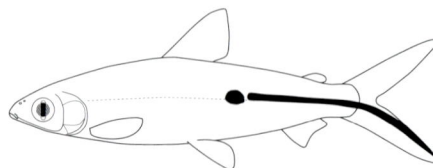

Orinoco basins. *Hemiodus* species can be divided into two groups. The first is composed of large-sized species (14–25 cm SL) with more than 85 perforated scales along lateral line, frequently with larger distributions along main river channels. The second is composed of small-sized species (7.0–16 cm SL) with fewer than 85 perforated scales along lateral line and occurring in the creeks and streams of the headwater portions of the basins in the Guiana and Brazilian crystalline shields; the small-sized species also have the most distinctive color patterns. **BIOLOGY** Common names from their ability to jump; "a group of *voadores* leaping away from predators or over a seine net to safety is an impressive sight" (Roberts 1971). Diet includes a wide variation of items, such as detritus, filamentous algae, microcrustaceans, plant parts; this broad trophic diversity could explain the relative success of the genus in reservoirs (Santos et al. 2004, da Silva et al. 2008). Garrone Neto and Carvalho (2011) report a nuclear-follower feeding association among characiform fishes as followers (including species of *Hemiodus)* and *Potamotrygon orbignyi*, a freshwater stingray species (Potamotrygonidae), as nuclear.

Micromischodus (17 cm SL)

Characterized by: teeth very small, unicuspid and pedicellate; a single tooth row in the premaxillary and two rows in the dentary. **SPECIES** One, *M. sugillatus*; for details about its identification see Roberts (1971). **DISTRIBUTION AND HABITAT** Main river channels, pools, and creeks of the Amazon basin. Typically in places with slow-flowing, black or crystalline water (Roberts 1971). **BIOLOGY** Generalist benthic omnivores, including small insect larvae (corixid and dipteran) and fish feces (Roberts 1971).

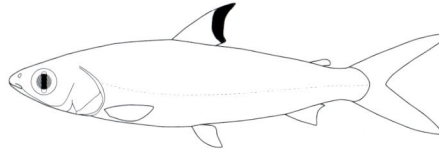

FAMILY IGUANODECTIDAE—IGUANODECTID CHARACIFORMS

— PETER VAN DER SLEEN and CRISTIANO R. MOREIRA

DIVERSITY Iguanodectinae has traditionally been recognized as a subfamily within the Characidae, containing two genera, *Iguanodectes* (8 species) and *Piabucus* (3 species). In a recent molecular phylogenetic study, Oliveira et al. (2011a) found *Bryconops* (19 species) related to Iguanodectinae (*Iguanodectes* and *Piabucus*). As a consequence, Oliveira et al. (2011a) proposed to unite these three genera into the family Iguanodectidae, with 30 species, including 27 species in the AOG region. **COMMON NAMES** *Piaba alongado* (Brazil); *Mojarita alargado* (Peru). **GEOGRAPHIC DISTRIBUTION** Orinoco, Amazon, São Francisco, and Paraná-Paraguay basins, and most coastal rivers flowing to the Atlantic Ocean in Venezuela, the Guianas, and northern Brazil. All three genera are widely distributed in the AOG region. **ADULT SIZES** Small to medium body sizes, from 3.1 cm SL in *Bryconops durbini* from the Tapajós River, Brazil, to 13 cm SL in *P. dentatus* from coastal drainages of the Guianas and Gulf of Pariá to the mouth of the Amazon River and the lower Amazon basin.

DIAGNOSIS OF FAMILY Supraorbital bone present; elongate (smelt-shaped) body; long anal fin with more than 20 (unbranched and branched) fin rays; multicuspid teeth; and few (3 or fewer) maxillary teeth. Additional characters of internal morphology are also shared by these genera (Vari 1977, Moreira 2003). **SEXUAL DIMORPHISM** Sexually mature males develop hooks on the pelvic fin (only *Bryconops*) and anal-fin rays. In *Iguanodectes* (except *I. geisleri*) and *Piabucus* a pair of lappets is present on the anterior 4–6 anal-fin rays of mature males. **HABITATS** Occur in a variety of habitats, but typically found in small streams. **FEEDING ECOLOGY** *Bryconops* species are omnivorous or insectivorous, while *Iguanodectes* and *Piabucus* are mainly herbivores. **BEHAVIOR** Species usually occur in schools. Both *Piabucus* and *Bryconops* species have been reported to jump out of the water to catch insects (Costa-Pereira and Severo-Neto 2012, Mol 2012b).

KEY TO THE GENERA

1a. Posterior margin of maxilla not surpassing anterior margin of eye **2**
1b. Posterior margin of maxilla surpassing anterior margin of eye ....*Bryconops*

2a. Pectoral keel present ....*Piabucus*
2b. Pectoral keel absent...*Iguanodectes*

GENUS ACCOUNTS

Bryconops (3.1–12 cm SL)

Distinguished from other Characidae by: ventral edge of maxilla curves sharply posteriorly, almost at 90°, extending beyond the quadrate socket of the articular; antorbital (AOR) with developed sensory canal; and supraorbital (SOR) sensory canal extending onto nuchal scales (Chernoff and Machado-Allison 1999). Body spindle-shaped. Males bear hooks on the anal- and pelvic-fin rays. **SPECIES** 19, including 16 species in the AOG region. Partial key to the species in Chernoff and Machado-Allison (2005). Species are divided into two subgenera, *Bryconops* and *Creatochanes* (Chernoff and Machado-Allison 1999). **COMMON NAMES** *Piaba* (Brazil); *Yaya sadine* (French Guiana); *Denton* (Peru). **DISTRIBUTION AND HABITAT** Widely distributed in the cis-Andean lowlands of South America from the Orinoco to the Paraná-Paraguay basins, including most coastal basins of the Guianas and Brazil. Most species are found in the upper water column in habitats with swift-flowing waters, which are typically acidic and transparent. Many species inhabit areas that are dominated by the moriche palm (*Mauritia* sp.) (Chernoff and Machado-Allison 2005). **BIOLOGY** Omnivorous, consuming plant material, terrestrial insects, and small fish. *Bryconops caudomaculatus* has been reported to jump out of the water to catch flies (Costa-Pereira and Severo-Neto 2012). Usually in schools; also spawns in schools (Planquette et al. 1996, Chernoff and Machado-Allison 2005).

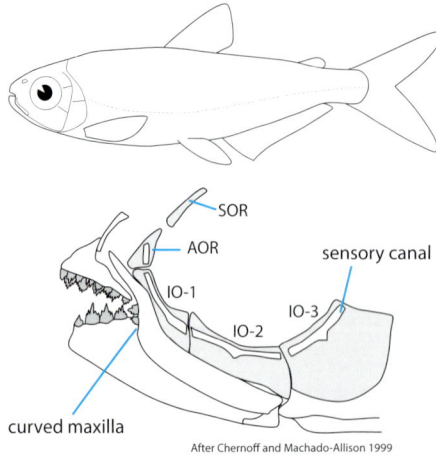

After Chernoff and Machado-Allison 1999

Iguanodectes (4.6–10 cm SL)

Shares with *Piabucus* an elongate (smelt-like) body; multicuspid teeth with constricted bases; gill membranes united and free from the isthmus; the posterior end of the maxilla not extending to the eye; dorsal-fin origin generally posterior to the middle of the body; and a long anal fin (Géry 1977, Vari 1977). Differs from *Piabucus* by a shorter pectoral fin, and absence of a well-developed pectoral keel (vs. a long pectoral fin and a well-developed pectoral keel in *Piabucus*) (Moreira 2003). Most species have a dark lateral band and the caudal-fin upper lobe or middle rays pigmented. Sexually mature males develop hooks and lappets on the anterior anal-fin rays. **SPECIES** Eight, all in the AOG region. Overview of genus in Géry (1993). **DISTRIBUTION AND HABITAT** Throughout the Amazon and Orinoco basins and the coastal rivers in the Guianas. Most species inhabit small streams and spend most time close to the substrate, feeding on periphyton. Midwater and surface-dwelling are more commonly observed among species inhabiting larger rivers such as *I. spilurus* and *I. purusi* (J. Zuanon pers. comm.). **BIOLOGY** Primarily herbivorous (Knöppel 1970, Goulding et al. 1988, Galvis et al. 2006).

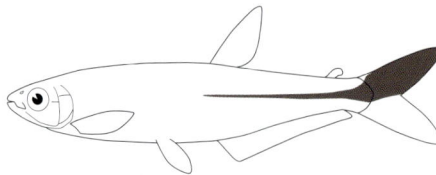

Piabucus (10–13 cm SL)

Shares with *Iguanodectes* an elongate (smelt-shaped) body; multicuspid teeth with contracted bases; gill membranes united and free from the isthmus; the posterior end of the maxilla not extending to the eye; dorsal-fin origin posterior to the middle of the body; and a long anal fin (Géry 1977, Vari 1977). Differs from *Iguanodectes* by a longer pectoral fin, and a well-developed pectoral keel (vs. a short pectoral fin and lacking a well-developed keel in *Iguanodectes*) (Moreira 2003). Species have a brilliant longitudinal band, sometimes overlying a darker lateral band. Sexually mature males develop hooks and lappets on the anterior anal-fin rays. **SPECIES** Three, all in the AOG region. **DISTRIBUTION AND HABITAT** *P. caudomaculatus* (9.6 cm SL) from the Mamoré basin in Bolivia, *P. dentatus* (13 cm SL) from coastal drainages from Gulf of Paría (Venezuela) to the mouth of the Amazon River and lower Amazon basin, and *P. melanostomus* (12 cm SL) from the upper Madeira and Paraguay basins. Species inhabit the upper zone of the water column (Planquette et al. 1996). **BIOLOGY** Primarily herbivorous.

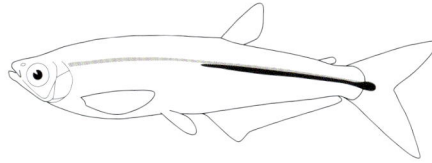

FAMILY LEBIASINIDAE—PENCILFISHES
— *ANDRÉ LUIZ NETTO-FERREIRA*

DIVERSITY 77 species in seven genera, including 59 species in the AOG region. The subfamily Lebiasininae includes *Lebiasina* (27 species), and the subfamily Pyrrhulininae includes *Copeina* (2 species), *Copella* (8 species), *Derhamia* (1 species), *Nannostomus* (20 species), and *Pyrrhulina* (19 species) (Weitzman and Weitzman 2003).

COMMON NAMES Copeína, Peixe-lápis, Pirá-tan-tan (Brazil); Chalquoque, Choro-coque, Flechita, Urquisho (Peru).

GEOGRAPHIC DISTRIBUTION Lebiasinids are distributed from southern Costa Rica to the La Plata basin. Does not occur naturally in the São Francisco basin, the eastern and southeastern Brazilian coastal basins, or Chile.

ADULT SIZES 1.6 cm TL in *Nannostomus anduzei* from the upper Orinoco and Ereré rivers (Negro basin), Venezuela, to >16 cm SL in *L. erythrinoides* from the Llanos of Colombia and Venezuela and Lake Maracaibo in Venezuela.

DIAGNOSIS OF FAMILY Distinguished from other characiform fishes by the large, thick scales on surface of the body and posterior portion of the head, reduced or absent laterosensory lines on the surface of the body and head, and the absence of cranial fontanels.

SEXUAL DIMORPHISM Characters observed only in males constitute an important source of taxonomic and systematic information to help diagnose most lebiasinid species. Most lebiasinid species are sexually dimorphic, with males exhibiting thicker and longer anal-fin rays and membranes, larger pterygiophores and pterygiophore muscles, and thick dermal membrane on the last anal-fin ray (Netto-Ferreira 2012, Netto-Ferreira and Marinho 2013). Many species also exhibit sexual differences in body coloration pattern. Males of most *Nannostomus* species show more intense stripes and red coloration on the anal and caudal fins. In *Copella* and most *Pyrrhulina* species the pelvic, anal, dorsal and caudal fins have much longer rays in males than in females, and the rays of the dorsal fin and dorsal lobe of the caudal fin may also be highly elongate and filamentous (Géry 1977). Breeding tubercles are present on the ventral portion of the head of a few species of *Nannostomus* (Wiley and Collette 1970), and on the lateral scales and fins of *Lebiasina*. Fertilization occurs externally, and some species engage in parental care (Vazzoler and Menezes 1992).

HABITATS Lebiasinids inhabit small forest streams with slow water flow and standing water in forest pools.

FEEDING ECOLOGY Members of *Copeina*, *Copella*, and *Pyrrhulina* are insectivores, feeding on allochthonous arthropods (mainly ants) near the surface, whereas relatives of *Nannostomus* feed on periphyton and benthic invertebrates associated with the vegetation. Lebiasinines are generalists, feeding mainly on insects, and also small crustaceans, juvenile fishes, flowers, and fruits (Breder 1927, Zaret and Rand 1971, Taphorn and Lilyestrom 1980, Taphorn 1992, Román-Valencia 1996, 1997, 2004)

BEHAVIOR Lebiasinids are gregarious, apparently not territorial, and show rudimentary parental care (except for *Copella arnoldi* in which males guard and moisten eggs spawned outside the water), forming schools of several individuals of mixed generations.

ADDITIONAL NOTES The spectacular coloration of some pyrrhulinine species and the unique reproductive behavior of *C. arnoldi* have made members of this family important in the global aquarium fish trade (Hoedeman 1954, Weitzman and Weitzman 2003).

KEY TO THE GENERA

1a. Mouth terminal or subterminal **2**
1b. Mouth distinctly superior **3**

2a. Body slender; anterior and posterior nares distinctly separated by a fleshy bridge; maxilla usually shorter than orbital diameter, slightly rounded; small-sized adult specimens (16–30 mm SL), some miniaturized; teeth multicuspid, distinctly spatulate or spoon shaped; color pattern usually consisting of two or more conspicuous longitudinal stripes and two oblique bands, but no humeral spot (if stripes absent, then several oblique bands exhibit); primary longitudinal stripe extending from tip of snout to at least end of caudal peduncle, usually onto caudal-fin median rays...*Nannostomus*
2b. Body robust; anterior and posterior nares adjacent to each other; maxilla usually longer than orbital diameter; medium- to large-sized adult specimens (5.0–30 cm SL); teeth tricuspid, nearly cylindrical, only slightly compressed; color pattern consisting of a humeral blotch, two (rarely three) longitudinal stripes and no oblique bands; primary stripe usually continuous from humeral blotch to tip of caudal peduncle, but a fragmented primary stripe may occur in Guiana Shield species...*Lebiasina*

3a. Dorsal fin located at middle of the body, dorsal-fin terminus not reaching vertical through anal-fin origin............................**4**
3b. Dorsal fin displaced posteriorly, dorsal-fin terminus reaching posterior to anal-fin origin **5**

4a. Primary stripe indistinct; snout distinctly blunt; body robust with head distinctly enlarged laterally, head nearly as long as wide; lower lip hypertrophied; single series of conical teeth on premaxilla; maxilla short, nearly rounded; opercular membrane united at the posterior portion of isthmus, near tip of pectoral girdle ....*Copeina*
4b. Primary stripe conspicuous, at least on anterior portion of head; snout pointed anteriorly; body relatively slender; head distinctly longer than wide; lower lip similar in size to upper lip; two series of conical teeth on premaxillary; maxilla long, not rounded; opercular membrane united near the anterior tip of the isthmus ....*Pyrrhulina*

5a. Presence of an adipose fin; anal-fin distal margin concave or nearly straight; anterior and posterior nares adjacent to each other; cephalic pores of lateral line system hypertrophied; short lateral line canal present on body, with 5–6 perforated scales; teeth tricuspid (Essequibo basin, Guyana) ....*Derhamia*
5b. Adipose fin absent; anal-fin distal margin distinctly convex; anterior and posterior nares distinctly separated by a fleshy bridge; cephalic pores of lateral line system reduced in size; lateral line absent from body; teeth conical..*Copella*

GENUS ACCOUNTS

Copeina (5.0–8.0 cm SL)

Characterized by: primary stripe indistinct; mouth superior; snout distinctly blunt; eyes large in comparison with head length; body robust with head distinctly enlarged laterally; head nearly as long as wide; lower lip hypertrophied; single series of several conical teeth on the premaxillary; maxilla short, nearly rounded; opercular membrane united to the isthmus posteriorly; adipose fin absent (Géry 1977). Sexual dimorphism typical of that of Lebiasinidae, involving mainly modifications of the anal fin in adult males. **SPECIES** Two, *C. guttata* and *C. osgoodi*, both in the Amazon basin. See Géry (1977) for distinguishing characters between both species. **COMMON NAMES** *Copeína rubi* (Brazil); *Urquisho rojo* (Peru). **DISTRIBUTION AND HABITAT** Both species inhabit lowland rivers (below c. 250 m elevation) of the central and western Amazon basin (Weitzman and Weitzman 2003). A population of *C. guttata* inhabits the eastern Amazon (Anapu basin, Brazil), although its taxonomic status is uncertain. *Copeina* species differ from *Copella* and *Pyrrhulina* by inhabiting the lower layer of the water column. **BIOLOGY** *Copeina* are diurnal, nonmigratory species that feed on insects. Spawning takes place in a depression in the sand and the male guards the eggs (Géry 1977).

Copella (3.0–5.0 cm SL)

Characterized by: primary longitudinal stripe of variable extent; very elongate body; filamentous rays on dorsal and caudal fins and an S-shaped maxilla in males; anterior and posterior nares separated by a fleshy bridge; a single series of few conical teeth on the premaxillary; opercular membrane united to the isthmus posteriorly; posteriorly displaced dorsal fin with rear end reaching beyond vertical through anal-fin origin; and absence of an adipose fin (Géry 1977). Along with the S-shaped maxilla of males, the sexual dimorphism pattern observed in the genus is typical of that of other Lebiasinidae, involving mainly modifications of the anal fin in mature males, although elongation of the pelvic-, dorsal-, and caudal-fin rays is also observed. In *C. compta*, *C. eigenmanni*, *C. nattereri*, *C. nigrofasciata*, and *C. vilmae* the dorsal and ventral caudal-fin procurrent rays of males are also densely covered by dark pigments (Netto-Ferreira 2010). **SPECIES** Seven, all in the AOG region. An outdated key to the species is available at Zarske and Géry (2006). **COMMON NAMES** *Ti-yaya* (French Guiana); *Urquisho a rayas* (Peru); Splashing tetra (English). **DISTRIBUTION AND HABITAT** Widespread throughout the Amazon, Orinoco, and Tocantins-Araguaia basins, and the adjacent coastal rivers draining from the Brazilian and Guiana shields, occupying mainly the lowland sections of rivers. **BIOLOGY** Insectivorous fishes inhabiting the upper portion of the water column near the surface. Most species spawn on submerged leaves overhanging from adjacent vegetation. *Copella arnoldi* is the species with the most unusual breeding behavior: spawning occurs outside the water and the male keeps guarding and moistening the eggs until embryos hatch. Diurnal, nonmigratory species.

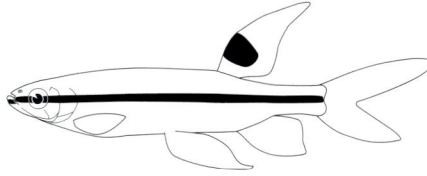

Derhamia (6.0 cm SL)

Distinguished from other lebiasinids by: a superior mouth; a posterior position of the dorsal fin with the rear end reaching beyond a vertical through the anal-fin origin; presence of an adipose fin; anal-fin distal margin concave or nearly straight; anterior and posterior nares adjacent to each other; cephalic pores of lateral line system hypertrophied; lateral line onto body short; and tricuspid teeth cylindrical (Géry and Zarske 2002). Juveniles can be easily overlooked and misidentified as small *Lebiasina* or *Copella*. **SPECIES** One, *D. hoffmannorum*. **DISTRIBUTION AND HABITAT** Inhabits the Kamarang, Kukui, Mazaruni, Membaru, and Waruma rivers in Guyana. Specimens are considered "shy" and occupy the upper portions of the water column, hiding under floating leaves or other floating objects (Géry and Zarske 2002). **BIOLOGY** Breeding behavior not reported. Apparently a diurnal species.

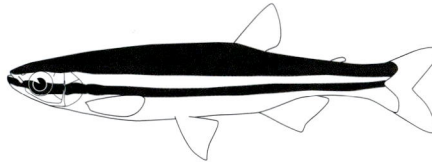

Lebiasina (6.0–16 cm SL)

Characterized by: medium adult body sizes; robust body; position of anterior and posterior nares adjacent to one another; maxilla usually longer than orbital diameter; tricuspid teeth slightly compressed and nearly cylindrical; a color pattern consisting of a humeral blotch with usually 2 (rarely 3 or 4) longitudinal stripes and no oblique blotches; primary stripe usually continuous from humeral blotch to end of caudal peduncle (fragmentation of the stripe sometimes occurs in Guiana Shield species and longitudinal blotches in Andean species) (Géry 1977, Netto-Ferreira 2010). Mature male specimens of *Lebiasina* usually exhibit conspicuous sexual dimorphism (Netto-Ferreira et al. 2011, Netto-Ferreira 2012, Netto-Ferreira et al. 2013), with longer and thicker anal-fin rays, thickened

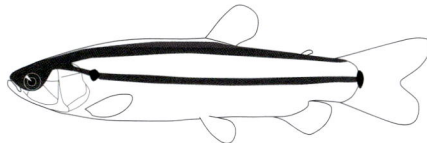

fin membranes, breeding tubercles on variable portions of head, fins, and body, and primary longitudinal pigment stripe being more intensely marked in male specimens in comparison with females in certain species from the Guiana Shield. SPECIES 27, including 12 endemic to the AOG region. Species-level identification in Netto-Ferreira (2010). Species of *Lebiasina* without an adipose fin were formerly ascribed to *Piabucina*, here treated as a junior synonym. **DISTRIBUTION AND HABITAT** Inhabit the upper courses of streams with rocky or sandy bottoms and well-oxygenated waters. The only exceptions are the lowland trans-Andean species *L. bimaculata*, *L. boruca*, and *L. festae*, which may occur in standing waters near the shore. *Lebiasina intermedia* was presumably described from Santarém in the lower Amazon (Géry 1977, Weitzman and Weitzman 2003), but no specimens matching an original description by Meinken (1936) were ever collected in the lower Amazon. **BIOLOGY** Apparently diurnal, nonmigratory species. Diet includes essentially allochthonous insects (Ardila-Rodríguez 1978), but small fish, flowers, fruits, and other alimentary items are also observed, although in small proportions (Breder 1927, Zaret and Rand 1971, Taphorn and Lilyestrom 1980, Taphorn 1992, Román-Valencia 1996, 1997, 2004).

Nannostomus (1.6–6.5 cm SL)

Includes the only truly miniaturized species (*sensu* Weitzman and Vari 1988) among the Erythrinoidea (*sensu* Buckup 1998). Characterized by: a slender body; anterior and posterior nares distinctly separated by a fleshy bridge; maxilla slightly rounded and usually shorter than orbital diameter; teeth multicuspid, distinctly spatulate and spoon-shaped; color pattern usually consisting of two or more conspicuous longitudinal stripes and two oblique bands with no humeral blotch, or if stripes are absent, then several oblique bands (e.g., *N. espei*); and with a primary stripe extending from tip of snout to at least end of caudal peduncle, usually onto caudal-fin median rays (Weitzman and Cobb 1975, Weitzman 1978). Nocturnal color pattern for most species different from diurnal coloration, and usually composed of two or three comma-shaped blotches. The sexual dimorphism of most species is similar to other Lebiasininae, usually with more intensely marked stripes and red coloration on the anal and caudal fins, and thicker anal-fin rays and membranes in mature males, although a few species have less developed sexual dimorphism. Breeding tubercles are sometimes present on the anteroventral portion of the head in males (Wiley and Collette 1970, Netto-Ferreira 2012). SPECIES 18, all in the AOG region. See Weitzman (1978) and Fernandez and Weitzman (1987) for species identifications. **DISTRIBUTION AND HABITAT** Geographic distribution largely overlaps with *Copella*, being widespread in lowland rivers of the Amazon, Orinoco, and Tocantins-Araguaia basins, and adjacent coastal rivers of the Brazilian and Guiana Shields. **BIOLOGY** Species feed on benthic invertebrate, and are diurnal and nonmigratory.

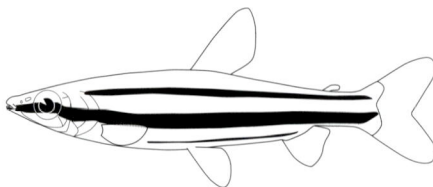

Pyrrhulina (3.5–8.0 cm SL)

Characterized by: superior mouth; a primary stripe that extends onto the anterior portion of head; anteriorly oriented snout; relatively slender body; head distinctly longer than wide; lower lip similar in size to the upper lip; two series of conical teeth on the premaxilla; maxilla long and straight (not rounded); opercular membrane united to the isthmus anteriorly; dorsal fin located at middle of the body; dorsal-fin rear end not reaching vertical through anal-fin origin; adipose fin absent; third postcleithrum absent (Géry 1977). Sexual dimorphism in *Pyrrhulina* is similar to that of other Lebiasinidae, involving elongation of the fin rays in all fins of mature males, except *P. australis* and *P. marilynae*. SPECIES 19, all in the AOG region. Genus is in need of a taxonomic review. See

Géry (1977) and Netto-Ferreira and Marinho (2013) for limited aid on species identifications. **DISTRIBUTION AND HABITAT** Most widespread geographic distribution of all Lebiasinidae, being known from throughout the Amazon, Orinoco, Tocantins-Araguaia basins, and adjacent coastal rivers of the Brazilian and Guiana shields, and unlike *Copella* and *Nannostomus*, is also present in upland waters (above c. 250 m elevation). All species inhabit the upper portions of the water column near the surface. **BIOLOGY** Species are all diurnal, insectivorous, and nonmigratory. Spawning can occur on flat rocks or submerged leaves (Géry 1977).

FAMILY PARODONTIDAE—SCRAPETOOTHS

— *PETER VAN DER SLEEN and JAMES S. ALBERT*

DIVERSITY 32 species in three genera: *Apareiodon* (15 species), *Parodon* (14 species), and *Saccodon* (3 species), with a total of 9 species in the AOG region.

TAXONOMIC NOTE Roberts (1974) already pointed out that "*Parodon, Saccodon* and *Apareiodon* are so poorly defined in the literature that their distinctness may be questioned." Traditionally *Parodon* is separated from *Apareiodon* and *Saccodon* by the presence of teeth on the lower jaw (vs. absent) and this is still the case (*Saccodon* is further distinguished from *Apareiodon* by two unbranched pectoral-fin rays, vs. only one in *Apareiodon*). However, absence or presence of teeth is difficult to see in young specimens, and the teeth are relatively small, weak, and easily broken in adults (Pavanelli 2003). In addition, Ingenito and Buckup (2005) assigned their new species, *P. moreirai* from the upper Paraná system, to *Parodon* even though some adults lacked dentary teeth.

COMMON NAMES Canivete, Duro-duro (Brazil); Mazorca, Tuzo (Colombia); Yulilla fluyendo (Peru); Marranito (Venezuela); Pongo characin (English).

GEOGRAPHIC DISTRIBUTION Throughout much of tropical South America and southern Panama, including Andean foothills and upland shields in the AOG region (Pavanelli 2003). *Saccodon* is found only in trans-Andean South America from Peru to Panama.

ADULT SIZES 5.6 cm SL in *Apareiodon cavalcante* from the Tocantins-Araguaia basin in Brazil, to 15 cm SL in *Parodon carrikeri* from the Bermejo and Pilcomayo rivers, in the Paraguay basin of Argentina.

DIAGNOSIS OF FAMILY All species have a fusiform body, no cranial fontanel, and an inferior mouth with a poorly developed, or absent, upper lip. There are commonly 4 (rarely 2) spatulate premaxillary teeth, which have a straight or cusped cutting border; dentary and maxillary teeth occur in some species; gill membranes are joined together and free of the scaly isthmus (Pavanelli 2003). Color patterns are variable, but in many species include a broad, dark, longitudinal stripe, in addition to other stripes, bars, and blotches.

KEY TO THE GENERA

1a. Dentary teeth absent ....*Apareiodon*
1b. Dentary teeth present .... *Parodon*

GENUS ACCOUNTS

Apareiodon (5.6–14 cm SL)

Differentiated from *Parodon* by absence of dentary teeth (vs. present in *Parodon*). Nuptial males of some species develop tubercles, mainly on the side of the snout and internasal regions (Wiley and Collette 1970, Londoño-Burbano et al. 2011), in yet other species tubercles develop in both sexes (Pavanelli 2006). **SPECIES** 15, including at least 5 species in the AOG region. Key to the species in the Tocantins-Araguaia basin in Pavanelli and Britski (2003). **DISTRIBUTION AND HABITAT** Upland portions of cis-Andean South America, except in coastal Atlantic basins south of Bahia state in Brazil, Patagonia, and the Amazon River channel (Pavanelli 2003). Generally found in fast-flowing waters. **BIOLOGY** Diurnal grazers on algae and small benthic invertebrates. They use their pectoral fins to grip the substrate while

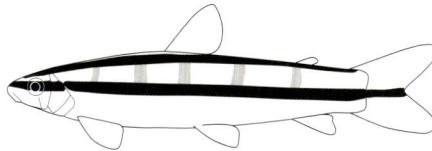

grazing (Sazima 1980). Some species form large schools (Starnes and Schindler 1993, Taphorn et al. 2008). At night, fish congregate in the shallows and in fast-flowing water pack closely among the rocks (Sazima 1980). Spawning in schools and marked seasonally (e.g., Sazima 1980).

Parodon (10–15 cm SL)

Differentiated from *Apareiodon* by the presence of dentary teeth (vs. absent in *Apareiodon*). Nuptial males of some species develop tubercles, mainly on the side of the snout and internasal regions (Wiley and Collette 1970, Londoño-Burbano et al. 2011). **SPECIES** 14, including at least 6 species in the AOG region. Review and key to the species in Colombia in Londoño-Burbano et al. (2011). **DISTRIBUTION AND HABITAT** Atrato, Cauca, Magdalena, Lake Maracaibo, Orinoco, upper Amazon, São Francisco, and Paraná-Paraguay basins. Generally found in rapidly flowing waters with a rocky bottom. **BIOLOGY** Most species feed on algae and aquatic insects. They use their pectoral fins to grip the substrate while grazing. *Parodon guyanensis* has been reported to be nocturnal or crepuscular (Planquette et al. 1996). Territorial behavior of *P. nasus* has been studied by Silva et al. (2009b).

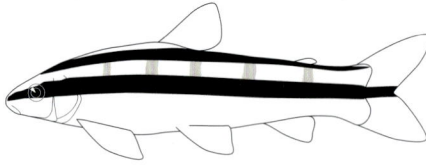

FAMILY PROCHILODONTIDAE—FLANNEL MOUTH CHARACIFORMS

— *BRUNO F. MELO and BRIAN L. SIDLAUSKAS*

DIVERSITY 21 species in three genera: *Ichthyoelephas* (2 species), *Prochilodus* (13 species) and *Semaprochilodus* (6 species) (Castro and Vari 2004), including 10 species in two genera in the AOG region.

COMMON NAMES *Curimatã pacu*, Curimbatá (Brazil); *Sábalo* (Peru); *Bocachico, Copoco, Coporo* (Venezuela).

GEOGRAPHIC DISTRIBUTION Widespread and abundant in aquatic ecosystems throughout South America on both sides of the Andes, and in all South American countries except Chile (Castro and Vari 2004). Most prochilodontid species are native to a single drainage basin, where their extensive migrations lead them to exhibit low levels of population structure, and little to no isolation-by-distance within each drainage (Sivasundar et al. 2001). Some species (e.g., *P. lineatus*) have been introduced into Brazilian drainages to which they are not native, and into parts of Asia and Oceania, such as Papua New Guinea, China, and Vietnam (Kalous et al. 2012).

ADULT SIZES From 24 cm SL in *P. britskii* from the Apiacá River in Mato Grosso, Brazil, to 80 cm SL in *P. lineatus* from the Paraná-Paraguay and Paraíba do Sul basins in Argentina, Brazil, and Paraguay (Castro and Vari 2003, 2004).

DIAGNOSIS OF FAMILY Easily distinguished by a disk-like, evertible mouth with multiple rows of tiny, spoon shaped teeth attached to fleshy lips (Castro and Vari 2004). These unusual jaw morphologies assist them in scraping algae and organic debris from the river bottom and other submerged surfaces. Prochilodontids share more than 50 specialized features of the oral jaws, gill arches, teeth, infraorbital bones, and opercular bones as well as other osteological and soft anatomical systems (Castro and Vari 2004).

SEXUAL DIMORPHISM The anterior ribs of male prochilodontids are expanded relative to the female condition, presumably to provide an increased area of attachment for muscles that produce sound via the gas bladder (Schaller 1971, 1974). Female *I. longirostris* appear deeper-bodied than males (Castro and Vari 2004).

HABITATS Large rivers, floodplains, oxbow lakes, and moderately sized streams throughout tropical South America (Castro and Vari 2003, 2004). Because of their moderate to large body sizes, adults do not typically occur in small creeks.

FEEDING ECOLOGY Specialized feeding on hyperabundant detritus could explain the large population sizes of many prochilodontid species. Detritus feeding also makes the prochilodontids a critical link in the riverine and floodplain food webs and nutrient cycling (Bowen et al. 1984, Flecker 1996, Taylor et al. 2006).

BEHAVIOR Notable for extensive feeding and spawning migrations of up to 1,500 km (Duque et al. 1998). When not migrating they often feed on loosely attached periphyton and sift detritus from the river bottom in large schools. Males of some prochilodontids produce sounds by drumming their muscles on the anteriormost ribs together with the anterior portion of the gas bladder (Schaller 1971, 1974, Castro and Vari 2004).

ADDITIONAL NOTES Because of their large bodies and large population sizes, prochilodontids support important commercial and artisanal fisheries in many parts of South America (Goulding et al. 1988, Castro and Vari 2003). Recent dam constructions threaten these species by blocking migration routes and fragmenting their populations (Hatanaka et al. 2006, Barroca et al. 2012).

KEY TO THE GENERA FROM CASTRO AND VARI (2004)

1a. Anal and caudal fins with dark stripes in life and in preservative, with up to 5 stripes on anal fin and up to 14 stripes on caudal fin; a black collar-like mark behind the opercular opening; modified scales in the median line between the dorsal and adipose fins, with each scale having a well-developed, posteriorly rounded, fleshy flap that significantly overlaps the dorsal surface of the succeeding scale ....*Semaprochilodus*

1b. Anal and caudal fin without dark stripes, but with hyaline caudal-fin lobes or, in some species, with 2–8 irregular vertical bars or wavy lines formed of small dark spots of chromatophores; absence of a black collar-like mark behind the opercular opening; scales in the median line between the dorsal and adipose fins with only a very slightly developed fleshy margin .... *Prochilodus*

GENUS ACCOUNTS

Prochilodus (24–80 cm SL)

Distinguished from *Semaprochilodus* by: absence of large stripes on the anal and caudal fins; absence of a black collar-like mark behind the opercular opening; and lacking modified scales in the median line between the dorsal and adipose fins. The fins are commonly hyaline or with irregular vertical bars or wavy lines. **SPECIES** 13, four species in the AOG region. Identification key to the species and discussion of their phylogenetic relationships in Castro and Vari (2004). **COMMON NAMES** *Sábalo* (Bolivia), *Curimatã* (Brazil); *Bocachico, Boquichico* (Peru, Venezuela). **DISTRIBUTION AND HABITAT** *P. mariae* is endemic to the Orinoco basin in Colombia and Venezuela; *P. rubrotaeniatus* is present in the rivers of the Guianas and the Branco and Marauiá rivers (tributaries of Negro River) in northern Brazil; *P. nigricans* is widely distributed throughout the Amazon basin; and *P. britskii* is a rare and endangered species endemic to the Apiacá River, a tributary of the Tapajós basin in the Brazilian Shield. **BIOLOGY** All species are detritivorous. Individual *Prochilodus* migrate great distances downstream to feed and upstream to reproduce. Details of migration behavior and ecology in Carolsfeld et al. (2003). Their larval and juvenile stages are associated with marginal lagoons along rivers where food is abundant. These marginal lagoons have an important role in the maintenance of genetic variability among populations of *Prochilodus* (Melo et al. 2013). *Prochilodus* species represent an important food resource for humans in all basins where they occur.

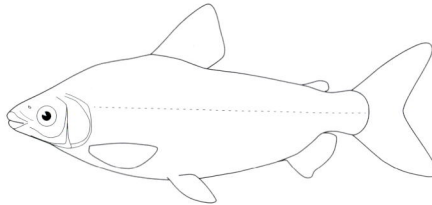

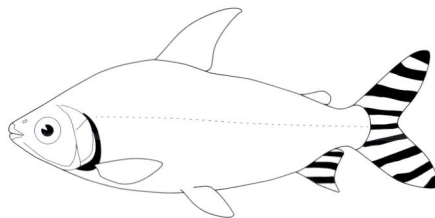

Semaprochilodus (24–44 cm SL)

Readily differentiated from *Prochilodus* by: dark, colored stripes on the anal and caudal fins (but absent or very faint in adults of *S. brama*); a black collar-like mark behind the opercular opening; and modified scales in the median line between the dorsal and adipose fins, with each scale having a well-developed, posteriorly rounded, fleshy flap that significantly overlaps the dorsal surface of the succeeding scale. **SPECIES** Six, all present in the AOG region. Identification key to the species and discussion of their phylogenetic relationships in Castro and Vari (2004). **COMMON NAMES** *Sabalín, Sabalina* (Bolivia), *Jaraqui* (Brazil); *Yaraqui* (Peru); *Copoco* (Venezuela). **DISTRIBUTION AND HABITAT** *S. brama* is restricted to the Tocantins-Araguaia and Xingu basins; *S. insignis* and *S. taeniurus* are abundant in most parts of the Amazon basin; *S. kneri* and *S. laticeps* inhabit the Orinoco basin; and *S. varii* occurs in the Marowijne basin at the border of French Guiana and Suriname. **BIOLOGY** All species are detritivorous and migrate seasonally for feeding and reproduction (Fernandes 1997). Ribeiro and Petrere Jr. (1990) present details on their ecology. *Semaprochilodus* species support one of the most important fisheries in the AOG region (Petrere Jr. 1985, Ribeiro and Petrere Jr. 1990).

FAMILY SERRASALMIDAE—PIRANHAS AND PACUS

— LEO G. NICO, MICHEL JÉGU, and MARCELO C. ANDRADE

DIVERSITY The Serrasalmidae, formerly often treated as a subfamily (i.e., Serrasalminae) within the Characidae, is a morphologically and ecologically diverse group of South American freshwater fishes consisting of about 91 valid species in 16 genera: *Acnodon* (3 species), *Catoprion* (1), *Colossoma* (1), *Metynnis* (15), *Mylesinus* (3), *Myleus* (5), *Myloplus* (12), *Mylossoma* (3), *Ossubtus* (1), *Piaractus* (2), *Pristobrycon* (6), *Pygocentrus* (3), *Pygopristis* (1), and *Serrasalmus* (26), *Tometes* (6), and *Utiaritichthys* (3). In addition, the fossil record indicates at least one extinct genus, †*Megapiranha* (Cione et al. 2009). The most species-rich genera are *Serrasalmus* (26 species), *Metynnis* (15 species), and *Myloplus* (12 species). Three living genera are monotypic: *Catoprion*, *Colossoma*, and *Ossubtus*. Additional species are in the process of being described as new, and a few undescribed forms likely remain undiscovered in the wild and in museum collections.

COMMON NAMES *Pacu*, *Piranha* (Brazil); *Palometa* (Peru); *Caribe* (Venezuela).

TAXONOMIC NOTE The history of serrasalmid systematics and nomenclature is complicated and fraught with confusion and instability (Eigenmann 1915b, Norman 1929, Gosline 1951, Géry 1972a, 1976, Machado-Allison 1982a, b, 2002, Machado-Allison and Fink 1995, Jégu 2003). Contributing factors include a shortage of comparative material, few distinct or reliable external features for distinguishing some genera and species, wide intraspecific morphological variation, little morphological differentiation among species, substantial morphological changes during ontogeny in many taxa, poorly known geographic ranges, superficial original descriptions of many nominal species (in some cases coupled with absence or loss of type material), probable existence of species complexes, and numerous synonymies (e.g., Fink 1993, Machado-Allison and Fink 1996, Machado-Allison 2002, Jégu 2003, Freeman et al. 2007). As a consequence, many names used in past publications (including identification keys), museum collections, and databases (e.g., GenBank) may be suspect or incorrect.

Fortunately, understanding of serrasalmid interrelationships has improved due to ongoing anatomical and genetic analyses (e.g., Freeman et al. 2007, Ortí et al. 2008, Thompson et al. 2014). Recent nomenclatural changes have revived some older generic names and reshuffled species among certain genera. The genus *Myleus* previously included 13 species but is now

represented by 5 species (Jégu and Santos 2002), with most of the species reassigned to *Myloplus* (Andrade et al. 2016b). The "piranhas" are still regarded as a natural group including four genera (i.e., *Pristobrycon*, *Pygopristis*, *Pygocentrus*, and *Serrasalmus*), and now also including *Catoprion* and *Metynnis*. The remaining serrasalmids are a diverse assemblage composed of at least three separate groups: the "pacus" (large herbivores of the genera *Colossoma*, *Piaractus*, and *Mylossoma*), the "*Myleus*-like" rheophilic (rapids-loving) pacus, and other non-piranhas mainly of the genus *Myloplus*.

GEOGRAPHIC DISTRIBUTION Inhabit most freshwater drainages of tropical and subtropical South America from Argentina to Venezuela (about 35°S to 10°N latitude), from sea level to about 300 m elevation. Most diverse in the Amazon and Orinoco basins, and moderately diverse in the La Plata and the Guianas drainages. Some relatively large but isolated basins have fewer serrasalmid species; the São Francisco basin of Brazil has only four native species (Britski et al. 1988), and the Lake Maracaibo basin in Venezuela only one species (Taphorn and Lilyestrom 1984a). Although the Magdalena River of Colombia has no living native serrasalmids (Maldonado-Ocampo et al. 2008), they were formerly present during the Miocene Epoch (Lundberg et al. 2010), and large pacus have recently been introduced by humans (Barletta et al. 2016). The geographic distributions of genera and species are quite variable. For example, some serrasalmids (e.g., *Serrasalmus rhombeus*) have very broad geographic ranges and may be species complexes. Other species inhabit a single drainage or tributary system.

ADULT SIZES Some of the smaller serrasalmids, such as *Catoprion mento* and certain *Metynnis* species, mature at about 6.0 cm SL. The largest member of the family, *Colossoma macropomum*, matures at about 60 cm SL (Loubens and Panfili 1997) and attains a maximum size greater than 1 m SL (Rapp Py-Daniel et al. 2015).

DIAGNOSIS OF FAMILY Teeth in both jaws, body covered with scales, relatively large eyes, nonprotractile (usually terminal) mouth, complete lateral line, single continuous dorsal fin consisting of rays (without spines), an adnexed (flag-like) adipose fin between dorsal fin and tail, an emarginate or forked caudal fin, and silver base body color in many species. Distinguished from other South American freshwater fishes by: (1) medium to large adult body sizes; (2) body laterally compressed and deep, with ratio of body depth (BD) to SL 45–100% (*S. elongatus* as low as 38%); (3) upper jaw with one or two rows of teeth, all restricted to the premaxillary bones (except in *Piaractus*); lower jaw with a complete set of outer row teeth, and in most non-piranhas also a pair of small inner teeth at the mandibular symphysis; (4) abdominal serrae; the number, coverage, and prominence of abdominal serrae vary, depending on taxa and age of fish; (5) single predorsal spine in most species, projecting anteriorly from the first pterygiophore bone immediately anterior to the dorsal-fin origin (absent in *Colossoma*, *Piaractus*, and *Mylossoma*); (6) small scales, with high (63–128) lateral line scale counts; (7) dorsal fin of moderate length, with 2–4 unbranched rays and 12–26 branched rays; (8) anal fin long, with 2–4 unbranched rays and 21–44 branched rays extending from near anus to caudal peduncle; and (9) adipose-fin base ranging from short (most species) to long (*Metynnis*).

SEXUAL DIMORPHISM Differences in the outward appearance of adult males versus females may be obvious, subtle, absent, or unknown, depending on taxa. Sexual dimorphism is most conspicuous in members of eight genera: *Acnodon*, *Metynnis*, *Mylesinus*, *Myleus*, *Myloplus*, *Ossubtus*, *Tometes*, and *Utiaritichthys*. In these genera, mature males develop a second lobe on branched anal-fin rays, sometimes with small, stiff hooklets at ray tips. Their dorsal-fin rays sometimes with long filaments, and the body and fins are beautifully colored (especially during the breeding season). In contrast, mature females have a single-lobed anal fin that is markedly elongate and typically falcate-shaped and, although often colorful, body colors that are generally less striking than those of males (Norman 1929, Gosline 1951, Jégu et al. 2004, Andrade et al. 2013, 2016a, Pereira and Castro 2014). In many pacus and piranhas, the females are slightly larger than males (Loubens and Panfili 1997, Villacorta-Correa and Saint-Paul 1999, Duponchelle et al. 2007, Gomes et al. 2012); a major exception is *Ossubtus xinguense*, in which the females tend to be smaller (although heavier) than mature males (Andrade et al. 2015).

HABITATS Common in a wide diversity of habitats, including the margins of large river channels, the channels of tributary rivers and streams, floodplain lakes, marshes, seasonally flooded forests, and artificial lakes or reservoirs (Goulding 1981, Nico 1991, Taphorn 1992, Mol 2012b). They are most abundant in lowland or floodplain areas, although some are found in high-gradient rivers. They are generally associated with shallower waters, such as the nearshore areas of large rivers, and frequently congregate near mats of submerged and floating vegetation. Many serrasalmid species are found in all three of the major riverine water types (white, black, and clear) and are common in both forested and savanna regions; however, others are closely associated with particular water types and some are only known from forested environments (Goulding 1980, Nico 1991, Taphorn 1992). Absence from a particular environment may be due to either ecological or biogeographical factors

(Goulding et al. 1988). Members of six genera, *Acnodon*, *Mylesinus*, *Ossubtus*, *Tometes*, *Utiaritichthys*, and some species of *Myloplus*, are primarily rheophilic. Most of these species spend much or all of their lives in high-gradient streams with moderate to strong currents, habitats where the rocky substrate is densely covered with aquatic plants of the family Podostemaceae (riverweeds), a source of food and shelter (Jégu et al. 2002a, b, Jégu 2004, Jégu and Keith 2005, Mol 2012b, Andrade et al. 2013, 2016a, b).

FEEDING ECOLOGY Diverse diets and feeding behaviors, including predators, scavengers, herbivores, and omnivores. Many species are opportunistic feeders, and a few species are relatively specialized. The dentition and intestine length of adult serrasalmids are closely correlated with diets. For instance, the intestine is short (<1.5 in SL) in highly carnivorous species, but relatively long in the more herbivorous taxa (Leite and Jégu 1990, Nico 1991). As with most fishes, the very young of nearly all serrasalmids feed heavily on small invertebrates (Nico and Taphorn 1988, Jégu et al. 1989, Nico 1991, Araujo-Lima and Goulding 1997).

Pygocentrus piranhas and some of the larger *Serrasalmus* are the most carnivorous serrasalmids, usually biting chunks of flesh from large fish, sometimes from other vertebrates, and also frequently taking bits of fin or consuming small fish whole (Nico and Taphorn 1988, Sazima and Machado 1990). Scavenging by piranhas on dead animals is also normal behavior (Sazima and Machado 1990). Many young and some adult *Serrasalmus* regularly clip off and consume pieces of fins; *Serrasalmus elongatus* feeds heavily on both fins and scales, and *Catoprion mento* is a scale-eating specialist (Roberts 1970, Goulding 1980, Nico and Taphorn 1988, Sazima 1988). Although many piranhas are highly carnivorous, the adults of most *Pristobrycon* and certain *Serrasalmus* species consume considerable numbers of seeds. These piranhas are considered seed predators because they masticate all or most seeds before swallowing (Goulding 1980, Nico 1991).

Species in at least 11 of the 16 serrasalmid genera are primarily herbivorous. These species feed most heavily on fruits, seeds, leaves, and occasionally flowers, with the types and amounts of exploited plant matter varying by fish species, habitat, season, and water level (Goulding 1980, Nico 1991, Araujo-Lima and Goulding 1997). On occasion, some opportunistically take animal prey (Goulding 1980). The smaller herbivorous serrasalmids, such as *Mylossoma*, eat mostly smaller seeds, whereas larger species, especially *Colossoma* and *Piaractus*, eat mostly larger fruits and seeds (Goulding 1980). Field and laboratory studies indicate the smaller or younger fish tend to destroy all or most seeds ingested, using their teeth to crush or masticate the food item before it is swallowed. In contrast, older and larger-sized individuals, especially *Colossoma* and *Piaractus*, frequently eat only the fleshy parts of a fruit while swallowing and defecating large numbers of seeds whole, thereby functioning as highly effective seed dispersers for a wide variety of floodplain tree and liana species (Goulding 1980, 1983b, Kubitzki and Ziburski 1994, Galetti et al. 2008, Anderson et al. 2009, Correa et al. 2015a).

Among the herbivorous members of the family, many of the rheophilic species (represented by 6 genera) have evolved a unique and intimate relationship with species in the riverweed family (Podostemaceae), a largely tropical group of aquatic herbaceous angiosperms specialized to grow in rocky rapids and waterfalls. Juvenile rheophilic serrasalmids forage on small aquatic animals that inhabit mats of these submerged plants, with their diets shifting as they grow into feeding more heavily, some almost exclusively, on the Podostemaceae plants themselves, clipping and consuming the leaves, flowers, and seeds (Jégu et al. 1989, 2002a, Andrade et al. 2013, Pereira and Castro 2014).

BEHAVIOR Both carnivorous and herbivorous serrasalmids are drawn to noises and splashing. Piranhas respond to disturbances because of the possibility of encountering disadvantaged prey; in contrast, vegetarian species are attracted to sounds corresponding to dropping fruits. Subsistence fishers are well aware of this behavior and often strike the water or the side of their dugout canoes with their rods or paddles so as to imitate the sounds of fruits or nuts hitting water (Goulding 1980, Sazima and Machado 1990, Araujo-Lima and Goulding 1997; LG Nico and MC Andrade pers. obs.).

Although a few serrasalmids forage as solitary individuals, most species are relatively social and usually occur in small groups or shoals, and a few form large schools (Géry 1972a, Nico and Taphorn 1986, 1988, Sazima and Machado 1990). Some young fin-eating piranhas shoal with characins of similar size and color, an apparent form of aggressive mimicry that allows attack from close range (Nico and Taphorn 1988, Sazima and Machado 1990). Certain vegetarian serrasalmids join large aggregates of single or mixed species while making long riverine migrations to reach spawning or feeding grounds. Such riverine migrations are well documented for *Colossoma*, *Piaractus*, and *Mylossoma*, and the behavior has also been reported for

certain species within the genera *Myleus*, *Myloplus*, *Tometes*, and *Acnodon* (Lowe-McConnell 1964, Goulding 1980, 1983a, Novoa 1989, Araujo-Lima and Goulding 1997, Carolsfeld et al. 2003, Agostinho et al. 2007, Mol 2012b). Piranha are not known to make long migrations in rivers, but many piranhas and other serrasalmids usually move between river channels and adjacent floodplain habitats, such lateral migrations apparently in response to changes in water levels and food resources (Goulding 1980, Taphorn 1992, Araujo-Lima and Goulding 1997).

ADDITIONAL NOTES Large pacus like *Piaractus* and *Colossoma* are economically important to commercial fisheries and aquaculture (Araujo-Lima and Goulding 1997, Roubach et al. 2003). Piranhas are generally less valued, although they are commonly consumed by subsistence fishers and frequently sold for food in local markets. Various serrasalmids have long been popular in the aquarium fish trade industry, with usually available ornamentals being *Metynnis*, *Myloplus*, and *Mylossoma*. Piranhas also occasionally appear in the ornamental fish trade, although they are prohibited in many parts of the world due to fear of introduction into the wild. In South America, piranhas occasionally bite bathers and swimmers, but serious injuries are rare and the threat to humans largely exaggerated (Goulding 1980, Nico and Taphorn 1986, Sazima and Andrade-Guimarães 1987, Haddad Jr. and Sazima 2003, Mol 2006). However, piranhas are a costly nuisance to commercial, subsistence, and sport fishers because they steal bait, mutilate catch, damage nets and other gear, and may bite when handled (Goulding 1980, Nico and Taphorn 1986, Agostinho et al. 1997).

Those serrasalmids that seasonally migrate are seriously affected where dams prevent fish from reaching critical spawning and feeding grounds (Agostinho et al. 2008, Barletta et al. 2016). The rheophilic serrasalmids, particularly populations endemic to just a few tributaries, are especially vulnerable and face possible elimination if their free-flowing, high-gradient streams are replaced or disrupted by dams and reservoirs (Jégu and Keith 2005, Andrade et al. 2015). In contrast, some of the more common but less desirable carnivorous piranhas often invade and flourish in artificial lakes (Mol et al. 2007a, Trindade and Jucá-Chagas 2008).

GUIDANCE FOR HANDLING AND EXAMINING SERRASALMID SPECIMENS

Because many serrasalmids are large and have jaws and teeth adapted for crushing hard nuts or for clipping out large chunks of flesh, handling of live specimens can be dangerous. Serious pacu bites can be avoided by simply not placing fingers in the mouth of the fish. Much greater attention must be devoted to handling piranhas. Indeed, many a careless fisherman and field biologist, even persons experienced with piranhas, have been bitten at least once. Often, bites occur during the process of removing live fish from nets or hooks. In the confines of a dugout canoe there is added risk, especially if toes are exposed, because landed piranhas have the irksome habit of flopping about while simultaneously snapping their jaws and teeth together.

Jaw teeth

Tooth shape, relative size, number, and spatial arrangement vary widely within the family and consequently are important traits for distinguishing different genera. However, the teeth of many serrasalmids, including those of piranhas, are often partially hidden by lips and gum tissues so some manipulation of the mouth or cutting may be necessary. When dealing with live piranhas, it is generally best to sedate or sacrifice the specimen. If the desire is to preserve the fish first and then examine the teeth later in the laboratory, it is recommended the fish be preserved with mouth fully open (such as by placing a wood wedge in the mouth). Otherwise, the specimen will preserve with jaws locked shut or only slightly open. In such cases, any follow-up attempt to view teeth may then require cutting of jaw muscles and some prying, techniques that may damage both jaws and teeth.

Ectopterygoid teeth

The presence and shape of these small teeth are important in distinguishing between certain piranha genera. If present, the teeth are in single rows on the ectopterygoids, a pair of small elongate bones which form part of the complex of bones on the inner roof of the mouth. The ectopterygoids roughly parallel the premaxillary bones. In early literature on piranhas, the ectopterygoid teeth are commonly referred to, incorrectly, as palatine teeth. To verify presence of these teeth, it is usually necessary to widely open a specimen's mouth. If the ectopterygoid teeth are not clearly visible, a probe or finger can be passed along the roof of the mouth to confirm their presence. If absent, the roof of the mouth is noticeably smooth to touch.

Predorsal spine

This structure, sometimes referred to as the predorsal-fin spine, is an anteriorly directed spine just before the dorsal fin, and represents an extension of the first pterygiophore bone. It is immovable and continuous with the first pterygiophore (Machado-Allison 1982b). All serrasalmids have a predorsal spine except *Mylossoma*, *Colossoma*, and *Piaractus*. The size and shape of the predorsal spine vary among taxa, and it may be prominent and easily observable or slightly hidden beneath the skin. Presence of a predorsal spine can often be confirmed with a fingertip, feeling for the sharp point of the spine. The structure is very evident in radiographs. Serrasalmids taken in gill nets are commonly snagged by their predorsal spine.

Preanal spine

This spine is a small bony element situated posterior to the anal opening at the anal-fin origin. If not detected visually, it can usually be felt by a finger as a sharp process. In some specimens the structure may be difficult to find but its presence/absence can be confirmed by radiograph (Fink and Machado-Allison 1992). A preanal spine is present in most piranhas, including all *Pygocentrus* and *Serrasalmus* species, and three of the six *Pristobrycon* species, including *P. aureus*, *P. calmoni*, and *P. eigenmanni* (Machado-Allison and Fink 1995, Jégu and Santos 2001). The structure is absent in *Pygopristis denticulata* and in *Pristobrycon careospinus*, *P. maculipinnis*, and *P. striolatus* (Machado-Allison and Fink 1995).

Abdominal serrae

All serrasalmids possess a series of hard, sharp-pointed, bony serrae or scutes along the midventral portion of their abdomen. In the literature, the abdominal serrae of serrasalmids are sometimes referred to as ventral spines, scutes, or belly serrae. The number, distribution, prominence, shape, and size of these serrae vary among taxa and even within certain species. All serrasalmids have abdominal serrae between the anal opening forward to near the pelvic-fin insertion, and in many species these serrae extend to near the isthmus or throat of the fish. In many serrasalmids, the abdominal serrae form a sharp keel and are clearly visible to the naked eye. In other serrasalmids, the serrae may be relatively small and less obvious. With small juveniles, it is usually possible to determine if abdominal serrae are present by touch, running a fingertip along the ventral margin. Although the total number of abdominal serrae is often reported, positive identification may require separately counting abdominal serrae anterior to the pelvic-fin insertion point (i.e., prepelvic serrae) versus postpelvic serrae. The number of serrae with double spines, located near the anal opening, is also sometimes reported. In those serrasalmids that have few or no prepelvic serrae, the prepelvic abdomen is typically smooth and rounded or flattened.

Fins

Although fin shape and size are useful traits to separate some genera and species, wild-caught fish are often missing parts of their fins due to predators, often fin-nipping piranhas. Specimens and their fins are also often easily damaged during capture and handling, during preservation in buckets and barrels, and during transport to museums.

Adipose fin

The length of the adipose-fin base is commonly reported and the measurement is important in separating *Metynnis* from other serrasalmid genera. How to precisely measure the base length of this fin is sometimes challenging because in some species or specimens the hyaline portion of the fin is continuous with, or sits atop, a fleshy ridge. In the current work, only the hyaline portion is measured.

KEY TO THE GENERA ILLUSTRATIONS IN KEY BY LEO NICO

NOTICE This key is primarily for identification of adult specimens, although some characters are evident in juveniles. As far as possible, the more obvious external features have been used in separating taxa. In a number of cases, however, it has been necessary to use characters that are less easily discerned.

1a. Predorsal spine absent (fig. 1a) **2**
1b. Predorsal spine present, indicated by the forward-directed spine at base of dorsal fin (fig. 1b).......................**4**

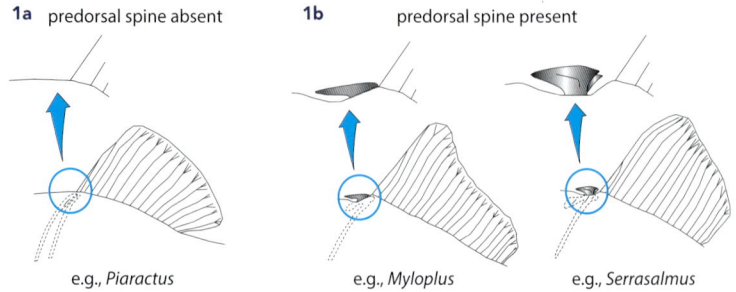

1a predorsal spine absent   **1b** predorsal spine present

e.g., *Piaractus*   e.g., *Myloplus*   e.g., *Serrasalmus*

2a. Body very deep, typically rounded in lateral profile; anal fin with 28–36 branched rays; anal fin with rounded or pointed distal margin, with longest rays within central portion of fin (fig. 2a); anal fin densely covered, over basal half or more, with minute scales (>25 rows) (fig. 2a) ......**Mylossoma**
2b. Body moderately deep and more elongate, elliptical in lateral profile; anal fin with 20–24 branched rays; anal-fin distal margin slightly concave in juveniles and nearly straight in adults, with rays near anterior portion of fin the longest (fig. 2b); anal fin scaled only near base (fig. 2b) **3**

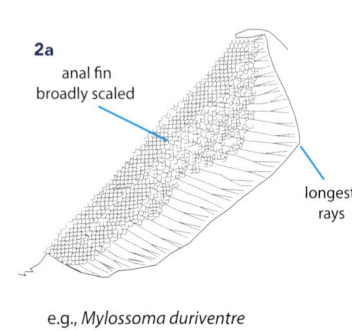

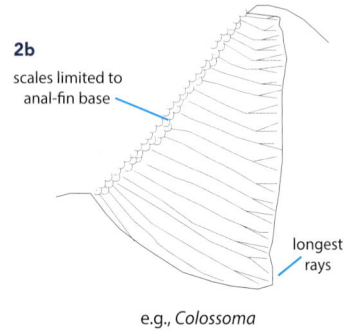

2a anal fin broadly scaled

2b scales limited to anal-fin base

longest rays

longest rays

e.g., *Mylossoma duriventre*   e.g., *Colossoma*

3a. Outer and inner rows of front upper jaw teeth in contact (fig. 3a); maxillary bone (corner of mouth) without teeth; adipose fin has pronounced and ossified rays in individuals >5.5 cm standard length (SL); opercle bone elongate, its maximum horizontal width is 48–57% of bony postorbital distance (posterior edge of orbit to posterior edge of bony opercle, see fig. 3c); number of gill rakers on first branchial arch generally >100 in specimens >15 cm SL..............................**Colossoma***

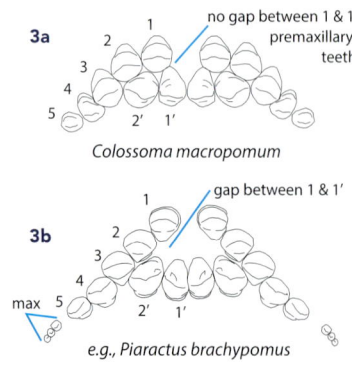

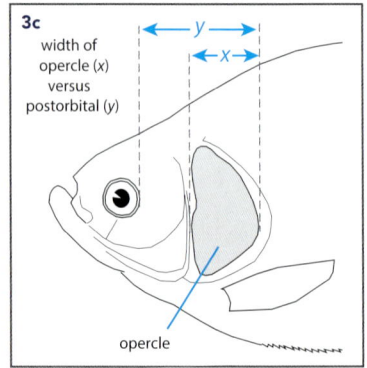

3a no gap between 1 & 1' premaxillary teeth

Colossoma macropomum

3b gap between 1 & 1'

max

e.g., *Piaractus brachypomus*

3c width of opercle (x) versus postorbital (y)

opercle

3b. Outer and inner rows of front upper-jaw teeth separated by a medial space or "hiatus" (fig. 3b); 1–3 small teeth on maxillary (max, fig. 3b); adipose fin without pronounced rays (faint, ray-like structures present in some specimens); opercle bone moderately elongate, its maximum horizontal width is 40–46% of bony postorbital distance; <40 gill rakers on first branchial arch ....**Piaractus***

* Hybrid *Colossoma* × *Piaractus*: the majority are artificially cultivated, and may display intermediate characters.

4a. Lower jaw very prominent (but narrow and weak), projects far forward of upper jaw (prognathous) (fig. 4a); mouth strongly slanting upward (oblique) when closed; jaw teeth widely separated, not touching, with anterior teeth projecting forward, external of mouth (fig. 4aa); general tooth shape mammiliform (nipple-like) or thorn-like, consisting of a central somewhat conical projection atop a lower, broad base (often hidden by lip and gum tissues) (fig. 4aaa); maximum size less than about 15 cm SL ....*Catoprion*

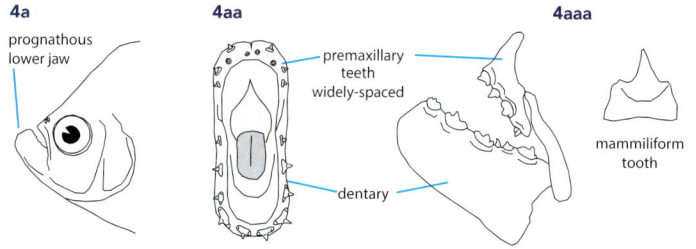

4a prognathous lower jaw

4aa premaxillary teeth widely-spaced
dentary

4aaa mammiliform tooth

4b. Lower jaw ranging from very short and undershot to moderately prominent in that it only slightly extends forward of upper jaw; closed-mouth does not slant upward (except in *Tometes lebaili* whose mouth is obliquely directed upwards); in most taxa the jaw teeth are not widely spaced and in the majority of taxa none of the teeth projects external of closed mouth; teeth not mammiliform; adults of many species much larger than 15 cm SL **5**

5a. Adipose fin very long, substantially longer than high, with length of base equal or longer than the distance between posterior end (insertion) of dorsal-fin base to adipose-fin origin (fig. 5a)...............................*Metynnis*

5b. Adipose fin short, length of base shorter, typically much shorter, than the distance between dorsal-fin insertion to adipose-fin origin (fig. 5b)...........................**6**

5a y x $x > y$
e.g., *Metynnis guaporensis*

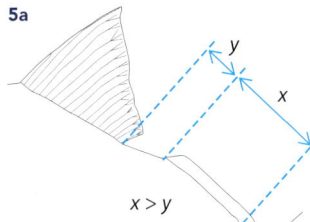

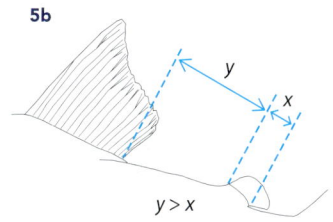

5b y x $y > x$
e.g., *Serrasalmus gouldingi*

6a. Two rows of teeth on upper jaw (fig. 6a); teeth shape best described as molariform, incisiform, or spatulate, or some combination of these ....**7** (*Myleus* group and relatives)

6b. One row of teeth on upper jaw (fig. 6b), although some piranhas have small teeth on inner roof of mouth on ectopterygoid bone; teeth incisor-like (thin and blade-like), with pointed cusps, adjacent teeth closely set and interlocking to form a continuous, saw-like, sharp cutting edge adapted for rapid puncture and shearing ....**13** (true or traditional piranhas)

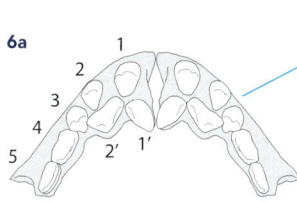

6a two rows versus single row of teeth on premaxillaries
e.g., *Utiaritichthys esguiceroi*

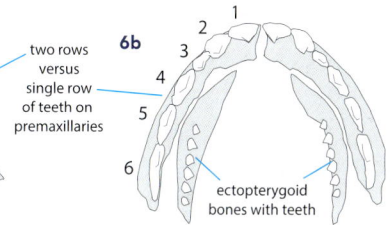

6b ectopterygoid bones with teeth
e.g., *Serrasalmus rhombeus*

7a. Abdomen flattened and always lacking prepelvic abdominal serrae (fig. 7a); lateral profile of body elongate and diamond-shaped; dorsal fin with 15–18 branched rays; flanks of adults with many vertical gray bars that taper from upper to lower flanks; adult males have a bilobed anal fin, but the secondary or posterior lobe is relatively small and rounded, and situated near posterior portion of fin at unbranched rays 20–25 (fig. 7a); dorsal-fin ray without long filaments, although first few rays may be elongate (fig. 7a) ......*Acnodon*

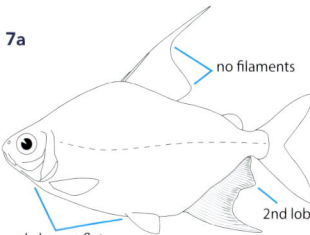

7a no filaments
2nd lobe
abdomen flat, no prepelvic serrae,
e.g., *Acnodon normani* - male

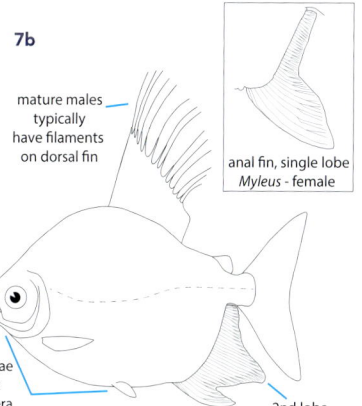

7b mature males typically have filaments on dorsal fin
anal fin, single lobe *Myleus* - female
abdomen with keel or rounded, & with prepelvic serrae present in most of the 6 genera
2nd lobe
e.g., *Myleus setiger* - male

7b. Abdomen rounded or with ventral keel, with or without abdominal serrae (fig. 7b); body lateral profile round or oval; dorsal fin with ≥18 branched rays; no vertical gray bars on flanks; adult males typically have a well-developed bilobed anal fin, the secondary lobe situated on central part of fin from unbranched rays 14–18 (fig. 7b); dorsal-fin rays of mature males with long filaments (fig. 7b) **8**

8a. Outer and inner rows of upper-jaw teeth abutting, not separated by a space (fig. 8a); jaw teeth mainly adapted for clipping or grazing, their general form incisiform (i.e., anteroposteriorly flattened and with relatively thin base)...........................**9**
(to the 4 genera that make up the *Myleus* Clade of rheophilic fishes)

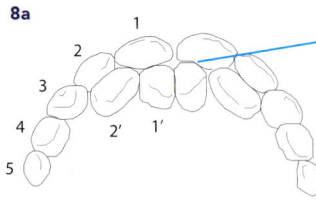

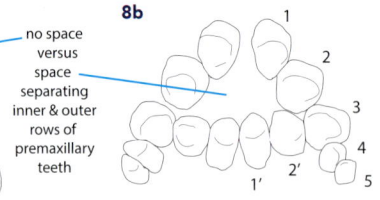

e.g., *Myleus setiger*

e.g, *Myloplus rhomboidalis*

8b. Outer row of upper-jaw teeth separated from inner row by a space (fig. 8b); jaw teeth generally adapted for crushing, basic shape is molariform, those on lower jaw (dentary) characterized by a more rounded base or by being anteroposteriorly wide **12**

9a. All premaxillary teeth on outer row of upper jaw approximately the same size and shape (fig. 9a); last 3 teeth on outer row of upper jaw aligned and their crown ridges form a nearly continuous cutting edge; jaw teeth relatively fragile, weakly attached to jaws; incisiform jaw teeth markedly slender, very flattened anteroposteriorly **10**

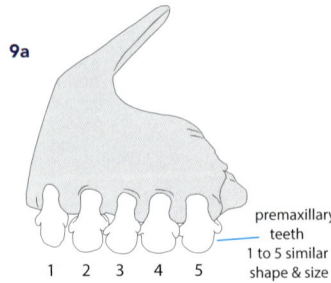

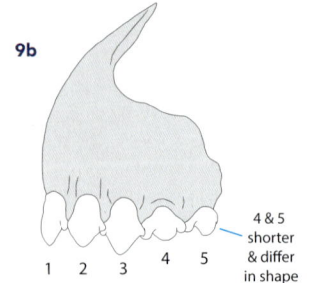

9b. Two posteriormost teeth (4 and 5) on each side of outer row of upper jaw shorter than the other premaxillary teeth (fig. 9b); jaw teeth robust, firmly attached to jaws; last two teeth in outer row of upper jaw with crown ridges that form an S-shaped cutting edge; incisiform teeth only somewhat flattened anteroposteriorly **11**

10a. Mouth ventral (in individuals >50 mm SL) (fig. 10a); snout strongly rounded; only 4 teeth on each side of lower jaw .... *Ossubtus*

10b. Mouth terminal to subterminal (fig. 10b); snout tapered; 7 or more teeth on each side of lower jaw................... *Mylesinus*

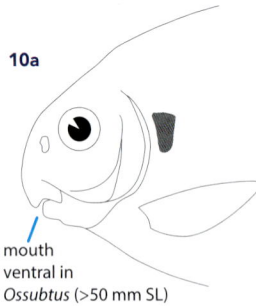

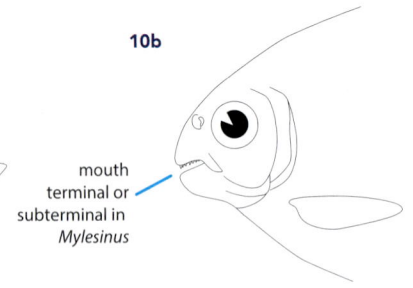

Ossubtus xinguense

e.g., *Mylesinus paraschomburgkii*

11a. Two middle front teeth on outer row of upper jaw close together, in contact or nearly so (fig. 11a); outer-row teeth on each side of upper jaw form a continuous, unbroken row, with no gaps between adjacent teeth; inner row of teeth on lower jaw consists of a single pair of very small conical teeth (always present) close to the symphysis, both are generally hidden from view because they are lower than crowns of median teeth of the outer row..*Myleus*

11b. The two middle frontal teeth on outer row of upper jaw are separated from each other by a broad space, the space greater than tooth width (fig. 11b); in some species, there is also a spatial gap separating upper-jaw tooth numbers 1 and 2 (fig. 11bb); inner row of teeth on lower jaw consists of a single pair of large conical teeth (absent in some specimens of *T. makue*), close to the symphysis, and whose tips are clearly visible, not hidden by crowns of the two median teeth of the outer row ....*Tometes*

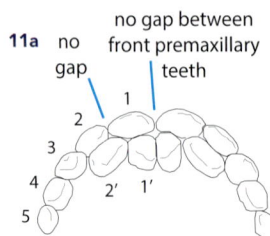

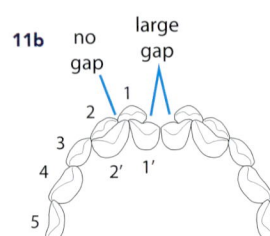

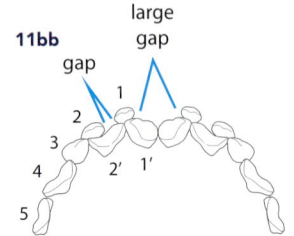

12a. Body lateral profile ranges from rounded to moderately elongate; body depth usually >50% SL (fig. 12a).......................*Myloplus*

12b. Body lateral profile elliptical, very elongate; body depth usually <50% SL (fig. 12b) .... *Utiaritichthys*

13a. Teeth on jaws with 5 or more cusps, the middle cusp typically only slightly larger than the other cusps (fig. 13a); preanal spine always absent .... *Pygopristis*

13b. Teeth on jaws typically with 2 or 3 cusps, the more central cusp substantially larger than others and their shape is broadly triangular, especially those more forward in mouth (fig. 13b), preanal spine present in most taxa.............................**14**

14a. Dorsal profile between mouth and dorsal-fin origin moderately to strongly convex (fig. 14a); adults are typically large, with blunt snouts and powerful, rather massive jaws; head is notably large and robust, and broader (width usually >15% of SL in individuals >10 cm SL) than in other piranhas at similar body sizes; ectopterygoid teeth absent (i.e., no teeth on inner roof of mouth), except in small juveniles (1 cm SL or smaller).............................*Pygocentrus*

14b. Dorsal profile between mouth and dorsal-fin origin with slight concavity (fig. 14b-bb); adults of this group include small, moderately sized, and large piranhas, with snout shape among these taxa ranging from very pointed to blunt; head typically moderately slim and narrow (width ranging from about 8% to about 14% of SL in big individuals of the largest species); ectopterygoid teeth generally present, although commonly reduced in number or even absent in larger specimens **15**

15a. Adults and most juveniles characterized by relatively sharp or pointed snouts, the lateral profile in front of nostrils somewhat rounded to pointed (fig. 15a)..*Serrasalmus* (in part)

15b. Adults and some juveniles characterized by relatively short, blunt snouts, the lateral profile in front of nostrils markedly or somewhat squarish relative to upper jaw (fig. 15b).. **16**

12a *Myloplus*: BD usually >50% SL

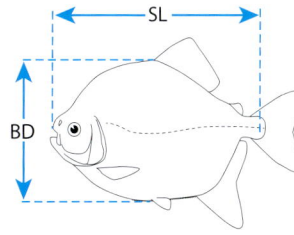

12b *Utiaritichthys*: BD usually <50% SL

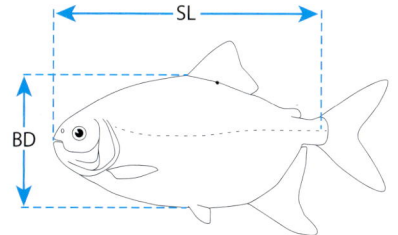

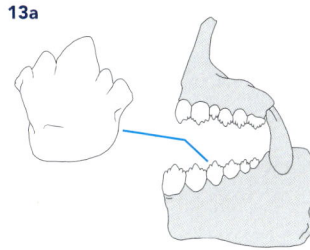

13a

13b

Pygopristis denticulatus

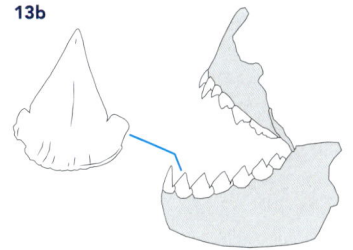

e.g., *Serrasalmus rhombeus*

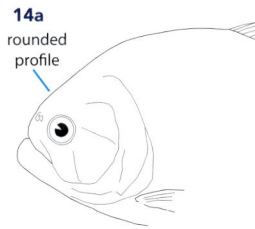

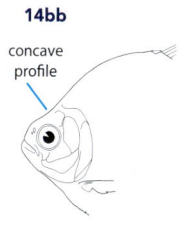

14a rounded profile

14b concave profile

14bb concave profile

e.g., *Pygocentrus nattereri*

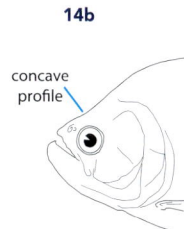

e.g., *Serrasalmus rhombeus*

e.g., *Pristobrycon calmoni*

15a snout sharp or pointed

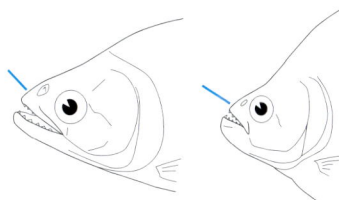

15b snout short & blunt

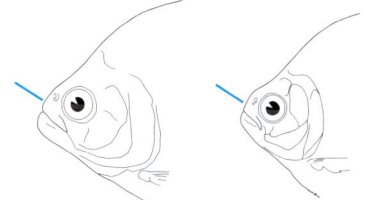

e.g., *Serrasalmus elongatus*   e.g., *Serrasalmus altuvei*

e.g., *Serrasalmus gouldingi*   e.g., *Pristobrycon calmoni*

16a. Preanal spine always present (fig. 16a) **17**

16b. Preanal spine absent (fig. 16b) ... **Pristobrycon**
(in part, includes 3 of 6 members of the genus: *P. careospinus*, *P. maculipinnis*, and *P. striolatus*)

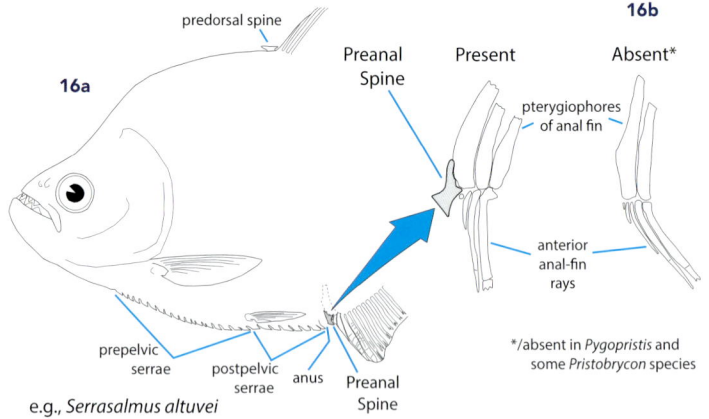

17a. Piranhas of small to moderate size (generally less than about 18 cm SL); adult body typically not very stout, head and body not especially broad; adult body depth about 55–68% of SL; ectopterygoid teeth absent or present, but when present generally few in number, shape blunt or squarish, not similar to jaw teeth (fig. 17a); cheek of adult specimens not extensively armored with bone, the third infraorbital bone relatively undeveloped so there is a relatively large naked area separating the third infraorbital and the preopercle bone (fig. 17aa).................................. **Pristobrycon**
(in part: *P. calmoni*, *P. aureus*, and *P. eigenmanni*)

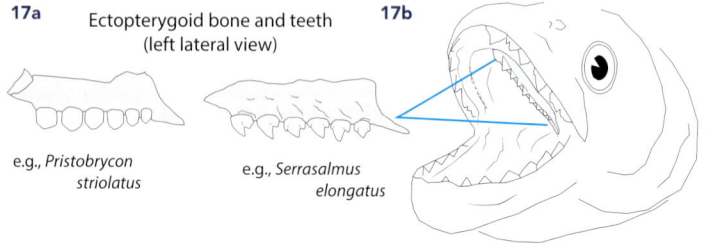

17b. Piranhas ranging from small to large-bodied (many species surpass 18 cm SL and some reach 40 cm SL or more), adult body typically stout, with relatively broad head and body; adult body depth generally <60% SL; ectopterygoid teeth, when present, strongly triangular, smaller but similar in shape to the jaw teeth (fig. 17b); cheeks of adults typically well armored, the third infraorbital bone (IO3) is relatively large and, consequently, the naked area separating the IO3 from the preopercle bone is small, or even absent in largest specimens (fig. 17bb) .... **Serrasalmus** (in part)

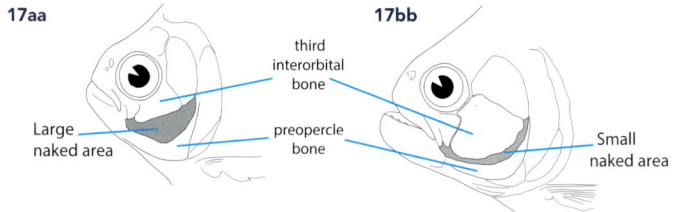

16a

predorsal spine

16b

Preanal Spine   Present   Absent*

pterygiophores of anal fin

anterior anal-fin rays

prepelvic serrae   postpelvic serrae   anus   Preanal Spine

e.g., *Serrasalmus altuvei*

*/absent in *Pygopristis* and some *Pristobrycon* species

17a   Ectopterygoid bone and teeth (left lateral view)   **17b**

e.g., *Pristobrycon striolatus*   e.g., *Serrasalmus elongatus*

17aa   **17bb**

third interorbital bone

Large naked area   preopercle bone   Small naked area

GENUS ACCOUNTS

Acnodon (13–20 cm SL)

Characterized by: silvery, rhomboidal-shaped body of medium size; flanks of adults with many vertical, vermiculate gray bars; distinctly forked caudal fin; a moderately long adipose fin; preanal spine present; mouth of adult subterminal (1 species) or terminal (2 species); abdominal area in front of pelvic fins relatively flat and without spines (i.e., no prepelvic serrae); abdominal postpelvic serrae includes 5–9 spines (i.e., between pelvic-fin origin and anus); upper jaw with 5 (sometimes 4) teeth per side on outer row and 1 (rarely 2) teeth per side on inner row; lower jaw without small pair of conical teeth behind the front mandibular teeth (different from many other *Myleus*-like serrasalmids). Distinguished from larger-bodied

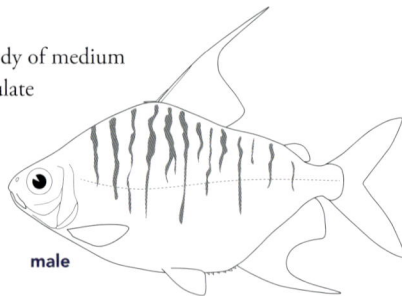

male

curimatids in the presence of oral teeth (vs. absent in curimatids). Sexual dimorphism: adult males have a bilobed anal fin; the secondary lobe is relatively small and rounded and situated near posterior portion of fin at unbranched rays 20–25 (fig. 7a). Juveniles: young *A. oligacanthus* have a conspicuous eyespot (ocellus) centered on lateral line just above pectoral fin. The lateral body profile of smaller juveniles is highly elongate but it becomes progressively more deep-bodied with increased age (body depth to SL ranging from 28% at 2.5 cm long to about 50% at 18 cm long) (Gosline 1951, Géry 1972a, Jégu and Santos 1990). **SPECIES** Three; *A. oligacanthus* (type species), *A. normani*, and *A. senai* (Jégu and Santos 1990, Jégu 2003). **COMMON NAMES** *Pacu branquinha* (Brazil); Sheep-pacu (English). **DISTRIBUTION AND HABITAT** Two species from the Amazon basin: *A. normani* (Tocantins, Xingu basins) and *A. senai* (Jari basin); *A. oligacanthus* in northern Guiana Shield drainages (Jégu and Santos 1990, Jégu 1992b, Mol 2012b). Rheophilic, inhabiting rivers with moderate to strong currents, usually in or around rocky rapids (Planquette et al. 1996, Mol 2012b). Adult *A. oligacanthus* migrate from deeper waters to small tributaries to reproduce during the late rainy season or early dry season; juveniles congregate in sandy, shallow waters, either in near-shore sites or in tributary streams (Géry 1972a, Planquette et al. 1996, Mol 2012b). **BIOLOGY** Primarily herbivorous, feeding heavily on vascular plant material, mainly fruits, seeds, flowers, and leaves, including parts of aquatic plants of the family Podostemaceae (Leite and Jégu 1990, Planquette et al. 1996, Mol 2012b). *A. normani* consume vascular plant matter, insects, shrimp, gastropods, and fish scales (Leite and Jégu 1990, Pereira et al. 2007b). Geographically restricted distributions and dependence on high-gradient streams make these fishes especially vulnerable to the effects of dams.

Catoprion (10–15 cm SL)

Characterized by: small adult body sizes; highly compressed head and body, discoid or deep body shape; silvery color; cheek usually marked with reddish orange; preanal spine present. Further distinguished from other serrasalmids by: prognathous lower jaw extending far anterior to upper jaw when mouth is closed (fig. 4a); lower jaw highly evertible, easily opened to a vertical position with slight pressure; teeth widely separated (not touching) in both jaws (fig. 4aa); most teeth with a rounded base and narrower tip, usually described as mammiliform or tuberculate (i.e., nipple or thorn-like); lower jaw with total of 12 teeth (6 per side); upper jaw with a total of 10 teeth (5 per side) consisting of 4 large forward-directed (antrorse) teeth on "outer" row and 6 much smaller teeth forming an "inner" row (10 premaxillary teeth described as "a single, irregular row"; see fig. 4); large oral valve or skin-flap inside mouth immediately behind arch of lower-jaw teeth (generally less prominent in other serrasalmids) adipose-fin base moderately long; pre- and postventral abdominal serrae present, with about 32–34 total spines. Sexual dimorphism: not reported. Both males and females may have long filaments extending from the first rays of the dorsal and anal fins. Juveniles: slightly less deep-bodied but otherwise closely resembles adults (Eigenmann 1915b, Gosline 1951, Géry 1972a, Taphorn 1992). **SPECIES** One; *C. mento*. Closer examination of widely separated geographic populations may reveal subspecific differences (Taphorn 1992). **COMMON NAMES** *Catirini, Piranha mucura, Piranha queixuda* (Brazil); *Palometa caribe* (Venezuela); Whipple piranha (English). **DISTRIBUTION AND HABITAT** Widely distributed in AOG region and upper Paraguay basin. Common in savannas where it inhabits pools, marshes, small creeks and streams, and artificial ponds (Lowe-McConnell 1964, Sazima 1988, Nico 1991, Maldonado-Ocampo and Prada-Pedreros 1999). Also known from reservoirs, floodplain lakes, and lagoons, and some forested regions (Ferreira 1984, Siqueira-Souza and Freitas 2004; LG Nico pers. obs.). Typically associated with black- and

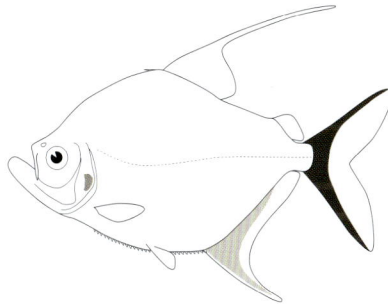

clearwaters, primarily in sites with abundant aquatic vegetation (Nico 1991, Taphorn 1992).
BIOLOGY One of only a few characoids (including *Roeboides*) that feed predominantly on scales
(i.e., lepidophagous), although occasionally consuming small fish, crustaceans, insects, and,
perhaps inadvertently, plant matter (Roberts 1970, Vieira and Gery 1979, Sazima 1983, Nico
and Taphorn 1988, Nico 1991). Generally solitary, commonly stalking and ambushing prey
using plants as cover. Typical attack is a short chase culminating in a high-speed, open-mouth
collision, whereby numerous scales are knocked loose and the free-floating scales then consumed
(Sazima 1983, 1988, Janovetz 2005). Territorial, responding agonistically to conspecifics that
approach its clump of vegetation (Sazima 1988).

Colossoma (60–110 cm SL)

The largest serrasalmid and the second-largest
scaled, freshwater fish in South America,
exceeded in size only by *Arapaima gigas*.
Characterized by: rather elongate body shape
as adults; absence of a predorsal spine (shared
with *Piaractus* and *Mylossoma*, distinguishing
these three genera from all other serrasalmids,
fig. 1); molariform and multicuspid teeth in oral jaws as subadults and adults; adults with
distinctive countershading, ventral area markedly darker than the dorsal, although colors vary
with water type. Distinguished from *Piaractus* by: (1) outer and inner rows of front upper jaw
(premaxillary) teeth in contact (vs. separated by medial space or "hiatus" in *Piaractus*; fig. 3a-b);
(2) maxilla lacking teeth (vs. 1–3 small teeth in *Piaractus*; fig. 3b); (3) pronounced, ossified rays
in the adipose fin of individuals larger than about 5.5 cm SL (vs. adipose not rayed in *Piaractus*);
(4) opercle bone elongate, maximum horizontal width 48–57% bony postorbital distance
(vs. moderately elongate, maximum horizontal width 40–46% bony postorbital distance in
Piaractus); and (5) gill rakers on first branchial arch generally >100 in specimens >15 cm SL
(vs. <40 in *Piaractus*). Sexual dimorphism: females slightly larger than males (Loubens and
Panfili 1997, Villacorta-Correa and Saint-Paul 1999). Juveniles: individuals less than about
10 cm SL have a round eye-spot (ocellus) on midbody below the dorsal fin similar to juvenile
Piaractus and *Mylossoma*. Young are rhomboidal in shape, gradually changing to elongate by
about 30 cm TL. Hybrids of *C. macropomum* and *Piaractus* species are common, produced
by the aquaculture industry (Saint-Paul 1992, Hashimoto et al. 2011), and are occasionally
observed in the wild in the Amazon (Araujo-Lima and Goulding 1997). These hybrids display
intermediate morphological characteristics of the parent species (Goulding 1981, Machado-
Allison 1982a, b, 1983, 1986, 1987, Araujo-Lima and Goulding 1997). **SPECIES** One, *Colossoma
macropomum*. **COMMON NAMES** *Curumim, Tambaqui* (Brazil), *Gamitana* (Peru), *Ruelo, Cachama*
(Venezuela); Black pacu (English). **DISTRIBUTION AND HABITAT** Naturally widespread in the
Amazon and Orinoco basins where it coexists with *Piaractus brachypomus*; its presence in
other drainages is the result of introductions, generally either from stocking or escapes from
aquaculture (Araujo-Lima and Goulding 1997, Loubens and Panfili 1997, Mojica et al. 2002,
Carolsfeld et al. 2003, Ortega et al. 2007). Miocene fossil record indicates *C. macropomum*
was present in the Magdalena River of Colombia during the Miocene Epoch (Lundberg et al.
1986); current reproducing populations in that drainage are due to recent human introductions
(Mojica et al. 2002). Typically inhabits river and stream channels in lowland areas and associated
floodplain lakes, and tends to be closely associated with whitewater (muddy) rivers, although
it also occurs in deeper sections of clearwater rivers. Schools often take refuge in woody areas
(Goulding 1981). *Colossoma* typically disperse into flooded areas during high-water periods,
then congregate in floodplain lakes and rivers during the low-water season (Loubens and Panfili
1997). **BIOLOGY** Adult *C. macropomum* feed extensively on fleshy fruits and seeds, using their
heavy molar-like dentition and strong jaws to crush even some of the hardest nuts (Araujo-Lima

and Goulding 1997). *Colossoma* swallow many fruits and seeds whole, and serve as an important dispersal vector for many floodplain tree species (Goulding 1980, 1983b, Kubitzki and Ziburski 1994, Galetti et al. 2008, Anderson et al. 2009). A single 10 kg *C. macropomum* feeding in the flooded forest can contain >1 kg of seeds in its gut (Goulding 1988). *Colossoma* is one of several herbivorous serrasalmids known to migrate long distances for spawning and feeding purposes (Araujo-Lima and Goulding 1997, Carolsfeld et al. 2003). In the wild, *C. macropomum* mature at about 60 cm SL (Loubens and Panfili 1997), and many individuals live to 40 years, with reported maximum longevity of about 65 years (Loubens and Panfili 1997). The largest *C. macropomum* on record, from Brazil's Japurá River system, measured 1.1 m SL and weighed 44 kg (Rapp Py-Daniel et al. 2015). Because of its large size and desirability as food for humans, *C. macropomum* has long been one of the most important commercial and subsistence fish resources in the Amazon and Orinoco regions (Novoa 1989, Araujo-Lima and Goulding 1997). Unfortunately, over-exploitation and loss of natural floodplain habitats have led to declines in many wild populations, including reduction in total numbers and substantially fewer large-sized individuals (Araujo-Lima and Goulding 1997, Reinert and Winter 2002). Artificially produced *C. macropomum* have been widely stocked in reservoirs and other waters to replenish diminishing native populations and to introduce the species to new areas (Saint-Paul 1992).

Metynnis (8–15 cm SL)

Characterized by: silvery coloration; small to moderate adult body sizes, adults typically much smaller than other serrasalmids; very long adipose-fin base, equal to or longer than distance between dorsal-fin insertion and adipose-fin origin (fig. 5a); and a relatively small mouth. *Metynnis* includes species with a body profile that is very deep and discoid (e.g., *M. luna*, *M. lippincottianus*) and those which superficially resemble young *Mylossoma* or *Myleus*. In contrast, certain other *Metynnis* species have a slightly elongate body (e.g., *Metynnis maculatus* that resembles young *Piaractus*). Sexual dimorphism: females larger than males; anal-fin margin of adult male slightly lobed or sinuous in some species, versus straight margin in females. Juveniles: body shape of young *Metynnis* similar to adults, with high ontogenetic variability (Taphorn 1992, Machado-Allison and Fink 1995, Zarske and Géry 2008, Pavanelli et al. 2009, Gomes et al. 2012, Queiroz et al. 2013). **SPECIES** 15, including type species *M. hypsauchen* (Müller and Troschel, 1844) from Guyana. Most recently described is *M. cuiaba* from Paraguay basin, Brazil (Zarske and Géry 2008, Pavanelli et al. 2009). **COMMON NAMES** *Pacu marreca* (Brazil); *Palometa moteada* (Peru, Venezuela); Silver Dollar (English). **DISTRIBUTION AND HABITAT** Widespread throughout the AOG region and Paraguay basins (Jégu 2003). A few have naturally broad distributions and others with restricted geographic ranges. Some species have been introduced within and outside their native ranges (Magalhães et al. 2002, Gomes et al. 2012). Inhabit a wide variety of habitats, in both still and flowing waters, including rivers, small streams and creeks, floodplain lakes, pools, marshes, flooded forests, and reservoirs; they are known from all three of the major water types (Lowe-McConnell 1964, 1991, Sazima 1986, Goulding et al. 1988, Nico 1991, Taphorn 1992, Pavanelli et al. 2009, Gomes et al. 2012). **BIOLOGY** Generalized omnivores or herbivores. Browse using specialized, shearing teeth to bite off small pieces of plants, taking small seeds, fruits, flowers, and leaves of macrophytes, some filamentous algae and periphyton, and various small invertebrates, including both aquatic and terrestrial insects, microcrustaceans, and freshwater sponges (Smith 1981, Santos et al. 1984, Sazima 1986, Goulding et al. 1988, Nico 1991, Taphorn 1992, Machado-Allison and Fink 1995). Masticate and destroy seeds during consumption, and are considered seed predators rather than dispersers (Goulding 1983b). Tend to move in small groups, although juveniles may form

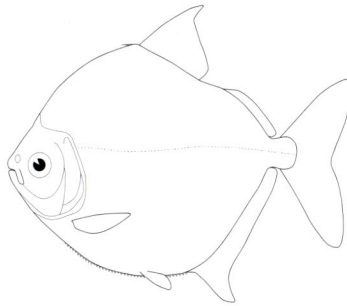

large shoals and some join schools of other plant-eating fishes (Sazima 1986, Machado-Allison and Fink 1995). Some species form large schools and make seasonal migrations (Goulding et al. 1988), although remaining within the clearwater or blackwater tributaries in which they live (Horn et al. 2011). Their deep body shape provides protection from predators that swallow prey whole (Lowe-McConnell 1964). Some species are popular in the aquarium trade, and *M. maculatus* is widely stocked in reservoirs and rivers of Brazil for sport fishing purposes (Gomes et al. 2012), although some occurrences are attributed to escapes from ornamental fish farms (Magalhães et al. 2002). At least one native population in northeastern Brazil was extirpated by predation from an introduced *Cichla* (Alves et al. 2007).

Mylesinus (22–40 cm SL)

Characterized by: medium to large adult body sizes; elliptical or oval body shape; mouth terminal to subinferior and snout tapered; outer and inner rows of upper-jaw teeth abut and not separated by a space; upper jaw with 5 teeth per side in outer row and 2 per side on inner row; lower jaw with 7 or more teeth on each side; teeth relatively fragile, weakly attached to jaws, used for clipping or grazing, incisiform with a spatulate edge (i.e., anteroposteriorly flattened and with relatively thin base); all teeth on upper jaw's outer row approximately the same size and shape; the posterior three teeth in outer row aligned and their crown ridges forming a nearly continuous cutting edge; abdomen rounded, lacking a marked keel, due to presence of only a few prepelvic spines, but with strong abdominal serrae in postpelvic area; dorsal branched rays 18–23; anal branched rays 26–32. Sexual dimorphism: adult males with well-developed bilobed anal fin, secondary lobe situated on central part of fin, with small stiff hooks on tip of branched anal-fin rays, and their dorsal-fin rays with long filaments. Mature females with a single-lobed anal fin forming a long keel. Juveniles become deep-bodied with size up to about 8 cm SL (Jégu and Santos 1988, Jégu et al. 1989, Santos et al. 1997, Andrade et al. 2016a, b). **SPECIES** Three: *M. paucisquamatus*, *M. paraschomburgkii*, and *M. schomburgkii* (type species). **COMMON NAMES** *Curupité*, *Pacu curupité*, *Pacu borracha* (Brazil); *Pacu* (Guyana), *Surapire* (Venezuela). **DISTRIBUTION AND HABITAT** Rivers of the Guiana and Brazilian shields (Jégu 1992b, Pagezy and Jégu 2002). Two of the three *Mylesinus* species are from the lower Amazon basin, with *M. paucisquamatus* endemic to the Tocantins-Araguaia basin, and *M. paraschomburgkii* in left bank tributaries of the Amazon, from the Uatumã River to the Araguari River. *Mylesinus schomburgkii* inhabits northern drainages of Guiana Shield, including the Essequibo (Jégu 1992b, Pagezy and Jégu 2002), Orinoco, Caroní, Caura, Ventuari, Bita, and upper Orinoco basins (Hurtado-Sepulveda 1984, Machado-Allison and Fink 1995, Lasso et al. 2004). Rheophilic fishes inhabiting rapids in the main tributaries of the Shield drainages (Jégu and Santos 1988, Jégu et al. 1989, Andrade et al. 2013, 2016a, b). **BIOLOGY** Adult *M. paraschomburgkii* feed mostly on rupestrian aquatic plants (Podostemaceae) (Santos et al. 1997). Young juveniles feed mainly on insect larvae associated with these same plants. Between 5 and 10 cm SL, young *Mylesinus* shift to a predominantly herbivorous diet. Intestines grow longer with age, reaching four times relative SL as adults (Jégu et al. 1989). *Mylesinus* intestines often contain many parasites, for example the nematode *Rondonia rondoni*, although heavily parasitized fish appear to be in good health (Santos et al. 1997). Among the more important food fishes for Amerindian people of the Guiana Shield region (Mol 2012b).

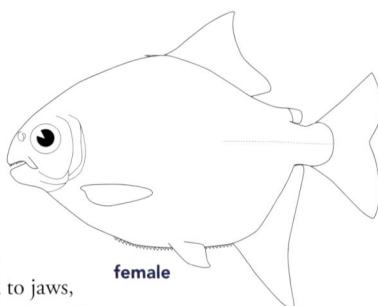

female

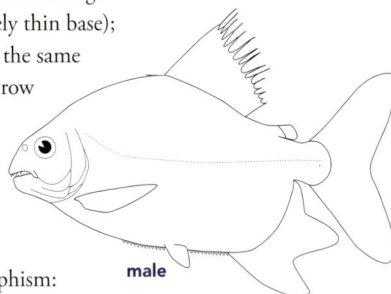

male

Myleus (12–42 cm SL)

Characterized by: adult body shape round or oval in lateral profile; predorsal spine present; abdominal serrae present in prepelvic area; abdomen somewhat rounded with a small keel extending from the mid-abdomen to pelvic-fin origin and strong abdominal serrae in postpelvic area; dorsal branched rays 17–24; anal branched rays 30–34; mouth terminal and snout tapered. Also distinguished from related genera by a unique dentition, with two rows of teeth in upper jaw, with the two middle front teeth on outer row of upper jaw close together, either in contact or nearly so (fig. 11a); outer-row teeth on each side of upper jaw form a continuous, unbroken row, with no gaps between adjacent teeth. The teeth are incisiform with acute edges and relatively robust (i.e., anteroposteriorly flattened and with relatively robust base), strongly attached to jaws, and adapted mainly for clipping or grazing. The inner row of teeth on the lower jaw consists of a single pair of very small conical teeth close to the symphysis, both of which are generally hidden from view because they are situated under the crowns of the median teeth of the outer row. Sexual dimorphism: adult males have a well-developed bilobed anal fin with small stiff hooks near tip of lobes; the secondary lobe is situated on central part of fin, and dorsal-fin rays are ornamented with long filaments. Mature females have a keel-shaped, single-lobed anal fin. Juveniles: morphology is not well documented. Small young *Myleus*, similar to those of *Tometes*, are generally cryptically colored (e.g., Machado-Allison 1983: fig. 9). Juvenile *Myleus* become progressively more deep-bodied up to 8 cm SL (Britski et al. 1988, Jégu and Santos 1988, 2002, Andrade 2013, Queiroz et al. 2013, Andrade et al. 2016b). **SPECIES** Five: *M. setiger* (type species), *M. knerii, M. altispinnis, M. micans,* and *M. pacu.* Other species previously assigned to this genus have been transferred to *Myloplus* (Jégu and Santos 2002, Jégu et al. 2003, Ortí et al. 2008, Andrade et al. 2016a). **COMMON NAMES** *Pacu dente-seco* (Brazil); *Palometa, Pámpano* (Peru, Venezuela); Pacu (English). **DISTRIBUTION AND HABITAT** Portions of the Amazon, Orinoco, and São Francisco basins, and drainages of the Guianas (Britski et al. 1988, Jégu and Santos 2002, Queiroz et al. 2013, Andrade et al. 2016a, b). Mostly restricted to rivers and streams of the Guiana and Brazilian shields, and *M. setiger* also in the Madeira and western Amazon basins, the most widespread of all the rheophilic serrasalmid species. *Myleus knerii* was described from the Maroni River in French Guiana and *M. pacu* was described from the Essequibo River, Guyana. *Myleus altipinnis* and *Myleus micans* are endemic to the São Francisco basin. Inhabit black- and clearwater river reaches with moderate to fast currents, often associated with rapids. **BIOLOGY** Feed on fruits and other plant parts, and insects (Jégu and Santos 2002). *M. setiger* from the Orinoco basin consumes seeds, fleshy fruit, leaves, flowers, and occasionally invertebrates, and the flesh, fins, and scales of other fish (Nico 1991). Reported to make upstream migrations as river levels rise at the start of the rainy season, with movements suspected to be related to spawning (Lowe-McConnell 1964). Important food fishes for humans, especially Amerindians (Figueiredo-Silva et al. 2012, Mol 2012b, Santos and Nóbrega-Alves 2016).

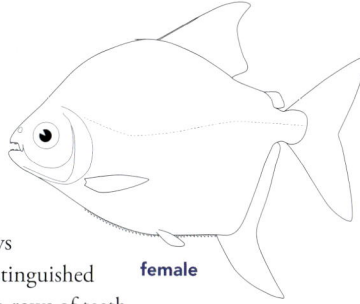

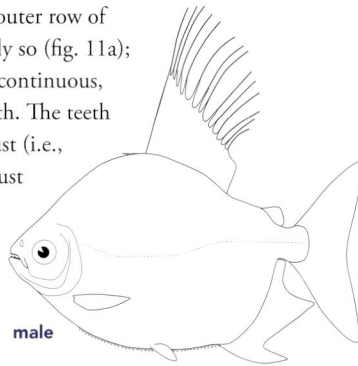

Myloplus (16–56 cm SL)

Characterized by: medium to large adult body sizes; snout rounded and the mouth terminal to somewhat superior; predorsal spine present; body lateral profile ranges from rounded to moderately elongate, body depth usually >50% of standard length (fig. 12a); abdominal serrae on pre- and postpelvic area clearly exposed in most species, forming a marked prepelvic, median keel; branched dorsal-fin rays range from 18 to 26 (or more). Separated from most members of the *Myleus* Clade by dentition. In *Myloplus* (and *Utiaritichthys*), outer row of upper-jaw teeth separated from the inner row by a space (fig. 8b); jaw teeth molariform, used for crushing. Teeth on lower jaw characterized by a rounded base or by being anteroposteriorly wide. Lower jaw with 5 teeth on each side and also a big pair of thorn teeth at symphysis behind main tooth row. *Myloplus* differs from *Utiaritichthys* by having a deeper body, approximately 60% SL (vs. <50% SL), and an abdomen with a conspicuous keel in most species. Sexual dimorphism: adult males have a well-developed bilobed anal fin (with or without stiff hooks near tip of lobes), the secondary lobe situated on central part of fin, and dorsal-fin rays ornamented with long

female

male

filaments. Mature females have a single-lobed anal fin forming a long keel. Adults, especially males, with vivid colors during the breeding season. Juveniles: young *M. rhomboidalis* marbled with brown colors (Mol 2012b). (Goulding 1980, Jégu et al. 2003, 2004, Mol 2012b, Andrade et al. 2016a, b). **SPECIES** 12, including *M. asterias* (type species). Most species until recently assigned to genus *Myleus* (Andrade et al. 2016a). **COMMON NAMES** *Pacu* (Brazil and Venezuela), *Watau* (French Guiana), *Kumalu*, *Kumaru* (Suriname). **DISTRIBUTION AND HABITAT** Broadly distributed in the AOG region (Mol 2012b, Queiroz et al. 2013, Andrade et al. 2016a, b), with one species in the La Plata (Paraguay-Paraná) basin (Correa et al. 2015a). Usually inhabit black- and clearwater rivers and streams, less common in whitewaters. Normally occur in main channels in slow and fast waters, including rapids, areas with dense aquatic vegetation, and old stream meanders isolated from main channels during low-water periods (Goulding 1980, Boujard et al. 1990, Nico 1991, Taphorn 1992, Carolsfeld et al. 2003). **BIOLOGY** Primarily herbivorous, with seeds as a main food source, but occasionally take small aquatic animals (Goulding 1980, Boujard et al. 1990, Nico 1991, Correa 2012, Mol 2012b). Other plant items commonly found in stomachs include fruits, leaves, flowers, and filamentous algae, and terrestrial and aquatic invertebrates such as beetles, caterpillars, termites, locusts, shrimp, crabs, copepods, snails, fish scales, and mammal excrement. *Myloplus rhomboidalis* from French Guiana consume 45 different kinds of seeds, including those of the açaí palm *Euterpe oleracea*, and occasionally parts of the rapids-dwelling aquatic Podostemonaceae plants (Boujard et al. 1990). In the Amazon, *Myloplus* feed on seeds of riverside *Amanoa* (Phyllanthaceae), and subsistence fishermen use these seeds as bait (Goulding 1980). *Myloplus* are mainly seed predators, but do act as seed dispersers for certain plants with small-sized seeds (Goulding 1980, Correa et al. 2015a). Reportedly make seasonal migrations for reproductive purposes (Carolsfeld et al. 2003). Most populations remain in home tributaries for their entire life cycle (Goulding 1980). Many species are important food for humans, exploited by both subsistence and commercial fishers; juveniles of some species are popular in the ornamental fish trade (Machado-Allison 1987, Carolsfeld et al. 2003, Mol 2012b, Andrade et al. 2016a). The largest record is a *M. planquettei* measuring 56 cm SL and weighing 6 kg (Mol 2012).

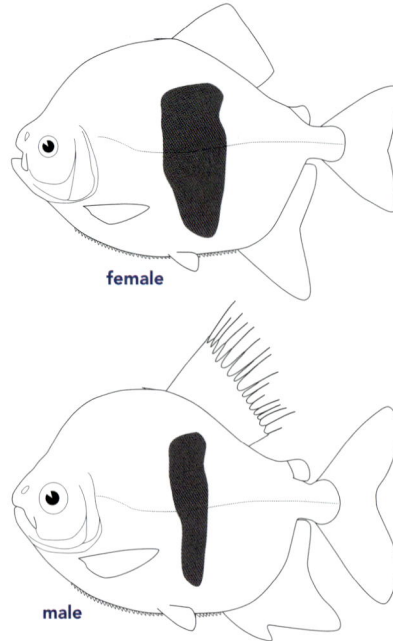

Mylossoma (20–30 cm SL)

Characterized by: moderate adult body sizes; mostly silvery coloration; a very deep body shape, rounded in lateral profile; a short adipose fin; and a small projecting head. *Mylossoma*, *Colossoma*, and *Piaractus* differ from other serrasalmid genera by absence of a predorsal spine (fig. 1). *Mylossoma* is readily distinguished from *Colossoma* and *Piaractus* by features of the anal fin (fig. 2): (1) 28–36 branched rays in *Mylossoma* (vs. 20–24 branched rays in *Colossoma* and *Piaractus*); (2) margin rounded or pointed, with longest rays within central portion of fin (vs. fin margin concave or straight), and (3) broadly covered with minute scales (vs. scaled only near fin base). Teeth generally molariform, with the same broad bases as those of *Colossoma* and *Piaractus*, but cusps are higher and sharper, somewhat incisiform, and used more for cutting and slicing seeds than crushing. Two rows of teeth in upper jaw, separated by a space somewhat similar to *Piaractus*. Sexual dimorphism: no obvious external differences. Juveniles: young and adults with similar body shape. Body color and degree of cryptic coloration of young vary with age and among species. Individuals from 1.2 to 7.0 cm SL with an ocellus on midbody below dorsal fin, most conspicuous in young *M. duriventre* and *M. acanthogaster*. Young *M. aureum* with a darkly patterned body and fins, and small and inconspicuous ocellus (absent in *M. aureum*) (Goulding 1980, Machado-Allison 1987, López and Nass 1989, Machado-Allison and Castillo 1992, Taphorn 1992, Britski et al. 1999). **SPECIES** Three: *M. acanthogaster*, *M. aureum*, and *M. duriventre* (type species). Key to species and other information in Schultz (1944b), Machado-Allison and Castillo (1992), Taphorn (1992), Queiroz et al. (2013).

COMMON NAMES *Pacu-manteiga*, *Pacu-toba* (Brazil), *Palometa* (Venezuela), *Garopa* (Colombia).

DISTRIBUTION AND HABITAT Distributed throughout the Amazon, Orinoco, La Plata (Paraguay-Paraná), and Maracaibo basins (Machado-Allison and Castillo 1992, Britski et al. 1999, Mojica et al. 2002). *Mylossoma duriventre* is the most widespread species (including *M. paraguayensis* and *M. orbignyanum*, both junior synonyms of *M. duriventre*). *Mylossoma aureum* occurs in the Amazon and Orinoco basins (Jégu 2003). *Mylossoma acanthogaster*, endemic to a few streams that drain to Lake Maracaibo, is vulnerable to extinction because of its restricted distribution and low numbers (Mojica et al. 2002). *Mylossoma* inhabit a variety of river and floodplain habitats, including the main channels of large and small rivers, streams, and flooded forests, floodplain lakes, and lagoons; they occur in all major water types, but tend to be most common in nutrient-rich whitewater systems (Goulding 1980, Goulding et al. 1988, Barbarino-Duque and Taphorn 1995, Carolsfeld et al. 2003). **BIOLOGY** Primarily herbivorous, feeding on a wide variety of seeds, fruits, leaves, and sometimes flowers and filamentous algae. Also occasionally consume animal matter, including aquatic and terrestrial/arboreal invertebrates such as spiders, ants, beetles, and cockroaches, and are also known to eat feces, presumably of monkeys (Goulding 1980, Nico 1991). Like *Myloplus*, *Mylossoma* eat mostly smaller seeds and serve as seed predators, since a large proportion of seeds are masticated or crushed. However, they do act as seed dispersers for certain plants with small-sized seeds, their importance as seed dispersers augmented with increased age and size (Goulding 1980, Correa et al. 2015a). *Mylossoma* are migratory, with complex and incompletely understood migration patterns. In the Amazon, adults migrate from floodplain lakes to main rivers during the beginning of the flood season, generally forming large schools that move upriver a few hundred kilometers for reproduction. After spawning, adults return to the floodplain and disperse into flooded forests to feed until water levels drop and they move back to the more permanent rivers and lakes (Goulding 1980, Smith 1981, Carolsfeld et al. 2003). *Mylossoma duriventre* populations in the Orinoco Llanos move into main rivers at the onset of the dry season, traveling upriver to more permanent creeks

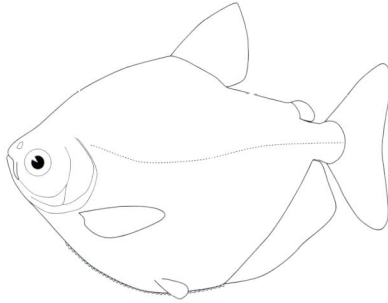

in the piedmont and high savannas (Taphorn 1992). In contrast, *M. acanthogaster* migrate upstream at the beginning of the rainy season and back downstream at the beginning of the dry season (Mojica et al. 2002). *Mylossoma* are important commercially as a food fish (Goulding 1980, 1981, Taphorn 1992, Carolsfeld et al. 2003) and are also common in the ornamental fish trade. Sport fishers targeting *Mylossoma* commonly use small hooks baited with a ball of mash cornmeal (Barbarino-Duque and Taphorn 1995).

Ossubtus (10–23 cm SL)

Characterized by: moderately large adult body sizes; deep body shape; forked caudal fin; short adipose fin; coppery brown body coloration, with sides often irregularly mottled or blotched with darker pigment. Differ from other members of Serrasalmidae by: strictly ventral position of the mouth in individuals >5 cm SL (fig. 10a); and only 4 teeth in each side of lower jaw (dentary). Also characterized by: a strong predorsal spine; lack of prepelvic serrae; incisiform teeth, markedly slender, and relatively fragile, weakly attached to jaws; 19–21 unbranched rays in dorsal-fin range; and 22–25 unbranched rays in anal fin. Sexual dimorphism: adult males with a bilobed anal fin, and adult females, a single-lobed, falcate-shaped, anal fin. Unique among many serrasalmids, mature females are smaller in body length (although heavier) than mature males. Juveniles: position of mouth shifts with age, being terminal in postlarval stage, slightly downturned in specimens of around 3.0 cm SL, and inferior in individuals >5.0 cm SL. Young *Ossubtus* with a black, triangular-shaped, vertically elongate, humeral spot (Jégu 1992a, Jégu and Zuanon 2005, Andrade et al. 2015, 2016b). **SPECIES** One: *O. xinguense* (see species information in Jégu 1992a, 2003, Andrade et al. 2013). **COMMON NAMES** *Pacu-capivara* (Brazil), Eagle-beak pacu (English). **DISTRIBUTION AND HABITAT** Known only from the Volta Grande (Big Bend) region of the lower Xingu basin in the lower Amazon basin, Brazil. Adults and juveniles strictly rheophilic, known only from river reaches with rocky rapids, and rocky shallow sites populated by rupestral seedlings of Podostemaceae (in areas with moderate to strong currents). Young *Ossubtus* up to 4.0 cm SL form schools of 20–30 individuals and take shelter under broad stones in the rapids (Jégu 2003, Jégu and Zuanon 2005, Andrade et al. 2013, 2015, Sabaj-Pérez 2015). **BIOLOGY** Diet includes aquatic macrophytes and filamentous algae (Jégu 1992a), although feeding behavior in the wild has not been observed (Jégu and Zuanon 2005). Highly territorial in aquaria, and strongly aggressive toward congeners. Wild-caught specimens parasitized by a host-specific isopod of the family Cymothoidae (Thatcher 1995). Because of its highly restricted distribution, dependence on rapids as its primary or only habitat, and apparent rarity, *Ossubtus xinguense* is on the Red List of Threatened Brazilian Fauna and its survival considered highly threatened, mainly by loss of lotic habitat associated with ongoing and planned constructions of dams (Jégu and Zuanon 2005, Andrade et al. 2015, Sabaj-Pérez 2015).

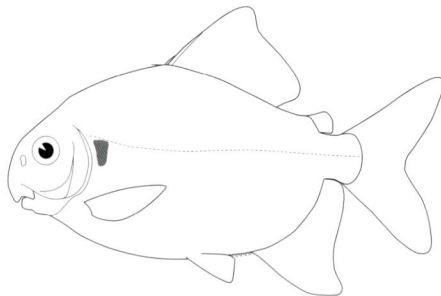

Piaractus (85 cm SL)

Characterized by: large adult body size, second only to *Colossoma* among serrasalmids; adult body shape somewhat elongate and elliptical; absence of a predorsal spine, shared only with *Colossoma* and *Mylossoma* among serrasalmids (fig. 1). Separated from *Colossoma* as described above. Adult *P. brachypomus* are typically bluish gray or

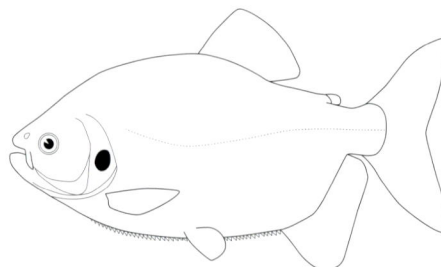

brownish gray on the dorsum and sides, with a pale abdomen. A few individuals have orange-red areas on the throat, cheek, abdomen, and lower fins. Older specimens are darker, usually dark brown or even black. *Piaractus mesopotamicus* tends to be evenly colored, typically gray or dark, with throat and abdomen that is usually yellow. Some *Piaractus* specimens show countershading (possibly indicating some degree of introgressive hybridization with *Colossoma*). Adipose fin reduced or absent in older adults. Sexual dimorphism: no marked sexual dimorphic differences, although adult females tend to be slightly larger than adult males (Loubens and Panfili 2001, Costa and Mateus 2009). Juveniles: smaller young have a round eye-spot (ocellus) on midbody below dorsal fin, also present in juvenile *Colossoma* and *Mylossoma*. Older juveniles are silver with red-orange on the lower fins, belly, and cheek, and various black markings on body and fins. Those about 8.0 cm SL have a body shape and color pattern remarkably similar to young piranhas of the genus *Pygocentrus*. A black opercular spot is present in specimens up to about 10 cm SL. In contrast to adults, juvenile *Piaractus* are deep-bodied, almost discoid. **HYBRIDS** Both *Piaractus* species are commonly hybridized with *Colossoma* for aquaculture and stocking purposes (Saint-Paul 1992, Hashimoto et al. 2011), and the progeny of these crosses display intermediate morphological characteristics (Britski 1977, Goulding 1980, Machado-Allison 1982a, b, 1983, 1986, Géry 1986, Taphorn 1992, Britski et al. 1999). **SPECIES** Two: *P. brachypomus* (type species; synonyms include *Colossoma bidens*) and *P. mesopotamicus* (synonyms include *Colossoma mitrei*); see Géry (1986) for key to the species. **COMMON NAMES** *P. brachypomus*: Pirapitinga, *Caranha* (Brazil), *Morocoto* (Venezuela), *Paco* (Colombia, Peru), Red-bellied pacu (English). *P. mesopotamicus*: Pacu, Caranha (Brazil), Small-scaled pacu (English). **DISTRIBUTION AND HABITAT** *P. brachypomus* is widespread in the Amazon and Orinoco basins, and *P. mesopotamicus* is widespread in the La Plata (Paraná-Paraguay) basin (Taphorn 1992, Britski et al. 1999). *Piaractus* species occur mainly in lowland portions of larger rivers and their floodplains; habitats typically include main channels of rivers, streams, and occasionally creeks; they are seasonally present in flooded forests and floodplain lakes (Goulding 1980, Taphorn 1992, Loubens and Panfili 2001). *Piaractus* have become much less common in parts of their native range because of overexploitation by commercial and recreational fishers, and damming of rivers (Carolsfeld et al. 2003, Costa and Mateus 2009, Correa et al. 2015a). Aquaculture-reared individuals of both species, including hybrids with *Colossoma*, have been introduced to areas outside their native ranges within and outside South America (Fuller et al. 1999, Carolsfeld et al. 2003, Ortega et al. 2007). A curious case involves presence and possible establishment of *P. brachypomus* in the Sepik-Ramu basin of Papua New Guinea, where it has been given the name "ball cutter" by locals based on claims that these non-native fish bite the testicles of wading fishermen (Correa et al. 2014). **BIOLOGY** Attain 85 cm SL and 20 kg; together with *Colossoma*, they are the largest of the living characins (Goulding 1980, 1981). Similar to *Colossoma*, adult *Piaractus* are primarily herbivorous and feed extensively on fleshy fruits and seeds. Because many seeds are swallowed whole, these fish play an important role as seed-dispersal agents for a wide assortment of floodplain plants (Goulding 1980, 1983b, Kubitzki and Ziburski 1994, Galetti et al. 2008, Anderson et al. 2009, Correa et al. 2015a). In Peru, *Piaractus* and *Colossoma* disperse large quantities of seeds from up to 35% of the trees and lianas that fruit during the high-water period (Anderson et al. 2009). *Piaractus* also consume large amounts of leaves, and occasionally flowers, insects, crabs, mollusks, bivalves, small fish, and mammal feces (Canistri 1970, Goulding 1980, Nico 1991, Taphorn 1992, Carolsfeld et al. 2003). *Piaractus* schools make seasonal migrations of many hundreds of kilometers up rivers for purposes of tracking food resources or for reproduction (Carolsfeld et al. 2003, Makrakis et al. 2007a, Costa and Mateus 2009). Young *Piaractus brachypomus* and *Pygocentrus cariba* are similar in external appearance and sometimes shoal together, suggesting Batesian mimicry (Mago-Leccia 1978) or Mullerian (Machado-Allison 1982a).

Pristobrycon (11–26 cm SL)

Pristobrycon is an artificial genus, poorly separated from *Serrasalmus* (Machado-Allison et al. 1989, Fink and Machado-Allison 1992, Freeman et al. 2007, Ortí et al. 2008, Thompson et al. 2014). Characterized by: relatively small adult body sizes, generally <20 cm SL, with only *P. maculipinnis* and *P. striolatus* to 24 and 26 cm SL; adult body shape not very stout, head and body not especially broad, body depth ranging from about 55 to 68% of SL; short, blunt snout, somewhat squarish lateral profile in front of nares (fig. 15b); tricuspid teeth in oral jaws with a large median cusp so that individual teeth are approximately triangular in profile (fig. 13b) (shared with *Pygocentrus* and *Serrasalmus*, but different from *Pygopristis* which has pentacuspid teeth, fig. 13a); ectopterygoid teeth either absent or few in number, with a blunt or squarish shape, dissimilar to the jaw teeth (fig. 17a; also found in many *Serrasalmus* in which ectopterygoid teeth are more numerous, often 4–5 per side and strongly triangular, roughly similar in shape to their jaw teeth); cheeks of adults not extensively armored with bone, the third infraorbital bone relatively undeveloped with a relatively large naked area separating the third infraorbital (IO3) border from the preopercle bone (fig. 17aa; cheeks of adult *Serrasalmus* typically well armored; the naked area separating the IO3 from the preopercle bone is small, or even absent in largest specimens, fig. 17bb). Genus is divided into two groups: those with a preanal spine (*P. calmoni*, *P. aureus*, and *P. eigenmanni*, see fig. 16a), and those without (*P. careospinus*, *P. maculipinnis*, and *P. striolatus*). A preanal spine is present in all *Pygocentrus* and *Serrasalmus* species, and is absent in all *Pygopristis* species. *Pristobrycon* can be distinguished from non-piranha serrasalmids by their dentition; all piranhas (i.e., *Serrasalmus*, *Pygocentrus*, and *Pygopristis*) have a terminal, slightly prognathous mouth, with a single row of teeth on upper and lower jaws (fig. 6b), with incisor-like (thin and blade-like) teeth, and adjacent teeth closely set and interlocking to form a continuous, saw-like sharp cutting edge adapted for rapid puncture and shearing. Sexual dimorphism: little information on external differences has been documented. According to Mol (2012b), male piranhas in Suriname tentatively identified as *Pristobrycon eigenmanni* were reported to have a falciform or sickle-shaped anal fin, suggesting a possible difference from females. In contrast, males of *Pristobrycon striolatus* in Trombetas basin have a bilobed anal fin, with the second lobe on middle rays, similar to that of many herbivorous serrasalmids (MCA pers. obs.). In most species, the morphology of young individuals is approximately similar to that of adults (Eigenmann 1915b, Machado-Allison et al. 1989, Fink and Machado-Allison 1992, Jégu and Santos 2001, Freeman et al. 2007, Ortí et al. 2008, Mol 2012b). **SPECIES** Six, including *P. calmoni* (type species). Some *Pristobrycon* in this chapter are treated as *Serrasalmus* in other recent literature (e.g., *Serrasalmus eigenmanni*), and vice versa (e.g., *Pristobrycon serrulatus*) (see key to 4 of the 6 species in Machado-Allison and Fink 1996). **COMMON NAMES** Piranha, Pirambeba (Brazil), *Caribito* (Venezuela). **DISTRIBUTION AND HABITAT** Throughout the AOG region (Fink and Machado-Allison 1992, Taphorn 1992, Machado-Allison and Fink 1996, Mol 2012b). *Pristobrycon striolatus* is the most widespread member (Taphorn 1992, Machado-Allison and Fink 1996, Mol 2012b). Inhabits a wide range of habitats, from large rivers to small streams, flooded forests, flooded savannas, lagoons, oxbow lakes, and marshes (Nico and Taphorn 1988, Nico 1991, Taphorn 1992). *Pristobrycon striolatus* inhabits all three major water types, but *P. maculipinnis* is known only from blackwater systems (Nico 1991). **BIOLOGY** Juvenile and adult *Pristobrycon* are omnivorous, opportunistic feeders (Goulding 1980, Machado-Allison and Garcia 1986, Nico and Taphorn 1988, Nico 1991). Small juveniles consume a variety of aquatic insects and small seeds. In the Orinoco Llanos, young piranhas tentatively identified as *P. striolatus* were found to have stomachs full of fins of other fishes (Nico and Taphorn 1988). Adult *P. striolatus* consume

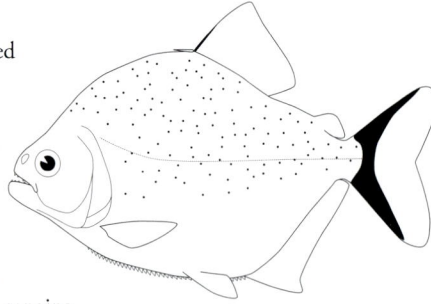

seeds, fruits, flowers, and other plant fragments, as well as various aquatic insects, shrimp, other invertebrates, and small fish (Goulding 1980, Nico and Taphorn 1988, Nico 1991, Mol 2012b). *P. maculipinnis* and *P. striolatus* inhabit flooded forests and are important seed predators during certain seasons (Goulding 1980, Nico 1991).

Pygocentrus (28–34 cm SL)

Medium- to large-sized piranhas and generally viewed as the true piranhas. Distinguished from non-piranha serrasalmids by their dentition, characterized by a single row of teeth on upper and lower jaws (fig. 6b). Teeth in oral jaws incisor-like (thin and blade-like), adjacent teeth closely set and interlocking to form a continuous, saw-like sharp cutting edge used to puncture and shear. Oral teeth tricuspid with a large median cusp, individual teeth approximately triangular in profile (fig. 13b), shared with *Pristobrycon* and *Serrasalmus*. Preanal spine present (also in *Serrasalmus* and some *Pristobrycon* species, but absent in *Pygopristis* and some *Pristobrycon*). Head of larger juveniles and adult *Pygocentrus* blunt and wide relative to other piranhas (10–20% [mean = 15%] SL). Can be distinguished from *Serrasalmus* and *Pristobrycon* by: (1) Dorsal profile between mouth and dorsal-fin origin moderately to strongly convex (fig. 14a); (2) adults large, with blunt snouts and powerful, massive jaws; (3) head large and robust, and broader (width >15% SL in individuals >10 cm SL) than other piranhas at similar body sizes; and (4) absence of ectopterygoid teeth, except in small juveniles (≤1.0 cm SL). Color variable, depending on age, reproductive state, and water conditions. Young adults silvery, with red-orange ventral region, larger adult males and females dark purple or nearly black. Sexual dimorphism: females slightly larger than males (Duponchelle et al. 2007). Juveniles: with numerous rounded or oval spots on body. Older juveniles silver with red-orange on the lower fins, belly, and cheek, and black markings on body and fins. Small juvenile *P. cariba*, up to 8.0 cm SL similar to young *Piaractus brachypomus* (see *Piaractus* account). Although absence of ectopterygoid teeth helps distinguish adults, small juveniles (<1.0 cm SL) can have 6 or more minuscule, unicuspid teeth on the ectopterygoid bone (Fink and Machado-Allison 1992, Fink 1993, Machado-Allison and Fink 1996, Machado-Allison 2002). **SPECIES** Three: *P. piraya* (type species), *P. cariba* (synonyms include *P. notatus* and *P. caribe*), and *P. nattereri* (see key in Fink 1993). A fourth, *P. palometa*, listed as valid by Eschmeyer et al. (2016), is of doubtful taxonomic status (Géry 1976, Fink 1993). **COMMON NAMES** *Piranha, Piranha vermelha, Piranha caju* (Brazil); *Paña roja* (Peru); *Caribe colorado, Capaburro* (Venezuela). **DISTRIBUTION AND HABITAT** *P. nattereri* widespread in the AOG region and La Plata basins, *P. cariba* in the Orinoco basin, and *P. piraya* in the São Francisco basin (Fink 1993). Inhabits savannas and forested regions in a variety of habitats, including small rivers and streams, floodplain lakes and swamps, reservoirs, roadside ditches, and natural and artificial ponds, lagoons, and marshes (Lowe-McConnell 1964, Nico 1990, 1991). Juveniles prefer vegetated habitats (Nico 1990, 1991). Common in lowland, whitewater systems, where they are often the most abundant piranha and most common predatory fish. Generally uncommon or absent in black- and clearwaters (Goulding 1980, Nico and Taphorn 1988, Nico 1991). Widely introduced outside their native ranges in South America and elsewhere, although reproducing non-native populations are known only from drainages in South America (Fuller et al. 1999, Latini and Petrere Jr. 2004, Trindade and Jucá-Chagas 2008). **BIOLOGY** Aggressive and highly carnivorous predators. Adults feed mainly on fish, often other characins (including other piranhas), either by biting pieces of flesh from large fish or taking small fish whole. Also consume fish fins and scales, whole frogs, flesh or bones of lizards, young caiman, ducks, rodents, small crustaceans, insects, spiders, and plant debris (Nico and Taphorn 1988, Nico 1991, Carvalho et al. 2007, Ferreira et al.

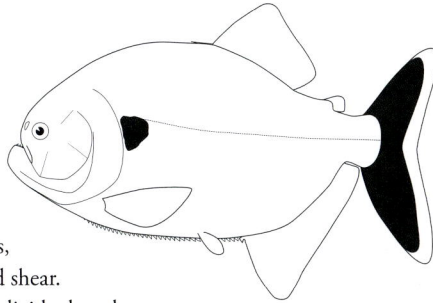

2014). Young and adult *Pygocentrus* are opportunistic scavengers and will attack and consume dead or dying fish and other vertebrate animals (Sazima and Andrade-Guimarães 1987, Nico and Taphorn 1988). Small juvenile *Pygocentrus* commonly feed on small aquatic insects and microcrustaceans, and sometimes fish fins, fish scales, and even small seeds; and larger, older juveniles include more fish in their diets (Machado-Allison and Garcia 1986, Nico 1990, 1991).

Pygopristis (18 cm SL)

A small- to medium-sized piranha whose adult morphology suggests an intermediate condition between a *Metynnis* and a *Pristobrycon*. Jaws weakly developed for a piranha. Distinguished from non-piranha serrasalmids by their dentition, characterized by a single row of sharp, blade-like teeth on upper and lower jaws (fig. 6b). Distinguished from other piranhas (i.e., *Serrasalmus*, *Pristobrycon*, and *Pygocentrus*) by their teeth, most of which are relatively symmetrical and have 5 (or more) cusps, with the middle cusp typically only slightly larger than the other cusps (fig. 13a). Teeth of other piranhas have 2 or 3 cusps, with one cusp (the central cusp in teeth with 3 cusps or the anterior cusp in teeth with 2 cusps) substantially larger so that each tooth is broadly triangular, especially those more forward in mouth (fig. 13b). Also characterized by: lack of preanal spine (shared with some *Pristobrycon*); lack of ectopterygoid teeth; ≤2 rows of scales on the anal-fin base (anal fin broadly scaled in other piranhas); adipose-fin base length varies from short to moderately long; body silvery to olive green and sometimes dark; parts of cheeks, belly, and most fins yellow to red-orange. Sexual dimorphism: females larger than males (Lowe-McConnell 1964). Anal fin of adult males bilobed, with a second lobe that is similar to but smaller than that of males of non-piranha serrasalmids (Taphorn 1992, Planquette et al. 1996). Juveniles: young are silvery, median fins partly reddish orange. From about 2.5 cm SL, sides marked by a series of about 10 brownish vertical bars, narrow and somewhat irregularly shaped; the bars become faint with age (Machado-Allison 1985, Jégu and Santos 1988, Nico 1991, Taphorn 1992, Machado-Allison and Fink 1996, Mol 2012b). **SPECIES** One, *P. denticulata* (synonyms include *Serrasalmus denticulatus*, *Pygopristis fumarius*, and *Serrasalmus punctatus*). **COMMON NAMES** *Piranha-amarela*, *Piranha-mafurá* (Brazil), *Pilin*, *Piray*, *Pireng* (French Guiana), *Caribe palometa* (Venezuela), Silver dollar piranha (English). **DISTRIBUTION AND HABITAT** Throughout the AOG region (Géry 1972a, Machado-Allison and Fink 1996, Lasso et al. 2004, Mol 2012b). A relatively rare piranha, most commonly found in blackwater environments. Known from savanna and forested regions in rivers, streams, creeks, marshes, lagoons, swamps, oxbow lakes, pools, and rice fields (Lowe-McConnell 1964, Nico 1991, Taphorn 1992, Jégu and Keith 1999, Mol 2012b). **BIOLOGY** Adults are opportunistic omnivores, consuming seeds, aquatic insects (mainly larvae), crustaceans, fish flesh, fish fins, and plant debris (Nico 1991, Machado-Allison and Fink 1996, Jégu and Keith 1999). Juveniles consume seeds (masticated) and other plant materials, aquatic insects, microcrustaceans, various other invertebrates, and fish fins (Nico 1991, Machado-Allison and Fink 1996). A young wild-caught *Pygopristis* reared in aquaria was observed (at about 4 cm SL) stalking and chasing other fishes and clipping out pieces of their fins (Nico 1991). *Pygopristis* is regarded as harmless to humans (Mol 2006).

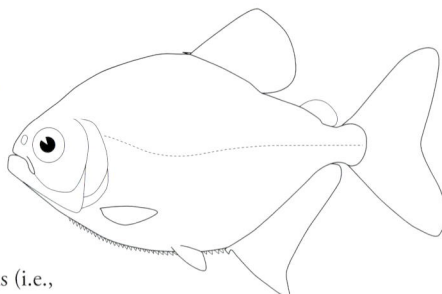

Serrasalmus (11–42 cm SL)

Morphologically diverse genus of small- to large-sized piranhas, some very similar in appearance to *Pristobrycon* (see *Pristobrycon* account for traits useful in separating the two genera). Relative body depth varies widely, ranging from the elongate-bodied *S. elongatus* (BD/SL <40%) to deep-bodied *S. altuvei* (50–66%) and *S. geryi* (to 60%). Distinguished from non-piranha serrasalmids by their dentition, characterized by a single row of sharp, blade-like teeth on upper and lower jaws (fig. 6b). Similar to *Pygocentrus* and *Pristobrycon*, teeth triangular (especially anterior teeth) with 2 or 3 cusps, one cusp substantially larger (fig. 13b). Similar to *Pygocentrus* and some *Pristobrycon*, preanal spine present (fig. 16a). Snout shape variable, depending on species and body size, ranging from narrow to broad, and short and blunt (e.g., *S. maculatus*) to long and pointed (fig. 15). Lower jaw extends well beyond upper jaw when the mouth is closed. Dorsal fin moderately long, with 2–3 unbranched rays and 12–17 branched rays; anal fin long, with 2–4 unbranched rays and 25–33 branched rays. Length of intestine short compared with SL. Body color silvery, often with reddish orange on the cheeks, belly, and some fins; however, large adults of several species are dark, almost grayish black to dense black (e.g., *S. gouldingi*, *S. manueli*, *S. rhombeus*), darkest individuals from blackwaters. Sexual dimorphism: anal fin pale green in males and yellow with red hues in females of *S. rhombeus* (Braga 1956). No other cases of sexual dichromatism in piranhas are known (Leão 1996). During the breeding season, the body and fins of many adult piranhas, such *S. rhombeus* and *P. cariba*, become noticeably darker and red hues tend to disappear, but both male and female undergo the same color change (Nico and Taphorn 1986, Leão 1996). Some piranhas maintain dark coloration following their first reproductive season, whereas others revert to their normal coloration during the nonbreeding period (Taphorn 1992). Juveniles: young *Serrasalmus* with a single row of small, well-developed, triangular teeth on inner roof of mouth on the ectopterygoid (up to about 8 per side and similar in shape to jaw teeth though much smaller). Ectopterygoid teeth reduced or lost with size (fig. 17) (Machado-Allison 1985, Machado-Allison et al. 1989, Jégu et al. 1991b, Nico 1991, Fink and Machado-Allison 1992, Taphorn 1992, Machado-Allison and Fink 1996, Jégu and Santos 2001). **SPECIES** 26. **COMMON NAMES** *Piranha*, *Pirambeba* (Brazil); *Paña* (Peru); *Caribe* (Venezuela). *Serrasalmus rhombeus* (type species): *Piranha-branca*, *Piranha-preta* (Brazil), *Caribe pinche* (Venezuela), Red eye piranha (English). *Serrasalmus elongatus*: *Piranha-mucura* (Brazil), *Caribe Pinche* (Venezuela). **DISTRIBUTION AND HABITAT** The most widespread serrasalmid genus, present throughout the AOG region, as well as the São Francisco and La Plata basins (Britski et al. 1988, Machado-Allison and Fink 1996, Britski et al. 1999, Jégu 2003, Mol 2012b, Queiroz et al. 2013). Greatest diversity in the Amazon (17 species) and Orinoco (8 species) basins, with four species distributed across both of these basins. *Serrasalmus rhombeus* has the broadest geographic range of all piranha species (perhaps of all serrasalmid species), and is one of the most abundant piranhas, and may represent a species complex. *Serrasalmus neveriensis* is known only from two small coastal drainages of Venezuela and represents the sole member of the genus naturally occurring in a Caribbean drainage (Machado-Allison et al. 1993). *Serrasalmus* piranhas occur in a wide diversity of habitats. Representative species are known from still and flowing waters of both savanna and forested regions in habitats that include large and small rivers, streams, and creeks, flooded forests, floodplain and oxbow lakes, lagoons, reservoirs, among others (Paiva 1958, Lowe-McConnell 1964, Goulding 1980, Nico 1991, Taphorn 1992). *Serrasalmus rhombeus* is known

from all three major waters types, but is most common in black- and clearwater environments. Other species are restricted to, or at least more common in whitewater habitats (Goulding 1980, Nico 1991, Taphorn 1992). Some *Serrasalmus* species have been introduced outside their native ranges. For instance, a reproducing population (likely *S. rhombeus*) survived for more than 10 years in a small body of water in Florida (USA) until all were eradicated with rotenone (Fuller et al. 1999). **BIOLOGY** Highly carnivorous (generally piscivorous), opportunistically consuming fruits or seeds when available. Larger *Serrasalmus* attack and bite out chunks of flesh from large fish, sometimes from other vertebrates, and also frequently take bits of fin or consume small fish whole (Nico and Taphorn 1988, Sazima and Machado 1990). Many young and some adults regularly clip off and consume pieces of fins, and *S. elongatus* feeds heavily on both fins and scales (Roberts 1970, Goulding 1980, Nico and Taphorn 1988, Nico 1991). Stomach contents include remains (typically only chunks) of frogs and toads, reptiles (e.g., teiid water lizards), birds, mammals (e.g., rodents), crabs, insects, snails, rubber tree seeds, palm fruits, various flowers, leaves, and plant resins (Goulding 1980, Nico 1991). Juvenile fin-eating *S. irritans* use aggressive mimicry to approach other fish (Nico and Taphorn 1986, 1988). Small juveniles identified as *S. gouldingi* have a prominent black spot on the dorsal fin, providing an advantage in approaching and nipping the fins of small characins with similar markings (Nico 1991). *Serrasalmus maculatus* actively feed on leaves of aquatic plants while scanning for aquatic insects and small crustaceans (Sazima and Machado 1990). *Serrasalmus rhombeus* do not form large schools but travel in small groups of about 5–20 individuals (Goulding 1980). The largest members of genus include *S. rhombeus* (to 41.5 cm SL), *S. manueli* (to 36 cm SL), *S. elongatus* (to 30 cm SL), and *S. gouldingi* (to about 28 cm SL). *Serrasalmus rhombeus*, along with *Pygocentrus* and some of the other larger and more aggressive *Serrasalmus*, are responsible for all or most of the more serious attacks on human bathers and swimmers (Haddad Jr. and Sazima 2003, Mol 2006).

Tometes (28–60 cm SL)

Medium- to large-sized fishes characterized by a predorsal spine and a terminal mouth, except *T. lebaili*, whose mouth is obliquely directed upward. One of four genera in the *Myleus* Clade (along with *Myleus*, *Mylesinus*, and *Ossubtus*) distinguished from other serrasalmids by: (1) prepelvic midventral abdominal area somewhat rounded in cross section, not forming a marked keel; (2) teeth incisiform in shape; and (3) outer and inner rows of upper-jaw teeth abutting, not separated by a space (fig. 8a). *Tometes* most similar in appearance to *Myleus*, sharing: two posteriormost teeth (4 and 5) on each side of outer row of upper jaw shorter than other premaxillary teeth (fig. 9b); teeth robust and firmly attached to jaws; last two teeth of outer row of upper jaw with crown ridges forming an S-shaped cutting edge; and incisiform teeth somewhat flattened anteroposteriorly (unlike *Ossubtus* and *Mylesinus* whose incisiform teeth are very slender and flat). Distinguished from *Myleus* by: two middle frontal teeth on outer row of upper jaw separated from each other by a broad space, greater than tooth width (fig. 11b), versus middle front teeth close together in *Myleus* (fig. 11a); spatial gap separating upper-jaw teeth numbers 1 and 2 (fig. 11bb); inner row of teeth on lower jaw a single pair of large conical teeth, close to the mandibular symphysis, the tips clearly visible (vs. symphyseal teeth small and hidden by the crowns of the median teeth of the outer row in *Myleus*). Sexual dimorphism: adult males with a well-developed bilobed anal fin, secondary lobe situated on central part of fin, with small stiff hooks on tip of branched anal-fin rays, and dorsal-fin rays with long filaments. Mature females with a single-lobed anal fin that forms a long keel. Juveniles: marbled brown coloration changing to faint, reticulated bars with growth, with a small humeral

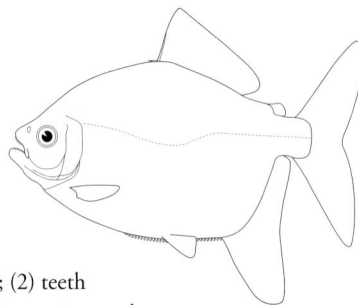

spot, faint in some species (Jégu et al. 2002a, b, Jégu and Keith 2005, Andrade et al. 2016b).
SPECIES Six, including *T. trilobatus* (type species) (Andrade et al. 2016b). Before its revalidation,
Tometes was synonymized with *Myletes* or *Myleus*. **COMMON NAMES** *Curupité, Pacu-curupité,
Pacu-borracha* (Brazil). **DISTRIBUTION AND HABITAT** Throughout the AOG region. Greatest
diversity in the Amazon basin, with four **SPECIES** *T. camunani* in the Trombetas basin;
T. kranponhah and *T. ancylohynchus* in the Xingu basin, *T. ancylohynchus* in the Tocantins-
Araguaia basin; and *T. makue* in the middle and upper Negro basin, including the Casiquiare.
Tometes makue inhabits the Orinoco basin. Non-Amazonian species include: *T. lebaili* in coastal
drainages of French Guiana and Suriname, and *T. trilobatus* in northeastern rivers of the Guiana
Shield. All species are rheophilic, known only from rocky rapids (Jégu et al. 2002a, Jégu 2003,
Andrade et al. 2013). **BIOLOGY** Adults strictly herbivorous, a main food source being
Podostemaceae, the rupestral plants characteristic of rapids (Jégu and Keith 2005, Andrade et al.
2013). Submerged Podostemaceae plants function as a nursery and foraging area for juveniles
(Jégu and Keith 2005). Occur in sympatry with other rheophilic serrasalmids and often are
social. Juvenile *T. kranponhah* and *T. ancylohynchus* form mixed schools, occasionally also with
young *Myleus*, suggesting protective mimicry (Andrade et al. 2016b). *Tometes kranponhah*
> 10 cm SL often swim in strong currents or rest behind rocks after traversing vortex zones,
where they are vulnerable to capture by cast nets. Specimens >30 cm SL commonly taken with
harpoons or by hooks baited with araçá guava fruit. *Tometes* are important food fishes for local
populations of Amerindians (Jégu and Keith 2005). The largest *Tometes* reported, a *T. lebaili*
captured in Guyana, was estimated to be at least 60 cm SL (Lord et al. 2007). Because of their
extreme habitat specificity, *Tometes* populations are especially vulnerable to dam construction
and impoundments, and other habitat-destructive practices such as gold mining (Jégu and Keith
2005, Andrade et al. 2013).

Utiaritichthys (20–29 cm SL)

Medium- to large-sized fishes characterized by a
predorsal spine, terminal mouth, and elliptical,
relatively elongate body. The abdominal serrae
are present but not especially well developed;
therefore the abdominal keel is present but not
obvious because of the reduced size and/or
number of prepelvic spines. Total number of
abdominal prepelvic spines varies among species,

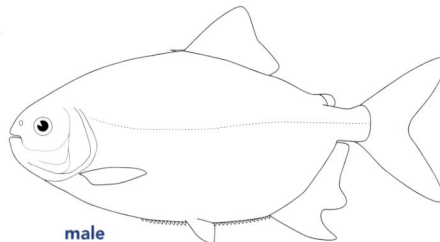

male

from 9–10 in one species to 26–31 in another species. *Utiaritichthys* together with *Myloplus* are
distinguished from members of the *Myleus* Clade (i.e., *Myleus, Mylesinus, Ossubtus,* and *Tometes*)
by: molariform teeth, used for crushing (vs. incisiform teeth in the *Myleus* Clade); an outer row
of premaxillary teeth separated from inner row by an internal gap (vs. absence of gap between
premaxillary tooth rows); and ascending process of premaxilla wide from its base to the tip (vs.
tapering from its base to the tip). *Utiaritichthys* is distinguished from *Myloplus* by: more slender
body, BD <50% SL (fig. 12b) (vs. >50% SL in *Myloplus*; fig. 12a). Superficially resembles
Piaractus, with a somewhat elongate body and gap separating the inner and outer rows of their
upper teeth, but *Piaractus* lacks a predorsal spine, and anal-fin shape of mature male and female
Piaractus are the same. Sexual dimorphism: adult males with a well-developed bilobed anal fin,
the secondary lobe situated on central part of fin. Adult females with anteriormost anal-fin rays
distinctly longer than more posterior rays, forming a single, long and distinct lobe. Juveniles: no
information available (Jégu et al. 1992, 2002a, b, Jégu and Keith 2005, Pereira and Castro 2014,
Andrade et al. 2016a, b). **SPECIES** Three: *U. sennaebragai* (type species), *U. esguiceroi,* and
U. longidorsalis (see key in Pereira and Castro 2014). **COMMON NAMES** *Pacu* (Brazil and
Venezuela). **DISTRIBUTION AND HABITAT** Known only from right-bank tributaries of the Amazon
basin (Jégu et al. 2003, Pereira and Castro 2014). *Utiaritichthys sennaebragai* and *U. esguiceroi* in

the Tapajós basin, and *U. longidorsalis* in the Madeira basin. Reports of *U. sennaebragai* in the Orinoco, Xingu, and Tocantins basins are probably misidentifications of *Tometes* species. *Utiaritichthys* are rheophilic fish and closely associated with fast-flowing reaches of major tributaries (Jégu et al. 1992, 2003, Pereira and Castro 2014, Andrade et al. 2016a, b).

BIOLOGY Juvenile *U. esguiceroi* inhabit river margins and feed on aquatic and terrestrial insects and organic matter, whereas adults inhabit rapids and riffle areas where they feed almost exclusively on Podostemaceae macrophytes and filamentous algae (Pereira and Castro 2014). The ontogenetic shifts in habitat and diets are similar to those reported by Jégu et al. (1989) for *Mylesinus paraschomburgkii*.

FAMILY TRIPORTHEIDAE—ELONGATE HATCHETFISHES AND RELATIVES
— *PETER VAN DER SLEEN* and *ANGELA ZANATA*

DIVERSITY 23 species in five genera (Oliveira et al. 2011a, Mariguela et al. 2016): *Agoniates* (2 species), *Clupeacharax* (1 species) *Engraulisoma* (1 species), *Lignobrycon* (1 species), and *Triportheus* (18 species). Twenty species in four genera in the AOG region.

GEOGRAPHIC DISTRIBUTION *Triportheus* are found in most of the major river drainages of South America; *Agoniates* is widespread in the Amazon basin; *Clupeacharax anchoveoides* and *Engraulisoma taeniatum* have more restricted distributions in the Western Amazon and Paraná-Paraguay basins; *Lignobrycon myersi* occurs in the Cachoeira and Contas basins, northeastern Brazil.

ADULT SIZES From 4.2 cm in *Engraulisoma taeniatum* from the western Amazon and upper Paraguay basins, to 24 cm in *Triportheus elongatus* from throughout the AOG region.

DIAGNOSIS OF FAMILY The five genera in the Triportheidae were grouped based on genetic data (Oliveira et al. 2011a, Mariguela et al. 2016). Zanata (2000) also showed 16 morphological characters shared by *Agoniates*, *Lignobrycon*, and *Triportheus*, including features of the infraorbital series,

pectoral and pelvic girdle, vertebrae, and ribs. Triportheidae is easily characterized by: a laterally compressed body; coracoids expanded to form a distinct thoracic keel (except *Engraulisoma taeniatum*); clupeiform (herring-like) body shape; and silvery coloration (but having an adipose fin, which is absent in the Clupeiformes). The expanded coracoids are, however, not unique among Characiformes and also found in members of Gasteropelecidae, Cynodontidae, and several other genera (e.g., *Gnathocharax* and *Piabucus*).

SEXUAL DIMORPHISM Females possibly larger than males in *Triportheus* (Martins-Queiroz et al. 2008).

HABITATS Typically found near the water surface of lakes, flooded forests, and streams.

FEEDING ECOLOGY *Triportheus* species vary from omnivorous to carnivorous, depending on the species and environment inhabited (e.g., Almeida 1984, Braga 1990, Catella and Petrere Jr. 1998, Gama and Caramaschi 2001, Pereira et al. 2011). *Agoniates* species are invertivorous to piscivorous (Goulding et al. 1988, Hawlitschek et al. 2013); *Engraulisoma taeniatum* is apparently insectivorous (Ohara 2012).

KEY TO THE GENERA

1a. Dentary with tricuspid teeth anteriorly, followed by strong conical teeth..*Agoniates*
1b. Dentary teeth uni- or multicuspid, but without strong conical teeth...**2**

2a. Coracoids distinctly expanded to form a thoracic keel; presence of a relatively large inner tooth proximate to dentary symphysis; 3 rows of premaxillary teeth (except a few species such as *T. magdalenae*)......................*Triportheus*
2b. Coracoids not or slightly expanded; absence of inner tooth proximate to dentary symphysis; 1 or 2 series of premaxillary teeth **3**

3a. Premaxillary teeth of the outer row directed outward; coracoids not expanded; maxilla edentulous; anal-fin origin posterior to the dorsal-fin origin...*Engraulisoma*
3b. Premaxillary teeth of the outer row ventrally directed; coracoids at least slightly expanded; maxilla with 1 to about 20 teeth; anal-fin origin anterior to the dorsal-fin origin...*Clupeacharax*

GENUS ACCOUNTS

Agoniates (15–22 cm SL)

Medium-sized fishes, characterized by an elongate body shape, resembling clupeiforms (e.g., sardines) in overall body shape and coloration. They possess a combination of unusual features, such as conic and canine teeth alternating on the dentary and tricuspid teeth on the premaxilla (Zarske and Géry 1997, Lima and Zanata 2003, Dagosta and Datovo 2013). Additional features that can help to recognize the genus: mouth oblique; maxilla elongate, with conical teeth along its entire border; belly with only a weakly developed keel in the coracoids region in *A. halecinus*, or with a well-developed keel from the head to the origin of the anal fin in *A. anchovia*; well-developed circum-orbital bones of nearly the same width around the eye; 5 branchiostegal rays; lateral line complete, sharply decurved on its first 5 scales; dorsal-fin origin behind the middle of the body; pectoral fin large; pelvic fin small (Zarske and Géry 1997). **SPECIES** Two, *A. anchovia* and *A. halecinus*, both in the Amazon basin. See review of genus in Zarske and Géry (1997). **COMMON NAMES** *Apapai, Maiaca* (Brazil); *Mojarita anchova* (Peru). **DISTRIBUTION AND HABITAT** Both species in the Amazon basin; *A. halecinus* predominates in clear- and blackwaters and *A. anchovia* in whitewaters. **BIOLOGY** Invertivorous to piscivorous (Goulding et al. 1988, Hawlitschek et al. 2013).

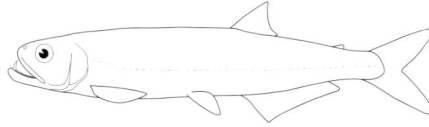

Clupeacharax (6.6 cm SL)

Recognized by a striking combination of features: elongate body; expanded coracoid bone; midventral keel; long anal fin with its origin anterior to the dorsal fin; and the dentary and premaxilla with tri- to pentacuspid teeth (Lima 2003b). It is similar to *Triportheus* but differs by having the dentary teeth in a single row (vs. 2 rows in *Triportheus*, with a single conical symphyseal tooth on the inner row) and by having 2 premaxillary tooth rows (vs. 3 rows, except two rows in a few species, such as *T. magdalenae*). **SPECIES** One, *C. anchoveoides*. **DISTRIBUTION AND HABITAT** Upper Amazon, Madeira River, and Paraná-Paraguay basins. **BIOLOGY** No data available.

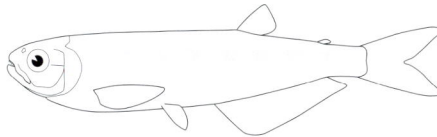

Engraulisoma (4.2 cm SL)

Easily recognized by: upper jaw longer than lower; two series of multicuspid teeth on the premaxilla; maxilla without teeth; anterior teeth on dentary multicuspid and posterior ones conical; long anal fin (but initiating after dorsal-fin origin) mostly with unbranched rays; lateral line low and interrupted; and a silvery lateral band (Castro 1981). **SPECIES** One, *E. taeniatum*; species description in Castro (1981). **DISTRIBUTION AND HABITAT** Napo, upper Madeira, and upper Paraguay basins. Reported to occur in small forest streams and the margins of larger rivers such as the Aripuanã, Manicoré, and Jaciparaná. Most abundant in shallow streams with warm and slightly acidic water (Ohara 2012). **BIOLOGY** Insectivorous (Ohara 2012).

Triportheus (11–24 cm SL)

Easily recognized by: laterally compressed body; expanded coracoids forming a prominent anteroventral keel; and elongate pectoral fins. In addition, most species are also characterized by: two or three series of premaxillary teeth; a second (inner) tooth row on each dentary represented by a large symphyseal tooth; and an anteriorly located quadrate-articular joint (Malabarba 2004). Females possibly larger than males (Martins-Queiroz et al. 2008). **SPECIES** 18, including 16 species in the AOG region. Review of the genus and key to the species in Malabarba (2004). **COMMON**

NAMES *Sardinha papuda, Sardinha chata* (Brazil); *Pechón* (Ecuador); *Sardina larga* (Peru); *Arenca* (Venezuela). **DISTRIBUTION AND HABITAT** Magdalena, Orinoco, Essequibo, Amazon, Tocantins, São Francisco, Parnaíba, and Paraguay-Paraná basins. Generally found in the upper layer of the water column.

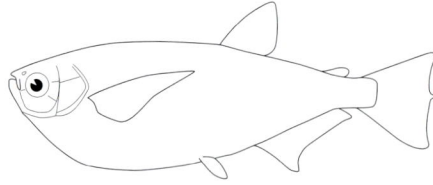

BIOLOGY Omnivorous to carnivorous, depending on the species and environment (e.g., Almeida 1984, Braga 1990, Catella and Petrere Jr. 1998, Gama and Caramaschi 2001, Pereira et al. 2011). In poorly oxygenated water, *Triportheus* are able to develop barbel-like expansions on the lower lip that help to direct oxygenated water at the surface into the mouth (Malabarba 2004). Reproduction takes place during the rainy season (Planquette et al. 1996, Martins-Queiroz et al. 2008). The largest species, *T. elongatus* (24 cm SL), is important in commercial and subsistence fisheries, and is used especially when larger and more appreciated species are scarce (e.g., Barthem and Goulding 2007).

FAMILY ARIIDAE—SEA CATFISHES
— *ALEXANDRE P. MARCENIUK and PETER VAN DER SLEEN*

DIVERSITY 153 species in 30 genera, including 15 species in 6 genera in the AOG region (Betancur-R et al. 2007, Marceniuk and Menezes 2007, Marceniuk et al. 2012b).

COMMON NAMES *Bagre bandeira* (Brazil); *Bagre marino, Pez crucifijo* (Spanish); Crucifix fish (English).

GEOGRAPHIC DISTRIBUTION Tropical and warm-temperate regions around the world (Atlantic, Indian, and Pacific oceans). As their name suggests, sea catfishes inhabit coastlines and estuaries, but also tidal rivers.

ADULT SIZES Species in the AOG range from 30 cm SL in *Cathorops arenatus* to 120 cm SL in *Sciades parkeri*.

DIAGNOSIS OF FAMILY Distinguished from other Siluriformes by: maxillary and mental barbels usually present, nasal barbels absent; anterior and posterior nostrils close together, the posterior with a valve; lateral ethmoid and frontal bones connected medially and laterally delimiting a fontanel; strong spines in dorsal and anal fins; anal-fin and adipose-fin bases short; and caudal fin deeply forked. Sea catfishes also share many characters of the internal anatomy, including a unique shape of the bones on the ventral surface of the skull that resembles a man on a cross.

SEXUAL DIMORPHISM Males practice oral incubation of the eggs. As a consequence, males have a longer and wider head than females. In females, the accessory tooth plates are larger than in males, while in males these tooth plates are generally covered by epithelial tissue during the breeding season (Marceniuk et al. 2009). Additionally, the pelvic fin is longer and wider at base in females than in males.

HABITATS Preferentially inhabit turbid waters, over muddy bottoms and low salinity.

FEEDING ECOLOGY Omnivorous to carnivorous, feeing on detritus, fish, and crustaceans (Le Bail et al. 2000).

BEHAVIOR Spawning period is associated with the wet monsoon, when anadromous movements, associated with breeding, are reported for several estuarine and marine species. Eggs can be few but relatively large in some species (e.g., 30–35 eggs of 10–15 mm in *Amphiarius rugispinis*), or much smaller eggs but in higher quantities in other species (e.g., 100–165 eggs of 2 mm in *Sciades couma*) (Le Bail et al. 2000). After eggs are laid, the male in many species incubates them in its mouth and may continue to protect the fry after hatching.

KEY TO THE GENERA

1a. Maxillary barbel compressed, tape-like; a single pair of mental barbels ....*Bagre*
1b. Maxillary barbel round in cross section; two pairs of mental barbels..2

2a. Adipose fin very short, its base shorter than 50% of anal-fin base length; accessory tooth plates small and oval-shaped, perpendicularly disposed (fig. 1D).. *Cathorops*
2b. Adipose fin moderately long to very long, its base >50% of anal-fin base length; accessory tooth plates large, round to subtriangular, or longitudinally elongate, or transversely elongate and parallel to the premaxillary (fig. 1A,B,E,F,G)...3

3a. Adipose-fin base moderately long, longer than half of anal-fin base; lateral ethmoid and frontal limiting a very small fenestra (fig. 2D), not visible under the skin .... *Sciades*
3b. Adipose-fin base very long, as long or longer than anal-fin base; lateral ethmoid and frontal limiting a wide fenestra (fig. 2A,B,C), visible under the skin 4

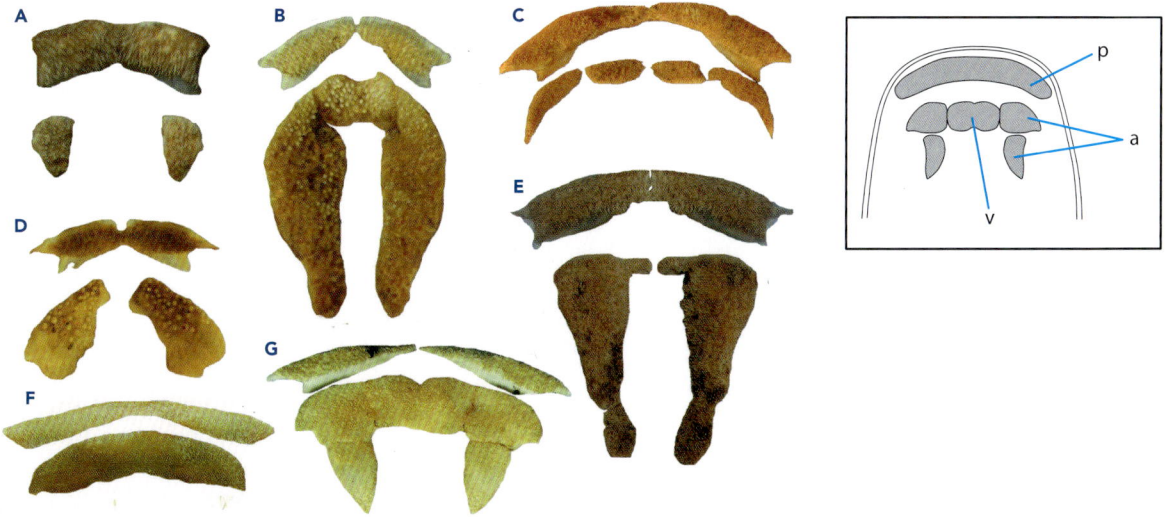

FIGURE 1 Premaxillary, vomerine, and accessory tooth plates in ventral view. (A) *Amphiarius rugispinis*, (B) *Aspistor quadriscutis*, (C) *Bagre bagre*, (D) *Cathorops agassizii*, (E) *Notarius grandicassis*, (F) *Sciades passany*, (G) *Sciades couma*, all from Pará state in Brazil. Inset: illustration of roof of the mouth with generalized position of the premaxillary (p), vomerine (v), and accessory (a) tooth plates.

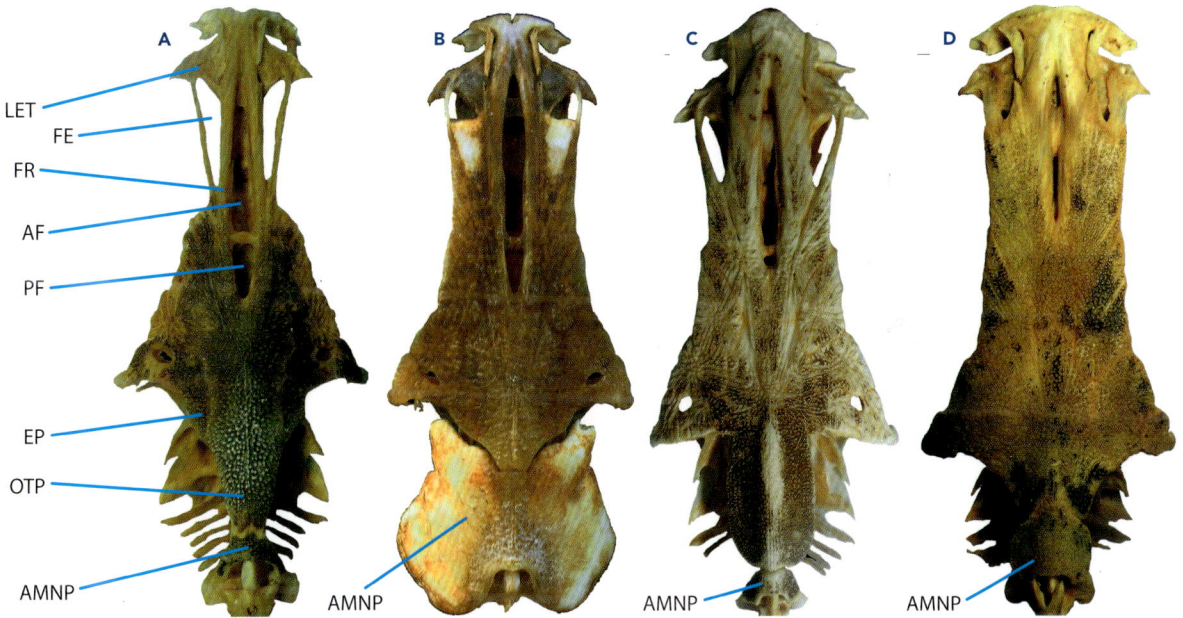

FIGURE 2 Neurocranium in dorsal view of (A) *Amphiarius rugispinis*, (B) *Aspistor quadriscutis*, (C) *Notarius grandicassis*, and (D) *Sciades couma*, all from Pará state in Brazil. Codes indicate: lateral ethmoid (LET), fenestra (FE), frontal (FR), anterior fontanel (AF), posterior fontanel (PF), epioccipital (EP), occipital process (OTP), and anterior and median nuchal plates (AMNP).

4a. Nuchal plate very wide, quadrangular or butterfly-shaped (fig. 2B)...*Aspistor*
4b. Nuchal plate small, V-shaped (fig. 2A,C)..**5**

5a. Tooth plate on vomer absent (fig. 1A); accessory tooth plate small and round (fig. 1A); occipital process
without a constriction at the base (fig. 2A) ....*Amphiarius*
5b. Tooth plate on vomer present (fig. 1E); accessory tooth plate wide and posteriorly elongate (fig. 1E),
occipital process with a constriction at the base (fig. 2C)... *Notarius*

GENUS ACCOUNTS

Amphiarius (20–35 cm SL)

Characterized by: cephalic shield granulated, visible under the skin; a large fenestra limited by frontal and lateral ethmoid bones (fig. 2A), clearly visible under the skin; medial groove of neurocranium absent; a large posterior cranial fontanel (fig. 2A); epioccipital invading or not into dorsal portion of cephalic shield; occipital process funnel-shaped and long (fig. 2A); anterior and median nuchal plates fused, indistinct, forming a structure of semilunar aspect (fig. 2A); vomerine tooth plates absent; accessory tooth plates round to oval-shaped (fig. 1A), with acicular (needle-shaped) teeth; maxillary barbel fleshy and cylindrical; two pairs of mental barbels; base of adipose fin very long, as long as anal-fin base; lateral line not bifurcated at caudal region, reaching base of caudal-fin upper lobe (Marceniuk and Menezes 2007). **SPECIES** Two species in the Neotropics, both in the AOG region: *A. phrygiatus* (*Bagre mucuro*, Kukwari sea catfish, 30 cm SL) and *A. rugispinis* (*Bagre tumbeló*; Softhead sea catfish, 35 cm SL). **DISTRIBUTION AND HABITAT** Both species occur in low parts of coastal rivers, estuaries, and coastal waters from Venezuela to the mouth of the Amazon River, found on shallow muddy bottoms. **BIOLOGY** Females lay relatively few but large eggs, which the males guard and mouth-brood until hatching (Le Bail et al. 2000).

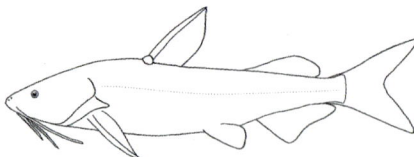

Aspistor (25–120 cm TL)

Characterized by: granulated cephalic shield visible under the skin; a moderately developed fenestra limited by frontal and lateral ethmoid bones (fig. 2B), visible under the skin; medial groove of neurocranium absent; moderately developed posterior cranial fontanel, oval-shaped (fig. 2B); epioccipital invading into dorsal portion of cephalic shield; occipital process very short and wide at base, round-shaped (fig. 2B); anterior and median nuchal plates fused and indistinct, forming a large butterfly-shaped structure (fig. 2B); vomerine tooth plates present; accessory tooth plates longitudinally elongate (fig. 1B), with molar-like teeth; maxillary barbel fleshy and cylindrical; two pairs of mental barbels; base of adipose fin very long, as long as anal-fin base; lateral line not bifurcated at caudal region, reaching base of caudal-fin upper lobe (Marceniuk and Menezes 2007). **SPECIES** Two species in the Neotropics, one (*A. quadriscutis*) in the AOG region. **COMMON NAMES** *Caiacoco*, *Cangatá* (Brazil); *Bagre bresú* (Spanish); Bressou sea catfish (English). **DISTRIBUTION AND HABITAT** *A. quadriscutis* inhabits coastal muddy shallow waters, but especially juvenile individuals can be found in the lower parts of coastal rivers, from Venezuela to the mouth of the Amazon River. **BIOLOGY** Females lay relatively large eggs, which the males mouth-brood. Diet includes bottom-living invertebrates (Le Bail et al. 2000).

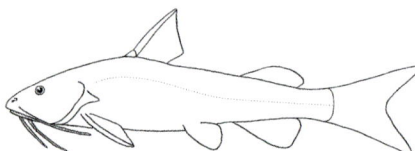

Bagre (50–70 cm SL)

Characterized by: filamentous dorsal and pectoral fins; cephalic shield smooth, covered by muscle and scarcely visible under the skin; a well-developed fenestra limited by frontal and lateral ethmoid bones visible under the skin, but sometimes obliterated by superficial ossifications; medial groove of neurocranium not very distinct; posterior cranial fontanel very reduced; epioccipital not invading dorsal portion of cephalic shield; occipital process short, progressively narrower toward its posterior part; anterior and median nuchal plates fused and indistinct (except in *B. bagre*), forming a structure

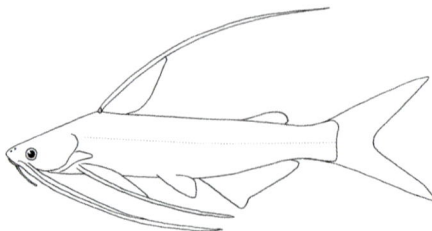

of semilunar aspect (as in fig. 2A,C); vomerine tooth plates present and transversely elongate (fig. 1C); accessory tooth plates narrow and transversely elongate (fig. 1C), bearing conical teeth; base of maxillary barbel bony, fleshy for remaining part of its length and compressed in form of tape; a single pair of mental barbels; base of adipose fin very short, less than one-half length of anal-fin base; lateral line bifurcated at caudal region, reaching base of caudal-fin upper and lower lobes (Marceniuk and Menezes 2007). **SPECIES** Four species in the Neotropics, including two species in the AOG region: *B. bagre* (Coco sea catfish) and *B. marinus* (Gaff-topsail catfish). **COMMON NAMES** *Bagre bandeira* (Brazil); *Chivo chinchorro* (Colombia). **DISTRIBUTION AND HABITAT** Both species are chiefly marine, but young specimens also enter the brackish waters of estuaries from Venezuela to the mouth of the Amazon River. Especially, *Bagre marinus* is rare or absent from the region influenced by the Atlantic Plume (Le Bail et al. 2000), having a greater preference for more saline waters. **BIOLOGY** Diet includes small fish and invertebrates.

Cathorops (20–30 cm SL)

Characterized by: granulated cephalic shield visible under the skin; lateral ethmoid and frontal limiting a wide fenestra, very conspicuous and visible under the skin; medial groove of neurocranium distinct; posterior cranial fontanel very reduced; epioccipital not invading dorsal portion of cephalic shield; occipital process funnel shaped and moderately long, its posterior part considerably narrower than its base (as in fig. 2A); anterior and median nuchal plates fused and indistinct, forming a structure of semi-lunar aspect (as in fig. 2A,C); vomerine tooth plates absent; accessory tooth plates small, oval, and perpendicularly disposed (fig. 1D), bearing molar-like teeth; maxillary barbel fleshy and cylindrical; two pairs of mental barbels; base of adipose fin very short, less than one-half length of anal-fin base; lateral line not bifurcated at caudal region, reaching base of caudal-fin upper lobe (Marceniuk and Menezes 2007). **SPECIES** At least 20 species in the Neotropics, including 4 species in the AOG region: *C. agassizii* (Gaviota sea catfish), *C. arenatus* (Yellow sea catfish), *C. nuchalis* (Orinoco sea catfish), and *C. spixii* (Madamango sea catfish). Review of genus and key to the species in the AOG in Marceniuk et al. (2012a). **DISTRIBUTION AND HABITAT** Estuaries and freshwaters from Venezuela to the mouth of the Amazon River. Especially *A. agassizii* and *A. nuchalis* are common in freshwater. **BIOLOGY** Diet includes invertebrates.

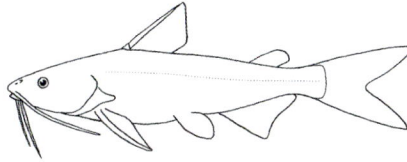

Notarius (40–60 cm SL)

Characterized by: granulated cephalic shield visible under the skin; lateral ethmoid and frontal bones limiting a moderately developed fenestra visible under the skin (fig. 2C); medial groove of neurocranium absent (except in *N. planiceps*); posterior cranial fontanel moderately developed, oval-shaped (fig. 2C); epioccipital not invading dorsal portion of cephalic shield; occipital process with a constriction at the base (only in *N. grandicassis*; fig. 2C); anterior and median nuchal plates fused and indistinct, forming a structure of semilunar aspect (fig. 2C); tooth plates associated with vomer present or absent, when present round (fig. 1E); accessory tooth plates well developed, triangular-shaped, bearing conical teeth (fig. 1E); maxillary barbel fleshy and cylindrical; two pairs of mental barbels; base of adipose fin very long, as long as anal-fin base; lateral line not bifurcated at caudal region, reaching base of caudal-fin upper lobe (Marceniuk and Menezes 2007). **SPECIES** 12 species in the Neotropics, including one species in the AOG region (*N. grandicassis*). **COMMON NAMES** *Bagre-papai*, *Yurupiranga* (Brazil); *Grondé* (French Guiana); *Tampoco* (Guyana); *Bagre Tomás* (Spanish); Thomas sea catfish (English). **DISTRIBUTION AND HABITAT** *N. grandicassis* prefers marine habitats, but juvenile specimens can be

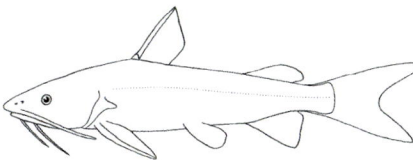

found in lower parts of coastal rivers and estuaries from Venezuela to the mouth of the Amazon River. **BIOLOGY** Females lay 20–30 eggs with a diameter of 10–12 mm, which the males guard for 10–12 days until hatching (Le Bail et al. 2000). Diet includes fish and shrimp.

Sciades (30–150 cm SL)

Characterized by: cephalic shield conspicuously granulated and visible under the skin; lateral ethmoid and frontal limiting a small fenestra (fig. 2D), not visible under the skin; medial groove of neurocranium distinct or indistinct; posterior cranial fontanel always closed (except in *S. platypogon*); epioccipital not invading dorsal portion of cephalic shield; occipital process triangular, its length and width variable, progressively narrower toward its posterior part (fig. 2D); anterior and median nuchal plates fused and indistinct, forming a structure of semilunar aspect or broad and in the form of a shield (fig. 2D); tooth plates associated with vomer round, often fused, forming a single large plate (fig. 1F,G); accessory tooth plates large, transversely disposed (fig. 1F) or subtriangular (fig. 1G), bearing conical teeth; maxillary barbel fleshy and cylindrical; two pairs of mental barbels; base of adipose fin moderately long, about half as long as anal-fin base; lateral line not bifurcated at caudal region, reaching caudal-fin upper lobe (with exception of *S. couma*) (Marceniuk and Menezes 2007). **SPECIES** Six species in the Neotropics, including five in the AOG region: *S. couma*, *S. herzbergii*, *S. parkeri*, *S. passany*, and *S. proops*. **COMMON NAMES** *Bagre catinga* (Brazil); *Couman-couman* (French Guiana); *Bagre cuma* (Spanish); Couma sea catfish (English). **DISTRIBUTION AND HABITAT** AOG species inhabit turbid waters of estuaries and the lower parts of coastal rivers from Venezuela to the mouth of the Amazon River. *Sciades herzbergii* and *S. passany* are especially common in tidal rivers. **BIOLOGY** Some species reach sexual maturity after 1.5–2 years (*S. parkeri* and *S. proops*). Females lay their eggs in a sandy depression, which in some species are many but relatively small eggs, whereas others produce few but large eggs. Males mouth-brood the eggs. Food items include fish, shrimp, and other invertebrates (Le Bail et al. 2000).

FAMILY ASPREDINIDAE—BANJO CATFISHES

— *JOHN P. FRIEL and TIAGO CARVALHO*

DIVERSITY 41 species in 13 genera, including 29 species in the AOG region. The subfamily Aspredininae includes *Aspredo* (1 species), *Aspredinichthys* (2 species), and *Platystacus* (1 species), and the subfamily Bunocephalinae includes *Acanthobunocephalus* (1 species), *Amaralia* (1 species), *Bunocephalus* (13 species), *Pseudobunocephalus* (6 species), *Pterobunocephalus* (2 species), *Xyliphius* (7 species), *Hoplomyzon* (3 species), *Dupouyichthys* (1 species), *and Ernstichthys* (3 species) (Friel 2003, 2008, Cardoso 2010, Figueiredo and Britto 2010, Carvalho et al. 2015). **COMMON NAMES** *Rabeca* (Brazil); *Croncron* (French Guiana); *Banjaman, Banjo-man* (Guyana); *Guitarrero* (Paraguay); *Sapo cunshi* (Peru); *Guitarillo* (Venezuela). **GEOGRAPHIC DISTRIBUTION** Tropical rivers of South America, including the Orinoco, Amazon, São Francisco, and Paraná-Paraguay basins, a few rivers west of the Andes (Atrato, San Juan, Patia, and Magdalena), and some brackish and marine waters between the Orinoco and Amazon river deltas. **ADULT SIZES** From 1.6 cm SL in *Micromyzon akamai* to 38 cm SL in *Aspredo aspredo*.

DIAGNOSIS OF FAMILY Easily distinguished from other siluriform fishes by several features including: overall body shape with a broad flattened head and relatively narrow body and caudal peduncle, roughened skin covered by keratinized tubercles often arranged in parallel rows along the body, opercular openings restricted to small ventral slits, lack of a rigid dorsal spine in most species, lack of an adipose fin, and 10 or fewer caudal-fin rays (Friel 1994, 2003). **SEXUAL DIMORPHISM** In most species mature females are larger than males. This pattern is reversed in *Hoplomyzon sexpapillostoma*. Furthermore, in *Aspredo* and *Platystacus*, the first dorsal-fin ray is elongate as a filament and much longer in males than in females. **HABITATS** Inhabit shallow backwaters to deep river channels to tidal estuaries. In general, most species are cryptically pigmented, benthic, and sluggish unless disturbed. Many are semifossorial during the day, often resting just beneath the substrate surface. **FEEDING ECOLOGY** Most aspredinid species are generalized omnivores and their stomachs often contain aquatic invertebrates, terrestrial insects, and organic debris. A notable exception is *Amaralia* that feeds on the eggs of other catfishes (Friel 1994, Roberts 2015).

BEHAVIOR Aspredinids display some very unusual behaviors and features. Their skin is completely keratinized and covered with tubercles. Periodically the entire outer layer of skin is shed like that of amphibians and reptiles (Friel 1994, 2003). While aspredinids can swim by typical undulatory movements, they can also use jets of water thrust from their restricted gill openings to skip along the substrate (Gradwell 1971, Farina et al. 2015). When agitated, some species produce audible stridulatory sounds by repeatedly abducting and adducting their pectoral spines. Parental care is known in *Dupouyichthys* (Miles 1945), *Pterobunocephalus*, *Platystacus*, *Aspredo*, *Aspredinichthys* (Friel 1994, 2003), and some species of *Bunocephalus*

(Carvalho et al. 2015). In all cases, adults carry their developing embryos attached to the ventral surface of their bodies. In some *Bunocephalus* and *Pterobunocephalus*, the developing eggs are directly attached to the body (Friel 1994, Carvalho et al. 2015) whereas in *Platystacus*, *Aspredo*, and *Aspredinichthys* they are attached to fleshy stalks, called cotylephores, which grow out from the female (Friel 1994, 2003). These develop seasonally and may function in the exchange of materials between the parent and her developing embryos (Wetzel et al. 1997). Also, specimens of the genus *Amaralia* are taught to mimic bean seeds by folding their tails and caudal fin around one side of their body and then remaining motionless whenever disturbed (Roberts 2015), this last behavior being more widespread within the family (e.g., *Bunocephalus*).

KEY TO THE GENERA

1a. >50 anal-fin rays; caudal peduncle with a continuous bony dorsal ridge **2**
1b. <20 anal-fin rays; caudal peduncle without a continuous bony dorsal ridge **4**

2a. 9 principal caudal-fin rays; tubercle rows well developed ....*Platystacus*
2b. 10 principal caudal-fin rays; small barbel at base of maxillary barbel...**3**

3a. Maxillary barbels connected to lateral surface of head by large skin flap; enlarged humeral process; tubercle rows on body highly reduced; 7 pectoral-fin rays..*Aspredo*
3b. Maxillary barbels without large skin flap connected to head; additional pairs of barbels on ventral surface of body; 8 pectoral-fin rays; antorbitals and mesethmoid with hook-like processes ....*Aspredinichthys*

4a. Lower lip equal to upper lip; outermost caudal-fin rays shortened; posterior margin of last anal and dorsal-fin ray not attached to body by a membrane; lateral line truncated at level of dorsal fin **5**
4b. Lower lip subequal to upper lip; posterior margin of last anal- and dorsal-fin ray attached to body by a membrane; lateral line complete **6**

5a. Rigid, locking dorsal spine and spinelet present ....*Acanthobunocephalus*
5b. Dorsal spine flexible; no dorsal spinelet ....*Pseudobunocephalus*

6a. No bony plates on body **7**
6b. Dorsal, ventral, and lateral series of bony plates present on body **10**

7a. Lower lip with numerous papillae; no premaxillary teeth ....*Xyliphius*
7b. Lower lip without papillae; premaxillary teeth present..**8**

8a. 2 or 3 dorsal-fin rays; caudal peduncle deep, laterally compressed...*Amaralia*
8b. >4 dorsal-fin rays; caudal peduncle slender, not laterally compressed **9**

9a. Head and body extremely depressed; skull ornamentation highly reduced or absent; anal fin with 10–20 rays ....*Pterobunocephalus*
9b. Head and body deep or moderately depressed; skull ornamentation typically developed; anal fin with 5–10 rays .... *Bunocephalus*

10a. Papillae present on upper lip; maxillary barbels extensively attached to lateral surface of head; dentary teeth present; 3–4 sets of paired pre-anal-fin plates (fig. 10a; after Stewart 1985) .... *Hoplomyzon*
10b. No papillae present on upper lip; maxillary barbels slightly attached to lateral surface of head; dentary teeth absent; 1–2 sets of paired pre-anal-fin plates...**11**

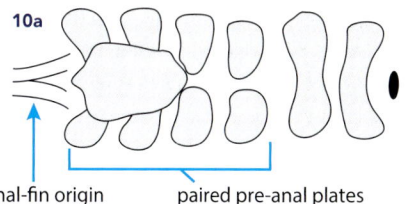

10a

anal-fin origin   paired pre-anal plates

11a. Eyes absent; pigmentation reduced .... *Micromyzon*
11b. Eyes present; heavily pigmented 12

12a. One set of paired pre-anal-fin plates (fig. 12a; after Stewart 1985); pectoral spines only slightly longer than first branched ray; skull ornamentation well developed .... *Dupouyichthys*
12b. Two sets of paired pre-anal-fin plates (fig. 12b; after Stewart 1985); pectoral spines much longer than first branched ray; skull ornamentation reduced..*Ernstichthys*

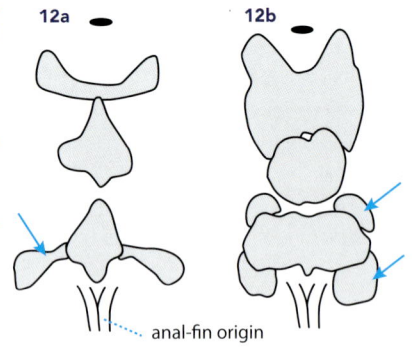

anal-fin origin

GENUS ACCOUNTS

Acanthobunocephalus (2.0 cm SL)

Distinguished from other aspredinids by: dorsal spinelet and rigid dorsal spine that can be locked in position, and absence of procurrent caudal-fin rays (Friel 1995). **SPECIES** One, *A. nicoi*. **DISTRIBUTION AND HABITAT** Edges of clear, tannin-stained rivers in flooded plains, primarily grasses and sedges along upper Orinoco basin, including the Sipapo and Casiquiare rivers in Venezuela and also upper portions of Negro River near São Gabriel da Cachoeira, Brazil. **BIOLOGY** Unknown.

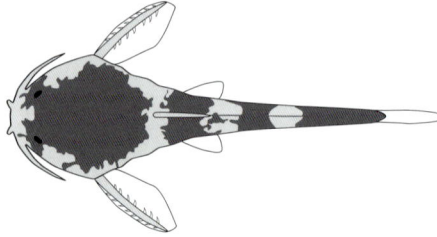

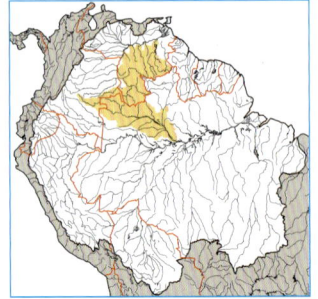

Amaralia (12 cm SL)

Distinguished from other aspredinids by: a highly reduced dorsal fin with only 2 or 3 rays; deep, laterally compressed caudal peduncle; and thickened S-shaped procurrent caudal-fin rays. Other characters not unique to this genus but useful for identification include: anterior margin of pectoral-spine serrations reduced or absent and only 9 caudal-fin rays. **SPECIES** One, *A. hypsiura*. **DISTRIBUTION AND HABITAT** Inhabits the Amazon, Essequibo, and Paraguay-Paraná basins. **BIOLOGY** Feeds on eggs and developing embryos of other catfishes (Friel 1994).

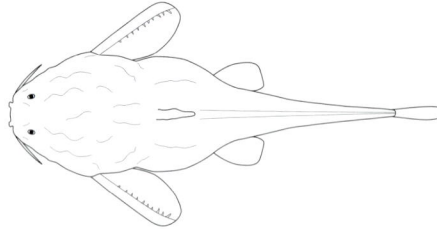

Aspredinichthys (22–24 cm SL)

Distinguished from other aspredinids by: hook-like processes developed on mesethmoid and antorbital bones; 8 pectoral-fin rays; and several pairs of accessory mental barbels present. **SPECIES** Two, both in the AOG region: *A. filamentosus* (22 cm SL) and *A. tibicen* (23.5 cm SL) (Mees 1987). **DISTRIBUTION AND HABITAT** Sandy-muddy littoral waters, estuaries, and flooded zones between northern Brazil and Orinoco River delta. **BIOLOGY** Peculiar mode of egg incubation wherein the adult carries the eggs firmly attached to the underside of the body via fleshy structures called cotylephores (Wetzel et al. 1997, Friel 2003). Tolerant of marine waters, found up to 160 km off the coast of Brazil in the Amazon River mouth.

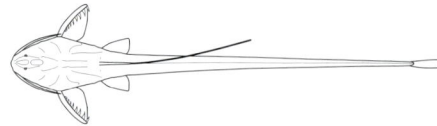

Aspredo (38 cm SL)

Distinguished from other aspredinids by expanded humeral processes, often contacting the posterior coracoid processes in large specimens. Other characters not unique to this genus but useful for identification include the following: mouth broad; maxillary barbel extensively attached to lateral surface of head; one pair of accessory barbels; body pigmentation uniform without any pattern of dark saddles; and unculiferous tubercle rows on the body highly reduced. **SPECIES** One, *A. aspredo*. **DISTRIBUTION AND HABITAT** Benthic fishes that live in sandy-muddy littoral waters, estuaries, and flooded zones of lower portions of rivers between northern Brazil and Orinoco River delta. **BIOLOGY** Peculiar mode of egg incubation wherein the adult carries the eggs firmly attached to the underside of the body via fleshy structures called cotylephores (Wetzel et al. 1997, Friel 2003).

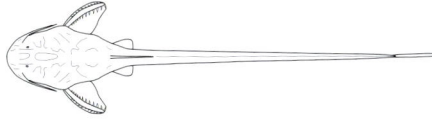

Bunocephalus (4.0–12 cm SL)

Characters not unique to this genus but useful for identification include: mouth subterminal; dorsal surface of head conspicuously tuberous; anal fin with 5–10 rays; and caudal peduncle depressed. **SPECIES** 12, including 7 in the AOG region. **DISTRIBUTION AND HABITAT** Widespread in the AOG, Paraguay-Paraná, Laguna dos Patos system and some rivers west of Andes, living in slow-flowing waters. **BIOLOGY** Generalized omnivores, feeding on terrestrial insects, larvae of aquatic insects, small fish, leaves, and flowers (Mérigoux and Ponton 1998, Melo et al. 2004).

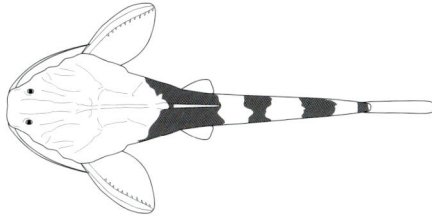

Dupouyichthys (3.0 cm SL)

Distinguished from other aspredinids by having only one set of paired pre-anal-fin plates. Other characters not unique to this genus but useful for identification include: bony ornamentation of skull better developed than in any other members of the Hoplomyzontini. **SPECIES** One, *D. sapito*. **DISTRIBUTION AND HABITAT** Lake Maracaibo basins. Possibly in the Madeira basin (Ohara and Zuanon 2013). **BIOLOGY** Can carry developing eggs attached to the pelvic fins (Miles 1945).

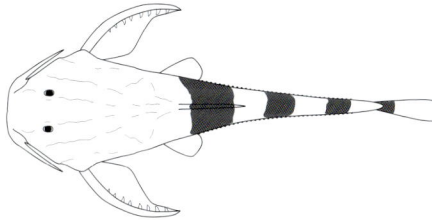

Ernstichthys (3.5–6.7 cm SL)

Distinguished from other aspredinids by: tip of parapophysis of fifth vertebrae expanded; 2 sets of paired pre-anal-fin plates; and pectoral spine long and strongly recurved. **SPECIES** Three, all in the AOG region: *E. anduzei*, *E. intonsus*, and *E. megistus*. See Stewart (1985) for species identification. **DISTRIBUTION AND HABITAT** Western portions of Orinoco and Amazon basins and also in the Essequibo basin in medium- to large-sized rivers. **BIOLOGY** Unknown.

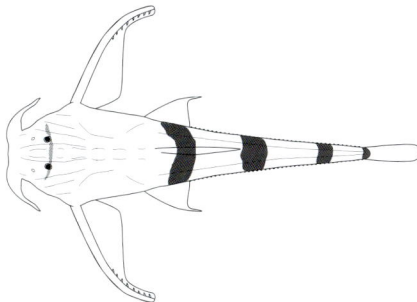

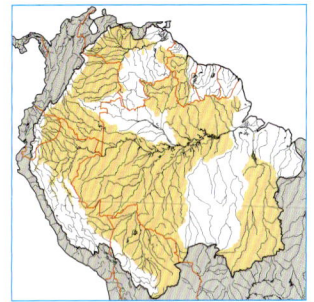

Hoplomyzon (1.7–3.2 cm SL)

Distinguished from other aspredinids by: each premaxilla with 2 bony knobs superficially covered by fleshy papillae; dorsal and ventral armor plates do not overlap; and 2 or 3 sets of paired pre-anal-fin plates. Other characters not unique to this genus but useful for identification include: maxillary barbel attached to head and pectoral spine <25% of standard length. **SPECIES** Three, including two in the AOG region: *H. papillatus* and *H. sexpapilostoma*. See Stewart (1985) for species identification. **DISTRIBUTION AND HABITAT** Western portions of Orinoco and Amazon tributaries in the Andean piedmont, occurring in fast-flowing riffles (Taphorn and Marrero 1990). **BIOLOGY** Unknown.

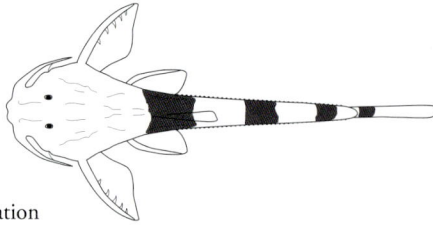

Micromyzon (1.6 cm SL)

Distinguished by: absence of eyes; anterior cranial fontanel highly reduced and posterior cranial fontanel absent; Weberian complex highly reduced or absent; premaxillae extremely reduced; lateral line ossicles hypertrophied to form an armor of overlapping crescent-shaped plates with dorsal and ventral limbs tilted anteriorly; absence of rows of large tubercles along lateral line and posterior portion of body; postcleithral processes short, rounded; posterior coracoid processes very short, not extending past anterior limit of basipterygia; and typical banding pattern of other hoplomyzontines reduced. **SPECIES** One, *M. akamai*. **DISTRIBUTION AND HABITAT** Lower portions of Amazon River and its tributary Madeira River from about 5–20 m in channels of whitewater rivers on sandy substrates (Friel and Lundberg 1996, Ohara and Zuanon 2013). **BIOLOGY** Unknown.

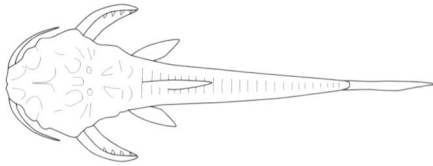

Platystacus (32 cm SL)

Distinguished from other aspredinids by: 4 + 5 caudal-fin rays; absence of accessory maxillary barbels; and well-developed unculiferous tubercles rows. **SPECIES** One, *P. cotylephorus*. **DISTRIBUTION AND HABITAT** Benthic fishes that live in sandy-muddy estuaries and flooded zones of lower portions of rivers between Amazon and Orinoco rivers. **BIOLOGY** Adults carry developing embryos firmly attached to the underside of the body via fleshy structures called cotylephores (Wetzel et al. 1997, Friel 2003).

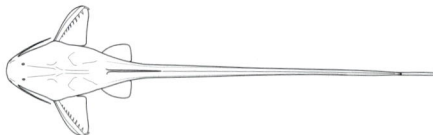

Pseudobunocephalus (3.0–5.3 cm SL)

Distinguished from other aspredinids by: dentary teeth restricted to broad tooth patch near symphysis of lower jaw; metapterygoid does not contact quadrate; posterior end of autopalatine forked; fourth pharyngobranchial absent; gill rakers absent on all gill arches; and parapophysis of fifth vertebrae oriented anteriorly. Characters not unique but useful for identification include: jaws equal; lateral line truncated at approximately the level of the dorsal-fin origin; dorsal- and ventralmost principal caudal rays much shorter than others; and dorsal- and anal-fin membranes not adnate with body. **SPECIES** Six, including four species in the AOG region: *P. amazonicus*, *P. bifidus*, *P. lundbergi*, and *P. quadriradiatus*. See Friel (2008) for species identifications. **DISTRIBUTION AND HABITAT** Widespread in the Orinoco, Amazon, Paraguay-Paraná, and Laguna dos Patos systems and some records for coastal streams near Rio de Janeiro (Mees 1987, Friel 2008) and Mearim basin in northern Brazil. Inhabit slow-flowing waters. **BIOLOGY** Unknown.

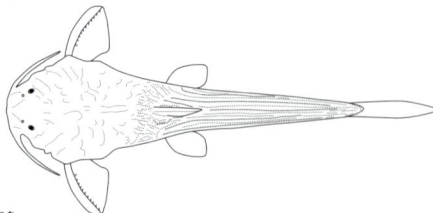

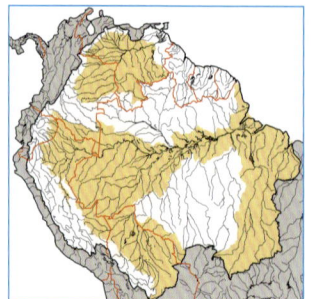

Pterobunocephalus (7.3–8.9 cm SL)

Distinguished from other aspredinids by: head and body extremely depressed; head ornamentation highly reduced or absent; often a distinct notch in upper jaw; 10–20 anal-fin rays. **SPECIES** Two, both in the AOG region: *P. depressus* and *P. dolichurus*. **DISTRIBUTION AND HABITAT** Widespread in the Orinoco, Amazon, Essequibo, and Paraguay-Paraná river systems and typically occurring in waters >5 m in depth. During the dry season can also be found in sandy/muddy beaches in rivers of the Amazon basin (J. Zuanon pers. comm.). **BIOLOGY** Adults carry developing embryos directly attached to the underside of the body (Friel 2003).

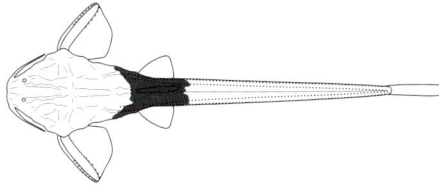

Xyliphius (8.0–15 cm SL)

Distinguished from other aspredinids by: eyes highly reduced; premaxillae toothless and displaced lateral to mesethmoid; row of fleshy papillae projecting anteriorly off lower lip; unculi and unculiferous tubercles flattened. Other characters not unique to this taxon but useful for identification include: openings of anterior nares with papillae; anterior portion of pectoral-fin spine with no serrations; and no dark saddles on body. **SPECIES** Seven, including three species in the AOG region: *X. anachoretes*, *X. lepturus*, and *X. melanopterus*. **DISTRIBUTION AND HABITAT** Tributaries of the western Orinoco and Amazon, upper Tocantins, and Paraguay-Paraná basins, inhabiting deep river channels. **BIOLOGY** Unknown.

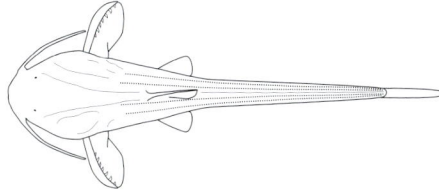

FAMILY ASTROBLEPIDAE—ANDEAN HILLSTREAM OR CLIMBING CATFISHES
— *PETER VAN DER SLEEN* and *JAMES S. ALBERT*

Family includes 58 species in a single genus, *Astroblepus*. found in the Andean mountains at middle and high elevations.

Astroblepus (2.5–30 cm SL)

Characterized by: fleshy, sucker-like mouth with expanded lips, naked body, and a dorsal opening to the gill chamber between the dorsal opercle margin and ventral edge of the pterotic (Schaefer 2003a). Head is broad and depressed; body is somewhat compressed and deep; first unbranched ray of the dorsal, anal, pectoral and pelvic fins thickened and robust, bearing enlarged odontodes, or dermal teeth; first pelvic-fin ray greatly thickened; adipose fin is present or absent, when present, adipose spine is present or absent, often embedded within the fin; dorsal fin short, with 6 or 7 branched rays; pectoral fin broad, with 9–12 rays; pelvic fin with 3 or 4 rays; anal fin with 4–6 rays; caudal fin with 11 branched rays; upper jaw is movably articulated with the skull, broad and flat, bearing several rows of teeth; teeth vary in size and shape within an individual and include large, unicuspid teeth, both large and small symmetrically bifid teeth, and asymmetrically bifid teeth (Schaefer 2003a). Males are sexually dimorphic, with an enlarged genital papilla and modified anal-fin rays (Buitrago and Galvis 1997). **SPECIES** 58. Many species were described before 1950, and no taxonomic review of genus exists since Regan (1904). Species-level taxonomy may be over-split and the number of valid species lower than currently recognized (Schaefer 2003a, Schaefer et al. 2011). **COMMON NAMES** *Guapucha* (Colombia); *Preñadilla* (Ecuador); *Bagre de torrente* (Peru). **DISTRIBUTION AND HABITAT** High-elevation streams in the Andes, from Panama,

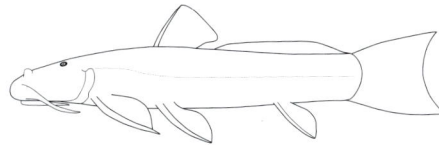

Colombia, Venezuela, Ecuador, Peru, and Bolivia. Most *Astroblepus* species have restricted geographical distributions, being limited to portions of single river drainage basins at elevations above 1,000 m (Schaefer 2003a). **BIOLOGY** Unique among catfishes in possessing a specialized pelvic-fin anatomy that enables them to grasp surfaces and, in association with a ventral sucker-like mouth, to climb vertical rocks or waterfalls (Shelden 1937, de Crop et al. 2013). They feed on aquatic invertebrates.

FAMILY AUCHENIPTERIDAE—DRIFTWOOD CATFISHES

— *JOSÉ L. O. BIRINDELLI and ALBERTO AKAMA*

DIVERSITY At least 123 species in 21 genera and 2 subfamilies, Auchenipterinae and Centromochlinae, including 86 species in the AOG region (Birindelli 2014, Calegari et al. 2014, Vari and Calegari 2014, Sarmento-Soares and Birindelli 2015, Walsh et al. 2015). The Centromochlinae comprises only 4 genera and around 30 species that are diagnosed by having the anal fin of mature males short and entirely modified into an intromittent organ. The remaining species belong to Auchenipterinae, diagnosed by having a long anal fin with only the last unbranched and first branched rays of mature males modified to facilitate insemination.

COMMON NAMES *Hocicón, Mandubé, Mapará, Mararate* (Brazil); *Pez barbudito* (Ecuador); *Maparate* (Peru).

GEOGRAPHIC DISTRIBUTION Small and large rivers throughout the Amazon and Orinoco basins, and in the coastal drainages of the Guianas and Brazil. At least one representative of each auchenipterid genus occurs in the AOG region, except *Pseudotatia*, which includes only a poorly known species endemic to the São Francisco basin.

ADULT SIZES From miniature species of *Gelanoglanis* at 2.5 cm SL (Rengifo et al. 2008), to *Ageneiosus* at more than half a meter in length (Ferraris Jr. 2003b).

DIAGNOSIS OF FAMILY Distinguished from other catfishes by: a naked body (without dermal plates or scutes); a well-developed cephalic shield, large and continuous with the dorsal-fin spine; suborbital groove present; and urogenital opening at base or tip of anterior anal-fin rays in males (Birindelli 2014).

SEXUAL DIMORPHISM Associated with elaborate reproductive behavior (see below), mature males of all species exhibit modified anal-fin rays and urogenital opening forming an inseminating organ. In addition, mature males of several species undergo impressive modifications in the dorsal surface of the head, dorsal-fin spine, and maxillary bone during the breeding period (Ferraris Jr. 2003b). Gonadal modifications are also present: males having testis, seminal vesicles, and pouch distinct from other catfishes, all related to the modifications leading to the production and storage of spermatozoa, with the spermatozoa having long and thin nucleus, tightly compacted in bundles; females having differentiated structures to favor sperm storage (Loir et al. 1989, Reno et al. 2000).

HABITATS Driftwood catfishes are found in distinct aquatic habitats. Centromochlinae species are usually adapted for swimming close to the water surface at night or dusk, often in the middle of river channels or lakes. Auchenipterinae members have more diversified habits, some like *Auchenipterus, Epapterus,* and *Pseudopapteus* form schools and are pelagic in river channels of any size; other members of the subfamily are usually found in flooded forests and marginal lakes, usually associated with submerged plants, driftwood, and litter (Sarmento-Soares et al. 2013).

FEEDING ECOLOGY The dietary range among the species is unusually broad. Some species eat insects near the surface, as do many species of Centromochlinae in crepuscular hours. Large-bodied species of *Ageneiosus, Asterophysus,* and *Trachycorystes* are piscivorous, but with different strategies; *Ageneiosus* are active midwater predators whereas *Asterophysus* and *Trachycorystes* are stalking predators. Pelagic species like *Auchenipterus, Epapterus, Pseudopaterus,* and *Entomocorus* are usually planktivorous. Some species are invertivores but can also forage on algae and other plant materials, and other species forage mainly on fruits and seeds (*Tocantinsia* and *Auchenipterichthys*).

BEHAVIOR Unique reproductive biology among catfishes in that all species have insemination, and the females carry mature, unfertilized eggs and packets of sperm inside the reproductive organs for an extended period of time (Ferraris Jr. 2003b). It is unknown what triggers fertilization and even if that occurs inside the female body (Zuanon pers. comm.). Some species migrate moderate to long distances (Lin and Caramaschi 2005). Most driftwood catfishes are nocturnal, hiding under sunken trees and rocks during the day and coming out to feed at night, during which most species swim close to the surface. Some species of auchenipterids (e.g., *Auchenipterichthys*) produce groaning sounds, by the friction of the pectoral-fin spine against the pectoral girdle (Ladich 2001), and by the contraction of muscles that move the parapophyses of the fourth vertebra (= Müllerian ramus) and gas bladder, in a special arrangement named the elastic-spring apparatus (Birindelli et al. 2012a). Some of the smaller species, like *Entomocorus gameroi*, have an annual life cycle, reaching sexual maturity in one year and dying soon after.

KEY TO THE GENERA MODIFIED FROM BIRINDELLI (2010)

1a. 7–16 rays in the anal fin (total) **2**
1b. 18–60 rays in the anal fin (total) **8**

2a. 5 branched rays in the pelvic fin; 3–5 branched
rays in the dorsal fin; males with strongly angled
anal fin (more or less 45° in relation to body axis),
and with urogenital opening at the base of the
anal-fin rays (fig. 2a)..**3**
2b. 6 or more branched rays in the pelvic fin; 6
branched rays in the dorsal fin; males with anal
fin little angled in relation to body axis and with
urogenital opening located in a small tube that
runs parallel to the anterior anal-fin rays to half of
their length (fig. 2b)...**6**

3a. 12–14 rays in the anal fin (total) .... *Glanidium*
3b. 7–11 rays in the anal fin (total) **4**

4a. One pair of mental barbels; small sized (<4 cm SL); small eyes (<10% of the length of the head) .... *Gelanoglanis*
4b. Two pairs of mental barbels; medium sized (>4 cm SL); big eyes (20–40% of the length of the head)................................**5**

5a. Anal fin relatively long (>8% of SL); caudal peduncle relatively low (<10% of SL) .... *Centromochlus*
5b. Anal fin relatively short (between 3.3% and 8% of SL); caudal peduncle low (between 10.1 and 18.6% of SL) .... *Tatia*

6a. Mouth extremely large, with a diagonal opening; 9–10 branched rays in the pelvic fin..................................... *Asterophysus*
6b. Mouth normal, with a more or less horizontal opening; 6 branched rays in the pelvic fin...**7**

7a. Adipose fin with long base (approximately as long as
the anal fin); cephalic shield from above the eyes with
angle (in lateral view; fig. 7a)............................. *Liosomadoras*
7b. Adipose fin with short base (shorter than anal fin);
cephalic shield from above the eye more or less
horizontal (in lateral view; fig. 7b) .... *Tocantinsia*

8a. 18–34 rays in anal fin (total)...**9**
8b. Anal fin with 35–60 rays (total) **16**

9a. Mental barbels absent; pelvic fin with 6 branched rays...**10**
9b. Mental barbels present; pelvic fin with 5, 7, or 8 branched rays **11**

10a. Pseudotympanum large, similar in size to eye diameter (fig. 10a)... *Tympanopleura*
10b. Pseudotympanum small, much smaller than eye diameter.................. *Ageneiosus*

11a. Inferior lobe of caudal fin with 9–12 branched rays.....................................**12**
11b. Inferior lobe of caudal fin with 8 branched rays..**13**

12a. Pectoral- and dorsal-fin spines with 4 rows of serrations.................... *Spinipterus*
12b. Pectoral- and dorsal-fin spines with 2 rows of serrations............. *Trachelyopterus*

13a. Pelvic fin with 5 branched rays... *Entomocorus*
13b. Pelvic fin with 7–10 branched rays **14**

14a. Dorsal fin with 5 branched rays... *Trachycorystes*
14b. Dorsal fin with 6 branched rays...**15**

15a. Pelvic fin with 7 branched rays.. *Pseudauchenipterus*
15b. Pelvic fin with 8–10 branched rays .... *Auchenipterichthys*

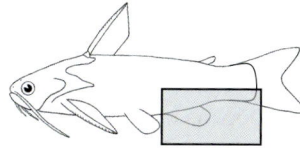

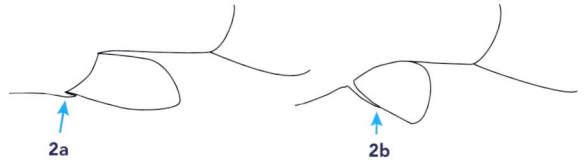

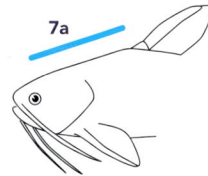

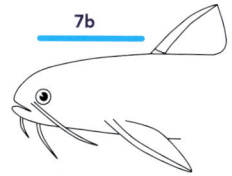

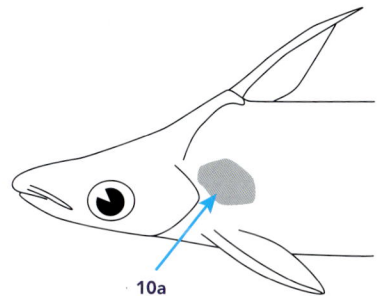

16a. Mental barbels absent or one pair only; pelvic fin with 6 branched rays **17**
16b. Two pairs of mental barbels; pelvic fin with 9–16 branched rays **18**

17a. One pair of mental barbels; pectoral fins with 9–11 branched rays; ventral lobe of caudal fin with 9–12 branched rays; all fins mostly dark brown .... *Tetranematichthys*
17b. Mental barbels absent; pectoral fins with 6 branched rays; ventral lobe of caudal fin with 8 branched rays; fins mostly hyaline .... *Ageneiosus*

18a. Adipose fin absent **19**
18b. Adipose fin present **21**

19a. Dorsal fin with 5 branched rays; ventral lobe of caudal fin with 9–12 branched rays .... *Trachelyichthys*
19b. Dorsal fin with 4 branched rays; ventral lobe of caudal fin with 8 branched rays **20**

20a. Pectoral fin with 6–9 branched rays; premaxilla and dentition with teeth; dorsal fin normally developed; mental barbels located distinctly posterior to mouth margin .... *Trachelyopterichthys*
20b. Pectoral fin with 8–12 branched rays; premaxilla and dentition without teeth; dorsal fin reduced in size (except in mature males); mental barbels located near mouth margin .... *Epapterus*

21a. Dorsal fin normally developed, with 6–7 branched rays .... *Auchenipterus*
21b. Dorsal fin reduced in size, with 3–5 branched rays .... *Pseudepapterus*

GENUS ACCOUNTS

Ageneiosus (12–59 cm SL)

Characterized by: urogenital opening at tip of anteriormost anal-fin rays in mature males; adipose fin present; 21–47 branched rays in anal fin; maxillary barbel short, not reaching pupil of eye; mental barbels absent; pseudotympanum small (much smaller than eye diameter and hardly visible externally); gas bladder reduced in size. Males undergo extreme changes in external morphology during the reproductive period, including elongation of the dorsal-fin spines, which rotate more than 90° and develop extra serrations at base and on top, ossification of the maxillary barbels, and development of spines and swellings of the anteriormost anal-fin rays. **SPECIES** 12, including 6 species in the AOG region. **COMMON NAMES** *Fidalgo, Mandubé, Mapará, Palmito, Perna-de-moça Ximbé* (Brazil); *Cunshi novia* (Peru); *Doncella* (Venezuela); Bottlenose catfish (English). **DISTRIBUTION AND HABITAT** Widespread in the AOG region. *Ageneiosus inermis* grows to 59 cm SL and prefers vegetated backwaters with weak current, where it feeds on small fishes and crustaceans. **BIOLOGY** Many larger-bodied *Ageneiosus* species undertake reproductive migrations (Goulding 1980). Fished by local people for large size (to 2.5 kg) and fine flavor (Boujard et al. 1997).

Asterophysus (25 cm SL)

Characterized by: urogenital opening at tip of anteriormost anal-fin rays in mature males; adipose fin present; 10 to 13 branched anal-fin rays; upper and lower jaws extremely long, mouth gape extremely wide; gas bladder almost completely covered by diverticula. **SPECIES** One, *A. batrachus*. **COMMON NAMES** *Mamyacú* (Brazil). **DISTRIBUTION AND HABITAT** Negro basin in Brazil and upper Orinoco basin in Venezuela. **BIOLOGY** Voracious piscivores that can eat very large prey, up to 80% of their own size (Zuanon and Sazima 2005b).

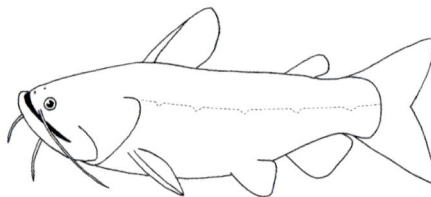

Auchenipterichthys (10–15 cm SL)

Characterized by: urogenital opening at tip of anteriormost anal-fin rays in mature males; adipose fin present; 18–27 branched anal-fin rays; lateral surface of body with several vertically oriented rows of pale spots above the lateral line; 8 branched pelvic-fin rays. **SPECIES** Four species, all in the AOG region. Key to the species in Ferraris Jr. et al. (2005). **COMMON NAMES** *Carawatouí* (Brazil); *Leguia* (Peru). **DISTRIBUTION AND HABITAT** Amazon, Orinoco, and Essequibo basins, including *A. coracoideus* (10 cm SL) in the western Amazon basin, Peru; *A. longimanus* (15 cm SL) in the Amazon basin, Brazil; *A. punctatus* (15 cm SL) in the Negro basin, Brazil; and *A. thoracatus* (11 cm SL) in the Amazon basin, Peru. **BIOLOGY** Mostly insectivorous, although *A. longimanus* also consumes and disperses seeds (Mannheimer et al. 2003).

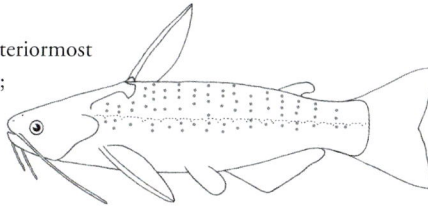

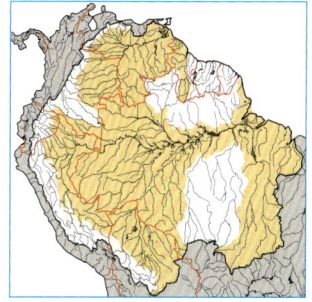

Auchenipterus (8.4–27 cm SL)

Characterized by: urogenital opening at tip of anteriormost anal-fin rays in mature males; adipose fin present; 32–50 branched anal-fin rays; 10–14 branched pelvic-fin rays; 10–14 branched pectoral-fin rays; 6 or 7 branched dorsal-fin rays; mature males with almost completely ossified maxillary barbel, elongate dorsal-fin spine, and hook on third unbranched ray of anal fin. **SPECIES** 11, including 10 species in the AOG region. Key to the species in Ferraris Jr. and Vari (1999). **COMMON NAMES** *Hocicón*, *Carataí*, *Mapará*, *Maparate* (Brazil); *Mararate* (Peru). **DISTRIBUTION AND HABITAT** Widespread in the Amazon, Essequibo, and Orinoco basins and smaller drainages in the Guianas and Suriname. **BIOLOGY** More active at dusk, feeding mainly on insects and small crustaceans. *Auchenipterus* species have pelagic habits and form medium to large schools.

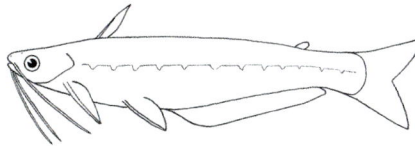

Centromochlus (2.8–10 cm SL)

Characterized by: urogenital opening at base of anteriormost anal-fin rays in mature males; adipose fin present in all species included herein; 5–8 branched anal-fin rays; anal fin oblique relative to the body axis in mature males; caudal peduncle oval in cross section. **SPECIES** 17, including 15 species in the AOG region. **COMMON NAMES** *Aceitero*, *Pirillo* (Peru). **DISTRIBUTION AND HABITAT** Widespread in the Amazon, Essequibo, and Orinoco basins and smaller drainages in the Guianas and Suriname. **BIOLOGY** More active at dusk, feeding mainly on insects and small crustaceans. Many species stay sheltered in logs or crevices in rocks during daytime, sometimes in small groups.

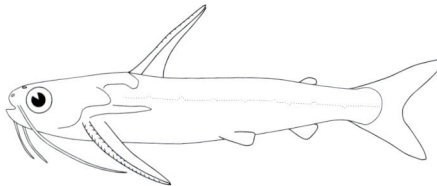

Entomocorus (5.9–7.0 cm SL)

Characterized by: urogenital opening at approximately half length of anteriormost anal-fin rays and located on a tube free from margin of anal-fin rays in mature males; adipose fin present; 14–22 branched anal-fin rays; 5 branched pelvic-fin rays; 8 branched pectoral-fin rays; dermal tubercles on dorsal-fin spine, abdomen, and isthmus in mature males. Sexual dimorphic features of *Entomocorus* males include: dorsal-fin spine elongate, first pelvic-fin ray extremely elongate, maxillary barbel enlarged and rigid for its proximal one-half, and a unique pattern of unculiferous tubercles on dorsal, pectoral, and pelvic fins and dorsal surface of cranium and

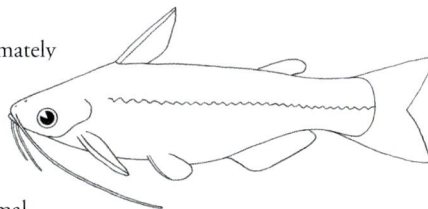

isthmus (see thorough descriptions in Akama and Ferraris Jr. 2003 and Reis and Borges 2006). **SPECIES** Four, including three species in the AOG region. Key to the species in Reis and Borges (2006). **DISTRIBUTION AND HABITAT** *E. benjamini* (7.0 cm SL) from the upper Madeira basin, Bolivia; *E. gameroi* (7.0 cm SL) from the Apure and upper Orinoco basins, Venezuela; and *E. melaphareus* (5.9 cm SL) from the lower Amazon, Brazil. Usually inhabit open environments such as river channels and bays. **BIOLOGY** *E. gameroi* has an annual life cycle, reaching sexual maturity in one year and dying soon after (Reis and Borges 2006).

Epapterus (8.5–13 cm SL)

Characterized by: urogenital opening at tip of anteriormost anal-fin rays in mature males; adipose fin absent; 46–57 branched anal-fin rays; 9–16 branched pelvic-fin rays; 8–12 branched pectoral-fin rays; 4 branched dorsal-fin rays; teeth absent on dentary and premaxilla. Mature males with almost completely ossified maxillary barbel, elongate dorsal-fin spine, and gap on anterior portion of anal fin. **SPECIES** Two, both in the AOG region. Key to the species in Vari and Ferraris Jr. (1999). **DISTRIBUTION AND HABITAT** *E. blohmi* (8.5 cm SL) from the Orinoco basin and Tuy River of the Caribbean coast of Venezuela, and *E. dispilurus* (13 cm SL) from the central and western Amazon basin and Paraguay basin in Paraguay, northern Argentina, and southern Brazil. More abundant in the lower portions of larger rivers. **BIOLOGY** More active at dusk, feeding mainly on insects and small crustaceans when available, and filamentous algae and other plant material during the dry season (Burgess 1989).

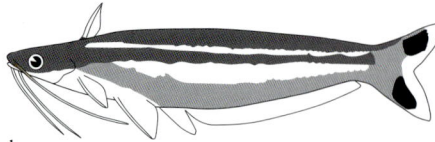

Gelanoglanis (2.2–4.0 cm SL)

Characterized by: small body sizes; urogenital opening at base of anteriormost anal-fin rays or on tube free from margin of anal-fin rays in mature males; adipose fin present; 5 or 7 branched anal-fin rays; anal fin oblique relative to the body axis in mature males; a single pair of mental barbels; premaxillary tooth patches separated at midline; mandibular branch of sensory canal external of lower jaw bones; mouth gap large and sinuous. **SPECIES** Four, all in the AOG region. Key to the species in Calegari et al. (2014). **DISTRIBUTION AND HABITAT** Occurs in disjunct areas of the Amazon and Orinoco basins, rare in collections; *G. nanocticolus* (2.2 cm SL) is from the upper Orinoco and Negro basins in Venezuela; *G. pan* (2.5 cm SL) from the Teles Pires River in the Tapajós basin, Brazil; *G. stroudi* (3.6 cm SL) from the Meta, Apure, and Masparro rivers in the Orinoco basin in Venezuela; and *G. travieso* (2.8 cm SL) from the Marañón River in the Amazon basin in Peru. All species inhabit fast-flowing waters with sand or gravel bottoms and are usually collected in the upper layer of the water column, frequently near rapids. **BIOLOGY** Species are presumably feeding on allochthonous items such as flying insects that fall on the water. X-ray images have shown that females carry no more than a dozen large eggs.

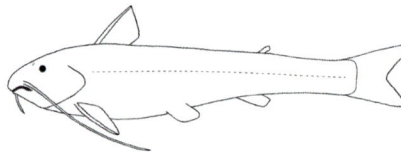

Glanidium (4.8–19 cm SL)

Characterized by: urogenital opening at base of anteriormost anal-fin rays in mature males; adipose fin present in all species included herein; 7–10 branched anal-fin rays; anal fin oblique relative to the body axis in mature males; caudal peduncle oval in cross section; body marbled with dark blotches. Differs from similar-looking *Tatia* and *Centromochlus* species by having a higher number of anal-fin rays (12–14 in total, vs. 7–11 in the latter two). **SPECIES** Eight, including one in the AOG. **DISTRIBUTION AND HABITAT** *G. leopardum* (11 cm SL) inhabits coastal rivers of Guianas. **BIOLOGY** Often captured alone, suggesting solitary or territorial behaviors (Le Bail et al. 2000).

Liosomadoras (6.0–10 cm SL)

Characterized by: urogenital opening at tip of anteriormost anal-fin rays in mature males; adipose fin present; 10–12 branched anal-fin rays; conspicuous spines on the posterior cleithral process; pectoral-fin and dorsal-fin spines strong with serrations on anterior and posterior faces. **SPECIES** Two, both in the AOG region. See species descriptions in Birindelli and Zuanon (2012). **COMMON NAME** Jaguar catfish, from their distinctive color pattern. **DISTRIBUTION AND HABITAT** *L. morrowi* (10 cm SL) from the western Amazon, Purus, and Ucayali rivers, in Peru and Brazil; and *L. oncinus* (6.0 cm SL) from the Negro, Branco, and upper Orinoco basins in Brazil and Venezuela (Birindelli and Zuanon 2012). Often found in cavities of submerged tree trunks (*L. oncinus*) or in association with macrophyte mats (*L. morrowi*) (Birindelli and Zuanon 2012). **BIOLOGY** No data available.

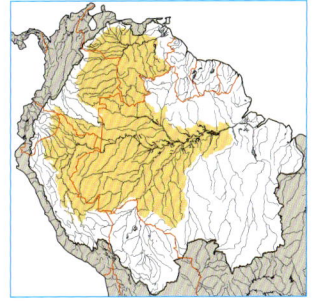

Pseudauchenipterus (22 cm SL)

Characterized by: urogenital opening at tip of anteriormost anal-fin rays in mature males; adipose fin present; 15–20 branched anal-fin rays; pelvic fin with 7 branched rays; soft epidermal papillae over tympanal area; anterior anal-fin rays extending beyond the base of the last anal-fin ray in sexually mature males; a pouch in the tip of gonopodium; a keel on the anterior margin of dorsal-fin spine. **SPECIES** Four, including one in the AOG region, *P. nodosus* (Cocosoda catfish). **DISTRIBUTION AND HABITAT** Lower reaches of rivers and estuaries from Venezuela to Brazil, at least to Bahia, including southern Trinidad. Inhabit muddy bottoms in brackish river mouths. **BIOLOGY** Omnivorous, with tendency toward a detritivorous diet. Commonly found searching for food near village waste waters. Pelvic fins and mandible turn pink during reproduction in December, when male and female mature gonads account for 10% and 25% of body weight, respectively (Le Bail et al. 2000). Lay eggs on mangroves tree roots in the tidal zone that are exposed during low tide.

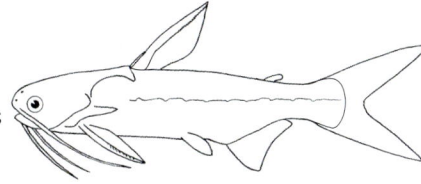

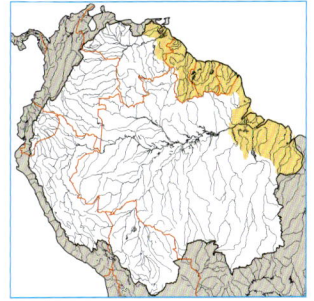

Pseudepapterus (5.9–10 cm SL)

Characterized by: urogenital opening at tip of anteriormost anal-fin rays in mature males; adipose fin present; 44–57 branched anal-fin rays; 9–16 branched pelvic-fin rays; 10–14 branched pectoral-fin rays; 3 or 5 (rarely 6) branched dorsal-fin rays; teeth absent on dentary and premaxilla; mature males with almost completely ossified maxillary barbel, elongate dorsal-fin spine. **SPECIES** Three, all in the AOG region. Key to the species in Ferraris Jr. (2000). **DISTRIBUTION AND HABITAT** *P. cucuhyensis* (5.9 cm SL) from the Amazon and Negro basins, Brazil; *P. gracilis* (8.1 cm SL) from the Caura basin, Venezuela; and *P. hasemani* (10 cm SL) from the Amazon and Tocantins rivers and their larger tributaries. Prefer deep-water habitats (Ferraris Jr. 2000). **BIOLOGY** No data available.

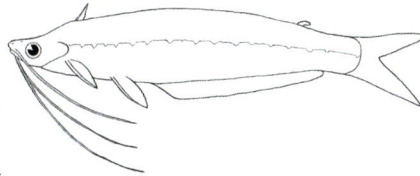

Spinipterus (4.0 cm SL)

Characterized by: urogenital opening at tip of anteriormost anal-fin rays in mature males; adipose fin present; 16 branched anal-fin rays; pectoral and dorsal fin with 4 rows of serrations; lateral margin of skull roofing bones with single row of spines; and a groove along dorsal midline posterior to dorsal fin. **SPECIES** One, *S. acsi*. Species information in Akama and Ferraris Jr. (2011). **DISTRIBUTION AND HABITAT** Known only from the Nanay River near

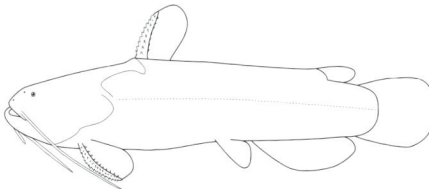

Iquitos, Peru. Additional (yet undescribed) species inhabit the Negro, Madeira, and Branco basins. Cryptobiotic habits, often hidden in wood and leaves (Akama and Ferraris Jr. 2011). **BIOLOGY** No data available.

Tatia (2.3–16 cm SL)

Many species with conspicuous color patterns, incl densely spotted bodies. Characterized by: urogenita opening at base of anteriormost anal-fin rays in mature males; adipose fin present; 5–8 branched anal-fin rays; anal fin oblique relative to the body axis in mature males; anal-fin base of adult males

reduced (anal-fin base length 3.3–8.0% SL); and caudal peduncle compressed and deep (caudal-peduncle depth 10.1–18.6% SL), with a middorsal keel posterior to adipose fin (Sarmento-Soares and Martins-Pinheiro 2008). **SPECIES** 17, including 14 species in the AOG region. Outdated key to the species in Sarmento-Soares and Martins-Pinheiro (2008); 5 new species have been described since (Pavanelli and Bifi 2009, Vari and Ferraris Jr. 2013, Vari and Calegari 2014). **DISTRIBUTION AND HABITAT** Widespread in the Amazon, Orinoco, Essequibo basins and coastal drainages of the Guianas and Suriname (Sarmento-Soares and Martins-Pinheiro 2008). **BIOLOGY** Commonly swim in groups making large circles close to the water surface when foraging at dusk or at night, in the middle of river channels or lakes (Sarmento-Soares et al. 2013, Birindelli et al. 2015).

Tetranematichthys (16–21 cm SL)

Characterized by: urogenital opening at tip of anteriormost anal-fin rays in mature males; adipose fin present; 35–42 branched rays in anal fin; a single pair of mental barbels; head, body, and fins dark brown; mature males with ossified maxillary barbel, elongate dorsal-fin spine and arched cephalic shield; tip of mental barbel bearing projections or serrations.

SPECIES Three, all in the AOG region. Key to the species in Peixoto and Wosiacki (2010). **DISTRIBUTION AND HABITAT** *T. barthemi* (18 cm SL) from the eastern Amazon between the mouth of the Negro and Trombetas rivers in Brazil; *T. quadrifilis* (16 cm SL) has a restricted distribution in upper Madeira basin (including the Guaporé River), and *T. wallacei* (21 cm SL) from the Amazon, Tocantins, and upper Orinoco basins (Vari and Ferraris Jr. 2006, Peixoto and Wosiacki 2010). Species are typically associated with leaf litter in smaller tributaries. **BIOLOGY** Feed primarily on insect larvae and small crustaceans.

Tocantinsia (30 cm SL)

Characterized by: caudal fin forked; urogenital opening at tip of anteriormost anal-fin rays in mature males; adipose fin present; 9 or 10 branched rays in anal fin; posterior process of cleithrum elongate, reaching

vertical through base of pectoral-fin spine; gas bladder with a pair of digitiform posterior diverticula. **SPECIES** One, *T. piresi*. **COMMON NAMES** *Jaú-de-loca* (Brazil). **DISTRIBUTION AND HABITAT** Tocantins, Xingu, Tapajós, and Jari rivers in the Brazilian Amazon. **BIOLOGY** More active at dusk, feeding mainly on insects and small crustaceans (Carvalho and Resende 1984), but also consuming small fish.

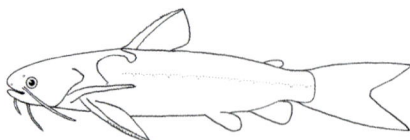

Trachelyichthys (8.0 cm SL)

Characterized by: urogenital opening at tip of anteriormost anal-fin rays in mature males; adipose fin absent; 33–35 branched anal-fin rays; rounded fontanel; body with numerous somewhat rectangular blotches. **SPECIES** Two, both in the AOG region. **DISTRIBUTION AND HABITAT** *T. decaradiatus* (8.0 cm SL) from the Rupununi basin, Guyana, and lower portion of the Negro River, Brazil; and *T. exilis* (8.0 cm SL) known only from the Nanay River near Iquitos, Peru. **BIOLOGY** Cryptic fish with nocturnal or crepuscular habits, usually collected hiding in the substrate or in aquatic plant roots.

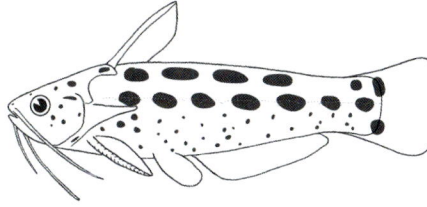

Trachelyopterichthys (14–15 cm SL)

Characterized by: urogenital opening at tip of anteriormost anal-fin rays in mature males; adipose fin absent; 9–16 pelvic-fin branched rays; four branched dorsal-fin rays; body with 2 or 3 dark longitudinal stripes; 39 (*T. anduzei*) or 52–58 (*T. taeniatus*) anal-fin branched rays. **SPECIES** Two, both in the AOG region. **DISTRIBUTION AND HABITAT** *T. anduzei* (14 cm SL) from upper Orinoco River; and *T. taeniatus* (15 cm SL), the Striped woodcat from the Amazon basin. **BIOLOGY** Cryptic fishes often collected associated with substrate or in holes or cracks in submerged trunks.

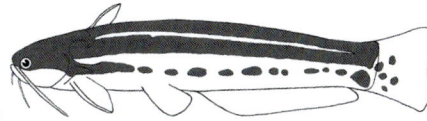

Trachelyopterus (13–28 cm SL)

Characterized by: urogenital opening at tip of anteriormost anal-fin rays in mature males; adipose fin present in most species and absent in *T. coriaceus*; 21–30 anal-fin branched rays; caudal fin obliquely truncate or oblique rounded; prognathous mouth; body marbled brown. **SPECIES** 16, including 4 in the AOG region. **DISTRIBUTION AND HABITAT** *T. ceratophysus* (27 cm SL) from the Guaporé, Branco, and Negro basins in Brazil; *T. coriaceus* (18 cm SL) from the Amazon basin and coastal rivers of Brazil and the Guianas; *T. galeatus* (22 cm SL) is widely distributed in the Amazon and Orinoco basins and coastal rivers of Brazil and the Guianas; and *T. porosus* (13 cm SL) from the Amazon basin. Often collected in floating vegetation. **BIOLOGY** More active at night, and hiding in the vegetation during the day. Omnivorous diet, feeding on invertebrates, small fish, and fruits.

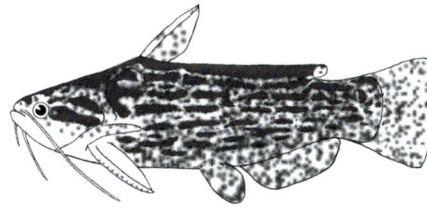

Trachycorystes (6.1–35 cm SL)

Characterized by: urogenital opening at tip of anteriormost anal-fin rays in mature males; adipose fin present; 16–19 anal-fin branched rays; gill rakers reduced or absent; last infraorbital expanded into plate; jaws equal to slightly prognathous; body uniformly dark brown to black. **SPECIES** Four, including two in the AOG region. **DISTRIBUTION AND HABITAT** *T. trachycorystes* (35 cm SL) is widely distributed in Amazon and Orinoco basins; *T. menezesi* (20 cm SL) restricted to the Aripuanã drainage, Madeira basin. **BIOLOGY** Piscivorous, feeding mainly on small fish. *Trachycorystes trachycorystes* (Black catfish) is an ambush predator, and *T. menezesi* an active predator that lives in rocky-bottomed stretches above Dardanelos falls (Britski and Akama 2011).

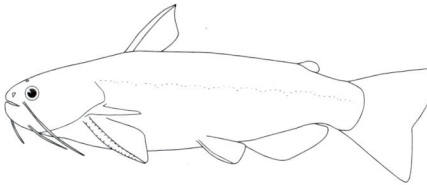

Tympanopleura (4.8–16 cm SL)

Characterized by: urogenital opening at tip of anteriormost anal-fin rays in mature males; adipose fin present; 22–40 branched rays in anal fin; maxillary barbel short, not surpassing pupil; mental barbels absent; pseudotympanum large (larger that eye diameter); gas bladder large (Walsh et al. 2015). Males undergo extreme changes in the external morphology during the reproductive period, including elongation of the dorsal-fin spines, which rotate more than 90° and develop extra serrations at base and on top, ossification of the maxillary barbell, and development of spines, in addition to the swelling of the anteriormost anal-fin rays. **SPECIES** Six species, all in the AOG region. Review of genus and key to the species in Walsh et al. (2015). **DISTRIBUTION AND HABITAT** Widespread in the Amazon, Essequibo and Orinoco basins. **BIOLOGY** No data available.

FAMILY CALLICHTHYIDAE—CALLICHTHYID ARMORED CATFISHES

— *ROBERTO E. REIS and PETER VAN DER SLEEN*

DIVERSITY 204 species in 8 genera and 2 subfamilies, with 154 species in 7 genera in the AOG region. At least 160 species in the genus *Corydoras*, including about 121 species in the AOG region.

GEOGRAPHIC DISTRIBUTION All major river drainages of cis-Andean South America, from the Province of Buenos Aires in Argentina, north to the Orinoco basin, including most Atlantic coastal drainages, and in trans-Andean Colombia and Panama. The highest species diversity inhabits the upper reaches of the Amazon drainage and the Guiana Shield. All genera except *Scleromystax* are found in the AOG region.

ADULT SIZES Small to medium-sized fishes, from 2.5 cm SL in *Corydoras cochui* (Barred-tail corydoras) from the upper Araguaia basin in Brazil, to 24 cm SL in *Hoplosternum littorale* from throughout most of the AOG region.

DIAGNOSIS OF FAMILY Distinguished from other catfish families by their bony armor, which is composed of two rows of plates that almost completely cover the whole body from head to tail.

SEXUAL DIMORPHISM Males often bigger and with thickened pectoral-fin ray. In some species, males also develop distinct coloration during the breeding season.

HABITATS Most waterways in tropical America, from small closed-canopy rainforest streams to the floodplains and margins of larger rivers, including swamps and other waters with very low oxygen levels. All callichthyids can swallow air from the surface, and they absorb oxygen from the air in their intestines. Air breathing is also used for hydrostatic balance, and therefore callichthyids also continuously breathe air in oxygen-rich waters (Gee and Graham 1978). Most are bottom dwellers; the two *Dianema* species and some small-sized *Corydoras* species swim in midwater.

FEEDING ECOLOGY Aquatic invertebrates and organic detritus.

BEHAVIOR Species of Corydoradinae (*Corydoras* and *Aspidoras*) are all substrate brooders, depositing their eggs on rocks, sunken trees, or leaves. Adult males of Callichthyinae (*Callichthys*, *Dianema*, *Hoplosternum*, *Lepthoplosternum*, and *Megalechis*) build floating nests during the time of high water, e.g., in flooded swamps. Nests are made by the males in the center of their territories of plant debris and oxygen-rich foam and defended vigorously (using pectoral spines). Several females might deposit their eggs in a nest, but females are selective, as up to half the nests do not acquire spawns (Mol 1996). Males strongly guard the eggs and often continue to guard newly hatched larvae for several days. The function of the floating nest is to provide oxygen to the eggs by lifting them above the water surface and keeping them moist (Hostache and Mol 1998). This function might be especially important in the hypoxic waters of tropical swamps. *Megalechis*, *Callichthys*, and *Corydoras* species are known to produce sounds. *Corydoras* produces sounds in at least four different behavioral contexts, mostly during courtship (Kaatz and Lobel 1999). These sounds are made, as in other catfishes, by minute bony ridges located on the proximal end of the pectoral-fin spine that are rubbed against the wall of the spinal fossa. In many *Corydoras* species, males and females assume a T-position during courtship, in which the female attaches her mouth to the male's genital opening and directly drinks his semen, which passes through her intestine and is discharged together with eggs into the "pouch" formed by her pelvic fins. In doing so, eggs are mixed with fresh undispersed sperm in an enclosed space, ensuring effective insemination (Kohda et al., 1995).

ADDITIONAL NOTES Many *Corydoras* and *Aspidoras* species are popular aquarium fishes; some species are harvested as a food fish (e.g., *Hoplosternum littorale*)

KEY TO THE GENERA AFTER REIS (1998)

1a. Maxillary barbel long, usually passing the gill opening; dentary with teeth; snout depressed (subfamily Callichthyinae) **2**
1b. Maxillary barbel short, usually not passing the eye; dentary without teeth; snout compressed (subfamily Corydoradinae) .. **6**

2a. Head very depressed, <75% of cleithral width, coracoids not exposed ventrally; infraorbital bones
covered with skin .... *Callichthys*
2b. Head moderately depressed, >75% of cleithral width; coracoids exposed ventrally; infraorbital bones exposed **3**

3a. Caudal fin forked or emarginated **4**
3b. Caudal fin truncated or convex **5**

4a. Dorsal-fin spine as long as the first branched ray .... *Dianema*
4b. Dorsal-fin spine about half the length of the first branched rays .... *Hoplosternum*

5a. Dorsal fin with 1 short spine, 1 unbranched ray, and 7–8 branched rays; anal fin with 2 unbranched and
5–6 branched rays; mature males with a very elongate pectoral-fin spine and with curved distal tip .... *Megalechis*
5b. Dorsal fin with 1 spine and 7 branched rays; anal fin with 1 unbranched and 5 branched rays; mature
males with pectoral-fin spines thickened but not especially elongate .... *Lepthoplosternum*

6a. Two cranial fontanels, the anterior (frontal) fontanel round
or slightly oval in shape, the posterior (supraoccipital)
fontanel is closed in adult specimens, leaving a small
roundish shallow pit (on parieto-supraoccipital) (fig. 6a;
after Nijssen and Isbrücker 1976); ossified portion of
pectoral and dorsal-fin spines short, about one-half length
of first branched ray .... *Aspidoras*
6b. A single, large and elongate, open (frontal) fontanel (fig. 6b; after
Nijssen and Isbrücker 1976); no shallow pit on parieto-supraoccipital;
ossified portion of pectoral and dorsal-fin spines long, about as long
as the first branched ray .... *Corydoras*

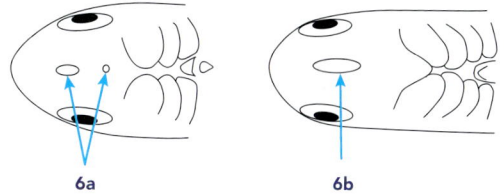

6a   6b

GENUS ACCOUNTS

Aspidoras (2.5–5.0 cm SL)

Characterized by a distinctive body shape
with a foreshortened snout and slender body.
Distinguished from *Corydoras* by a reduced
frontal fontanel; a roundish fossa in the
parieto-supraoccipital; and a short ossified
portion of the pectoral and dorsal-fin spines,
being about one-half the length of first
branched ray. **SPECIES** 20, all in the Amazon basin. Outdated key to the species in Nijssen and
Isbrücker (1976). **COMMON NAMES** *Comboeiro* (Brazil). **DISTRIBUTION AND HABITAT** Only known
from Brazil. Most species are found on the Brazilian Shield and about 9 occur in the Xingu
and Tocantins basins, with only one species (*A. pauciradiatus*) widespread in the Amazon basin.
BIOLOGY Most species are active during the day. They have omnivorous or detritivorous feeding
habits, and some are important in the ornamental fish market. They are substrate brooders and
deposit eggs on rocks, sunken trees, or leaves.

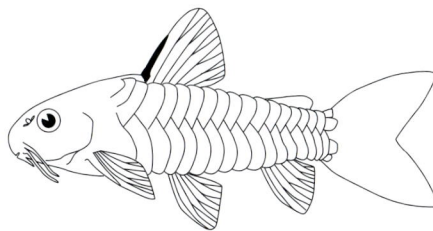

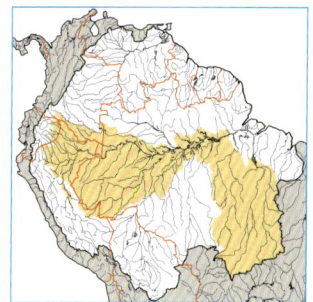

Callichthys (13–17 cm SL)

Distinguished from other callichthyids by: a very depressed head (<75% of the cleithral width,
except sometimes in mature males); eyes set on midlateral portion of head; coracoids covered
by skin and not exposed ventrally; and infraorbital bones also covered by skin. **SPECIES** Four,

including two species in the Amazon basin. Key to the species in Lehmann and Reis (2004). **COMMON NAMES** *Caboje, Tamatá, Tamboatá* (Brazil); *Shirui corto* (Peru); *Curito, Cochinito* (Venezuela). **DISTRIBUTION AND HABITAT** *C. callichthys* from throughout most of tropical and subtropical South America; *C. serralabium* from the upper Orinoco and lower Negro basins. Usually found in slow-moving rivers, pools, and swamps with muddy bottoms and use their intestines for additional air breathing in these low-oxygen habitats. **BIOLOGY** Feed at night on minute fish, insects, and plant matter. Juveniles feed on rotifers, in addition to the microcrustaceans and aquatic insect larvae they find when digging into the substrate (Le Bail et al. 2000). Males make a floating bubble nest during reproduction period and strongly guard it after the female lays down her eggs (Mol 1996). They can produce sounds, e.g., grunts and squeaks. Illustration after Mol (1995).

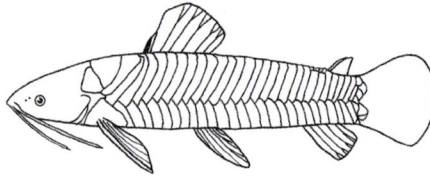

Corydoras (2.5–8.8 cm SL)

Easily identified by their distinctive body shape, with a compressed snout and deep body with a high back. Differentiated from *Aspidoras* by having elongate frontal fontanel; absence of a roundish fossa in the parieto-supraoccipital; and a long ossified portion of the pectoral- and dorsal-fin spines, about as long as the first branched ray. All species have strong spines in the dorsal and pectoral fins. **SPECIES** The most species-rich catfish genus, with at least 160 species, including at least 121 species in the AOG region (Alexandrou et al. 2011, Mariguela et al. 2013a). Photographic guides to **SPECIES** e.g., Glaser et al. (1996) and Fuller and Evers (2005). **COMMON NAMES** *Camboatazinho, Corredeira, Coridora, Papa-areia, Tamuatá* (Brazil). **DISTRIBUTION AND HABITAT** Throughout most of tropical South America east of the Andes to the Atlantic coast, and from Trinidad to the Río de la Plata drainage in northern Argentina. Typically found in smaller streams, river margins, and marshes and ponds. **BIOLOGY** Usually found in schools and active during the day. Most species are bottom dwellers, with diets including insects, larvae, worms, and vegetable matter. They are substrate brooders, and deposit eggs on rocks, sunken trees, or leaves. Many species are popular aquarium fishes.

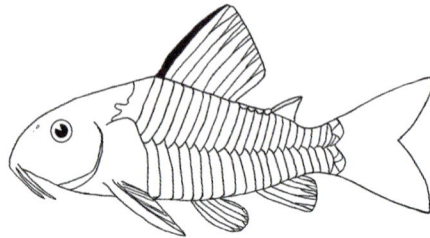

Dianema (8.2–8.4 cm SL)

Characterized by: maxillary barbel long, usually passing the gill opening; dentary with teeth; snout depressed; head moderately depressed, more than 75% of cleithral width; coracoids exposed ventrally; infraorbital bones exposed; caudal fin forked or emarginated; dorsal-fin spine as long as first branched ray. **SPECIES** Two, *D. longibarbis* (Porthole catfish) and *D. urostriatus* (Flagtail catfish). **DISTRIBUTION AND HABITAT** Both species from the Amazon basin. *Dianema* species swim in the midwater (whereas other callichthyids are in general bottom dwellers). **BIOLOGY** Omnivorous. Males make a floating bubble nest during reproduction period and guard the eggs during incubation.

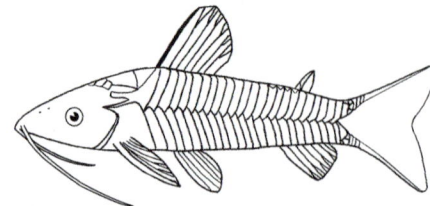

Hoplosternum (7.7–24 cm SL)

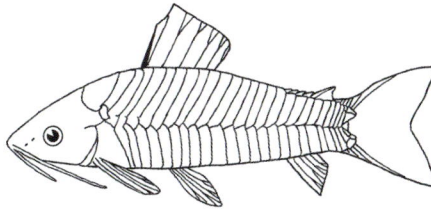

Characterized by: long maxillary barbels, usually passing the gill opening; dentary with teeth; head moderately depressed, >75% of cleithral width; coracoids exposed ventrally; infraorbital bones exposed; caudal fin bifurcated or concave; dorsal-fin spine half the length of the first branched ray. **SPECIES** Three, including one nominal species, *H. littorale*, in the AOG region. Key to the species in Reis (1997). **COMMON NAMES** *Camboatá, Tamboatá* (Brazil); *Shirui alargado* (Peru). **DISTRIBUTION AND HABITAT** *H. littorale* from most of the Amazon and Orinoco basins, and coastal rivers of the Guianas and northern Brazil. Lives mostly in swamps and in stagnant floodplain waters of whitewater rivers, and uses its intestines for additional air breathing in these low-oxygen habitats. **BIOLOGY** Males make a floating bubble nest during reproduction period and guard the eggs during incubation (Mol 1996, Hostache and Mol 1998). Mainly nocturnal, feeding on benthic invertebrates and detritus. *Hoplosternum littorale* is harvested as a food fish throughout its range. Illustration after Mol (1995).

Lepthoplosternum (4.6–6.0 cm SL)

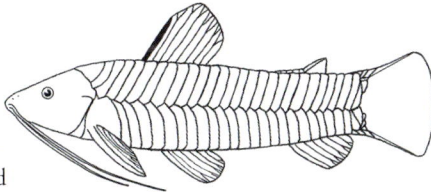

Characterized by: lower lip with a deep medial notch and an additional small lateral notch, forming fleshy projections on each side (Reis and Kaefer 2005); maxillary barbel long, usually passing the gill opening; dentary with teeth; head moderately depressed, >75% of cleithral width; coracoids exposed ventrally; infraorbital bones exposed; caudal fin truncated or convex; dorsal fin with one spine and 7 branched rays; anal fin with one unbranched and 5 branched rays; and body often with dark spots. Mature males have a thickened pectoral-fin spine but not very elongate (different from *Megalechis*). **SPECIES** Six, all in the AOG region. Key to the species in Reis and Kaefer (2005). **DISTRIBUTION AND HABITAT** Widely distributed in cis-Andean South America south of the Orinoco basin. Species typically inhabit slow-flowing waters with marginal or floating vegetation. One species has been found in hypoxic conditions (*L. ucamara*), whereas another was also found in a well-oxygenated stream (*L. stellatum*). **BIOLOGY** Males make a floating bubble nest during reproduction period and guard the eggs during incubation.

Megalechis (12–16 cm SL)

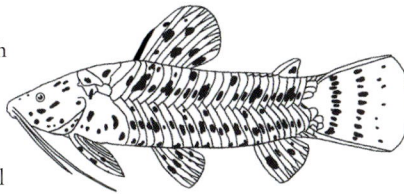

Characterized by: caudal fin convex; dorsal fin with one short spine, one unbranched ray and 7–8 branched rays; anal fin with 2 unbranched and 5–6 branched rays; body usually with many dark spots; maxillary barbel long, usually passing the gill opening; dentary with teeth; head moderately depressed, >75% of cleithral width; coracoids exposed ventrally; and infraorbital bones exposed. Mature males have a thickened pectoral-fin spine which is very elongate and with an upward-curved tip. **SPECIES** Two, both in the AOG region, including *M. picta* (Spotted hoplo). Species information in Reis et al. (2005). **DISTRIBUTION AND HABITAT** Both species inhabit the Amazon and Orinoco basins, and coastal rivers of the Guianas and northern Brazil. *Megalechis thoracata* is also found in the upper Paraguay basins. They are found in ponds, swamps, flooded savannas, and creeks. **BIOLOGY** Intestinal breathing enables *Megalechis* species to adapt to extreme hypoxic conditions. In the dry season, when swamps dry, they burrow in the mud at a depth of 15–25 cm and remain there until the first rains (Le Bail et al. 2000). *Megalechis thoracata* have territories in which they take care of the brood. Males defend their territories strongly and emit sounds at their nest sites, especially during spawning.

FAMILY CETOPSIDAE—WHALE CATFISHES

— *PETER VAN DER SLEEN and JAMES ALBERT*

DIVERSITY 39 species in 5 genera and 2 subfamilies; Cetopsinae with 4 genera and 38 species, and Helogeninae with 1 genus (*Helogenes*) and 4 species (Vari et al. 2005, De Pinna et al. 2007). All genera (except *Paracetopsis*) and 34 species in AOG region.
COMMON NAMES *Candirú, Candirú-açu* (Brazil); *Canero* (Peru); *Bagre ciego* (Spanish).
GEOGRAPHIC DISTRIBUTION Species in the subfamily Cetopsinae inhabit northern and central South America, on both sides of the Andes. *Paracetopsis* is found exclusively on the Pacific slope of Ecuador and northwestern Peru. *Helogenes* occurs throughout much of the Amazon and upper Orinoco basins, and coastal rivers of the Guianas.
ADULT SIZES From 1.8 cm SL in *Denticetopsis royeroi* from small rainforest streams in the upper Negro basin of Venezuela, to 27 cm SL in Cetopsis coecutiens from larger rivers throughout the AOG region.
DIAGNOSIS OF FAMILY Whale catfishes have smooth bodies without bony plates; an anal fin with a long base; absence of a nasal barbel; absence of a free orbital margin; absence of spines in the pectoral and dorsal fins, other than in a few species of the subfamily Cetopsinae; adipose fin is absent in adults (except in some adults of the genus *Helogenes*); and usually very reduced eyes, or even lacking eyes in *Cetopsis oliveirai* (Vari and Ferraris Jr. 2003). The name Whale catfish is derived from the robust bodies and relatively blunt heads of some of the first described species in this family, hence vaguely resembling whales.
SEXUAL DIMORPHISM Adult males in the genera *Cetopsidium*, *Cetopsis*, and *Paracetopsis* have an elongate filament extending from the tip of the first rays of the dorsal and pectoral fins (Vari et al. 2005). In *Denticetopsis* and *Helogenes* no sexual dimorphism has been observed.
HABITATS From large rivers (*Cetopsis*) to small forest streams (*Cetopsidium, Denticetopsis*, and *Helogenes*). *Helogenes* are "shy" or cryptobiotic fishes, hiding in submerged leaf litter, with *H. marmoratus* having strong morphological and color resemblances to dead leaves.
FEEDING ECOLOGY Most species feed largely on terrestrial insects (Vari and Ferraris Jr. 2003). Some species have been said to be parasitic, but none are. Several of the larger species in the genus *Cetopsis* are strongly carnivorous (mostly carrion feeders) and are notorious for their feeding habits, occasionally attacking humans (e.g., Barthem and Goulding 1997).

KEY TO THE GENERA BASED ON VARI ET AL. (2005)

1a. Dorsal fin clearly located on anterior half of the body; adipose fin absent in adults **2**
1b. Dorsal fin located halfway or on posterior half of the body; adipose fin usually present..*Helogenes*

2a. First pectoral-fin ray spinous for basal one-half of its length..**3**
2b. First pectoral-fin ray not spinous... *Cetopsis*

3a. Mental barbels extending beyond the rear margin of the gill cover; deeply forked caudal fin............................ *Cetopsidium*
3b. Mental barbels not extending beyond the rear margin of the gill cover; shallowly forked or obliquely truncate caudal fin..*Denticetopsis*

GENUS ACCOUNTS

Cetopsidium (3.6–5.8 cm SL)

Characterized by: dorsal fin clearly located on anterior half of the body; adipose fin absent in adults; dorsal fin with first ray spinous and with an ossified spinelet anterior of dorsal-fin spine; first pectoral-fin ray spinous for the basal one-half of its length; mental barbels passing beyond the gill cover; and a deeply forked caudal fin (Vari et al. 2005). Sexual dimorphism includes filaments on the first rays of the dorsal and pectoral fins, and a broadly convex anal-fin margin in mature males (vs. filaments lacking and a straight or nearly straight anal-fin margin in females and immature males). **SPECIES** Seven, all in the AOG region. Key to species and species information in Vari et al. (2005), except for *C. soniae* (Vari and Ferraris Jr. 2009). **DISTRIBUTION AND HABITAT** Madeira basin, Guiana Shield area and Orinoco basin. *Cetopsidium* species typically inhabit streams and smaller rivers. **BIOLOGY** No data available.

Cetopsis (3.6–27 cm SL)

Characterized by: dorsal fin clearly located on anterior half of the body; adipose fin absent in adults; absence of a spinelet associated with the dorsal fin; absence of spines on the dorsal and pectoral fins; and the possession of a single row of teeth on the vomer (Vari et al. 2005). Mature males with filaments on the first rays of the dorsal and pectoral fins that are proportionally more elongate than the extensions on those rays in females and immature males. **SPECIES** 21, including 16 species in the AOG region. Key to the species and species information in Vari et al. (2005). **COMMON NAMES** *Candirú, Candirú-açu* (Brazil); *Canero* (Peru); *Bagre ciego* (Spanish). **DISTRIBUTION AND HABITAT** Throughout the Amazon and Orinoco basins. Most species live in large rivers. **BIOLOGY** Larger-bodied species like C. coecutiens (27 cm SL) are strong carnivores and scavengers, notorious for their voracious feeding habits. Others like *C. oliveirai* (3.6 cm SL), a blind species without pigments, feed on terrestrial arthropods, including spiders, ants and dipterans.

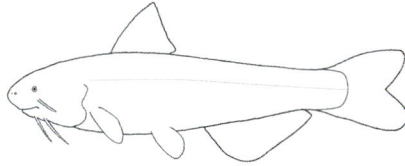

Denticetopsis (1.8–6.7 cm SL)

Characterized by: dorsal fin clearly located on anterior half of the body; adipose fin absent in adults; dorsal fin with first ray spinous in some, but not all species; dorsal fin without ossified spinelet anterior of dorsal-fin spine; first pectoral-fin ray spinous for the basal one-half of its length; mental barbels not passing beyond the gill cover; a conspicuous w-shaped constriction behind the head; and a shallowly forked or even obliquely truncate caudal fin with the outmost rays no more than 1.5 times the length of the innermost rays (vs. 1.7–2.0 times that length in other genera of Cetopsinae; Vari et al. 2005). **SPECIES** Seven, all in the AOG region. Key to the species and species information in Vari et al. (2005). **DISTRIBUTION AND HABITAT** Amazon basin and coastal rivers in the Guianas. Frequently found sheltered amid submerged dead palm leaves in small forest streams during the day (J. Zuanon pers. comm.). **BIOLOGY** Stomach content of *D. praecox* contained insect larvae (Ferraris Jr. and Brown 1991).

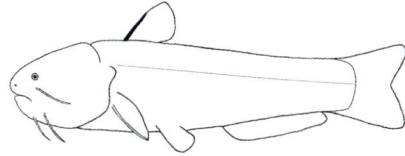

Helogenes (4.3–7.3 cm SL)

Characterized by the position of dorsal fin at or posterior to midbody (vs. anterior half of body in Cetopsinae). In addition, some (but not all) adult *Helogenes* have an adipose fin, which is absent in adult Cetopsinae (although sometimes present in juveniles). **SPECIES** Four, all in the AOG region. Revision of the genus and key to the species in Vari and Ortega (1986). **DISTRIBUTION AND HABITAT** H. castaneus (4.7 cm SL) from the upper Orinoco, Guaviare, and Meta rivers in the Orinoco basin; H. gouldingi (4.7 cm SL) from the Madeira basin; H. marmoratus (7.3 cm SL) widespread throughout the Amazon, Orinoco, and Guianas; and H. uruyensis (4.3 cm SL) from the Uruyén (or Yurwan) River near Angel Falls in the pantepui region of Venezuela. Species typically inhabiting smaller blackwater and clearwater forest streams. **BIOLOGY** Feed mainly on terrestrial insects, e.g., ants (Vari and Ortega 1986). *Helogenes marmoratus* is well camouflaged and can easily be mistaken for a dead leaf or a piece of wood. It is nocturnal and hides in plant debris and leaf litter during the day (Sazima et al. 2006), and often swims on one side in undulating movements.

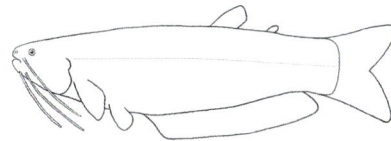

FAMILY DORADIDAE—THORNY CATFISHES

— *JOSÉ L. O. BIRINDELLI and LEANDRO M. DE SOUSA*

DIVERSITY 94 species (93 extant and one fossil) in 31 genera, including 82 species in 26 genera in the AOG region (Arce et al. 2013, Birindelli 2014, Sabaj Pérez et al. 2014). Approximately 30% of all valid doradid species have been described since 2005. The family is composed of three subfamilies: Wertheimerinae, Astrodoradinae, and Doradinae (Birindelli 2014). The first group is composed exclusively of species endemic to drainages of eastern Brazil: *Wertheimeria maculata*, *Kalyptodoras bahiensis*, and possibly *Franciscodoras marmoratus*. The second group is composed of relatively small fishes with depressed head and body and bulged pectoral girdle, encompassed in the genera *Amblydoras*, *Anadoras*, *Astrodoras*, *Hypodoras*, *Physopyxis*, and *Scorpiodoras* (Higuchi et al. 2007). The Doradinae comprise the majority of species in the family, easily diagnosed from other doradids by the fimbriate barbels. Nevertheless, there is still controversy on the phylogenetic hypothesis of doradids inferred from morphological and molecular data, especially concerning the relationships of *Acanthodoras*, *Agamyxis*, and *Franciscodoras* (Arce et al. 2013, Birindelli 2014).

COMMON NAMES *Abotoados, Armados* (Brazil); *Bagres espinosos, Sierras* (Spanish).

GEOGRAPHIC DISTRIBUTION Inhabit all large rivers of the Amazon and Orinoco basins, and coastal drainages of the Guianas (Sabaj and Ferraris 2003). Only nine species are not found in the AOG region: *Franciscodoras marmoratus*, *Kalyptodoras bahiensis*, and *Wertheimeria maculata* restricted to eastern Brazil, *Amblydoras nheco* and *Ossancora eigenmanni* from the Pantanal region of the Paraguay basin, *Rhinodoras dorbignyi* from the Paraná, Paraguay, and Uruguay basins, and *Centrochir crocodili*, *Doraops zuloagai*, and *Rhinodoras thomersoni* endemic to trans-Andean drainages in Colombia and Venezuela.

ADULT SIZES From 2.2 cm SL in *Physopyxis ananas* (Sousa and Rapp Py-Daniel 2005) to 100 cm SL in *Oxydoras niger* (Sabaj and Ferraris 2003).

DIAGNOSIS OF FAMILY Characterized by a series of bony scutes along the side of the body (fig. 1a), each of which has a robust, backward-directed, medial thorn and sometimes smaller additional thorns and serrations; a well-developed cephalic shield, large and continuous until the dorsal-fin spine (fig. 1b); a subterminal mouth; a large, exposed posterior cleithral process (fig. 1c); strongly serrated dorsal and pectoral-fin spines; and a ventrally flattened body (Sabaj and Ferraris 2003).

SEXUAL DIMORPHISM The only dimorphic feature observed in doradids is an elongation of the dorsal-fin spine in mature males of some species, including those of the genera *Hassar*, *Hemidoras*, *Nemadoras*, *Tenellus*, and *Trachydoras* (Rapp Py-Daniel and Cox Fernandes 2005, Birindelli et al. 2011, Sabaj Pérez et al. 2014; pers. obs.).

HABITATS Thorny catfishes inhabit distinct aquatic habitats. Most of the large-sized species are typically found in the main channels of big rivers. Small-sized species with large eyes and deep head (e.g., fimbriate-barbel doradids) are commonly collected in sand beaches as well as in the main channel of big rivers (Sabaj-Pérez and Birindelli 2008, Birindelli et al. 2011, Sabaj Pérez et al. 2014). Small-sized species with small eyes and depressed head (e.g., Astrodoradinae) usually inhabit flooded forests and marginal lakes, preferably associated with submerged plants and litter (Higuchi et al. 2007, Sousa and Birindelli 2011).

FEEDING ECOLOGY Two of the largest species, *Lithodoras dorsalis* and *Megalodoras uranoscopus*, feed on snails as young and on seeds and fruits as adults (Miranda Ribeiro 1911, Starks 1913, Eigenmann 1925, Goulding 1980). Additional malacophagous species are *Franciscodoras marmoratus* (von Ihering and Azevedo 1934) and *Rhinodoras dorbignyi* (Veitenheimer and Mansur 1975). *Doraops zuloagai* and *Kalyptodoras bahiensis* feed on mollusks and crabs (Schultz 1944a, Higuchi et al. 1990; pers. obs.). Species of *Platydoras* and *Hassar* are omnivores feeding on insects, crabs, and vegetal items (Menezes and Menezes 1948, Menezes 1949). The long-snouted species are generally considered benthivores. For example, *Oxydoras niger* is a detritivore (Goulding 1980), *Leptodoras* cf. *cataniai* feeds on shallowly buried chironomid larvae ingested with detritus and grains of sand (Sabaj 2005), and *Trachydoras paraguayensis* is an omnivore feeding on microcrustaceans, protozoans, nematodes, aquatic mites, insect larvae, and algae, also ingested with detritus and sand (Hahn et al. 1991, Fugi et al. 2001). The only predominantly herbivorous species are *Pterodoras granulosus* and *Lithodoras dorsalis* (Hahn et al. 1992, Ferriz et al. 2000, Devincenzi and Teague 1942).

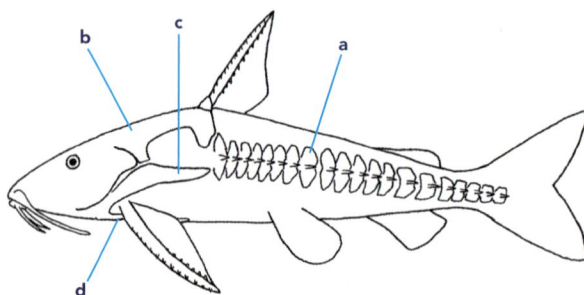

FIGURE 1 A generalized doradid species showing (a) the series of lateral bony scutes, (b) the cephalic shield, and the pectoral girdle formed by the paired cleithrum (c) and coracoid bones (d). The cleithrum is extended posteriorly, forming the posterior process of the cleithrum (c), which is large and exposed in all doradids.

BEHAVIOR Information on the behavior and reproduction of doradids is scarce. Some species are known to migrate moderate to long distances, fertilize externally, and have no parental care (Agostinho et al. 2003, Sato et al. 2003). In fish tanks, some doradids (e.g., Astrodoradinae and *Platydoras*) show very little activity, whereas others (e.g., fimbriate-barbel doradids) are much more active. Most (if not all) species are predominantly nocturnal. Some species often burry themselves in sand, or look for shelter among roots of floating plants (e.g., *Physopyxis* and *Amblydoras*) or inside logs (e.g., *Platydoras* and *Acanthodoras*). Doradids are commonly known as talking catfishes because of their ability to produce groaning sounds by the friction of the pectoral-fin spine against the pectoral girdle (Ladich 2001), and by the contraction of muscles that move the parapophyses of the fourth vertebra (= Müllerian ramus) and gas bladder, in a special arrangement named the elastic-spring apparatus. Among doradids, species with smaller gas bladders have relatively more sensitive hearing (Zebedin and Ladich 2013).

KEY TO THE GENERA MODIFIED FROM BIRINDELLI (2010) AND SOUSA (2010)

1a. Maxillary barbels simple (fig. 1a); lateral mental barbels much longer than medial ones (fig. 1aa); midlateral scutes perpendicular to long axis of body **2**
1b. Maxillary barbels branched, with fimbriae (fig. 1b); mental barbels similar in length; midlateral scutes oblique to long axis of body **18**

2a. 5–7 branched rays on ventral lobe of caudal fin **3**
2b. 8 branched rays on ventral lobe of caudal fin **10**

3a. Coloration of head and body dark brown or black, with lighter spots or stripes over entire body; dorsal-fin spine with serrations on its lateral face; pectoral-fin spine with serrations on its dorsal face; caudal-fin rays with spines; 5 branched rays in the pelvic fin..**4**
3b. Coloration of head and body beige or light brown, with darker spots or stripes over entire body; dorsal-fin spine without serrations on its lateral face; pectoral-fin spine without serrations on its dorsal face; caudal-fin rays without spines; 6 branched rays in the pelvic fin...**5**

4a. Procurrent rays of caudal fin modified to plates (fig. 4a); caudal fin truncate; 6 branched rays in dorsal lobe of caudal fin; 6 branched rays in dorsal fin; body with lighter spots not forming a midlateral stripe.. *Agamyxis*
4b. Procurrent rays of caudal fin normal, not modified to plates; caudal fin rounded; 7 branched rays in dorsal lobe of caudal fin; 5 branched rays in dorsal fin; body with lighter spots forming (in general) a midlateral stripe .... *Acanthodoras*

5a. A dark longitudinal stripe on each lobe of the caudal fin .... *Anadoras*
5b. Dark dots or spots on entire area of the caudal fin **6**

6a. Posterior process of coracoid elongate, distal tip more posterior than that of cleithrum (fig. 6a); small size (specimens <3 cm SL)............................. *Physopyxis*
6b. Posterior process of coracoid short, distal tip more anterior than that of cleithrum (fig. 6b); medium size (specimens reaching >3 cm SL)... **7**

7a. Body relatively flattened (height at dorsal-fin origin <60% of width at cleithral bulge)..**8**
7b. Body relatively high (height at dorsal-fin origin >65% of width at cleithral bulge) **9**

8a. A bony scute immediately anterior to adipose fin; 6 branched rays in the dorsal lobe of the caudal fin .... *Hypodoras*
8b. No scutes anterior to adipose fin; 7 branched rays in the dorsal lobe of the caudal fin .... *Astrodoras*

9a. Caudal fin bifurcate .... *Scorpiodoras*
9b. Caudal fin truncate .... *Amblydoras*

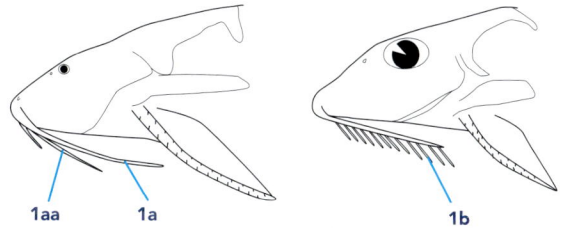

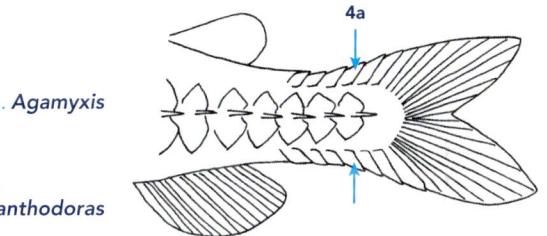

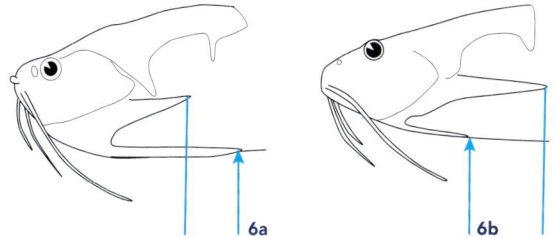

10a. Procurrent rays of caudal fin modified to plates (fig. 4a) **11**
10b. Procurrent rays of caudal fin not modified to plates..**14**

11a. Mouth subterminal; mental barbels with separated bases; anterior
nuchal plate well developed (fig. 11a) **12**
11b. Mouth formed as a tweezers and ventrally oriented, mental barbels
with joint bases; anterior nuchal plate reduced (in smaller fishes) or
absent (in larger fishes).. *Rhynchodoras*

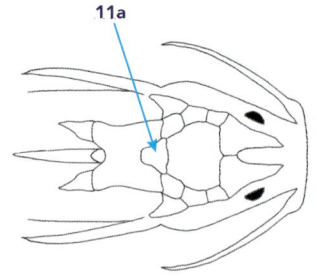

12a. Lips large, with lateral expansions (fig.
12a); posterior process of cleithrum
high and triangular (fig. 12aa); dermal
tubercles well developed **13**
12b. Lips thin, without lateral expansions;
posterior process of cleithrum slender
and long (fig. 12b); dermal tubercles
little developed .... *Platydoras*

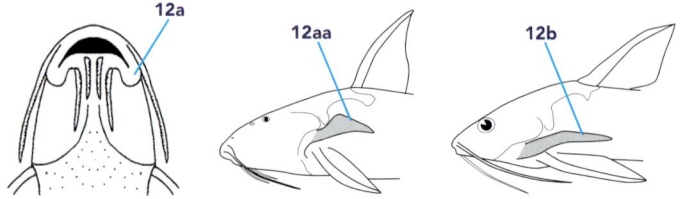

13a. Head and body with black ground color and a longitudinal clear stripe (yellow in life); barbels
without transversal grooves ....*Orinocodoras*
13b. Head and body with tan ground color and brown saddles, blotches and spots; barbels with
transversal grooves (straie)..*Rhinodoras*

14a. Head compressed (higher than wide); lips large; upper and lower jaw without teeth.............................*Oxydoras*
14b. Head depressed (wider than high); lips thin; upper and lower jaw with teeth **15**

15a. Medium sized (<40 cm SL); 30–40 midlateral scutes ....*Centrodoras*
15b. Large sized (>40 cm SL); 15–25 midlateral scutes **16**

16a. Adipose fin with small lobe, but with long base; 15–17 midlateral scutes; body with mosaic of
light and dark spots...*Megalodoras*
16b. Adipose fin with large lobe and long base; 16–25 midlateral scutes; body uniformly colored or with small dark spots...........**17**

17a. Body completely covered with bony plates (in individuals >15 cm); 16–20 midlateral scutes; body uniformly
cream- to beige-colored ....*Lithodoras*
17b. Body not covered with body plates; 19–25 midlateral scutes; body brownish with small dark spots ....*Pterodoras*

18a. Maxillary and mental barbels united by the lip membrane; 33–46 midlateral scutes **19**
18b. Maxillary and mental barbels free (not united by the lip membrane); 26–37 midlateral scutes.................................**20**

19a. Maxillary and mental barbels united by a membrane in their entire extension; 33–46 lateral plates....................*Leptodoras*
19b. Maxillary and mental barbels united by a membrane at base only; 39–40 lateral plates................................*Anduzedoras*

20a. Paired nuchal foramina present (fig. 20a)..**23**
20b. Nuchal foramina absent **21**

21a. Posterior process of coracoid long (about as long as posterior process
of cleithrum); upper and lower jaw generally with >15 teeth; multiple
pores ventral of posterior process of cleithrum absent .... *Ossancora*
21b. Posterior process of coracoid usually short (shorter than posterior process
of cleithrum); upper and lower jaw generally with <15 teeth; multiple
pores ventral of posterior process of cleithrum usually present.............................**22**

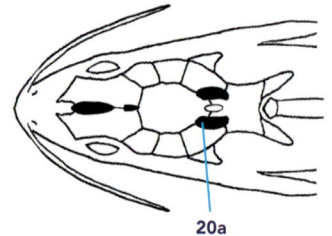

22a. Snout rounded and short (eye diameter >50% of snout length); maxillary barbel short (not reaching gill opening); posterior fontanel present (fig. 22a) ....*Trachydoras*
22b. Snout long and pointed (eye diameter <50% of snout length); maxillary barbel long (passing gill opening); posterior fontanel absent.........*Doras*

23a. Dorsal fin with dark spot at the distal end of the branched rays; anterior midlateral scutes reduced .... *Hassar*
23b. Dorsal fin hyaline or with dark spot at the proximal portion of first rays; anterior midlateral scutes generally well developed (except in *Tenellus leporhinus*)...**24**

24a. Dorsal fin with a dark spot at the proximal portion of first rays and one in the middle of each lobe of the caudal fin ....*Tenellus*
24b. Fins hyaline and uniformly colored..**25**

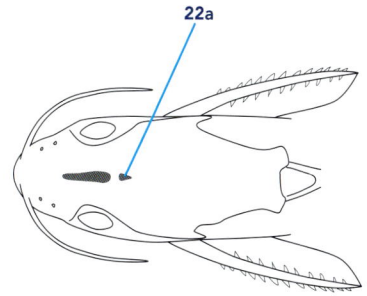

25a. Posterior process of cleithrum triangular, with posterior extremity pointed (fig 25a); multiple pores ventral of posterior process of cleithrum absent; bony scutes between dorsal and adipose fin sometimes present (mostly in large specimens); fimbriae of mental barbels well developed (fig. 25aa)..*Hemidoras*
25b. Posterior process of cleithrum trapezoid, with posterior extremity truncate (fig. 25b); multiple pores ventral of posterior process of cleithrum usually present; bony scutes between dorsal and adipose fin always absent; fimbriae of mental barbels little developed (fig. 25bb).....................*Nemadoras*

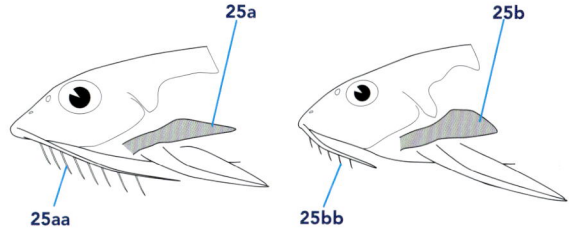

GENUS ACCOUNTS

Acanthodoras (8.1–14 cm SL)

Characterized by: dark background color with clear spots and stripes along the body; simple maxillary and mental barbels; pectoral-fin spine dorsally serrated in adults; dorsal-fin spine laterally serrated in adults; caudal fin rounded; 7 branched rays on lower lobe of caudal fin; 24–27 midlateral scutes; midlateral scutes vertically oriented; procurrent rays not expanded; infraorbital 1 bearing spines. **SPECIES** Three species, all in the AOG region. **COMMON NAMES** *Bacu pedra* (Brazil); *Bagre hueso* (Ecuador); *Lechero* (Peru). **DISTRIBUTION AND HABITAT** *A. cataphractus* (12 cm SL) inhabits the Amazon, Orinoco, Essequibo, and smaller drainages of Guianas and Suriname, *A. spinosissimus* (14 cm SL) in the Amazon and Essequibo basins, and *A. depressus* (8.1 cm SL) in the Amazon and Negro basins. **BIOLOGY** Omnivorous diet. Active during the night, hiding in submerged macrophytes, sunken wood or roots during the day. Secrete white milky substance from an axillary pore, bitter to the human taste, that discourages predation. Possible cleaning behavior of juveniles has been observed by the authors: juvenile approached an adult *Asterophysus batrachus*, who lay down with mouth open. Clear longitudinal stripes (as present in *Acanthodoras*) are a common visual signal in marine fishes with cleaning behavior (Carvalho et al. 2003).

Agamyxis (15 cm SL)

Characterized by: dark background color with clear dots along the body; simple maxillary and mental barbels; pectoral-fin spine dorsally serrated in adults; dorsal-fin spine posteriorly serrated; dorsal-fin spine laterally serrated in adults; anteriormost rays of pelvic fin bearing spines; caudal fin forked; 6 branched rays in the upper lobe of caudal fin; 7 branched rays on lower lobe of caudal fin; 24–27 midlateral scutes; midlateral scutes vertically oriented; procurrent rays laterally expanded; infraorbital 1 bearing spines. **SPECIES** Two, both from the AOG region. **COMMON NAMES** *Dorita, Espinosísimo, Rego rego* (Peru); *Sierra pintada* (Venezuela). **DISTRIBUTION AND HABITAT** *A. albomaculata* (15 cm SL) from the Orinoco basin in Venezuela, *A. pectinifrons* (15 cm SL) from the Amazon basin in Brazil. **BIOLOGY** Generalist omnivores, active during the night, hiding in wood and roots during the day.

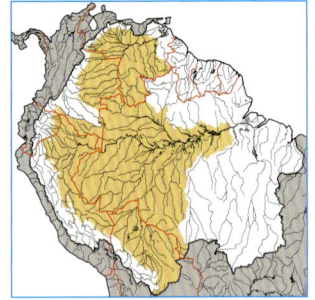

Amblydoras (10 cm SL)

Characterized by: simple maxillary and mental barbels; dorsal-fin spine usually serrated anteriorly, sometimes only in basal portion; 7 branched rays on lower lobe of caudal fin; 23–27 midlateral scutes; midlateral scutes vertically oriented; black saddles below midlateral scutes usually connected longitudinally, forming a stripe; infraorbital 1 bearing spines; suture between contralateral bones of pectoral girdle ventrally exposed (visible). **SPECIES** Five, including four species in the AOG region. **DISTRIBUTION AND HABITAT** *A. affinis* (10 cm SL) has an apparently widespread distribution in the Amazon and Essequibo basins, and is possibly senior synonym of *A. monitor* and *A. nauticus* (Sousa, 2010); *A. bolivarensis* (10 cm SL) inhabits the Orinoco basin. **BIOLOGY** Generalist omnivores, hiding in vegetation and leaf litter packs or partially buried in sand during the day. Very abundant in some localities using seines or trawls over flooded beaches (Correa and Ortega 2010).

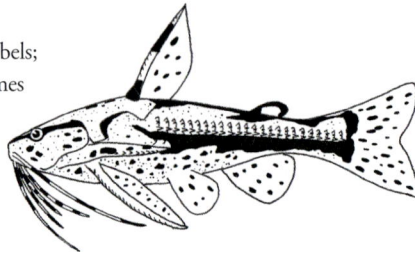

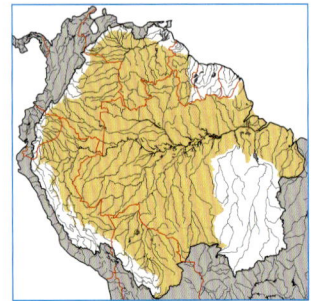

Anadoras (15 cm SL)

Characterized by: simple maxillary and mental barbels; dorsal-fin spine smooth anteriorly, not serrated; 7 branched rays on lower lobe of caudal fin; 23–28 midlateral scutes; midlateral scutes vertically oriented; procurrent rays laterally expanded; infraorbital 1 without spines. **SPECIES** Two species, both occurring in the AOG region. **COMMON NAMES** *Uarioroch, Quiri-quiri* (Brazil); *Rego rego* (Peru). **DISTRIBUTION AND HABITAT** *A. grypus* (15 cm SL) from the eastern Amazon and Madeira basins; and *A. weddellii* (15 cm SL) from western Amazon basin. **BIOLOGY** Generalist omnivores, hiding in vegetation during the day.

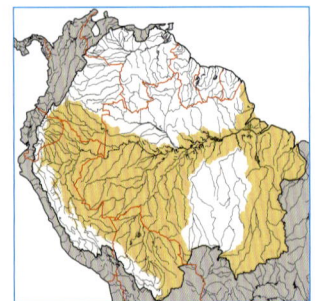

Anduzedoras (32 cm SL)

Characterized by: fimbriate maxillary barbels; 8 branched rays on lower lobe of caudal fin; 39 or 40 midlateral scutes; midlateral scutes obliquely oriented; barbels united by lip extension forming a relatively short oral hood; nuchal foramina reduced or absent in specimens >150 mm SL; 16–25 gill rakers; body generally pale or tan; short bony capsule covering anteriormost vertebrae. **SPECIES** One; *A. oxyrhynchus*.

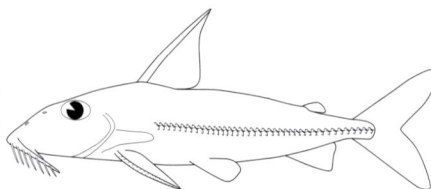

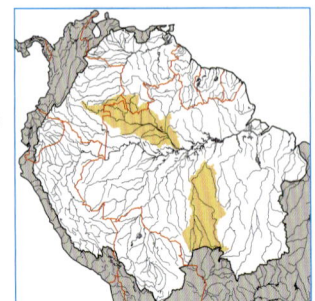

COMMON NAMES *Cuyu-cuyu* (Brazil). **DISTRIBUTION AND HABITAT** *A. oxyrhynchus* occurs in the blackwater rivers of the Negro, Trombetas, and upper Orinoco rivers, and also in the Tapajós basin. **BIOLOGY** Detritivorous feeding on invertebrates buried in the river bottom, as in most doradids with fimbriate barbels (Sabaj 2005).

Astrodoras (10 cm SL)

Characterized by: simple maxillary and mental barbels; depressed body; eyes protruding above overall head surface; dorsal-fin spine serrated anteriorly; 8 branched rays on lower lobe of caudal fin; 24–25 midlateral scutes; midlateral scutes vertically oriented; procurrent rays sometimes laterally expanded; caudal fin forked; infraorbital 1 bearing spines. **SPECIES** One, *A. asterifrons.* **DISTRIBUTION AND HABITAT** Widespread throughout the Amazon basin. **BIOLOGY** Generalist omnivores, hiding in vegetation during the day.

Centrodoras (21–41 cm SL)

Characterized by: simple maxillary and mental barbels; 8 branched rays on lower lobe of caudal fin; 30–40 midlateral scutes; midlateral scutes vertically oriented; infraorbitals tubular (not forming plates); last tympanal, infranuchal and first postinfranuchal midlateral scutes sutured to posterior nuchal plate; adipose fin with shallow anterior keel and a small posterior lobe; head and body generally tan; large and distinctly curved pectoral-fin spine. **SPECIES** Two, both in the AOG region. **DISTRIBUTION AND HABITAT** *C. brachiatus* (41 cm SL) is widespread in the Amazon basin, but not found in the Negro basin; *C. hasemani* (21 cm SL) is apparently endemic to the Negro basin. **BIOLOGY** Medium-sized species that possibly undertake long-distance migrations. Diets includes mollusks.

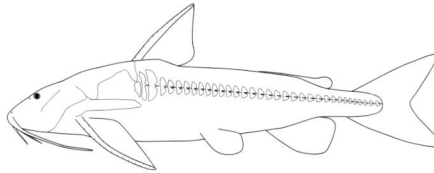

Doras (17–29 cm SL)

Characterized by: fimbriate maxillary barbel; 8 branched rays on lower lobe of caudal fin; 30–36 midlateral scutes; midlateral scutes obliquely oriented; mental barbel with papillae; pores on belly; nuchal foramina absent; posterior cranial fontanel absent; anterior nuchal plate large. **SPECIES** Five species, all in the AOG region. Key to the species in Sabaj-Pérez and Birindelli (2008). **COMMON NAMES** *Peixe botinho, Mandi-serra* (Brazil); *Pirillo* (Peru). **DISTRIBUTION AND HABITAT** *D. carinatus* (20 cm SL) inhabits tributaries of the lower Orinoco, Essequibo, and coastal rivers of Guianas and Surinam; *D. micropoeus* (29 cm SL) occurs in the Essequibo basin and coastal drainages of the Guianas and Suriname; *D. higuchii* (25 cm SL) inhabits the eastern Amazon, including the Xingu river; *D. phlyzakion* (20 cm SL) occurs in western Amazon basin, including the Negro River in Brazil; and *D. zuanoni* (17 cm SL) inhabits the Araguaia River. **BIOLOGY** Species include the largest-bodied doradids with fimbriate barbels; detritivorous, feeding on invertebrates buried in the river bottom (Sabaj 2005, Sabaj-Pérez and Birindelli 2008).

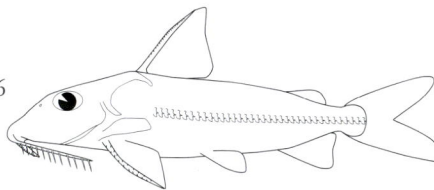

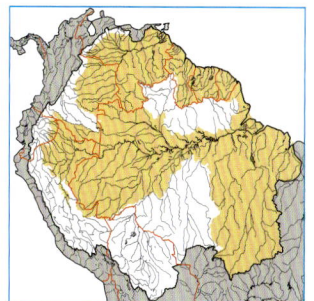

Hassar (17–25 cm SL)

Characterized by: fimbriate barbels; 8 branched rays on lower lobe of caudal fin; 30–34 midlateral scutes; midlateral scutes obliquely oriented; dark blotch on the distal area of anterior branched rays of dorsal fin; mental barbels with papillae; anteriormost infranuchal scutes reduced in size; pores beneath cleithral process numerous; well-developed nuchal foramina;

anterior nuchal plate reduced and enclosed by parieto-supraoccipital and middle nuchal plate. **SPECIES** Five, including four species in the AOG region. Key to the species in Birindelli et al. (2011). **DISTRIBUTION AND HABITAT** *H. gabiru* (16 cm SL) is apparently endemic to the Xingu River above the Volta Grande falls, *H. orestis* (25 cm SL) is widespread in the Amazon, Orinoco, and Essequibo basins (although absent in the Xingu and Tocantins-Araguaia), *H. wilderi* (19 cm SL) is endemic to the Tocantins-Araguaia basin. **BIOLOGY** Feed primarily on insects and to a lesser extent on other invertebrates and plant material, and form schools on sandy and muddy beaches of medium to large rivers (Birindelli et al. 2011).

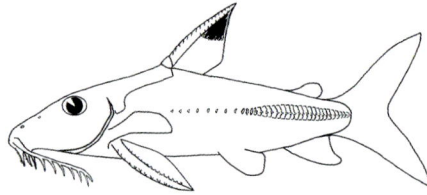

Hemidoras (13–15 cm SL)

Characterized by: fimbriate maxillary barbel; 8 branched rays on lower lobe of caudal fin; 30–35 midlateral scutes; midlateral scutes obliquely oriented; bony scutes/dermal plates sometimes present in the dorsal midline between dorsal and adipose fins (mostly in large specimens); mental barbels with 2 rows of fimbriae; cleithral process long and thin; few pores beneath posterior cleithral process. **SPECIES** Five, all in the AOG region. **DISTRIBUTION AND HABITAT** *H. boulengeri* (15 cm SL) occurs in the Amazon basin, *H. morei* (11 cm SL) inhabits the Negro River, *H. morrisi* (14 cm SL) occurs in the western Amazon basin (and it is possibly synonym of *H. boulengeri*, according to Sabaj 2003), *H. stenopeltis* (13 cm SL) is widespread in the Amazon basin, *H. stubelii* (11 cm SL) inhabits the Amazon and Orinoco basins. **BIOLOGY** Detritivorous, feeding on invertebrates buried in the river bottom, as in most doradids with fimbriate barbels (Sabaj 2005).

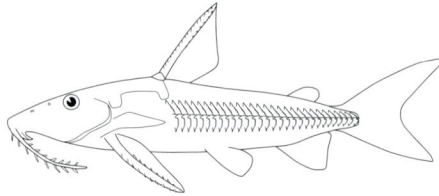

Hypodoras (11 cm SL)

Characterized by: simple maxillary and mental barbels; dermal plate immediately anterior to adipose fin; depressed body; dorsal spine smooth, not serrated anteriorly; 6 branched rays on lower lobe of caudal fin; 24 or 25 midlateral scutes; midlateral scutes vertically oriented; procurrent rays laterally expanded; caudal fin rounded; infraorbital 1 bearing spines. **SPECIES** One, *H. forficulatus*. **DISTRIBUTION AND HABITAT** Known only from the Nanay and Itaya rivers near Iquitos, Peru. Sabaj (2002) reports "one specimen from the Río Nanay was taken along shore in shallow water with little or no current over a bottom of mud, silt and sand." **BIOLOGY** No data available.

Leptodoras (10–29 cm SL)

Characterized by: distinctly elongate body in most species; fimbriate maxillary barbels; 8 branched rays on lower lobe of caudal fin; 33 or 46 midlateral scutes; midlateral scutes obliquely oriented; barbels united by lip extension forming a relatively long oral hood; 16–25 gill rakers; pelvic-fin origin located in the anteriormost half of the body; accessory lamellae on the branchial filaments; short bony capsule covering anteriormost vertebrae. **SPECIES** 12 species, all in the AOG region. Key to the species in

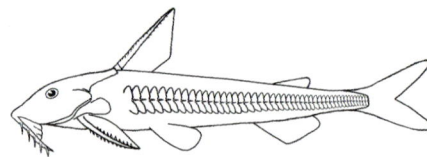

Sabaj (2005). **COMMON NAMES** *Bagre espinoso delgado* (Spanish); Slender thorny-catfish (English). **DISTRIBUTION AND HABITAT** Widespread in the Amazon, Orinoco and Essequibo basins (Sabaj 2005, Birindelli et al. 2008). **BIOLOGY** Most species are benthic, inhabiting deep swift-flowing waters of large rivers (Sabaj 2005). Some species migrate at dusk into shallow waters near shore to forage over beaches and shoals of sand or silt; *L. juruensis, L. myersi* restricted to deep channel habitats and best captured via bottom trawls at depths down to 50 m. Feed on chironomid larvae and detritus (Sabaj 2005).

Lithodoras (80 mm SL)

Characterized by: simple maxillary and mental barbels; 8 branched rays on lower lobe of caudal fin; 16–20 midlateral scutes; midlateral scutes vertically oriented; (non-midlateral) scutes on the body in specimens >15 cm SL (body completely covered by scutes in specimens >50 mm SL); penultimate and last infraorbitals expanded into plates; adipose fin with relatively short base (generally shorter than that of anal fin). **SPECIES** One, *L dorsalis*. **DISTRIBUTION AND HABITAT** Occurs in the Amazon basin. Most common in the lower stretches of the Amazon River, below Santarém. **BIOLOGY** Feeds on mollusks when young and on seeds and fruits as adult (Vono and Birindelli 2007).

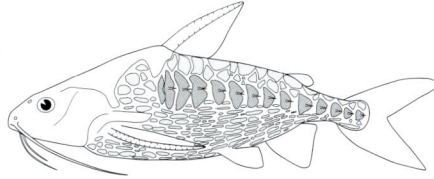

Megalodoras (53–60 cm SL)

Characterized by: simple maxillary and mental barbels; 8 branched rays on lower lobe of caudal fin; 15–18 midlateral scutes; midlateral scutes vertically oriented; head and body marbled with pale and dark; penultimate infraorbital forming a plate; adipose fin with shallow anterior keel and a small posterior lobe. **SPECIES** Two, both in the AOG region. **COMMON NAMES** *Cuiú cuiú amarelo*, *Dukeru*, *Mãe de caracóis* (Brazil); *Key-way-mamma* (Guyana), *Piro, Churero* (Peru). **DISTRIBUTION AND HABITAT** *M. guayoensis* (53 cm SL) in the Orinoco basin; *M. uranoscopus* (60 cm SL) in the Amazon and Essequibo basins. **BIOLOGY** *M. uranoscopus* is a migratory species that feeds predominately on mollusks (J. Zuanon pers. comm.).

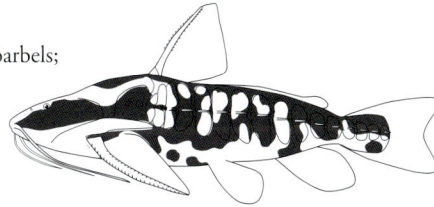

Nemadoras (17 cm SL)

Characterized by: fimbriate maxillary barbel; 8 branched rays on lower lobe of caudal fin; 30–33 midlateral scutes; midlateral scutes obliquely oriented; mental barbels with papillae; posterior fontanel large; medium to large nuchal foramina; weakly developed adipose eyelid; gas bladder with diverticulae abundant and extended on ventral and dorsal walls. Sexually mature males exhibit dorsal-fin spines prolonged by a thin flexible filament nearly as long as the spine itself, whereas mature females exhibit normal spines (Sabaj Pérez et al. 2014). **SPECIES** Three, all in the AOG region. Key to species in Sabaj Pérez et al. (2014). **DISTRIBUTION AND HABITAT** *N. elongatus* (17 cm SL) and *N. hemipeltis* (17 cm SL) occur widely in whitewater rivers of the middle and upper Amazon basin, and *N. humeralis* (17 cm SL) in the Amazon and Tocantins basins (Sabaj Pérez et al. 2014). **BIOLOGY** No data available.

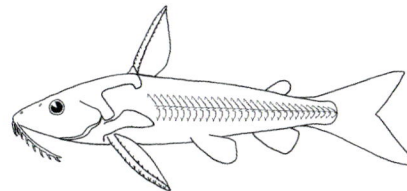

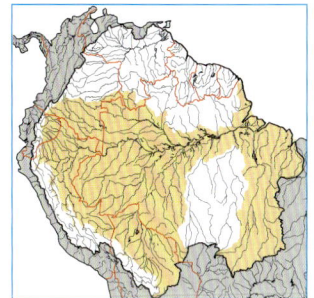

Orinocodoras (18 cm SL)

Characterized by: simple maxillary and mental barbels; 8 branched rays on lower lobe of caudal fin; 27–29 midlateral scutes; midlateral scutes vertically oriented; body dark with a pale midlateral stripe; adipose fin thick with base longer than base of anal fin; thick lips; procurrent caudal-fin rays expanded forming plates framing caudal peduncle. **SPECIES** One, *O. eigenmanni.* **DISTRIBUTION AND HABITAT** Orinoco basin in Venezuela. **BIOLOGY** Coloration is possibly related to complex cleaning behavior (Birindelli 2014).

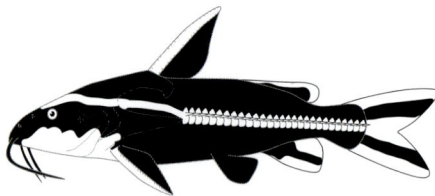

Ossancora (10–11 cm SL)

Characterized by: fimbriate maxillary barbel; 8 branched rays on lower lobe of caudal fin; 27–30 mid-lateral scutes obliquely oriented; mental barbels with fimbriae; numerous teeth on premaxilla and dentary; posterior cranial fontanel occluded; posterior coracoid process approximately as long as posterior cleithral process (forming U-shaped posterior margin of pectoral girdle in ventral view). **SPECIES** Four, three in the AOG region. Key to the species in Birindelli and Pérez (2011). **DISTRIBUTION AND HABITAT** *O. asterophysa* (11 cm SL), *O. fimbriata* (10 cm SL), and *O. punctata* (11 cm SL) are widespread in the middle and upper Amazon basin, above the mouth of the rio Curuá-Una (Birindelli and Pérez 2011). Inhabits lowland floodplain lakes and waterways below about 200 m. Records span a variety of black-, clear-, and whitewater habitats and include main river channels to depths of 22.5 m. **BIOLOGY** No data available.

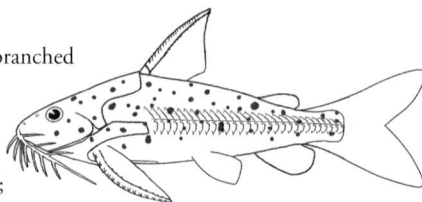

Oxydoras (60–100 cm SL)

Characterized by: simple maxillary and mental barbels; 8 branched rays on lower lobe of caudal fin; teeth absent on premaxillae and dentary; 20–26 midlateral scutes; midlateral scutes vertically oriented; gas bladder with a secondary chamber without inner septum. **SPECIES** Three, including two in the AOG region. **COMMON NAMES** *Peixe cujuba, Peixe tuiú* (Brazil); *Saurawari* (Guyana); *Turushuqui* (Peru); Ripsaw catfish (English). **DISTRIBUTION AND HABITAT** *O. niger* (100 cm SL) from the Amazon and Essequibo basins; and *O. sifontesi* (60 cm SL) from the Orinoco basin in Colombia and Venezuela. **BIOLOGY** Include the largest doradid species, and are benthivorous and detritivorous (Vono and Birindelli 2007).

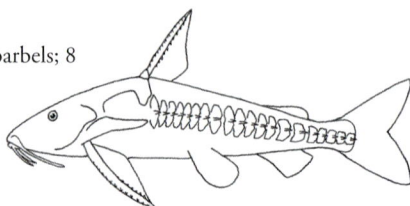

Physopyxis (3.0–4.0 cm SL)

Characterized by: small adult body sizes; simple maxillary and mental barbels; 5–27 midlateral scutes; lateral line complete or incomplete; dorsal-fin spine serrated anteriorly only in basal portion; muscles of pectoral girdle enclosed in bone ventrally; posterior coracoid process longer than posterior cleithral process; 5 branched rays on lower lobe of caudal fin; infraorbital 1 bearing spines. **SPECIES** Three, all in the AOG region. **DISTRIBUTION AND HABITAT** *P. ananas* (3.0 cm SL) in lowlands of entire Amazon basin and Essequibo basin; *P. cristata* (3.0 cm SL) in the middle portion of Negro basin; and *P. lyra* (4.0 cm SL) in the Ampiyacu River (and lowland portions of other tributaries to the upper Amazon in northeastern Peru) to the rio Uatumá

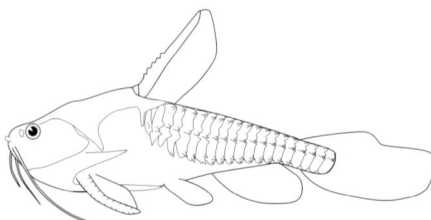

in Brazil. These miniaturized species inhabit low-energy systems where silt and fine organic matter accumulate, such as small root tangles and submerged litter. **BIOLOGY** Adults commonly collected schooling together with juvenile of *Amblydoras* spp. in sandy- or muddy-bottom river banks (Sousa and Rapp Py-Daniel 2005). Individuals frequently bury in the sand during the day (observations in aquaria; J. Zuanon pers. comm.).

Platydoras (15–24 cm SL)

Characterized by: simple maxillary and mental barbels; 8 branched rays on lower lobe of caudal fin; 28–30 midlateral scutes; midlateral scutes vertically oriented; body dark with a pale midlateral stripe (more conspicuous marked in juveniles); procurrent caudal-fin rays expanded, forming plates framing caudal peduncle; midlateral scutes deep and with serrated posterior margin; adipose fin with small posterior lobe, sometimes extended anteriorly by shallow keel. **SPECIES** Four species, three in the AOG region. Key to the species in Piorski et al. (2008). **DISTRIBUTION AND HABITAT** *P. armatulus* (20 cm SL) is widespread in the Amazon and Orinoco basins, *P. costatus* (24 cm SL) is endemic to the Essequibo basin and smaller coastal drainages of the Guianas and Suriname, and *P. hancockii* (15 cm SL) occurs in the Essequibo, upper Orinoco, and Negro drainages. **BIOLOGY** Nocturnal fishes that hide in holes of logs and rocks during the day and feed mainly on insects and other invertebrates (Piorski et al. 2008). Carvalho et al. (2003) report a cleaning behavior of *P. armatulus* (identified as *P. costatus*) on a large *Hoplias malabaricus* in a stream of the Araguaia drainage.

Pterodoras (50–70 cm SL)

Characterized by: simple maxillary and mental barbels; 8 branched rays on lower lobe of caudal fin; 19–25 midlateral scutes, which are vertically oriented; body generally grayish tan with dark blotches; penultimate and last infraorbitals expanded into plates; posterior cleithral process extremely thin; adipose fin with relatively short base (generally shorter than that of anal fin). **SPECIES** Two, both in the AOG region. **DISTRIBUTION AND HABITAT** *P. granulosus* (70 cm SL) occurs widely in the Amazon basin; *P. rivasi* (50 cm SL) in the Orinoco basin. **BIOLOGY** Some species are migratory, and most species are predominantly herbivores (Vono and Birindelli 2007).

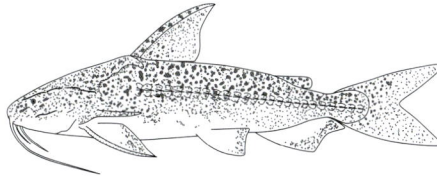

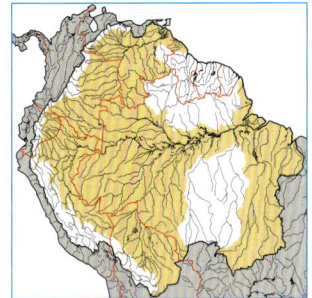

Rhinodoras (12–14 cm SL)

Characterized by: simple maxillary and mental barbels; 8 branched rays on lower lobe of caudal fin; 28–32 midlateral scutes; midlateral scutes vertically oriented; maxillary and mental barbels with striae; body tan with dark blotches sometimes forming saddle-like bands; adipose fin thick with base longer than base of anal fin; thick lips (considerably expanded at corners of mouth forming relatively thick and rounded flap-like extensions); procurrent caudal-fin rays expanded, forming plates framing caudal peduncle. **SPECIES** Five, including four species in the AOG region. Key to the species in Sabaj et al. (2008). **DISTRIBUTION AND HABITAT** *R. armbrusteri* (12 cm SL) inhabits the Essequibo basin, *R. boelhkei* (13 cm SL) occurs in the Amazon, *R. gallagheri* (12 cm SL) is known from the Orinoco basin, and *R. thomersoni* (14 cm SL) is endemic to the Maracaibo lake (Sabaj et al. 2008). Rare in museum collections; prefers the main channel of rivers with a rocky bottom. **BIOLOGY** Secretive fishes that hide in hollow logs or cavities in lateritic boulders during the day and forage at night. Diet includes fish, crustaceans, bivalves, aquatic insects, and vegetable matter (Sabaj et al. 2008).

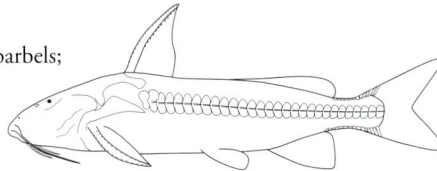

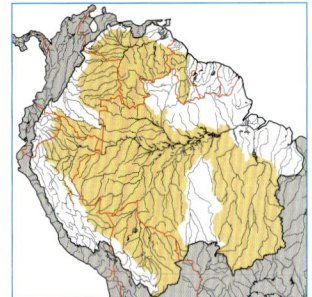

Rhynchodoras (7.0–11 cm SL)

Characterized by: simple maxillary and mental barbels;
8 branched rays on lower lobe of caudal fin;
33–37 midlateral scutes; midlateral scutes
vertically oriented; mental barbels united at
base; upper and lower jaw strongly modified
as a ventrally directed "forceps-like" structure;
adipose fin thick with base longer than base of anal fin; procurrent caudal-fin rays expanded
as plates framing caudal peduncle; anterior nuchal plate absent; gas bladder reduced in size.
SPECIES Three, all in the AOG region. **DISTRIBUTION AND HABITAT** *R. castilloi* (8.0 cm SL) in the
middle to lower Apure River (Orinoco basin); *R. woodsi* (11 cm SL) in the Amazon basin in
Brazil, Ecuador, and Peru and in the Essequibo in Guyana; and *R. xingui* (7.0 cm SL) in the
Xingu and Tocantins basins (Birindelli et al. 2007). Species are typically found in deep portions
of large river channels associated with rocky beds. **BIOLOGY** No data available.

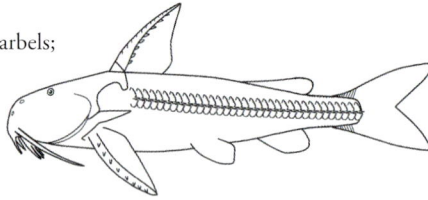

Scorpiodoras (16–17 cm SL)

Characterized by: simple maxillary and mental barbels;
dorsal spine serrated anteriorly; 8 branched rays
on lower lobe of caudal fin; 27–30 midlateral
scutes; midlateral scutes vertically oriented;
infraorbital 1 bearing spines. **SPECIES** Three,
all in the AOG region. **DISTRIBUTION AND
HABITAT** *S. calderonensis* (17 cm SL) occurs in the upper Amazon basin; *S. heckelii* (17 cm SL)
in the Orinoco and Negro basins and along the Amazon River at and below the mouth of the
Negro River; and *S. liophysus* (16 cm SL) in the Madeira basin (Sousa and Birindelli 2011).
BIOLOGY Generalist omnivorous, hiding in vegetation and leaf litter packs during the day.

Tenellus (11–21 cm SL)

Characterized by: fimbriate maxillary barbels; 8
branched rays on lower lobe of caudal fin; 29–37
midlateral scutes; midlateral scutes obliquely
oriented; mental barbels with papillae or
fimbriae; large posterior cranial fontanel;
anterior nuchal plate reduced and enclosed in
parieto-supraoccipital; middle nuchal plate reduced in size and diamond-shaped or absent; dorsal
fin usually with a dark spot at the proximal portion of first rays and one in the middle of each
lobe of the caudal fin. Sexually mature males exhibit dorsal-fin spines prolonged by a thin flexible
filament nearly as long as the spine itself, whereas mature females exhibit normal spines (Sabaj
Pérez et al. 2014). **SPECIES** Four species, all in the AOG region. Key to the species in Sabaj Pérez
et al. (2014). **TAXONOMIC NOTES** *Nemadoras cristinae* was described by Sabaj Pérez et al. (2014)
concomitantly with the description of the genus *Tenellus* by Birindelli (2014). *Nemadoras cristinae*
was not included in the phylogenetic analysis of Birindelli (2014) and, therefore, not assigned to
Tenellus by that author. However, *N. cristinae* was considered closely related to *Tenellus leporhinus*
and *T. ternetzi* in the molecular-based phylogeny of Arce et al. (2013), and is herein assigned to
Tenellus. **DISTRIBUTION AND HABITAT** *T. cristinae* (11 cm SL) in the middle and upper Amazon
basin and in the Orinoco basin, *T. leporhinus* (21 cm SL) in tributaries of the lower and
middle Amazon River, and in the Orinoco basin, *T. ternetzi* (13 cm SL) in the Amazon,
Essequibo, and Orinoco basins, and *T. trimaculatus* (12 cm SL) in the Amazon and Orinoco
basins. **BIOLOGY** Detritivorous, feeding mostly on insects buried in the river bottom
(Sabaj Pérez et al. 2014).

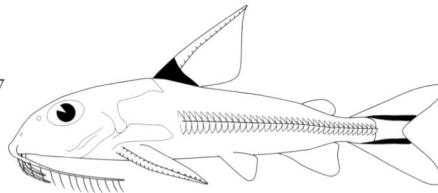

Trachydoras (8.0–10 cm SL)

Characterized by: fimbriate maxillary barbel; 8 branched rays on lower lobe of caudal fin; 29–35 midlateral scutes; midlateral scutes obliquely oriented; mental barbels with papillae or smooth; mouth small and ventral; adipose eyelid weakly developed; posterior fontanel large; anterior nuchal plate large; nuchal foramina usually absent (rarely small); exposed bones on the cheek area. **SPECIES** Five species, all in the AOG region. **COMMON NAMES** *Armado*, *Armadinho* (southern Brazil); *Corno*, *Maniquim* (Brazilian Amazon); *Bagre hueso* (Ecuador); *Pirillo* (Peru). **DISTRIBUTION AND HABITAT** *T. brevis* (10 cm SL), *T. nattereri* (10 cm SL), and *T. steindachneri* (10 cm SL) occur in the Amazon basin, *T. microstomus* (8.0 cm SL) is distributed in the Amazon, Orinoco and Essequibo basins, and *T. paraguayensis* (10 cm SL) occurs in the Madeira basin. **BIOLOGY** Detritivorous, feeding on invertebrates buried in the river bottom, as most fimbriate-barbel doradids (Sabaj 2005).

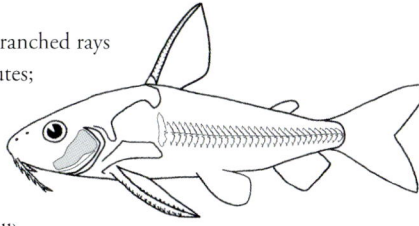

FAMILY HEPTAPTERIDAE—THREE-BARBELED CATFISHES

— *FLÁVIO A BOCKMANN and VERONICA SLOBODIAN*

ILLUSTRATIONS BY VERONICA SLOBODIAN

DIVERSITY 210 species allotted to 24 genera, with many undescribed forms currently known and awaiting formal description (see informal descriptions of eight genera below).

COMMON NAMES *Bagrinho*, *Mandizinho* (Brazil); *Bagrecito* (Peru).

TAXONOMIC NOTE Heptapteridae systematics is still in an early stage of development, with the following genera having been revised using phylogenetic approaches: *Brachyrhamdia*, *Gladioglanis*, *Mastiglanis*, *Myoglanis*, *Nemuroglanis*, *Phenacorhamdia*, *Rhamdella*, *Rhamdia*, and *Rhamdiopsis* (Lundberg and McDade 1986, Britski 1993, Bockmann 1994, Silfvergrip 1996, Bockmann and Ferraris Jr. 2005, Masson 2007, Bockmann and Miquelarena 2008, DoNascimiento and Milani 2008, Bockmann and Castro 2010, Slobodian 2013, Slobodian and Bockmann 2013). According to recent phylogenetic analyses based on sequences of nuclear and mitochondrial genes, Heptapteridae is a member of the superfamily Pimelodoidea, along with Pimelodidae and Pseudopimelodidae, as well as the enigmatic genera *Phreatobius* and *Conorhynchos* (Sullivan et al. 2006, Sullivan et al. 2013a).

GEOGRAPHIC DISTRIBUTION Throughout most of the humid Neotropics, occurring in Atlantic drainages from the Gulf of Mexico to central Argentina, and in Pacific drainages from southern Mexico to central Peru (Silfvergrip 1996, Bockmann and Guazzelli 2003). Despite this wide distribution, the family has several locally exclusive representatives in all areas of ichthyological endemism recognized in the Neotropics.

ADULT SIZES The smallest described heptapterid species to date is *Horiomyzon retropinnatus* at 2.4 cm SL, which inhabits the benthos of deep (>20 m; >66 ft) channels of large lowland Amazonian rivers (Stewart 1986a, Bockmann and Slobodian 2013). The largest heptapterid species belong to the genera *Goeldiella*, *Rhamdella*, *Rhamdia*, and *Pimelodella*, usually exceeding 17 cm SL. Among these, *Rhamdia quelen* commonly attains >30 cm SL, and the maximum reported size is 47 cm TL for a specimen from the upper Uruguay River in southeastern Brazil (Zaniboni Filho et al. 2004).

DIAGNOSIS OF FAMILY Recognized based on internal (mostly osteological) features, and difficult to separate from Pimelodidae and Pseudopimelodidae (Bockmann and Guazzelli 2003). Most heptapterid species can be distinguished from other South American siluriforms by: small adult body size, usually <20 cm SL (except some species of *Goeldiella*, *Heptapterus*, *Pimelodella*, *Rhamdella*, *Rhamdia*, and *Rhamdioglanis*); skin naked, with a few species having slightly enlarged lateral line ossicles; cephalic laterosensory canals simple, with single pores at the skin surface (a few species have multibranched laterosensory canals on the head and anterior trunk); nares well separated and lacking barbels; eyes varying from large to small (sometimes completely absent), with or without a free orbital margin; 3 pairs of barbels (maxillary, inner, and outer mentals); gill membranes free, branchial openings not restricted; dorsal fin usually with 6 branched rays; adipose fin large; dorsal and pectoral fins ranging from bearing an anterior pungent (sharp and hard) fin spine to a completely flexible and mostly segmented fin ray; caudal fin usually forked, but can also be emarginate, rounded, or lanceolate. Some heptapterid genera have a somewhat generalized catfish body shape (e.g., *Pimelodella*, *Rhamdia*), while other genera are highly elongate (e.g., *Heptapterus*). General reduction of body size is a common phenomenon among heptapterids, with the most extreme miniaturization in the genera *Gladioglanis*, *Horiomyzon*, and *Nemuroglanis*.

SEXUAL DIMORPHISM Not known in most heptapterids. Juveniles of most species are miniature replicas of adults (Bockmann and Guazzelli, 2003). In some *Pimelodella* species, males have elongate and filamentous first pectoral-fin rays and second dorsal-fin rays (Souza-Shibatta et al. 2013; F. B and V. S. pers. obs.). The reproductive cycle of heptapterids is poorly known, probably a result of their reduced importance in commercial fisheries and aquarium trade, except for *Brachyrhamdia* and *Pimelodella* that are more frequently used as ornamental fishes.

HABITATS Most heptapterid species inhabit small rainforest streams with a dense canopy cover and high water quality, so that species in this group serve as effective indicators of environmental quality (Bockmann and Guazzelli 2003). Most heptapterids are benthic, although *Brachyrhamdia* and *Pimelodella* sometimes swim up into the water column. Most heptapterids have cryptic habits and are most active at twilight and at night, sheltering among rocks or sand or leaf litter during the day (Bockmann 1998, Bockmann and Guazzelli 2003). Heptapteridae has among the largest diversity of subterranean species among catfish families worldwide, with 14 species (Muriel-Cunha and de Pinna 2005, Bockmann and Castro 2010, Bockmann and Slobodian 2013).

BEHAVIOR Tend to be solitary or to organize in small schools up to 10 individuals, e.g., species of *Acentronichthys*, *Brachyrhamdia*, *Heptapterus*, and *Pimelodella* (Bockmann and Guazzelli 2003; F. B and V. S. pers. obs.).

FEEDING ECOLOGY The dietary habits of only a few heptapterid species have been studied in detail, revealing that they are generalized carnivores or omnivores that consume chiefly arthropods in all stages, from autochthonous and allochthonous sources, as well as algae, organic debris, nematodes, and small fish (Bockmann and Guazzelli 2003).

KEY TO THE GENERA

NOTE Correctly identifying heptapterid genera is difficult. The majority of heptapterid genera are uniquely characterized by osteological features that are not easily seen from external observation, and many external features commonly utilized in keys are highly variable within genera. Some of the external traits valuable in identifying heptapterid genera are difficult to observe, such as the anteriormost rays of the anal fin that are generally embedded within a thick skin, and the counts of pores of the laterosensory system. Some heptapterid genera, such as *Pimelodella*, *Rhamdella*, and *Rhamdia*, are artificial as currently diagnosed and circumscribed. Other heptapterid genera, such as *Cetopsorhamdia*, *Chasmocranus*, and *Horiomyzon*, are easily characterized by external attributes, but include certain species with highly divergent morphologies. There are also many heptapterid genera and species that are not yet formally described, which will lead to substantial modifications of the generic classification. For these reasons, the key to heptapterid genera provided below employs characters shared by the large majority of species in each genus, with some exceptions as mentioned. In order to further reduce confusion, eight new genera are herein diagnosed, although not formally named.

1a. Posterior process of supraoccipital reaching the nuchal plate (in front of dorsal fin) (fig. 1a–c) **2**
1b. Posterior process of supraoccipital not reaching the nuchal plate (in front of dorsal fin) (fig. 1d,e) **4**

FIGURE 1 Heads in dorsal view of (A) *Goeldiella eques*, (B) *Pimelodella australis*, (C) *Brachyrhamdia thayeria*, (D) *Brachyglanis microphthalmus*, (E) *Imparfinis mirini*. Indicated are the anterior cranial fontanel (AF), posterior cranial fontanel, posterior process of supraoccipital bone (PPSO), and the nuchal plate (NP).

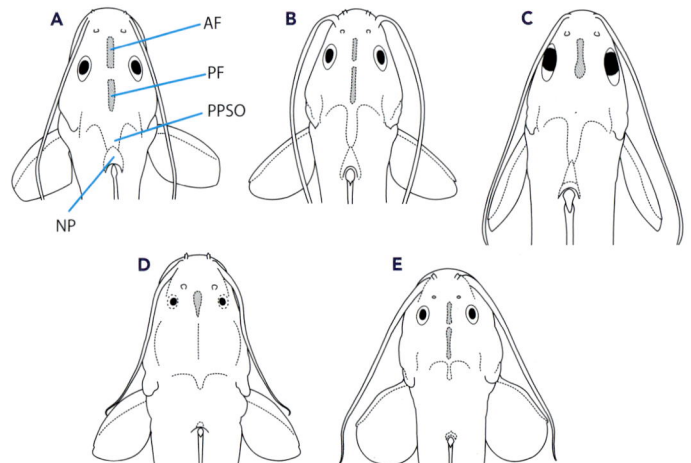

2a. Supraoccipital process triangular, with broad base (fig. 1a); cephalic laterosensory canals opening in skin into multiple pores; anterior margin of pectoral spine with well-defined antrorse dentations (tip of dentations oriented toward the apex of the spine; fig. 2a); an oblique dark stripe from pectoral-fin base to anterior region of dorsal-fin base, extending dorsally along the first dorsal-fin rays.................................... *Goeldiella*

2b. Supraoccipital process rectangular, with roughly constant width from base to tip (fig. 1b,c); cephalic laterosensory canals opening in skin into single pores; anterior margin of pectoral spine with antrorse, usually poorly developed dentations (tip of dentations oriented toward the apex of the spine) (fig. 2b); an oblique dark stripe from pectoral-fin base to first dorsal-fin rays absent **3**

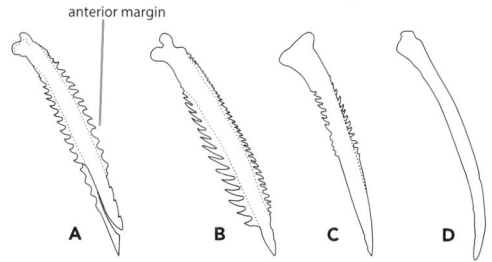

FIGURE 2 Left first pectoral-fin ray of (A) *Goeldiella eques*, (B) *Pimelodella serrata*, (C) *Leptorhamdia essequibensis*, (D) *Cetopsorhamdia insidiosa*.

3a. Head of medium size, 20–30% of standard length; posterior fontanel as a long opening from behind the eyes to the base of supraoccipital (fig. 1b); body elongate .... *Pimelodella*

3b. Head large, 31–37% of standard length; posterior fontanel as a small opening at base of supraoccipital (fig. 1c); body deep and compact .... *Brachyrhamdia*

4a. First ray of the pectoral fin with most of its length broad and rigid, forming a pungent spine, usually with dentations and serrations along its anterior and posterior margins (fig. 2c)... **5**

4b. First ray of the pectoral fin with its basal portion thin and slightly stiffened, forming a brittle structure without any distinct ornamentation along its margins (fig. 2d)... **11**

5a. Mandibular adductor muscles extending dorsally onto the midline of the head, giving it a soft texture (fig. 1d)............... **6**

5b. Mandibular adductor muscles restricted to the facial region, so that the top of head is hard, covered by skin only (fig. 1a,b,c,e)... **9**

6a. Anterior and posterior fontanels as long openings; postcleithral process with a conspicuous fleshy projection; lateral line very short, formed by 2 pores only... **new Genus A**

6b. Anterior and posterior fontanels as short openings, the anterior as a short anterior slit and the posterior as an oval opening at base of supraoccipital; postcleithral process without a fleshy projection; lateral line long, ending at middle of caudal-fin skeleton **7**

7a. Body very short and compact; anal fin with 11–12 rays.. *Brachyglanis*

7b. Body moderately long to very elongate; anal fin with 14–35 rays... **8**

8a. Premaxillary tooth plate formed as a compact triangular or rectangular structure, bearing a distinct posterolateral projection (fig. 3a); adipose fin very low, with its depth approximately constant for most of its length; adipose fin posteriorly not fused with caudal fin, leaving a distinct posterior lobe....... *Leptorhamdia*

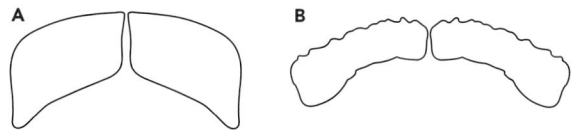

FIGURE 3 Premaxillary tooth plate of (A) *Leptorhamdia nocturna* and (B) *Myoglanis potaroensis*.

8b. Premaxillary tooth plate formed as an arch, lacking a posterolateral projection (fig. 3b); adipose as a continuous convex curve, with greatest depth at its midlength; adipose fin posteriorly fused with caudal fin so that its posterior limit is barely discernible .... *Myoglanis*

9a. Pectoral spine proportionally very large, dorsally curved; cephalic pores large; lateral line very short, with 3–4 pores anterior to dorsal-fin origin; caudal fin deeply forked, with ventral lobe much longer; dorsal lobe of caudal fin with 5–6 branched rays; ventral lobe of caudal fin with 4–5 branched rays .... *Gladioglanis*

9b. Pectoral spine moderately large or slender, straight along its dorsal surface; cephalic pores small; lateral line long, extending to middle of caudal-fin rays; caudal fin forked, lobes similar in length or only slightly different; dorsal lobe of caudal fin with 7 branched rays; ventral lobe of caudal fin with 8–9 branched rays **10**

10a. Posterior fontanel as a long slit..*Rhamdella*
10b. Posterior fontanel as small orifice at base of supraoccipital...*Rhamdia*

11a. Dorsal fin with 10 branched rays...**new Genus B**
11b. Dorsal fin with 6 branched rays, rarely 5 or 7...**12**

12a. First pectoral-fin ray and second dorsal-fin ray prolonged into very long filaments..........................*Mastiglanis*
12b. First pectoral-fin ray and second dorsal-fin ray not distinctly extending beyond the adjacent rays or, at
most, forming a short, discrete filament...**13**

13a. Anterior and posterior cranial fontanels short, separated from each other by a broad osseous bridge........*Cetopsorhamdia*
13b. Anterior and posterior cranial fontanels long, separated from each other by a narrow osseous bridge.............................**14**

14a. Trunk laterally with 4–5 dark oval spots **15**
14b. Trunk laterally without any oval spots **16**

15a. Mouth ventral; eye small, with rounded pupil; caudal fin with ventral lobe longer than dorsal one; ventral
lobe of caudal fin with 6 branched rays ....*Horiomyzon*
15b. Mouth subterminal; eye large, with comma-shaped pupil; caudal fin with dorsal lobe longer than ventral
one; ventral lobe of caudal fin with 8 branched rays ....**new Genus C**

16a. Pelvic-fin origin on vertical through the mid third of dorsal-fin base or posterior **17**
16b. Pelvic-fin origin on vertical through the dorsal-fin origin or anterior....................**24** (except *Chasmocranus brevior*)

17a. Mouth distinctly dorsal (prognathous); ventral lobe of caudal fin longer than the dorsal lobe ....*Phenacorhamdia*
17b. Mouth subterminal; ventral lobe of caudal fin as long as the dorsal lobe, or shorter.....................................**18**

18a. Dorsum of trunk lacking dark transverse bars...*Nemuroglanis*
18b. Dorsum of trunk with distinct dark transverse bars **19**

19a. Adipose fin broadly fused with caudal fin, with its posterior limit barely demarked; caudal fin rounded or
shallowly forked...**20**
19b. Adipose fin not fused with caudal fin, so that its posterior limit is distinctly demarked by a free lobe; caudal
fin distinctly forked..**21**

20a. Tip of distal extremity of maxillary barbel reaching at most the middle of pectoral fin or slightly beyond;
eyes significantly distanced from each other (24–29% of head length); anal fin from moderately long to
long, with 16–29 rays; posterior region of trunk with darker pigmentation along the myosepta; lateral line
incomplete, posteriorly fragmented ....*Heptapterus*
20b. Tip of distal extremity of maxillary barbel barely reaching the limits of gill slit; eyes very close to each other
(15–17% of head length); anal fin short, with 14–15 rays; posterior region of trunk without any distinct
pigmentation along the myosepta; lateral line complete, continuous, ending at base of region of caudal
skeleton ....**new Genus D**

21a. Body poorly pigmented; comma-shaped pupil ....**new Genus E**
21b. Body heavily pigmented; pupil rounded...**22**

22a. Body short and deep; snout conical in dorsal view; region from snout to dorsal-fin origin with large papillae ....**new Genus F**
22b. Body moderately elongate; snout convex in dorsal view; region from snout to dorsal-fin origin lacking large papillae **23**

23a. Eye of medium size, 15–20% of head length, with free orbital margin; adipose fin roughly triangular; caudal
fin deeply forked, with elongate lobes ....*Imparfinis*
23b. Eye small, 11–15% of head length, without free orbital margin; adipose fin roughly rectangular; caudal fin
shallowly forked, with attenuated lobes ....**new Genus G**

24a. Body with 5 broad, well-defined unpigmented zones covering dorsolaterally the body (one from snout tip
to region behind the orbits, one in the back of the head, one in front of the dorsal fin, covering its anterior
portion, one anterior to adipose fin, covering its anterior portion, and one around the caudal peduncle) ....*Nannoglanis*
24b. Body with background coloration mostly uniform brown...**25**

25a. Eye of medium size, 10–18% of head length; axillary pore present, large; lateral line continuous, ending at base of caudal fin .... *Chasmocranus*
25b. Eye small, 6.4–10.0% of head length; axillary pore absent; lateral line posteriorly fragmented, not reaching posteriorly the base of caudal fin **26**

26a. Mouth dorsal (prognathous); posterior portion of head with an unpigmented collar; region immediately anterior to dorsal fin with unpigmented mark; dorsal lobe of caudal fin slightly longer than ventral lobe .... *Pariolius*
26b. Mouth subterminal; body uniformly brown, without unpigmented collar around the posterior portion of head and without unpigmented mark immediately anterior to dorsal fin; dorsal lobe of caudal fin much more pronounced than ventral lobe ....**new Genus H**

GENUS ACCOUNTS

Brachyglanis (5.0–10 cm SL)

Characterized by: well-developed adductor mandibulae jaw muscles that extend from the cheeks onto the top of the head (also in *Leptorhamdia* and *Myoglanis*, which have more slender bodies). *Brachyglanis* attain relatively small adult body sizes and have a short and deep body shape in lateral profile. Also characterized by: snout flat or convex in dorsal view; mouth subterminal; eye small, without a free margin; premaxillary tooth plate wide and not bearing projection at posterolateral angle; maxillary barbel short, usually reaching to the middle of the pectoral fin; anterior fontanel present as a short slit, closed posteriorly; posterior fontanel present as small opening at base of supraoccipital; posterior supraoccipital process short, not contacting the first nuchal plate; posterior border of pectoral, dorsal, and anal fins convex (rounded); base of first pectoral-fin ray modified into strong spine, with distinct anteriorly oriented dentations on both anterior and posterior borders; pectoral fin usually with 6–7 branched rays; first dorsal-fin ray (spinelet) short and triangular; base of second dorsal-fin ray modified into spine; dorsal fin with 6–8 branched rays; pelvic-fin origin at a vertical through the middle to the last third of dorsal-fin base; adipose fin moderately long, with posterior lobe free, not connected to caudal fin; anal fin short, usually bearing 11–12 rays; caudal fin not deeply forked, with 7 branched rays on dorsal lobe and usually 8 branched rays on ventral lobe; lateral line continuous to middle of caudal-fin skeleton. **COLORATION** Brown or marbled coloration over the body surface, a patch of pale or unpigmented skin extending from the gular region dorsoposteriorly along the branchial margin, a dark pigment stripe at the base of caudal fin, and one or two irregular dark pigment stripes in the middle of the dorsal, pectoral, pelvic, anal, and caudal fins. **SPECIES** Five, all of which occur in the AOG region: *B. frenata* (type species), *B. magoi*, *B. melas*, *B. microphthalmus*, and *B. phalacra*. "*Brachyglanis*" *nocturnus* is actually a member of *Leptorhamdia* (Bockmann 1998). **DISTRIBUTION AND HABITAT** Restricted to northern South America, in north-bank affluents of the Amazon, Negro, Urubu, and Trombetas rivers, and in the Orinoco and Essequibo basins. *Brachyglanis* often live in small shoals in the leaf litter in backwaters of streams, but also among and under rocks in shallow riffles. **BIOLOGY** No data available.

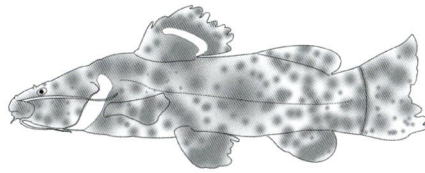

Brachyrhamdia (3.0–8.0 cm SL)

Characterized by: small, foreshortened, and deep body in lateral profile; snout convex in dorsal view; mouth subterminal; eye large, dorsolateral, with free margin; premaxillary tooth plate wide and not bearing projection at posterolateral angle; maxillary barbel very long (usually surpassing anal-fin origin); anterior fontanel as a long anterior slit; posterior fontanel as small opening at base of supraoccipital; mandibular adductor muscles restricted to facial region; posterior supraoccipital process long, contacting the first nuchal plate; posterior border of pectoral, dorsal, and anal fins convex; base of first pectoral-fin ray modified into a strong spine, with small serrations along its anterior margin, sometimes absent, and with conspicuous dentations along the basal portion of its posterior margin; pectoral fin usually with 7 branched rays; first dorsal-fin ray (spinelet) short and triangular; base of second dorsal-fin ray modified as a spine; dorsal fin with 6 branched rays; origin of pelvic fin on vertical through the region from end of dorsal-fin base to a point slightly posterior to it; adipose fin long, with posterior lobe free; anal fin short, bearing few rays (usually with 12–14 rays); caudal fin deeply forked, with lobes subequal in length, with 7 branched rays on dorsal lobe and 8 branched rays on ventral lobe; lateral line continuous to base of caudal-fin rays. *Brachyrhamdia* species are similar to *Pimelodella* species, from which they differ in being smaller (including fewer vertebrae, 32–37 vs. 38–46); larger heads (31–37% SL vs. 20–30%); posterior fontanel almost fully closed, except for a small anterior opening immediately posterior to the epiphyseal bar and a rounded opening at the base of the supraoccipital process (vs. posterior fontanel completely opened as a long slit); and dentations on posterior margin of unbranched pectoral-fin ray restricted to the basal portion (vs. extending considerably beyond its basal half, often approaching its tip) (Slobodian 2013, Slobodian and Bockmann 2013). **COLORATION** Highly variable among species, with marbled pattern or uniformly pigmented, dorsolateral band present or absent, lateral stripe present or absent, with dark transversal stripe across caudal peduncle present or absent, dark mark at caudal-fin base present or absent. All *Brachyrhamdia* species except *B. marthae* have a subocular blotch. **SPECIES** Six, all of which inhabit the AOG region: *B. heteropleura, B. imitator, B. marthae, B. meesi, B. rambarrani,* and *B. thayeria*. **DISTRIBUTION AND HABITAT** Throughout AOG region, in both white- and blackwater streams and rivers, especially associated with marginal vegetation or along sandy beaches in shallow streams (Slobodian and Bockmann 2013). **BIOLOGY** Some species have mimetic associations with syntopic *Corydoras* species. Popular among fish hobbyists (e.g., Slobodian and Bockmann 2013).

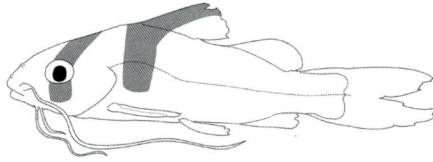

Cetopsorhamdia (7.2–12 cm SL)

Characterized by: small adult body sizes; a slender and slightly dorsoventrally depressed body in lateral profile, elliptical in cross section through the dorsal-fin origin. All species (except *C. boquillae*) with a conical snout; ventral mouth; posterior edge of the dorsal and pectoral fins truncate or concave; adipose fin short and distinctly triangular; and narrow lobes of the caudal fin. Also characterized by: eyes small, dorsolaterally placed, without a free rim; premaxillary tooth plate wide, not bearing projection at posterolateral angle; maxillary barbels short, reaching at most the origin of the dorsal fin, but usually shorter; anterior and posterior fontanels short, due to a broad ossification of the epiphyseal bar and adjacent regions; posterior border of pectoral, dorsal, pelvic, and anal fins concave to convex; mandibular adductor muscles restricted to facial region; posterior supraoccipital process short, not contacting the first nuchal plate; first pectoral-fin ray mostly flexible, except for its basal half to third; pectoral fin usually with 8–9 branched rays; first dorsal-fin ray (spinelet) absent; second dorsal-fin ray rigid on its basal third or completely flexible; dorsal fin with 6 branched rays; pelvic-fin origin usually at a vertical through posterior half of dorsal-fin base; adipose fin moderately short, with posterior lobe free; anal fin short, with few (usually 12–13) rays; caudal fin forked, usually the lower lobe with the longest fin rays, with 7 branched rays on the dorsal lobe and 8 branched rays on the ventral lobe; lateral line usually continuous to base of caudal-fin rays. **COLORATION** Mostly dark brown, with conspicuous V- or W-shaped vertical, dark band at the end of the caudal peduncle. **SPECIES** Five, including one species, *C. insidiosa*, in the AOG region. Several additional species remaining to be described (Bockmann and Reis 2011, Bockmann and Slobodian 2013). **DISTRIBUTION AND HABITAT** Magdalena, Maracaibo, Orinoco, Amazon, São Francisco, upper

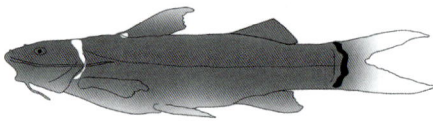

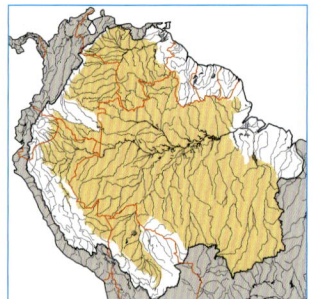

Paraná, and Uruguay basins (Bockmann and Guazzelli 2003; F. B. pers. obs.). *Cetopsorhamdia insidiosa* inhabits the Branco River in Brazil. Species typically inhabit the rocky bottoms of turbulent and well-oxygenated waters of small to medium-sized streams and rivers (Schultz 1944a, Schubart and Gomes 1959, Casatti and Castro 1998, Miranda and Mazzoni 2003; F. B. pers. obs.). **BIOLOGY** *Cetopsorhamdia* are active mainly at night, when they hunt their prey (mostly insect larvae) by excavating the substrate, using their mouths and barbels to turn over the sand-gravel substrate (Casatti and Castro 1998).

Chasmocranus (7.9–17 cm SL)

Characterized by: small to medium adult body sizes; a low and slender to moderately slender body shape in lateral profile; elliptical to quadrangular in cross section through the dorsal-fin origin; and possessing a large axillary pore. Also characterized by: snout convex in dorsal view; mouth subterminal or dorsal (prognathous); eye large, dorsal or dorsolateral, without free margin; premaxillary tooth plate wide and lacking projection at posterolateral angle, or narrow and bearing a distinct projection at posterolateral angle; maxillary barbel short (reaching at most the middle of pectoral fin or slightly beyond); anterior and posterior fontanels opened as long slits; mandibular adductor muscles restricted to facial region; posterior supraoccipital process short, not contacting the first nuchal plate; posterior border of pectoral, dorsal, pelvic, and anal fins convex; first pectoral-fin ray mostly flexible, except for its basal half; pectoral fin usually with 8–9 branched rays; first dorsal-fin ray (spinelet) absent; second dorsal-fin ray rigid on its basal half; dorsal fin with 6 branched rays; origin of pelvic fin at a vertical through the dorsal-fin origin or slightly posterior; adipose fin low, moderately long, with posterior lobe free; anal fin usually short, bearing few rays (12–15; single exception is *C. surinamensis*, with 19–22 anal-fin rays); caudal fin forked or rounded, with 7 branched rays on dorsal lobe and 8–9 branched rays on ventral lobe; lateral line continuous to base of caudal fin; species of *Chasmocranus* other than *C. brevior* have caudal fin shallowly forked, with lobes posteriorly rounded and equal in length, or rounded, with dorsal lobe slightly longer. **COLORATION** Mostly dark brown, usually with conspicuous unpigmented collar around the posterior portion of head and unpigmented mark anterior to dorsal fin. **SPECIES** Five, all of which are present in the AOG region: *C. brevior*, *C. chimantanus*, *C. longior*, *C. surinamensis*, and *C. tapanahoniensis*. Six other nominal species of "*Chasmocranus*" have been described that should be reassigned to other heptapterid genera, some of them new (Bockmann 1998, Bockmann and Guazzelli 2003; see below). Other *Chasmocranus* species remain to be described. **DISTRIBUTION AND HABITAT** Inhabit crevices formed in rocky bottom of pools of fast-flowing rivers draining the Brazilian and Guiana shields (Bockmann and Guazzelli 2003), and are common in rapids, where they live under stones and in dense growths of riverweed (Podostemacea) aquatic herbs (Mees 1967, Horeau et al. 1998). Smaller specimens occur in smaller rivers with faster currents (Mees 1967). **BIOLOGY** Feed mostly on aquatic insects (Horeau et al. 1998).

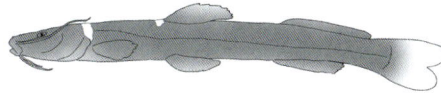

Gladioglanis (2.6–3.9 cm SL)

Includes species with the smallest body sizes of all heptapterids (Bockmann and Guazzelli 2003). Characterized by: massive, heavily serrated pectoral-fin spines, with distinct dentations on both anterior and posterior margins; and elongate, dorsoventrally depressed body, elliptical in cross section through the dorsal-fin origin (Ferraris and Mago-Leccia 1989, Lundberg et al. 1991b, Rocha et al. 2008b). Also characterized by: snout convex in dorsal view; mouth dorsal (prognathous); eye very small, dorsal, without free margin; premaxillary tooth plate wide, not bearing projection at posterolateral angle; maxillary barbel short, its tip reaching the region from the middle of pectoral fin to slightly posterior to the dorsal-fin origin; anterior and posterior fontanels as broad openings; mandibular adductor muscles restricted to facial region

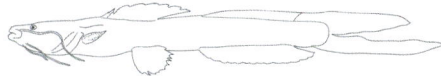

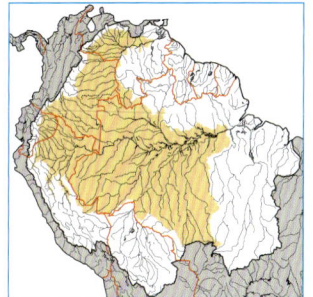

(with the exception of one undescribed species; see Bockmann and Slobodian 2013); posterior supraoccipital process short, not contacting the first nuchal plate; posterior border of pectoral, dorsal, pelvic, and anal fins convex; pectoral fin with 4 branched rays; first dorsal-fin ray (spinelet) short and triangular; second dorsal-fin ray forming a short spine (spinelet and second dorsal-fin ray missing in *G. anacanthus*); dorsal fin with 5–6 branched rays; origin of pelvic fin on vertical through the posterior half of dorsal-fin origin; adipose fin long, low, posteriorly confluent with caudal fin; anal fin moderately long to long, bearing 15–25 rays; caudal fin deeply forked (ventral lobe the longest), with 5–6 branched rays on dorsal lobe and 4–5 branched rays on ventral lobe; lateral line very short, bearing 3–4 pores. **COLORATION** Body uniformly brown. **SPECIES** Three, all of which are present in the AOG region: *G. anacanthus*, *G. conquistador*, and *G. machadoi*. At least six additional species remain to be described (Bockmann and Lundberg 2006, Bockmann and Slobodian 2013). **DISTRIBUTION AND HABITAT** Small, forested streams with large leaf-litter deposits in lowlands of the main tributaries of the Orinoco and Amazon, and Capim basins (Ferraris and Mago-Leccia 1989, Lundberg et al. 1991b, Bockmann 1998, Rocha et al. 2008b, Bockmann and Slobodian 2013). **BIOLOGY** Feeds on small invertebrates, including copepods, bosminid cladocerans, chironomids, and rotifers (Ferraris and Mago-Leccia 1989).

Goeldiella (10–30 cm SL)

Characterized by: relatively large adult body size; a pectoral spine with well-defined serrations, with anteriorly oriented barbs on the anterior margin, and posteriorly oriented barbs on the posterior margin; and a triangular supraoccipital process with a broad base. Also characterized by: body moderately long, deep, and elliptical in cross section through the dorsal-fin origin; snout convex in dorsal view; mouth subterminal; eye large, dorsolateral, with free margin; premaxillary tooth plate wide, not bearing projection at posterolateral angle; maxillary barbel very long (usually surpassing caudal-fin base); anterior and posterior fontanels as long slits; mandibular adductor muscles restricted to facial region; posterior supraoccipital process long, contacting the first nuchal plate; posterior border of pectoral, dorsal, and anal fins convex; pectoral fin usually with 8–9 branched rays; first dorsal-fin ray (spinelet) short and triangular; base of second dorsal-fin ray modified into a spine; dorsal fin with 6 branched rays; origin of pelvic fin at vertical through the end of dorsal-fin base; adipose fin long and deep, with posterior lobe free; anal fin short, bearing few rays (usually with 14–15 rays); caudal fin forked, with lobes rounded (the ventral lobe the longest), bearing 7 branched rays on dorsal lobe and 9 branched rays on ventral lobe; lateral line continuous to base of caudal-fin rays. **COLORATION** Immediately recognizable by a unique color pattern consisting of an oblique dark stripe extending from the pectoral-fin base to the anterior region of the dorsal-fin base and along the first dorsal-fin rays, a clear zone adjacent to the oblique dark stripe on the trunk, and a marbled color pattern on the trunk and adipose fin interrupted by an irregular dorsolateral stripe along the trunk. **SPECIES** One, *G. eques*. **DISTRIBUTION AND HABITAT** Throughout the Amazon basin and coastal rivers of the Guianas (Bockmann and Guazzelli 2003). **BIOLOGY** Feeds on small arthropods and fishes (Andrade-López and Machado-Allison 2009).

Heptapterus (4.7–21 cm SL)

Characterized by: elongate slender body, elliptical in cross section through the dorsal-fin origin, bearing a typical eel-like aspect. Also characterized by: snout convex in dorsal view; mouth subterminal; eye small, dorsal, without free margin; premaxillary tooth plate wide, not bearing projection at posterolateral angle; maxillary barbel short (its tip reaching at most the middle of pectoral fin or slightly beyond); anterior and posterior fontanels opened; mandibular adductor muscles restricted to facial region; posterior supraoccipital process short, not contacting the first nuchal plate; posterior border of pectoral, dorsal, pelvic, and anal fins convex; first pectoral-fin ray mostly flexible, except for its basal portion; pectoral fin usually with 6–8 branched rays; first dorsal-fin ray (spinelet) absent; second dorsal-fin ray flexible except for its basal portion; dorsal fin with 6 branched rays; origin of pelvic fin on vertical through the anterior half of dorsal-fin base; adipose fin deep, sloped, and very long, with posterior portion confluent with caudal fin, so that its posterior limit is barely noticeable; anal fin medium-sized to moderately long, bearing 16–29 rays; caudal fin rounded, with region corresponding to dorsal lobe more pronounced, and many more procurrent rays on ventral lobe of caudal fin than on dorsal lobe; dorsal caudal-fin lobe with 5–6 branched rays and ventral lobe with 4–7 branched rays; lateral line posteriorly fragmented, not reaching the base of caudal fin. **COLORATION** Body mostly light brown to grayish, with 4 transverse dark bars on back (one immediately posterior to head, one in front of dorsal-fin origin, one on posterior half of dorsal-fin base, and one on adipose-fin origin) and posterior region of trunk with darker pigmentation along the myosepta. **SPECIES** Six, only one of which occurs in the AOG region, an undescribed species from the headwaters of the Grande River, affluent to the Mamoré River, in the Madeira basin, Bolivia (Bockmann 1998). **DISTRIBUTION AND HABITAT** Well-oxygenated waters of small to medium-sized streams with moderate to rapid flow draining the Brazilian Shield, sheltering in crevices formed in the rocky bottom (Aguilera et al. 2011). Some species also occur in slow-flowing waters with a sandy bottom (Azpelicueta et al. 2011). **BIOLOGY** Individuals organize in small schools comprising several individuals (F. B. pers. obs.).

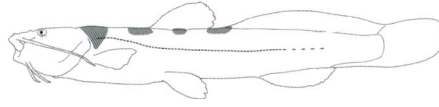

Horiomyzon (1.5–2.4 cm SL)

Characterized by: very small adult body sizes; ventrally oriented mouth; caudal peduncle very slender (5.0–6.0% SL); ventral lobe of caudal fin longer than dorsal lobe (Bockmann and Slobodian 2013); body moderately short and slender, elliptical in cross section through the dorsal-fin origin; snout conical or convex in dorsal view; eye small or very small, dorsal, without free margin; premaxillary tooth plate wide, not bearing projection at posterolateral angle; maxillary barbel moderately long (its tip surpassing the origin of the pelvic fin); anterior and posterior fontanels opened; mandibular adductor muscles restricted to facial region; posterior supraoccipital process short, not contacting the first nuchal plate; posterior border of pectoral, dorsal, pelvic, and anal fins convex; first pectoral-fin ray mostly flexible, except for its basal portion; pectoral fin usually with 7–8 branched rays; first dorsal-fin ray (spinelet) absent; second dorsal-fin ray flexible except for its basal portion; dorsal fin with 6 branched rays; origin of pelvic fin on vertical through origin of dorsal fin or slightly anterior; adipose fin deep, triangular, and moderately long, with posterior portion distinctly apart from caudal fin; anal fin short, bearing 11–12 rays; caudal fin forked, with 5 or 7 branched rays on dorsal lobe and 6 branched rays on ventral lobe; lateral line continuous to middle of caudal-fin skeleton. The type species *H. retropinnatus* is markedly modified (in opposition to the undescribed species; see Bockmann and Slobodian, 2013), being characterized by several characteristics, such as a very small eye, top of head with 2 rows of large papillae, small and distinctly ventral mouth, pseudotympanum largely visible, pectoral fin dorsally inclined, medial rays of pectoral fin longer than lateral-most rays, and most medial ray of pelvic fin unbranched. **COLORATION** Body ground color lightly pigmented, with 4–5 oval spots laterally on the trunk, sometimes fusing into an irregular lateral stripe. The spots are located immediately posterior to head, near the dorsal-fin base, between the dorsal-fin and adipose-fin bases, and at the adipose-fin origin and end. **SPECIES** Two, *H. retropinnatus* and an undescribed species (Bockmann and Slobodian 2013). **DISTRIBUTION AND HABITAT** Deep river channels (>20 m) of the Napo, Madeira, Solimões, and Amazon rivers (Stewart 1986a, Bockmann and Slobodian 2013). **BIOLOGY** No data available.

Imparfinis (3.0–12 cm SL)

Diagnosed by osteological characters (Bockmann 1998) that are not easily observed externally. Characterized by: moderately elongate body, oval in cross section through the dorsal-fin origin; snout convex in dorsal view; mouth subterminal; eye of medium size, dorsolateral, with free margin; premaxillary tooth plate wide, not bearing projection at posterolateral angle; maxillary barbel moderately long (its extremity usually reaching the end of the pectoral fin, and often surpassing the pelvic-fin base); anterior and posterior fontanels opened; mandibular adductor muscles restricted to facial region; posterior supraoccipital process short, not contacting the first nuchal plate; posterior border of pectoral, dorsal, pelvic, and anal fins convex; first pectoral-fin ray mostly flexible, except for its basal portion; pectoral fin usually with 8–10 branched rays; first dorsal-fin ray (spinelet) absent; second dorsal-fin ray flexible except for its basal portion; dorsal fin with 6 branched rays; origin of pelvic fin at vertical through the middle of dorsal-fin base or slightly posterior; adipose fin low, sloped, and moderately long, with posterior lobe free, not confluent with caudal fin; anal fin short, bearing 11–15 rays; caudal fin deeply forked (usually the dorsal lobe the longest), with 7 branched rays on dorsal lobe and 8 branched rays on ventral lobe; lateral line usually continuous to base of caudal-fin rays. **COLORATION** Brown to grayish, with 7 transverse dark bars on back (one immediately posterior to head, one in front of dorsal-fin origin, one on posterior half of dorsal-fin base, one between dorsal-fin base and adipose-fin base, one on adipose-fin origin, one on adipose-fin end, and one on base of procurrent rays of caudal fin); usually the first 4 are the most conspicuous. A lateral stripe is normally present. **SPECIES** At least 15 species, of which 13 inhabit the AOG region. **DISTRIBUTION AND HABITAT** Coastal drainages of Guianas, Amazon, Orinoco, São Francisco, and San Juan basins; Paraná, Uruguay, and Paraguay basins; and along eastern Brazilian coastal basins (Bockmann and Guazzelli 2003, Almirón et al. 2007, Ortega-Lara et al. 2011). *Imparfinis* typically occur in environments of small water volume and bottom with stones and gravel (e.g., Moraes and Braga 2011). **BIOLOGY** Individuals of *Imparfinis* are solitary, active during the night, hunting their prey (mostly insect larvae) by excavating the substrate, using their mouths and barbels to turn over the sand-gravel substrate, while actively swimming among the bottom rocks or burrowing in sand (Sazima and Pombal Jr. 1986, Casatti and Castro 1998, Moraes and Braga 2011).

Leptorhamdia (6.0–16 cm SL)

Characterized by: well-developed cheek muscles (adductor mandibulae) that extend dorsally toward the head midline; medium body sizes; short to very long body shape; and a quadrangular body shape in cross section through the dorsal-fin origin. Also characterized by: snout convex in dorsal view; mouth subterminal; eye small, without free margin; premaxillary tooth plate narrow and with a distinct projection at posterolateral angle; maxillary barbel moderately long (usually reaching the origin of pelvic fin); anterior fontanel as a short anterior slit (posteriorly closed); posterior fontanel as small opening at base of supraoccipital; posterior supraoccipital process short, not contacting the first nuchal plate; posterior border of pectoral, dorsal, and anal fins convex; base of first pectoral-fin ray modified into strong spine, with small to moderate dentations oriented toward the base of spine on both anterior and posterior margins; pectoral fin usually with 7–9 branched rays; first dorsal-fin ray (spinelet) short and triangular; base of second dorsal-fin ray modified into spine; dorsal fin usually with 6 branched rays; origin of pelvic fin at vertical through the middle to the last third of dorsal-fin base; adipose fin long, very low, with constant height along its length, with posterior lobe free (not connected to caudal fin); anal fin moderately varying from short to long, bearing 14–35 rays; caudal fin shallowly forked or rounded, with 7 branched rays on dorsal lobe and 9–10 branched rays on ventral lobe; lateral line continuous

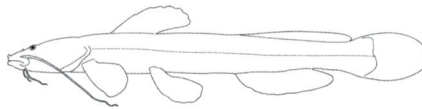

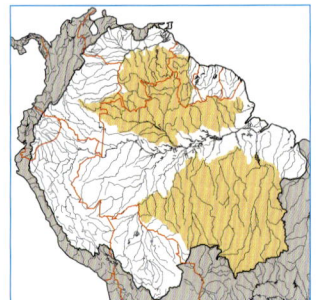

to middle of caudal-fin skeleton. **COLORATION** Dark brown, lacking an unpigmented collar or any unpigmented region at base of dorsal and adipose fins; dorsal, pectoral, pelvic, anal, and caudal fins lacking any stripe in their middle portions. **SPECIES** Three, all of which are present in the AOG region: *L. essequibensis*, *L. nocturna*, and *L. schultzi*. "*Myoglanis*" *aspredinoides* is also likely a species of *Leptorhamdia* (DoNascimiento and Lundberg 2005). **DISTRIBUTION AND HABITAT** Coastal Guyana drainages, Orinoco basin, and Xingu, Tocantins, Tapajós, and Negro basins. Species are typically found close to rapids and waters with rocky bottoms in fast-flowing rivers draining the Brazilian and Guiana shields (e.g., DoNascimiento and Lundberg 2005). **BIOLOGY** No data available.

Mastiglanis (4.0–7.0 cm SL)

Characterized by: the first pectoral-fin ray and first unbranched dorsal-fin ray prolonged as very long filaments; body weakly pigmented (translucent when alive); and a comma-shaped pupil (Bockmann 1994, Zuanon et al. 2006a). *Mastiglanis* has a moderately elongate body,

elliptical in cross section through the dorsal-fin origin; snout elliptical in dorsal view; mouth ventral; eyes large, dorsal, very close to each other, without free margin (except anterodorsally); maxillary barbel very long (its extremity extending beyond anal-fin origin); anterior and posterior fontanels opened; mandibular adductor muscles restricted to facial region; posterior supraoccipital process short, not contacting the first nuchal plate; posterior border of pectoral, dorsal, pelvic, and anal fins convex; first pectoral-fin ray stiffened basally (but friable), with its filament flexible; pectoral fin usually with 8–9 branched rays; first dorsal-fin ray (spinelet) absent; second dorsal-fin ray rigid (but friable) at its basal portion that corresponds to the length of first branched ray; dorsal fin with 6 branched rays; origin of pelvic fin on vertical through the posterior third of dorsal-fin base; adipose fin triangular and moderately long, with posterior lobe free, not confluent with caudal fin; anal fin short, bearing 9–11 rays; caudal fin deeply forked (lobes usually equal in length), with 7 branched rays on dorsal lobe and 8 branched rays on ventral lobe; lateral line continuous to base of caudal-fin rays. **COLORATION** Dorsum of trunk with 7 faint transverse dark bars on back: one immediately posterior to head, one in front of dorsal-fin origin, one on posterior half of dorsal-fin base, one between dorsal-fin base and adipose-fin base, one on adipose-fin origin, one on end of adipose-fin base, and one on base of procurrent rays of caudal fin. A poorly defined stripe is present along lateral line. **SPECIES** One, *M. asopos*. **DISTRIBUTION AND HABITAT** Throughout the AOG region and Capim River basins (Bockmann and Guazzelli 2003, Zuanon et al. 2006a). Inhabiting the superficial layers of sand patches (Zuanon et al. 2006a). **BIOLOGY** Forages on insects by using an exclusive "drift-trap" system made up of the extended barbels and filamentous first pectoral-fin rays (Zuanon et al. 2006a).

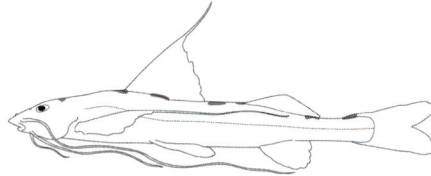

Myoglanis (6.0–15 cm SL)

Characterized by: well-developed cheek muscles that extend dorsally toward the head midline (Bockmann 1998); medium to small adult body sizes; elongate body; elliptical

in cross section through the dorsal-fin origin; a first dorsal-fin ray (spinelet) very reduced, not observable externally; base of second dorsal-fin ray soft, not modified into a spine; a moderately long anal fin, bearing 18–29 rays; and a very deep and long adipose fin, forming an ascending arch and fused posteriorly with caudal fin (so that its posterior limit is barely discernible). Also characterized by: snout convex in dorsal view; mouth subterminal; eye small, without free margin; premaxillary tooth plate wide and not bearing projection at posterolateral angle (except

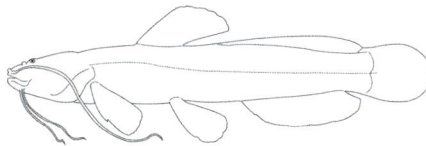

in *M. koepckei*); maxillary barbel of moderate size to long (its tip reaching from middle to pectoral fin to origin of pelvic fin); anterior fontanel as a short anterior slit (posteriorly closed); posterior fontanel as small opening at base of supraoccipital; posterior supraoccipital process short, not contacting the first nuchal plate; posterior border of pectoral, dorsal, and anal fins convex; base of first pectoral-fin ray modified into strong spine, with dentations oriented toward the base of spine; dentations on anterior margin and almost straight dentations on posterior margin; pectoral fin usually with 6–8 branched rays; dorsal fin usually with 6 branched rays; origin of pelvic fin on vertical through the middle to the last third of dorsal-fin base; caudal fin usually rounded or forked, with shallow lobes, with 5–8 branched rays on dorsal lobe and 5–10 branched rays on ventral lobe; lateral line complete, continuous to the middle of caudal-fin skeleton, or incomplete, with its end at vertical through the dorsal-fin origin. **COLORATION** Dark brown, an unpigmented collar present or absent, but unpigmented region at base of dorsal and adipose fins or dark transverse bars are absent; dorsal, pectoral, pelvic, anal, and caudal fins lacking any stripe in their middle portions. **SPECIES** Two: *M. potaroensis* and *M. koepckei*. *Myoglanis aspredinoides* (DoNascimiento and Lundberg 2005) is probably a member of *Leptorhamdia*. **DISTRIBUTION AND HABITAT** Coastal drainages of the Guianas, Orinoco, Madeira, Tapajós, Negro, and Amazon basins (Masson 2007; F. B and V. S. pers. obs.). *Myoglanis koepckei* inhabits leaf litter packs in small terra firme forest streams (Lima et al. 2005). **BIOLOGY** No data available.

Nannoglanis (4.5 cm SL)

Characterized by: a short and relatively slender body; elliptical in cross section through the dorsal-fin origin; snout convex in dorsal view; mouth subterminal; eyes small, dorsal, very close to each other, without free margin; premaxillary tooth plate wide, not bearing projection at posterolateral angle; maxillary barbel short (its extremity reaching the region from the external border of branchiostegal membrane to the first third of pectoral fin); anterior and posterior fontanels opened; mandibular adductor muscles restricted to facial region; posterior supraoccipital process short, not contacting the first nuchal plate; posterior border of pectoral, dorsal, pelvic, and anal fins convex; first pectoral-fin ray mostly flexible, except for its basal portion; pectoral fin usually with 8–9 branched rays; first dorsal-fin ray (spinelet) absent; second dorsal-fin ray flexible except for its basal portion; dorsal fin with 6 branched rays; origin of pelvic fin on vertical through the middle of dorsal-fin base or slightly posterior; adipose fin rectangular, low, and moderately long, with posterior lobe free, not confluent with caudal fin; anal fin short, bearing 10–11 rays; caudal fin shallowly forked, almost truncated (the dorsal lobe slightly longer), with 6 branched rays on dorsal lobe and 6 branched rays on ventral lobe; lateral line posteriorly fragmented, reaching the vertical through the adipose-fin origin. **COLORATION** A unique color pattern among heptapterids, consisting of 5 broad, well-defined unpigmented zones covering the body dorsolaterally: one from snout tip to region behind the orbits, one in the back of the head, one in front of the dorsal fin, covering its anterior portion, one anterior to adipose fin, covering its anterior portion, and one around the caudal peduncle. **SPECIES** One, *N. fasciatus*. **DISTRIBUTION AND HABITAT** Western and central Amazon from the Napo to Japurá basins (Bockmann and Guazzelli 2003; F. B. pers. obs.), where it inhabits small (0.5–1.5 m wide) moderately fast-flowing streams with pebbles and gravel on the bottom (Saul 1975). **BIOLOGY** No data available.

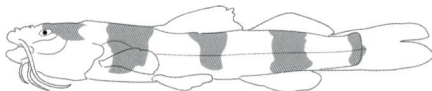

Nemuroglanis (3.3–4.7 cm SL)

Characterized by: small adult body sizes; a pectoral fin with few rays (I + 6–7); a ventral caudal-fin plate distinctly smaller than the dorsal one, supporting 7–3 rays; a ventral caudal-fin lobe with 9–19 rays, 6–3 of which are branched; a dorsal caudal-fin lobe much longer than the ventral lobe (usually >35% SL); and a lateral line fragmented and reaching at maximum the vertical through middle of adipose-fin base (Bockmann and Ferraris Jr. 2005, Ribeiro et al. 2011). Also characterized by: body short to elongate, ovoid in cross section through the dorsal-fin origin; snout convex in dorsal view; mouth subterminal or slightly dorsal; eye small, dorsolateral, without free margin; premaxillary tooth plate wide, not bearing projection at posterolateral angle; maxillary barbel long (its distal tip surpassing pelvic-fin origin); anterior and posterior cranial fontanels long and wide; mandibular adductor muscles restricted to facial region; posterior supraoccipital process short, not contacting the first nuchal plate; posterior border of pectoral, dorsal, pelvic, and anal fins convex; first pectoral-fin ray distally flexible; dorsal fin broad, with 6 branched rays; first dorsal-fin ray (spinelet) absent; second dorsal-fin ray flexible except for its basal portion; origin of pelvic fin on vertical through anterior third of dorsal-fin base; adipose fin moderately long, forming a continuous arch, with posterior region basally fused to caudal fin; anal fin short, bearing 10–15 rays; caudal fin deeply forked or lanceolate, with 4–7 branched rays on dorsal lobe and 3–7 branched rays on ventral lobe; lateral line fragmented, with its posteriormost limit usually reaching region beneath dorsal fin (may reach adipose fin in *N. furcatus*). **COLORATION** Mostly light brown on its dorsal half, with conspicuous lateral stripe extending to below adipose fin or further; absence of dark bars and stripes on back of trunk, as well as absence of dark band at end of the caudal peduncle. **SPECIES** Four, all in the AOG region: *N. furcatus*, *N. lanceolatus*, *N. mariai*, and *N. pauciradiatus*. At least five additional species remain to be described. **TAXONOMIC NOTE** *Nemuroglanis* is the senior synonym of *Medemichthys* and *Imparales* (Bockmann and Ferraris Jr. 2005). **DISTRIBUTION AND HABITAT** Throughout lowlands of the Orinoco, upper reaches of Amazon, and Madeira, Purus, and Negro basins (Bockmann and Ferraris Jr. 2005, Ribeiro et al. 2011; F. B. pers. obs.). *Nemuroglanis* is a typical inhabitant of leaf litter accumulated in backwaters. **BIOLOGY** No data available.

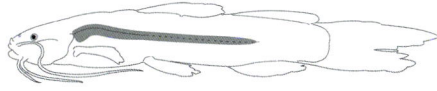

Pariolius (3.2–3.8 cm SL)

Characterized by: small, slender, and short bodies; quadrangular in cross section through the dorsal-fin origin; unpigmented collar behind head; snout short, and convex in dorsal view; sometimes a transversal unpigmented stripe in front of eyes and an unpigmented mark on adipose-fin origin are present; mouth dorsal (prognathous); eyes very small, dorsal, without free margin, and distant from each other; premaxillary tooth plate wide and lacking projection at posterolateral angle; maxillary barbel short (its tip usually reaching from middle of pectoral fin to its posterior margin); anterior and posterior fontanels opened as long slits; mandibular adductor muscles restricted to facial region; posterior supraoccipital process short, not contacting the first nuchal plate; posterior border of pectoral, dorsal, pelvic, and anal fins convex; first pectoral-fin ray mostly flexible, except for its basal half; pectoral fin usually with 6–8 branched rays; first dorsal-fin ray (spinelet) absent; second dorsal-fin ray rigid on its basal portion; dorsal fin with 6 branched rays; origin of pelvic fin slightly anterior to vertical through the dorsal-fin origin; adipose fin low, short, basally fused with caudal fin, but with posterior lobe distinguishable; anal fin short, usually with 12–13 rays; caudal fin rounded or shallowly forked, sometimes with truncate aspect (the dorsal lobe the longest), with 5 branched rays on dorsal lobe and 5 branched rays on ventral lobe; lateral line posteriorly fragmented, its terminus reaching beneath the dorsal-fin base. **COLORATION** Dark brown, with conspicuous unpigmented collar around the posterior region of head and unpigmented mark in front of dorsal fin. **SPECIES** Three, all in the AOG region, including *P. armillatus* and two undescribed species. **DISTRIBUTION AND HABITAT** Lowlands along the upper Amazon, Japurá River, and western parts of Negro River (Bockmann and Guazzelli 2003, Lima et al. 2005; F. B. pers. obs.). *Pariolius* lives in sand and gravel-bottomed channels of creeks, in areas devoid of aquatic vegetation, but often with overhanging marginal vegetation (Saul 1975). Lima et al. (2005) also reported *Pariolius* in both rivers and streams, living on the bottom, associated with rocks and leaf litter. **BIOLOGY** *Pariolius* is omnivorous, feeding on plant material, insects, shrimps, and small fishes (Saul 1975, Lima et al. 2005).

Phenacorhamdia (2.5–7.0 cm SL)

Diagnosed by osteological features (Bockmann 1998, DoNascimiento and Milani 2008). Externally characterized by a moderately to
very elongate body; distinctly prognathous mouth; first pectoral-fin ray usually longer than the second one, prolonged as a short filament; and caudal fin with ventral lobe markedly longer than the dorsal lobe. Also characterized by: snout convex in dorsal view; eyes of small size, dorsal, distant from each other, and without free margin; premaxillary tooth plate narrow, bearing projection at posterolateral angle; body elliptical or quadrangular in cross section through the dorsal-fin origin; maxillary barbel of moderate size, tips usually reaching to the last third of the pectoral fin or slightly surpassing the posterior border of pectoral fin; anterior and posterior fontanels opened; mandibular adductor muscles restricted to facial region; posterior supraoccipital process short, not contacting the first nuchal plate; posterior border of pectoral, dorsal, pelvic, and anal fins convex; first pectoral-fin ray mostly flexible, except for its basal third; pectoral fin usually with 6–9 branched rays; first dorsal-fin ray (spinelet) absent; second dorsal-fin ray flexible except for its basal portion; dorsal fin with 6 branched rays; origin of pelvic fin at vertical through the middle of dorsal-fin base or slightly anterior; adipose fin low, rectangular, and moderately long, not confluent with caudal fin, leaving a posterior lobe free; anal fin short to moderately long, bearing 12–18 rays; caudal fin deeply forked, usually with 7 branched rays on dorsal lobe and 8 branched rays on ventral lobe; lateral line continuous to base of caudal-fin rays. **COLORATION** Uniformly brown, lacking dorsal bars, unpigmented regions, or a midlateral stripe. **SPECIES** 10, including 8 in the AOG region (Ferraris Jr. 2007, DoNascimiento and Milani 2008). **DISTRIBUTION AND HABITAT** Throughout the Orinoco, Amazon, Paraná-Paraguay, and São Francisco basins, and coastal drainages of the Guianas. Found in small riparian streams, where it lives in crevices in the rocky or gravel bottom or in woody debris. **BIOLOGY** Hunts prey (mostly insect larvae) by excavating the substrate (Casatti et al. 2010).

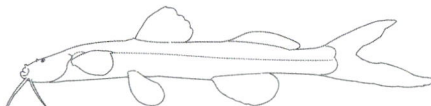

Pimelodella (7.0–30 cm SL)

Characterized by: very long barbels with tips usually reaching between the end of pelvic fin and the caudal-fin insertion; a membrane between the most medial caudal-fin rays extending
onto the basal half of the fin; pungent (sharp) pectoral- and dorsal-fin spines; pectoral spine bearing conspicuous dentations oriented toward the base of the spine on its posterior margin, and less conspicuous dentations on the anterior margin; 38–46 vertebrae; a medium-sized head (20–30% SL); and a dark lateral stripe along the body. Shares with *Goeldiella*, *Rhamdella*, *Rhamdia*. and *Rhamdioglanis* a relatively large adult body size compared with that of other heptapterids (≤20 cm SL). Also characterized by; snout convex in dorsal view; mouth subterminal; eyes of medium to large size, moderately distant from each other, with free margin; premaxillary tooth plate rectangular, not bearing a projection at posterolateral angle; body elliptical in cross section through the dorsal-fin origin; anterior and posterior fontanels opened; mandibular adductor muscles restricted to facial region; posterior supraoccipital process long, contacting the first nuchal plate; posterior border of pectoral, dorsal, pelvic, and anal fins convex; pectoral fin usually 7–8 branched rays; first dorsal-fin ray (spinelet) present; second dorsal-fin ray ossified as a pungent spine, sometimes bearing small retrose dentations (i.e., tip of dentitions oriented toward base of spine) on its posterior margin; dorsal fin with 6 branched rays; origin of pelvic fin usually at vertical through the posterior half of dorsal-fin base; adipose fin usually high, may be moderately short to long, not confluent with caudal fin, leaving a posterior lobe free; anal fin short, bearing 12–16 rays; caudal fin deeply forked, with 7 branched rays on dorsal lobe and 8 branched rays on ventral lobe; lateral line continuous to base of caudal-fin rays. Some *Pimelodella* species are sexually dimorphic, with males having the first pectoral-fin rays and second dorsal-fin rays as

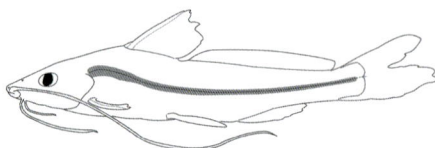

elongate filaments (Souza-Shibatta et al. 2013; F. B and V. S. pers. obs.). **COLORATION** Variable, although usually with a dark stripe along the lateral body surface, extending from snout or just posterior to the opercle to caudal-fin base or along its central rays. **SPECIES** *Pimelodella* is the most species-rich genus of Heptapteridae, with 77 species currently recognized as valid, with about 30 species in the AOG region. However, *Pimelodella* as currently circumscribed is an artificial genus, and these species numbers will probably change. In addition, there are many species assignable to *Pimelodella* that remain to be described. **COMMON NAMES** *Mandí chorão* (Brazil); *Bagre cunshi* (Peru). **DISTRIBUTION AND HABITAT** Throughout the Neotropical region, from Panama to the south of Peru and central Argentina. **BIOLOGY** Omnivorous and mostly active during the twilight and night. School in groups of up to ten individuals. As with species of *Brachyrhamdia*, some species of *Pimelodella* are appreciated as ornamental fishes.

Rhamdella (10–21 cm SL)

Diagnosed by several osteological characters (Bockmann and Miquelarena 2008), but can be identified by a combination of externally visible characters: eye medium to large, dorsolateral, with a free margin; first pectoral-fin ray ossified for two-thirds its base and flexible at its distal third; posterior supraoccipital process short, not reaching the first nuchal plate. *Rhamdella* (along with *Goeldiella*, *Pimelodella*, *Rhamdia* and *Rhamdioglanis*) can exceed the small body size common for most other heptapterids (≤20 cm SL). Also characterized by; body long, elliptical in cross section through the dorsal-fin origin; snout convex in dorsal view; mouth subterminal; premaxillary tooth plate rectangular, not bearing projection at posterolateral angle; maxillary barbel usually long, reaching at least the insertion of pelvic fin; anterior and posterior fontanels opened; mandibular adductor muscles restricted to facial region; posterior supraoccipital process not reaching the first nuchal plate; posterior border of pectoral, dorsal, and anal fins convex; first pectoral-fin ray having its basal two-thirds ossified into a spine, usually with dentations oriented toward the base of spine; dentations at its posterior margin; pectoral fin usually with 8–9 branched rays; first dorsal-fin ray (spinelet) short and triangular; basal two-thirds of second dorsal-fin ray modified into a spine; dorsal fin with 6–7 branched rays; origin of pelvic fin at vertical through the posterior end of dorsal-fin base; adipose fin long and deep, with posterior lobe free; anal fin moderately short, bearing few rays (usually with 16–18 rays); caudal fin forked, with 7 branched rays on dorsal lobe and 8 branched rays on ventral lobe; lateral line continuous to base of caudal-fin rays. **COLORATION** Similar to *Pimelodella*, with a dark lateral stripe present or not. **SPECIES** Six (Bockmann and Miquelarena 2008, Reis et al. 2014), including only one in the AOG region: *R. rusbyi* (Bockmann and Miquelarena 2008, Reis et al. 2014). The taxonomic position of "*R.*" *montana* in the family Heptapteridae is uncertain, and is here included provisionally in *Rhamdella* following Bockmann (1998) and Bockmann and Miquelarena (2008). **DISTRIBUTION AND HABITAT** Most diverse in the Southern Brazilian coastal basins and the Uruguay basin. One *Rhamdella* species in the AOG region, *R. rusbyi*, in the upper Beni basin of Bolivia (Bockmann and Guazzelli 2003). Typically found over rocky bottoms in clear water streams, showing a tendency for gregarity, grouping under large stones (Bockmann and Miquelarena 2008). **BIOLOGY** No data available.

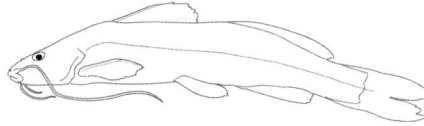

Rhamdia (4.5–47 cm SL)

Attain the largest adult body sizes of the family. Characterized by the following unique combination of characters: posterior fontanel almost entirely closed, with just a small opening at supraoccipital remaining; posterior process of the supraoccipital short, not reaching the first nuchal plate; body long, relatively deep, elliptical in cross section through the dorsal-fin origin; snout convex in dorsal view; mouth subterminal; eye medium to large, dorsolateral, with free margin; premaxillary tooth plate rectangular, not

bearing projection at posterolateral angle; maxillary barbel usually long, reaching at least the insertion of pelvic fin; anterior fontanel opened, as a long slit; mandibular adductor muscles restricted to facial region; posterior border of pectoral, dorsal, and anal fins convex; pectoral fin usually with 8–9 branched rays; first dorsal-fin ray (spinelet) short and triangular; basal two-thirds of second dorsal-fin ray modified into a spine; dorsal fin with 6–7 branched rays; origin of pelvic fin at vertical through the end of dorsal-fin base; adipose fin long and deep, with posterior lobe free; anal fin moderately short, bearing few rays (9–17 rays, usually 12–15); caudal fin forked, with 7 branched rays on dorsal lobe and 8–9 branched rays on ventral lobe; lateral line continuous to base of caudal-fin rays. **COLORATION** Usually with a marbled dark-and-light coloration pattern, and sometimes a dark lateral stripe along the lateral line. **SPECIES** 21, including about six in the AOG region. In the most recent taxonomic revision of the genus, Silfvergrip (1996) recognized only eleven species, ascribing the rest of named species as part of the natural variation within species. **DISTRIBUTION AND HABITAT** Rivers and streams of forested areas throughout the humid Neotropics, from southern Mexico to central Argentina (Silfvergrip 1996), and are present to abundant in most rivers of the AOG region. **BIOLOGY** Feed on small benthic aquatic animals (like insect larvae) and small fishes.

Heptapteridae Genus A (3.4–4.1 cm SL)

Diagnosed by: well-developed jaw muscles that extend dorsally toward the head midline (also present in *Brachyglanis*, *Leptorhamdia*, and *Myoglanis*); and an unusually well-developed postcleithral process with a fleshy projection, covered with dark pigmentation. Species of Genus A are all of small adult body size, and the body is compact or moderately elongate. Also characterized by: snout convex in dorsal view; mouth prognathous; eye small, without free margin; premaxillary tooth plate wide and not bearing projection at posterolateral angle; maxillary barbel short (usually barely surpassing the posterior border of pectoral fin); anterior fontanel and posterior fontanels broadly opened; posterior supraoccipital process short, not contacting the first nuchal plate; posterior border of pectoral, dorsal, and anal fins convex; base of first pectoral-fin ray modified into strong spine, with strong dentations oriented toward the base of spine; dentations on both anterior and posterior borders; pectoral fin usually with 5 branched rays; first dorsal-fin ray (spinelet) very reduced, not observable externally; second dorsal-fin ray absent; dorsal fin with 6 branched rays; origin of pelvic fin at a vertical through the last two-thirds of the dorsal-fin base; adipose fin moderately long, with posterior region connected with caudal fin, with its posterior limit demarked by a shallow concavity; anal fin long, bearing 21–29 rays; caudal fin forked, with 5 branched rays on dorsal lobe and usually 5–6 branched rays on ventral lobe; lateral line very short, with 2 pores only. **COLORATION** Uniformly brown, lacking unpigmented regions, dark dorsal bars, or lateral stripe. **SPECIES** One, undescribed and not previously identified in other genera. **DISTRIBUTION AND HABITAT** Endemic to the upper Negro basin. Individuals of Genus A inhabit small blackwater streams with abundant leaf litter and a muddy bottom. **BIOLOGY** No data available.

Heptapteridae Genus B (3.4–4.0 cm SL)

Diagnosed by osteological characters (Stewart 1985, Bockmann 1998). Recognized by several external traits: body deep, distinctly compressed; snout conical in dorsal view; region from top of head to dorsal-fin origin with large papillae; dorsal fin with 10 branched rays; first ray of pectoral fin and second ray of dorsal fin prolonged into short filaments. Also characterized by: mouth subterminal; eye of medium size, dorsolaterally placed on the head, and without a free margin; premaxillary tooth plate wide, not bearing projection at posterolateral angle; maxillary barbel very long with a tip usually reaching beyond the anal-fin base; anterior

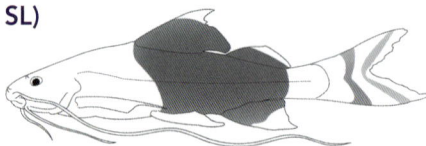

and posterior fontanels opened; mandibular adductor muscles restricted to facial region; posterior supraoccipital process short, not contacting the first nuchal plate; posterior border of pectoral, dorsal, pelvic, and anal fins concave; first pectoral-fin ray with basal third rigid and distal portion flexible; pectoral fin with 9–10 branched rays; first dorsal-fin ray (spinelet) absent; second dorsal-fin ray flexible except for its basal third; origin of pelvic fin on vertical through the first third of dorsal-fin base; adipose fin of moderate size, triangular, deep, with posterior lobe free, not confluent with caudal fin; anal fin short, bearing 11–13 rays; caudal fin deeply forked (ventral lobe the longest), with 7 branched rays on dorsal lobe and 8 branched rays on ventral lobe; lateral line usually continuous to base of caudal fin skeleton. **COLORATION** Midportion of trunk region between the dorsal fin and the end of adipose-fin base obliquely crossed by a wide, well-defined dark area, interspersed with a broad anterior and posterior clear areas, dorsal fin almost completely dark, except for distal portion of the first rays. Transverse dark bars on back and lateral stripe absent. **SPECIES** Two, both in the AOG region, one previously identified as "*Cetopsorhamdia*" *phantasia*, and one as yet undescribed (Bockmann 1998). **DISTRIBUTION AND HABITAT** Napo and Madeira basins (Bockmann 1998, Bockmann and Slobodian 2013). Individuals of "*C.*" *phantasia* were collected in sectors where the water current was very fast and the bottom of sandy composition, with leaf litter (Stewart 1985). **BIOLOGY** No data available.

Heptapteridae Genus C (4.0–4.8 cm SL)

Diagnosed by having a large eye with a comma-shaped pupil, and a caudal fin with the dorsal lobe much longer than the ventral lobe. Also characterized by: a moderately elongate body, elliptical in cross section through the dorsal-fin origin; snout convex in dorsal view; mouth subterminal; eye large, dorsolateral, without free margin; premaxillary tooth plate wide, not bearing projection at posterolateral angle; maxillary barbel long to very long (its extremity usually reaching the origin of pelvic and usually surpassing the anal-fin base); anterior and posterior fontanels opened; mandibular adductor muscles restricted to facial region; posterior supraoccipital process short, not contacting the first nuchal plate; posterior border of pectoral, dorsal, pelvic, and anal fins concave; first pectoral-fin ray stiffened basally (but friable), with a short distal filament flexible; pectoral fin usually with 8–10 branched rays; first dorsal-fin ray (spinelet) absent; second dorsal-fin ray rigid (but friable) at its basal portion and distally flexible, as a short filament; dorsal fin with 6 branched rays; origin of pelvic fin on vertical through the middle of dorsal-fin base or slightly posterior; adipose fin low, roughly triangular, and moderately long, with posterior lobe free, not confluent with caudal fin; anal fin short, bearing 10–11 rays; caudal fin deeply forked (usually the dorsal lobe the longest), with 7 branched rays on dorsal lobe and 8 branched rays on ventral lobe; lateral line usually continuous to base of caudal-fin rays. **COLORATION** Body weakly pigmented and translucent when alive, with 7 transverse dark bars on back (one immediately posterior to head, one in front of dorsal-fin origin, one on posterior half of dorsal-fin base, one between dorsal-fin base and adipose-fin base, one on adipose-fin origin, one on adipose-fin end, and one on base of procurrent rays of caudal fin; usually the first 4 are the most conspicuous) and 4 dark, oval spots on the lateral surface of the trunk (one above the pectoral fin, one below the dorsal fin, one below the anterior portion of the adipose fin, and one below posterior region of adipose fin), and a dark ventral lobe of the caudal fin. A lateral stripe is either faint or absent. **SPECIES** Three, all in the AOG region, one previously identified as "*Nannorhamdia*" *stictonotus*, another as "*Imparfinis*" *pseudonemacheir*, and one as yet undescribed. **DISTRIBUTION AND HABITAT** Broadly distributed along lowlands of Amazon, Orinoco, and upper Paraguay basins, as well as rivers of the Guyana region, but confined to sandy bottom habitats (Bockmann and Guazzelli 2003; F. B. pers. obs., Bockmann and Slobodian 2013). **BIOLOGY** No data available.

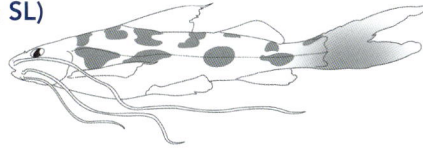

Heptapteridae Genus D (13–23 cm SL)

Diagnosed by: medium to large body adult sizes; and an elongate eel-like aspect, elliptical in cross section. Snout convex in dorsal view; mouth subterminal; eyes large, dorsal, without free margin, very close to each other (15–17% of head length); maxillary barbel very short (its end barely reaching the limits of gill slit); anterior and posterior fontanels opened; mandibular adductor muscles restricted to the facial region; posterior supraoccipital process short, not contacting the first nuchal plate; posterior border of pectoral, dorsal, pelvic, and anal fins convex; first pectoral-fin ray mostly flexible, except for its basal third to half; pectoral fin usually with 7–9 branched rays; first dorsal-fin ray (spinelet) absent; second dorsal-fin ray flexible except for its basal portion; dorsal fin with 6 branched rays; origin of pelvic fin on vertical through the anterior third of dorsal-fin base; adipose fin deep, sloped, and very long, with posterior portion confluent with caudal fin, so that its posterior limit is barely noticeable; anal fin short, bearing 14–15 rays; caudal fin almost rounded, with lobes attenuated; region corresponding to dorsal lobe usually more pronounced, sometimes forming a pointed projection; dorsal caudal-fin lobe with 6–7 branched rays and ventral lobe usually with 5–7 branched rays; lateral line continuous, ending at base of region of caudal skeleton. **COLORATION** Body mostly light brown to grayish, with 4 transverse dark bars on back (one immediately posterior to head, one in front of dorsal-fin origin, one on posterior half of dorsal fin, and one on adipose-fin origin). **SPECIES** Six, all in the AOG region, including one previously identified as "*Chasmocranus*" *brachynema*, one as "*Imparfinis*" *borodini*, one as "*Pariolius*" *hollandi*, and at least three species not yet described. **DISTRIBUTION AND HABITAT** Distributed predominantly in more meridional basins, occurring in the upper Paraná, upper São Francisco, and upper Tocantins, and marginally in the upper reaches of Iguaçu and Paraguay. Genus D lives in protected areas of medium to large rivers, with less turbulent waters, usually associated with marginal vegetation, fallen logs, sometimes interspersed among rooted weeds and stones (Caramaschi 1986). In the upper rio Tocantins, state of Goiás, Brazil, individuals of this genus have also been reported in a stretch with riffles, rapids, and rocky bottom, sometimes intercalated with pools of sandy bottom (Miranda and Mazzoni 2003). In that stretch, most of the riverbed was devoid of vegetal coverage. **BIOLOGY** Hahn et al. (1997) reported an invertivorous diet for a species of this genus, with a predominace of microcrustaceans.

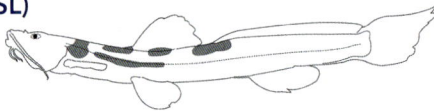

Heptapteridae Genus E (3.8 cm SL)

Diagnosed by a very short body, with 33 vertebrae; an elliptical snout; a comma-shaped pupil; and a poorly pigmented body, translucent when alive (Bockmann 1998, Zuanon et al. 2006a). Also characterized by: body elliptical in cross section; a poorly defined stripe present along lateral line; mouth ventral; eyes of medium size, dorsal, very close to each other, without free margin; maxillary barbel short (its extremity extending to base of pectoral fin); anterior fontanel anteriorly closed; posterior fontanel entirely opened; mandibular adductor muscles restricted to facial region; posterior supraoccipital process short, not contacting the first nuchal plate; posterior border of pectoral, dorsal, pelvic, and anal fins convex; first pectoral-fin ray mostly flexible, except for its basal portion; pectoral fin usually with 8–9 branched rays; first dorsal-fin ray (spinelet) absent; second dorsal-fin ray flexible, except for its basal portion; dorsal fin with 6 branched rays; origin of pelvic fin on vertical through the middle of dorsal-fin base; adipose fin short, triangular, with posterior lobe free, not confluent with caudal fin; anal fin short, bearing 10 rays; caudal fin deeply forked (lobes usually equal in length), with 6 branched rays on dorsal lobe and 7 branched rays on ventral lobe; lateral line continuous to base of caudal-fin rays. **COLORATION** Back of trunk with 7 transverse dark bars: one immediately posterior to head, one in front of dorsal-fin origin, one on posterior half of dorsal-fin base, one between dorsal-fin base and adipose-fin base, one on

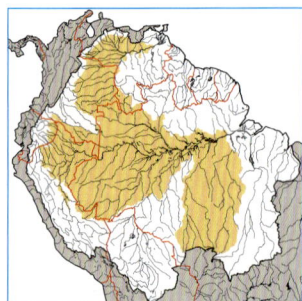

adipose-fin origin, one on end of adipose-fin base, and one on base of procurrent rays of caudal fin. **SPECIES** One, a species formerly ascribed to "*Imparfinis*" *pristos* (Mees and Cala 1989). **DISTRIBUTION AND HABITAT** Throughout the Orinoco, Madeira, Negro, Purus, Xingu, and Tapajós river basins (Bockmann and Guazzelli 2003, Zuanon et al. 2006a). Individuals of this genus are psammophilous (sand-dwelling), inhabiting the superficial layers of silica sand patches. **BIOLOGY** Feed mainly on invertebrates (Zuanon et al. 2006a).

Heptapteridae Genus F (3.0–3.6 cm SL)

Externally identified by a short and deep body, elliptical in cross section through the dorsal-fin origin; region from snout to dorsal-fin origin with large papillae; first ray of pectoral fin and second ray of dorsal fin distally prolonged into short filaments. Also characterized by: snout convex in dorsal view; mouth subterminal; eye of medium size, dorsolateral, and without free margin; premaxillary tooth plate wide, not bearing projection at posterolateral angle; maxillary barbel of moderate length (its extremity usually reaching the end of the pectoral fin border); anterior and posterior fontanels opened; mandibular adductor muscles restricted to facial region; posterior supraoccipital process short, not contacting the first nuchal plate; posterior border of pectoral, dorsal, pelvic, and anal fins shallowly concave; first pectoral-fin ray distally flexible, with its basal two-thirds stiffened (but friable); pectoral fin usually with 8–10 branched rays; first dorsal-fin ray (spinelet) absent; second dorsal-fin ray flexible except for its basal half; dorsal fin with 6–7 branched rays; origin of pelvic fin on vertical through the posterior half of dorsal-fin base; adipose short, triangular, with posterior lobe free, not confluent with caudal fin; anal fin short, bearing 12–13 rays; caudal fin deeply forked (the ventral lobe sometimes the longest), with 7 branched rays on dorsal lobe and 8 branched rays on ventral lobe; lateral line continuous to base of caudal skeleton. **COLORATION** A peculiar color pattern comprising an unpigmented collar and dorsal and lateral regions of body with three unpigmented or weakly pigmented zones intercalated by dark areas: one below dorsal-fin base, one below the region between dorsal and adipose fins, and one immediately posterior to adipose fin; sometimes melanistic forms are present, with the whole body darkly pigmented. These dark zones extending ventrally from dorsum to ventral surface result from expansion and fusion of the transverse dark bars present on dorsum of other heptapterid genera (e.g., *Horiomyzon*, *Imparfinis*). Lateral stripe absent. **SPECIES** Four, including three in the AOG region, one previously identified as "*Cetopsorhamdia*" *molinae*, one as "*C.*" *shermani*, and at least two species yet to be described. **DISTRIBUTION AND HABITAT** Disjunct distribution in the Magdalena and Orinoco basins, and in the upper Tocantins basin (Bockmann and Guazzelli 2003). **BIOLOGY** No data available.

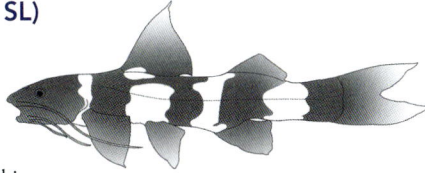

Heptapteridae Genus G (4.5–9.9 cm SL)

Diagnosed by a moderately elongate body, elliptical in cross section through the dorsal-fin origin; small eye (11.0–15% head length), without free orbital margin; maxillary barbel short (its tip reaching the region from end of branchial slit to the second third of pectoral); and adipose fin roughly rectangular. Also characterized by: snout convex in dorsal view; mouth subterminal; eye dorsolateral; premaxillary tooth plate wide, not bearing projection at posterolateral angle; anterior and posterior fontanels opened; mandibular adductor muscles restricted to facial region; posterior supraoccipital process short, not contacting the first nuchal plate; posterior border of pectoral, dorsal, pelvic, and anal fins convex; first pectoral-fin ray mostly flexible, except for its basal half; pectoral fin usually with 8–10 branched rays; first dorsal-fin ray (spinelet) absent; second dorsal-fin ray flexible except for its basal portion; dorsal fin with 6 branched rays; origin of pelvic fin on vertical through the anterior two-thirds of dorsal-fin base; adipose fin low, sloped, and moderately long, with

posterior lobe free, not confluent with caudal fin; anal fin short, usually bearing 10–12 rays; caudal fin forked (with lobes attenuated subequal), with 6–7 branched rays on dorsal lobe and 7–8 branched rays on ventral lobe; lateral line usually continuous to base of caudal-fin rays. **COLORATION** Background body coloration brown to grayish. Coloration variable among species. Some species have 6 transverse dark bars on back extending ventrally (one immediately posterior to head, one below dorsal-fin base, one between dorsal-fin base and adipose-fin base, one on adipose-fin origin, one on adipose-fin end, and one on base of procurrent rays of caudal fin) and 2 or 4 dark dashes along lateral line. In some species, dorsal bars expand ventrally and laterally, so that 3 dark zones are demarked: the largest from posterior region of head to adipose-fin origin, one below adipose-fin base, and one on caudal peduncle. **SPECIES** Five, all in the AOG region, including one previously described as "*Cetopsorhamdia*" *orinoco*, one as "*Chasmocranus*" *quadrizonatus*, one as "*Chasmocranus*" *rosae*, and at least two other species as yet undescribed. **DISTRIBUTION AND HABITAT** Broadly distributed along the Orinoco and upper Amazon (Napo, Marañón, Huallaga, and Ucayali basins) (Bockmann and Guazzelli 2003; F. B. pers. obs). **BIOLOGY** No data available.

Heptapteridae Genus H (5.9–14 cm SL)

Diagnosed by several osteological attributes (Bockmann 1998). Members of this genus have a small to medium adult body size; a slender, elongate (eel-like) body that is elliptical in cross section through the dorsal-fin origin; a very small eye (6.0–10.0% head length); and caudal-fin lobes separated by a shallow concavity in which the dorsal lobe is more pronounced and the ventral lobe more rounded. Also characterized by: snout convex in dorsal view; mouth subterminal; premaxillary tooth plate wide and lacking at posterolateral angle; eye dorsolateral, well distanced from each other, without free margin; maxillary barbel short (reaching at most the middle of pectoral fin or slightly beyond); anterior and posterior fontanels opened as long slits; mandibular adductor muscles restricted to facial region; posterior supraoccipital process short, not contacting the first nuchal plate; posterior border of pectoral, dorsal, pelvic, and anal fins convex; first pectoral-fin ray mostly flexible, except for its basal third; pectoral fin usually with 6–7 branched rays; first dorsal-fin ray (spinelet) absent; second dorsal-fin ray rigid on its basal half; dorsal fin with 6 branched rays; origin of pelvic fin anterior to vertical through the dorsal-fin origin; adipose fin low and very long, with posterior portion fused to caudal fin so that its posterior limit demarked by a small free lobe or barely perceivable; anal fin usually short to moderately long, bearing 13–25 rays; caudal fin with 5–9 branched rays on dorsal lobe and 6–8 branched rays on ventral lobe; lateral line posteriorly fragmented, not reaching the base of caudal fin. **COLORATION** Uniformly brown to grayish, lacking any dark dorsal bars, unpigmented regions, and midlateral stripe. **SPECIES** Three, all in the AOG region, one previously described as "*Heptapterus*" *bleekeri*, one as "*Imparfinis*" *microps*, and at least one other species as yet undescribed (Bockmann 1998). **DISTRIBUTION AND HABITAT** Meta (Orinoco basin), Marowijne, Trombetas river drainages in Amapá state (Brazil), and Marañón (Bockmann 1998; F. B. pers. obs.).

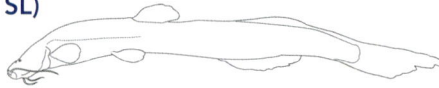

FAMILY LORICARIIDAE—SUCKERMOUTH ARMORED CATFISHES
— *JONATHAN ARMBRUSTER, PETER VAN DER SLEEN, and NATHAN LUJAN*

DIVERSITY With 924 recognized species and possibly hundreds of undescribed species, the Loricariidae is the fifth-most species-rich vertebrate family on Earth, and the second-most species-rich family in the AOG region after Characidae. Loricariid species are distributed in 99 genera and 6 subfamilies (Lujan et al. 2015a, Eschmeyer et al. 2016): Lithogeninae (1 genus; 3 species), Delturinae (2 genera; 7 species), Rhinelepinae (3 genera; 6 species), Loricariinae (31 genera; 239 species), Hypoptopomatinae (including the Neoplecostomini; 22 genera; 205 species) and Hypostominae (40 genera; 464 species). Of this diversity, 405 loricariid species inhabit the AOG region, allotted to 70 genera and 5 subfamilies. Only the subfamily Delturinae does not occur in the AOG region, being instead restricted to Atlantic coastal streams of southeastern Brazil (Reis et al. 2006). The family is hypothesized to have originated more than 85 million years ago (Lundberg et al. 2007), so much of the family's species richness results from a long evolutionary history, and also the philopatric tendencies of most species. The family as a whole is characterized by specialized oral jaws and teeth used to scrape algae and organic debris from hard surfaces (Adriaens et al. 2009, Lujan and Armbruster 2012). This specialized feeding system is composed of a distinctive tripartite jaw apparatus, consisting of bilaterally independent lower jaws and a functionally independent upper jaw (Schaefer and Lauder 1996, Lujan and Armbruster 2012). The biomechanically unique tripartite jaw apparatus is thought to have contributed to the ecomorphological diversity of the family as a whole (Lujan et al. 2011, 2012).

COMMON NAMES *Acarí, Bodó, Cascudo* (Brazil); *Carachama* (Ecuador, Peru); *Chorrosco, Corroncho* (Venezuela).

GEOGRAPHIC DISTRIBUTION Loricariids are distributed throughout the Atlantic slope of Central and South America north of Buenos Aires to central Panama and the Pacific slope of South America north of Peru to southern Costa Rica.

ADULT SIZES Maximum adult sizes range from 2.2 cm SL (*Nannoplecostomus eleonorae*) to approximately 60 cm SL (*Acanthicus hystrix* and *Panaque schaeferi*), with most species being around 7.5–15. cm SL (Lujan et al. 2010).

DIAGNOSIS OF FAMILY Loricariids are easily distinguished from other fishes by having bodies covered in ossified dermal plates, an abundance of extraoral integumentary teeth on the body's surface known as odontodes (Garg et al. 2010), and a ventral oral disk that facilitates surface attachment and feeding (Geerinckx et al. 2011). Generalized loricariid anatomy is summarized in fig. 1.

SEXUAL DIMORPHISM The only consistent, externally visible, primary sexual characteristic of most loricariid species is the morphology of the urogenital papilla, with females exhibiting a relatively broader, flatter urogenital papilla more closely associated with the anus, and males often exhibiting an only slightly more elongate and tubular urogenital papilla more clearly distinct from the anus (e.g., Lujan et al. 2015a). Secondary sexual characteristics vary from absent to pronounced, depending on genus and species. The following traits indicate a sexually mature male, either alone or in combination: elongation of odontodes on the snout, cheek, body, and/or fins, enlargement of the myodome and dilator operculi muscle that erects the cheek odontodes, extension of the lower lip, fleshy tentacles on the snout dorsum, enlargement of the snout, elongation of branched and/or unbranched fin rays, and development of fleshy folds along unbranched fin rays.

HABITATS Loricariid species have colonized nearly all Neotropical freshwater habitats below approximately 1,500 m elevation, including torrential waters flowing from the Andes, quiet brackish waters of some estuaries, black and acidic waters of the Guiana Shield, floodplains and deep channels of large lowlands rivers, and subterranean watercourses and caverns (Covain and Fisch-Muller 2007). They are more speciose in lotic environments, and are generally absent from temporary pools.

FEEDING ECOLOGY Most species use their oral disk to attach themselves to substrates like rocks or submerged trees and branches. They are obligate surface-feeders and consume an often indistinguishable mix of detritus and algae (Buck and Sazima 1995). However, diverse sympatric assemblages may partition benthic food resources, and several lineages show morphological specializations for the consumption of specific foods, including wood, seeds, and macroinvertebrates (Lujan and Armbruster 2012). Some species, notably in the *Hypostomus cochliodon* group, *Panaqolus*, and *Panaque*, are specialized for scraping the surfaces of submerged wood, which they ingest but likely do not digest, assimilating instead the microbes (primarily fungi but also bacteria) that colonize the surface of the wood (German 2009, Lujan et al. 2011). Because their dense skeletons and ossified dermal plates are made up largely of calcium phosphate, loricariids have the highest body

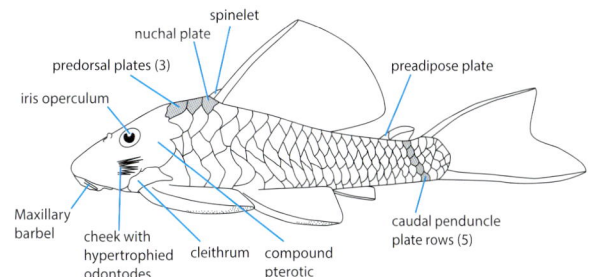

FIGURE 1 Generalized loricariid morphology indicating some of the main attributes in the key and diagnoses. Head bones are not shown, but the area of the compound pterotic is indicated.

phosphorus (P) concentration of any fish measured, and the demand for dietary P may restrict growth rates of some populations during the dry season when resources are scarce (Hood et al. 2005).

BEHAVIOR Most species are nocturnal, sedentary, and territorial, although some members of the Rhinelepinae migrate long distances (i.e., *Rhinelepis* spp., Makrakis et al. 2007), and other species, like *Otocinclus*, are social and live in large shoals. Many species in tropical lowland rivers have modifications of the digestive tract to air-breathe during hypoxic conditions (Armbruster 1998a). Males of many Hypostominae species defend nest sites and territory by forcefully and aggressively erecting enlarged cheek odontodes via a mechanical apparatus derived from the operculum (Geerinckx and Adriaens 2006).

Reproductive strategies are highly variable, including cavity spawning, attachment of eggs under rocks, and egg-carrying by males. When present, parental care is usually limited to males guarding eggs and sometimes larvae (e.g., Sabaj et al. 1999). Many Hypostominae species produce a harsh grinding alarm sound when captured, which is produced via abduction of the pectoral fin and interaction of opposing stridulatory ridges on the pectoral-fin spine condyle and socket (Alexander 1965, Heyd and Pfeiffer 2000). This behavior may be co-opted for nonthreat communication in some species.

ADDITIONAL NOTES Many species are boldly patterned or distinctively shaped and are exported in large numbers to the international ornamental fish trade (Prang 2007), where individual fish may fetch hundreds to thousands of dollars.

KEY TO THE AMAZONIAN SUBFAMILIES

1a. Body behind head incompletely plated dorsally and laterally......**Lithogeninae** (*Lithogenes*; p. 286)
1b. Body behind head completely plated dorsally and laterally (plates may be absent ventrally)....................**2**

2a. Ventral surface of the pectoral girdle exposed (i.e., supporting odontodes) mesial to the coracoid strut (fig. 2a) ....**Hypoptopomatinae** (p. 254)
2b. Ventral surface of the pectoral girdle covered in skin or plates mesial to the coracoid strut (fig. 2b; coracoid strut may be exposed; plates may cover the pectoral girdle, but the odontodes are supported by the plates and not the girdle)....................**3**

3a. Caudal peduncle long and depressed, dorsoventrally flattened; rectangular in cross section; adipose fin absent.................**Loricariinae** (p. 287)
3b. Caudal peduncle robust, not dorsoventrally flattened; oval, round, or triangular in cross section (if caudal peduncle is depressed, it is never rectangular); adipose fin absent or present...**4**

4a. Anus well-separated from first anal-fin ray; eye generally with dorsal flap of iris (*iris operculum*); if iris operculum is absent, the body lacks plates on the abdomen; adipose fin generally present, if absent, it is replaced by a low ridge of medial, unpaired plates; abdomen plated or unplated ....**Hypostominae** (p. 259)
4b. Anus almost in contact with first anal-fin ray; eye without dorsal flap of iris (iris operculum); adipose fin absent, not replaced by low ridge of medial, unpaired plates; abdomen plated...**Rhinelepinae** (*Pseudorinelepis genibarbis*; p. 298)

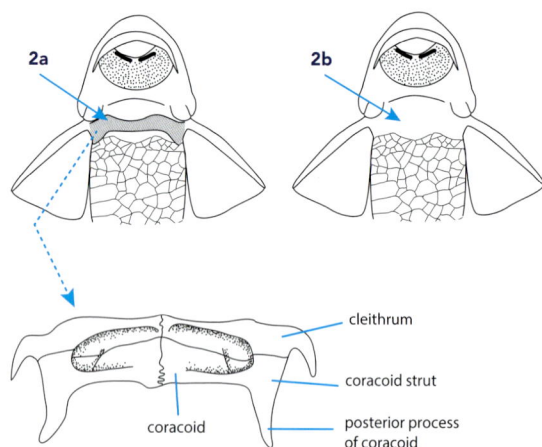

SUBFAMILY HYPOPTOPOMATINAE—OTOS AND RELATIVES

— *ROBERTO E. REIS*

DIVERSITY 205 species in 22 genera, with at least 56 species in 9 genera in the AOG region.

COMMON NAMES *Cascudinho limpa vidros* (Brazil); Dwarf sucker, Oto (English).

GEOGRAPHIC DISTRIBUTION Hypoptopomatine catfishes are broadly distributed in cis-Andean South America, from Colombia and Venezuela to northern Argentina, with most of the

diversity occurring on the Brazilian Shield, especially in eastern and southeastern Brazil.

ADULT SIZES Small-sized fishes (2.0–6.0 cm SL), but some *Hypoptopoma* reaching to about 10 cm.

DIAGNOSIS OF SUBFAMILY Members of this subfamily are distinguished from other loricariids by the morphology of their pectoral-fin girdle, where both the cleithrum and the coracoid bear ventral laminar expansions that partially or completely cover the fossa for the pectoral-fin erector muscles. In addition, the pectoral girdle is totally or partially exposed in the ventral surface and bears odontodes, contrary to other loricariids where the pectoral girdle is always covered with thick skin or dermal plates. Also, the odontodes on the ventral surface of the pelvic-fin spine are curved and turned medially (this feature also present in the neoplecostomine *Kronichthys*), instead of aligned with the main axis of the spine.

SEXUAL DIMORPHISM Sexual dimorphism occurs in the presence of a small, conical urogenital papilla immediately posterior to the anus and a conspicuous skin flap on the dorsal surface of the first pelvic-fin ray of males. Also, larger nostrils and a swirl of odontodes on the posteroventral portion of the caudal peduncle of males have been reported for some species.

HABITATS Most species live in close association with marginal vegetation of creeks and rivers, while few, usually more basal species, are found between rocks in the river bottom.

FEEDING ECOLOGY All are herbivorous, feeding on algae and periphyton.

BEHAVIOR Mostly diurnal.

KEY TO THE GENERA

1a. Dermal plates extremely reduced, leaving trunk almost completely naked...*Gymnotocinclus*
1b. Body typically entirely covered by dermal plates **2**

2a. Caudal peduncle depressed, horizontally oval in cross section...**3**
2b. Caudal peduncle compressed, vertically oval in cross section **5**

3a. Anterior margin of snout expanded in conspicuous spatulate projection (fig. 3a); caudal fin with 10–12 rays.........*Acestridium*
3b. Anterior margin of snout blunt, not expanded; caudal fin with 14 rays **4**

4a. Eye in lateral position, visible from both dorsal and ventral views..*Oxyropsis*
4b. Eye in dorsolateral position, only visible dorsally ....*Niobichthys*

5a. Eye in lateral position, visible from both dorsal and ventral views; nuchal plate wide, its width at least twice that of base of dorsal-fin spine (fig. 5a; after Aquino and Schaefer 2010)...*Hypoptopoma*
5b. Eye in dorsolateral position, only visible dorsally; nuchal plate narrow, its width equal or slightly surpassing that of base of dorsal-fin spine **6**

6a. Adipose fin present; cheek canal plate elongate posteroventrally and contacting the cleithrum ....*"Parotocinclus"*
6b. Adipose fin absent; cheek canal plate not elongate posteroventrally and not contacting the cleithrum **7**

7a. Abdomen covered by >5 longitudinal series of small dermal plates (fig. 7a); lateral abdominal plates absent (fig. 7a); dorsal-fin locking mechanism not functional...........*Corumbataia*
7b. Abdomen covered by <5 longitudinal series of dermal plates; lateral abdominal plates present (fig. 7b; after Schaefer 1997); dorsal-fin locking mechanism functional **8**

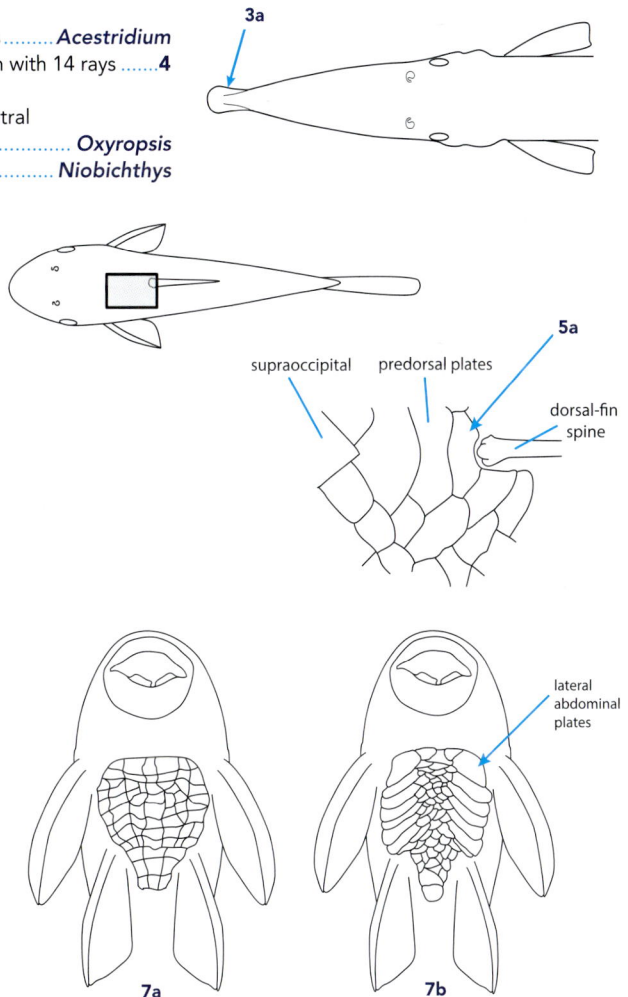

3a

5a

supraoccipital   predorsal plates

dorsal-fin spine

lateral abdominal plates

7a   7b

8a. Mid-dorsal lateral plate series with >9 plates (fig. 8; after Schaefer 1997); preopercle reduced, without sensory canal; fourth infraorbital not expanded ventrally.........*Otocinclus*

8b. Mid-dorsal lateral series of plates with <9 plates; preopercle large, with branch of sensory canal; fourth infraorbital expanded ventrally.......................*Curculionichthys*

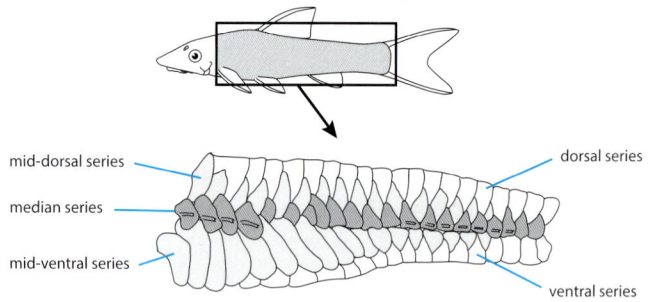

GENUS ACCOUNTS

Acestridium (4.0–6.0 cm SL)

Readily diagnosed from other hypoptopomatines by: very elongate and slender body and snout; expanded spatulate projection on the snout tip; and 10–12 rays in the caudal fin (vs. 14 in remaining hypoptopomatines). **SPECIES** Seven, all in the AOG region. Key to the species in Reis and Lehmann (2009). **DISTRIBUTION AND HABITAT** Endemic to two broad areas: the Orinoco and Negro basins (*A. dichromum*, *A. discus*, *A. colombiensis*, and *A. martini*), and lower Madeira and lower Tapajós region (*A. gymnogaster*, *A. scutatum*, and *A. triplax*) (Rodriguez et al. 2015). Commonly found in swift-flowing stretches of clearwater creeks with dense vegetation, where they hang on leaves and stalks using their pelvic fins and suctorial lips. **BIOLOGY** Herbivorous, feeding on algae and periphyton.

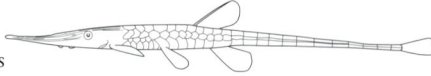

Corumbataia (3.0–4.0 cm SL)

Diagnosed from other members of the Hypoptopomatinae by: a broadly rounded snout; abdomen covered by several longitudinal series or irregularly arranged small dermal plates; lacking lateral abdominal plates; and rostral border of the snout devoid of dermal plates and covered by skin only. **SPECIES** Four, including two in the AOG region. Can be identified based on Britski (1997) and Carvalho (2008). **DISTRIBUTION AND HABITAT** Amazon tributaries of the Brazilian Shield: *C. tocantinensis* (upper Araguaia and Tocantins basins), and *C. veadeiros* (upper Tocantins basin). Usually inhabit creeks with fast-flowing water and bottom covered with stones and pebbles. **BIOLOGY** Herbivorous, feeding on algae and periphyton.

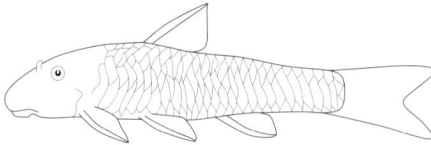

Curculionichthys (3.0–5.0 cm SL)

Recently described genus for species from central Brazil previously assigned to *Hisonotus* but not closely related to *Hisonotus* species from south and southeastern Brazil (Roxo et al. 2015). Distinguished from other members of Hypoptopomatinae (except *Otocinclus*) by a V-shaped dorsal-fin spinelet (see fig. 2 in Silva et al. 2014) and the functional dorsal-fin locking mechanism. From *Otocinclus*, by having <9 plates in the mid-dorsal series of lateral plates (vs. >9 plates); a large preopercle with a branch of the sensory canal (vs. small preopercle without a sensory canal); and the fourth infraorbital bone expanded ventrally (vs. fourth infraorbital not expanded). **SPECIES** About 13 species, 6 in the AOG region. **DISTRIBUTION AND HABITAT** Amazon basin in the western Brazilian Shield: *C. acuen* and *C. sabaji* in the Xingu basin (Silva et al. 2014); and *C. bockmanni*, *C. chromodontus*, and *C. luteofrenatus* in the upper Tapajós River (Britski and Garavello 2007, Carvalho and Datovo 2012). These species are usually found in clearwater creeks of slow to moderately fast current, generally associated with marginal vegetation. **BIOLOGY** Herbivorous, feeding on algae and periphyton.

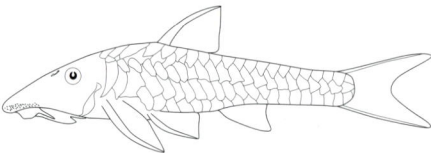

Gymnotocinclus (4.0 cm SL)

Distinguished from other hypoptopomatines by the extreme reduction of body dermal plates, having an almost completely naked body. Additional anatomical traits diagnosing this genus can be found in Carvalho et al. (2008). **SPECIES** One: *G. anosteus*. **DISTRIBUTION AND HABITAT** Tocantinzinho River (upper Tocantins basin). Known only from two small creeks with fast-flowing waters and bottom covered in stones and pebbles. **BIOLOGY** Herbivorous, feeding on algae and periphyton.

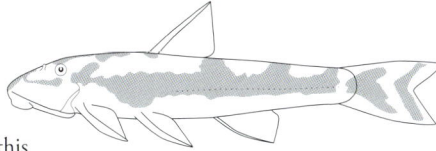

Hypoptopoma (2.0–10 cm SL)

Distinguished from all other hypoptopomatines by the wide nuchal plate, its width at least twice that of the base of the dorsal-fin spine (vs. nuchal plate narrow, its width equal or slightly surpassing that of the base of the dorsal-fin spine). It can also be distinguished from most hypoptopomatines by the presence of an adipose fin (but frequently lost in adult individuals), and by having the eye laterally placed, being visible both dorsally and ventrally (also in *Oxyropsis*). From *Oxyropsis*, *Hypoptopoma* can be distinguished by having the caudal peduncle compressed, vertically oval in cross section (vs. caudal peduncle depressed, horizontally oval in cross section). Secondary sexual dimorphism includes the typical urogenital papilla, a skin flap on the first pelvic-fin ray, and a patch of tightly arranged small odontodes sometimes oriented as a swirl on first plates of the ventral series, lateral to the urogenital papilla of males. **SPECIES** 15, including 14 species in the AOG region. Key to the species and diagnoses in Aquino and Schaefer (2010). Includes two species with small adult body sizes originally described as *Nannoptopoma*. **DISTRIBUTION AND HABITAT** 14 species in the Amazon, Orinoco, or the Guianas. One species, *H. inexspectatum*, in the Paraná-Paraguay basin. Typically inhabits large, open habitats like slow-flowing rivers and lakes, always associated with marginal or floating vegetation. **BIOLOGY** Herbivorous, feeding on algae and periphyton.

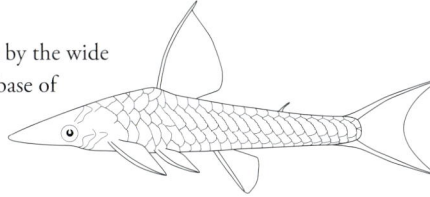

Niobichthys (7.0 cm SL)

The elongate and depressed trunk, merging into a narrow and shallow caudal peduncle (depth 9–17% of body depth) distinguishes *Niobichthys* among the Hypoptopomatinae (except from *Oxyropsis* and *Acestridium*). From *Acestridium* it can be distinguished by lacking the expanded spatulate projection on the snout tip, and from *Oxyropsis* by the dorsolaterally positioned eye, only visible dorsally. Reported sexual secondary traits are the urogenital papilla behind the anal opening and skin flaps on the unbranched and branched pelvic-fin rays (Schaefer and Provenzano 1998). **SPECIES** One, *N. ferrarisi*; see species description in Schaefer and Provenzano (1998). **DISTRIBUTION AND HABITAT** Baria River, tributary to Casiquiare Canal; collected over a cobble-pebble riffle section of the river. **BIOLOGY** Herbivorous, feeding on algae and periphyton.

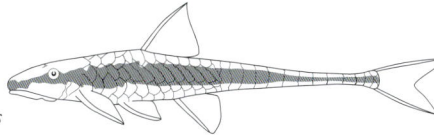

Otocinclus (2.0–4.0 cm SL)

Distinguished from other hypoptopomatines by: head and snout not very depressed; eye large and not visible from below; and the dorsal-fin locking mechanism functional. In addition, species of *Otocinclus* are diagnosed by the possession of a specialized accessory gas bladder formed by an esophageal diverticulum, which helps in fish buoyancy. Mature males of *Otocinclus* sometimes have a contact organ composed of a swirl of odontodes on the ventrolateral caudal-fin base.

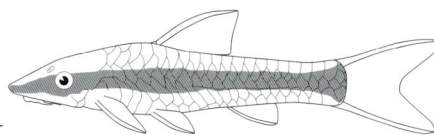

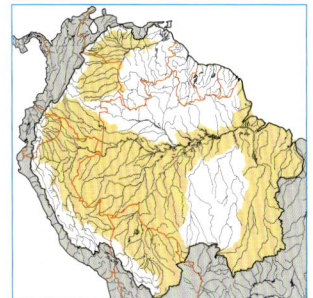

Other sexual dimorphism includes the urogenital papillae and skin flap on the first pelvic-fin ray in mature males, and a larger (usually 10–20%) body size in females (Schaefer 1997). **SPECIES** 19, including 13 species in the Amazon and Orinoco basins. Species identifications in Schaefer (1997), Britto and Moreira (2002), Reis (2004), Lehmann (2006), Lehmann et al. (2010a,b). **DISTRIBUTION AND HABITAT** Widely distributed in most major cis-Andean basins of South America, from Venezuela to Argentina, but absent in the Guianas. Inhabit a variety of habitats, including small forest creeks to large rivers and even lakes, but always associated with marginal or floating vegetation. **BIOLOGY** Herbivorous, feeding on algae and periphyton. Diurnal and often found in shoals.

Oxyropsis (4.0–8.0 cm SL)

Distinguished from most hypoptopomatines (except *Niobichthys* and *Acestridium*) by a depressed, shallow caudal peduncle. From *Acestridium* it can be distinguished by lacking the expanded spatulate projection on the snout tip, and from *Niobichthys* by the laterally positioned eye, visible both dorsally and ventrally. *Oxyropsis* is also distinguished from all hypoptopomatines by possessing a single row of enlarged odontodes laterally, along the trunk midline, adjacent and immediately dorsal to the lateral line canal (Aquino and Schaefer 2002). Males are distinguished from females by the possession of a urogenital papilla and by a cluster of enlarged odontodes on second, third, and/or fourth plates of the ventral series of lateral plates. **SPECIES** Three, all in the AOG region. Species identifications in Aquino and Schaefer (2002). **DISTRIBUTION AND HABITAT** Western and central Amazon, Negro, and upper Orinoco basins. Commonly found in swampy areas and in large, open habitats like slow-flowing river and lakes, always associated with marginal or floating vegetation. **BIOLOGY** Herbivorous, feeding on algae and periphyton.

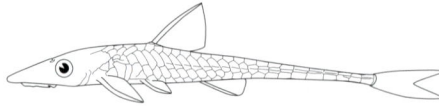

"*Parotocinclus*" (2.0–5.0 cm SL)

Parotocinclus species in the AOG basins are not closely related to *Parotocinclus* species in eastern Brazil (from where the type-species of the genus comes), and for this reason are being treated as "*Parotocinclus*" herein. These species, including several undescribed ones, will likely be transferred to a new genus in the near future. Characterized by the possession of an adipose fin, a trait unique among hypoptopomatines, but also present in a few *Hypoptopoma* species. They are easily distinguished from *Hypoptopoma* by having dorsolaterally positioned eyes, only visible dorsally. The possession of a urogenital papilla and a skin flap on the first pelvic-fin ray distinguish males from females. **SPECIES** 28, including 8 species in the AOG region, which can be diagnosed among loricariids by having the cheek canal plate elongate posteroventrally and contacting the cleithrum. Species of "*Parotocinclus*" in the Amazon, Orinoco, and the Guianas can be identified using keys provided by Schaefer and Provenzano (1993) and Lehmann et al. (2014, 2015). **DISTRIBUTION AND HABITAT** Most species occur in the eastern and southeastern Brazilian coastal basins, but eight species are distributed in the Amazon, Orinoco, or Guianas. Commonly found in creeks and small rivers, typically with slow to moderate water current and sand or gravel in the bottom. Individuals are usually caught on the vegetation or associated with the gravel of the river bottom. **BIOLOGY** Herbivorous, feeding on algae and periphyton.

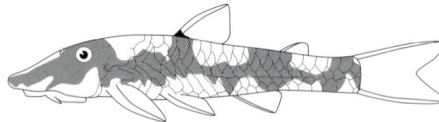

SUBFAMILY HYPOSTOMINAE—PLECOS AND RELATIVES
— *JONATHAN ARMBRUSTER, PETER VAN DER SLEEN, and NATHAN LUJAN*

DIVERSITY Hypostominae is the most species-rich subfamily of Loricariidae, with about 464 species in 40 genera, including at least 232 species in 36 genera in the AOG region. The classification used here follows Lujan et al. (2015a) in recognizing nine clades: Ancistrini (traditionally recognized as a separate subfamily, with 10 valid genera and ~97 species), Hypostomini (with 2 described genera and ~170 species), *Chaetostoma* Clade (with 6 genera and 62 species), *Pseudancistrus* Clade (with 1 genus and 6 species), *Lithoxus* Clade (with 2 genera and 10 species), '*Pseudancistrus*' Clade (with 1 genus and 2 species), *Acanthicus* Clade (with 4 genera and 14 species), *Hemiancistrus* Clade (with 6 described genera and 21 species), and *Peckoltia* Clade (with 9 described genera and 52 species).

COMMON NAMES *Acari, Barbadinho, Bodó, Bodozinho, Cascudo* (Brazil); *Aletón carachama, Cascudo, Shitari* (Peru); *Pez palito* (Ecuador); *Agujeta, Barbón, Corroncho* (Venezuela); Pleco, Sailfin pleco, Whiptail catfish (English).

GEOGRAPHIC DISTRIBUTION Widespread throughout South America, from Costa Rica to Peru on the Western side of the Andes and from Panama to Argentina on the Eastern side.

ADULT SIZES Ranges from 4.2 cm SL in *Micracanthicus vandragti* from near the confluence of the Ventuari and upper Orinoco rivers, to approximately 53 cm SL in *Acanthicus hystrix* and 60 cm SL in *Panaque schaeferi*, both from the Amazon basin. Most species grow to around 7.5–15 cm SL (Lujan et al. 2010).

DIAGNOSIS OF SUBFAMILY Most Hypostominae can be diagnosed from all other loricariids by being able to at least partially evert the cheek plates (>30°) versus the cheek plates unable to be flexed away from the head. Hypostomines can be separated from the Lithogeninae by having the sides posterior to the head fully plated, from the Hypoptopomatinae (excluding Neoplecostomini) by having maximally the coracoid strut (far lateral portion of pectoral girdle) exposed and supporting odontodes (vs. having regions medial to the strut exposed and supporting odontodes), from the Neoplecostomini and most of the rest of the Hypoptopomatinae by having the spinelet V-shaped and the dorsal-fin spine lock functional (vs. spinelet rectangular and the dorsal-fin spine lock not functional); from the Loricariinae by generally having an adipose fin (vs. always absent) and by having a robust caudal peduncle (vs. very dorsoventrally flattened), and from the Rhinelepinae by generally having an adipose fin, generally having an iris operculum, by having the anus distant from the anal fin (vs. almost contacting one another), and by lacking ordered rows of odontodes on the compound pterotic.

SEXUAL DIMORPHISM Pronounced in many species, including often more numerous and enlarged odontodes on the snout, pectoral fins, and/or body plates of males. *Ancistrus* males also have large fleshy tentacles on their snout, and sexually mature males of the *Chaetostoma* Clade (*Andeancistrus, Chaetostoma, Cordylancistrus, Dolichancistrus, Leptoancistrus, Transancistrus*) may exhibit an enlarged and fleshy anterior snout region (genus *Chaetostoma* only) or elongation of various paired-fin regions, sometimes accompanied by longitudinal dermal folds along the dorsal or lateral ridges of fin rays

HABITATS Generally restricted to freshwater, except for *Hypostomus watwata*, which lives in the estuarine brackish waters of coastal rivers in the Guianas. Most species live in the bottom and banks of sandy and rocky rivers with highest diversities in rapids habitats of clearwater rivers of the Brazilian and Guiana shields. Most species are active at night, while staying in their woody or rocky refuges during the day. Some species of *Ancistrus*, and possibly some *Chaetostoma*, are cave-specialized (i.e., eyeless, depigmented; Hoese et al. 2015).

FEEDING ECOLOGY Most species are generalized grazers of algae and detritus. *Parancistrus* species and *Ancistrus ranunculus* are specialized for the consumption of flocculent detritus; species in the *Lithoxus* Clade, *Leporacanthicus, Hypancistrus, Scobinancistrus*, and *Panaqolus albomaculatus* are invertivores; *Megalancistrus* and *Pseudoqolus koko* eat sponges, and most species of *Panaque, Panaqolus*, and members of the *Hypostomus cochliodon* group consume wood, although these species do not directly digest wood but rather assimilate microbes living on the wood surface (German 2009, Lujan et al. 2011).

BEHAVIOR Parental care is well developed in many species, in which males excavate nest cavities and guard eggs and sometimes larvae. The snout tentacles of male *Ancistrus* produce a nutrient-rich mucus to feed larvae when under paternal care (Yan 2009). Individuals of most or all species are philopatric, moving only short distances within their home territory (Power 1984).

ADDITIONAL NOTES Hypostominae include many of the most boldly patterned and colorful loricariid catfishes, many of which are small-sized, making them internationally popular in the aquarium fish hobby. Many species in the genera *Acanthicus, Ancistrus, Baryancistrus, Hypancistrus, Leporacanthicus, Megalancistrus, Peckoltia, Panaque, Parancistrus, Pseudacanthicus, Pterygoplichthys, Scobinancistrus*, and *Spectracanthicus*, for example, are marketed throughout the world, with rare species fetching hundreds to thousands of dollars each. Wild fish stocks supplying this demand, mostly in Brazil, Colombia, Peru, and Venezuela, are largely unmonitored, and there have been no published studies on the sustainability of these commercial harvests. Members of *Ancistrus, Hypostomus*, and *Pterygoplichthys* have become invasive in tropical and subtropical regions of the world, with the most common introduced species appearing to be a hybrid of *P. disjunctivus* and *P. pardalis* (JWA, pers. obs.).

KEY TO THE GENERA

1a. 5 rows of plates in the region below the dorsal fin **2**
1b. 3 rows of plates below the dorsal fin ....***Nannoplecostomus***

2a. Portions of anterodorsal snout surface naked, having a marginal crescent or medial patches of skin lacking both plates and odontodes (fig. 1A–D) **3**
2b. Entire snout covered with ossified plates bearing odontodes.......**6**

3a. No fleshy tentacles or digitate papillae on snout, 5 rows of plates on caudal peduncle; 7–10 branched dorsal-fin rays......**4**
3b. Fleshy tentacles on snout (best developed in males, but females typically possess at least small tentacles or digitate papillae along outer margin of snout); 3 rows of plates on caudal peduncle; usually 7 branched dorsal-fin rays (except up to 10 in *A. dolichopterus*) ....***Ancistrus***

4a. 8–10 dorsal-fin rays ....***Chaetostoma*** (all species except *Chaetostoma platyrhynchus*, which has a fully plated snout)
4b. 7 dorsal-fin rays **5**

5a. Naked areas on snout restricted to oval patches on either side of the mesethmoid, column of plates and odontodes along dorsal surface of mesethmoid, plates present on snout margin (fig. 1C); fleshy keel on preadipose plate and slightly anterior; body mottled tan and brown; restricted to the Mazaruni River in northwest Guyana..............***Paulasquama callis***
5b. Naked area on snout includes snout margin and a narrow medial strip along mesethmoid (fig. 1D); body with thin, irregular light yellow to gold bars; restricted to rivers draining the southwestern slope of Mount Duida in the upper Orinoco of southern Venezuela..........***Soromonichthys***

6a. Postdorsal ridge of 3 or more median, unpaired preadipose plates present (fig. 2B).......**7**
6b. Postdorsal ridge absent or restricted to usually 1 (occasionally 0 or 2) median, unpaired preadipose plates (fig. 2A)...................**8**

7a. Cheek plates strongly evertible (>75° from head) and possessing hypertrophied odontodes .... ***Neblinichthys*** (part)
7b. Cheek plates not strongly evertible (~30° from head) and lacking hypertrophied odontodes .... ***Corymbophanes***

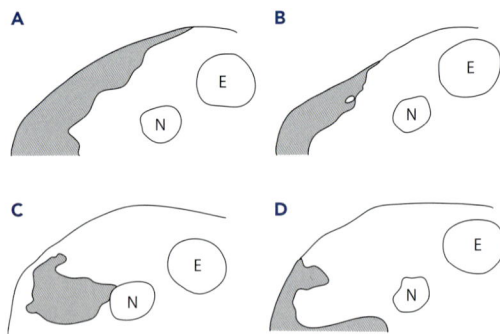

FIGURE 1 Detail of the right anterior portion of the snout (dorsal view) of some Hypostominae loricariids showing the distribution of naked areas (i.e., without plates) in gray: (A) *Chaetostoma anomalum*, (B) *Ancistrus pirareta* (female, males would have a wider region without plates), (C) *Paulasquama callis*, and (D) *Soromonichthys stearleyi*. White regions have plates or odontodes and the circles indicate the eye (E) and naris openings (N). In *Paulasquama*, the plates on the edge and tip of the snout are deeply embedded, and anterior ones do not have odontodes. Figure redrawn from Lujan and Armbruster (2011).

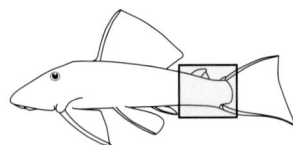

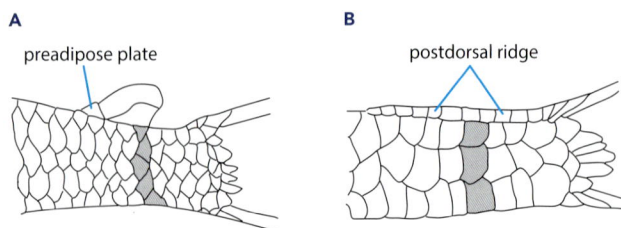

FIGURE 2 Caudal peduncles showing (A) the presence of a normal adipose fin with a single preadipose plate, a spine, and a membrane and 5 rows of plates on the caudal peduncle, and (B) the adipose fin replaced by a postdorsal ridge of azygous plates and 3 rows of plates on the caudal peduncle.

8a. 8 or more branched dorsal-fin rays **9**
8b. 7 branched dorsal-fin rays **17**

9a. >60 teeth per jaw ramus; nuchal plate covered by plates and dorsal-fin spinelet covered by skin, both rarely supporting odontodes; abdomen lacking plates; usually no strong keels on lateral plates (*Andeancistrus platycephalus* develops keels on the plates, but they are not well developed) **10**
9b. <60 teeth per jaw ramus; nuchal plate and dorsal-fin spinelet exposed, supporting odontodes; usually with plates on abdomen (plates may develop late ontogenetically); if specimen is large (>50 mm SL) and has no plates on abdomen, lateral plates have strong keels formed by long, stout, sharp odontodes **13**

10a. One or two cheek odontodes extending past posterior margin of cleithrum; high-gradient rivers in Colombia or Venezuela **11**
10b. All cheek odontodes of approximately the same length, none extending past posterior margin of cleithrum **12**

11a. Outermost snout margin lacking enlarged odontodes; cheek odontodes extending past posterior margin of cleithrum but not beyond first plate in midventral plate series; pectoral-fin spines of sexually mature males similar in length to those of females and juveniles of both sexes .... *Cordylancistrus*
11b. Outermost snout margin having enlarged odontodes; typically at least one cheek odontode extending past posterior margin of first plate in midventral plate series; pectoral-fin spines of sexually mature males much longer than those of females and juveniles of both sexes .... *Dolichancistrus*

12a. Cheek odontodes with straight tips; 9–10 branched dorsal-fin rays; black spot at the anteroventral corner of the dorsal fin; restricted to the upper Aguarico basin of northern Ecuador and the upper Caquetá basin of southern Colombia .... *Chaetostoma platyrhynchus*
12b. Cheek odontodes distally hooked; 8 branched dorsal-fin rays; restricted to the upper Pastaza and Santiago basins of Ecuador and the Marañón drainage of northern Peru between approximately 300 and 1200 m above sea level .... *Andeancistrus*

13a. Dorsal laminae of plates of ventral series on caudal peduncle strongly concave, accentuating a ventrolateral keel; cluster of papillae located medially along internal, anterodorsal surface of dentary (fig. 3a); cheek odontodes of mature males extending beyond base of pectoral fin; main channels of the upper Orinoco River and Casiquiare Canal in southern Venezuela ...*'Pseudancistrus' pectegenitor*
13b. Mouth lacking dentary papillae (fig. 3b); caudal peduncle ventrum rounded, lacking ventrolateral keel; cheek odontodes not extending past posterior margin of cleithrum **14**

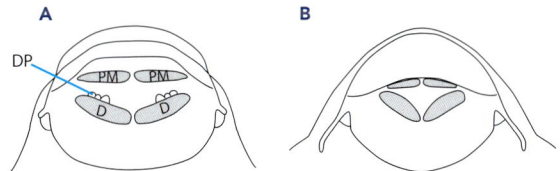

FIGURE 3 Mouths, ventral view showing position of dentary papilla(e). There may be one to several papillae depending on individual and species. D: Dentary, DP: dentary papillae, PM: premaxilla.

14a. Adipose fin absent; adult size >50 cm SL; widespread in main river channels of the Amazon and Orinoco river drainages .... *Acanthicus*
14b. Adipose fin present **15**

15a. Usually >10 teeth per jaw ramus .... *Pterygoplichthys*
15b. Usually ≤10 per jaw ramus **16**

16a. Lower lip oval (fig. 4A); fimbriae absent above upper lip (fig. 4A); supraoccipital not raised into a crest; usually >4 teeth per premaxilla; teeth moderately long ....*Pseudacanthicus*
16b. Lower lip round (fig. 4C); fimbriae present above upper lip, looking like tentacles (fig. 4C); supraoccipital raised into a crest; only 1–4 teeth per premaxilla that are extremely long .... *Leporacanthicus*

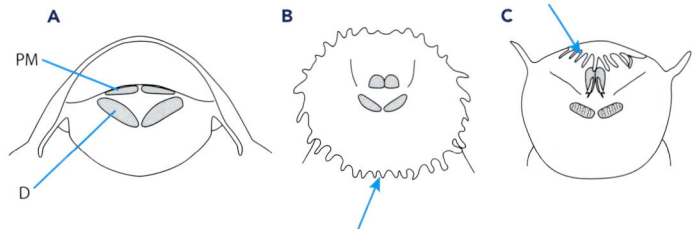

FIGURE 4 Lip shapes and mouth fimbriae: (A) standard oval-lipped hypostomine, (B) round-lipped *Exastilithoxus* with arrow indicating lower lip fimbriae, and (C) round-lipped *Leporacanthicus* with arrow indicating upper lip fimbriae). D = Dentary, PM = premaxilla.

17a. Dorsal fin with membrane of last ray not expanded posteriorly, not attaching to preadipose plate or the adipose spine (fig. 5A)... **21**
17b. Dorsal fin with membrane of last ray expanded posteriorly, attaching to preadipose plate or the adipose spine (fig. 5B,C) **18**

18a. Gill openings restricted (fig. 6A); skin at base of dorsal fin and pectoral fin without fleshy, fimbriate folds.................. **19**
18b. Gill openings large (fig. 6B); skin at base of dorsal-fin and pectoral-fin spine textured with fleshy, fimbriate folds (at least in nuptial males).. *Parancistrus*

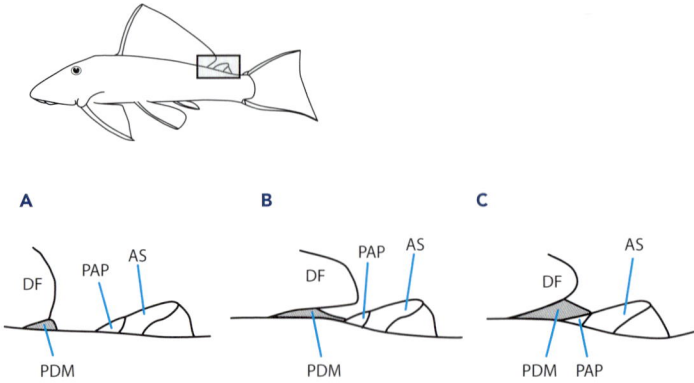

FIGURE 5 Conditions of the posterior dorsal-fin membrane (PDM): (A) well separated from adipose fin, (B) contacting base of preadipose plate (PAP), and (C) contacting all of preadipose plate to adipose spine (AS). DF = dorsal fin.

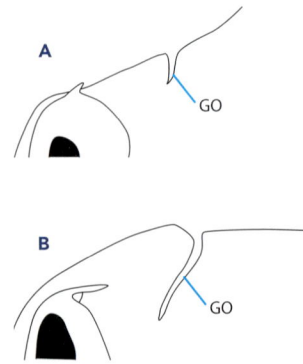

FIGURE 6 (A) The restricted gill opening (GO) of most loricariids vs. (B) the expanded gill openings of *Parancistrus*, a condition also present in *Ancistrus ranunculus*.

19a. Premaxillae long, nearly forming oblique angle at union (fig. 7A); >26 teeth per jaw ramus **20**
19b. Premaxillae short, forming acute angle at union (fig. 7B); ≤26 per jaw ramus .... *Spectracanthicus*

20a. Body green with gold spots or entirely dark (blue-black in life); Orinoco basin .... *'Baryancistrus'*
20b. Body dark with either gold spots wwor gold edging to fins; Amazon basin .... *Baryancistrus*

FIGURE 7 Relative dentary angles: (A) oblique (>120°), and (B) acute (≤90°). Angle of these examples given. Figure redrawn from Armbruster (2008).

21a. Teeth large, spoon-shaped with little or no stalk (fig. 8B) **22**
21b. Teeth villiform (thread-like; fig. 8A) or large with long stalk (fig. 8C) **24**

22a. Cheek without elongate, evertible odontodes (fig. 9A) .... *Hypostomus cochliodon* group
22b. Cheek with elongate, evertible odontodes (fig. 9B)................................... **23**

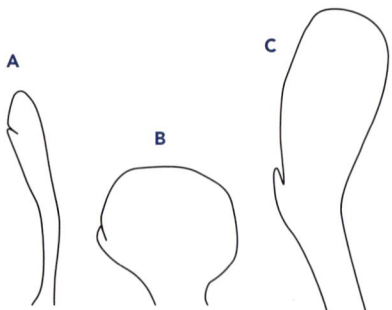

FIGURE 8 Tooth shapes: (A) villiform tooth, (B) spoon-shaped tooth, and (C) spatulate tooth. Redrawn from Armbruster (2004a).

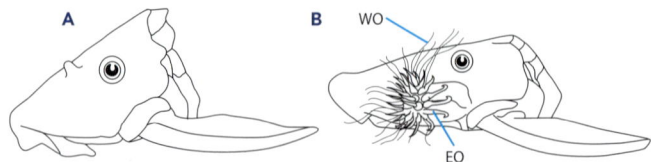

FIGURE 9 Head, left side, lateral view showing (A) *Hypostomus plecostomoides* and (B) *Lasiancistrus* sp. with the evertible cheek plates and associated odontodes (EO) as well as the whisker-like odontodes (WO). Drawings after M. H. Sabaj in Armbruster (2004a).

23a. Absence of a keel on the lateral surface of the caudal peduncle; preopercle with deep lateral groove located near the reflected lateral margin of the bone; orbital margin round, lacking posterodorsal notch (fig. 10A); adults typically <15 cm SL..................................*Panaqolus*
23b. Presence of a keel on the lateral surface of the caudal peduncle; preopercle without a deep lateral groove located near the reflected lateral margin of the bone; orbit with posterodorsal notch (fig. 10B); adults typically >25 cm SL...*Panaque*

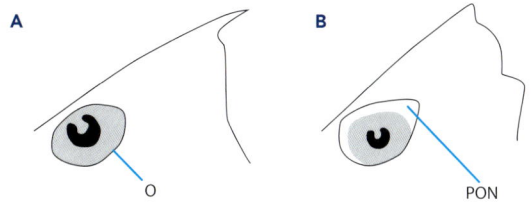

FIGURE 10 Eyes (gray) with orbital rim (O) indicated, (A) based on *Panaqolus changae*, (B) based on *Panaque nigrolineatus* with distinctly depressed postorbital notch (PON) indicated.

24a. Cheek plates not evertible (not loosely embedded in connective tissue; firmly attached to suspensorium), if hypertrophied odontodes are present on cheek, their supporting plates are fairly immobile; cheek plates covering anteroventral corner of the opercle do have slight mobility, but can be held only to about a 30° angle from head **25**
24b. Cheek plates evertible (loosely embedded in connective tissue, not firmly attached to suspensorium), usually with hypertrophied odontodes; evertible cheek plates and odontodes can be held at an angle >75° from the head (and usually <90°)...**27**

25a. Hypertrophied odontodes present along the snout of males and females, never any hypertrophied odontodes on the lateral plates; plates absent on abdomen of adults...**26**
25b. Hypertrophied odontodes either absent along the snout or present only on nuptial males and then only with hypertrophied odontodes on the lateral plates; plates generally present on abdomen in adults **49**

26a. A single papilla or group of papillae located medially along dentary behind teeth (fig. 3A); Orinoco basin in Colombia and Venezuela ....*Lithoxancistrus*
26b. No papilla located medially along dentary behind teeth (fig. 3); Orinoco and Amazon basin, and coastal rivers in the Guianas...*Pseudancistrus sensu stricto*

27a. Median row of odontodes on each lateral plate forming keels...**28**
27b. Median row of odontodes on each lateral plate not forming sharp keels...**29**

28a. Body with large dark spots; 5 rows of plates on the caudal peduncle (fig. 2A); body deep; Maroni River, Suriname, and French Guiana ....*Hemiancistrus medians*
28b. Body dark with light spots or tan with dark saddles and/or bars; 3 rows of plates on the caudal peduncle (fig. 2B); body dorsoventrally flattened; upper Orinoco and Amazon drainages...*Dekeyseria*

29a. <10 teeth per jaw ramus; dentaries short, forming acute angle (≤90°) at union (fig. 7B) **30**
29b. >10 teeth per jaw ramus; dentaries variable, may be short or long, and forming an oblique angle (≥120) to nearly straight line at union **33**

30a. Lower lip oval (fig. 4A); 5 rows of plates on the caudal peduncle (fig. 2A) **31**
30b. Lower lip round (fig. 4B); 3 rows of plates on the caudal peduncle (fig. 2B) **32**

31a. Dentary teeth long and strong, crowns consisting of one large and one small lobe, spatulate (fig. 8C); with gold spots in life; large adult size >100 mm ....*Scobinancistrus*
31b. Dentary teeth moderately long; crowns consisting of 2 nearly equally sized lobes, not spoon-shaped ....*Hypancistrus*

32a. Lower lip fimbriate (fig. 4B)...*Exastilithoxus*
32b. Lower lip not fimbriate ....*Lithoxus*

33a. Cheek generally with only 3 odontodes (can be 1–5, but usually 3); cheek odontodes stout, strongly hooked anteriorly (fig. 11), and remarkably movable forward ....*Hopliancistrus*
33b. Cheek with 4 or more odontodes; cheek odontodes not particularly stout and relatively straight to moderately hooked...**34**

FIGURE 11 Head of *Hopliancistrus tricornis*, dorsal view, showing the 3 stout, anteriorly curved and highly evertible odontodes of each cheek.

34a. 4–5 rows of lateral plates on caudal peduncle (fig. 2A) **35**
34b. 3 rows of lateral plates on caudal peduncle (fig. 2B) **46**

35a. Opercle not supporting odontodes in adults (maximally 1 or 2) **36**
35b. Opercle supporting several odontodes in adults.. **48**

36a. Body with dorsal saddles and no light edges to fins ....*Peckoltia* (part)
36b. Body with dark or light spots or mottling, some with light edges to fins.................................. **37**

37a. Body with light spots on a dark gray or green background **38**
37b. Body with dark spots, mottled dark and light, or entirely dark **39**

38a. Dorsal fin reaching adipose-fin spine when adpressed; Orinoco drainage of Venezuela
and Colombia..*'Hemiancistrus' subviridis*
38b. Dorsal fin not reaching adipose-fin spine when adpressed; Tocantins drainage, Brazil .... *Baryancistrus longipinnis*

39a. Dentaries forming an acute angle (≤90°; fig. 7B)...*Peckoltia* (part)
39b. Dentaries forming an oblique angle (>120°) or straight (fig. 7A) **40**

40a. Eyes and body pigmentation reduced; upper Amazon in Peru.................................*Peckoltia pankimpuju*
40b. Eyes and body pigmentation well developed **41**

41a. Spots on caudal peduncle larger than eye diameter in adults **42**
41b. Spots on caudal peduncle equal to eye diameter or less in adults, or absent **43**

42a. Dorsal fin does not reach adipose fin when adpressed; postanal length 34.3–38.8% of SL; Orinoco,
Amazon, and Essequibo of Guyana, Colombia, Venezuela, and Brazil ....*Peckoltia sabaji*
42b. Dorsal fin reaches adipose fin when adpressed; postanal length 28.4–32.4% of SL; Xingu River,
Brazil..*Ancistomus feldbergae*

43a. Widened pelvic-fin spines; enlarged plates on the abdomen; eye low on the head, dorsal rim of orbit lower
than interdorsal space; Amazon basin, Colombia, Ecuador, Peru, and Brazil ....*Peckoltichthys bachi*
43b. Pelvic-fin spines narrow; plates on abdomen small or absent; eyes set high on the head; dorsal rim of orbit
higher than interdorsal space... **44**

44a. Body mottled with light to red band at edge of caudal fin ....*'Hemiancistrus' guahiborum*
44b. Body with small to large spots on head and at least anterior trunk or with body entirely dark, no light band
at edge of caudal fin **45**

45a. Spots on body smaller than eye diameter, body base color light gray; southern tributaries of the lower
Amazon, Brazil...*Ancistomus* (part)
45b. Spots on body equal to eye diameter or body entirely dark, body base color tan to dark gray;
Tocantins River, Brazil ....*'Hemiancistrus' cerrado*

46a. Shorter, whisker-like odontodes anterior to elongate cheek odontodes (fig. 9B); tentacles on edge of snout
of nuptial male longer than associated odontodes ....*Lasiancistrus*
46b. No whisker-like odontodes anterior to elongate cheek odontodes (all cheek odontodes thick); if tentacles
present along snout in nuptial male, they are not longer than their associated odontodes.................. **47**

47a. No extremely elongate, anteriorly oriented odontodes on snout of nuptial male; hypertrophied odontodes
along snout margin present in adult males and females; pectoral-fin spine reaches anus in adults................*Pseudolithoxus*
47b. Extremely elongate, anteriorly oriented odontodes on snout of nuptial male; hypertrophied odontodes
along snout margin present only in adult males; pectoral-fin spine does not reach anus in adults....................*Neblinichthys*

48a. Dorsal laminae of plates of ventral series on caudal peduncle strongly concave, accentuating a ventral
keel; white to gold spots on base of lower lobe of caudal fin; Orinoco of Venezuela ....*'Pseudancistrus' sidereus*
48b. Dorsal laminae of plates of ventral series on caudal peduncle flat to slightly convex; alternating dark and
light bands on lower lobe of caudal fin or caudal fin uniformly dark; Guiana Shield of Suriname, French
Guiana, and Brazil ....*Guyanancistrus*

49a. Base color dark tan to brown; dorsal fin separated by 2 or fewer plates from preadipose plate when adpressed or surpassing preadipose plate; no hypertrophied odontodes on heads or lateral plates of nuptial males ....*Hypostomus*
49b. Base color off-white to tan; dorsal fin separated by 3 or more plates from preadipose plate when adpressed; hypertrophied odontodes on head and lateral plates of nuptial males... **50**

50a. Caudal peduncle does not become greatly elongate with size, and is oval in cross section; small dark spots (less than half plate diameter) on a light tan background; Orinoco, Amazon, and Essequibo drainages of Bolivia, Brazil, Colombia, Ecuador, Guyana, Peru, and Venezuela..*Aphanotorulus*
50b. Caudal peduncle gets proportionately longer with size, and becomes round in cross section; spots almost as large as lateral plates on a nearly white background; Guayas basin of western Ecuador and the middle Orinoco basin of Colombia and Venezuela...*Isorineloricaria*

GENUS ACCOUNTS

Acanthicus (21–53 cm SL)

Differs from other loricariids by: very large adult body sizes; absence of an adipose fin; large swim bladder capsule extending from cranium to dorsal fin (visible externally by large pterotic bone measured as distance from posterior margin of eye to end of skull greater than interorbital width). Color is typically black or gray, occasionally with white spots on the entire body. Four or more asymmetrical predorsal plates (including nuchal plate); 5 rows of plates on the caudal peduncle; abdomen fully plated; cheek odontodes long (extending past base of pectoral spine); caudal fin forked, lower lobe slightly longer; principal caudal-fin rays forming elongate filaments; pectoral-fin spines extremely long and armed with sharp odontodes; the entire dorsal surface of the head is covered in stout, sharp odontodes; the odontodes form a sharp keel on the lateral plates and, in juveniles, there are few to no odontodes on the plates above and below the keel rows (Armbruster 1997). Most similar to *Leporacanthicus* and *Pseudacanthicus*, but the jaws are long, nearly forming a straight line at union (vs. short, forming an acute angle at union), the adipose fin is absent, and the pterotics are enlarged. Males with more and longer cheek odontodes and greatly elongate odontodes on the pectoral-fin spine. **SPECIES** Two, both in the Amazon basin. **DISTRIBUTION AND HABITAT** *A. adonis* (21 cm SL) from the lower Tocantins basin, and *A. hystrix* (53 cm SL) from the Amazon basin, both are found in large rivers. **BIOLOGY** Algivores and detritivores.

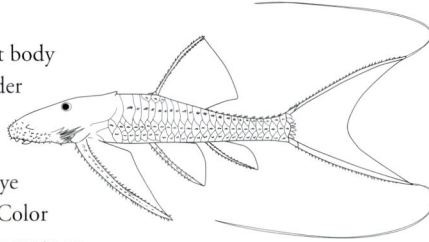

Ancistomus (12–15 cm SL)

Species now considered part of *Ancistomus* have historically been placed in either *Peckoltia* or *Hemiancistrus*. Species assigned to *Ancistomus* (Armbruster et al. 2015) have characteristics in common with *Peckoltia*, like hypertrophied odontodes on the lateral plates of nuptial males (not reported in all species); however, *Peckoltia* species with hypertrophied body odontodes also have acutely angled dentaries (≤90°), whereas *Ancistomus* has obliquely angled (>120°) dentaries. *Ancistomus* can further be separated from most *Peckoltia* (except *P. sabaji*), *Hemiancistrus*, and the 'H.' *guahiborum* group, by having black spots (vs. white to gold spots or by being mottled); from *Peckoltia sabaji* by lacking spots on the dorsal and caudal fins or only present proximally (vs. spots throughout both fins); from 'H.' *cerrado* by having a forked caudal fin (vs. emarginate); from *Baryancistrus*, *Parancistrus*, and *Spectracanthicus* by lacking an expanded posterior membrane of the dorsal fin and by having black spots (vs. dorsal-fin membrane contacting preadipose plate

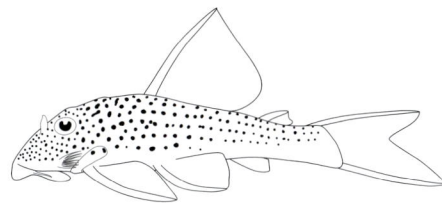

or adipose spine in most species and body either dark or with white spots); and from *Panaque*, *Peckoltia* Group 2 (see above), and *Scobinancistrus* by having obliquely angled dentaries (vs. acutely angled). Three predorsal plates (including nuchal plate); 5 rows of plates on the caudal peduncle; abdomen from almost completely unplated to almost completely plated; cheek odontodes long (reaching beyond base of pectoral spine in some); caudal fin forked, lower lobe longer. Males of some species known to have hypertrophied odontodes on the lateral plates and pectoral-fin spines. **SPECIES** Four, all in the AOG region. **DISTRIBUTION AND HABITAT** Tocantins, Tapajós, and Xingu basins of Brazil in shallow rapids. **BIOLOGY** Stomach contents of *A. feldbergae* included mostly diatoms as well as sponge spicules (de Oliveira et al. 2012).

Ancistrus (5.0–20 cm SL)

Readily distinguished from other loricariids by tentacles on the snout. In females and juveniles, the tentacles are short and located only along the snout margin; however, in breeding males, the tentacles of most species are quite long and located on the top of the snout. Color pattern typically dark gray to black, often with white, gold, or brown spots or reticulations; abdomen naked; caudal fin straight, angled posteroventrally, to slightly emarginate with the lower lobe longer; snout is naked along the edge in females and juveniles and, maximally, to the nares in adult males; usually 4 predorsal plates; 3 rows of plates on the caudal peduncle; the nuchal plate exposed or covered in plates (Armbruster 1997). In addition to the conspicuous fleshy tentacles on the snout, males have larger and more numerous odontodes on the cheek and pectoral-fin spines than females, and an enlarged myodome dilator operculi muscle (Geerinckx and Adriaens 2006). **SPECIES** 66, including at least 26 species in the AOG region. **COMMON NAMES** *Barbadinho*, *Bodozinho* (Brazil); *Barbón* (Venezuela). **DISTRIBUTION AND HABITAT** Widespread, from Panama to Río de la Plata in Argentina. Inhabit a wide variety of habitats, from fast-flowing rivers to floodplain lakes. **BIOLOGY** *Ancistrus* can breathe air under hypoxic conditions (Gee 1976). Males defend nests in cavities where females lay large clumps of eggs, and continue to defend young after hatching until they have absorbed their yolk sacs and have pigment (Burgess 1989). Several clutches in various stages of development, from eggs to free-swimming larvae, can be found in one nest (Sabaj et al. 1999). Because females in some fish groups prefer laying eggs in the nest of males that already have eggs, the tentacles of males may act as larval mimics and trick the female into laying eggs in a nest. However, Yan (2009) examined histological sections of the tentacles and snout skin of *Ancistrus triradiatus* and found they contain many goblet cells. These cells secrete a nutrient-rich mucus and are absent in females. The author therefore hypothesized that the snout tentacle secretions provide nutrients to fish larvae when they are under paternal care.

Andeancistrus (12–14 cm SL)

Like other members of the *Chaetostoma* Clade, *Andeancistrus* has five rows of plates on the caudal peduncle, 8 branched dorsal-fin rays, and wide, straight jaws with many small teeth (typically 90–125 teeth per premaxillary ramus, 100–185 teeth per dentary). *Andeancistrus* can be distinguished from most *Chaetostoma* (all except *C. platyrhynchus*) by having a fully plated snout, from *C. platyrhynchus* by having distally hooked cheek odontodes (vs. straight), and from *Cordylancistrus*, *Dolichancistrus*, and *Leptoancistrus* (not in AOG) by lacking cheek odontodes that extend posteriorly beyond the cleithrum. Color patterns consist of a dull gray to dark black base color with small white, yellow-gold, or light blue spots distributed across the head, body, and fins. **SPECIES** Two, both in the Amazon basin: *A. eschwartzae* and *A. platycephalus*. Another similar species has been reported from northern Peru (Salcedo and Ortega

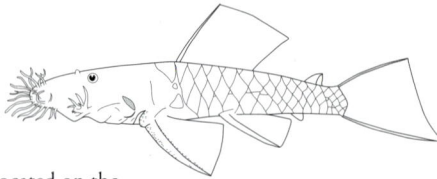

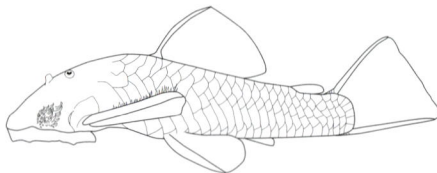

2015); however, the relationships of this species to other members of the *Chaetostoma* Clade are unresolved. **DISTRIBUTION AND HABITAT** High-gradient habitats in the respective headwaters of the Pastaza and Santiago river drainages in Ecuador, between approximately 300 and 1,200 m above sea level. **BIOLOGY** No data available, although presumably similar to other members of the *Chaetostoma* Clade (i.e., algivorous and territorial nest builders).

Aphanotorulus (11–51 cm SL)

Range from the small piedmont species of *Aphanotorulus sensu stricto* to larger species of the Amazon, Essequibo, and Orinoco formerly assigned to *Squaliforma*. Differs from other AOG hypostomines, except *Corymbophanes, Hypostomus, Isorineloricaria, Pseudancistrus, Pseudorinelepis*, and some *Spectracanthicus*, by lacking highly evertible cheek plates. *Aphanotorulus* can be separated from all the above taxa by having a light tan base color (vs. brown to dark gray in most and nearly white in *Isorineloricaria*), by having hypertrophied odontodes on the lateral plates of nuptial males, and from all but *Isorineloricaria* by having either an enlarged buccal papilla or many buccal papillae (vs. a small or absent buccal papilla). *Aphanotorulus* can be further distinguished from *Corymbophanes* and *Pseudorinelepis* by having an iris operculum; from *Corymbophanes* by having plates on the abdomen (vs. abdomen naked); from *Isorineloricaria* by having the spots smaller than the lateral plates (vs. nearly as large as the plates) and by having the caudal peduncle not changing in proportions as the fish get larger and remaining oval in cross section (vs. greatly elongating and becoming round in cross section); from *Pseudancistrus* by having plates on the abdomen; 5 rows of plates on the caudal peduncle (vs. 4), having only hypertrophied odontodes on the cheeks (if present) in nuptial males (vs. all individuals); from *Pseudorinelepis* by having an adipose fin and by lacking regular rows of odontodes on the compound pterotic; and from *Spectracanthicus* by lacking connection between the dorsal fin and adipose spine. Color is light tan, almost golden in life, with small black spots on all surfaces except ventrally; abdomen white. (Armbruster and Page 1996). Three predorsal plates (including nuchal plate); 5 rows of plates on the caudal peduncle; abdomen plated; cheek odontodes absent except in some nuptial males; caudal fin forked, lower lobe longer. Nuptial males exhibit hypertrophied odontodes on the lateral plates, dorsal and adipose spines, dorsal- and caudal-fin simple rays, and occasionally on the cheek. Nuptial males of *A. ammophilus* and *A. unicolor* have unicuspid (instead of bicuspid) teeth. **SPECIES** Six, all in the AOG region. **TAXONOMIC NOTE** Originally diagnosed by the presence of numerous small papillae inside the buccal cavity; however, the group is very similar to some of the species placed in *Squaliforma* and one species (*Plecostomus phrixosoma*) is intermediate between these *Squaliforma* and *Aphanotorulus sensu stricto*, so Lujan et al. (2015a) recognized *Squaliforma* as a synonym of *Aphanotorulus*. However, some species placed in *Squaliforma* by Isbrücker and Michels (in Isbrücker et al. 2001) were moved to *Isorineloricaria* by Ray and Armbruster (2016), who revised both *Aphanotorulus* and *Isorineloricaria*. **DISTRIBUTION AND HABITAT** Lowland and piedmont streams throughout the Amazon, Essequibo, and Orinoco systems. Species are typically found in areas of slow to moderately fast-flowing water over sand or gravel (Armbruster and Page 1996). **BIOLOGY** No data available.

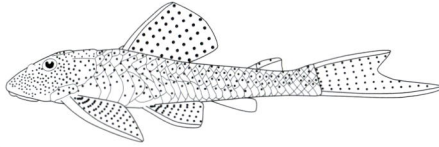

Baryancistrus (8.0–24 cm SL)

Colorful fishes with a dark olive, brown, or black base color and either yellow or white spots or yellow to orange edging of the dorsal fins. Species described in *Baryancistrus* include four species from Brazil and two from Venezuela (Rapp Py-Daniel et al. 2011), although this result is disputed by other studies which restrict *Baryancistrus* sensu stricto to only those species of the Brazilian Shield (Armbruster 2004a; Lujan et al., 2015a). Characterized by: a heavy and robust head and

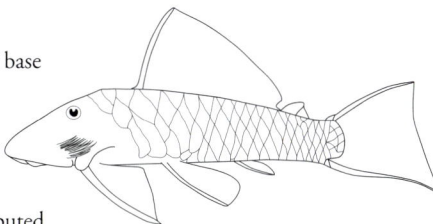

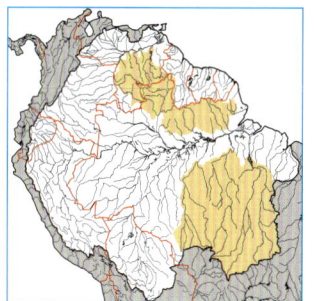

body; strong odontodes on the cheek plates; numerous teeth; large and parallel dentary tooth rows; and almost complete abdominal plating (Lujan et al. 2009, Rapp Py-Daniel et al. 2011). All these features are present in many other loricariids, from which *Baryancistrus* is distinguished by a hypertrophied membrane posterior to the last branched dorsal-fin ray, this membrane reaching the supporting plate of the adipose fin except in *B. longipinnis*. This is different from the genera *Parancistrus* and *Spectracanthicus* and '*B.*' *beggini* in which the dorsal fin is completely connected to the adipose fin, but the same as in '*B.*' *demantoides*. *Baryancistrus* can be separated from '*B.*' *demantoides* by having a dark background body color (vs. bright green) and by having gold spots (when present) on the lower half of the caudal peduncle (vs. maximally on the dorsal quarter of the caudal peduncle). Three predorsal plates (including nuchal plate); 5 rows of plates on the caudal peduncle; abdomen almost completely plated; cheek odontodes long (beyond base of pectoral spine in adults); caudal fin emarginate, lower lobe longer. **SPECIES** Four, all in the AOG region. Key to the species in Rapp Py-Daniel et al. (2011). **DISTRIBUTION AND HABITAT** Southern Guiana and northern Brazilian shields. Habitats include granite rocks and submerged boulders in flowing water. Juveniles of some species form groups of several individuals and are found in different habitats than adults, including marginal areas of rapids or under flat rocks at the bottom of shallow rapid stretches (Rapp Py-Daniel et al. 2011). Species in the Xingu basin may partition habitats, with *B. chrysolomus* preferring heavily sedimented backwaters, and *B. xanthellus* preferring higher-gradient habitats with rocky substrates generally lacking sediment (L. Sousa pers. comm.). **BIOLOGY** Predominantly detritivorous (Lujan et al. 2009, Rapp Py-Daniel et al. 2011).

Upper Orinoco '*Baryancistrus*' + '*Hemiancistrus*' *guahiborum* and '*H.*' *subviridis* (8.0–15 cm SL)

A group of four species from the Orinoco basin ('*Baryancistrus*' *beggini*, '*B.*' *demantoides*, '*Hemiancistrus*' *guahiborum* and '*H.*' *subviridis*) share the following traits (Lujan et al. 2015a): 3 predorsal plates (including nuchal plate); 5 rows of plates on the caudal peduncle; abdomen without plates or partially plated; cheek odontodes moderate to long (extending beyond gill opening in most and beyond base of pectoral spine in '*B.*' *beggini*); cheek odontodes long (beyond base of pectoral spine in adults); caudal fin straight in '*B.*' *beggini* and forked in the rest, lower lobe longer. Within this group '*Hemiancistrus*' *guahiborum* and '*H.*' *subviridis* have the dorsal and adipose fins well separated, '*B.*' *demantoides* has the posterior edge of the dorsal fin contacting the preadipose plate, and '*B.*' *beggini* has the two fins connected. Color ranges from black with a blue sheen in '*B.*' *beggini*, gray with a green or blue hue with gold spots in '*B.*' *demantoides* and '*H.*' *subviridis*, and mottled brown in '*H.*' *guahiborum*. '*Baryancistrus*' can be separated from all other loricariids except *Baryancistrus*, *Spectracanthicus*, and *Parancistrus* by an enlarged posterior membrane contacting at least the preadipose plate. '*Baryancistrus*' *beggini* can be separated from *Baryancistrus* by having the dorsal-fin membrane connected to the adipose-fin spine, from *Spectracanthicus* by having ≥15 teeth per jaw ramus (vs. ≤15 in the dentary and ≤10 in the premaxilla), from *Parancistrus* by having small gill openings (vs. large), and from *Parancistrus* and *Spectracanthicus* by having the pectoral spine reaching the anus when adpressed ventral to the pelvic fin (vs. distant from the anus). '*Hemiancistrus*' *subviridis* differs from all other hypostomines except '*B.*' *demantoides* by having a bright-green background color (some Meta river specimens perhaps belonging to '*H.*' *subviridis* have the base color a dark blue). '*Hemiancistrus*' *guahiborum* differs from most loricariids by having a mottled body with a reddish band on the edge of the dorsal fin (such bands are normally seen either in species with saddles or in species with a dark gray or black base color. **SPECIES** Four, all in the AOG region. See species information in Werneke et al. (2005a,b) and Lujan et al. (2009). **DISTRIBUTION AND HABITAT** Guiana Shield tributaries of the upper Orinoco and Casiquiare River of Venezuela. All species are found on granite in flowing waters. **BIOLOGY** All feed on detritus and algae scraped from epilithon.

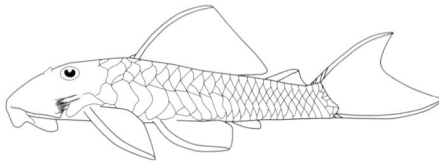

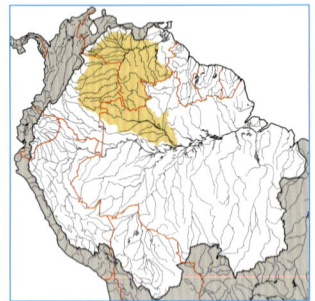

Chaetostoma (5.0–25 cm SL)

Characterized by: absence of plates on the abdomen; 5 rows of plates on the caudal peduncle; 8 or more branched dorsal-fin rays; and wide and straight jaws with many small teeth (typically 50–166 teeth per premaxillary ramus, 70–220 teeth per dentary). All but one species (*C. platyrhynchus*) have a wide margin of naked skin without plates around the margin of the snout (Armbruster 1997). Live color patterns generally consist of a gray to black base color with a wide range of spotting and banding patterns that are helpful in distinguishing species. These patterns fade substantially in preserved individuals. The caudal-fin margin can vary from slightly emarginate to obliquely straight, with the dorsal lobe shorter than the ventral lobe, and often has a white margin or white tips dorsally and ventrally. Predorsal plates number 4 or more with the nuchal plate covered by other plates so that it appears absent; the dorsal-fin spinelet is present and triangular but covered in skin and appears absent because it generally lacks odontodes. Lower jaws often have papillae that are slightly larger than the main buccal papilla situated just behind the teeth; some species have a very small black, fleshy keel behind the supraoccipital. Live, sexually mature males can develop exceptionally contrasting color patterns, take on an overall golden yellow hue, or become dark with decreased pattern contrast. Other male sexual dimorphisms can include enlargement of the snout, elongation of the medial pelvic-fin rays into a lobe, enlargement of a fleshy fold along the posterodorsal surface of the pectoral-fin spine, and elongation of unbranched pectoral and anal-fin rays. **SPECIES** 48, including 20 species in the Amazon and Orinoco basins. A recent taxonomic revision by Lujan et al. (2015b) provides a guide to *Chaetostoma* species of the Amazon basin; however, many species of *Chaetostoma* remain difficult to distinguish from each other, and some currently recognized species are likely invalid. **DISTRIBUTION AND HABITAT** Distributed across Brazil (1 species), Colombia (17 species), Ecuador (7 species), Panama (1 species), Peru (10 species), and Venezuela (12 species). Within these countries, distributed across the Amazon (20 species), Orinoco (5 species), and Magdalena (9 species) basins, Caribbean coastal drainages (6 species), and Pacific coastal drainages (13 species). Most species live in piedmont habitats of the Andes or coastal mountains, although two described and a few undescribed species are known from localities around the Guiana and Brazilian shields. **BIOLOGY** Scrape algae and detritus from rocks (Saul 1975, Power 1984, Ortaz 1992, Kramer and Bryant 1995a, b, Hood et al. 2005). Males defend nests under flat rocks where the eggs are laid in a single layer on the roof of the nest (Page et al. 1993).

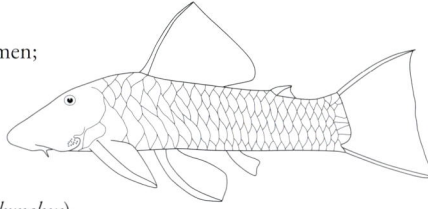

Cordylancistrus (7.0–13 cm SL)

Shares with the *Chaetostoma* Clade absence of odontodes on the nuchal plate; absence of a dorsal-fin spinelet; and 8 or more dorsal-fin rays. Differs from *Chaetostoma* by having plates on the snout and generally longer cheek odontodes that extend past the posterior margin of the cleithrum but not beyond the first midventral plate, from *Dolichancistrus* by lacking one or more enlarged cheek odontode extending to beyond the first midventral plate and by having pectoral-fin rays in breeding males that are similar in length to those of females and juveniles of both sexes (Armbruster 1997), and from *Leptoancistrus* (not in AOG) by having an adipose fin. **SPECIES** Three, including one from the Orinoco basin; *Co. torbesensis* from the Torbes drainage of southeastern Venezuela. Key to the species in Provenzano and Milani (2006). **DISTRIBUTION AND HABITAT** High montane streams and rivers of a few Andean drainages in Venezuela and Colombia. **BIOLOGY** No data available, although likely similar to other members of the *Chaetostoma* Clade (i.e., algivorous and territorial nest builders).

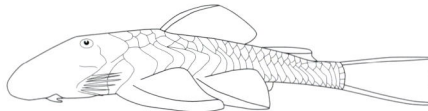

Corymbophanes (6.5–8.5 cm SL)

Readily distinguished from all other loricariids by: a low elongate postdorsal ridge formed by 13–17 raised median unpaired plates; absence of an adipose fin; absence of evertible cheek odontodes; and absence of an iris operculum (Armbruster et al. 2000). Among loricariids, the only species with a postdorsal ridge and no adipose-fin membrane are species of *Leptoancistrus* (not in the AOG), *Chaetostoma carrioni* (from upper Amazon piedmont rivers in northern Peru and southeastern Ecuador), and an undescribed *Neblinichthys* from the upper Ireng River—all of which have clusters of evertible cheek odontodes. *Neblinichthys roraima* has a similar ridge, but the adipose spine is strongly adnate and there is often a short membrane. *Chaetostoma carrioni* and *Neblinichthys* can be separated from *Corymbophanes* by evertible cheek plates and hypertrophied odontodes, and a dorsal flap of the iris (iris operculum). *Chaetostoma carrioni* can be further separated from *Corymbophanes* by having 5 rows of plates on the caudal peduncle (vs. at least one column of 3 plates) and the complete loss or reduction of the anal fin (Armbruster 1997). *Delturus* and *Upsilodus* have a postdorsal ridge, but an adipose-fin membrane is present. Sexual dimorphism in development of odontodes has not been observed, although recent specimens of *C.* cf. *kaiei* from the Kuribrong River of Guyana included a nuptial male with very long, anteriorly directed odontodes on the snout and shorter odontodes on the rest of the head, which is similar to conditions seen in *Neblinichthys*. **SPECIES** Two, *C. andersoni* and *C. kaiei*, both in the AOG region. Key to the species and species information in Armbruster et al. (2000). **DISTRIBUTION AND HABITAT** Known only from the Potaro River and tributaries above Kaieteur Falls in Guyana, and from above Amaila Falls in the Kuribrong River (a tributary of the Potaro). Inhabits flowing pools to swift riffles. **BIOLOGY** No data available.

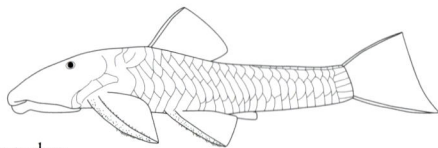

Dekeyseria (10–21 cm SL)

The median rows of slightly enlarged, sharp odontodes that form keels on the lateral plates and highly evertible cheek plates separate *Dekeyseria* from all other hypostomines except *Acanthicus*, *Leporacanthicus*, *Megalancistrus*, and *Pseudacanthicus* (the *Acanthicus* Clade). In *Dekeyseria*, however, there are regular rows of odontodes above and below the keels, whereas in the *Acanthicus* Clade there are usually few odontodes above or below the keels in all but the largest adults (>20 cm SL) and no complete rows. Also, there are only 7 dorsal-fin rays in *Dekeyseria* (vs. 8 or more), 3 rows of plates on the caudal peduncle (vs. 5), and no plates on the abdomen. Two basic body colors are present in Dekeyseria **SPECIES** dark gray (almost black) with white spots or mottling, and dark brown dorsal saddles on a tan to red-brown background. Additional characters that might aid identification: 3 or 4 predorsal plates; species are flattened more than most other Ancistrini; tentacles are associated with odontodes on the pectoral-fin spine and snout, but are shorter than the supporting odontodes. *Dekeyseria* is most similar to *Lasiancistrus* but lacks whisker-like odontodes on the cheek, and tentacles are longer than the supporting odontodes on the snout of breeding males. Breeding males develop elongate odontodes along the snout margin anterior to the evertible cheek odontodes and on the pectoral-fin rays. **SPECIES** Six, all in the AOG region. **DISTRIBUTION AND HABITAT** Known from the Negro basin (Brazil), floodplain lakes in the middle and upper Amazon basin (Brazil), and the upper Orinoco basin (Venezuela). **BIOLOGY** Reported as an air-breather (Val and de Almeida-Val 1995).

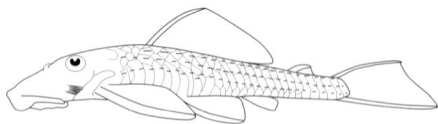

Dolichancistrus (9.0–13 cm SL)

Differs from *Andeancistrus*, *Cordylancistrus*, *Chaetostoma*, and *Leptoancistrus* by the much longer pectoral-fin spines in adult males than in adult females and juveniles of both sexes (vs. pectoral spines of equal length); from *Andeancistrus*, *Cordylancistrus*, and *Chaetostoma* by having one or two enlarged cheek odontodes extending beyond the first midventral plate and by having enlarged odontodes along the outermost snout margin, from all species of *Chaetostoma* except for *C. platyrhynchus* by having a plated snout (vs. naked), and from *Leptoancistrus* (not in AOG) by having an adipose-fin spine and membrane (Ballen and Vari 2012). *Dolichancistrus* further differs from *Chaetostoma platyrhynchus* in having distally hooked cheek odontodes that extend past the posterior margin of the cleithrum and are evertible to >90º (vs. straight odontodes not reaching the anterior margin of the cleithrum and being evertible only to approximately 30–40º; Ballen and Vari 2012). **SPECIES** Four, two in AOG region. A fifth (*D. setosus*) from the Colombian Andes is tentatively assigned to this genus. Review of genus and key to the species in Ballen and Vari (2012). **DISTRIBUTION AND HABITAT** Both sides of the Andes in Colombia and Venezuela. Species inhabit high-altitude montane, torrential streams. **BIOLOGY** No data available, although likely similar to other members of the *Chaetostoma* Clade (i.e., algivorous and territorial nest builders).

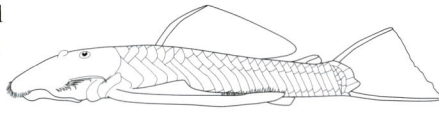

Exastilithoxus (5.0–6.0 cm SL)

Distinguished from other Hypostominae by fimbriae along the posterior margin of the lower lip; however, *Lithoxus* also has projections on the lower lip, but they are much smaller than in *Exastilithoxus*. *Exastilithoxus* is a small, dorsoventrally depressed loricariid, with 4 or more predorsal plates (including nuchal plate); 3 rows of plates on the caudal peduncle; abdomen without plates; cheek odontodes very small (not extending beyond gill aperture); caudal fin slightly forked, lower lobe longer (Armbruster 1997). Color pattern is generally mottled and dark brown with paler areas under and just posterior to the dorsal fin; the abdomen is white; fins are mottled; ventral surface of the caudal peduncle is colored like the sides, but somewhat lighter. **SPECIES** Two, in the Amazon and Orinoco basin. **DISTRIBUTION AND HABITAT** *E. fimbriatus* in the Caroni and Matacuni basins in Orinoco basin, Venezuela; *E. hoedemani* in the Marauiá River in upper Negro basin, Brazil. **BIOLOGY** Feed on insect larvae (Armbruster 1997).

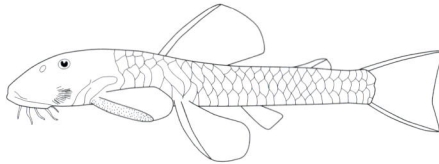

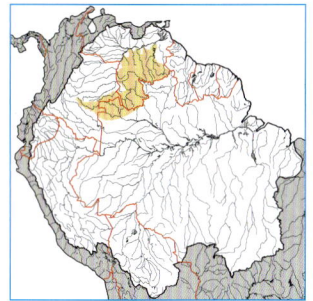

Guyanancistrus (10–16 cm SL)

Similar to species of *Pseudancistrus* and *Lithoxancistrus*, but with short or no hypertrophied odontodes on the snout. Differ from all other members of the Ancistrini *sensu* Lujan et al. (2015a) except *Lithoxancistrus* and *Paulasquama* by having 5 rows of plates on the caudal peduncle (vs. 3). Color generally dark brown to gray with either mottling, white spots, or dark spots on the body and fins and occasionally bands on the fins. Three or four predorsal plates; abdomen without plates; short cheek odontodes (not extending much past the gill aperture) in *G. brevispinnis* and *G. niger* and long in *G. longispinnis*; cheek plates fully evertible; short hypertrophied odontodes occasionally present along snout edge with longest at the corners of the snout. *Guyanancistrus* can be separated from most other Ancistrini genera mainly by the lack of characters of those genera; from *Ancistrus* by having plates on the snout and lacking tentacles; from *Lasiancistrus* by lacking whisker-like odontodes; from *Corymbophanes* and *Neblinichthys* by lacking hypertrophied odontodes on the top of the snout of nuptial males; from *Pseudolithoxus* by having short or no hypertrophied odontodes along the snout margin; from *Paulasquama* and *Soromonichthys* by having the snout fully plated (vs. having some naked areas); from *Hopliancistrus* by having numerous thin, short, almost straight cheek odontodes (vs. 3–5 stout, strongly hooked cheek odontodes); from *Corymbophanes* by having an iris operculum; from *Dekeyseria* by lacking keeled lateral plates and by having the odontodes above and below the keel rows random (vs. keels and

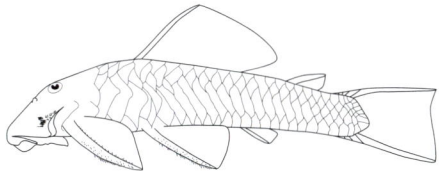

regular rows of odontodes above and below them); and from *Lithoxancistrus* by lacking a large papilla or set of papillae on the dentary posterior to the teeth. In addition, *Guyanancistrus* can be separated from *Pseudancistrus* by having 5 rows of plates on the caudal peduncle (vs. 4), fully evertible cheek plates (≥75° vs. <30°), and by having short (vs. long) or no hypertrophied odontodes along the snout edge. Also similar to '*Pseudancistrus*' but has straight laminae of the ventral plates posterior to the pelvic fin (vs. strongly concave, accentuating the ventral keel). **SPECIES** Three, all in the AOG region: *G. brevispinis*, *G. longispinnis*, and *G. niger*. **DISTRIBUTION AND HABITAT** Coastal drainages of Suriname, French Guiana, and Amapá in Brazil and have apparently crossed the divide into the Amazon in northern Brazil (Cardoso and Montoya-Burgos 2009; pers. obs.). Prefer rapids where the depth is shallow and the substrate is rocky (Le Bail et al. 2000). **BIOLOGY** No data available.

Hemiancistrus (19 cm SL)

This genus has long been used as a repository for species with unclear relationships. Lujan et al. (2015a) restricted *Hemiancistrus* to just the type species, *H. medians*. *Hemiancistrus* is a relatively large hypostomine with a robust, deep body and large eyes. Like *Dekeyseria*, the lateral plates have keels (though not as large as in *Dekeyseria*) and the

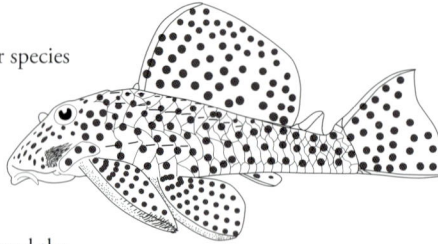

odontodes above and below the keels are arranged in ordered rows; however, there are 5 rows of plates on the caudal peduncle (vs. 3) and the body is much more robust than in *Dekeyseria*. No other hypostomines with strongly evertible cheek plates have such keels and well-ordered rows of odontodes. Three predorsal plates (including nuchal plate); 5 rows of plates on the caudal peduncle; abdomen without plates when young and becoming fully plated as an adult; cheek odontodes moderate (extending beyond gill opening); caudal fin emarginate, lower lobe longer. **SPECIES** One, *H. medians*. **TAXONOMIC NOTE** Other species formerly assigned to *Hemiancistrus* are recognized as species groups in '*Hemiancistrus*' in single quotes until further study. Three species groups are recognized (Armbruster et al. 2015): '*H.*' *chlorostictus* group (with 6 species), '*H.*' *guahiborum* group (including *H. guahiborum* and *H. subviridis* described with the Orinoco *Baryancistrus* above), and '*H.*' *landoni* group (including one trans-Andean species). These species groups lack diagnostic characteristics. Information on *Hemiancistrus medians* in Le Bail et al. (2000) and Mol (2012b). For the Amazonian species in the '*Hemiancistrus*' *chlorostictus* group (*H. cerrado*) see De Souza et al. (2008); and for the species in the '*Hemiancistrus*' *guahiborum* see Werneke et al. (2005a, b). The AOG species groups are described above and below. **DISTRIBUTION AND HABITAT** Maroni basin in French Guiana and Suriname, generally preferring flowing-water habitats in medium to large rivers. **BIOLOGY** A generalist algivore/detritivore.

'Hemiancistrus' chlorostictus group

'*Hemiancistrus*' *cerrado* was tentatively placed in the '*H.*' *chlorostictus* group (Armbruster et al. 2015), but the remainder of the '*H.*' *chlorostictus* group is from southeastern Brazil and Uruguay, very distant from the range of '*H.*' *cerrado* in the Tocantins. '*Hemiancistrus*' *cerrado* differs from

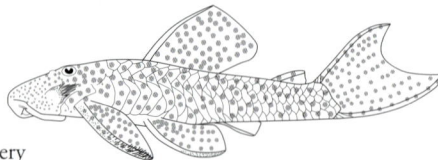

H. medians by lacking keels on the lateral plates and has unordered odontodes above and below the keels (vs. odontodes ordered in rows), and by having the dorsal fin not reaching the preadipose plate when adpressed (vs. reaching). '*Hemiancistrus*' *cerrado* differs from most *Peckoltia* by having spots on the body and fins versus saddles, and from *Peckoltia* with spots by having the dentaries straight (vs. forming an acute angle) or by having the spots on the caudal peduncle smaller than the plates (vs. larger). '*Hemiancistrus*' *cerrado* is sympatric with species of *Ancistomus* from which it differs by lacking hypertrophied odontodes on the lateral plates in nuptial males, by having spots

nearly as large as the lateral plates (vs. much smaller), and by having the dorsal fin not reaching the adipose fin when adpressed (vs. reaching). Three predorsal plates (including nuchal plate); 5 rows of plates on the caudal peduncle; abdomen with plates on the throat and below the pectoral girdle and slightly posterior; cheek odontodes moderate (extending beyond gill opening); caudal fin emarginate, lower lobe longer. **SPECIES** Seven, including one species in the AOG region: '*H.*' *cerrado*, see De Souza et al. (2008). **DISTRIBUTION AND HABITAT** '*H.*' *cerrado* inhabits rocky riffles of second order streams in tributaries of the Araguaia River, Tocantins basin, Brazil (De Souza et al. 2008); the other species live in coastal streams of southeastern Brazil and Uruguay. **BIOLOGY** A generalist algivore/detritivore.

Hopliancistrus (10 cm SL)

Separated from all other hypostomines by: usually 3 (1–5) stout, strongly anteriorly curved and highly evertible odontodes in the cheek mass (vs. thinner with only a slight anterior bend at the tips). Body color is dark and mottled in the preserved specimens with a slight hint of white spots, but in life, the white spots are much more distinct. Three predorsal plates; 5 rows of plates on the caudal peduncle; abdomen without plates; cheek odontodes short (not extending posterior to gill aperture), but stout and strongly hooked anteriorly, capable of being everted anteriorly to the point of contacting the snout; short hypertrophied odontodes present in nuptial males at the corner of the snout, present in at least 2 rows. *Hopliancistrus* is similar to *Lasiancistrus*, but has 5 rows of plates on the caudal peduncle (vs. 3 in *Lasiancistrus*). **SPECIES** One described species, *H. tricornis*, and at least two undescribed ones. **DISTRIBUTION AND HABITAT** Tapajós and Xingu basins in Brazil. **BIOLOGY** No data available.

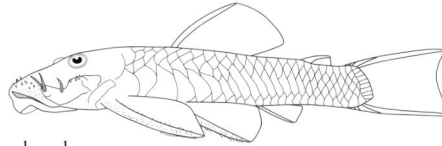

Hypancistrus (5.0–13 cm SL)

Hypancistrus is not readily separated from other hypostomines except that most species (except *H. inspector* adults) have the dentary teeth about twice as long as the premaxillary teeth (Armbruster 2002, Armbruster et al. 2007). Coloration is typically some combination of dark brown to black base color with white markings ranging from stripes to spots to squiggles, and can be quite dramatic as in the case of the stripes of *H. zebra*. The genus *Micracanthicus* was described for a small species with similarities to the *Acanthicus* group, including >3 randomly arranged predorsal plates; however, molecular phylogenetic (Lujan et al. 2015a) and mitochondrial haplotype (M. Tan pers. comm.) analyses show that *Micracanthicus* is closely related to Orinoco *Hypancistrus* and nested within *Hypancistrus*. *Micracanthicus* is therefore recognized here as a synonym of *Hypancistrus*. *Hypancistrus* species are most similar to *Panaqolus* and some species of *Peckoltia*, which share acutely angled (≤90°) dentaries. *Hypancistrus* can be separated from *Peckoltia* with acutely angled dentaries (Group 2 as defined below) and from *Panaqolus* by having only slightly hypertrophied odontodes on the lateral plates of nuptial males or no hypertrophied odontodes (vs. very long, longer than the supporting plates, odontodes in nuptial males); from *Peckoltia* with acutely angled dentaries by having no plates on the central part of the abdomen behind the pectoral girdle (vs. abdomen usually fully plated); from *Panaqolus* by having thin teeth (vs. spoon-shaped); and from *Peckoltichthys* by having the eyes high on the head and the supraorbital ridge taller than the interorbital space (vs. eyes low and supraorbital ridge below the interorbital space). Some of the *Panaqolus* from the distribution of

Hypancistrus in the Orinoco lack the spoon-shaped teeth diagnostic for *Panaqolus* and can be difficult to separate from *Hypancistrus*, but those *Panaqolus* have the dorsal fin reaching the preadipose spine (instead of just the preadipose plate) when adpressed and have smaller eyes. Three predorsal plates in most species (including nuchal plate; *H. vandragti* has 4 or 5 erratically arranged plates); 5 rows of plates on the caudal peduncle; abdomen plated only below pectoral fin and along sides; cheek odontodes moderate (reaching base of pectoral spine); caudal fin emarginate, lower lobe longer. Breeding males have larger odontodes on the pectoral-fin spines and on the cheek (Isbrücker and Nijssen 1991). In some species, breeding males also have hypertrophied odontodes on the lateral plates but not to the extreme seen in some *Peckoltia* or *Panaqolus* (Armbruster et al. 2007). **SPECIES** Eight, all in the AOG region. Outdated key to the species in Armbruster et al. (2007). Many still undescribed species are distributed throughout mainly the lower Amazon and its tributaries and many members of the genus are popular in the ornamental fish trade. **DISTRIBUTION AND HABITAT** Negro and upper Orinoco basins, and southern drainages of the Amazon (Tan and Armbruster 2016). Generally inhabits areas with slight or moderate flow along bedrock outcrops. **BIOLOGY** Omnivorous, diet includes green algae, diatoms, and aquatic insects (Armbruster et al. 2007). *Hypancistrus inspector* also consumes seeds (Armbruster 2002). Female deposits the eggs in caves; male defends the nest and tends the clutch.

Hypostomus (4.0–70 cm SL)

AOG species differ from all other AOG hypostomines except *Aphanotorulus*, *Isorineloricaria*, *Pseudorinelepis*, and some *Spectracanthicus* by lacking highly evertible cheek plates (marginally evertible to approximately 30° from the head), and by never having hypertrophied cheek odontodes, even in nuptial males (occasionally, nuptial males have a very slight increase in cheek odontode length, but this is not very discernible). Within the AOG *Hypostomus* can be separated from *Aphanotorulus* and *Isorineloricaria* by having a brown base color (vs. almost white to tan), lacking hypertrophied odontodes on the lateral plates, and having a small, single buccal papilla (vs. large, single papilla or many); from *Pseudorinelepis* by having an adipose fin (*H. levis* lacks one), by having an iris operculum, anal fin i + 4 (vs. i + 5), and by lacking well-developed rows of odontodes on the compound pterotic (Armbruster and Hardman 1999); and from *Spectracanthicus* by lacking an expanded posterior membrane of the dorsal fin contacting the adipose spine. In addition, *Hypostomus* is similar to *Pterygoplichthys* (some of which have evertible cheek plates but no hypertrophied odontodes on them) and can be separated by having 7 branched dorsal-fin rays (vs. 9 or more). Body coloration in the AOG is dark brown with black spots on most of the body (inverted in *H. faveolus* from Tocantins and Xingu basins, which have pale spots); the abdomen may or may not be spotted; some species have dorsal saddles, and some can change color between being spotted and saddled (Armbruster 2003). Three predorsal plates (including nuchal plate); 5 rows of plates on the caudal peduncle; abdomen naked to fully plated; caudal fin forked, lower lobe longer. In most species, males develop hypertrophied odontodes on the leading edge of the pectoral-fin spine and the distal tip of the spine may become swollen. Nuptial males of some species of the *H. cochliodon* group develop wider, more widely spaced odontodes on the lateral plates. **SPECIES** The most species-rich genus of Loricariidae, with 141 species, including at least 34 species in the AOG region, mostly in coastal rivers of the Guianas (Montoya-Burgos 2003, Armbruster 2004a, Lujan et al. 2015a). *Hypostomus* reaches its zenith of diversity in southeastern Brazil, the Paraguay basin, and Suriname. Several subgroups are recognized within the genus, including around 20 wood-eating species in the *H. cochliodon* group (or subgenus *Cochliodon*), diagnosed by the loss of a notch between the metapterygoid and hyomandibula, strongly angled dentaries, and spoon-shaped teeth, although the last character is not present in all members of the group (Armbruster 2004a). Overview of species from the

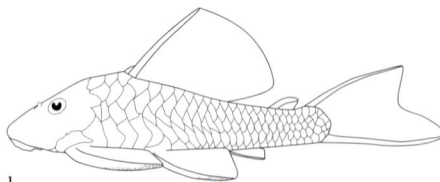

Guianas in Weber et al. (2012). Outdated key to the species in the *H. cochliodon* group in Armbruster (2003). **COMMON NAMES** *Acari*, *Bodó*, *Cascudo* (Brazil); *Carachama comun* (Peru); *Corroncho* (Venezuela). **DISTRIBUTION AND HABITAT** Throughout tropical and temperate South America east of the Andes, in nearly all aquatic habitats including some brackish water estuaries. Most species are lowland, sluggish stream and lake dwellers usually found associated with submerged wood; however, many species are found among rocks or wood in piedmont to mountain streams with moderate to swift flow. Members of *Hypostomus* occupy substrates ranging from mud and detritus, to gravel and cobble and boulders, to sand. *Hypostomus* has been widely introduced outside its native range, and over the past few decades non-native populations have established reproducing populations in inland waters of Texas (Pound et al. 2011), Sri Lanka (Bambaradeniya 2002), and China (Xu et al. 2012). **BIOLOGY** Mainly herbivorous or detritus-eating fishes (Montoya-Burgos 2003). Many species are able to breathe air and use a modified stomach for gas exchange (Armbruster 1998a). Members of the *H. cochliodon* group have large, spoon-shaped teeth like *Panaque* and *Panaqolus* and consume wood (Armbruster 2003, Lujan et al. 2011).

Isorineloricaria (21–52 cm SL)

Originally described for *I. spinosisssima*, a species from the Guayas drainage of Ecuador, and named for the elongate caudal peduncle that is similar to that in Loricariinae (Isbrücker 1980). Juveniles of *I. spinosissima* are shaped almost like a regular *Hypostomus*, but the caudal peduncle increases in relative length with size and goes from being oval to almost round in cross section. A similar growth pattern is seen in *Plecostomus tenuicauda* from the Magdalena and *P. villarsi* from the Maracaibo basin as well as a recently described species (*I. acuarius*) from the Orinoco basin, and these species should be placed in *Isorineloricaria* (Ray and Armbruster 2016). Additionally, these species are unusual in that nuptial males gain hypertrophied odontodes over all dorsolateral surfaces (vs. usually just postcranially in *Aphanotorulus* and other members of the *Peckoltia* Clade or just the pectoral-fin spine in most other loricariids). The elongate, transversely round caudal peduncle separates *Isorineloricaria* from other members of the *Peckoltia* Clade, and the distribution of hypertrophied odontodes on all the dorsolateral surfaces of nuptial males separates *Isorineloricaria* from all other hypostomines. In addition, *Isorineloricaria* can be separated from all other AOG hypostomines except *Corymbophanes*, *Hypostomus*, *Isorineloricaria*, *Pseudancistrus*, *Pseudorinelepis*, and some *Spectracanthicus* by lacking highly evertible cheek plates; from all the above taxa by having a nearly white base color (vs. brown to dark gray in most and tan in *Aphanotorulus*); from *Corymbophanes* and *Pseudorinelepis* by having an iris operculum; from *Corymbophanes* by having plates on the abdomen (vs. abdomen naked); from *Aphanotorulus* by having the spots nearly the size of the lateral plates (vs. smaller than the plates); from *Pseudancistrus* by having plates on the abdomen, 5 rows of plates on the caudal peduncle (vs. 4), and hypertrophied odontodes on the cheeks (if present) only in nuptial males (vs. all individuals); from *Pseudorinelepis* by having an adipose fin and by lacking regular rows of odontodes on the compound pterotic; and from *Spectracanthicus* by lacking a membranous connection between the dorsal fin and adipose spine. Three predorsal plates (including nuchal plate); 5 rows of plates on the caudal peduncle; abdomen plated; cheek odontodes absent except in nuptial males; caudal fin forked, lower lobe longer. Nuptial males have hypertrophied odontodes on all dorsolateral surfaces including fin spines/simple rays, head, cheek, and snout. **SPECIES** Four, one in the Orinoco basin (*I. acuarius*). **DISTRIBUTION AND HABITAT** Lowland and piedmont streams of trans-Andean drainages as well as the Orinoco basin of Venezuela and Colombia. The undescribed species generally inhabits streams with slow to moderate flow. **BIOLOGY** No data available.

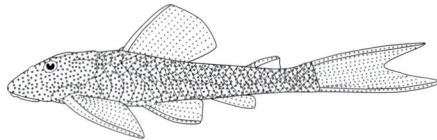

Lasiancistrus (10–15 cm SL)

Adult *Lasiancistrus* can be separated from all other loricariids by whisker-like odontodes on the cheek (all species) and at the corner of the snout (all except *L. tentaculatus*). Three predorsal plates; 3 rows of plates on the caudal peduncle; abdomen without plates; cheek odontodes long (longest extending beyond base of pectoral spine); caudal fin emarginate to forked with lower lobe longer. *Lasiancistrus* can be further separated from other loricariids (except Ancistrini and *Pterygoplichthys*) by evertible cheek odontodes; from *Pterygoplichthys* by 3 rows of plates on the caudal peduncle (vs. 5), a modified, bar-shaped opercle (vs. a triangular opercle), and <10 dorsal-fin rays (vs. ≥10); from all hypostomines (except *Ancistrus, Corymbophanes, Dekeyseria, Exastilithoxus, Lithoxus, Neblinichthys,* and *Pseudolithoxus*) by having 3 rows of plates on the caudal peduncle (vs. 4 or 5); from all but *Ancistrus* by the presence of tentacules on the pectoral-fin spines in nuptial males that are longer than their associated odontodes (tentacules is the name Sabaj et al., 1999, use for short tentacles still attached at the base to odontodes); from *Ancistrus* by having plates along the edge of the snout (vs. snout plates absent), and by maximally having translucent tentacules on the snout that have odontodes associated with them (vs. larger tentacles without associated odontodes colored the same as the head; even female and juvenile *Ancistrus* have some tentacles along the snout margin); from *Dekeyseria* by lacking keels on the lateral plates; from *Exastilithoxus* and *Lithoxus* by having >30 teeth per jaw ramus (vs. <10) and by having oval lips (vs. round); from *Neblinichthys* by lacking hypertrophied odontodes on top of the head in nuptial males; and from *Pseudolithoxus* by lacking long, bristle-like odontodes on the leading edge of the pectoral-fin spine (Armbruster 2005). Sexual dimorphism includes an almost square snout in nuptial males, versus rounded in females and juveniles. Adult males of most species (except *Lasiancistrus tentaculatus*) have whisker-like odontodes at the anterolateral corner of snout; males of *L. tentaculatis* have tentacules instead of whisker-like odontodes along anterior margin of snout (Armbruster 2005). Nuptial males with tentacules longer than their associated odontodes on the pectoral-fin spine (Sabaj et al. 1999). **SPECIES** Six, including four in the AOG region: *L. heteracanthus, L. saetiger, L. schomburgkii,* and *L. tentaculatus*. Key to the species in Armbruster (2005). **DISTRIBUTION AND HABITAT** Amazon, Essequibo, Orinoco, Magdalena, San Juan, Atrato, Tuyra, and Bayano basins and the Lake Maracaibo drainage. Most common in lower piedmont streams. **BIOLOGY** The functions of the tentacules and whisker-like odontodes are unknown. Presence of whisker-like odontodes on the cheek is not sexually dimorphic, although male odontodes may be longer. See discussion on function of tentacules in *Ancistrus*.

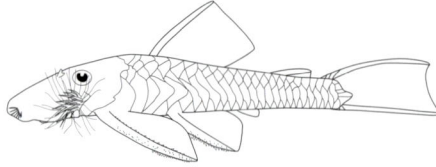

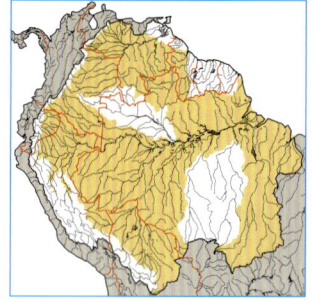

Leporacanthicus (10–25 cm SL)

Readily identifiable from all other Ancistrini by the development of usually only 2 teeth in each premaxilla, the inner teeth being very long (all other hypostomines have >2 teeth in the premaxilla); by having a well-raised supraoccipital crest (the supraoccipital in most of the Hypostominae is flat to slightly raised); and by having numerous fimbriate papillae along the margin of the upper lip (Armbruster 1997). *Leporacanthicus* looks most similar to some *Panaque, Spectracanthicus, Megalancistrus,* and *Pseudacanthicus* from which it can also be identified by having a round lower lip (vs. oval) and fimbriae on the upper lip. In addition, *Spectracanthicus* has the dorsal fin connected to the adipose fin spine (well separated in *Leporacanthicus*), and *Panaque* has a wide, rounded head (vs. pointed and narrow in *Leporacanthicus*) and short, spoon-shaped teeth (vs. long and acute). Color pattern is generally dark gray to black with white to golden spots or a light gray with medium-sized black spots. Three or more asymmetrical predorsal plates (including nuchal plate); 5 rows of

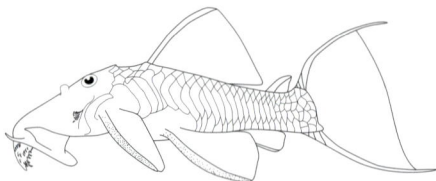

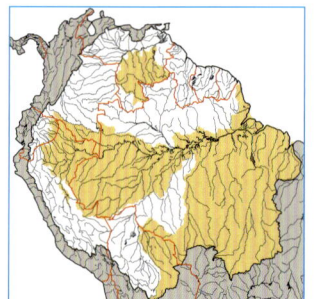

plates on the caudal peduncle; abdomen fully plated; cheek odontodes long (extending past base of pectoral spine); caudal fin forked, lower lobe slightly longer; principal caudal-fin rays forming short filaments; pectoral-fin spines extremely long and armed with enlarged odontodes; the entire dorsal surface of the head is covered in stout, sharp odontodes; the odontodes form a sharp keel on lateral plates; in juveniles, there are few to no odontodes on lateral plates besides those forming the keel; cheek odontodes fairly thin, but numerous, up to 100 or more (Armbruster 1997). **SPECIES** Four, all in the AOG region. Called the vampire plecostomus in the aquarium literature in reference to the large teeth in the upper jaw. **DISTRIBUTION AND HABITAT** Upper Orinoco basin, southern Amazon tributaries, and the Tocantins basin. *L. triactis* inhabits coarse woody debris and holes in muddy banks (L. Nico pers. comm.), whereas the sympatric *L.* cf. *galaxias* is more commonly collected from rocks in main river channels (NKL, pers. obs.). **BIOLOGY** Enlarged teeth of the upper jaw used to remove snails from their shells (Burgess 1994). This has been observed in *L. joselimai*, although specimens of *L.* cf. *galaxias* from Venezuela had guts containing pyralid and caddisfly larvae as well as freshwater sponges (Armbruster 1997).

Lithoxancistrus (10 cm SL)

Similar to *Pseudancistrus*, and sharing the presence of hypertrophied odontodes along the snout margin. *Lithoxancistrus* can be distinguished from *Pseudancistrus* by large papillae located along the dentary posterior to the teeth (Isbrücker et al. 1988, Silva et al. 2015). Body is dark to slightly mottled, with bands or spots on the fins. Four predorsal plates; 5 rows of plates on the caudal peduncle; abdomen without plates; cheek odontodes moderately short (not extending much posterior to base of pectoral spine); short hypertrophied odontodes present in both sexes along snout edge from tip of snout to evertible cheek odontodes. Can be separated from all other hypostomines except *Chaetostoma* and '*Pseudancistrus*' *pectegenitor* by having a large papilla or a set of papillae located along the dentary posterior to the teeth, and from *Chaetostoma* and '*Pseudancistrus*' *pectegenitor* by having 7 dorsal-fin rays (vs. 8 or more). **SPECIES** Three, all in the Orinoco basin: *L. coquenani*, *L. orinoco*, and *L. yekuana*. **DISTRIBUTION AND HABITAT** Inhabits swift waters of the Orinoco basin in Colombia and Venezuela. **BIOLOGY** No data available.

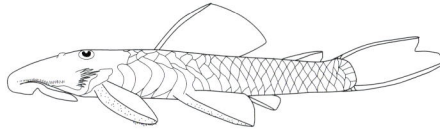

Lithoxus (4.0–8.5 cm SL)

Characterized by: highly dorsoventrally flattened body; lower lip large and round as in *Exastilithoxus* with the edge sometimes frilled, but not with the longer fimbriae seen in *Exastilithoxus*; jaws reduced with very few teeth (<6 per premaxilla); premaxillary teeth longer than the dentary teeth (Armbruster 1997). Color is slate gray to tan with a few lighter markings on the body and occasionally bands in the pectoral and caudal fins; ventral surface ranges from white to slightly lighter than the sides. Four or more predorsal plates (including nuchal plate); 3 rows of plates on the caudal peduncle; abdomen without plates; cheek odontodes short (extending just beyond gill aperture); caudal fin emarginated, lower lobe longer. *Lithoxus* can be separated from all other hypostomines except *Leporacanthicus* and *Exastilithoxus* by having a round (vs. oval) lower lip. *Lithoxus* can be separated from *Leporacanthicus* by lacking tentacles on the upper lip and by having 3 (vs. 5) rows of plates on the caudal peduncle, and from *Exastilithoxus* by lacking short fimbriae along the lower lip (the posterior rim of the lower lip is smooth in *Lithoxus*). Breeding males develop extremely long odontodes on the leading edge of the pectoral fin spine (Armbruster 1997). **SPECIES** Eight, all in the AOG region. **DISTRIBUTION AND HABITAT** Throughout the Guianas, the upper Negro River in Brazil, and the upper Orinoco basin in Venezuela. Inhabits fast-flowing waters of principal rivers or creeks where the

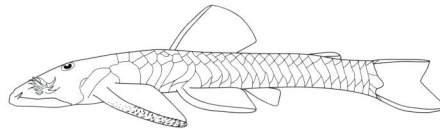

substrate consists of rock and sand, including stony rivulets and rapids. **BIOLOGY** *Lithoxus* (and *Exastilithoxus*) are unusual among loricariids in that they subsist on a diet consisting largely of invertebrates (a diet shared to varying degrees with *Hypancistrus*, *Leporacanthicus*, *Pseudacanthicus*, *Scobinancistrus*, *Spatuloricaria*, *Spectracanthicus*, and perhaps *Panaqolus albomaculatus*). *Lithoxus planquettei* is a pioneer species capable of colonizing environments which are nutritionally poor (Le Bail et al. 2000). The large size and small number of eggs suggest that parental care is well developed, but nothing is known of the breeding habits of *Lithoxus*.

Nannoplecostomus (2.2 cm SL)

Characterized by: very small adult body size, and a unique lateral dermal plate arrangement, in which the mid-dorsal and median plates are greatly reduced in number and elements. As a consequence, most of the body is covered by only 3 series of plates (the dorsal, midventral, and ventral plate series) posteriorly from the abdomen to caudal peduncle (vs. typically 5 lateral dermal plates present, with the plate series decreasing to 3 series only at the caudal peduncle in most other loricariidae genera; Ribeiro et al. 2012). Other characters that might aid identification: oral disk elliptical, broad, reaching anterior limit of pectoral girdle; lower lip covered with numerous similar-sized papillae; oral disk with fringed margins, formed by well-developed, elongate papillae; buccal papilla present, moderately developed; teeth slender, bifid, with larger medial cusp and smaller lateral cusp minute and pointed; body entirely covered by plates, except for ventral surface of head, abdomen, and insertions of pectoral and pelvic fins; plates and odontodes of head and body not arranged in crests, keels, or conspicuous rows; head and body plates covered by similarly sized odontodes; and body color brown, with 3 dark brown vertical bars; the intensity of which may depend on habitat (Ribeiro et al. 2012). Females with an enlarged, swollen urogenital opening; males possess a small genital papilla just posterior to the anus, which is absent in females. Additionally, males also possess a thickened flap of skin on the dorsal surface of the pelvic-fin spine, which is more developed proximally (flap absent in females). **SPECIES** One, *N. eleonorae*, see species description in Ribeiro et al. (2012). **DISTRIBUTION AND HABITAT** Known only from small tributaries of the São Domingos and from the das Pedras rivers, both tributaries of the Paraná River, upper Tocantins basin, Goiás state, Brazil. **BIOLOGY** Individuals were observed attached to the rock surfaces, presumably scraping periphyton.

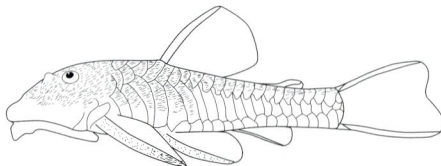

Neblinichthys (5.0–9.0 cm SL)

An extremely unusual loricariid, very similar in appearance to *Lasiancistrus*, but males with extremely elongate, anteriorly oriented odontodes on the snout (a so-called snout brush) and fairly elongate odontodes on the rest of the head, especially over the orbits (Taphorn et al. 2010). Some species have shorter odontodes in the snout brush (*N. peniculatus*), although the odontodes develop only during the breeding season. Color is generally a rich brown with lighter bands in the fins; the elongate odontodes of the snout brush are orange, and the abdomen is dark but lighter than the sides. Four or more predorsal plates (not arranged symmetrically); nuchal plate and dorsal fin spinelet are exposed; at least one column of 3 rows of plates on the caudal peduncle; abdomen without plates; cheek odontodes fairly long (longest extending to base of pectoral spine or beyond); caudal fin straight, angled posteroventrally to slightly emarginated with the lower lobe longer (Taphorn et al. 2010, Armbruster and Taphorn 2013). **SPECIES** Six, all in the AOG region. Key to the species in Armbruster and Taphorn (2013). **DISTRIBUTION AND HABITAT** Occurs throughout the Western Guiana Shield at midaltitude locations, and in small creeks and medium-sized rivers along the flanks of tepuis and other granite outcrops (Armbruster and Taphorn 2013). **BIOLOGY** No data available.

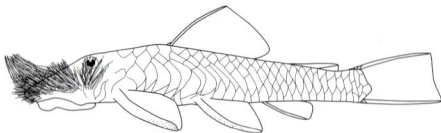

Panaqolus (7.0–14 cm SL)

Very similar to *Panaque*, but smaller-bodied (adults <15 cm SL, vs. >25 cm SL) with a relatively smaller head. *Panaqolus* and *Panaque* both possess acutely angled rows of generally short, robust teeth with large spoon-shaped cusps (also shared with the *Hypostomus cochliodon* group); however, in some shield-draining tributaries of the Orinoco, *Panaqolus maccus* has normal, villiform teeth, and the upper Amazon species *Panaqolus albomaculatus* has rows of more elongate dentary teeth that are nearly parallel with the body midline. Characteristics used to distinguish *Panaqolus* include: a rounded caudal peduncle in cross section, lacking a ventrolateral keel (vs. squared and ventrolateral keel present in *Panaque*); preopercle with deep lateral groove located near the reflected lateral margin of the bone (vs. groove absent in *Panaque* and other ancistrines); and absence of the posterior orbital notch (vs. notch present in *Panaque*) (Schaefer and Stewart 1993, Lujan et al. 2013). *Panaqolus maccus* with villiform teeth can be separated from *Hypancistrus* by having the adpressed dorsal fin reaching the adipose spine (vs. preadipose plate) and by having smaller eyes, and from *Peckoltia* with acutely angled dentaries by maturing at a much smaller size and by completely lacking a buccal papilla (vs. very small buccal papilla). Three predorsal plates (including nuchal plate); five rows of plates on the caudal peduncle; abdomen plated; cheek odontodes moderate (reaching base of pectoral spine); caudal fin emarginate, lower lobe longer. Males with hypertrophied odontodes on posterior margin of lateral plates (Chockley and Armbruster 2002). **SPECIES** Nine, all in the AOG region. See key to the species in Cramer (2014), except for *Panaqolus albivermis* (see Lujan et al. 2013). **TAXONOMIC NOTE** Previously recognized as the *Panaque dentex* species group (Schaefer and Stewart 1993). Fisch-Muller et al. (2012) tentatively assigned the species *Pseudoqolus koko* to genus *Panaqolus* because it has spoon-shaped tooth cusps and other general characteristics of the *Peckoltia* Clade. Molecular phylogenetic evidence confirms the placement of this species within the *Peckoltia* Clade but provides little support for its placement within genus *Panaqolus* (Lujan et al. 2015a). Close examination of the teeth of *P. koko* reveals that they are longer and flatter than most other *Panaqolus*, being intermediate between this genus and *Scobinancistrus*, and examination of *P. koko* gut contents has revealed that they eat sponges (NKL pers. obs.). **DISTRIBUTION AND HABITAT** Widespread in the Amazon and Orinoco basins, and one species in the Atlantic Coastal drainages of the Guianas. Locally uncommon throughout much of their range, although this might be a sampling artifact. Some *Panaqolus* species inhabit rocky bottoms of large rivers in rapids stretches, making them difficult to collect (J. Zuanon pers. comm.). **BIOLOGY** See discussion on trophic partitioning among species of *Panaqolus*, *Panaque*, and *Hypostomus pyrineusi* in Lujan et al. (2011). The specialized teeth and oral jaws of *P. albomaculatus* are used to pick invertebrates from wood crevices (Lujan et al. 2011).

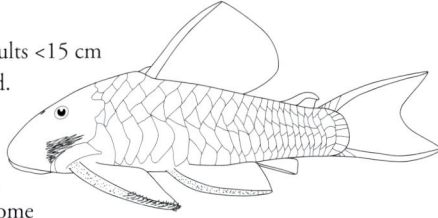

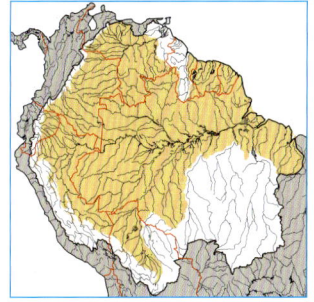

Panaque (28–60 cm SL)

Panaque (as well as *Panaqolus* and the *Hypostomus cochliodon* species group) possess acutely angled rows of generally short, robust, spoon-shaped teeth. Additional characteristics used to distinguish *Panaque* include: clusters of straight, elongate, and highly evertible cheek odontodes (vs. absent in *Hypostomus cochliodon* group); absence of odontodes on the preopercle; a larger body size (adults typically >25 cm; vs. <15 cm for *Panaqolus*); a relatively large swim bladder and overlying externally visible compound pterotic bone (absent in *P. bathyphilus*); presence of a strongly angled ventrolateral keel on the caudal peduncle (vs. absent in *Panaqolus*); preopercle without a deep lateral groove located near the reflected lateral margin of the bone (vs. groove present in *Panaqolus*); and a

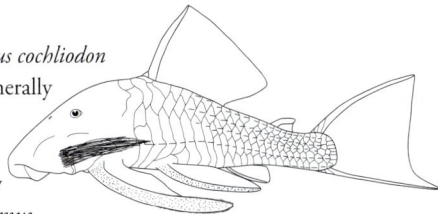

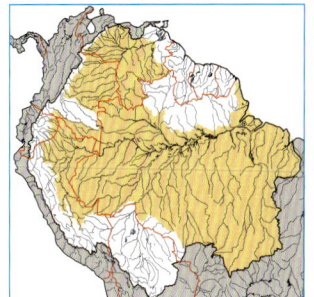

posterior orbital notch (vs. notch absent in *Panaqolus*) (Schaefer and Stewart 1993, Lujan et al. 2010, 2013). Internally, *Panaque* also differs from *Panaqolus* by having relatively larger adductor mandibulae jaw muscles and an enlarged coronoid flange on the dentary where this muscle inserts (Lujan and Armbruster 2012). **SPECIES** Seven, including five species in the AOG region. Revision of genus in Lujan et al. (2010), but for *Panaque bathyphilus* see Lujan and Chamon (2008). **DISTRIBUTION AND HABITAT** Lake Maracaibo drainages and the Magdalena, Orinoco, and Amazon basins, including the Araguaia-Tocantins. See species distribution maps in Lujan et al. (2010). **BIOLOGY** Consume large amounts of wood both in aquaria and in the wild, although this wood is not directly assimilated. *Panaque* are nutritionally dependent on environmental microbes that naturally colonize the wood surface (German 2009, Lujan et al. 2011). The spoon-shaped teeth of *Panaque* and reinforced jaws and hypertrophied musculature are specializations for gouging wood and consuming a subset of wood distinct from that consumed by sympatric *Panaqolus* and *Hypostomus cochliodon* group members. Native fishermen locate these species by listening for the sounds of them scraping submerged wood (Nelson et al. 1999).

Parancistrus (18–19 cm SL)

Characterized by: dorsoventrally depressed but stout body; dorsal-fin membrane connected to the adipose-fin spine. Expanded gill openings are present only in *Parancistrus, Ancistrus ranunculus,* and *Rhinelepis*; these are much larger than the very restricted openings of most loricariids. Unique to *Parancistrus* is the presence of fimbriate folds in the epidermis around the dorsal fin (Armbruster 1997), which may be a nuptial male characteristic. Color typically slate gray to black, occasionally with small white or light blue spots, large white blotches, or entirely yellow-white; abdomen completely plated in adults (*P. nudiventris* from Xingu River). *Parancistrus* differs from all loricariids except '*Baryancistrus*' *beggini* and *Spectracanthicus* by having the dorsal-fin membrane attached to the adipose spine. *Parancistrus* can be separated from *Spectracanthicus* by having much larger gill openings, a wider, more rounded head (vs. slender and pointed), premaxillae forming nearly a straight line at their union (vs. forming an acute angle), and thin teeth (vs. stout); and from '*B.*' *beggini* by having the pectoral-fin spine reaching just beyond pelvic-fin base when adpressed (vs. to anus), and by being dorsoventrally flattened (vs. body deep). Three predorsal plates (including nuchal plate); 5 rows of plates on the caudal peduncle; cheek odontodes moderate (extending to pectoral-spine base); caudal fin largely straight, lower lobe longer. Breeding males have elongate odontodes on the pectoral-fin spines and the posterolateral plates. In addition, the fleshy, fimbriate folds at the dorsal- and pectoral-fin insertions may be a breeding male trait (Armbruster 1997, Rapp Py-Daniel and Zuanon 2005). **SPECIES** Two, both in the AOG region: *P. aurantiacus* and *P. nudiventris*. **DISTRIBUTION AND HABITAT** Tocantins and Xingu basins in Brazil. *P. aurantiacus* is collected in rapids, under large flat rocks with some deposited sediments; *P. nudiventris* inhabits rocky-bottom areas subjected to moderate to strong current (Rapp Py-Daniel and Zuanon 2005). **BIOLOGY** *P. nudiventris* is nocturnal and feeds by grazing over rock surface, browsing the epilithon (Rapp Py-Daniel and Zuanon 2005) and flocculent detritus (Zuanon 1999). These characteristics are morphologically and ecologically convergent with the sympatric but distantly related species *Ancistrus ranunculus*, suggesting that their specialized jaw and branchial morphologies are adaptations for feeding on flocculent detritus.

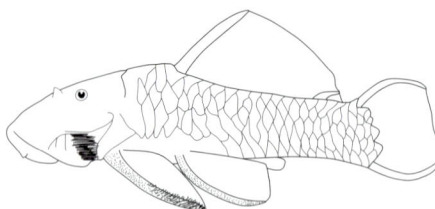

Paulasquama (5.0 cm SL)

Readily distinguished from other hypostomines by three characters: (1) Absence of plates in two large oval areas on top of snout lateral to mesethmoid. It may appear as if entire anterior snout except for its edge is devoid of plates because the skin is thick and odontodes on the plates over the mesethmoid may not penetrate in smaller specimens (vs. completely plated, with plates absent from a broad anterior margin of the snout in *Ancistrus* and *Chaetostoma*). (2) Plates in the dorsal series lateral to dorsal fin reduced in size to less than half that expected for a loricariid of its size (vs. plates in the dorsal series lateral to dorsal fin large). (3) Fleshy keel along and slightly anterior to the preadipose plate extending to dorsal caudal-fin simple ray along dorsal procurrent rays (vs. preadipose plate and dorsal procurrent rays covered in odontodes along medial surface). Further distinguished by: 6–7 asymmetrically arranged predorsal plates (including skin-covered nuchal plate); 5 rows of plates on the caudal peduncle; abdomen without plates; cheek odontodes very short (not reaching beyond gill aperture); caudal fin weakly emarginated, lower lobe longer. Sexual dimorphism includes hypertrophied odontodes along snout, longer odontodes on pectoral spine, posterior rays of pectoral fin with thicker skin in nuptial males (Armbruster and Taphorn 2011). **SPECIES** One, *P. callis*. **DISTRIBUTION AND HABITAT** Known only from the Waruma River, a tributary of the Mazaruni River in northwest Guyana. Collected from a riffle in a shallow side channel over polished cobble and gravel. **BIOLOGY** No data available.

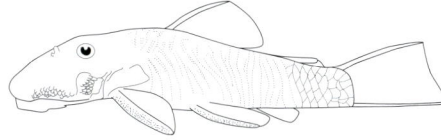

Peckoltia (6.0–17 cm SL)

Peckoltia, sensu Lujan et al. (2015a) and Armbruster et al. (2015), lacks a diagnosis. Species in this genus generally have pigment saddles, separating them from most of the other species of the *Peckoltia* Clade; however, some species are spotted and saddles are faint or absent. Most species of *Peckoltia* have dentary tooth rows angled ≤90º, but this character is variable within the genus. Just within the Orinoco basin, *P. wernekei* has acutely angled dentaries, whereas its sister, *P. lujani*, has straight dentaries. Some species have hypertrophied odontodes on the lateral plates of nuptial males (as in *Ancistomus* and *Panaqolus*), but some do not. Given the importance of dentary tooth row angle as a taxonomic character, it is useful to first divide species of *Peckoltia* into two groups: those with nearly straight dentaries (Group 1), and those with acutely angled dentaries (Group 2).

Group 1 can be separated from *Panaqolus, Peckoltichthys, Hypancistrus,* and *Scobinancistrus* by having straight dentaries (vs. acutely angled); from most *Panaqolus* by having villiform teeth (vs. stout teeth with a spoon-shaped cusp), from *Peckoltichthys* by having the eyes high on the head and the supraorbital ridge taller than the interorbital space (vs. eyes low and supraorbital ridge below the interorbital space), and from *Hypancistrus* by having the dentary and premaxillary teeth the same length (vs. dentary teeth twice as long); from *Ancistomus* by generally having dorsal saddles (vs. spots; only *P. sabaji* has spots and it differs by having spots throughout the dorsal and caudal fins, vs. spots present only proximally on fins if at all in *Ancistomus*); from *Hemiancistrus* and the 'H.' guahiborum group generally by having saddles (vs. dark spots, light spots, or mottled color; only *P. sabaji* has dark spots, and it can be separated by having those dark spots larger than the plates on the caudal peduncle; vs. smaller than plates or absent); from *Baryancistrus*, 'Baryancistrus', *Parancistrus,* and *Spectracanthicus* by lacking an expansion of the posterior dorsal-fin membrane and always lacking white spots (vs. expansion generally reaching preadipose plate or adipose spine and, when not, having white spots); and from *Aphanotorulus, Hypostomus,* and *Isorineloricaria* by having evertible cheek plates.

Group 2 *Peckoltia* can be separated from all except the *H. cochliodon* group, *Exastilithoxus, Leporacanthicus, Lithoxus, Panaqolus, Panaque, Peckoltichthys, Pseudacanthicus,* and *Spectracanthicus* by having acutely angled (≤90°) dentaries (vs. dentaries forming an oblique angle or straight); from the *H. cochliodon* group, *Panaqolus,* and *Panaque* by having villiform

teeth (vs. spoon-shaped); from *Exastilithoxus* and *Lithoxus* by having an oval lower lip (vs. round) and by having at least some plates on the abdomen (vs. abdomen naked); from *Peckoltichthys* by having the eyes high on the head such that the supraorbital ridge is higher than the interorbital area (vs. eyes low on the head, supraorbital ridge below interdorsal area) and by having small plates on the abdomen (vs. relatively large); from *Leporacanthicus* and *Pseudacanthicus* by lacking keels on the lateral plates (vs. keels of generally very sharp odontodes present) and by having 7 dorsal-fin rays (vs. 8 or more); from *Leporacanthicus* by having >2 teeth per premaxilla (vs. only 2); and from *Spectracanthicus* by not having the dorsal and adipose fins connected. Some specimens of *Panaqolus maccus* have villiform teeth, and Group 2 *Peckoltia* can be separated from them by having a very small buccal papilla (vs. buccal papilla absent).

Other characters of *Peckoltia* species include: 3 predorsal plates (including nuchal plate); 5 rows of plates on the caudal peduncle; abdomen ranging from unplated to completely plated (in general, Group 1 has an unplated to partially plated abdomen and Group 2 has a plated abdomen); caudal fin forked, lower lobe longer. Nuptial males with hypertrophied odontodes on sides of body from the head posteriorly including the pectoral and adipose spines, and unbranched and some branched rays of the dorsal lobe of the caudal fin. **SPECIES** 19, all in the AOG region. Partial key to the species in Armbruster (2008). **TAXONOMIC NOTE** Based on the phylogenetic analysis of Lujan et al. (2015a), we also recognize the enigmatic species *Hemiancistrus pankimpuju* and *Etsaputu relictum* among Group 1 *Peckoltia*. *Peckoltia pankimpuju* is a species with reduced eyes, elongate caudal-fin filaments, and variable pigment: from no pigments to uniformly dark black. *Peckoltia relictum* is unusual in that it has the cheek plates only slightly evertible and very short cheek odontodes, along with many other primitive osteological characteristics that would be reversals based on the molecular phylogeny of Lujan et al. (2015a). **DISTRIBUTION AND HABITAT** Amazon, Essequibo, and Orinoco systems and coastal drainages north of the Amazon to French Guiana. Inhabits a variety of habitats including swift riffles, rapids, small forest creeks, and slow-flowing muddy streams. **BIOLOGY** Most species hide inside cavities in submerged logs, rocks, or lateritic conglomerates during the day and forage on epilithic algae at night.

Peckoltichthys (14 cm SL)

Differs from all other hypostomines by having eyes very low on head (reminiscent of condition in *Hypoptopoma*), such that the supraorbital ridge is ventral to the highest point of the interorbital space (vs. eyes set high in the head such that the supraorbital ridge is generally higher than the highest point on the interorbital space or at least that there is a concavity medial to the eye). Similar to *Peckoltia*, but differentiated by: large pigment spots and mottling on head, and sometimes body (vs. saddles and occasionally spots); pelvic-fin spines widened and can be adducted ventral to the abdominal surface of the body; and plates on the abdomen rather large, about twice as large as in most other hypostomines (Armbruster 2008). Three predorsal plates (including nuchal plate); 5 rows of plates on the caudal peduncle; cheek odontodes moderate (reaching base of pectoral spine); caudal fin forked, lower lobe longer. Nuptial males with hypertrophied odontodes on sides of body and posterior part of head, but absent from caudal- and adipose-fin spines (Armbruster 2008). **SPECIES** One, *P. bachi*; see species information in Armbruster (2008). **DISTRIBUTION AND HABITAT** Upper Amazon River and its tributaries in Brazil, Colombia, Ecuador, and Peru. Occurs at the edge of medium to large rivers among submerged twigs and grasses, usually in flowing stretches (Armbruster 2008). **BIOLOGY** Hypertrophied pelvic muscles and widened pelvic-fin spines used to grasp grasses and twigs (Armbruster 2008).

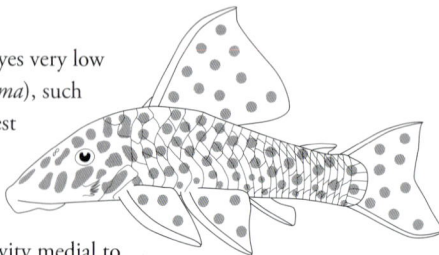

Pseudacanthicus (15–90 cm SL)

Large-bodied, spiny loricariids with a diverse range of color patterns: from light to dark gray, often with black spots; the fins and body may have red sections or a red wash. Four or more asymmetrical predorsal plates (including nuchal plate); 5 rows of plates on the caudal peduncle; abdomen completely covered in small plates in adults; cheek odontodes short (extending just beyond gill aperture); caudal fin mostly straight to forked, lower lobe slightly longer; principal caudal-fin rays occasionally elongate as relatively thick filaments; the odontodes form a sharp keel on the lateral plates (Armbruster 1997). *Pseudacanthicus* can be separated from all hypostomines except the *Acanthicus* group, *Pterygoplichthys*, the *Chaetostoma* Clade, and '*Pseudancistrus*' *pectegenitor* and *Ancistrus dolichopterus* by having 8 or more dorsal-fin rays, and from the *Chaetostoma* group and *Pseudancistrus* by having sharp odontodes forming keels on the lateral plates (vs. maximally weak keels). *Pseudacanthicus* can be separated from *Acanthicus* by having an adipose fin, and by dentaries forming an angle of 90° or less (vs. almost straight) and a small pterotic (vs. a very large pterotic and associated increase in the size of the swim bladder); from *Pterygoplichthys* by having 8 or 9 dorsal-fin rays (vs. 10 or more, rarely 9), a crescent-shaped opercle (vs. triangular), and by having the keel odontodes stout (vs. long if present; Armbruster 1997); and from *Ancistrus dolichopterus* by lacking tentacles on the snout (vs. present). *Pseudacanthicus histrix* has highly elongate odontodes that form a brush on the anterior margin of the pectoral-fin spine in breeding males (Burgess 1989), but dimorphism has not been reported for the other species. **SPECIES** Five, all in the AOG region. **DISTRIBUTION AND HABITAT** Swift waters of the Orinoco, the Guianas, the Negro, and the lower Amazonian tributaries. According to fishermen in Guyana, *P. serratus* lives in the deep and rocky zones of the main riverbed (Le Bail et al. 2000). **BIOLOGY** Feeds on macroinvertebrates and perhaps sponges.

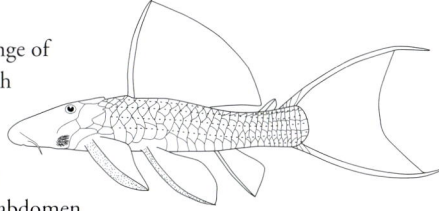

Pseudancistrus (4.3–24 cm SL)

Characterized by the extreme development of hypertrophied odontodes on snout in breeding males (Darwin 1882). Both males and females do have hypertrophied snout odontodes although they are longer in males. Armbruster (2004a, b) recognized a broad concept of the genus *Pseudancistrus* based on several characters, but molecular studies find these similarities to result from convergent evolution (Covain and Fisch-Muller 2012, Lujan et al. 2015a). *Guyanancistrus* and *Lithoxancistrus* were resurrected, and new genera were proposed for '*P*.' *pectegenitor* + '*P*.' *sidereus* and '*P*.' *genesetiger*. '*Pseudancistrus*' *genesetiger* was found to not even belong to the Hypostominae. *Pseudancistrus* is now limited to a group with fairly large adult body sizes, extreme development of the hypertrophied cheek odontodes, and 4 rows of plates on the caudal peduncle. As such, *Chaetostomus megacephalus*, which was moved to *Pseudancistrus* by Armbruster (2004a) and Armbruster (2004b) is not a member of that genus, and is here recognized as '*Pseudancistrus*' *megacephalus–incertae sedis* within the Hypostominae. Three predorsal plates (including nuchal plate); abdomen without plates; cheek odontodes long (extending beyond base of pectoral spine); cheek odontodes weakly evertible (to ~30° from head); caudal fin emarginate, lower lobe longer. The weakly evertible cheek plates, 4 rows of plates on the caudal peduncle, and elongate odontodes along the snout margin separate *Pseudancistrus* from all other hypostomines. It is further separated from *Lithoxancistrus* by lacking a large papilla posterior to the dentary teeth. Males generally with longer snout odontodes and possessing a papilla posterior to urogenital opening (absent in females) (Armbruster 2004b, Silva et al. 2015). **SPECIES** 18, including 10 species in the AOG region. **DISTRIBUTION AND HABITAT** Drainages of the Guiana Shield and northern Brazilian Shield

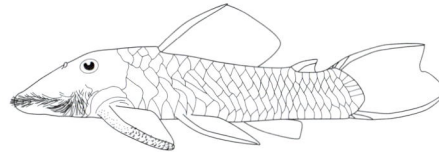

in Brazil, Colombia, French Guiana, Guyana, Suriname, and Venezuela. Inhabits swiftly flowing water, with gravel, cobble, and boulders, as well as rapids and zones of plunging waters. **BIOLOGY** Spawns in cavities; males provide parental care; possibly seasonal spawning (Lujan et al. 2007). The length of the male snout bristles might be related to hierarchic level of the individual within the population (Le Bail et al. 2000). Frequently in mixed-species groups with other depressed-bodied loricariids (J. Zuanon pers. comm.).

Pseudolithoxus (6.5–13 cm SL)

Similar to *Lasiancistrus* but more dorsoventrally flattened and with longer pectoral fins. Like *Pseudancistrus* and *Lithoxancistrus*, *Pseudolithoxus* has hypertrophied odontodes along the snout regardless of season or sex. Identified by: evertible cheek plates; a dorsoventrally flattened body; extremely hypertrophied odontodes on elongate pectoral-fin spines and along the snout margin; cheek odontodes long (extending beyond base of pectoral spine); 3 rows of plates on the caudal peduncle; 3–4 predorsal plates (including skin-covered nuchal plate); abdomen without plates; caudal fin weakly emarginated, with lower lobe longer (Armbruster and Provenzano 2000). Differs from *Lasiancistrus* by lacking thin, whisker-like odontodes on the evertible cheek plates and from all other ancistrines except *Ancistrus, Dekeyseria, Exastilithoxus, Lithoxus,* and *Neblinichthys* by having 3 rows of plates on the caudal peduncle (vs. 4–5). *Pseudolithoxus* further differs from *Ancistrus* by having plates along the anterior margin of the snout (vs. anterior part of snout naked) and by lacking tentacles on the snout; from *Dekeyseria* by lacking well-keeled lateral plates; from *Exastilithoxus* by lacking fimbriae on the lower lip; from *Exastilithoxus* and *Lithoxus* by having many more than 20 teeth per jaw ramus; and from *Neblinichthys* by lacking elongate odontodes on the top of the snout of breeding males (Armbruster and Provenzano 2000). **SPECIES** Five, all from the AOG region. Key to the species and species information in Armbruster and Provenzano (2000) except for *P. kelsorum* (see Lujan and Birindelli 2011). **DISTRIBUTION AND HABITAT** Known only from rocky main channel habitats in the Casiquiare Canal and tributaries of the Orinoco River draining the western and northern slopes of the Guiana Shield highlands. Species distribution maps in Lujan and Birindelli (2011). **BIOLOGY** No data available.

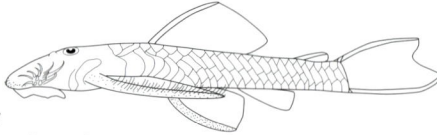

Pterygoplichthys (20–70 cm SL)

Readily diagnosed from nearly all other hypostomines in the AOG by: usually 10 (rarely 9) or more branched dorsal-fin rays (vs. 7 in most and usually 8–9, rarely 10, in the *Chaetostoma* and *Acanthicus* groups). *Pterygoplichthys* can be separated from the *Chaetostoma* group (*Andeancistrus, Chaetostoma, Cordylancistrus, Dolichancistrus,* and *Transancistrus*) by having a fully plated abdomen (vs. plates absent); and from *Acanthicus,* and *Pseudacanthicus* by having moderate keels (vs. keel odontodes stout and sharp); and from *Leporacanthicus* by having many more than 2 teeth per premaxilla. *Pterygoplichthys* further differs from the *Chaetostoma* group by having 3 (occasionally 2) plates between the suprapreopercle and the exposed opercle (vs. 1), an exposed nuchal plate (vs. covered by plates), and an exposed spinelet (vs. covered by skin; nuptial male *Dolichancistrus* do have odontodes on the spinelet). Color pattern is generally dark brown with either darker spots or lighter spots or vermiculations. Three predorsal plates (including nuchal plate); 5 rows of plates on the caudal peduncle; cheek odontodes short to moderate (extending to gill openings to pectoral-spine base) or absent; cheek plates evertible even in species without hypertrophied cheek odontodes to ≥75° from head; caudal fin strongly emarginated, lower lobe longer (Armbruster 1997). **SPECIES** 16, including 10 species in the AOG region. Key to the species in Armbruster and Page (2006). **COMMON NAMES** *Acari, Bodó, Cascudo* (Brazil); *Aletón carachama* (Peru); Sailfin

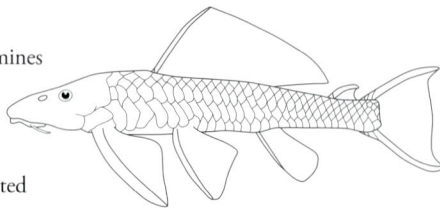

plecos (English). **DISTRIBUTION AND HABITAT** Orinoco, Amazon, Magdalena, Maracaibo, Paraná, and São Francisco systems. Several *Pterygoplichthys* species have been widely introduced outside their native ranges, and over the past few decades non-native populations have established reproducing populations in inland waters of North and Central America, the Caribbean, Hawaii, and Asia (Nico et al. 2009). **BIOLOGY** Males excavate tunnels into mud banks where eggs are laid; they guard the fertilized eggs (Burgess 1989, Galvis et al. 1997). *Pterygoplichthys* are lowland specialists and have a modified stomach for breathing air (Armbruster 1998a, da Cruz et al. 2013). The common plecostomus in the pet trade is often labeled *Hypostomus plecostomus* but is typically a *Pterygoplichthys*, most likely a hybrid of *P. pardalis* and *P. disjunctivus*. They are large-bodied species, commonly found alive in fish markets; their respiratory stomachs allow them to survive for long time periods out of water (Val and de Almeida-Val 1995).

Scobinancistrus (24–25 cm SL)

Readily diagnosed from other hypostomines by the few, stout, bicuspid teeth with an elongate and broadly flat or convex principal cusp, echoing aspects of dentition that are variously characteristic of *Leporacanthicus, Panaque,* and *Panaqolus*. These elongate, spatulate teeth separate *Scobinancistrus* from all other hypostomines. The species have a black base color with gold spots and gold to red edging on the fins. Like *Panaque, Panaqolus,* and *Peckoltia* Group 2, *Scobinancistrus* has acutely angled dentaries. *Scobinancistrus* can be separated from all of these by having 3 or 4 dentary and premaxillary teeth (vs. >4 teeth) and from all but *Panaqolus albomaculatus* by having white to gold spots (Lujan et al. 2010). Three predorsal plates (including nuchal plate); 5 rows of plates on the caudal peduncle; abdomen completely plated; cheek odontodes short (reaching maximally to base of pectoral spine); caudal fin strongly emarginate, lower lobe longer. **SPECIES** Two, both in the AOG region: *S. aureatus* and *S. pariolispos*. **DISTRIBUTION AND HABITAT** *S. aureatus* inhabits the Xingu basin; and *S. pariolispos* the Tocantins and Tapajós basins. **BIOLOGY** Burgess (1994) and Lujan et al. (2012) suggested these species feed on snails or insects.

Soromonichthys (3.0 cm SL)

Readily diagnosed from other hypostomines by a unique pattern of plate loss on the snout (see fig. 1D). Only three other Hypostominae genera have a loss of snout plates: *Ancistrus, Chaetostoma,* and *Paulasquama*. In *Soromonichthys*, the plate loss occurs from the front of the snout and along the mesethmoid to approximately the nares, whereas in *Ancistrus* females and *Chaetostoma* it occurs in a narrow band all along the snout; in *Ancistrus* males it is a broad band along the entire snout, and in *Paulasquama* it is in oval patches lateral to the mesethmoid (which does have plates above it). *Soromonichthys* is further distinguished from these genera by having a body coloration consisting of thin irregular light yellow to gold bars on a base color of irregularly mixed moderate to dark green (vs. coloration lacking thin, light-colored bars); from *Ancistrus* by lacking tentacles emergent from unplated regions of the snout; and from *Chaetostoma* by having 7 dorsal-fin rays (vs. 8–10; Lujan and Armbruster 2011). Three predorsal plates (including nuchal plate); 3 rows of plates on the caudal peduncle; abdomen without plates; cheek odontodes moderate (extending to base of pectoral spine); caudal fin weakly emarginate, lower lobe longer. **SPECIES** One, *S. stearleyi*. **DISTRIBUTION AND HABITAT** Known only from the lower reaches of Soromoni Creek, a clearwater tributary of the upper Orinoco draining the southwestern slope of Mount Duida, a tepui at the western margin of the Guiana Shield. Collected from shallow riffles and runs over clean cobble substrate interspersed with sand and patches of a rooted, moss-like aquatic macrophyte (Lujan and Armbruster 2011). **BIOLOGY** A benthic herbivore and detritivore (Lujan and Armbruster 2011).

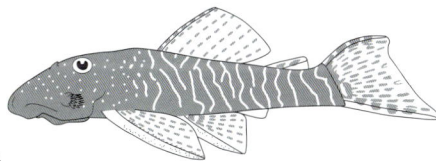

Spectracanthicus (6.0–13 cm SL)

Separated from all other loricariids except
'Baryancistrus' beggini and Parancistrus by:
posteriormost dorsal-fin ray entirely adnate
and connected to adipose spine via a low
fleshy membrane; however, unlike other
species with the dorsal–adipose fin connection,
Spectracanthicus has fewer teeth (<25 vs. >25) and the dentaries form an acute angle. It can be
further separated from Parancistrus by having restricted gill openings, reaching up to one-third
of cleithrum length (vs. large gill opening, reaching nearly one-half of cleithrum length) and
from 'B.' beggini by having the pectoral-fin spine reaching just beyond pelvic-fin base when
adpressed (vs. to anus) (Armbruster 2004a, Chamon and Py-Daniel 2014). Three predorsal
plates (including nuchal plate); 5 rows of plates on the caudal peduncle; abdomen naked to
almost fully plated; cheek odontodes short (to gill opening) or absent; cheek plates evertible
in some species, but not in S. murinus; caudal fin largely straight, lower lobe longer. Chamon
and Py-Daniel (2014) note that considerable diversity in tooth number and jaw morphology
is encompassed within Spectracanthicus punctatissimus. **SPECIES** Six, all in the AOG region.
Review of genus and key to the species in Chamon and Py-Daniel (2014). **DISTRIBUTION AND
HABITAT** Tapajós, Xingu, and Tocantins basins. **BIOLOGY** S. zuanoni is nocturnal, feeding on
epilithic algae and invertebrates (Chamon and Py-Daniel (2014). It occurs individually or in
groups often sympatric with S. punctatissimus, under shelter boulders (adults) and spaces beneath
rocks (juveniles). Spectracanthicus punctatissimus and S. zuanoni are among the most common
and abundant loricariids throughout the Xingu River, Brazil (NKL pers. obs.).

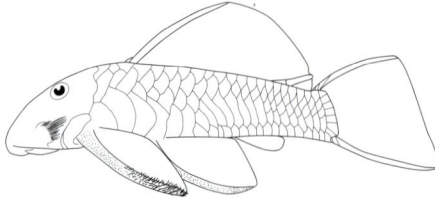

SUBFAMILY LITHOGENINAE—CLIMBING ARMORED CATFISHES

— JONATHAN ARMBRUSTER, PETER VAN DER SLEEN, and NATHAN LUJAN

Subfamily includes one genus (Lithogenes), with three species.
Lithogenes is the sister group to the rest of the Loricariidae
(Schaefer 2003b, Lujan et al. 2015a).

Genus Lithogenes

Unusual among armored loricariid catfishes
in lacking plates over most of the body
surface. The dermal plates that usually encase
the head and body of most loricariids are
restricted in Lithogenes to 3 paired series of small plates on the posterior trunk region (between
a vertical through the anal-fin origin and the caudal-fin base), plus a set of 2–5 dermal plates
on either side of the lateral cheek between the opercle and maxilla (Schaefer 2003b, Schaefer
and Provenzano 2008). All other loricarioids either lack plates entirely (Nematogenyidae,
Trichomycteridae, Astroblepidae), have dermal plates on the trunk arranged in 2 paired series
only (Callichthyidae, Scoloplacidae), or have 3 or more plate series along the full extent of the
trunk from the pterotic to the caudal fin (all other Loricariidae). Another distinctive feature
in Lithogenes is the modification of the pelvic fin: the first ray is branched, divided to near its
base, greatly thickened, flattened, and with a broad fleshy pad on the ventral surface that bears
numerous large odontodes (no odontodes in L. wahari). In addition to the ventral sucker-shaped
mouth (as in other loricariids), the modifications of the pelvic fin and associated musculature
function to assist in adhesion as well as in vertical propulsion, or climbing (Schaefer and
Provenzano 2008). Sexual dimorphism has been reported for L. wahari and includes differences
in the head, anal fin, and urogenital papilla (Schaefer and Provenzano 2008). **SPECIES** Three, all
in the AOG region; however, one of these (L. valencia) is likely extinct (Provenzano et al. 2003).

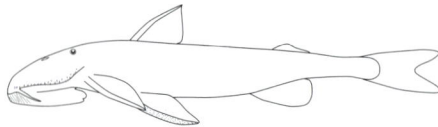

Key to the species and species information in Schaefer and Provenzano (2008). **DISTRIBUTION AND HABITAT** *L. villosus* inhabits the uplands of the western Guiana Shield in habitat dominated by rapids formed by bedrock outcrops, and is known only from the Potaro River (Hardman et al. 2002). *Lithogenes wahari* is known only from the middle section of the Cuao River, a clearwater tributary of the Sipapo River in the upper Orinoco basin in southwestern Venezuela (Schaefer and Provenzano 2008). It inhabits moderately high-gradient forest streams with clear water, swift current, and exposed bedrock substratum. *Lithogenes valencia* is known from only six specimens collected from a single locality in northern Venezuela sometime in the 1970s. This location is situated in a heavily populated and industrialized region of the Lago Valencia drainage in northern Venezuela and, despite extensive survey work in the area, no additional specimens of this species have been collected (Provenzano et al. 2003). Species distribution map in Schaefer and Provenzano (2008). **BIOLOGY** Poorly known. *Lithogenes wahari* is rheophilic with many specialized traits for living in fast-moving water, including robust oral disk and pelvic fins used to adhere to rocky substrates and climb rocks in powerful currents. *Lithogenes wahari* consumes primarily vegetable matter (Schaefer and Provenzano 2008). Based on the low number of ova counted in specimens of *L. wahari*, a high degree of parental care is suspected but unconfirmed (Schaefer and Provenzano 2008).

SUBFAMILY LORICARIINAE—LORICARIINE ARMORED CATFISHES
— RAPHAEL COVAIN and PETER VAN DER SLEEN

DIVERSITY 239 species in 31 genera and two tribes; Hartiini and Loricariini, the latter divided into two subtribes, Farlowellina and Loricariina, and with 108 species in 23 genera in the AOG region (Covain and Fisch-Muller 2007, Rodriguez et al. 2008, 2011, Covain et al. 2012, 2016).

GEOGRAPHIC DISTRIBUTION Distributed widely throughout tropical and subtropical South America, from La Plata estuary in Argentina to Pacific coastal rivers of Colombia, Ecuador, and Panama, and one species from the Caribbean drainages of Costa Rica (Ferraris Jr. 2003a). Some species, particularly in Harttiini, can be locally very abundant. Other species possess a very restricted distribution, such as *Harttiella* species that are restricted to a single mountain creek, making them highly vulnerable to extirpation or extinction.

ADULT SIZES Most species grow to 10–20 cm SL. *Harttiella parva* from small forest streams in French Guiana grows to about 3.1 cm SL, and *Spatuloricaria euacanthagenys* from the Caquetá River in Colombia reaches more than a half meter in length.

DIAGNOSIS OF SUBFAMILY Most diagnostic characters can be observed only in skeletal preparations. In general, characterized by a long and depressed caudal peduncle, absence of an adipose fin and, often, a depressed snout (Ferraris Jr. 2003a, Covain and Fisch-Muller 2007). They show dramatic variation in body shape, lip morphology, and dentition.

SEXUAL DIMORPHISM Often pronounced and expressed through the hypertrophy of odontodes on the pectoral-fin rays, on the snout margin, and sometimes on the predorsal area of mature males. Certain genera also show sexual differences in lip and tooth structures (Covain and Fisch-Muller 2007). Males are often larger than females, but for several species females can be larger than males, for example, *Loricaria* gr. *cataphracta* from Paraguay, *Hemiodontichthys acipenserinus* and *Spatuloricaria tuira* (Fichberg et al. 2014). Males of the *Loricariichthys* group (i.e., *Loricariichthys*, *Hemiodontichthys*, *Furcodontichthys*, *Limatulichthys*, and *Pseudoloricaria*) develop enlarged lower lips to carry fertilized eggs attached to their lip until the larvae hatch (Ferraris Jr. 2003a).

HABITATS Loricariinae occupy many habitats, from fast-flowing waters in rapids and falls to quiet and muddy areas of estuaries. Rheophilic species (e.g., Harttiini) live on rocky substrates made of stones, boulders, gravels, and sand, whereas limnophilic species preferring standing or slowly moving water (e.g., *Loricariichthys*) live over sand, mud, and decaying organic litter.

FEEDING ECOLOGY Members of Harttiini and Farlowellina possess numerous, long, and pedunculate teeth organized in a comb-like manner and feed by grasping algae growing on submerged wood and stones. Other loricariines with few and poorly differentiated teeth feed on small invertebrates on or in the sediment.

BEHAVIOR Loricariines are generally cryptobiotic, although many species are territorial when breeding.

KEY TO THE GENERA ADAPTED FROM COVAIN AND FISCH-MULLER (2007)

ILLUSTRATIONS IN KEY AFTER COVAIN AND FISCH-MULLER (2007)

1a. Caudal fin with i + 12 + i or i + 11 + i rays; teeth pedunculated, bicuspid, numerous (≥10 per premaxilla), organized in a comb-like manner and weakly differentiated (fig. 1a); sometimes with filamentous extensions on pectoral, dorsal, upper, and/or lower caudal-fin spines **2**

1b. Caudal fin with i + 12 + i or i + 10 + i rays; teeth straight, bicuspid, spoon-shaped (fig. 1b), not numerous (≤20 per premaxilla), strongly differentiated, sometimes reduced in size or absent; often with a more or less strong whip on upper caudal-fin spine..**10**

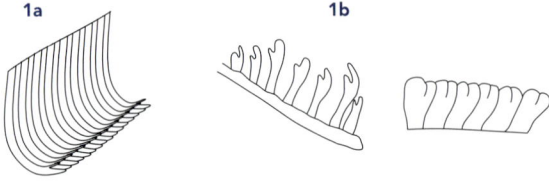

2a. Mouth shape elliptical (fig. 2a) **3**
2b. Mouth shape horseshoe-like (fig. 2b); with three buccal papillae, lateral ones trilobate; teeth small and not numerous (≈10 per premaxilla) ....*Metaloricaria*

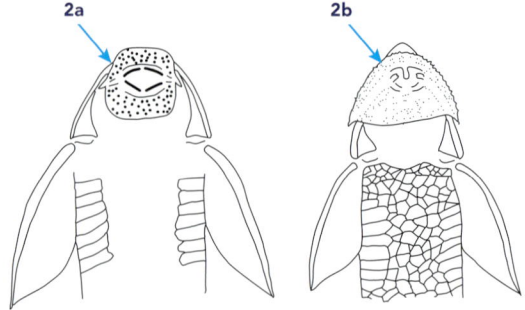

3a. Caudal peduncle strongly depressed, elliptical in transverse section (on average, the minimal depth of the caudal peduncle represents 1–3% of the SL) **4**
3b. Caudal peduncle weakly depressed, more or less circular in transverse section (on average, the minimal depth of the caudal peduncle represents 5% of the SL); abdomen naked; body covered in numerous short and dense odontodes, conferring a velvety aspect, species of small size (≈5 cm SL)...............................*Harttiella*

4a. Rostrum present (fig. 4a) **5**
4b. Rostrum absent (fig. 4b) **7**

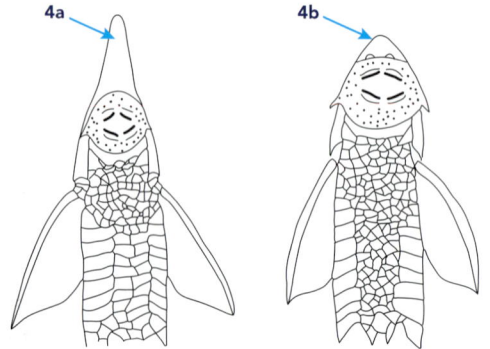

5a. Dorsal fin originating more or less in front of the anal-fin origin **6**
5b. Dorsal fin originating more or less in front of the pelvic-fin origin; abdominal cover complete and weakly structured in 2–3 rows ....*Sturisoma*

6a. Teeth not numerous (≈20 per premaxilla); 2–3 rows of abdominal plates; general appearance slender and resembling a woody stick....................................*Farlowella*
6b. Teeth numerous (≈100 per premaxilla); 3 rows of abdominal plates....................................*Aposturisoma*

7a. Pectoral fins with i + 6 or 1 + 7 rays; when i + 7, caudal, pectoral, and dorsal spines without filamentous extensions...........**8**
7b. Pectoral fins with i + 7 rays; caudal, pectoral, and dorsal spines often with filamentous extensions ....*Lamontichthys*

8a. Tip of snout naked **9**
8b. Tip of snout covered by plates; eye diameter small (on average ≈10% of head length); with filamentous extensions on pectoral, upper, and lower caudal-fin spines....................................*Pterosturisoma*

9a. Abdominal cover often incomplete, made of small granular platelets without particular organization, and rarely extending until pectoral girdle; caudal fin with a black basicaudal blotch (fig. 9a)...............................*Harttia*
9b. Abdominal cover complete, made of small rhombic platelets without particular organization, and extending until pectoral girdle; caudal fin with a black crescent (fig. 9b) ....*Cteniloricaria*

10a. Lower lip bilobate with a median furrow (figs. 10a, 11a); surface of this lip more or less smooth or weakly papillose; presence of a double abdominal keel (fig. 10a); throat never covered (fig. 10a); whip on upper caudal-fin spine weak or absent... **11**

10b. Absence of such a combination of characters; lower lip more often strongly papillose or filamentous (fig. 10b) **15**

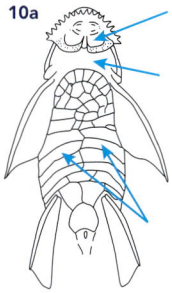

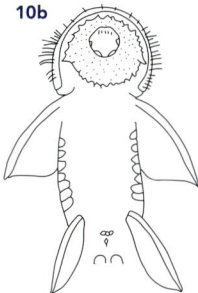

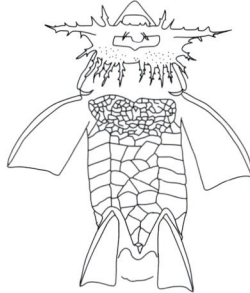

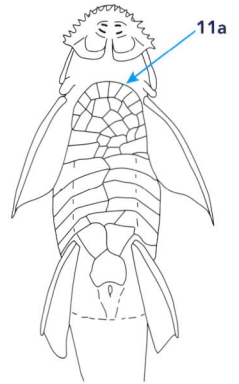

11a. Presence of a secondary structure in the organization of the abdominal cover forming a perfect elliptical area at the level of the pectoral girdle (fig. 11a).. *Loricariichthys*

11b. Without such structure **12**

12a. With conspicuous lines of odontodes on head and snout (fig. 12a)................................ **13**

12b. Without lines of odontodes on head and snout **14**

13a. Rostrum strongly pronounced and with a discoid tip (as in fig. 12a); maxillary barbels short; premaxillary teeth absent; abdomen covered by large rectangular plates organized in 3 rows ....*Hemiodontichthys*

13b. Rostrum weakly pronounced; maxillary and fringed barbels conspicuous and gathered in series at the lip corners; premaxillary teeth present; abdomen covered by large plates organized in 2 rows.............................*Furcodontichthys*

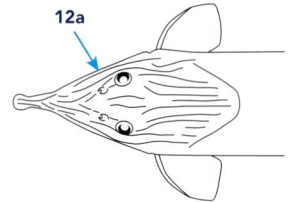

14a. Abdomen covered by small plates without particular organization; adults with pelvic-fin spine longer than last pelvic-fin rays; juveniles with a conspicuous basicaudal spot...*Pseudoloricaria*

14b. Abdomen covered by medium-sized plates, weakly structured in 2–3 rows; adults with last pelvic-fin rays longer than pelvic-fin spine; juveniles without a conspicuous basicaudal spot...*Limatulichthys*

15a. Lips papillose; fringed barbels of lower lip absent or inconspicuous **16**

15b. Lips generally filamentous or smooth; fringed barbels of lower lip generally conspicuous.................................... **18**

16a. Abdomen partially to completely covered by small to medium-sized contiguous plates... **17**

16b. Abdomen covered by very small plates, not contiguous; mouth circular and thick; postorbital notches weak (for comparison see deep notches in fig. 17a); teeth few (≈4 per premaxilla); body depth about 12% of SL; presence of a long whip on upper caudal-fin spine; predorsal keels strong ....*Spatuloricaria*

17a. Premaxillary teeth strong bicuspid, of same size as dentary teeth; postorbital notch deep (fig. 17a); predorsal keels more or less pronounced; caudal fin with i + 10 + i rays....................................*Rineloricaria*

17b. Premaxillary teeth much reduced in size when not missing; no postorbital notch; predorsal keels weak; caudal fin with i + 12 + i rays...*Fonchiiloricaria*

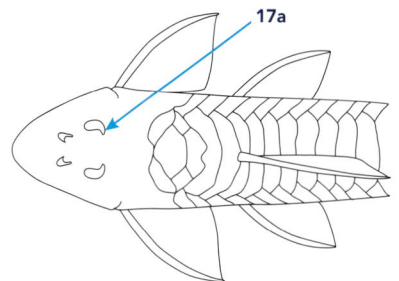

18a. Mouth opening without particular shape; predorsal keels most often strong; body generally weakly depressed ....*Loricaria*
18b. Mouth opening trapezoidal (fig. 18b); predorsal keels weak; body strongly depressed **19**

19a. Rostrum weakly pronounced or absent **20**
19b. Rostrum strongly pronounced (as in fig. 4a); premaxillary teeth absent; dentary teeth numerous (≈15 per dentary) and reduced in size; lips smooth; maxillary barbels short; abdominal cover complete, consisting of small plates without particular organization; throat covered ....*Reganella*

20a. Abdominal cover complete **21**
20b. Abdominal cover incomplete, most often consisting of a double median row of plates (fig. 20b) **22**

21a. Maxillary barbels inconspicuous, not reaching gill opening; teeth very difficult to observe, invisible in normally preserved specimens ....*Dentectus*
21b. Maxillary barbels conspicuous, reaching gill opening; teeth visible; head large; body strongly depressed ....*Pseudohemiodon*

22a. Premaxillary teeth present; head generally triangular; maxillary barbels often conspicuous, reaching beyond gill opening ....*Rhadinoloricaria*
22b. Premaxillary teeth absent; head rounded; eyes small; maxillary barbels conspicuous, reaching beyond pectoral-fin origin ....*Planiloricaria*

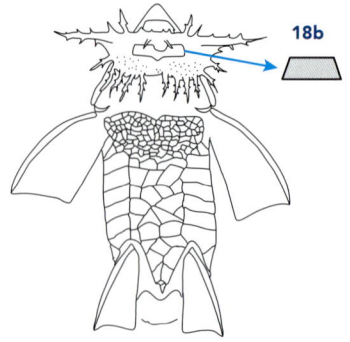

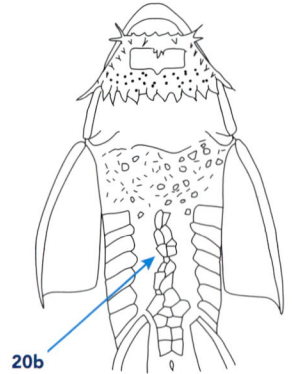

GENUS ACCOUNTS

Aposturisoma (20 cm SL)

Morphologically similar to *Farlowella*, with which it shares: numerous predorsal plates and dorsal-fin origin anterior to anal-fin origin. *Aposturisoma* differs from *Farlowella* in having a larger mouth, a much deeper and wider body, and a much thicker caudal peduncle, which can be interpreted as an adaptation to stream habitats (Covain and Fisch-Muller 2007). **SPECIES** One, *A. myriodon*. Species information in Isbrücker et al. (1983). **DISTRIBUTION AND HABITAT** Aguaytia basin, particularly in the Huacamayo River (a tributary of the Aguaytía River) in the western Amazon in Peru. Found in shallow clearwater rivers with a swift current, and a substrate consisting of stones, shingles, gravel, and sand (Isbrücker et al. 1983). **BIOLOGY** No data available.

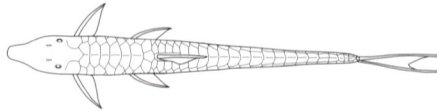

Cteniloricaria (13–20 cm SL)

Characterized by: abdomen completely covered with medium-sized rhombic plates, these plates becoming more numerous and decreasing in size toward the head; abdominal cover reaching gill opening, not organized in rows, and complete around 70 mm SL; presence of a black crescent in the caudal fin (Covain et al. 2012). **SPECIES** Two, both in the AOG region. Key to the species and species information in Covain et al. (2012). **DISTRIBUTION AND HABITAT** *C. napova* in the Paru de Oeste River, Suriname-Brazil border; and *C. platystoma* in Suriname and French Guiana. Found in main stream of lowland rivers, as well as in creeks and rapids where the vegetation is poor and water is clear and well-oxygenated. **BIOLOGY** No data available.

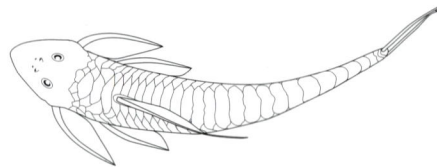

Dentectus (14 cm SL)

Characterized by: upper and lower lip with numerous long, filamentous barbels in several rather irregular rows; externo-lateral base of filaments of the upper lip and almost the entire margin of the maxillary barbel with dermal ossifications bearing odontodes; barbels along the margin of the lower lip reaching to the height of the branchiostegal membrane, usually these barbels have enlarged papillae laterally, whereas the other barbels (except for the maxillary barbel) are smooth; premaxillae very small, widely separated from each other and from the maxillary; up to 3 small, elongate teeth with expanded crowns present in each premaxilla; dentaries likewise widely separated from each other, distally including up to 3 teeth each, which teeth are similar to the ones in the premaxillary (Martín Salazar et al. 1982). The teeth and jaws are not visible without clearing and staining. *Dentectus* also has plates along the outer margin of the maxillary barbels, and a unique mouth structure. **SPECIES** One, *D. barbarmatus*. Species information in Martín Salazar et al. (1982). **DISTRIBUTION AND HABITAT** Western Orinoco basin, Venezuela. **BIOLOGY** No data available.

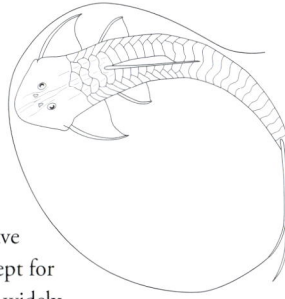

Farlowella (10–27 cm SL)

Characterized by a unique body shape that resembles a thin stick of wood. The body is slender and elongate, often with a pronounced rostrum and a brownish color with two lateral dark stripes beginning at the tip of the rostrum, passing over the eyes and ending at the tail, and regularly interrupted on the caudal peduncle (Covain and Fisch-Muller 2007). Sexual dimorphism includes hypertrophied odontodes along the sides of the rostrum or the head (in species with a short rostrum) in mature males. **SPECIES** 27, including 26 in the AOG region. Key and information to 25 valid species of *Farlowella* in revisionary study of Retzer and Page (1997). An additional species from the Beni River in Bolivia has been described by Retzer (2006), and a first trans-Andean species by Ballen and Mojica (2014). **COMMON NAMES** *Acarí, Acarí-cachimbo* (Brazil); *Pez palito* (Ecuador); *Shitari aguja* (Peru); *Agujeta* (Venezuela); Whiptail catfish (English). **DISTRIBUTION AND HABITAT** Broadly distributed in Amazon, Orinoco, Paraná, and coastal rivers of the Guiana Shield. Species distribution maps in Retzer and Page (1997). Absent from coastal rivers of the Brazilian shield. Species inhabit areas of gently flowing water in submerged dead leaves and sticks, among which it blends remarkably (Covain and Fisch-Muller 2007). Sometimes found in swift current over rocks and submerged wood. Uncommon in the wild, possibly due to their mimicry (Le Bail et al. 2000, Evers and Seidel 2005). **BIOLOGY** Species morphologically and behaviorally mimics sticks. Open-water brooders. Eggs laid on vertical surfaces such as submerged vegetation or rocks, in a single layer and guarded by the male (Covain and Fisch-Muller 2007).

Fonchiiloricaria (18 cm SL)

Distinguished from other genera of Loricariinae by usually possessing 1–3 premaxillary teeth (although these are often missing) that are much reduced in size, particularly in comparison to the dentary teeth (Rodriguez et al. 2011). Also differentiated from other Loricariinae by: lips with globular papillae on surface, except for some areas close to the opening of the mouth where the papillae are prolonged and digitiform; distal margin of lower lip with short, triangular filaments; premaxilla very reduced; abdomen covered by plates, medial plates small and rhombic between lateral abdominal plates; caudal fin with 14 total rays (12 branched); orbital notch absent; 5 lateral series of plates; dorsal-fin spinelet absent; preanal plate present, large and solid, and of irregular, polygonal shape; trunk and caudal peduncle becoming more compressed

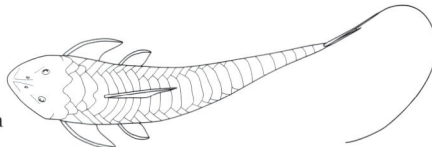

posteriorly for last 7–10 plates (Rodriguez et al. 2011). Sexual dimorphism includes the presence of weakly hypertrophied odontodes on sides of head (also in ventral view) and on dorsal surface of pectoral-fin rays, and a thickened unbranched pectoral-fin ray in males. **SPECIES** One *F. nanodon*. **DISTRIBUTION AND HABITAT** Huallaga River, Peru (600–700 m above sea level), in the vicinity of Tingo Maria, in the main stream of the río Huallaga and its tributaries, in swift current, over rocky substrata of stones, shingle, gravel, and sand (Rodriguez et al. 2011). **BIOLOGY** No data available.

Furcodontichthys (10 cm SL)

Characterized by the conspicuous fringed barbels at the lip corners, unique among the Loricariinae. These barbels have branching patterns comparable to those of the *Pseudohemiodon* group. **SPECIES** One, *F. novaesi*. Species information in Rapp Py-Daniel (1981). **DISTRIBUTION AND HABITAT** Middle Amazon at Lake Tefé and from the upper Juruá basin in the Solimões basin. *Furcodontichthys* inhabits sandy substrates. **BIOLOGY** Mature males of *Furcodontichthys* develop hypertrophied lips for brooding eggs. Eggs are laid in a mass and held by the male in the fold made by its lips, which they provide with ventilation during movement (J. Zuanon, pers. comm.).

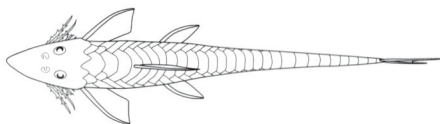

Harttia (6.0–18 cm SL)

Characterized by: abdomen partially to wholly covered by very small, granular plates without particular organization; abdominal plating sometimes restricted to preanal and lateral abdominal plates; large plates surrounding the anus; body large, flattened, and covered by very short odontodes giving a rather smooth aspect to the species; snout rounded; subpreopercle exposed; absence of keels on lateral plates, but lateral plates slightly keeled and coalescing toward the end of caudal peduncle; abrupt narrowing of caudal peduncle between the eighth and fourteenth postdorsal plates (Boeseman 1971, Covain et al. 2012, Oyakawa et al. 2013a). Sexual dimorphism includes hypertrophied odontodes on the pectoral spines and along the margins of the snout in mature males. **SPECIES** 23, including at least 5 species in the AOG region. *Harttia* is in need of revision. Partial keys are available for species occurring in Atlantic coastal drainages (Oyakawa 1993, Langeani et al. 2001), Amazon and Guianas drainages (Rapp Py-Daniel and Oliveira 2001), and Guianas drainages (Covain et al. 2012). **DISTRIBUTION AND HABITAT** Broad distribution in South America, with highest diversity in the rivers draining the Guiana Shield, coastal rivers in northeastern Brazil, and the Amazon basin. These rheophilic fishes are found in the upper courses of rivers over rocky and sandy bottoms. *Harttia* should be able to exploit areas with the strongest current, because of its extremely depressed body and long caudal peduncle, compared with other species (Casatti and Castro 2006). This proposition was also empirically noted by Le Bail et al. (2000). **BIOLOGY** Open-water brooders (Dotzer and Weiner 2003). *H. kronei* forages mainly at night and grazes on microscopic algae, mostly diatoms and green algae growing on rugged and light-colored rocks, and submersed vegetation (Buck and Sazima 1995). Feed also on chironomid and simuliid larvae, and microcrustaceans.

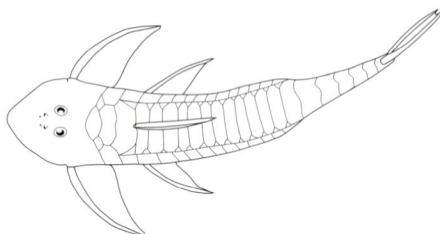

Harttiella (3.1–5.2 cm SL)

Characterized by: abdomen naked with exception of lateral abdominal plates and, rarely, preanal plates; small body sizes; body densely covered by odontodes; subpreopercle not exposed; lateral plates not keeled (Covain et al. 2012). Sexual dimorphism includes a larger head and thickened pectoral spines bearing hypertrophied odontodes in males. **SPECIES** Seven, all in the Guianas.

Key to the species and species information in Covain et al. (2012). **DISTRIBUTION AND HABITAT** Suriname and French Guiana. Each species has a very restricted geographic distribution. *Harttiella* species inhabit small creeks in mountainous areas where the water is cooler. Species distribution map in Covain et al. (2012). **BIOLOGY** Mol et al. (2007b) characterized the feeding habit of *H. crassicauda*. The species is detrivorous, feeding mostly on fine detritus, filamentous red algae, and epiphytic diatoms.

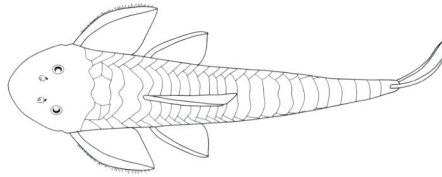

Hemiodontichthys (13 cm SL)

Characterized by: lower lip bilobate with a median furrow; surface of the lip more or less smooth or weakly papillose; a double abdominal keel; throat never covered; whip on upper caudal spine weak or absent; rostrum strongly pronounced; maxillary barbels short; premaxillary teeth absent; abdomen covered by large rectangular plates organized in 3 rows (Covain and Fisch-Muller 2007). Mature males develop a huge labial veil and teeth with spoon-shaped crown (vs. pointed ones in females and juveniles). They lack well-developed odontodes on the snout and pectoral fins (Le Bail et al. 2000). **SPECIES** One, *H. acipenserinus*. **DISTRIBUTION AND HABITAT** Widely distributed in the Amazon, Essequibo, Oyapock, and Paraguay basins (Bolivia, Brazil, French Guiana, Guyana, and Peru). Sand dweller that lives partially buried in the substrate, its cryptic coloration providing efficient protection. **BIOLOGY** Mature males develop hypertrophied lips for brooding eggs. Eggs are laid in a mass and held by the male in the fold of his lips, which they provide with ventilation during movement. About one week after hatching, the fry leave their parent's protection (Le Bail et al. 2000, Covain and Fisch-Muller 2007). *Hemiodontichthys acipenserinus* shows a remarkable feeding tactic: foraging alone at night and supported by its pectoral and pelvic fins, it projects its body forward and sinks the oral disk into the substrate, then the fish re-suspends and sucks the particles into the oral chamber (where particulate organic matter and small organisms are selected), expelling small clouds of sediments through the opercular openings (Brejão et al. 2013). This feeding tactic (digging) is also found in some cichlid species and unusual for loricariids that are generally grazers.

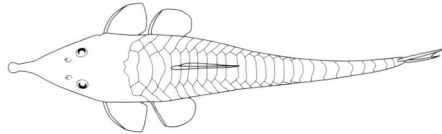

Lamontichthys (7.0–24 cm SL)

Distinguished from other loricariids in the AOG region by 7 branched rays in the pectoral-fin (vs. 6 in all others; Isbrücker and Nijssen 1978, Paixão and Toledo-Piza 2009). Sexual dimorphism includes hypertrophied odontodes on the pectoral spines in mature males. **SPECIES** Six, including five in the AOG region. Review of genus and key to the species in Paixão and Toledo-Piza (2009). **DISTRIBUTION AND HABITAT** *L. filamentosus* in the upper and middle Amazon basin; *L. llanero* in the Orinoco basin; *L. stibaros* in the western Amazon; *L. avacanoeiro* in the upper Tocantins basin; and *L. parakana* in the lower Tocantins basin. Live mainly in the mainstream of rivers, on rocky and sandy bottoms (Taphorn and Lilyestrom 1984b). **BIOLOGY** Open-water brooder, with eggs laid on the surface of rocks, submerged wood, or plants, and generally exposed to the current. Females lay a few large-sized (1.4–1.8 mm dia.) yellowish eggs during each spawning event (Taphorn and Lilyestrom 1984b).

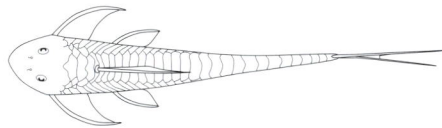

Limatulichthys (15–18 cm SL)

Similar to *Pseudoloricaria*, with *Limatulichthys* being distinguished by: abdomen covered by medium-sized plates weakly structured in 2–3 rows (vs. abdomen covered by small plates without any particular organization in *Pseudoloricaria*); last pelvic-fin ray longer than pelvic-fin spine (vs. shorter in *Pseudoloricaria*); and absence of a conspicuous basicaudal spot in juveniles (vs. present in *Pseudoloricaria*) (Isbrücker 1979, Covain and Fisch-Muller 2007). However, the difference in abdominal plate arrangement between *Limatulichthys* and *Pseudoloricaria* might not always be present (Londoño-Burbano et al. 2014). Male dimorphism related to the hypertrophy of odontodes on lateral portions of the head is lacking in both *Limatulichthys* and *Pseudoloricaria*. **SPECIES** Two, both in the AOG region. Notes on species differences in Londoño-Burbano et al. (2014). **DISTRIBUTION AND HABITAT** *L. punctatus* is widely distributed in the Amazon, Tocantins, Essequibo, western Orinoco, and Parnaiba basins; *L. nasarcus* from the middle Ventuari and lower Caura rivers in the Orinoco basin. Both species are sand dwellers. **BIOLOGY** Males develop enlarged lower lips to carry fertilized eggs, which remain attached to the lower lip until the larvae hatch.

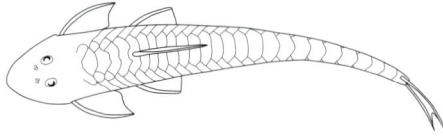

Loricaria (7.5–30 cm SL)

Readily distinguished from other loricariine genera by the elongate, slender filaments on the lips and a low number of bicuspid premaxillary teeth (usually 3–4 per side) that are about twice the length of the dentary teeth (Isbrücker 1979, Thomas and Rapp Py-Daniel 2008). Sexual dimorphism includes hypertrophied development of the pectoral spines, blunt odontodes on the pelvic and anal fin spines, and tooth crowns becoming shortened and rounded in mature males (Isbrücker 1979, 1981). **SPECIES** 17, including at least 6 species in the AOG region. See revision by Isbrücker (1981) who recognized 11 species (10 of which are still valid species), and 7 new species described subsequently; e.g., Thomas and Rapp Py-Daniel (2008) and Thomas and Pérez (2010). **DISTRIBUTION AND HABITAT** Distributed east of the Andes on nearly the entire subcontinent, including the Amazon, Orinoco, Paraguay-Paraná basin and smaller coastal rivers draining the Guiana and Brazilian shields. Species occur in a variety of habitats, from the main flow of rivers on sandy and rocky bottoms to flooded areas and lakes over muddy and sandy bottoms (Covain and Fisch-Muller 2007, Thomas and Rapp Py-Daniel 2008). **BIOLOGY** Lip brooders: males develop enlarged lower lips to carry fertilized eggs, which remain attached to the lower lip until the larvae hatch. Analyzed stomach contents included aquatic insect larvae, organic detritus, and sand (Thomas and Rapp Py-Daniel 2008).

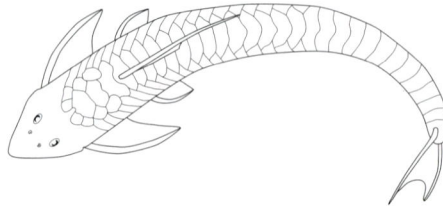

Loricariichthys (11–46 cm SL)

Identified by two unique characters of the upper and lower lips: (1) the upper lip completely fused to the premaxillary region medially, never having a free margin with barbel-like fringes along the transverse, medial portion; the fringes, when present, are restricted to the outer, more lateral portion of the lip; (2) the lower lip of immature males and females has two thick, cushion-like structures, covered with small papillae and with irregular fringes along the posterior edge (Reis and Pereira 2000). Additional diagnostic characters are discussed in Reis and Pereira (2000). Sexual dimorphism includes hypertrophied development of the lips in sexually mature males. **SPECIES** 17, including at least 8 species in the AOG region. A key to the species in southern South America is available in Reis and Pereira (2000). **DISTRIBUTION AND HABITAT** Widely distributed in the Amazon basin, the Paraná

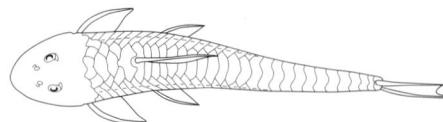

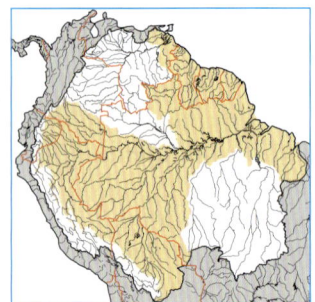

system, and coastal rivers of the Guiana and Brazilian shields. Species occur in a large diversity of habitats over sandy and muddy bottoms. **BIOLOGY** Lip brooders: males develop enlarged lower lips to carry fertilized eggs, which remain attached to the lower lip until the larvae hatch. Males of *L. castaneus* have also been recorded carrying larvae attached to the anterior ventral surface of the body (Gomes et al. 2011). Species are likely detritivorous-omnivorous (Ferreira et al. 2013).

Metaloricaria (27–30 cm SL)

Characterized by: reduction of the number of caudal-fin rays (i + 11 + i); low number of teeth (≈10 per premaxilla) and reduced tooth size; mouth shape horseshoe-like, with 3 buccal papillae, the lateral ones trilobate (Isbrücker and Nijssen 1982, Covain and Fisch-Muller 2007). Sexual dimorphism includes hypertrophied development of odontodes arranged in brushes along the sides of the head and on the spine and rays of the pectoral fins in mature males. Females also possess such brushes along sides of the head, but do not develop pectoral-fin enlarged odontodes (Covain and Fisch-Muller 2007). **SPECIES** Two, both in the Guianas. Revised by Isbrücker and Nijssen (1982), in which a key to the two species is available. **DISTRIBUTION AND HABITAT** Known only from French Guiana and Suriname. Species occupy an ecological niche similar to that of *Harttia* and inhabit primarily streams over rocky and sandy substrates (Le Bail et al. 2000). **BIOLOGY** No data available.

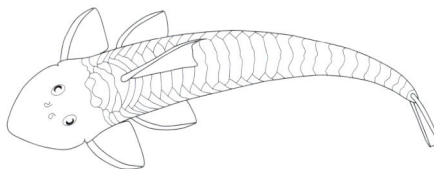

Planiloricaria (30 cm SL)

Characterized by: reduction in size and number of teeth; premaxillary teeth absent; a circular head shape; eyes reduced in size and without iris operculum. Evers and Seidel (2005) characterized sexual dimorphism by the shape of the genital area, with the genital area in males elongate and narrow compared with the large and roundish area of females. **SPECIES** One, *P. cryptodon*. **DISTRIBUTION AND HABITAT** Western Amazon basin: Ucayali, Purus, Madeira, and Mamoré basin (Bolivia, Brazil, and Peru). Species inhabits sandy substrates in the main streams of large rivers (Evers and Seidel 2005). **BIOLOGY** Reproductive ecology is unknown but could be reminiscent of those of other representatives of the *Pseudohemiodon* group. Species is frequently found in stomach contents of giant catfishes of the genus *Brachyplatystoma* (J. Zuanon pers. comm.).

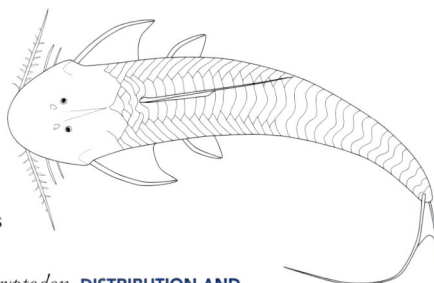

Pseudohemiodon (9–30 cm SL)

Characterized by: head large and triangular; rostrum weakly pronounced or absent; body strongly depressed; lips generally filamentous or smooth; fringed barbels of lower lip generally conspicuous; maxillary barbels conspicuous, reaching gill opening; mouth opening trapezoidal; teeth straight, bicuspid, spoon-shaped, not numerous (≤20 per premaxilla), strongly differentiated, sometimes reduced in size or absent; abdominal cover complete; predorsal keels weak; caudal fin with i + 10 + i rays; often with a more or less strong whip on upper caudal-fin spine (Covain and Fisch-Muller 2007). Sexual dimorphism is unknown. **SPECIES** Seven, including four species in the AOG region. A partial key to the species is available in Isbrücker (1975). **DISTRIBUTION AND HABITAT** Distributed in the Amazon, Orinoco, and Paraná basins. Like other members of the *Pseudohemiodon* group, *Pseudohemiodon* occurs primarily over sandy substrates. This ecological specialization is reflected in the dramatic

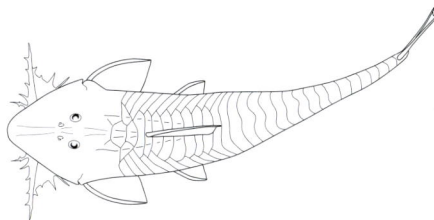

dorsoventral depression of the body and pelvic fins that are used mainly for locomotion on sand. **BIOLOGY** Species are abdomino-lip brooders; the very large eggs are incubated by the male (Covain and Fisch-Muller 2007).

Pseudoloricaria (31 cm SL)

Characterized by: lower lip bilobate with a median furrow; surface of this lip more or less smooth or weakly papillose; teeth straight, bicuspid, not numerous (≤20 per premaxilla); abdomen covered by small plates without particular organization; throat never covered; caudal fin with i + 10 + i rays; whip on upper caudal-fin spine weak or absent; adults with pelvic-fin spine longer than last pelvic-fin rays; juveniles with a conspicuous basicaudal spot (Covain and Fisch-Muller 2007). Similar to *Limatulichthys*, with *Pseudoloricaria* being distinguished by having the abdomen usually covered in small plates without any particular organization (vs. usually covered in medium-sized plates weakly structured in 2–3 rows); last pelvic-fin ray shorter than pelvic-fin spine (vs. longer) and a conspicuous basicaudal spot in juveniles (vs. absent). Sexual dimorphism includes hypertrophied development of the lower lip in males. **SPECIES** One, *P. laeviuscula*. Species information in Isbrücker and Nijssen (1976). **DISTRIBUTION AND HABITAT** Lower and middle Amazon basin, including Negro and Branco rivers. Species has been collected over sandy bottoms, in clear waters, in shallow beaches along the main river, and in neighboring temporary ponds (Covain and Fisch-Muller 2007). **BIOLOGY** Hypertrophied development of the lower lip in sexually mature males suggests that *Pseudoloricaria* is a lip brooder.

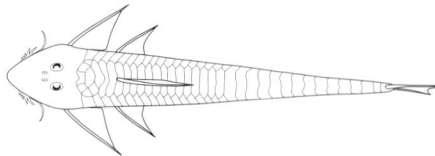

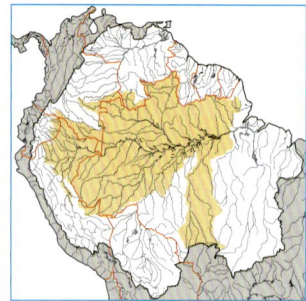

Pterosturisoma (16 cm SL)

Similar to *Lamontichthys* from which it differs primarily in the number of pectoral-fin rays (i + 6 in the former vs. i + 7 in the latter). These two genera share with *Sturisoma* a similar body depth at dorsal-fin origin, filamentous extensions on caudal-fin spines, and complete abdominal plate cover extending to the lower lip margin. Males do not possess hypertrophied odontodes either on the sides of head or on pectoral fin spines. Sexes distinguished by the width of a naked trapezoidal area framed by 4 bony plates in the genital region, broader in females and longer and narrower in males (Evers and Seidel 2005). **SPECIES** One, *P. microps*. **DISTRIBUTION AND HABITAT** Western Amazon in Peru. Rheophilic species frequently collected with bottom trawl nets in the main channel of whitewater rivers in Brazilian Amazon. **BIOLOGY** Open-water brooder (Evers and Seidel 2005).

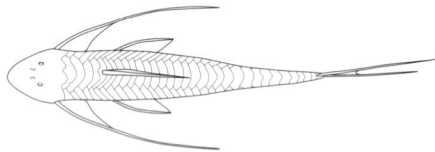

Reganella (11 cm SL)

Characterized by: body strongly depressed; rostrum strongly pronounced; mouth opening trapezoidal; premaxillary teeth absent; dentary teeth numerous (≈15 per dentary) and reduced in size; lips smooth; maxillary barbels short; fringed barbels of lower lip very short or absent; abdominal cover complete, consisting of little plates without particular organization; throat covered; predorsal keels weak; caudal fin with i + 10 + i rays; a basidorsal spot (Covain and Fisch-Muller 2007). **SPECIES** One, *R. depressa*. **DISTRIBUTION AND HABITAT** Negro, Branco, Xingu, Trombetas, and Tapajós basins, Brazil. The dorsoventrally flattened body suggests that *Reganella* inhabits flowing waters over sandy substrates. **BIOLOGY** No data available.

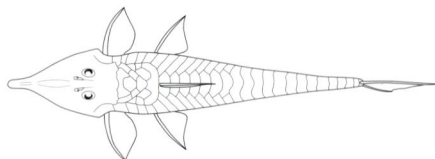

Rhadinoloricaria (14 cm SL)

Characterized by: body strongly depressed; lips strongly filamentous; maxillary barbels conspicuous, reaching beyond pectoral-fin origin; conspicuous fringed barbels on lower lip; mouth opening trapezoidal; teeth straight, bicuspid, spoon-shaped, not numerous (≤20 per premaxilla), strongly differentiated, sometimes reduced in size or absent; iris operculum generally present; predorsal keels weak; abdominal cover generally incomplete, most often consisting of a double median row of plates; caudal fin with i + 10 + i rays; often with a more or less strong whip on upper caudal-fin spine (Covain and Fisch-Muller 2007). Sexual dimorphism is apparent through differentiated lip structure. The lip surfaces of the male are rather papillose while those of the female are filamentous (Nijssen and Isbrücker 1988). Their body is strongly depressed and the pelvic fins are used for locomotion, enabling these fish to "walk" on the substrate. **SPECIES** Seven; a partial key to the species is available in Nijssen and Isbrücker (1988). Recent molecular works showed that *Apistoloricaria* and cis-Andean *Crossoloricaria* are members of *Rhadinoloricaria* (Covain et al. 2016). **DISTRIBUTION AND HABITAT** Upper Amazon and Orinoco basin; Essequibo and Tocantins drainages. Known to occur over sandy substrates. **BIOLOGY** Abdomino-lip brooders: eggs are laid in a single-layered mass, then attached to the surface of the lower lip and abdomen of the male. Stomach contents of *R. bahuaja* from the Madre de Dios basin in Peru included larvae of aquatic insects, small seeds, and debris (Chang and Castro 1999).

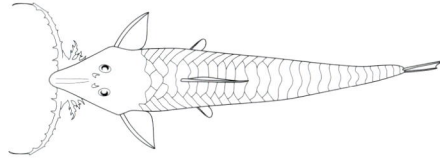

Rineloricaria (5.5–36 cm SL)

Characterized by: postorbital notch present; lower lip with short rounded papillae; premaxilla with 7–15 teeth on each series; dentary teeth strong, bicuspid; coloration of dorsal region with dark brown bars or blotches; abdomen with a conspicuous polygonal preanal plate, usually bordered by 3 other large trapezoidal plates (Fichberg and Chamon 2008). Some sexually dimorphic features of mature males also help identify *Rineloricaria*: numerous hypertrophied odontodes along the sides of the head and the dorsal surface of pectoral fin in some species (generally thick, short, and curved odontodes); and well-developed odontodes all over the predorsal area (generally thin, long, and erect or depressed odontodes) (Fichberg and Chamon 2008). **SPECIES** Most species-rich genus in Loricariinae, with 66 species, including at least 16 species in the AOG region. **DISTRIBUTION AND HABITAT** Widely distributed on nearly the entire subcontinent, from Costa Rica to Argentina, on both slopes of the Andes. The species inhabit an extremely diverse array of environments, including large rivers, streams, and lagoons; associated with sand or rock bottom, sometimes found in marginal vegetation. **BIOLOGY** Cavity brooders. Numerous eggs (often >100) are laid attached to one another in single layer masses on the cavity floor, and are brooded by males (Covain and Fisch-Muller 2007).

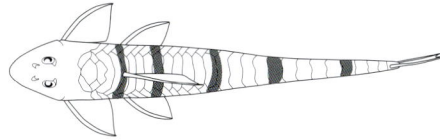

Spatuloricaria (11–52 cm SL)

Characterized by: abdomen covered by small not contiguous plates; mouth circular and thick, with lips strongly papillose; postorbital notch weak; predorsal keels strong; teeth few; body deep, a long and strong supracaudal whip (Covain and Fisch-Muller 2007). Sexual dimorphism includes hypertrophied development of claw-like odontodes along the sides of the head and on the pectoral spines in mature males. **SPECIES** 13, including at least 2 species in the AOG region. Genus is in need of revision, as species boundaries and distributions are poorly known (Isbrücker 1979, Fichberg et al. 2014).

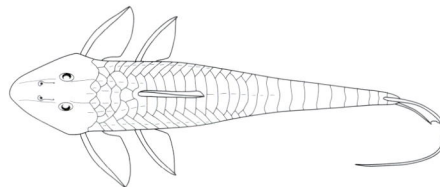

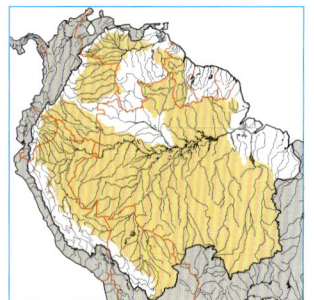

DISTRIBUTION AND HABITAT Distributed in northwestern South America, in drainages of the Pacific and Atlantic slopes of the Andes. *Spatuloricaria euacanthagenys* (52 cm SL) from the Caquetá River, Colombia, and *S. puganensis* (22 cm SL) from the Marañón River, Peru. Inhabit bottoms of medium- to large-sized rivers, in association with rocky or sandy substrates in fast-flowing sectors (Fichberg et al. 2014). **BIOLOGY** *S. evansii* from the Paraguay basin feeds on simuliid fly larvae (Rapp Py-Daniel and Py-Daniel 1984).

Sturisoma (13–28 cm SL)

Characterized by: rostrum present; mouth shape elliptical; teeth pedunculate, bicuspid, ≥10 per premaxilla, organized in a comb-like manner and weakly differentiated; abdominal cover complete and weakly structured in 2 to 3 rows; dorsal fin originating more or less in front of the pelvic-fin origin; caudal peduncle strongly depressed, elliptical in transverse section; caudal fin with i +12 + i rays; sometimes with filamentous extensions on upper and/or lower caudal-fin spines (Covain and Fisch-Muller 2007). Sexual dimorphism includes hypertrophied odontodes on the sides of the head of males. **SPECIES** 15, including 14 in the AOG region. **COMMON NAME** *Shitari* (Peru). **DISTRIBUTION AND HABITAT** Widely distributed in the Amazon, Orinoco, and Paraná basins. *Sturisoma* inhabits gently to swiftly flowing whitewaters where submerged wood is abundant in the main flow of rivers (Evers and Seidel 2005). Recent molecular results show *Sturisoma* is restricted to the cis-Andean region (Covain et al. 2016), with trans-Andean species being transferred to *Sturisomatichthys*. **BIOLOGY** Open-water brooders; male tends the clutch.

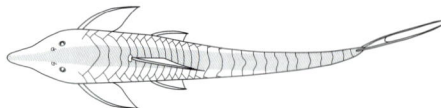

SUBFAMILY RHINELEPINAE—RHINELEPINE PLECOS
— *JONATHAN ARMBRUSTER, PETER VAN DER SLEEN, and NATHAN LUJAN*

Subfamily includes three genera: *Pogonopoma* with three species, *Pseudorinelepis* with one species, and *Rhinelepis* with two species (Armbruster 1998b, Lujan et al. 2015a). Five of the six species in Rhinelepinae are restricted to southeastern Brazil, northern Argentina, Paraguay, and Uruguay, with only one species (*Pseudorinelepis genibarbis*) distributed more broadly across the southern and western Amazon basin (Brazil and Peru).

Pseudorinelepis (35 cm SL)

Readily identified by: a round iris (versus a bilobed, omega-shaped iris observed in other large-bodied, sympatric loricariids); absence of an adipose fin; 5 branched anal-fin rays; a single medium-sized plate posterior to the compound pterotic (vs. many small plates in other rhinelepines and no plates in other loricariids); a patch of non-evertible, elongate odontodes on the cheek; well-developed ridges on the compound pterotic (vs. ridges not as well developed); and strongly keeled lateral plates with well-developed bony ridges and odontodes above and below the keels (Armbruster 1998b, Armbruster and Hardman 1999). Color patterns in *Pseudorinelepis* are variable and can change to match substrate. Males have longer odontodes on the cheek that are more dense and numerous than females; males often with orange patches on cheeks and on dorsal- and caudal-fin spines (Armbruster and Hardman 1999). **SPECIES** One, *P. genibarbis*. Species information in Armbruster and Hardman (1999). **COMMON NAMES** *Bodó sem costela* (Brazil), *Carachama sin costilla* (Peru). **DISTRIBUTION AND HABITAT** Amazon River and its major tributaries: Madeira and Negro rivers in Brazil and Marañón, Napo, and Ucayali rivers in Peru. Typically live in small sluggish streams, floodplain lakes, and large rivers. Often found on large submerged logs near the water surface (Armbruster and Hardman 1999). **BIOLOGY** Facultative air breather. *P. genibarbis* in aquaria exhibit more midwater behavior than other loricariids, and are able to maintain neutral buoyancy (Armbruster and Hardman 1999).

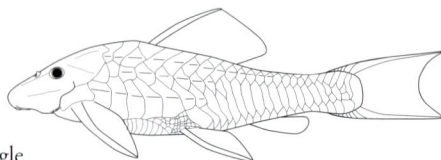

FAMILY PIMELODIDAE—LONG-WHISKERED CATFISHES

— *MARCELO SALLES ROCHA*

DIVERSITY 109 species in 30 genera; at least 59 species in 22 genera in the AOG region.

COMMON NAMES *Barbado, Cachara, Dourada, Filhote, Mapará, Mandubé, Mandi, Pintado, Piraíba, Piracatinga, Pirambucu, Piramutaba, Pirarara, Jaú, Surubim* (Brazil); *Cunchimama, Cunshi, Doncella, Maparate, Mota, Shiripira, Tigre zungaro, Zungarito, Zúngaro* (Peru); *Dormilón, Dorado, Paletón, Dorado, Valentón, Zamurito* (Venezuela).

GEOGRAPHIC DISTRIBUTION Distributed throughout most of the humid Neotropics, from Panama to La Plata basin, with highest diversity in the Amazon and Orinoco basins. Some species are endemic to the Magdalena, Maracaibo, and southeastern basins of Brazil.

ADULT SIZES From 20 cm SL in *Exallodontus aguanai* to >200 cm SL in *Brachyplatystoma filamentosum*, one of the largest freshwater fish species in South America (Goulding 1981, Barthem and Goulding 1997).

DIAGNOSIS OF FAMILY Distinguished from other Siluriformes by a dendritic arrangement of lateral line tubes in the skin of the snout, cheek, and nape; a uniquely shaped and elongate articulation of the lateral ethmoid and palatine bones; a bifurcate dorsolateral process on the premaxilla; infraorbital 1 contacting the lateral ethmoid by means of a cartilage; and a deep suture in the joint between fifth and sixth vertebral centra (Lundberg and Littmann 2003, Rocha 2012). Most pimelodids also have strong and serrated pectoral- and dorsal-fin spines; a long and fleshy adipose fin; 3 pairs of long barbels (2 pairs on the chin and 1 pair from the upper jaw); and complete lack of external bony plates. Pimelodids exhibit a very generalized catfish body form, with body colors ranging from uniform gray to elaborate patterns of stripes and spots. Juveniles of most species are miniature replicas of adults, although juveniles of many large-bodied species have elongate (filamentous) dorsal- and caudal-fin leading rays.

SEXUAL DIMORPHISM Scarcely developed. Pimelodids are externally fertilizing and not known to practice parental care (Lundberg and Littmann 2003).

HABITATS Pimelodids inhabit a wide range of environments, although most commonly found in major rivers and their larger tributaries. Pimelodids also inhabit high-gradient upland or mountain streams, but are rare or absent in small forest streams and stagnant swamps. Some species are abundant in natural lakes or those formed by construction of dams, like *Pimelodus, Hypophthalmus,* and *Pinirampus.* In the Amazon and Orinoco systems, pimelodid catfishes are found in all three waters types (i.e., white, clear, and black), with many species entering flooded forests and floodplain lakes during the high-water season.

FEEDING ECOLOGY Some pimelodids are carnivorous, and several larger-bodied species are the top predators of the Amazonian aquatic food web. The diet of some species includes fruits (*Pimelodus blochii* and *Phractocephalus hemioliopterus*), whereas others (*Pinirampus pirinampu, Pimelodus maculatus*) have an omnivorous diet. *Calophysus* is a flesh-biting scavenger equipped with incisiform teeth. *Hypophthalmus* are microphagous zooplanktivores. Most are benthic or bottom oriented, but *Hypophthalmus* and *Platynematichthys* are distinctly pelagic (Lundberg and Littmann 2003).

BEHAVIOR Some species of Pimelodidae, including *Brachyplatystoma, Platynematichthys, Pseudoplatystoma, Sorubimichthys,* and *Zungaro,* grow to large body sizes and undertake long breeding migrations. Barthem and Goulding (1997) showed that at the beginning of the high-water season, immature adults of *Brachyplatystoma rousseauxii* leave the lower Amazon River near the estuary and move toward headwaters in the western Amazon, a distance that can reach up to 5,500 km. Such long-distance migration is, however, uncertain for other species, although individuals of several species form large schools at certain times of the year and undertake short migrations for spawning during flood season.

ADDITIONAL NOTES Most of the larger species (e.g., *Brachyplatystoma* and *Pseudoplatystoma*) are important commercial fishes, although studies show overfishing of some populations in the Pantanal (Mateus and Penha 2007) and Amazon (Sant'Anna et al. 2014). Barthem and Goulding (1997) showed that 95% of the exploited catfishes in the Amazon basin are Pimelodidae with *B. rousseauxii* (the *dourada*) and *B. vaillantii* (the *piramutaba*) representing around 60% of the total catch by biomass.

KEY TO THE GENERA

1a. Bright red caudal fin; anterior and middle nuchal plates fused and expanded, forming a massive bilobed bone (fig. 1a); skull roof and nuchal plate almost completely ornamented by reticulating ridges and pits, with few or no elongate, parallel ridges and sulci .... *Phractocephalus*

1b. Caudal fin without red color; anterior and middle nuchal plates small, not fused (fig. 1b); skull roof and nuchal plate smooth or ornamented by ridges **2**

1a   1b

2a. Body laterally compressed; anal fin very elongate, >50% of body length; >50 very long gill rakers .... *Hypophthalmus*

2b. Body less laterally compressed to fairly depressed; anal fin short, occupying about a quarter or less of the body length; gill rakers generally short and few in number (exception: *Sorubim maniradii*) **3**

3a. Dorsal fin with 9–11 branched rays .... *Leiarius*

3b. Dorsal fin with 6 or 7 branched rays.. **4**

4a. Premaxilla much longer than dentary; premaxillary plate teeth exposed in ventral view **5**

4b. Premaxilla slightly longer than dentary; small portion of premaxillary teeth sometimes exposed in ventral view (except *Hemisorubim*).. **6**

5a. Body roughly circular in cross section; pigmentation pattern formed by small black spots and elongate dark spots on the body, particularly evident on the head; middle caudal-fin rays clear .... *Sorubimichthys*

5b. Body compressed laterally in the posterior portion; pigmentation pattern comprises a median longitudinal black stripe along the body extending onto the middle caudal-fin rays .... *Sorubim*

6a. Jaw prognathous, with dentary conspicuously longer than the premaxilla.. *Hemisorubim*

6b. Premaxilla longer than dentary with a subterminal mouth **7**

7a. First pectoral-fin ray flexible (pectoral spine absent)... **8**

7b. First pectoral-fin ray strongly ossified and pungent, transformed into pectoral spine.. **12**

8a. Premaxilla with 1 row of incisiform teeth (with an enlarged distal end), forming a cutting edge .... *Calophysus*

8b. Premaxilla with a patch of teeth of variable width.. **9**

9a. Snout narrow, tapered; mouth clearly subterminal; body gray with the pectoral and pelvic fins yellowish............ *Pimelodina*

9b. Snout broad; mouth terminal to slightly subterminal; body gray or brown; fins hyaline, sometimes with dark spots........... **10**

10a. Mouth approximately terminal; adipose fin very long, its distance from last dorsal-fin ray less than the length of dorsal-fin base; barbels flat in cross section; premaxilla with a median indentation in front view, forming a V with the vertex upward... *Pinirampus*

10b. Mouth subterminal; adipose fin short, its distance from last dorsal-fin ray equivalent to the dorsal-fin base length; barbels ovoid in section; premaxilla without median indentation, forming a steady arc in front view...................... **11**

11a. Upper lip narrow; presence of 2 dark, narrow spots at the base of the caudal-fin lobes (sometimes evident only in the upper lobe)... *Megalonema*

11b. Upper lip wide and fleshy; no dark spots on the base of the caudal-fin lobes.. *Aguarunichthys*

12a. Mouth narrow and fully ventral, with relatively fleshy lips; upper lip with a deep pocket on each side which extends medially almost to midline; teeth absent in adults; pectoral-fin spine thin and relatively flexible .... *Cheirocerus*

12b. Mouth wide and subterminal, with thin lips; no evident fold of skin at the corners of the mouth; teeth present in adults; pectoral-fin spine strong and well ossified... **13**

13a. Pectoral-fin spine short, about two-thirds the subsequent branched ray length; 2 tooth patches on vomer confluent at the midline .... *Duopalatinus*

13b. Pectoral-fin spine long, similar in size or slightly smaller than the subsequent branched ray; 2 tooth patches on vomer, when present, separated at midline... **14**

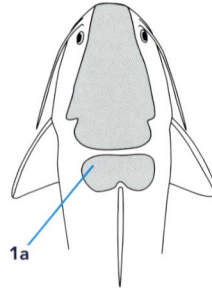

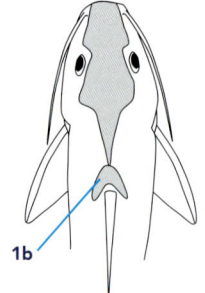

14a. Premaxillary teeth relatively large and tapered, arranged in 2–4 rows; adipose fin very long, covering almost entire length between the dorsal fin and the caudal peduncle **15**
14b. Premaxillary teeth arranged in several rows of narrow conical teeth; adipose fin relatively short and clearly separated from the posterior part of the dorsal fin **16**

15a. Eye small, >5 times in head length; premaxilla with 2 rows of strong teeth .... *Exallodontus*
15b. Eye large, <5 times in head length; premaxilla with 3–8 (irregular) rows of thin conical teeth .... *Propimelodus*

16a. Maxillary barbel very long and ossified (rigid) at its basal portion; ≥2 round black spots on the anterior part of body; a conspicuous black stripe on lower caudal-fin lobe **17**
16b. Maxillary barbel variable in length, flexible and without ossification; body color different from that mentioned above **18**

17a. Snout with rounded trapezoidal anterior margin, extending slightly ahead from the insertion point of the maxillary barbel (fig. 17a); premaxillary teeth plate partially exposed in ventral view.................................... *Platysilurus*
17b. Snout with triangular anterior margin, pointed, very long, extending much further from the insertion point of the maxillary barbel (fig. 17b); premaxillary teeth plate completely exposed, forming a large triangular area in ventral view .... *Platystomatichthys*

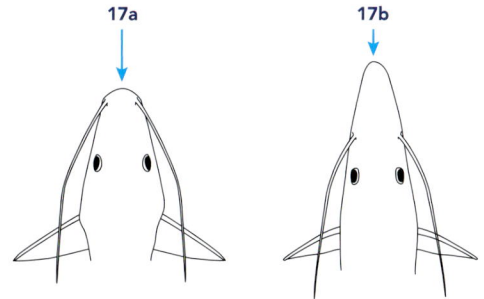

17a   17b

18a. Caudal fin emarginate; head short and wide; eyes small; interorbital space wide and flat; palatine tooth plate forming a narrow cross band with distal ends bent back; general body color gray to yellow, with numerous darker spots.................................. *Zungaro*
18b. Caudal fin usually forked; head relatively long; eye size and interorbital space variable; palatine tooth plate not forming a cross band; color pattern variable, but never as above... **19**

19a. Caudal fin weakly forked to emarginate with rounded lobes; body color gray or olive, with alternating narrow black vertical stripes with very narrow white bands or irregularly anastomosing to form various designs on the flanks; black spots on the head and lower half of the flanks; head flat with straight profile; snout strongly depressed...*Pseudoplatystoma*
19b. Caudal fin deeply forked with pointed lobes; body color uniform, or with stripes or spots; head not flat and with a straight to convex profile; snout tapered.. **20**

20a. Caudal fin shallowly to deeply forked with pointed lobes, sometimes with long filaments; premaxillary and palatine tooth plates well developed, with many slender teeth; supraoccipital process narrow and weakly attached to the nuchal plate (fig. 20a)...................................*Brachyplatystoma*
20b. Caudal fin deeply forked with pointed lobes but lacking filaments; premaxillary tooth plates not well developed, with few rows of slender teeth; palatine tooth plates not well developed when present, with only a few teeth; supraoccipital process wide and strongly attached to the nuchal plate (fig. 20b) **21**

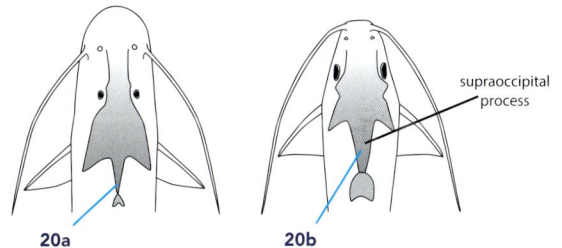

supraoccipital process

20a   20b

21a. Maxillary and mental barbels wide and flat in cross section; maxillary barbels short, not reaching the origin of the pelvic fin; first dorsal-fin ray with a long filament; body coloration formed by numerous dark spots on a gray background; caudal fin with the middle rays and part of the lower lobe blackened; adipose fin short, its length >3 times in the distance between dorsal fin and caudal peduncle .... *Platynematichthys*
21b. Maxillary and mental barbels ovoid in cross section; maxillary barbels long, surpassing the origin of anal fin and usually reaching the caudal peduncle; dorsal fin lacking a long filament; body coloration ranging from uniform to presence of spots and/or stripes; caudal fin hyaline or with dark dots; adipose fin medium to large, its length <3 times in the distance between the dorsal fin and the caudal peduncle .... *Pimelodus*

GENUS ACCOUNTS

Aguarunichthys (32–42 cm SL)

Characterized by: a distinctive spotted color pattern; robust body; ventrally positioned mouth; relatively long, conical, fleshy snout; premaxillary teeth in a broad band with posterolateral extensions; small eyes; and an adipose fin of medium length (Stewart 1986b). Also possesses a characteristic swim bladder with posterolateral margin with several finger-like projections. **SPECIES** Three, all in the AOG region. Key to the species in Zuanon et al. (1993). **COMMON NAMES** *Jundiá* (Brazil); *Achara* (Peru). **DISTRIBUTION AND HABITAT** *A. inpai* from the Madeira River and Solimões River in the region of Manaus and Manacapuru; *A. tocantinsensis* from the Tocantins River in the rapids above Marabá, Pará state; and *A. torosus* from the Cenepa basin in the Peruvian Amazon (Zuanon et al. 1993). Inhabit deeper portions of the main channel of large rivers (J. Zuanon pers. comm.). **BIOLOGY** Little is known about their biology. These species are rare in museum collections.

Brachyplatystoma (60–250 cm SL)

Large-bodied catfishes, with *B. filamentosum* being the largest Neotropical catfish species, reaching lengths of >250 cm SL. Also characterized by: strong pectoral-fin spines; small eyes; large tooth plates on premaxilla and palate; and long filaments of upper and lower lobes of the caudal fin in young and subadults (Lundberg and Akama 2005). **SPECIES** Seven, all in the AOG region, including the recently described *B. capapretum*, previously misidentified as *B. filamentosum* for almost 200 years (Lundberg and Akama 2005). Illustrations of species in Lundberg and Akama (2005). **COMMON NAMES** *Dourada, Piraíba, Filhote* (Brazil); *Zúngaro* (Peru); *Dorado, Valentón* (Venezuela). *Brachyplatystoma filamentosum*: *Piratinga* (Argentina, Brazil), *Piraíba, Filhote* (Brazil), *Blanco pobre* (Ecuador, Venezuela), *Saltón* (Peru), Kumakuma (English). *Brachyplatystoma juruense*: *Sorubim flamengo* (Brazil), *Zúngaro alianza* (Peru), Zebra catfish (English). **DISTRIBUTION AND HABITAT** Major rivers of Amazon and Orinoco basins and large rivers in the Guianas. *Brachyplatystoma filamentosum* and *B. vaillantii* also occur in the Parnaíba basin in northeastern Brazil. **BIOLOGY** *Brachyplatystoma rousseauxii* has one of the longest migrations of any freshwater fish in the world, spawning in headwater rivers of the Andean piedmont of Bolivia, Ecuador, Peru, and Colombia, and with a main nursery in the Marajó estuary close to the Atlantic Ocean. There is, however, some controversy regarding the possible occurrence of homing behavior (returning to spawn in the same locations where individuals were born) in this species. See Barthem and Goulding (1997) for more details. Juveniles of *Brachyplatystoma*, such as *B. juruense* and *B. tigrinum*, are commonly sold in the ornamental aquarium trade.

Calophysus (40 cm SL)

Similar to *Pinirampus pirinampu* with which it is sometimes collected; however, *C. macropterus* has spots along the sides of body (vs. absent in *P. pirinampu*); has maxillary barbels that are ovoid in cross section (vs. very flattened and large in *P. pirinampu*); and premaxillary teeth with sharp cusps arranged in a single row (vs. small villiform teeth arranged in a broad tooth plate in *P. pirinampu*). **SPECIES** One, *C. macropterus*. **COMMON NAMES** *Piracatinga* (Brazil); *Mota* (Peru); *Zamurito* (Venezuela). **DISTRIBUTION AND HABITAT** Major rivers of Amazon and Orinoco basins. **BIOLOGY** Consumes a variety of foods, sometimes scavenging bodies of dead animals. Fished in large numbers using bait made with the carcasses of large animals including caimans and dolphins, thereby adversely affecting local populations of these species (Brum et al. 2015).

Cheirocerus (13.8–21.5 cm SL)

Characterized by: a fully ventral mouth, with relatively fleshy lips; and an upper lip with a deep pocket on each side, extending medially almost to the midline. Stewart and Pavlik (1985) provided additional osteological and anatomical characters. **SPECIES** Three, two in the AOG region. **COMMON NAMES** *Mandi manteiga* (Brazil); *Bagre cunshi* (Peru). **DISTRIBUTION AND HABITAT** *C. eques* and *C. goeldii* from the Amazon basin; *C. abuelo* endemic to Maracaibo basin. **BIOLOGY** Diet includes small invertebrates (Stewart and Pavlik 1985).

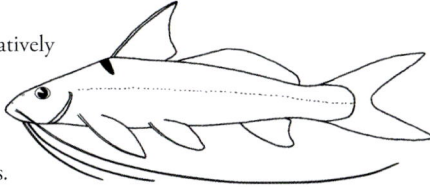

Duopalatinus (30–80 cm SL)

Characterized by: a premaxillary tooth plate with a posterolateral projection; two small tooth plates on vomer; one pair of tooth plates between the metapterygoid and vomer; and pectoral spine half the length of first branched pectoral-fin ray. **SPECIES** Two, one in the AOG region. **COMMON NAMES** *Mandi-açu* (Brazil); *Bagre cunshi* (Peru). **DISTRIBUTION AND HABITAT** *D. peruanus* has a wide distribution in major rivers in the Amazon and Orinoco basins, *D. emarginatus* inhabits the São Francisco basin (Lundberg and Parisi 2002, Parisi et al. 2006, Lundberg et al. 2011, Rocha 2012). **BIOLOGY** Little is known; some analyzed specimens had detritus and fragments of invertebrates in their stomach.

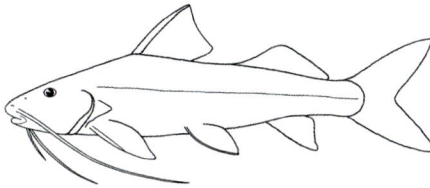

Exallodontus (20 cm SL)

Characterized by: a short head (~5.5 times in SL); long adipose fin; caudal fin deeply forked; and 2–3 rows of large teeth on premaxillary and dentary (Lundberg et al. 1991a). **SPECIES** One, *Exallodontus aguanai*. Species information in Lundberg et al. (1991a). **DISTRIBUTION AND HABITAT** Lower and middle Orinoco basin, in the mainstream of the Meta and Apure rivers, and in the Amazon mainstream from Iquitos (Peru) to Manaus (Brazil). Abundant in bottom trawls at about 3–35 m depth, commonly associated with other Pimelodidae species like *Propimelodus*. **BIOLOGY** Little information about diet and reproduction; some larger analyzed specimens had detritus and fragments of invertebrates in their stomachs.

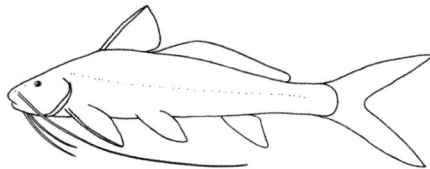

Hemisorubim (53 cm SL)

Characterized by: shape of the mouth, which presents a clearly prognathous lower jaw; conspicuous dark spots scattered on sides of body; and a dark spot at base of upper caudal-fin lobe. **SPECIES** One, *H. platyrhynchos.* **COMMON NAMES** *Mandubé pintadinho; Jurupoca* (Brazil); *Toa* (Peru); *Dormilón* (Venezuela); Porthole shovelnose catfish (English). **DISTRIBUTION AND HABITAT** Amazon, Maroni, Orinoco, Paraná, and Parnaíba basins. Found in whitewater (várzea) lakes of large tributaries and in main river channels. **BIOLOGY** Juveniles are common in the aquarium trade. Some specimens analyzed had small fishes in their stomachs.

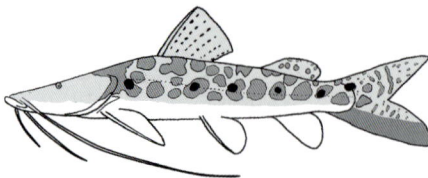

Hypophthalmus (50 cm SL)

Characterized by: laterally compressed body; eyes laterally on the head; absence of teeth; mouth wide; branchial aperture extending well forward on the ventral side of the head; very long and numerous gill rakers; anal fin long with 50–80 rays; and the complex structure of the swim bladder. **SPECIES** Four, with three species in the AOG: *H. edentatus, H. marginatus,* and *H. fimbriatus.* **COMMON NAMES** *Mapará* (Brazil); *Maparate* (Peru). **DISTRIBUTION AND HABITAT** Large rivers of the Amazon, Orinoco, and Paraná basins and some large rivers in Guyana, Suriname, and French Guiana. **BIOLOGY** Species are pelagic, using their long gill rakers to sieve planktonic crustaceans like cladocerans, copepods, and ostracods, following planktonic vertical movements during the daily cycle (Le Bail et al. 2000).

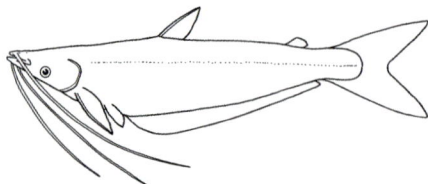

Leiarius (40–60 cm SL)

Large catfishes, easily recognizable by their marbled color pattern, and by a large dorsal fin with >10 rays. **SPECIES** Two, *L. marmoratus* and *L. pictus.* **COMMON NAMES** *Mandi de pedra, Jandiá* (Brazil); *Bagre achara* (Peru). **DISTRIBUTION AND HABITAT** *L. marmoratus* in the Amazon and Orinoco basins and *L. pictus* in Amazon, Orinoco, and Essequibo basins. Present, but uncommon in most Amazonian rivers, occurring in higher abundance in rivers with large rapids. **BIOLOGY** No data available.

Megalonema (10–30 cm SL)

Characterized by: a large mouth; medium-sized adipose fin; flexible spines in the dorsal and pectoral fins; a plain yellowish to tan coloration generally without prominent marks, except for an embedded dark spot in the base of the upper caudal-fin lobe, and sometimes also the lower lobe (Lundberg and Dahdul 2008). **SPECIES** Six, three species in the AOG region. **DISTRIBUTION AND HABITAT** *M. amaxanthum* in the Amazon basin, *M. orixanthum* in the Orinoco basin, and *M. platycephalum* in the Amazon, Essequibo, and Orinoco basins. Some species (*M. amaxanthum* and *M. orixanthum*) are found in the deep main channels of medium to large rivers, whereas *M. platycephalum* occurs in shallow sandy/muddy beaches (J. Zuanon pers. comm.). Not abundant in museum collections. **BIOLOGY** No data available.

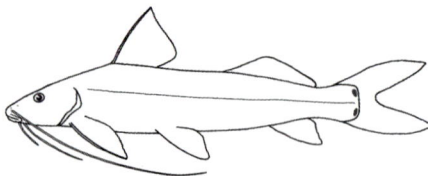

Phractocephalus (150 cm SL)

Readily identified by its color pattern, consisting of a brownish to dark gray back, with many dark spots, especially on the anterior half of the body, yellow sides, and characteristic orange-red dorsal and caudal fins. Also recognized by a large and thick, kidney-shaped predorsal plate and a large flat head with a rough surface. **SPECIES** One, *P. hemioliopterus*. **COMMON NAMES** *Pirarara* (Brazil); *Bagre cajano, Peje torre* (Peru); Red-tail catfish (English). **DISTRIBUTION AND HABITAT** Widely distributed in the Amazon and Orinoco basins. Found in large rivers, flooded forest, and várzea lakes. One of the few species of giant catfishes frequently found in flooded forests. **BIOLOGY** Consumes a large variety of food types, including fish, crabs, and fruits. Juveniles use floating macrophytes as a nursery and are commonly found in the aquarium trade. *Phractocephalus* has been interbred with *Pseudoplatystoma* to produce hybrids (*Cachapira*) for aquaculture purposes in Brazil

Pimelodina (40 cm SL)

Characterized by: a narrow snout; fully ventral mouth with thick lips; a long adipose fin; long barbels; and a bright greenish to olive color in life with spots over the body. The stomach is very small and muscular and apparently used to grind food. **SPECIES** One, *P. flavipinnis*. Species information in Stewart (1986b). **DISTRIBUTION AND HABITAT** Widespread in the Amazon and Orinoco basins, in the lower reaches of large tributaries such as the Negro, Madeira, and Purus (Stewart 1986b; pers. obs.). **BIOLOGY** Consumes aquatic insects and seeds (Stewart 1986b).

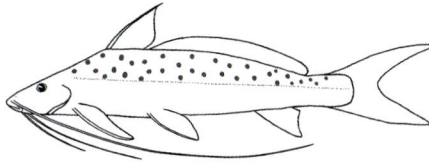

Pimelodus (10–40 cm SL)

Characterized by: its strong pectoral and dorsal-fin spines; adipose fin medium to large; dorsal surface of head rugose; and supraoccipital process triangular, strong and firmly attached to nuchal plate. **SPECIES** The most species-rich genus of Pimelodidae, with 33 species, 11 of which occur in the AOG (Lundberg et al. 2011, Rocha 2012). The most common species in the AOG region are *P. altissimus*, *P. blochii*, *P. ornatus*, and *P. pictus*. **COMMON NAMES** *Mandi amarelo, Mandi pintado* (Brazil); *Cunshi* (Peru); Pictus cat (English). **DISTRIBUTION AND HABITAT** The broadest distribution of the family, and occur from Panama (Tuira and Diablo basins) to Argentina (La Plata basin). Typically inhabit large rivers and lakes. **BIOLOGY** Omnivorous and opportunistic feeders. During the high-water season in the Amazon, *P. blochii* enters the flooded forest and feeds on fruits and seeds. Some species, especially *P. pictus* and *P. ornatus*, are common in the aquarium trade because of their attractive color patterns.

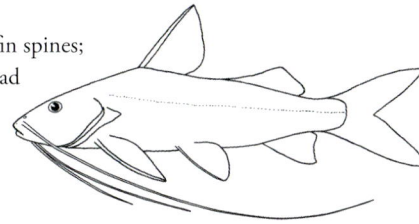

Pinirampus (75 cm SL)

Characterized by: a uniform silvery gray coloration; flat maxillary barbels; long adipose fin; dorsal fin with long filament; and premaxilla with a median indentation in front view, forming a V with the vertex upward. **SPECIES** One, *P. pirinampu*. **COMMON NAMES** *Barbado, Barbado branco* (Brazil); *Bagre mota blanca* (Peru); *Barba-chata* (Venezuela); Flat-whiskered catfish (English). **DISTRIBUTION AND HABITAT** A broad distribution in tropical South America, including the Amazon, Essequibo,

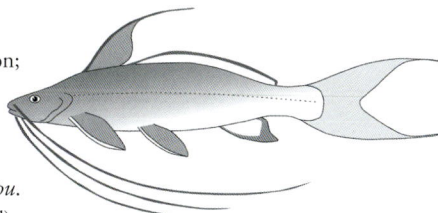

Orinoco, and Paraná basins, occurring in many environments. **BIOLOGY** *P. pirinampu* has an omnivorous diet and is migratory, moving upstream during the dry season to spawn at the onset of the rainy season (Dias et al. 2004, Peixer et al. 2006). Adults and juveniles drift downstream after spawning, using flooded areas or lakes for foraging and shelter against predation. *Pinirampus pirinampu* is an important commercial food fish in thwe Amazon, and is the target of sport fishers for its tasty meat, relatively large size, and fighting behavior. Reaches sexual maturity at ~60 cm SL.

Platynematichthys (80 cm SL)

Similar to *Brachyplatystoma*, from which it can be distinguished by a distinct color pattern with a black band on the lower lobe of the caudal fin; flat and wide maxillary barbels; and pelvic fins inserted posterior to the end of dorsal fin. A dorsal fin with a first filamentous ray is also a distinctive feature of this species. **SPECIES** One, *P. notatus.* **COMMON NAMES** *Coroatá, Pirá-tucandira* (Brazil); Lince catfish (English). **DISTRIBUTION AND HABITAT** Main river channel of the Amazon and Orinoco rivers and their major tributaries. **BIOLOGY** Pelagic and moderately common in the large Amazonian rivers.

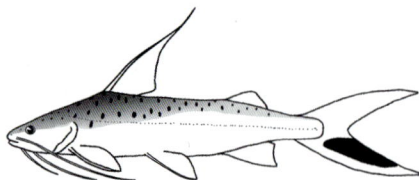

Platysilurus (20–40 cm SL)

Distinguished from other Pimelodidae by a very flat and strong head; snout protruding slightly over the lower jaw; ossified basal portion of the maxillary barbel; long maxillary barbel that surpasses the origin of anal fin; spots along the sides of the body; and a groove along the supraoccipital process. **SPECIES** Two, one in the AOG (*P. mucosus*). **COMMON NAMES** *Piramutaba* (Brazil); *Zungarito barbatus* (Peru). **DISTRIBUTION AND HABITAT** *P. mucosus* inhabits the main channel of large rivers in the Amazon and Orinoco basins; *P. malarmo* is endemic to the Maracaibo system. **BIOLOGY** Little is known; some specimens examined had small fishes in their stomachs. The species are usually sampled with bottom trawl nets and with hook-and-line in the river channel.

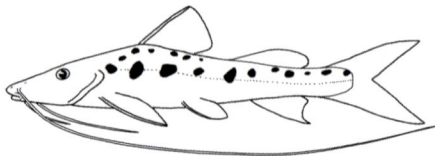

Platystomatichthys (40 cm SL)

Characterized by: a peculiar snout with a prolongation of the upper jaw exposing a triangular area of small teeth; long ossified maxillary barbels; and grooves along the supraoccipital process (also in *Platysilurus*). **SPECIES** One, *P. sturio.* **COMMON NAMES** *Pira-peuaua, Surubim mena* (Brazil); *Bagre sturion* (Peru). **DISTRIBUTION AND HABITAT** Main channels of large rivers in the Amazon basin. **BIOLOGY** Some analyzed specimens had small fishes in their stomachs. This species (like *Platysilurus*) is usually caught with bottom trawl nets and with hook-and-line in the river channel.

Propimelodus (25 cm SL)

Characterized by: a long adipose fin; strong pectoral and dorsal spines; relatively high back; long barbels, with the maxillary barbel extending beyond the caudal-fin base; mouth ventral; body sides without conspicuous spots or stripes; and caudal fin deeply forked with slender, pointed lobes. **SPECIES** Three, all in the AOG region: *P. araguayae, P. caesius,* and *P. eigenmanni* (Rocha et al. 2007). **DISTRIBUTION AND HABITAT** Widely distributed in large Amazonian rivers. *Propimelodus caesius* has the largest

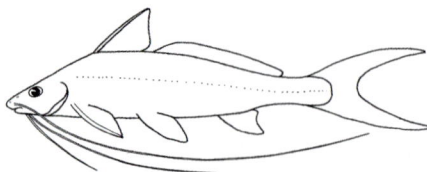

geographic range in the AOG, and has been collected from Peru to Brazil (Belém) in the Amazon mainstream and the lower reaches of its main tributaries. *Propimelodus eigenmanni* is distributed in the Amazon basin between the lower Madeira and Belém (close to the Amazon River estuary), as well as in some rivers in the Guianas. *Propimelodus araguayae* is endemic to the Araguaia River (Rocha et al. 2007). This genus together with *Exallodontus* are common species sampled by trawl nets in deep river channels. **BIOLOGY** Omnivorous.

Pseudoplatystoma (115 cm SL)

Large catfishes with snout and head strongly depressed; superior maxilla nonprominent; premaxillary teeth minute; vomero-palatine teeth in 4 patches (Buitrago-Suarez and Burr 2007). **SPECIES** Eight, with seven in the AOG region. Key to the species in Buitrago-Suarez and Burr (2007). **COMMON NAMES** *Cachara, Pintado, Pirambucu, Surubim tigre* (Brazil); *Bagre rayado* (Ecuador); *Doncella, Tigre zúngaro* (Peru); *Matafraile* (Venezuela); Tiger shovelnose catfish, Tiger sorubim (English). **DISTRIBUTION AND HABITAT** Throughout the Amazon and Orinoco basins, as well as coastal rivers in the Guianas. Adults typically occur in large rivers and lakes, while juveniles inhabit floating meadows of várzea lakes. **BIOLOGY** An important food resource; juveniles used as ornamental fish. Some aspects of its migration patterns were published by Goulding (1980).

Sorubim (25–42 cm SL)

Readily distinguished by: a very depressed head; long upper jaw that greatly projects over lower jaw; premaxillary tooth patch covered by minute, villiform teeth and exposed ventrally; eyes positioned laterally, usually visible from below; and black lateral stripe of highly variable width from snout to distal tip of median rays on caudal fin (Littmann 2007). **SPECIES** Five, including four species in the AOG region. Key to the species in Littmann (2007). **COMMON NAMES** *Bico de pato* (Brazil); *Paletón* (Colombia, Venezuela); *Shiripira* (Peru); Duckbill catfish, Lima shovelnose catfish (English). **DISTRIBUTION AND HABITAT** Amazon, Cauca, Essequibo, Lake Maracaibo, Magdalena, Orinoco, Paraná, and Sinú basins and in the Parnaíba basin in northeastern Brazil. Within the Amazon and Orinoco basins, individuals of *Sorubim* are more abundant in whitewater rivers and lakes. Juveniles commonly found amid aquatic macrophytes in várzea lakes. **BIOLOGY** A study by Goulding and Ferreira (1984) showed that shrimps formed the major part of the diet of *S. lima* and concluded that the greatly elongate upper jaw is an adaptation for pressing and holding prey against a substrate. More information about biology in Goulding (1980), Goulding and Ferreira (1984), and Littmann (2007).

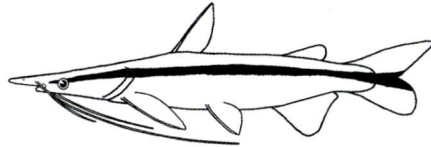

Sorubimichthys (150 cm SL)

Characterized by: large adult body size; distinctive body shape, with posterior body region round in cross section; very flat head; wide snout, premaxillary tooth plate exposed when mouth is closed; and a distinctive color pattern consisting of back and upper sides gray with scattered darker spots, lower portion of sides and ventral region mostly white with sparsely scattered dark spots. **SPECIES** One, *S. planiceps*. Lundberg et al. (1989) reviewed the taxonomy and ontogenetic development of *S. planiceps*. **COMMON NAMES** *Pirauacá; Surubim chicote; Surubim lenha* (Brazil); *Achacubo* (Peru); Firewood catfish (English). **DISTRIBUTION AND HABITAT** Amazon and Orinoco basins, where it typically inhabits shallow beaches of whitewater rivers. **BIOLOGY** Juveniles show remarkable ontogenetic changes in coloration and body morphology (Lundberg et al. 1989).

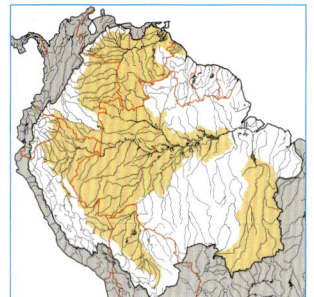

Zungaro (140 cm SL)

One of the largest Neotropical catfishes, weighing up to 50 kg. Mouth large; head massive and almost square-shaped (short and wide) in dorsal view; adipose fin short; strong dorsal and pectoral spines; large tooth plates on the palate. Color varies from light to dark gray with numerous dark spots and covered with a bright yellow mucus. **SPECIES** Two, including one in the AOG region (*Z. zungaro*). **COMMON NAMES** *Jaú* (Brazil), *Cunchimama* (Peru). **DISTRIBUTION AND HABITAT** *Z. zungaro* inhabits large tributaries of the Amazon and Orinoco basins. Often found in deep portions along the rivers, especially downstream of waterfalls and rapids. Juveniles frequently found in aquatic herbaceous banks. *Zungaro jahu* inhabits the Paraná-Paraguay basin. **BIOLOGY** Feeds mainly on fish. *Zungaro* has a high importance in regional commercial fisheries.

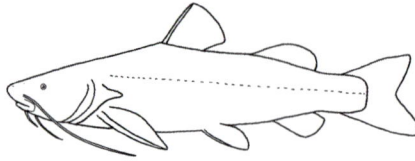

FAMILY PSEUDOPIMELODIDAE—BUMBLEBEE CATFISHES, DWARF-MARBLED CATFISHES

— OSCAR A. SHIBATTA and PETER VAN DER SLEEN

DIVERSITY 39 species in 6 genera, including 16 species in 4 genera in the AOG region.

COMMON NAMES *Bagrinho marmoreado* (Brazil); *Zungarito marmóreo* (Peru).

GEOGRAPHIC DISTRIBUTION Widely distributed in South America, with two genera occurring in Panama.

ADULT SIZES From 2.0 to 3.0 cm in some *Microglanis* species to ≤80 cm in *Batrochoglanis acanthochiroides* in the Catatumbo basin of Colombia draining into Lake Maracaibo.

DIAGNOSIS OF FAMILY Species are characterized by a wide mouth; small eyes without a free orbital margin; pectoral spine serrated anteriorly and posteriorly; and short barbels (Shibatta 2003a, b). Some species have beautiful patterns and coloration.

SEXUAL DIMORPHISM Not pronounced.

HABITATS All species are adapted to a benthic lifestyle.

FEEDING ECOLOGY Omnivorous to carnivorous.

KEY TO THE GENERA

1a. Premaxillary dental plate with rounded lateral margin; lateral line incomplete, not surpassing the vertical through beginning of the adipose fin; small-sized adults, SL <80 mm .... *Microglanis*

1b. Premaxillary dental plate with posteriorly pointed lateral margin; lateral line complete or almost complete; adults >80 mm SL 2

2a. Pectoral fin rays ≥7; caudal fin forked .... *Pseudopimelodus*

2b. Pectoral fin rays ≤6; caudal fin rounded or emarginated 3

3a. Mouth isognathous, lower and upper jaws equal; caudal fin rounded or emarginated; lateral line almost complete, not reaching the base of caudal fin .... *Batrochoglanis*

3b. Mouth prognathous, lower jaw slightly longer than upper; caudal fin rounded; lateral line complete .... *Cephalosilurus*

GENUS ACCOUNTS

Batrochoglanis (14–80 cm SL)

Characterized by: body shape rounded and wider than deep; a large head (rounded in dorsal view); almost complete lateral line; terminal mouth; short caudal peduncle, with the procurrent rays of the caudal fin close to the adipose and anal fins; emarginated caudal fin with rounded lobes, or completely rounded. Shares with *Microglanis* the following characters: pectoral fin spine with greatly enlarged anterior and posterior dentitions; absence of an axillary pore; and

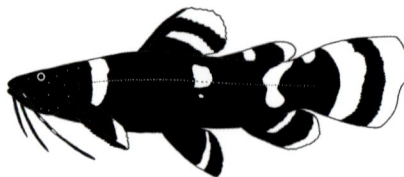

the first dark blotch broad, extending from the supraoccipital area to the end of the base of the dorsal fin (Shibatta 2003b). **SPECIES** Five, including two in the AOG region: *B. raninus* and *B. villosus*. *Batrochoglanis raninus* (20 cm SL) has a dark body color with several lighter blotches (as illustrated), and *B. villosus* (12 cm SL) is completely dark gray-brown, more or less mottled and lacks clear markings. **DISTRIBUTION AND HABITAT** *B. raninus* inhabits the Amazon basin and Guianas, and *B. villosus* is distributed throughout much of the AOG region. Map of species distributions in Shibatta and Pavanelli (2005). Often found in creeks and rivers with low currents. **BIOLOGY** *Batrochoglanis raninus* has been studied in French Guiana. It lies hidden under branches or rocks during the day and stalk-hunts at night, waiting for prey to pass. Juveniles start feeding on microcrustaceans and aquatic insect larvae, then shift diet to fishes of proportionally large size (Le Bail et al. 2000).

Cephalosilurus (8.5–35 cm SL)

Characterized by: depressed body, <26.2% of standard length; ramified gill rakers; a dark caudal fin, at least in the initial phase of life (Shibatta 2003b). Mostly dark-colored with many blotches and small streaks, except *C. apurensis* that is gray-brown mottled with black dots. **SPECIES** Four, including three species in the AOG region. **DISTRIBUTION AND HABITAT** *C. albomarginatus* (8.5 cm) in Guyana; *C. nigricaudus* (35 cm) in Suriname; and *C. apurensis* in the Arichuna basin, Apure state, Venezuela. *Cephalosilurus albomarginatus* occurs over sand and gravel partially covered with mud, leaves, and dead wood, and *C. nigricaudus* has been collected in pools and in the backwater of rapids and fast streams with sand, rocks, and decaying wood on the bottom (Le Bail et al. 2000). **BIOLOGY** Diet includes fish.

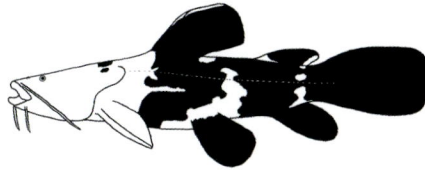

Microglanis (2.0–8.0 cm SL)

Similar to *Pseudopimelodus*, but with smaller adult body sizes, rarely exceeding 8 cm SL. Also characterized by: a wide mouth (gape equal to head width); short barbels (not surpassing the pectoral-fin base); small eyes, without a free orbital margin; incomplete lateral line; body with large dark brown blotches; axillary pore absent; premaxillary tooth plate with a rounded margin and thin mesocoracoid arch (Mattos et al. 2013, Shibatta 2014). Caudal fin variable in shape (emarginated, rounded, or forked). Species are often beautifully colored, with an overall orange to brown color and a characteristic light band running across the nape, and alternate light and dark blotches over the body. **SPECIES** 24, including 9 species in the AOG region. Key to the Amazonian species in Ruiz and Shibatta (2011), except for *M. lundbergi* (Jarduli and Shibatta 2013) and *M. maculatus* (Shibatta 2014). **DISTRIBUTION AND HABITAT** Amazon and Orinoco basins, Guianas, Paraná-Paraguay basin, and coastal rivers in southern Brazil. Inhabit calm water stretches of rivers, under rocks, submerged woods, and floating vegetation during the day. At least one species (*M. robustus*) has been found in rapids (Ruiz and Shibatta 2010). **BIOLOGY** Nocturnal foragers that consume algae and insect larvae (Alcaraz et al. 2008). Parental care has been reported for *M. iheringi* (Winemiller 1989b), but not for other species (Esguicero and Arcifa 2010).

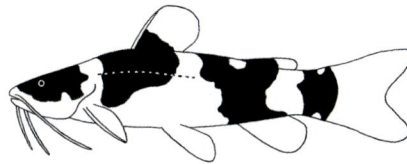

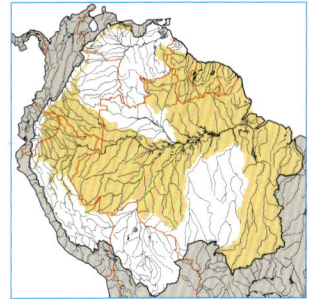

Pseudopimelodus (9.0–25 cm SL)

Characterized by: lateral line complete (but usually inconspicuous); caudal fin forked; vomer present; pectoral-fin spine covered by thick skin; pectoral-fin rays 7–8; color dark brown to blackish, with a pattern of pale, light brown or white bars and streaks, most distinctive in small specimens; fins often largely black, usually with white bands. **SPECIES** Five, including two species in the AOG region. **COMMON NAMES** *Jaú sapo* (Brazil); *Zungarito pulcher* (Peru). **DISTRIBUTION AND HABITAT** *P. bufonis* inhabits the Amazon and Orinoco basins and Guianas, and *P. pulcher* the western Amazon basin of Brazil and Ecuador. **BIOLOGY** Species are carnivorous, feeding on small aquatic animals (Galvis et al. 1997).

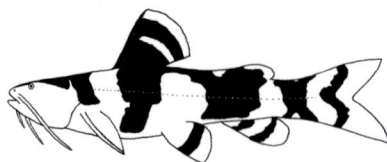

FAMILY SCOLOPLACIDAE—SPINY DWARF CATFISHES

— *PETER VAN DER SLEEN and MARCELO SALLES ROCHA*

Family includes a single genus with six species, including five species in the AOG region.

Scoloplax (1.0–2.0 cm SL)

Distinguished from other Siluriformes by very small body sizes and a conspicuous rostral plate bearing numerous large and curved odontodes (tooth-like structures). Also characterized by: two bilateral series of odontode-bearing plates and one midventral series of plates along the body; odontodes are also found on many other parts of the body; dorsal fin with a strong spine and 3–5 soft rays; anal fin with 5–6 soft rays; adipose fin absent (Schaefer et al. 1989). **SPECIES** Six, including five species in the AOG region. The genus was reviewed by Schaefer et al. (1989). Two new species have been described since (Rocha et al. 2008a, 2012). **COMMON NAMES** *Bagre anão espinhoso* (Brazil); *Bagre enano espinoso* (Peru). **DISTRIBUTION AND HABITAT** Inhabit terra firme streams, oxbow lakes, and backwater pools with stagnant water in the Amazon (Brazil, Bolivia, Peru, and Colombia) and Paraná-Paraguay basins. See species distribution map in Rocha et al. (2008a) and Rocha et al. (2012). *Scoloplax baileyi* inhabits Negro basin; *S. baskini* small tributaries of the Aripuanã River in the Madeira basin; *S. dicra* throughout the Amazon and Paraguay basins; *S. distolothrix* the Tocantins-Araguaia, Xingu, and Paraguay basins; and *S. dolicholophia* the Negro basin and Lake Amanã in the central Amazon, Brazil. **BIOLOGY** *Scoloplax* has an enlarged air-filled stomach, which is used to take up atmospheric oxygen (Armbruster 1998a). They hide during the day in plant debris, root tangles, or water plants and feed at twilight and night, during which they mostly sit and wait until something edible passes by (Sazima et al. 2000). Diet includes insect larvae and tiny worms. When disturbed they bury themselves in the sand, diving quickly headfirst and making lateral movements with body (observed for *S. empousa*) (Sazima et al. 2000). Species are likely inseminating (Spadella et al. 2008).

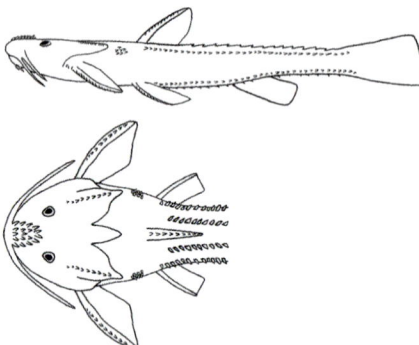

FAMILY TRICHOMYCTERIDAE—PENCIL CATFISHES, TORRENT CATFISHES, AND PARASITIC CATFISHES (CANDIRÚS)

— *LUIS FERNÁNDEZ*

DIVERSITY More than 240 species in 41 genera and 8 subfamilies, including 55 species in 24 genera and 6 subfamilies in the AOG region: Trichomycterinae, Sarcoglanidinae, Glanapteryginae, Tridentinae, Stegophilinae, and Vandelliinae (De Pinna and Wosiacki 2003, Ferraris Jr. 2007). The taxonomy of this great diversity is still poorly understood, with a majority of species assigned to the artificial subfamily Trichomycterinae and the artificial genus *Trichomycterus*.

COMMON NAMES *Candirú* (Brazil, Peru); *Canero* (Ecuador; Peru).

TAXONOMIC NOTE Trichomycteridae was first recognized as a family by Gill (1872), but Eigenmann and Eigenmann (1888, 1890) and Eigenmann (1918b) used the name Pygidiidae. Later, Tchernavin (1944) proposed maintaining the name Trichomycteridae rather than Pygidiidae. In the past, the Trichomycteridae included Cetopsinae, Nematogenyinae, and *Phreatobius* (Eigenmann 1918b, Myers and Weitzman 1966, De Pinna 1998).

GEOGRAPHIC DISTRIBUTION Trichomycterids inhabit freshwater systems throughout most of South America and southern Central America, from Costa Rica and Panama (10°N with *Trichomycterus*) to Patagonia (47°S with *Hatcheria* as the southernmost record) in Argentina and Chile, and within that range of latitudes, from the Atlantic forests of eastern Brazil to high-elevation (4,000 m) streams of the Andean Cordilleras. Only the subfamily Trichomycterinae is present throughout this whole range, and this subfamily includes the only representatives west of the Andes.

ADULT SIZES From *Trichomycterus anhanga* (1.3 cm SL) in the Amazon basin to *T. rivulatus* (38 cm SL) from lakes and streams in the High Andes.

DIAGNOSIS OF FAMILY Distinguished from other catfishes by a patch of odontodes (tooth-like structures; see figure below) on the interopercle (lost in *Glanapteryx*, *Pygidianops*, and *Typhlobelus*) and the opercle (lost in Copionodontinae, *Apomatoceros*, *Megalocentor*, *Glanapteryx*, *Pygidianops*, and *Typhlobelus*). Other external morphological characteristics (Baskin 1973, De Pinna 1989, 1998): a pair of maxillary barbels at the angle (or rictus) of the mouth (reduced in some

Stegophilinae and Vandelliinae); a nasal barbel on anterior nostril (reduced in Stegophilinae, Vandelliinae, and many Tridentinae); lack of pectoral- and dorsal-fin spines; 3–7 procurrent (short and unbranched) rays at the anterior portion of the dorsal fin; lack of a dorsal-fin spine-locking mechanism; the dorsal fin located on the middle or posterior half of the body; i + 4 pelvic-fin rays (except in the basal groups, Copionodontinae and Trichogeninae, with i + 6); and lack of an adipose fin (except in Copionodontinae).

SEXUAL DIMORPHISM Sexual dimorphism has been reported in body size and urogenital papillae size in some glanapterygine species. No known sexual dimorphism in body shape, head shape, or pigmentation/coloration.

HABITATS Trichomycterids occupy a diversity of habitats, showing a high potential for the colonization of extreme environments, such as high-elevation streams above 4,000 m (*T. yuska*), offshore islands (*T. gorgona*), subterranean drainages in caves (e.g., *T. chaberti*, *T. itacarambiensis*, *T. sketi*, *T. uisae*), phreatic (subterranean) waters (*Silvinichthys bortayro*), and warm thermal (*T. therma*) or cold spring waters (*Hatcheria macraei*).

FEEDING ECOLOGY Trichomycterids exhibit some of the most interesting feeding specializations observed among fishes. The subfamily Vandelliinae is exclusively hematophagous, sucking blood from the gills of other fishes. The Stegophilinae include mucus- and scale-eating from a variety of host fish species. Both vandelliines and stegophilines are known in the Brazilian Amazon as *candirú*, and some species (e.g., *Vandellia cirrhosa*) can accidentally penetrate the urethras of bathing humans, causing painful results (Gudger 1930a,b, Spotte et al. 2001, Spotte 2002). The parasitic catfishes can anchor themselves inside their host with the opercular and interopercular odontodes, making it painful to dislodge the fish by force. Despite these remarkable specializations, most species belonging to the big subfamily Trichomycterinae (150+ species) are nocturnal benthic predators of small aquatic macroinvertebrates.

BEHAVIOR The great majority of trichomycterids are nocturnal, bottom-dwelling species; a few species, such as some tridentines, freely swim in the water column and are active in the day. They are positively rheophilic, which means they tend to swim upstream. Trichomycterids generally are equipped with interopercles and opercles that support odontodes. They can evert these odontodes for protection and to provide friction when climbing waterfalls or to attach to the outer surface of the body of large fish (Zuanon and Sazima 2005a, Armbruster 2011).

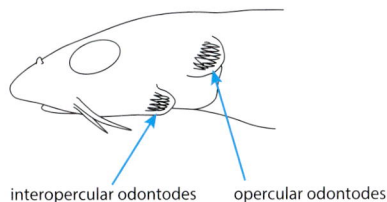

interopercular odontodes   opercular odontodes

KEY TO THE GENERA
BASED ON EIGENMANN (1918B), MYERS (1944), AND MIRANDA RIBEIRO (1951)

1a. Anal fin absent or short and with <12 rays **2**
1b. Anal fin long with 15–25 rays ....**6** (subfamily Tridentinae)

2a. Nasal barbel present **3**
2b. Nasal barbel absent **5**

3a. Anal fin present **4**
3b. Anal fin absent...**8** (subfamily Glanapteryginae)

4a. Anal-fin rays 8–12; premaxilla without a lateral process
(fig. 4a)...**10** (subfamily Trichomycterinae)
4b. Anal-fin rays ≤7; premaxilla with a toothless lateral process,
extending along the length of maxilla (fig. 4b).....**11** (subfamily Sarcoglanidinae)

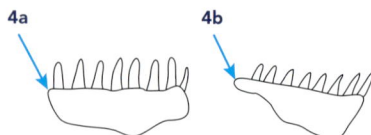

5a. Mouth wide, teeth many and in
1 or more rows (fig. 5a), rami of
mandible meeting anteriorly
(fig. 5aa)..............**13** (subfamily Stegophilinae)
5b. Mouth narrow, teeth few and large,
claw-like teeth on distal end of
premaxilla, enclosed in a
pocket of skin (fig. 5b), rami of
mandible separated anteriorly
(fig. 5bb) ....**23** (subfamily Vandelliinae)

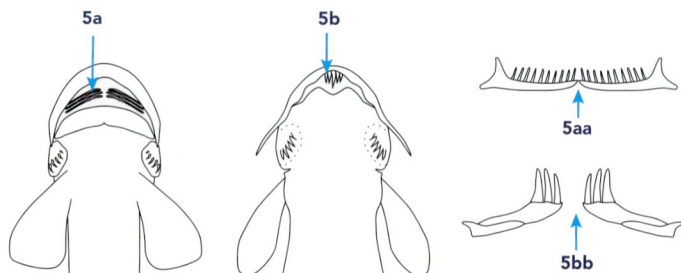

6a. Body greatly elongate; body depth 13 times in SL; head about 9 times in SL; barbels reduced;
interopercular odontodes reduced to three or four ....*Tridens*
6b. Body not greatly elongate or compact; body depth 4–8 times in SL; head 5.0 to 6.5 in SL; barbels
developed; interopercular odontodes 4–8 in number...**7**

7a. Opercular odontodes 10; interopercular odontodes 8–12; body depth 4.5–6.0 times in SL; 17–21 anal-fin rays ....*Tridentopsis*
7b. Opercular odontodes 6; interopercular odontodes 4–6; body depth 8 times in SL; 22–23 anal-fin rays..............*Tridensimilis*

8a. Eyes degenerate; caudal fin small but well developed..**9**
8b. Eyes functional; caudal fin reduced...*Glanapteryx*

9a. No externally visible eyes; snout shovel-shaped; pectoral fin present; body form compact and compressed......*Pygidianops*
9b. A reduced eye, visible as minute black dots; snout trowel-shaped; no pectoral fin; body very elongate and
smooth ....*Typhlobelus*

10a. Head usually wide and depressed; caudal peduncle generally
short and compressed; anterior portion of sphenotic-prootic-
pterosphenoid directed anteriorly (fig. 10a); supraoccipital fontanel
generally reduced to small round orifice on the posterior region
(fig. 10aa)... *Ituglanis*
10b. Head elongate and compressed; caudal peduncle long and
generally depressed; anterior portion of sphenotic-prootic-
pterosphenoid not directed anteriorly (fig. 10b); two
supraoccipital fontanels (fig. 10bb) .... *Trichomycterus*

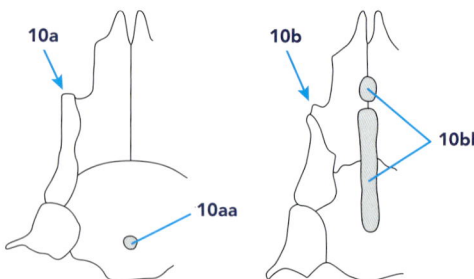

11a. Body with reduction of integument pigmentation, highly transparent
and nearly invisible in their habitat; interopercular and opercular odontodes reduced, with 3 interopercular
odontodes and 2 opercular odontodes...*Stauroglanis*
11b. Body with dark pigmentation forming a banded pattern or agglomerated chromatophores forming circular
patches; interopercular and opercular odontodes well developed, with 5–7 interopercular odontodes and 6
or 7 opercular odontodes **12**

12a. Body with dark chromatophores forming a banded pattern visible from both sides because of the transparency of the body in living fishes...*Ammoglanis*
12b. Body with dark pigmentation abruptly disappearing below lateral midline or agglomerated chromatophores forming circular patches twice the diameter of the eye...*Stenolicmus*

13a. Mouth subterminal, not sucker-like, one complete row of premaxillary teeth, lacking teeth embedded in the fleshy upper lip; eye diameter twice or more times snout length.....................................*Pareiodon*
13b. Mouth ventral, sucker-like, ≥4 rows of premaxillary teeth, ≥3 rows of teeth in the fleshy upper lip; eye diameter less than snout length...**14**

14a. Opercular odontodes absent and corresponding lack of the posterior process of the opercle **15**
14b. Opercular odontodes present and a posterior process of the opercle **16**

15a. Eyes dorsolateral, not visible in ventral view and a single large interopercular odontode ....*Megalocentor*
15b. Eyes lateral, visible in ventral view and numerous very small interopercular odontodes.................................*Apomatoceros*

16a. Eyes lateral, visible in ventral view, wide apart and not partially hidden by the cheeks when viewed from the side; head very flat and depressed; interorbital nearly as wide as head and almost perfectly flat.............*Haemomaster*
16b. Eyes dorso-ateral, not visible in ventral view, close together, and usually partly hidden by the cheeks when viewed from the side; interorbital narrow, usually concave **17**

17a. Accessory procurrent caudal rays very numerous and conspicuous, caudal tail tadpole-like ....*Ochmacanthus*
17b. Accessory procurrent caudal rays few and relatively inconspicuous, caudal not tadpole-like **18**

18a. Caudal fin deeply forked, the upper lobe prolonged and pointed; body with wide, dark, vertical bands.....*Pseudostegophilus*
18b. Caudal fin emarginate, the dorsal lobe not prolonged or obliquely rounded or truncate; body color plain or with small spots...**19**

19a. Deep suborbital groove dorsally separates the lateral portion of the upper lip from the rest of the head........*Schultzichthys*
19b. Suborbital groove absent...**20**

20a. Branchial-membranes united, free from the isthmus (fig. 20a)..*Acanthopoma*
20b. Branchial-membranes confluent with the isthmus (fig. 20b)................**21**

21a. Opercle with 2 or 3 odontodes, rarely 4................................*Henonemus*
21b. Opercle with ≥4 odontodes **22**

22a. Origin of pelvic fin distinctly closer to caudal-fin base than to snout tip...*Stegophilus*
22b. Origin of pelvic fin almost equidistant from snout tip and caudal-fin base ....*Homodiaetus*

20a   20b

23a. Branchiostegal membranes joined and free of the isthmus (as in fig. 20a) ....*Paracanthopoma*
23b. Branchiostegal membranes confluent with the isthmus (as in fig. 20b)...**24**

24a. Caudal-fin margin forked and mandible with a patch of minute teeth...*Plectrochilus*
24b. Caudal-fin margin straight or sometimes emarginated and mandible generally lacks teeth ....*Vandellia*

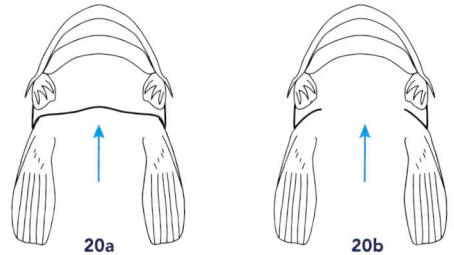

GENUS ACCOUNTS

Acanthopoma (12 cm SL)

Shares with other member of the subfamily Stegophilinae the absence of nasal barbels; two short pairs of maxillary barbels, located at the corner of the mouth; and a wide and inferior mouth, with numerous teeth arranged in several regular rows along the entire margin. Distinguished from other stegophilines by having the branchial-membranes united, free from the isthmus; eyes superior and close together; upper lip with fine, hair-like, movable teeth; superficial neuromasts on the caudal fin; anal fin short with 7–11 rays; procurrent caudal-fin rays few and

not conspicuous (Myers 1944, DoNascimiento and Provenzano 2006, DoNascimiento 2015). **SPECIES** One, *A. annectens*. **DISTRIBUTION AND HABITAT** Orinoco, Guyana, and upper and middle Amazon basin in Brazil and Peru. **BIOLOGY** Parasitic.

Ammoglanis (1.5–1.9 cm SL)

Shares with other members of the subfamily Sarcoglanidinae the reduction of external body pigments; translucent body tissues; inconspicuous or absent opercular and interopercular odontodes; maxilla longer than its associated barbel; and a conspicuous sac-like, fat-filled, adipose organ located dorsal to the pectoral fin. Distinguished from other Sarcoglanidinae by the dark chromatophores on the body surface forming a banded pattern visible from both sides because of the transparency of the body in living fish. Additional osteological characters include: a slender quadrate (fig. 5 in De Pinna and Winemiller 2000); expanded anterior tip of the interopercle; premaxilla posterior to mesethmoid cornua (fig. 6 in De Pinna and Winemiller 2000); short lateral process on the premaxilla; separate ossification of the anterior cartilage of autopalatine absent and the lack of dentary teeth. **SPECIES** Three, all in the AOG region: *A. diaphanous*, *A. pulex*, and *A. amapaensis*. **DISTRIBUTION AND HABITAT** Amazon basin (*A. diaphanous* and *A. amapaensis*) and Orinoco basin (*A. pulex*). *Ammoglanis* inhabit shallow streams where they burrow in sandbanks where the current is slow (Costa 1994, De Pinna and Winemiller 2000). **BIOLOGY** Feed on micro- and meiofauna, including protozoa, rotifers, microcrustaceans, and nematodes (De Pinna and Winemiller 2000).

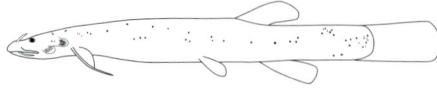

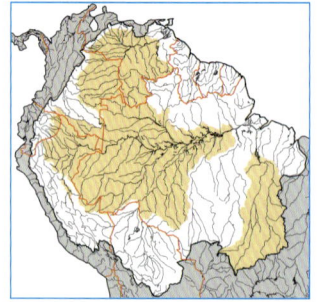

Apomatoceros (15 cm SL)

Shares with other member of the subfamily Stegophilinae the absence of nasal barbels; two short pairs of maxillary barbels, located at the corner of the mouth; and a wide and inferior mouth, with numerous teeth arranged in several regular rows along the entire margin. Distinguished from other stegophiles by the morphology of the opercle, which lacks odontodes and the corresponding dorsal posterior process, anal-fin rays partly spinous, black vertical stripe across basal portion of caudal fin (Eigenmann 1922, De Pinna and Britski 1991). Additional characters for identification: large eyes that are visible ventrally; wide eversible lips, when extruded extending backward in points behind corners of mouth, normally folded into mouth prolonged backward to near the tips of the interopercle patch (Eigenmann 1922, Myers 1944). **SPECIES** One, *A. alleni*. **DISTRIBUTION AND HABITAT** Upper and middle Amazon basin. **BIOLOGY** No data available; possibly (semi) parasitic.

Glanapteryx (6.3 cm SL)

Distinguished from other trichomycterids by: body eel-like; 3 barbels present; eyes small and embedded in the skin; a rudimentary pectoral fin with 3 rays; lack of dorsal and anal fins, and a reduced and diphycercal caudal fin (De Pinna 1989). Additional osteological features: triangle-shaped premaxilla with long and broad posterior process; extremely reduced and simplified pelvic bone (when present); pronounced interdigitations between frontals, sphenotics, and supraoccipital; high number of vertebrae (88 or 89); posterior nares mostly or totally mesial to eyes, adjoined to mesial ocular margin (De Pinna 1989). **SPECIES** Two, both in the AOG region: *G. anguilla* and *G. niobium*. **DISTRIBUTION AND HABITAT** Orinoco and Negro basins, and southern Guiana shield. The species spend most time burrowed in the sand. *Glanapteryx* species inhabit small forest streams with clear water, a slow current, and a sandy substrate (Nico and de Pinna 1996). **BIOLOGY** *Glanapteryx* is not parasitic, as indicated by its tiny mouth and weak, unspecialized dentition (Nico and de Pinna 1996).

Haemomaster (6.6 cm SL)

Shares with other members of the subfamily Stegophilinae the absence of nasal barbels; two short pairs of maxillary barbels, located at the corner of the mouth; and a wide and inferior mouth, with numerous teeth arranged in several regular rows along the entire margin. Distinguished from other stegophilines by a very flat and depressed head; lateral eye visible in ventral view, wide apart and not partially hidden by the cheeks when viewed from the side; interorbital nearly as wide as head; epiphyseal branch of supraorbital canal absent (S6 two pores, such as *Ochmacanthus* in Arratia and Huaquin 1995); mouth sucker-like; medial patch of larger teeth on the lip; patch of odontodes on the interopercle and opercle; color with a short fine band at peduncle and caudal fin (Myers 1944, Baskin 1973, DoNascimiento and Provenzano 2006). **SPECIES** One, *H. venezuelae*. **DISTRIBUTION AND HABITAT** Throughout the Amazon and Orinoco basins in Brazil and Venezuela. **BIOLOGY** No data available; possibly (semi) parasitic.

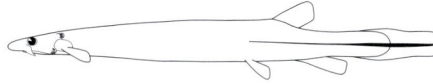

Henonemus (6.0–9.4 cm SL)

Shares with other members of the subfamily Stegophilinae the absence of nasal barbels; two short pairs of maxillary barbels, located at the corner of the mouth; and a wide and inferior mouth, with numerous teeth arranged in several regular rows along the entire margin. Distinguished from other stegophilines by a branchial-membrane that is confluent with the isthmus (not like a fold); opercle with 2 or 3 odontodes (rarely 4) and an epiphyseal branch of supraorbital canal, with one supraorbital pore. Additional characters that might aid identification: upper lip with 3 rows of teeth, posterior labial rows of teeth interrupted in the middle by a patch of larger teeth; premaxillary teeth arranged in 4 rows; superficial neuromasts aligned in 2 rows on the base of the middle rays of caudal fin; and lateral line long, reaching to or beyond a vertical through end of dorsal-fin base (DoNascimiento and Provenzano 2006, DoNascimiento 2015). **SPECIES** Five, all in the AOG region: *H. taxigistigma*, *H. punctatus*, *H. intermedius*, *H. macrops*, and *H. triacanthopomus*. **DISTRIBUTION AND HABITAT** Throughout the Amazon and Orinoco basins, and the Rupununi basin in Guyana. **BIOLOGY** The members of *Henonemus* feed on the mucus and scales of other large fishes (Winemiller and Yan 1989).

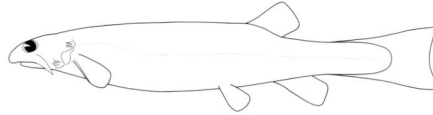

Homodiaetus (3.5–4.2 cm SL)

Shares with other members of the subfamily Stegophilinae the absence of nasal barbels; two short pairs of maxillary barbels, located at the corner of the mouth; and a wide and inferior mouth, with numerous teeth arranged in several regular rows along the entire margin. Very similar to *Acanthopoma*, *Henonemus*, and *Stegophilus*, from which it can be separated by: gill membrane confluent with the isthmus; ≥3 opercular odontodes; inferior lips not extending posteriorly; suborbital groove absent; epiphyseal branch of supraorbital canal absent, with 2 supraorbital pores; origin of pelvic fin almost equidistant from snout tip and caudal-fin base; emarginated or obliquely rounded caudal fin (Myers 1944, Baskin 1973, Koch 2002). **SPECIES** Four, including one as-yet unidentified species in the AOG region. Key to the species in Koch (2002). **DISTRIBUTION AND HABITAT** Upper Madeira, and Rupununi and Takutu rivers (Guyana). **BIOLOGY** Encountered in shallow rivers, free-swimming or attached to the back or sides of other large fish, such as *Pseudoplatystoma* species (Koch 2002).

Ituglanis (3.1–20 cm SL)

Very similar in external appearance to *Trichomycterus*, differing by having a wider and more dorsoventrally depressed head, and a shorter and deeper caudal peduncle. Additional osteological characters include: supraoccipital fontanel reduced to a small round orifice; autopalatine with a deep concavity on its medial margin and anterior portion of sphenotic-prootic-pterosphenoid directed anteriorly, levator internus IV muscle attached to the dorsal face of the posttemporal-supracleithrum (Costa and Bockmann 1993, Datovo 2014, Datovo and de Pinna 2014). **SPECIES** 23, including 5 species in the AOG region: *I. amazonicus, I. gracilior, I. guayaberensis, I. nebulosus,* and *I. parkoi.* **DISTRIBUTION AND HABITAT** Throughout the middle and lower Amazon basin, Tocantins, São Francisco, Paraná, and Paraguay drainages, Orinoco basin, and coastal rivers from Guyana to southeastern Brazil. *Ituglanis* stay hidden during the day under gravel, woody debris, and leaf litter. Five species from central Brazil are troglobitic (cave-dwelling) and one species from Brazil (*I. macunaima*) is a miniature, reaching 2.2–3.2 cm SL (Datovo and Landim 2005). **BIOLOGY** *Ituglanis* feed on benthic aquatic macroinvertebrates such as larvae of dipterans, coleopterans, ephemeropterans, trichopterans, plecopterans, and crustaceans.

Megalocentor (8.8 cm SL)

Shares with other members of the subfamily Stegophilinae the absence of nasal barbels; two short pairs of maxillary barbels, located at the corner of the mouth; and a wide and inferior mouth, with numerous teeth arranged in several regular rows along the entire margin. Distinguished from other stegophilines by a single hypertrophied interopercular odontode (De Pinna and Britski 1991). Some additional osteological features: autopalatine with a large fenestra, which makes its anterior portion ring-like; third hypobranchial of each side fused to each other at midline, but not to last basibranchial; narrow and elongate articulation between interopercle and preopercle; ascending process of median premaxilla broad and dorsally flat (De Pinna and Britski 1991). **SPECIES** One, *M. echthrus.* **DISTRIBUTION AND HABITAT** Orinoco and Western Amazon. **BIOLOGY** *Megalocentor* has an opportunistic parasitic behavior, eating mucus and bits of flesh of large catfishes (e.g., *Brachyplatystoma*) and taking scales when scaled prey are available (De Pinna and Britski 1991).

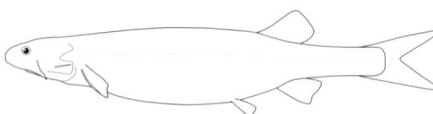

Ochmacanthus (3.2–4.6 cm SL)

Shares with other members of the subfamily Stegophilinae the absence of nasal barbels; two short pairs of maxillary barbels, located at the corner of the mouth; and a wide and inferior mouth, with numerous teeth arranged in several regular rows along the entire margin. Distinguished from other stegophilines by: procurrent caudal-fin rays very numerous and conspicuous; tail like that of tadpole and base of caudal fin narrow; head depressed; body spotted or plain and epiphyseal branch of supraorbital canal absent, with 2 supraorbital pores (Myers 1944, Baskin 1973, Santos 2014). **SPECIES** Five, including four in the AOG region. **DISTRIBUTION AND HABITAT** *O. alternus* (4.0 cm SL) from the Negro basin, Brazil; *O. flabelliferus* (3.5 cm SL) from Guyana and Venezuela; *O. orinoco* (4.6 cm SL) from Venezuela; and *O. reinhardtii* (4.5 cm SL) from French Guiana. The genus also occurs in Paraguay and Río de la Plata basins, Argentina. **BIOLOGY** *Ochmacanthus* feed at least partially on the mucus slime layer and scales of large fishes. Mucus provides a source of energy and amino acids, and feeding on mucus might be advantageous in lowland aquatic habitats where the availability of macroinvertebrate prey fluctuates seasonally (Winemiller and Yan 1989, Galvis et al. 2006).

Paracanthopoma (2.7 cm SL)

Shares with other members of the subfamily Vandelliinae a narrow mouth with small depressible and pointed teeth; few and large upper jaw teeth, placed in two discontinuous series; and the rami mandibulae are not in contact (Myers 1944, Baskin 1973, Schmidt 1993, De Pinna 1998). Distinguished from other vandelliines by an epiphyseal branch of the supraorbital canal (S6 one median pore); branchiostegal membranes joined and free of the isthmus; and an enlarged axillary gland (fig. 1 in Schmidt 1993) located near the base of pectoral fin that secretes mucus that promotes adhesion to the external surface of the host. **SPECIES** One, *P. parva*. Species information in Zuanon and Sazima (2005a). **DISTRIBUTION AND HABITAT** Throughout the Amazon basin and Essequibo basin. **BIOLOGY** Attached to the outer surface of larger fish even when not feeding, or sometimes lives within the gill cavities of other large fishes. Feeding habits of *P. parva* have not been documented; however, it possibly feeds on blood like others vandelliines (Schmidt 1993, De Pinna 1998, Zuanon and Sazima 2005a).

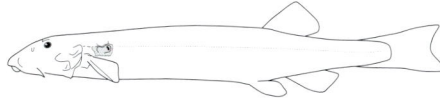

Pareiodon (15 cm SL)

Shares with other members of the subfamily Stegophilinae the absence of nasal barbels and two short pairs of maxillary barbels, located at the corner of the mouth. Distinguished from other stegophilines by: mouth subterminal; two complete rows of premaxillary teeth; absence of teeth embedded in the fleshy upper lips; premaxillary teeth short and robust; lack the median premaxilla bone (fig. 22 in Baskin 1973); eye small, its diameter fitting more than twice in snout length (vs. eye diameter less than twice in snout length); epiphyseal branch of supraorbital canal present, with one pore; branchial-membranes united with the isthmus; anal fin short; caudal fin deeply forked; and head rather deep (Myers 1944, Baskin 1973). **SPECIES** One, *P. microps*. **DISTRIBUTION AND HABITAT** Amazon basin, Brazil. **BIOLOGY** Specialized to take bites or chunks of flesh, especially from wounded or dead fish, in a manner similar to that reported for large cetopsids (such as *Cetopsis coecutiens*). *Pareiodon* has also been reported to feed on the scales of other fishes, but lacks the modified dentition for this behavior as is seen in other stegophilines (Baskin 1973, De Pinna 1998, Fernández and Schaefer 2009).

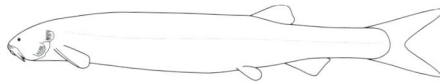

Plectrochilus (6.3–9.3 cm SL)

Shares with other members of the subfamily Vandelliinae a narrow mouth with small depressible and pointed teeth; few and large upper jaw teeth, placed in two discontinuous series; and the rami mandibulae are not in contact (Myers 1944, Baskin 1973, Schmidt 1993, De Pinna 1998). Distinguished by: two short pairs of maxillary barbels (rictal reduced); interopercle with a single greatly enlarged odontoid and very few tiny complementary ones; rudimentary opercular odontodes hidden beneath the skin; branchiostegal membranes confluent with the isthmus (vs. joined and free of the isthmus *Paracanthopoma*); caudal fin margin forked; more than 55 vertebrae (Schmidt 1993). **SPECIES** Three, all in the AOG region: *P. diabolicus*, *P. machadoi* and *P. wieneri*. **DISTRIBUTION AND HABITAT** Upper Amazon and possibly in Guyana. **BIOLOGY** Parasitic, like *Vandellia*, attacking the gills of large fishes and feeding on their blood with specialized teeth. Most species associate with the host only during feeding, leaving the larger fish immediately or shortly after filling their stomach. It has been well documented that the *candirú* or *carnero* of the Amazon region can accidentally penetrate the natural orifices of human bathers. Information on feeding ecology in Spotte (2002).

Pseudostegophilus (5.7–15 cm SL)

Shares with other members of the subfamily Stegophilinae the absence of nasal barbels; two short pairs of maxillary barbels, located at the corner of the mouth; and a wide and inferior mouth, with numerous teeth arranged in several regular rows along the entire margin. Distinguished from other stegophilines by: opercle with 4–12 odontodes; branchial membrane confluent with the isthmus; epiphyseal branch of supraorbital canal present, with one pore; anal fin short with 7–11 rays; superficial neuromasts on the caudal fin; and caudal fin strongly forked, the upper lobe prolonged in *P. nemurus*; body with 4 dark bands across the sides, the margins of the bands darkest; lower caudal-fin lobe and tip of the upper lobe black (Miranda Ribeiro 1951, Baskin 1973, DoNascimiento and Provenzano 2006). **SPECIES** Four, including two species in the AOG region: *P. haemomyzon* and *P. nemurus*. **DISTRIBUTION AND HABITAT** *P. haemomyzon* in the Orinoco basin; *P. nemurus* in the Amazon basin in Brazil, Bolivia, and Peru. **BIOLOGY** The ventral mouth with numerous minute teeth functions as a mucus scraper. The small interopercular odontodes and pectoral glands (secreting a sticky substance) promote adhesion to the external surface of the host.

Pygidianops (2.0–3.0 cm SL)

Distinguished from other trichomycterids by: absence of eyes; body compact and laterally compressed, body depth 6 times in standard length; snout flattened, shovel-shaped; nasal, rostral, and maxillary barbels all with a stiff core and membranous wings; rudimentary pectoral fin with just 1 fin ray; dorsal and pelvic fins absent; anal fin present (except in *P. magoi*); caudal fin well developed; gill openings below the origin of the pectoral fin and gill membranes forming a free fold across the isthmus, attached to the latter at a single median line; mouth a transverse slit, narrow and inferior, but without complicated lip structure or sucking disk; teeth comparatively large and conical, arranged in a single closely set series in each jaw; myomeres very conspicuous (Schaefer et al. 2005). Sexual dimorphism is observed in the shape of the urogenital papillae and in the body size (Carvalho et al. 2014). **SPECIES** Four, all in the AOG region: *P. amphioxus*, *P. cuao*, *P. eigenmanni*, and *P. magoi*. Key to the species in Schaefer et al. (2005), except *P. amphioxus* (see De Pinna and Kirovsky 2011). **DISTRIBUTION AND HABITAT** Lower Orinoco, Negro, and Amazon basins. Buried in sandy substrates at night (Carvalho et al. 2014). **BIOLOGY** Feeds on psammophilous microinvertebrates such as dipterans and crustaceans. Reproduces throughout the year, with more intense reproductive activity during the dry season to avoid disturbances caused by rains and flooding (Carvalho et al. 2014).

Schultzichthys (2.6–3.7 cm SL)

Shares with other members of the subfamily Stegophilinae the absence of nasal barbels; two short pairs of maxillary barbels, located at the corner of the mouth; and a wide and inferior mouth, with numerous teeth arranged in several regular rows along the entire margin; and the branchial membranes united, free from the isthmus. Very similar to *Acanthopoma*, *Henonemus*, *Homodiaetus*, and *Stegophilus*, from which it can be separated by a deep suborbital groove that separates the upper lip from the rest of the head (Baskin 1973, DoNascimiento and Provenzano 2006, DoNascimiento 2015). This characteristic is not present in any other stegophilines known. **SPECIES** Two, both species in the AOG region: *S. bondi* and *S. gracilis*. **DISTRIBUTION AND HABITAT** *S. bondi* in the Amazon and Orinoco basin; *S. gracilis* in the Guayabero River, Orinoco basin, Colombia. **BIOLOGY** No data available; possibly (semi) parasitic.

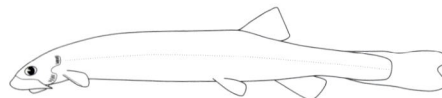

Stauroglanis (2.7 cm SL)

Shares with other members of the subfamily Sarcoglanidinae the reduction of external body pigments; translucent body tissues; inconspicuous or absent opercular and interopercular odontodes; maxilla longer than its associated barbel. Distinguished from other Sarcoglanidinae by: elongate body shape, roughly circular in cross section at the trunk region and gradually becoming laterally compressed toward caudal region; segmented muscle blocks (myomeres) conspicuous, readily visible along whole length of body; long lateral band of adipose tissue running from region of pectoral-fin attachment to posterior margin of anal-fin base. Other characters for identification: eyes large and conspicuous; nasal barbel very short; snout elongate; mouth subterminal not sucker-like; all teeth conical; no lateral sac-like adipose organ dorsoposterior to pectoral-fin insertion; anal fin reduced to 6 rays (De Pinna 1989). **SPECIES** One, *S. gouldingi*. Species information in Zuanon and Sazima (2004a) and Zuanon et al. (2006a). **DISTRIBUTION AND HABITAT** Amazon basin, Brazil, in patches of loose sand where the water flow forms ephemeral sand ripples. They are highly transparent and nearly invisible in their habitat. **BIOLOGY** Species is mostly a microcarnivore, foraging on immature aquatic insects. Feeding activity peaks at late morning and afternoon, and at night the fish remain completely buried. Sexual dimorphism is known in *Stauroglanis*, with the males having larger snouts and teeth (Carvalho et al. 2014).

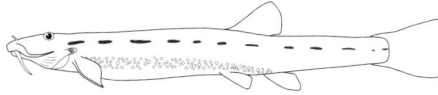

Stegophilus (4.1–4.4 cm SL)

Shares with other members of the subfamily Stegophilinae the absence of nasal barbels; two short pairs of maxillary barbels, located at the corner of the mouth; and a wide and inferior mouth, with numerous teeth arranged in several regular rows along the entire margin. Similar to *Henonemus*, from which it can be separated by 2 rows of teeth on the upper lip, while *Henonemus* has >3 rows (DoNascimiento and Provenzano 2006). Additional characters include: epiphyseal branch of supraorbital canal absent, with 2 pores; lower barbel at angle of mouth minute; 3 or 4 rows of premaxillary teeth; medial patch of larger teeth on the lip; branchial membrane confluent with the isthmus; opercle with ≥4 odontodes; anal fin short and with 7–11 rays (Baskin 1973, DoNascimiento and Provenzano 2006, DoNascimiento 2015). **SPECIES** Three, including two in the AOG region: *S. panzeri* and *S. septentrionalis*. **DISTRIBUTION AND HABITAT** *S. panzeri* inhabits the Amazon; *S. septentrionalis* the Orinoco basin. Distribution associated with the migration of their hosts (Zuanon and Sazima 2005a). **BIOLOGY** Feed on scales and mucus.

Stenolicmus (1.9–3.0 cm SL)

Shares with other members of the subfamily Sarcoglanidinae the presence of a conspicuous sac-like, fat-filled, adipose organ located dorsal to the pectoral fin. Distinguished from other Sarcoglanidinae by: an elongate body with heavy skin pigmentation or agglomerated chromatophores forming circular patches; autopalatine with an anterior ossification of its cartilaginous head (fig. 3 in De Pinna and Starnes 1990); 9 premaxillary teeth disposed in a single row; well-developed patch of opercular odontodes (6 or 7) and interopercular (5 to 7); 4 pelvic-fin rays; no fontanels on cranial roof; posterior process of dorsal expansion of quadrates straight and narrow; metapterygoid absent; expanded distal half of maxilla; numerous accessory caudal-fin rays (13 dorsal and 11 ventral); integument with extensive dark pigmentation (De Pinna and Starnes 1990). **SPECIES** Two, both in the AOG region: *S. sarmientoi* and *S. ix*. Species descriptions in De Pinna and Starnes (1990) and Wosiacki et al. (2011). **DISTRIBUTION AND HABITAT** *S. sarmientoi* in the Matos River, a tributary of the Apere River in the Mamoré basin, Bolivia; and *S. ix* in the Alenquer River, Pará state, Brazil. Both species are present in substrata sand with moderate water current. **BIOLOGY** Feed on microscopic invertebrates, given small body size and other morphological features related to the interstitial life.

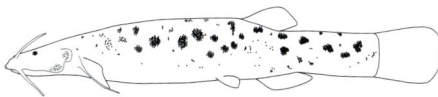

Trichomycterus (1.3–38 cm SL)

The genus *Trichomycterus* is artificial, being defined by absence of the diagnostic traits of the other trichomycterine genera (Baskin 1973, De Pinna 1989, 1998, Schaefer and Fernández 2009). *Trichomycterus* can be recognized by: body elongate and smooth; caudal peduncle compressed and deep; caudal fin truncate, slightly rounded or emarginated; anal fin short, with 7–12 rays; head depressed; interopercle with a variable number of odontodes (2–40) arranged in several rows, those in outer rows larger; opercle and interopercle with a similar pattern of odontodes; mouth moderate size and subterminal, with 2–7 rows of incisiform or conical teeth; 3 pairs of barbels; labial and vomerine teeth absent; very thick, rugose layer of fatty tissue between the skin and muscles, and absence of a large "pillow" of fatty tissue immediately posterior to the supraoccipital region; dorsal fin moderate (8 or 9 principal dorsal-fin rays); 15–20 ribs on each side; portion of the laterosensory canal system in the sphenotic present; with 2–4 supraorbital pores; elongate and compressed head; longer caudal peduncle; 2 supraoccipital fontanels; autopalatine not deeply concave on its medial margin; and anterior portion of sphenotic-prootic-pterosphenoid not directed anteriorly (Arratia 1990, De Pinna 1998, Fernández and Vari 2000). **SPECIES** More than 150 nominal species with at least seven species in the AOG region: *T. anhanga, T. barbouri, T. conradi, T. guianense, T. hasemani, T. johnsoni,* and *T. migrans*. **DISTRIBUTION AND HABITAT** Widely distributed in tropical South America. *Trichomycterus* species inhabit a remarkable variety of environments, including temporary streams, subterranean drainages in caves, high elevations, fossorial sandy settings, and warm thermal waters (Fernández and Vari 2000, Fernández and Schaefer 2003, Fernández and Miranda 2007). Some *Trichomycterus* species climb waterfalls using the opercular and interopercular odontodes to grip rocks and vegetation. Other species are burrowers that aestivate in times of drought until the coming of the next summer rains (some water remains in the burrow). **BIOLOGY** Range in adult body size from *T. anhanga* (1.3 cm SL) from the Amazon basin to *T. rivulatus* (38 cm SL) from high-altitude lakes and streams in the central Andes of western Bolivia, southern Peru, and northern Chile. Most *Trichomycterus* are benthic dwellers that stay hidden during the day by burrowing into the muddy or sandy substrate or hide under leaves, rocks, and other bottom material. In some species individuals aggregate into loosely organized schools. *Trichomycterus* feed on benthic aquatic macroinvertebrates, such as dipteran larvae, coleopterans, ephemeropterans, trichopterans, plecopterans, and crustaceans (Casatti 2003, Chará et al. 2006, Fernández and Vari 2012).

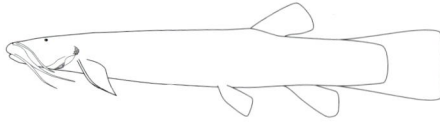

Tridens (2.7 cm SL)

Share with other members of the subfamily Tridentinae a long anal fin and a ventrally exposed eye. Distinguished from other tridentines by having a greatly elongate body (body depth 13 times in SL, head about 9 times in SL) and interopercular odontodes reduced to 3 or 4 (Myers 1944). Additional characters: head greatly depressed; interorbital area broad and flat; maxillary barbels reduced; opercle with 10 odontodes; series of fine labial teeth and stronger teeth in the jaws; branchial membrane united; pelvic fin small, nearer to the tip of snout than to base of caudal fin; anal fin long with 18–22 rays, inserted in front of the dorsal fin; last ray of anal fin positioned under the last ray of the dorsal fin; caudal fin rounded (Myers 1944, Baskin 1973). **SPECIES** One, *T. melanops*. **DISTRIBUTION AND HABITAT** Western and Central Amazon Basin. **BIOLOGY** No data available.

Tridensimilis (2.5–3.0 cm SL)

Share with other members of the subfamily
Tridentinae a long anal fin and a ventrally
exposed eye. Distinguished from other
tridentines by having a compact (not elongate) body, its depth 4–8 times in SL; head 5–6.5
times in SL; 4–6 interopercular odontodes; 6 opercular odontodes; eyes face more ventrally
than dorsally; anal fin with 22 rays; nasal barbel absent and rictal barbel not visible externally
(Myers 1944, Baskin 1973, Galvis et al. 2006). **SPECIES** Two, both in the AOG region: *T. brevis*
and *T. venezuelae*. **DISTRIBUTION AND HABITAT** *T. brevis* in the Amazon basin; *T. venezuelae* in the
Orinoco basin. **BIOLOGY** No data available.

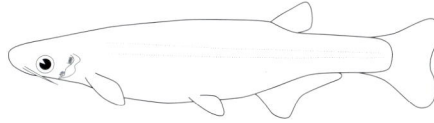

Tridentopsis (2.2–2.3 cm SL)

Share with other members of the subfamily
Tridentinae a long anal fin and a ventrally
exposed eye. Distinguished from other
tridentines by their compact (not elongate) body form, body depth 4–8 times in SL; head length
4.5–6.5 times in SL; 8–12 interopercular odontodes; 11–13 opercular odontodes; two maxillary
barbels developed and one nasal barbel (Myers 1944). Other characters include: nares widely
separated; interorbital broad and flat; first pectoral-fin ray enlarged; dorsal fin short with 7–10
rays; anal fin longer with 17–21 rays; pelvic fins developed; caudal fin deeply emarginated with
the upper lobe slightly longer (Myers 1944, Baskin 1973). **SPECIES** Three, including two in the
AOG region: *T. pearsoni* and *T. tocantinsi*. **DISTRIBUTION AND HABITAT** *T. pearsoni* from the upper
Madeira basin, Bolivia, and *T. tocantinsi* from the Tocantins basin in Brazil. Usually live in
shallow streams with clear water. **BIOLOGY** Diet is unknown, but tooth and mouth morphology
suggest a parasitic mucus- and scale-eating behavior similar to stegophilines.

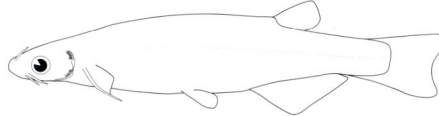

Typhlobelus (2.2–3.3 cm SL)

Resembles *Pygidianops* in many characters, but
differs as follows: eyes vestigial, visible as minute black dots; body greatly elongate and smooth;
depth 12–13 times in SL; snout elongate, trowel-shaped, not merging into the membranous
wings of the barbels; teeth more widely spaced in jaws; mouth a little anterior to insertion of
rostral barbels; no vestige of a pectoral fin; and caudal fin reduced (De Pinna and Zuanon 2013).
Schaefer et al. (2005) speculated on the existence of sexual dimorphism in *T. lundbergi* in the
form of variation in the size of the urogenital papillae and in the number of vertebrae of males
and females. **SPECIES** Five, all in the AOG region: *T. auriculatus*, *T. guacamaya*, *T. lundbergi*,
T. macromycterus, and *T. ternetzi*. Key to the species in De Pinna and Zuanon (2013).
DISTRIBUTION AND HABITAT Orinoco and Amazon basins, and Guianas. Buried in marginal
sandbanks, possibly never leaving the sand under normal conditions (De Pinna and Zuanon
2013). **BIOLOGY** Forages on small arthropods and organic debris (De Pinna and Zuanon 2013).

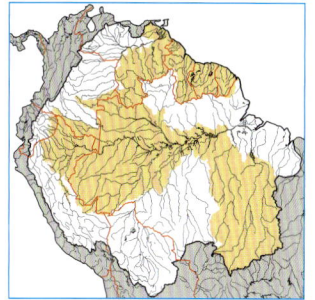

Vandellia (5.9–17 cm SL)

Shares with other members of the subfamily
Vandelliinae a narrow mouth with small depressible and pointed teeth; few and large upper
jaw teeth, placed in two discontinuous series; and the rami mandibulae are not in contact
(Myers 1944, Baskin 1973, Schmidt 1993, De Pinna 1998). Distinguished from other
vandelliines by its elongate and slender body with a smooth surface; mouth inferior and
narrow; teeth few and depressible, distributed in a single row on the upper jaw, dentary with
no teeth (except by *V. sanguinea* with minute teeth); branchial membranes are confluent with
the isthmus (Schmidt 1993); caudal-fin margin straight or occasionally emarginated; pelvic-
fin tip close to the urogenital opening; body elongate (62 vertebrae approximately). *Vandellia*
shares with *Plectrochilus* having more than 55 vertebrae. **SPECIES** Three, all in the AOG region:

V. beccari, *V. cirrhosa* (Toothpick fish), and *V. sanguinea* (Vampire fish). **DISTRIBUTION AND HABITAT** Throughout the AOG region. **BIOLOGY** All hematophagous, feeding mainly on blood in the gills of large fishes (Zuanon and Sazima 2004b, Adriaens et al. 2010). Vandelliines feed by lacerating a major vessel of the gill arch of a host using specialized teeth. Associated with the host only during feeding, leaving the larger fish immediately or shortly after filling their stomach. The ability of *V. sanguinea* to accidentally penetrate the natural orifices of human bathers is well documented (Eigenmann 1917, Gudger 1930a, b, Herman 1973). They attack fish hosts in shoals, producing fatal hemorrhages. Extensive information on ecology in Spotte et al. (2001), Spotte (2002), and Zuanon and Sazima (2004b).

SILURIFORM *PHREATOBIUS INCERTAE SEDIS* (4–6 cm SL)
— *PETER VAN DER SLEEN and JAMES ALBERT*

Species in the genus *Phreatobius* are worm-like subterranean catfishes that have been discovered in artificial wells. Because its members are so unusual, the genus has a long history of conflicting hypotheses about its phylogenetic placement, having been aligned at one time or another with six different siluriform families (De Pinna 1998, Sullivan et al. 2013a). As consensus on its position is still lacking, we include this genus here as *incertae sedis* pending future studies. *Phreatobius* species are small and slender and have: minute eyes, or lack eyes; bright red to pink body coloration; a paddle-shaped caudal region formed by a caudal fin continuous dorsally and ventrally by numerous large procurrent rays, and ventrally confluent with the anal fin; a strongly prognathous lower jaw; massively hypertrophied jaw muscles; and a soft (not spinous) first pectoral-fin ray (Fernandez et al. 2007, Shibatta et al. 2007).

SPECIES At least three species: *P. cisternarum* (see information in Muriel-Cunha and de Pinna 2005), *P. dracunculus* (Shibatta et al. 2007), and *P. sanguijuela* (Fernandez et al. 2007).

DISTRIBUTION AND HABITAT *P. cisternarum* from phreatic environments in region of the mouth of the Amazon River; *P. dracunculus* from an artificial well in the village of Rio Pardo (middle Madeira basin, Rondônia state, Brazil); and

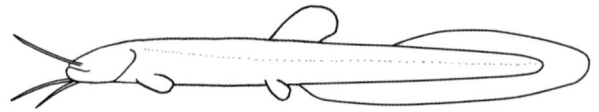

P. sanguijuela from artificial wells near río Paraguá, a tributary of the río Iténez (Bolivia).

BIOLOGY In captivity, *P. cisternarum* vigorously attacks earthworms and beetle larvae and is able to secure and ingest relatively large live prey items. The same fish were also able to survive long periods without food (Muriel-Cunha and de Pinna 2005).

FAMILY APTERONOTIDAE—GHOST KNIFEFISHES
— *MAXWELL J. BERNT and JAMES S. ALBERT*

DIVERSITY Apteronotidae is the most species-rich family of Gymnotiformes, with 91 species in 15 genera, of which 75 species in 14 genera occur in the AOG region. *Tembeassu* from the Paraná basin in Brazil is the only apteronotid genus not known from the AOG region. Apteronotids exhibit the greatest diversity of head, snout, and jaw morphologies among Gymnotiformes. Some groups exhibit long snouts with short jaws (*Sternarchorhamphus*, *Orthosternarchus*, *Sternarchorhynchus*), others long snouts and large jaws (*Platyurosternarchus*, *Parapteronotus*, some *Apteronotus*, *Compsaraia samueli*), short snouts with large jaws and robust dentition (*Sternarchella*,

Magosternarchus), or short snouts with small jaws and reduced dentition (*Adontosternarchus*, *Sternarchogiton*).

COMMON NAMES *Peixe-faca*; *Sarapó* (Brazil), *Macana fantasma* (Peru); *Cuchillo fantasma* (Venezuela).

GEOGRAPHIC DISTRIBUTION Widely distributed throughout the humid Neotropics, and most diverse in large rivers of the Amazon basin. Two species (*Apteronotus rostratus* and *A. spurellii*) are found in Pacific-draining streams in Panama and Colombia, and six species in three genera (including the monotypic *Tembeassu*) are endemic to the La Plata (Paraguay-Paraná) basin.

ADULT SIZES Body size in gymnotiform fishes is reported as maximum total length (cm TL) measured from tip of snout to tip of caudal fin or caudal appendage. Apteronotids exhibit a modest range of adult body sizes, from 15 cm TL in *Sternarchorhynchus gnomus* from the Orinoco River in Venezuela to a reported 129 cm TL in *Apteronotus magdalenensis* (Miles 1945) from the Magdalena River in Colombia, although Maldonado-Ocampo et al. (2011) disputed this later claim. In the AOG region, the largest recorded species are *Sternarchorhynchus mormyrus* (to 55 cm TL) from the Amazon and Negro rivers in Brazil, and *Apteronotus albifrons* (to 50 cm TL) from most of the AOG region.

DIAGNOSIS OF FAMILY Readily recognizable as the only gymnotiform fishes to possess an externally visible caudal fin and a fleshy dorsal organ, a strip of tissue along the posterior middorsum that functions in electroreception (Franchina and Hopkins 1996).

SEXUAL DIMORPHISM Many species within Apteronotidae exhibit pronounced sexual dimorphism in cranial morphology. Males within *Apteronotus*, *Compsaraia*, *Parapteronotus*, and *Porotergus* develop prominently elongate snouts and jaws at sexual maturity that are believed to function in agonistic jaw-locking behaviors. Mature male *Sternarchogiton nattereri* grow numerous large external teeth on the upper and lower jaws, also presumed to play a role in male-male aggression. In some species of *Sternarchorhynchus*, sexually mature males possess large teeth that project from the anterior tip of the lower jaw.

HABITATS The majority of apteronotids inhabit the bottom and margins of large lowland rivers, where they constitute a major portion of the benthic fish biomass in many localities (Lundberg and Lewis 1987, Crampton 1996). Other species, especially members of *Apteronotus* and *Platyurosternarchus* primarily inhabit small streams, while *Apteronotus lindalvae*, *Megadontognathus* spp., several species of *Sternarchorhynchus*, and *Sternarchogiton zuanoni* have been found exclusively in rapids or near waterfalls (Campos-da-Paz 1999, De Santana and Vari 2009, 2010a, b, De Santana and Fernandes 2012).

FEEDING ECOLOGY The ecology of most gymnotiform fishes is poorly known. Many apteronotids prey on benthic macroinvertebrates. The tube-snouted *Sternarchorhynchus* prefer chironomid larvae that burrow into the river bottom substrate. Members of *Sternarchella* and *Magosternarchus* possess robust jaws with numerous teeth. *Sternarchella* feed on small fishes and crustaceans, while *Magosternarchus* consume the tails of other electric fishes (Lundberg et al. 1996). Many apteronotids are important prey items for pimelodid catfishes in large river channels (Barthem and Goulding 1997). *Adontosternarchus* species are planktivores, and at least some *Sternarchogiton* species consume freshwater sponges.

BEHAVIOR Apteronotids produce a high frequency wave-type electric organ discharge that ranges in cycle frequency (numbers of discharges per second) from 421 Hz in *Orthosternarchus tamandua* to more than 2,000 Hz in *Sternarchella schotti*. The latter of these is the fastest and most stable biological oscillator known on Earth (Albert and Crampton 2005). The fastest electric organ discharge ever recorded was 2,179 Hz by a specimen of *S. schotti*, recorded by William Crampton at Tefé in the Central Amazon (Albert and Crampton 2005).

KEY TO THE GENERA AND SPECIES GROUPS

1a. Snout long, preorbital region more than one-third of head length; snout tubular, about as deep as wide at rictus; mouth small, rictus not posterior to anterior nares...2
1b. Snout short, preorbital region one-third of head length or less (except mature males of *Parapteronotus hasemani*, "*Apteronotus*" *bonapartii*, and *Compsaraia samueli*); snout laterally compressed, deeper than wide at rictus; mouth large, rictus extending posterior to posterior nares (except *Sternarchella*).........................5

2a. Mouth large, >20% head length; dorsal head profile straight, head downturned (fig. 1k); body surface with blotched or mottled pigment patterns; caudal peduncle thick and short.................... *Platyurosternarchus*
2b. Mouth small, <10% head length; dorsal head profile variable; body surface with uniform pigment patterns or no pigmentation; caudal peduncle narrow...3

3a. Snout highly recurved, convex dorsal margin in front of eyes (fig. 1q); uniform dark gray or brown coloration, some with pale middorsal stripe; most anal-fin rays branched; fleshy dorsal organ on posterior half of body.................... *Sternarchorhynchus*
3b. Snout not recurved, dorsal margin straight or concave in front of eyes (fig. 1h or 1p); body with pale ground color, no pale middorsal stripe; all anal-fin rays unbranched, fleshy dorsal organ extending posteriorly from nape.... 4

4a. Snout and head very long and straight (fig. 1h); pigments absent.................... *Orthosternarchus*
4b. Snout decurved, concave dorsal margin in front of eyes (fig. 1p). Pale ground color with darker dorsum, dark margins along pectoral and anal fins.................... *Sternarchorhamphus*

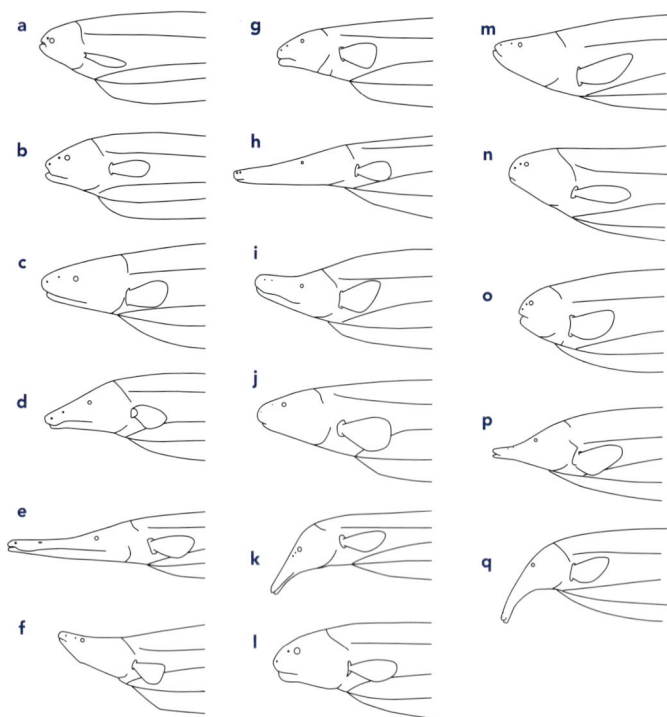

FIGURE 1 Head profiles of genera and some species groups (adapted from Albert 2001): (a) *Adontosternarchus sachsi*, (b) *Apteronotus albifrons*, (c) "*Apteronotus*" *bonapartii*, (d) *Apteronotus leptorhynchus*, (e) *Compsaraia samueli*, (f) *Magosternarchus raptor*, (g) *Megadontognathus kaitukaensis*, (h) *Orthosternarchus tamandua*, (i) *Parapteronotus hasemani*, (j) *Pariosternarchus amazonensis*, (k) *Platyurosternarchus crypticus*, (l) *Porotergus gimbeli*, (m) *Sternarchella schotti*, (n) *Sternarchella sima*, (o) *Sternarchogiton nattereri*, (p) *Sternarchorhamphus muelleri*, (q) *Sternarchorhynchus curvirostris*.

5a. Mouth V-shaped viewed end-on (fig. 5a); upper jaws inside a groove formed by lower jaws...*Adontosternarchus*
5b. Mouth not V-shaped; upper jaws located lateral to lower jaws................................. **6**

6a. Scales present on nape; snout elongate in mature males (>33% of head length)........**7**
6b. Scales absent on nape. Snout length moderate to short in mature males (≤33% of head length; except in *Compsaraia samueli* and *Magosternarchus raptor*).. **11**

7a. No teeth on anterior portion of dentary; 2–4 large conical teeth on posterior portion of dentary ....*Megadontognathus*
7b. Teeth arranged in two or more rows along dentary; no large conical teeth on posterior dentary (large teeth strongly recurved in *Apteronotus leptorhynchus* group) **8**

8a. Small scales, ≥9 above lateral line at midbody **9**
8b. Large scales, ≤8 above lateral line at midbody.. **10**

9a. Shape of mouth square in lateral view; uniform dark brown coloration; no markings at base of caudal fin; pale middorsal stripe sometimes present, but not extending onto head....................................*Parapteronotus*
9b. Mouth not square in lateral view; prominent pale (white or yellow) midsaggital stripe on head, body, and tip of lower jaw (fig. 9b); pale markings on caudal peduncle and caudal fin... **10**

10a. Two or more pale bars at caudal region; snout deep and not elongate, about one-third head length (~33% of head length; fig. 1b), except in mature males of some species ....*Apteronotus albifrons* group
10b. Single pale bar or spot at base of caudal fin; snout slender and elongate (>33% of head length) in all adult males and females (fig. 1d) ....*Apteronotus leptorhynchus* group

11a. Snout very long in mature males, more than one-third head length.. **12**
11b. Snout less than or equal to one-third head length in mature males.. **13**

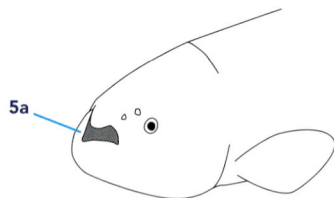

12a. Narrow, pale stripe anterior to eyes and lateral to nares; no pigments on anal or pectoral fins .... *Compsaraia*
12b. No pale stripe anterior to eyes; anal and pectoral fins with uniform or distal brown to
black pigmentation .... *"Apteronotus" bonapartii* group

13a. Head broad, with flattened ventral surface (fig. 13a); head
more than half as wide as long; large mandibular sensory
canal pores .... *Pariosternarchus*
13b. Head laterally compressed (fig. 13b); head less than half
as wide as long; small mandibular sensory canal pores **14**

14a. Gape large, rictus at or posterior to posterior nares **15**
14b. Gape small, rictus not extending to posterior nares **16**

15a. Head short and forehead convex in profile (fig. 1l or 1o);
dentary short and deep; mouth terminal or inferior **17**
15b. Forehead profile straight or concave (fig. 1f); mouth
strongly superior or terminal; dentary elongate; jaws
robust with numerous teeth .... *Magosternarchus*

16a. Forehead straight or slightly concave (fig. 1m); mouth terminal or superior .... *Sternarchella schotti* group
16b. Forehead rounded (fig. 1n); mouth inferior .... *Sternarchella sima* group

17a. Teeth present on premaxilla (except in *Porotergus duende*, which is recognizable by its light brown coloration ..... *Porotergus*
17b. Teeth absent on premaxilla (except in *Sternarchogiton preto* which is recognizable by its dark brown-
black coloration) .... *Sternarchogiton*

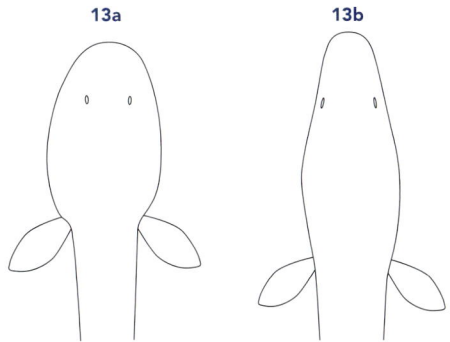

13a   **13b**

GENUS ACCOUNTS

Adontosternarchus (19–32 cm TL)

Characterized by: a beak-like mouth in which
the triangular upper jaw rests in the V-shaped
lower jaw. Teeth are absent from both jaws.
Unlike most other deep-channel apteronotids, species of *Adontosternarchus* possess dark brown,
black, and in some species yellow pigments over the body surface. These pigments are blotchy,
mottled, or speckled. No known sexual dimorphism. **SPECIES** Six, all in the AOG region. Species
descriptions in Mago-Leccia et al. (1985) and Lundberg and Cox Fernandes (2007). **COMMON
NAMES** *Sarapó granito* (Brazil); *Macana granito* (Peru). **DISTRIBUTION AND HABITAT** Exclusively
inhabits river channels in the Amazon and Orinoco basins. **BIOLOGY** All species are planktivores.
The deepest recorded collection for any gymnotiform is that of *A. devenanzii* at 84 m (276 ft)
in the Orinoco River, collected by John Lundberg and colleagues "downriver from Ciudad
Bolivar where the Shield rock comes up into the river bed" (J. Lundberg pers. comm.).
Adontosternarchus is a main prey item of the pimelodid catfish *Brachyplatystoma platynema* in
tributaries of the Orinoco River in Venezuela (Barbarino Duque and Winemiller 2003).

Distribution of all *Apteronotus* groups

Apteronotus (18–50 cm TL)

The most species-rich genus of Apteronotidae, with 27 described species, including 12 species in
the AOG region, assigned to one of the following three species groups:

Apteronotus albifrons group

Recognized by a characteristic color pattern of
black or dark brown ground color with a pale,
high-contrast mid-sagittal stripe extending from the
chin to the middorsum or beyond, at least two pale bands over the caudal region, one band at the
base of the caudal fin, and another band (sometimes broken into multiple bands or pale mottling

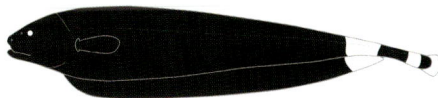

or merged with the posterior band) encircling the distal end of the anal fin anterior to the caudal peduncle (greatly reduced in *A. lindalvae*); snout rounded and deep, head bullet-shaped in lateral profile; scales small, arranged in 9–11 rows above the lateral line at midbody. At least two species (*A. camposdapazi* and *A. lindalvae*) are sexually dimorphic in head shape, with males exhibiting more elongate jaws and more robust snouts. Some populations of *A. albifrons* exhibit sexual size dimorphism, with males growing to a larger body size. **SPECIES** Eight, including four species in the AOG region: *A. albifrons*, *A. camposdapazi*, *A. lindalvae*, and *A. magoi*. **COMMON NAMES** *Ituí cavalo* (Southern Brazil); *Macana pero* (Peru); *Cuchillo fantasma* (Venezuela). Black ghost knifefish (English). **DISTRIBUTION AND HABITAT** Amazon and Orinoco basins as well as the Guianas, where they inhabit a wide range of habitats, from small rainforest streams and rapids to floodplain lakes and the margins of larger rivers. **BIOLOGY** Forage on a wide variety of prey items from benthic and midwater portions of the water column, including dipteran and coleopteran larvae, terrestrial arthropods, and small fishes (Winemiller and Adite 1997).

Apteronotus leptorhynchus group

Recognized by a light brown or chocolate brown ground color with a pale, high-contrast mid-sagittal stripe extending from the chin to the middorsum or beyond, one band at the base of the caudal fin, and 9–11 scales above the lateral line. The snout is long and moderately tapered. Most species in this group are sexually dimorphic, with males exhibiting elongate, slender snouts. **SPECIES** 14, including five species in the AOG region: *A. baniwa*, *A. galvisi*, *A. leptorhynchus*, *A. macrostomus*, and *A. pemon*. **COMMON NAMES** *Morenita* (Colombia, Panama); Brown ghost knifefish (English). **DISTRIBUTION AND HABITAT** Small rivers and streams in the Orinoco basin and Guianas. **BIOLOGY** Sexually mature males use their elongate oral jaws in aggressive interactions known as jaw-locking behavior (Albert and Crampton 2009).

"*Apteronotus*" *bonapartii* group

Recognized by: large scales, <7 above the lateral line at midbody; absence of scales on the anterior portion of the nape; brown to dark gray color dorsally with a pale underside; dark color at the tip of the pectoral and anal fins; a broad pale bar at the base of the caudal fin; robust and rounded snout. "*Apteronotus*" *bonapartii* is sexually dimorphic, with males possessing elongate snouts (Hilton and Cox Fernandes 2006). Specimens range in color from lighter to darker brown, depending on water type, and may change color from pale to dark brown in a matter of minutes after capture. **SPECIES** Four, including three species in the AOG region. **COMMON NAMES** *Sarrapu* (Brazil); *Macana murcielago* (Peru); Yellow ghost knifefish (English). **DISTRIBUTION AND HABITAT** Larger rivers in the Amazon, Orinoco, and Essequibo basins, and also the Paraná-Paraguay basin. **BIOLOGY** Consume aquatic arthropods.

Compsaraia (23–34 cm TL)

Distinguishable from other apteronotids by: a pale stripe anterior to eye, running laterally along the snout; inverted L-shaped supratemporal lateralis canal is visible on the dorsum immediately behind the head; long snout, greater than one-third head length; large gape, rictus posterior to eye; large scales; scales absent on anterior middorsum; body laterally compressed with gray pigmentation dorsally and semitranslucent ventrally and over the opercle. The Macana pekin *C. samueli* is sexually dimorphic, with males growing greatly elongate snouts and jaws (≤70% of head length). However, *C. samueli* from the central and western Amazon exhibit pronounced sexual dimorphism in head and cranial morphology, with elongation of the snout (preorbital) region in mature males, whereas *C.* cf. *samueli* from the upper Madeira and *C. compsa* from the Orinoco do not exhibit sexual dimorphism. **SPECIES** Two, both species in the AOG region:

C. samueli and *C. compsa*. See species information in Albert (2001) and Albert and Crampton (2009). **COMMON NAMES** *Macana pekin* (Peru); Elegant knifefishes (English). **DISTRIBUTION AND HABITAT** *C. samueli* from river channels in the Amazon basin, and *C. compsa* similar habitats in the Orinoco basin. **BIOLOGY** Sexually mature males of *C. samueli* use their elongate oral jaws in aggressive interactions known as jaw-locking behavior (Albert and Crampton 2009).

Magosternarchus (20–25 cm TL)

Distinguishable from other apteronotids by: large, elongate premaxillae, each containing two to three rows of large teeth, with 9–14 teeth in the outermost row; dentary longer than deep, with large teeth along its entire length; mouth terminal or strongly superior; and concave dorsal margin of opercle. Pigmentation is largely absent and color in life is pinkish white, though some specimens exhibit a fine speckling of dark chromatophores over the head and dorsum. No known sexual dimorphism. **SPECIES** Two, both species in the AOG region: *M. raptor* and *M. duccis*, see species descriptions in Lundberg et al. (1996). **COMMON NAMES** *Macana piraña* (Peru); Mago's knifefishes (English). **DISTRIBUTION AND HABITAT** Known only from deep channels of large rivers in the Amazon basin. **BIOLOGY** Analysis of gut content suggests this genus feed on the tails of other gymnotiform fishes (Lundberg et al. 1996). Whether or not electric fish tails (which can be regenerated) are the sole or primary food source of this group is not known.

Megadontognathus (20–24 cm TL)

Distinguishable from other apteronotids by: 2–4 large teeth on the posterior region of the dentary only. Body color is light or dark brown with no distinctive markings. Scales are small, with 14–15 above the lateral line at midbody. No known sexual dimorphism. **SPECIES** Two, both species in the AOG region: *M. cuyuniense* (Mago-Leccia 1994) and *M. kaitukaensis* (Campos-da-Paz 1999). **DISTRIBUTION AND HABITAT** *M. cuyuniense* is known from the Cuyuni River (Essequibo basin) in Guyana and the Caroni River (Orinoco basin) in Venezuela, and *M. kaitukaensis* from the Xingu River (Amazon basin), Brazil. Both species are collected from or near rapids. **BIOLOGY** Inhabit rapids of large rivers on the Guiana and Brazilian shields.

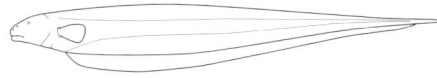

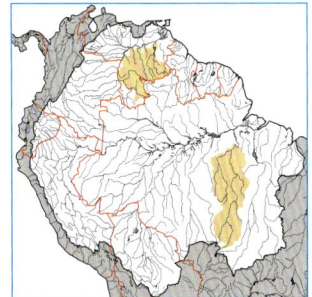

Orthosternarchus (45 cm TL)

Distinguishable from other apteronotids by: a long, straight snout and small mouth (gape not reaching beyond posterior nares). Dorsal organ begins immediately behind the head and extends the entire length of dorsum. Body laterally compressed. Dark pigments are absent, and color in life is white or pink due to subdermal blood vessels. Eyes are greatly reduced. No known sexual dimorphism. **SPECIES** One, *O. tamandua*. A thorough description can be found in Hilton et al. (2007). **COMMON NAMES** *Sarapó branco* (Brazil); *Macana blanco* (Peru); Tamandua knifefish (English). **DISTRIBUTION AND HABITAT** Restricted to the Amazon basin, where it inhabits the deep channels of white- and blackwater rivers. **BIOLOGY** Exhibits a wide range of characters associated with life in deep river channels, including strong lateral compression, a highly demineralized (reticulate) skeleton, and complete absence of skin pigments. The eyes are very small and laterally asymmetrical, a unique condition in Gymnotiformes. These characters suggest the ability to inhabit the deepest portions of large river channels, although this species also inhabits the same depths as other channel apteronotids and has not yet been collected at depths below 20 meters. Possesses the lowest electric organ discharge rate among apteronotids.

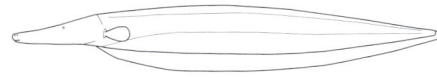

Parapteronotus (38 cm TL)

Distinguishable from other apteronotids by:
a large gape reaching the eye; terminal and
obliquely oriented square-shaped mouth in lateral view
on females and juveniles; body covered with fine brown to black chromatophores; scales small,
>8 above lateral line at midbody; pale stripe sometimes present on middorsum, not extending
onto head; caudal peduncle very short. Sexually dimorphic, with mature males possessing a
robust and elongate snout. **SPECIES** One, *P. hasemani*. **COMMON NAMES** *Sarapó pato* (Brazil);
Macana pato (Peru); Duckbill knifefish (English). **DISTRIBUTION AND HABITAT** Channels and
margins of large lowland rivers in the Amazon basin, including rootmats of floating vegetation.
BIOLOGY Sexually mature males use their elongate oral jaws in aggressive interactions known as
jaw-locking behavior (Albert and Crampton 2009).

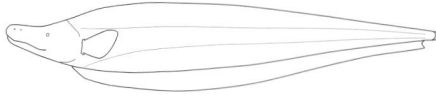

Pariosternarchus (18 cm TL)

Distinguishable from other apteronotids by:
a wide and flattened ventral surface of head, with
enlarged mandibular laterosensory canals, head somewhat globose when viewed from above, its
width >50% head length, small mouth in which the rictus does not extend to a vertical with
the eye, and reduced pigmentation, with pale white to pink coloration in life. No known sexual
dimorphism. **SPECIES** One, *P. amazonensis* (Albert and Crampton 2006). **DISTRIBUTION AND
HABITAT** Known only from the deep channels of large rivers in the Amazon basin. **BIOLOGY** The
flattened ventral surface of the head with enlarged laterosensory canals presumably functions in
enhanced vibration sensitivity to the region immediately below the head.

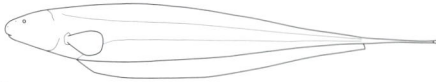

Platyurosternarchus (30–40 cm TL)

Distinguishable from other apteronotids
by: a distinctive mottled gray and brown coloration with larger blotches of dark brown or
black; a long, straight, and narrow snout; head downturned at a roughly 40° angle; large gape,
nearly reaching a vertical with the eye; anal fin lacking pigmentation, except for a dark patch
at the posterior end; caudal peduncle deep and not tapering; large, semicircular caudal fin.
No known sexual dimorphism. **SPECIES** Two, both species in the AOG region: *P. macrostomus*
and *P. crypticus*. Species are diagnosed in De Santana and Vari (2009). **COMMON NAMES** *Cara
de cavalo* (Brazil); *Macana caballo* (Peru); Flat-tail knifefish (English). **DISTRIBUTION AND
HABITAT** *P. macrostomus* (40 cm TL) widespread in the AOG region, and *P. crypticus* (30 cm
TL) from the upper Branco River near the headwaters of the Negro (Amazon) and Rupununi
(Essequibo) basins. Species typically inhabit shallow nearshore regions of rivers and forest
streams, particularly in areas of submerged woody debris and leaf litter. **BIOLOGY** Cryptic
inhabitants on vegetated benthos of slowly moving waters.

Porotergus (18–21 cm TL)

Distinguishable from other apteronotids by:
rounded head (vs. straight or concave) in
lateral profile (also rounded in *Sternarchogiton*
except *S. porcinum*); large mouth, reaching past posterior nares (vs. small mouth in
Adontosternarchus and *Sternarchella*); absence of scales from the nape (also in "*A.*" *bonapartii*
group, *Compsaraia*, *Magosternarchus*, *Pariosternarchus*, *Sternarchella*, and *Sternarchogiton*);
and dumbbell-shaped laterosensory canal ossicles (also in *Sternarchogiton*) (Albert 2001, De
Santana and Crampton 2010). Some *Porotergus* species are similar in external appearance to
Sternarchogiton, from which they can be identified by several internal characters: presence
(vs. absence) of premaxillary teeth (absent in *P. duende* and present in *Sternarchogiton preto*);
and maxillary descending blade rounded (vs. with a sharp angle). *Porotergus duende* can be

distinguished from *Sternarchogiton* by a light brown coloration (vs. little or no pigments, except *S. preto* with dark brown to black pigments). *P. gimbeli* is sexually dimorphic, with mature males possessing a moderately elongate snout; most mature *P. gimbeli* of both sexes with an enlarged chin pad. **SPECIES** Three, all in the AOG region. Key to the species and species information in De Santana and Crampton (2010). **DISTRIBUTION AND HABITAT** *P. gimbeli* and *P. duende* inhabit the channels of large rivers in the Amazon basin, while *P. gymnotus* is known only from Guyana. **BIOLOGY** Consume aquatic insect larvae and microcrustaceans.

Sternarchella (19–40 cm TL)

Distinguishable from other apteronotids by: a very short gape, less than twice the

diameter of the eye; numerous teeth on the dentary and premaxilla; maxilla crescent-shaped; gill rakers attached to gill arches. Pigmentation is greatly reduced and scales are absent from the anterior middorsum in all species. *Sternarchella* is divided into two species groups, the *S. schotti* and *S. sima* groups. Members of the *S. schotti* group possess a terminal to slightly superior mouth while those of the *S. sima* group possess distinctly subterminal or ventral mouths. No *Sternarchella* is known to have sexual dimorphism. **COMMON NAMES** *Sarapó pulso* (Brazil); *Macana muñeca* (Peru); Bulldog knifefish (English). **SPECIES** Six, all in the AOG region. The *S. schotti* group includes *S. schotti* (40 cm TL), *S. terminalis* (35 cm TL), and *S. calhamazon* (19 cm TL) from the Amazon basin, and *S. orthos* (30 cm TL) from the Orinoco basin. The *S. sima* species group includes *S. sima* (30 cm TL) from the Amazon and *S. orinoco* (30 cm TL) from the Orinoco basin. Descriptions and diagnoses are provided in Lundberg et al. (2013) and Ivanyisky and Albert (2014). **DISTRIBUTION AND HABITAT** Inhabit the main channels of large rivers in the Amazon and Orinoco basins. **BIOLOGY** Species with a terminal mouth consume mainly small-bodied animals ranging in size from zooplankton to small fishes; species with ventral mouths consume mainly benthic aquatic arthropods; *S. calhamazon* is the most abundant apteronotid species of the Brazilian and Peruvian Amazon (Lundberg et al. 2013).

Sternarchogiton (20–40 cm TL)

Distinguishable from other apteronotids by: a short, sharply sloping forehead; a

subterminal mouth; and absence of teeth on the premaxilla (except *S. preto*), and yellow-hyaline body color (semitranslucent in life) (dark purplish brown to black in *S. preto*). **SPECIES** Five, all in the AOG region: *S. nattereri* (25 cm TL), *S. labiatus* (35 cm TL), *S. porcinum* (30 cm TL), *S. preto* (40 cm TL), and *S. zuanoni* (18 cm TL). Review of genus and key to the species in De Santana and Crampton (2007), except for *S. zuanoni* (see De Santana and Vari 2010b). **COMMON NAMES** *Sarapó blanquita* (Brazil); *Macana blanquita* (Peru); White ghost knifefish (English). **DISTRIBUTION AND HABITAT** Most species inhabit large and deep white- and blackwater rivers in the lowland Amazon and Orinoco basins. *Sternarchogiton zuanoni* is known only from rapids in the lower Xingu River in Brazil (De Santana and Vari 2010b). A key to the species of *Sternarchogiton* is presented in De Santana and Crampton (2007). **BIOLOGY** *S. labiatus* possesses a fleshy 3-lobed structure projecting from the lower jaw, the function of which is unknown. Mature *S. nattereri* males grow large teeth on the outside of the upper and lower jaws, which are believed to function in male-male aggressive interactions.

Sternarchorhamphus (45 cm TL)

Distinguishable from other apteronotids by:
a long, tubular, decurved (upturned) snout.

The eye is very small, and pigmentation is greatly reduced. Body is strongly compressed laterally. Body coloration is white to pink, often with gray pigments on dorsum and over the snout. Pectoral and anal fins are black distally (Albert 2001, Hilton et al. 2007). No known sexual dimorphism. **SPECIES** One, *S. muelleri.* **COMMON NAMES** *Sarapó toupeira* (Brazil); *Macana pinocho* (Peru); Mole-nose knifefish (English). **DISTRIBUTION AND HABITAT** Large river channels in the Amazon and Orinoco basins. **BIOLOGY** *S. muelleri* is highly variable in coloration, with specimens collected together grading from a dark gray body with black margins on the pectoral and anal fins, to very pale with almost no pigmentation at all, and thus exhibiting a pink body color.

Sternarchorhynchus (15–54 cm TL)

Distinguishable from other apteronotids by:
a very long, tubular snout that is distinctly

recurved (turned down). The dentary is also highly elongate and gracile, but the mouth is very small, with a gape length of less than one-eighth of snout length. Body color is pale gray or brown to black. Scales are typically small, with 6–13 above the lateral line at midbody. Many species within this genus are sexually dimorphic. Mature males of some species develop a bulbous chin from which project numerous large teeth. **SPECIES** 32 species, with 31 in the AOG region. A review of genus and key to the species in De Santana and Vari (2010a). **COMMON NAMES** *Sarapó tapir* (Brazil); *Macana elefante* (Peru); Elephant-nose knifefish (English). **DISTRIBUTION AND HABITAT** Two species of *Sternarchorhynchus* (*S. freemani* and *S. galibi*) are endemic to the Guianas, while the remaining species inhabit rivers of the Amazon and Orinoco basins. Seven species (*S. caboclo, S. severii, S. inpai, S. mareikeae, S. jaimei, S. higuchii,* and *S. hagedornae*) inhabit large river rapids, while the remaining species inhabit river channels. **BIOLOGY** The strongly decurved (downturned) snout and small jaws allow *Sternarchorhynchus* to employ grasp-suction feeding for preying on benthic macroinvertebrates that burrow in sediments and crevices at the river bottom (Marrero and Winemiller 1993).

FAMILY GYMNOTIDAE—ELECTRIC EEL AND BANDED KNIFEFISHES

— *JACK M. CRAIG and JAMES S. ALBERT*

DIVERSITY 40 species in two genera (*Electrophorus* and *Gymnotus*), including 22 species in the AOG (Albert et al. 2005). *Gymnotus* of the AOG region are represented by five species groups: the *G. anguillaris* group (3 species), the *G. carapo* group (4 species), the *G. coatesi* group (8 species), the *G. tigre* group (1 species), and the *G. varzea* group 5 species). Two additional species groups (the *G. cylindricus* and *G. pantherinus* groups) are not present in the AOG region. Gymnotids exhibit substantial diversity among species in chromosome number and organization (Milhomem et al. 2007, 2008, Fonteles et al. 2008), and in the form of their electric signals (Crampton and Hopkins 2005, Crampton et al. 2008, Rodríguez-Cattáneo and Caputi 2009, Rodríguez-Cattáneo et al. 2013). The electric eel (*E. electricus*) reaches a substantially longer total length than other gymnotids (216 cm TL vs. 10–100 cm TL) and can be easily identified by its terminal mouth, dark brown uniform coloration, and yellow to red belly (Assunção and Schwassmann 1995).

COMMON NAMES *Morena, Poraquê, Tuvira* (Brazil); *Anguilla electrica, Chaviro* (Peru); *El Temblador, Gymnotus* (Venezuela).
GEOGRAPHIC DISTRIBUTION Most of the AOG region, with *E. electricus* and *G. carapo* present in every major basin (Albert et al. 2005). The *G. anguillaris* group is present only on the Guianas highlands. The *G. carapo* group and *G. coatesi* group are present throughout most of Amazonia (Crampton and Albert 2004). The *G. tigre* group and the *G. varzea* group are restricted to Amazonian lowlands.
ADULT SIZES From about 10 cm TL total length in *G. jonasi* and *G. melanopleura* from the floodplains of the Central Amazon basin, to 60 cm TL length in *G. carapo* in the AOG region, and 100 cm TL in *G. inaequilabiatus* from the Paraná-Paraguay basin. *Electrophorus electricus* grows to the largest body size of all gymnotiforms, reaching a maximum size of 216 cm TL.
DIAGNOSIS OF FAMILY Differentiated from other gymnotiforms by: cylindrical or subcylindrical body shape in cross section (except laterally compressed in the *G. carapo* group and *G. varzea*

group; also cylindrical in Hypopomidae) vs. laterally compressed in Apteronotidae, Rhamphichthyidae, and Sternopygidae; superior mouth in *Gymnotus* or terminal mouth in *Electrophorus* (also terminal or superior in the apteronotid *Magosternarchus* vs. terminal or subterminal); teeth present on premaxilla and dentary (vs. absent in Hypopomidae and Rhamphichthyidae); anal-fin origin posterior to pectoral-fin base (also in Hypopomidae); small (*Gymnotus*) or no (*Electrophorus*) caudal appendage, tail short, comprising 0–16% total length (vs. long caudal appendage in Apteronotidae, Hypopomidae, Rhamphichthyidae, and Sternopygidae). *Gymnotus* in Greater Amazonia display a series of 6–23 obliquely oriented dark pigment bands, and *Electrophorus* has a uniform dark brown coloration with a yellow to orange belly and ventral head surface.

SEXUAL DIMORPHISM No sexual dimorphism is known in *Gymnotus* morphology or electric organ discharges. In *E. electricus* males reach larger adult sizes (216 cm TL vs. 180 cm TL) and have darker orange or red (vs. yellow) coloration on their bellies at sexual maturity (Assunção and Schwassmann 1995).

HABITATS Inhabit most aquatic habitats in Greater Amazonia, with the notable exception of large and deep rivers. Some *Gymnotus* inhabit streams above 1,000 meters elevation in the Andean foothills and Guianas (JSA, pers. obs.), and riffles in the Gran Sabana of Venezuela (N. Lujan, pers. comm.). Large-bodied and geographically widespread species like *G. carapo* and *G. tigre* inhabit seasonally flooded forests, floodplains lakes, and river margins, but never the main river channels. The small-bodied *G. coropinae* is widespread in upland forest streams throughout much of the AOG region (excluding the Brazilian Shield). *Gymnotus curupira* is unique among gymnotiforms by inhabiting the moist leaf litter of isolated ephemeral forest pools and can survive for many hours completely out of water (Crampton et al. 2005). At least 13 *Gymnotus* species and *E. electricus* inhabit floodplains with seasonally hypoxic water.

FEEDING ECOLOGY All gymnotids are generalized predators (Silva et al. 2003). *Electrophorus* and species of the *G. carapo* group and *G. tigre* group grow to be large-bodied piscivores, whereas species of the smaller-bodied *G. anguillaris* and *G. coatesi* groups feed on crustaceans and insect larvae. Species of the *G. varzea* group feed on small fishes and macroinvertebrates.

BEHAVIOR All gymnotids are nocturnal and exhibit a pulse-type electric organ discharge used for navigation, foraging, and communication. Most, if not all, gymnotids are unusual among gymnotiforms in that the males are territorial; they defend nests and brood the eggs and larvae. Male *Electrophorus* build bubble nests at the water surface and fan their eggs and larvae to reduce risk of fungal or parasitic infections (Crampton and Hopkins 2005). Most if not all gymnotids gulp air and use the swim bladder for buoyancy, gas exchange, and sound detection (Johansen et al. 1968, Liem et al. 1984). *Electrophorus* is an obligate air breather, using a highly vascularized oral lining to extract oxygen from the water, and can drown in low oxygen water if prevented from obtaining air at the water surface (Johansen et al. 1968). *Electrophorus electricus* can live up to 15–20 years in aquaria and somewhat less in the wild.

ADDITIONAL NOTES Gymnotids have been historically important in the study of the biological production and control of electricity. The electric eel *E. electricus* was used in the original discovery of electricity as a force of nature, and the electric organs of gymnotids have been the subject of extensive studies into the genetic, physiological, developmental, neurobiological, and behavioral basis of biogenetic electrogenesis (see overview in Finger and Piccolino 2011). *Electrophorus* is the only gymnotiform fish with a strong electric organ discharge (up to 600 V in large adult fish). They use this electrical discharge to immobilize or kill prey items (Catania 2014). Older *Electrophorus* develop cataracts (clouded lenses in the eye) possibly from the accumulated effects of years of experiencing high-voltage shocks.

KEY TO THE GENERA

1a. Coloration dark brown dorsally and orange to red on ventral surface of head and abdomen; no pigment bands; no scales; terminal mouth; anterior nares on dorsal surface of head; fin rays on dorsal margin of tail; adults >100 cm TL and with strong electric discharges.................................. ***Electrophorus electricus***

1b. Body with numerous (6–23) vertical or oblique pigment bands; no orange or red coloration on ventral surface; scales present on body; superior mouth; anterior nares in gape; fin rays limited to ventral margin of tail; adults reach 10–60 cm TL and do not produce strong electric discharges........................... ***Gymnotus***

Key to *Gymnotus* Species Groups

1a. One pore at dorsoposterior corner of preopercle; needle-shaped teeth; small to intermediate adult body size (10–30 cm TL) **2**

1b. Two pores at dorsoposterior corner of preopercle; arrowhead-shaped teeth on dentary; intermediate to large adult body size (30–60 cm TL).. **3**

2a. Bands arranged in pairs with irregular, wavy margins; intermediate adult body size (≤30 cm TL) ....**G. anguillaris** group
2b. Bands not paired, band margins regular and sharp; small adult body size (<30 cm TL) ....**G. coatesi** group

3a. Head wide; head with irregular white pigment blotches; teeth conical with outwardly curved tips; anal fin with pigment stripes posteriorly; body subcylindrical (body depth / body width at anal-fin origin >75%) ....**G. tigre** group
3b. Head narrow; head without white blotches; anterior teeth arrowhead-shaped; anal fin not striped posteriorly; body laterally compressed (body depth / body width at anal-fin origin <75%) **4**

4a. Head short (head length <10% total length in specimens >10 cm TL); small to intermediate adult body size (17–28 cm TL); few arrowhead-shaped teeth on dentary; anal fin clear or evenly and lightly pigmented; sometimes with faint dark patch posteriorly in juveniles ....**G. varzea** group
4b. Head long (>10% total length); large adult body size (35–60 cm TL); many arrowhead-shaped teeth on dentary; anal fin dark with clear patch posteriorly in adults; posterior third of anal fin very dark in juveniles....**G. carapo** group

GENUS ACCOUNTS

Electrophorus (216 cm TL)

Readily distinguished from other gymnotids by: large adult body size; body shape cylindrical in cross section; coloration dark brown to olive green on dorsal surface with no pigment bands; yellow to red coloration on ventral surface of head and abdomen; anal fin with uniform dark pigmentation; no scales; large electric organs constituting more than half of total body mass; fin rays on dorsal margin of caudal appendage; capacity to produce powerful shocks of up to 600 volts (Coates and Cox 1945). Sexually dimorphic with males growing to a larger adult body size (180 cm TL vs. 216 cm TL in males), and getting a deep orange to red coloration on the abdomen and ventral head surface. **SPECIES** One, *E. electricus*. **COMMON NAMES** *Poraquê, Puraque* (Brazil); *Anguilla electrica, Chinkirma* (Ecuador, Peru), *Anguille électrique* (French Guiana), *El Temblador* (Venezuela), Electric eel (English). **DISTRIBUTION AND HABITAT** Widely distributed throughout the AOG, where it inhabits freshwater marshes, flooded forests and floodplain lakes, and small upland rainforest streams, but never the deep channels of large rivers. Its ability to breathe air allows it access to hypoxic habitats during the dry season. **BIOLOGY** Regularly grow to >2 meters (6.5 ft) total length, reaching sexual maturity at about 100 cm TL in males and 70 cm TL in females. Males build and guard bubble nests in thick vegetation at the water surface and mouth-brood the larvae (Assunção and Schwassmann 1995). Adults may be solitary or form large aggregates, sometimes accumulating under stands of the açaí palm *Euterpe oleracea* (Arecaceae) to consume the berries falling into the water. Omnivorous, feeding mainly on small aquatic animals (especially fishes) and fruits. Electric eels have three anatomically and physiologically distinct electric organs on each side of the body, using the anterior two-thirds of the main electric organ to produce strong electric discharges (up to hundreds of volts) and the posterior third of this organ along with portions of the other two organs to produce weak electric discharges (about 1 volt) for use in navigation and social communication. Electric eels are not widely consumed as a food fish except by some indigenous people, but the fish oils are sometimes extracted for use in alleviating muscle or joint pain.

Distribution of all *Gymnotus* groups

Gymnotus (10–100 cm TL)

The most species-rich genus of Gymnotiformes, with 39 described species, including 21 species in the AOG region, assigned to five species groups (Albert and Crampton 2001, Crampton and Albert 2003, Albert et al. 2005, Crampton et al. 2005, Maxime and Albert 2009, Maxime et al. 2011). Range substantially in body sizes and habits, from smaller-bodied members of the *G. coatesi* and *G. anguillaris* groups that consume insect larvae, live in leaf litter, and undercut banks of small rainforest streams, to larger-bodied members of the *G. carapo* and *G. tigre* groups that eat fish in floating vegetation along the margins of larger rivers.

Gymnotus anguillaris group (24–34 cm TL)

Distinguished from congeners by: larger adult body sizes than the *G. coatesi* group; cylindrical or almost cylindrical body shape in cross section; one pore at dorsoposterior corner of preopercle; dark pigment bands with irregular wavy margins (except *G. pedanopterus* with sharp margins) versus bands with regular sharp margins in the *G. coatesi* group and the *G. tigre* group; ovoid scales; 3 rows of electrocytes above caudal end of anal fin. **SPECIES** Four, all in the AOG region: *G. anguillaris* (30 cm TL), *G. cataniapo* (32 cm TL), *G. pedanopterus* (34 cm TL), and *G. tiquie* (24 cm TL). **COMMON NAMES** *Alapotaïni*, *Alapotanchiayi* (French Guiana); G2 group (English). **DISTRIBUTION AND HABITAT** Inhabit upland rainforest streams and margins of small rivers in the Guiana Shield.

Gymnotus carapo group (35–60 cm TL)

Distinguished from congeners by: larger adult body sizes; laterally compressed body; two pores at dorsoposterior corner of preopercle; 18–23 irregular wavy dark pigment bands or band pairs, often broken into spots above the lateral line (except *G. ucamara* with solid unbroken bands); black patch at the posterior of the anal fin; ovoid scales; 4 rows of electrocytes above caudal end of anal fin. Individual specimens attain adult body and head proportions at about 15 cm TL. Coloration also changes during growth from highly regular bands with high contrast margins in juveniles to broken bands or speckles in adults. The black pigment patch on the posterior third of the anal-fin membrane of juveniles fades with growth as well (Crampton and Albert 2003, Maxime and Albert 2009). **SPECIES** 11, including 4 in the AOG region: *G. arapaima* (55 cm TL) from the central Amazon lowlands, *G. capanema* from the eastern Amazon, *G. carapo* from throughout the AOG, and *G. ucamara* (35 cm TL) from the western Amazon. **COMMON NAMES** *Morena pintada*, *Tuvira* (Brazil); *Chaviro*, *Macana Zebra* (Peru). **DISTRIBUTION AND HABITAT** Widely distributed throughout AOG where they inhabit flooded forests and aquatic vegetation on river margins, floodplain lakes, and small upland rainforest streams, but never the deep channels of large rivers.

Gymnotus coatesi group (10–21 cm TL)

Distinguished from congeners by: small to moderate adult body sizes; cylindrical body; one pore at dorsoposterior corner of preopercle; 12–20 evenly pigmented and straight (not wavy) oblique pigment bands; anal fin free of pigmentation; ovoid scales; slender electric organ with 2 rows of electrocytes above caudal end of anal fin. **SPECIES** Seven, all in the AOG region. **COMMON NAMES** *Sarapó tigrinho* (Brazil); *Macana tigrito* (Peru); G1 group (English). **DISTRIBUTION AND HABITAT** *G. coropinae* is present in upland forest streams throughout the AOG, *G. stenoleucus* is restricted to the Orinoco basin, and other species are restricted to the Amazon basin (Crampton and Albert 2003). Most diverse in the whitewater floodplains (*várzeas*) with five species, including *G. coatesi* in the eastern Amazon of Brazil and *G. javari* in the western Amazon of Peru.

Gymnotus tigre group (100 cm TL)

Distinguished from its congeners by: large adult body sizes; elongate tube-shaped body with body depth about equal to head depth at the beginning of the neck; wide head with widely set eyes and a thick and bulbous chin; two pores at dorsoposterior corner of preopercle; 18–23 vertical or obliquely oriented dark pigment bands; striped patch of membrane at posterior end of anal fin; elongate scales on posterior portion of body; 4 rows of electrocytes above caudal end of anal fin. **SPECIES** Five, including one species in Amazonia, *G. tigre* (41 cm TL); *G. inaequilabiatus* from the Paraná-Paraguay basin is the largest species of the genus, growing to 100 cm TL. **COMMON NAMES** *Tuvirão* (Brazil); *Macana tigre* (Peru). **DISTRIBUTION AND HABITAT** *G. tigre* inhabits floodplain lakes, flooded forests, and marginal riverine vegetation in the western Amazon.

Gymnotus varzea group (17–28 cm TL)

Distinguished from its congeners by: small to intermediate adult body sizes; laterally compressed body; two pores at dorsoposterior corner of preopercle; 18–23 pigment bands or band pairs with irregular, wavy margins; black patch at the posterior of the anal fin; ovoid scales; 3 rows of electrocytes above caudal end of electric organ. **SPECIES** Six, including five species in the AOG region. **COMMON NAMES** *Tuvirinho* (Brazil); *Chavirito* (Peru). **DISTRIBUTION AND HABITAT** Distributed throughout the Amazon and Paraguay basins, and not known from the Orinoco basin or Guianas. All species inhabit floodplains with the exceptions of *G. capanema* and *G. curupira*, which inhabit rainforest streams.

FAMILY HYPOPOMIDAE—GRASS KNIFEFISHES

— *KORY M. EVANS and JAMES S. ALBERT*

DIVERSITY 34 species in 6 genera: *Akawaio*, *Brachyhypopomus*, *Hypopomus*, *Microsternarchus*, *Procerusternarchus*, and *Racenisia*, with 28 species in the AOG region. The genera *Hypopygus* and *Steatogenys* are herein recognized in the family Rhamphichthyidae. Four hypopomid genera (*Akawaio*, *Hypopomus*, *Microsternarchus*, *Procerusternarchus*) are known from a single species each (i.e., are monotypic). Three of these species (*Microsternarchus*, *Procerusternarchus*, *Racenisia*) have a small body size, not exceeding 14 cm TL, and a cylindrical or tube-shaped body, and are therefore called vermiform, or worm-like, knifefishes. Most hypopomid species are members of *Brachyhypopomus*, which is especially common in rooted and floating vegetation along river and stream margins. Several new species of *Brachyhypopomus* and *Microsternarchus* are known and await formal description. At least one *Brachyhypopomus* species has accessory electric organs on the opercle at the side of the head.

COMMON NAMES *Sarapó de grama* (Brazil); *Macana comun*, *Mayupita* (Peru).

GEOGRAPHIC DISTRIBUTION Hypopomids are distributed across most of tropical South America and southern Central America, from the semiarid savannas (pampas) of northern Argentina and Uruguay and Cerrado region of northeastern Brazil, to the humid Pacific coast from Ecuador to Costa Rica (Albert and Crampton 2003, Crampton 2011, Carvalho 2013).

ADULT SIZES Small to moderate body sizes, from *Microsternarchus brevis* reaching 5.3 cm TL, to *Brachyhypopomus pinnicaudatus* at 46 cm TL.

DIAGNOSIS OF FAMILY Recognizable among gymnotiform fishes by: a cylindrical or subcylindrical body that is roughly circular in cross section (slightly compressed laterally in some *Brachyhypopomus*; body also cylindrical or subcylindrical in many Gymnotidae) vs. laterally compressed in *Gymnotus carapo* group, Rhamphichthyidae, Sternopygidae, and Apteronotidae; snout (preorbital distance) about one-third head length (except *Akawaio*, where snout is more than one-third head length; snout also one-third head length in Gymnotidae) verus snout short in *Steatogenys* and *Hypopygus*, or snout long

in Rhamphichthyinae (snout length variable in Sternopygidae and Apteronotidae); small eyes covered by skin versus large eyes in most Sternopygidae (except some *Eigenmannia*) and not covered by skin in *Sternopygus* and *Archolaemus*; anal-fin origin posterior to the pectoral-fin base (also in Gymnotidae) versus anterior to branchial isthmus in Rhamphichthyidae, Sternopygidae, and Apteronotidae; thick fleshy semitransparent strip of tissue along base of anal fin versus absent in other gymnotiforms (except *Hypopygus*). All hypopomids (except most *Brachyhypopomus* species) have an even, mottled or reticulated pigmentation pattern on the body surface, whereas most *Brachyhypopomus* species have dark pigment saddles or vertical bars (saddles and bars also present in some *Gymnorhamphichthys*) versus obliquely angled and irregular dark pigment bands with wavy margins in *Steatogenys*, *Hypopygus*, *Rhamphichthys*, and many *Gymnotus*.

SEXUAL DIMORPHISM Most hypopomids for which electric organ discharge (EOD) are known exhibit sexual dimorphism in the shape and duration of EOD pulses (Hagedorn 1988, Curtis and Stoddard 2003, Crampton and Albert 2006, Salazar and Stoddard 2008). In several *Brachyhypopomus* species (e.g., *B. draco*, *B. pinnicaudatus*), males develop an expanded posterior tip to the caudal appendage, resembling a caudal fin but without fin rays or bony supports (Giora et al. 2008).

HABITATS Small streams in rainforests and savannas, oxbow lakes, and floating meadows, the vegetated margins of large river floodplains, and coastal marshes. Many *Brachyhypopomus* species on floodplains are capable of withstanding hypoxic environments, by trapping air bubbles in the gill chamber (*B. pinnicaudatus*) or by using aquatic surface respiration (Schaan et al. 2009).

FEEDING ECOLOGY Insect larvae and other small aquatic invertebrates (Cox-Fernandes et al. 2014, Giora et al. 2014).

BEHAVIOR Hypopomids like most other Gymnotiformes are nocturnal, hiding in substrate or leafy cover by day and venturing out at night when they use electroreception to locate insects and other small organisms. Little is known about the life cycle and mating behaviors of most species.

KEY TO THE GENERA

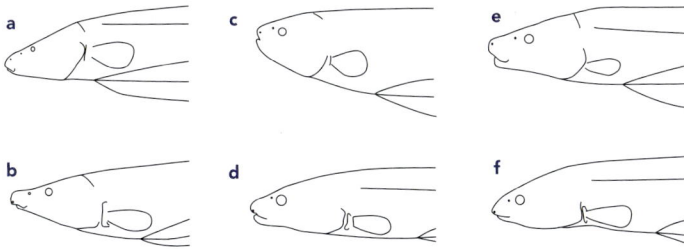

FIGURE 1 Head profiles of genera:
(a) *Akawaio penak*, (b) *Hypopomus artedi*,
(c) *Brachyhypopomus* sp., (d) *Microsternarchus bilineatus*, (e) *Procerusternarchus pixuna*, and
(f) *Racenisia fimbriipinna*.

1a. Snout more than one-third head length; posterior nares approximately equidistant between eye and snout tip; mouth strongly subterminal (fig. 1a,b) **2**
1b. Snout less than or equal to one-third head length; posterior nares closer to eye than tip of snout; mouth terminal or slightly subterminal (fig. 1c–f) **3**

2a. Opercle with convex (rounded) posterior margin; foramen absent on coracoids; reticulated or mottled pigmentation over most body surface; 17–22 pectoral-fin rays; <200 anal-fin rays; 15–16 precaudal vertebrae .... *Hypopomus*
2b. Opercle with concave posterior margin; foramen present on coracoids; reticulated or mottled pigmentation restricted to ventral body surface; 14 pectoral-fin rays; >200 anal-fin rays; 17 precaudal vertebrae ... *Akawaio*

3a. Adult body size 20–30 cm TL; body shape laterally compressed; pronounced pigment patterns (bars, stripes, or spots) usually present over most of body surface; no dorsal rami of posterior lateral line ....*Brachyhypopomus*
3b. Adult body size <13 cm TL; body shape cylindrical in cross section; no pronounced pigment patterns on body surface (except faint countershading or stripe over lateral line); dark pigmented dorsal rami of posterior lateral line on anterior-dorsal body surface (not readily visible in *Racenisia*) **4**

4a. Electric organ readily visible above most of anal-fin base; scales present on anterior portion of dorsum; 10 small scales above lateral line at midbody ....*Procerusternarchus*
4b. Electric organ not visible above most of anal-fin base; scales absent on anterior portion of dorsum; 3–5 large scales above lateral line at midbody **5**

5a. Parallel brown lines (dorsal branches of posterior lateral line nerve) visible on dorsolateral surface; body slightly deeper than wide, not greatly attenuated caudally; length of tail posterior to anal fin >16% total length; gill rakers absent ....*Microsternarchus*
5b. Parallel brown lines not readily visible on dorsal surface; body cylindrical, as deep as wide; girth strongly attenuated caudally; tail posterior to anal fin <16% total length; body uniformly dark brown; gill rakers present ......*Racenisia*

GENUS ACCOUNTS

Akawaio (21 cm TL)

Differs from other hypopomids by: long snout, more than one third head length (vs. snout less than one-third head length in most other hypopomid genera); posterior nares approximately equidistant between the eye and snout tip; mouth strongly subterminal; opercle with a concave posterior margin; reticulated or mottled pigmentation pattern restricted to ventral body surface; 14 pectoral-fin rays; 173–199 anal-fin rays and 17 precaudal vertebrae. **SPECIES** One, *A. penak* (Maldonado-Ocampo et al. 2014), known as *Logo-logo* by the Patamona indigenous communities in Guyana (N. Lujan pers. comm.). **DISTRIBUTION AND HABITAT** Known only from the Mazaruni River in the Essequibo basin of Guyana. **BIOLOGY** *Akawaio* feeds nocturnally over sandy banks no deeper than 1 meter, hiding during the day in woody debris and vegetation.

Brachyhypopomus (8.0–46 cm TL)

Discernible from other hypopomids by: larger adult body sizes; body shape subcylindrical to laterally compressed; prominent pigmentation patterns usually present over most of body surface, and absence of dorsal rami of the posterior lateral line. Additional characters of internal anatomy include: a gracile premaxilla with a curved anterior margin, a gracile dentary, and a body cavity with 16–17 precaudal vertebrae. Secondary sexual dimorphism has been observed in several species in which mature males grow a longer or thicker caudal appendage (Hopkins 1999, Sullivan et al. 2013b). **SPECIES** 28, with 22 species in the AOG region (Carvalho 2013), including 15 new species reported in Crampton et al. (2016b). **DISTRIBUTION AND HABITAT** Widely distributed in lowland freshwaters from southern Costa Rica to southern Uruguay. Abundant throughout most of the AOG region where they inhabit slowly moving streams, floodplains, and floating meadows, taking refuge in vegetation diurnally and foraging for invertebrates nocturnally (Crampton et al. 2016a). **BIOLOGY** *Brachyhypopomus* are cryptically pigmented, nocturnally active predators of small aquatic invertebrates and may be locally abundant in shallow and slow-flowing upland streams, swamps, and seasonal floodplains (Giora et al. 2014). *Brachyhypopomus occidentalis* from Panama, and *B. draco* from southern Brazil, breed during the transition from wet to dry season (Giora et al. 2008, Schaan et al. 2009).

Hypopomus (33 cm TL)

Differs from other hypopomids by: long snout, more than one-third head length (vs. snout less than one-third head length in most other hypopomid genera); posterior nares approximately equidistant between eye and snout tip; strongly subterminal mouth; opercle with convex (rounded) posterior margin; characteristic reticulated or mottled pigmentation over most body surface; 17–22 pectoral-fin rays; <200 anal-fin rays, and 15–16 precaudal vertebrae. **SPECIES** One, *H. artedi*, also known as *Logo-logo* by the Patamona indigenous communities in Guyana (N. Lujan pers. comm.). **DISTRIBUTION AND HABITAT** Small rivers and streams across most of the Guiana Shield, in Guyana, French Guiana, and Suriname. **BIOLOGY** Poorly known. Consumes primarily insect larvae (Mérigoux and Ponton 1998).

Microsternarchus (5.3–12 cm TL)

Differs from other hypopomids by: snout length moderate, not elongate or very short; preorbital region about one-third head length; nasal capsule near the eye, posterior nares closer to eye than tip of snout; mouth terminal or slightly subterminal; snout slender, longer than deep; absence of accessory electric organ near pectoral-fin base; small adult body sizes, <13 cm TL; body cylindrical in cross section; absence of pronounced pigment patterns on body surface (except faint countershading or stripe over lateral line); dark pigmented dorsal rami of posterior lateral line clearly visible on anterior-dorsal body surface; electric organ not visible above most of anal-fin base; scales absent on anterior portion of dorsum; 3–5 large scales above lateral line at midbody; body depth not greatly attenuated caudally; length of tail posterior to anal fin >16% total body length, and absence of gill rakers. **SPECIES** Two, both in the AOG region: *M. bilineatus* and *M. brevis*. **DISTRIBUTION AND HABITAT** Small forest streams throughout the Amazon and Orinoco basins, taking refuge during the day in aquatic vegetation and feeding nocturnally on small aquatic animals. **BIOLOGY** Individual *Microsternarchus* reported to exhibit distinct behavioral syndromes or "personalities" in their activity levels, aggression, and degree of behavioral flexibility (Berry 2011).

Procerusternarchus (14 cm TL)

Diagnosed by: snout length moderate, not elongate or very short, preorbital region about one-third total head length; nasal capsule near the eye; posterior nares closer to eye than tip of snout; terminal or slightly subterminal mouth; slender snout, longer than deep; absence of accessory electric organ near pectoral-fin base; small adult body size; a body shape cylindrical in cross section; absence of pronounced pigment patterns on body surface (except faint countershading or stripe over lateral line); presence of dark pigmented dorsal rami of posterior lateral line on anterior-dorsal body surface; electric organ readily visible above most of the anal-fin base; scales on anterior portion of dorsum; 10 small scales above lateral line at midbody. No known sexual dimorphism in either head or body proportions, or in electric signal waveform **SPECIES** One, *P. pixuna* (Cox-Fernandes et al. 2014). **DISTRIBUTION AND HABITAT** Swiftly flowing waters, small waterfalls, and river margins in Negro basin, Brazil. Refuge during the day in aquatic vegetation. **BIOLOGY** Feeds primarily on insect larvae. *Procerusternarchus pixuna* has the fastest and most stable electric organ discharge (EOD) of all hypopomids (and of all electric fishes with a pulse-type EOD), ranging from 100 Hz while at rest, to up to 144 Hz while actively foraging at night, and with coefficients of variation for the EOD repetition rates on the order of 10^{-4} (Cox-Fernandes et al. 2014).

Racenisia (12 cm TL)

Diagnosed by: snout length moderate, not elongate or very short; preorbital region about one-third head length; nasal capsule near eye; mouth terminal or slightly subterminal; snout slender, longer than deep; no accessory electric organ near pectoral-fin base; small adult body size (<13 cm TL); body cylindrical in cross section; no pronounced pigment patterns on body surface (except faint countershading or stripe over lateral line); body uniformly dark brown; pigmented dorsal rami of posterior lateral line on anterior-dorsal body surface present but not readily visible against dark background coloration; electric organ not visible above most anal-fin base; scales absent on anterior portion of dorsum; 3–5 large scales above lateral line at midbody; body cylindrical, about as deep as wide; girth strongly attenuated caudally; length of tail posterior to anal fin <16% total length; gill rakers present. **SPECIES** One, *R. fimbriipinna* (Mago-Leccia 1994). **DISTRIBUTION AND HABITAT** Small forest streams at the headwaters of the upper Negro and Orinoco basins. **BIOLOGY** No data available.

FAMILY RHAMPHICHTHYIDAE—PAINTED KNIFEFISHES, SAND KNIFEFISHES, AND TRUMPET KNIFEFISHES

— LESLEY Y. KIM and JAMES S. ALBERT

DIVERSITY 26 species in five genera and two subfamilies, Rhamphichthyinae (*Gymnorhamphichthys*, *Iracema*, and *Rhamphichthys*), and Steatogenae (*Hypopygus* and *Steatogenys*), and with 23 species in the AOG region. Rhamphichthyids exhibit a diversity of accessory electric organs on and around the head, including the chin (*Gymnorhamphichthys*, *Steatogenys*) and above the pectoral fin (in the painted knifefishes *Hypopygus* and *Steatogenys*). All rhamphichthyid species for which data are available have 2n = 50 chromosomes (Cardoso et al. 2011, da Silva et al. 2013).

COMMON NAMES *Sarapós de areia, Folheto, Pintado; Ituiterçado* (Brazil); *Macanas arena, Hojita, Cinturon, Sierra* (Peru).

GEOGRAPHIC DISTRIBUTION Most water bodies in lowland tropical South America east of the Andes, ranging from the La Plata estuary in Argentina to the Orinoco basin of Venezuela, and throughout most of Greater Amazonia.

ADULT SIZES A large range of body sizes, from 5.6 cm TL (including a very thin caudal appendage 1.5 cm TL long) in *Hypopygus hoedemani* from small rainforest streams draining the Guiana Shield region, to >100 cm TL in *Rhamphichthys rostratus* from the mainstream of the Amazon, Tocantins, Orinoco, and Essequibo rivers.

DIAGNOSIS OF FAMILY Recognizable among gymnotiform fishes by: laterally compressed body (also in the *Gymnotus carapo* group, Sternopygidae, and Apteronotidae versus a cylindrical or subcylindrical body that is roughly circular in cross section in Hypopomidae and several groups of Gymnotidae); terminal or subterminal mouth vs. superior mouth (in *Gymnotus*); anterior nares near margin of upper lip (or inside upper lip in *Hypopygus*) versus remote from upper lip in most gymnotiforms (also inside upper lip in *Gymnotus*); small eyes covered by skin vs. large eyes in most Sternopygidae (except some *Eigenmannia*) and not covered by skin in *Sternopygus* and *Archolaemus*; anal-fin origin anterior to branchial isthmus (also in Sternopygidae and Apteronotidae) versus posterior to the pectoral-fin base (in Hypopomidae and Gymnotidae); oral jaws lacking teeth (also in Hypopomidae and the apteronotid *Adontosternarchus*); absence of a thick fleshy semitransparent strip of tissue along base of anal fin (present in *Hypopygus*) versus present in Hypopomidae; obliquely angled and irregular dark pigment bands with wavy margins (also present in many *Gymnotus*; *Gymnorhamphichthys* with vertical pigment bars or saddles) versus even, mottled, or reticulated pigmentation pattern on the body surface in most Hypopomidae, or dark pigment saddles or vertical bars in most *Brachyhypopomus* species. Steatogenae are immediately recognizable by their short, deep head, and an accessory electric organ, appearing as a shiny smooth tissue patch, above the pectoral-fin base. Rhamphichthyinae are immediately recognizable by a long and tubular snout, much greater than one-third head length, and a very small mouth (also in some apteronotids).

SEXUAL DIMORPHISM Sexual differences in external morphology are not known in any rhamphichthyid species. Most, if not all, species differ in aspects of the electric organ discharge.

HABITATS Some rhamphichthyids move from shallow-water habitats (e.g., streams and floodplains) to deep river channels as they grow. Adults of some *Gymnorhamphichthys* inhabit the sandy bottoms of streams and small rivers in lowland Amazonian rainforests (Zuanon et al. 2006a). *Steatogenys elegans* and *S. ocellatus* inhabit white- and blackwater rivers, respectively, in the central Amazon, and *S. duidae* inhabits small rainforest streams throughout most of the Amazon and Orinoco basins (Crampton et al. 2004b, Crampton 2007). Some species of painted knifefishes (*H. lepturus*, *S. duidae*) are leaf-mimics, with a coloration that matches the leaf litter on the bottom of small rainforest streams (Schwassmann 1984, 1989, Triques 1996, 2005, 2007, Albert 2001, Lundberg 2005, Carvalho and Albert 2011, 2015, De Santana and Crampton 2011, Carvalho 2013).

FEEDING ECOLOGY Rhamphichthyines use their elongate tubular snouts and small mouths to forage on insect larvae and other small aquatic animals from river bottoms (Tesk et al. 2014). Steatogenes consume a variety of small aquatic animals, either from the water column in *S. elegans* and *S. ocellatus*, or from the leaf litter and aquatic vegetation in *S. duidae* (Crampton et al. 2004b, De Santana and Crampton 2011).

BEHAVIOR All rhamphichthyids are nocturnal and use a pulse-type EOD to forage and for sexual communication. Some species forage at night on flooded beaches and take refuge during the day in the deep river channels (Pimentel-Souza and Fernandes-Souza 1987, Schindler 1994).

KEY TO THE GENERA

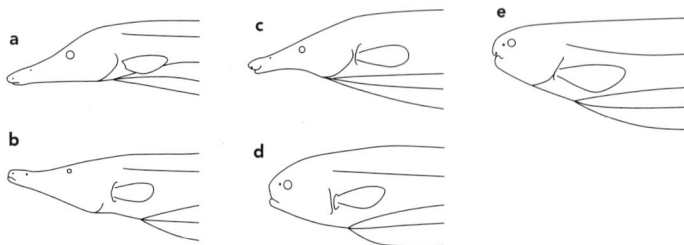

FIGURE 1 Head profiles of genera:
(a) *Gymnorhamphichthys bogardusi*,
(b) *Iracema caiana*, (c) *Rhamphichthys marmoratus*,
(d) *Hypopygus nijsseni*, and (e) *Steatogenys elegans*.

1a. Snout long and slender, more than one-third head length (fig. 1a–c); no accessory electric organ above pectoral-fin base ..**2**
1b. Snout short and deep, about one-third head length (fig. 1d,e); accessory electric organ (shiny smooth tissue patch) above pectoral-fin base **4**

2a. Anal fin relatively short, with 139–211 rays; pectoral fin with 10–13 rays; color pattern either present as vertical bars or saddles or absent; origin of the anal fin near the branchial opening ....***Gymnorhamphichthys***
2b. Anal fin relatively long, with ≥240 rays; pectoral fin with 15–22 rays; color pattern variable, never as vertical bars or saddles...**3**

3a. Anal-fin rays 240–257, pectoral-fin rays 15 or 16; series of dark rounded blotches on side of body over the lateral line; scales absent on anterior portion of body; anal-fin origin at vertical with brachial opening......................***Iracema***
3b. Anal-fin rays 304–470, pectoral-fin rays 17–22; dark oblique transverse bands on side of body; scales present on anterior portion of body; anal-fin origin anterior to vertical with brachial opening......................***Rhamphichthys***

4a. Paired filamentous mental accessory electric organs affixed to ventral head margin (fig. 4a; after Crampton et al. 2004b); posterior nares present; larger adult body sizes (21–41 cm TL) .... *Steatogenys*

4b. No accessory electric organs along ventral head margin; posterior nares absent; smaller adult body sizes (6.4–12 cm TL) .... *Hypopygus*

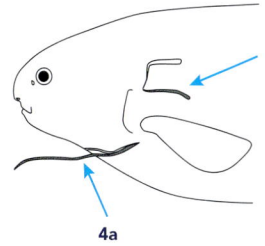

4a

GENUS ACCOUNTS

Gymnorhamphichthys (13–24 cm TL)

Distinguished from other genera of Rhamphichthyidae by: long snout, more than one-third head length (vs. less than one-third head length in *Hypopygus* and *Steatogenys*); absence of scales on anterior portion of body (vs. present in *Rhamphichthys*); a relatively low number of anal-fin rays (140–211) versus relatively higher number of rays (≥240) in *Iracema* and *Rhamphichthys*; anal-fin origin near branchial opening (vs. anterior to branchial opening in *Rhamphichthys*); pigments either as vertical bars or saddles or absent (vs. irregular oblique bands in *Rhamphichthys*). **SPECIES** Five, including four species in the AOG region. Review of genus and key to the species in Carvalho (2013). **COMMON NAMES** *Sarapó de areia* (Brazil); *Macana de arena* (Peru); Sand knifefishes (English). **DISTRIBUTION AND HABITAT** Channels and flooded beaches of medium to large lowland rivers like the Amazon, Orinoco, and Essequibo. *Gymnorhamphichthys rondoni* inhabits terra firme forest streams and smaller tributaries of the Guiana and Brazilian shields. **BIOLOGY** *G. rosamariae* undergo daily migrations from deeper portions of the river during the day onto flooded beaches at night in the Negro River (Schwassmann 1971); *G. rondoni* swim along the stream bottom in a head-down body posture, probing the substrate repeatedly with their long snouts (Brejão et al. 2013).

Hypopygus (6.4–15 cm TL)

Distinguished from other gymnotiform fishes by: a short snout, less than one-third the head length; a pointed snout in lateral profile; a terminal mouth; absence of posterior nares; an accessory electric organ over each pectoral fin that extends anteriorly onto the back of the head (vs. posteriorly over the pectoral fin in *Steatogenys*); absence of filamentous accessory electric organs on the ventral surface of the head; a semitranslucent (vs. opaque) body with pigment patches composed in obliquely oriented bands and saddles, never extending onto anal-fin membrane; and small adult body sizes (not more than 15 cm TL). **SPECIES** Nine, all in the AOG region. Review of genus and key to the species in De Santana and Crampton (2011). **COMMON NAMES** *Sarapó folheto* (Brazil); *Macana folleto* (Peru); Leaflet knifefishes (English). **DISTRIBUTION AND HABITAT** Widely distributed throughout the Amazon, Orinoco, and Essequibo basins, where it inhabits rootmats, undercut banks, and leaf litter of small rainforest streams and creeks. **BIOLOGY** *Hypopygus* consume small aquatic invertebrates, mostly aquatic insect larvae, and also ants and termites that fall into the water. Female *H. lepturus* place eggs on the surface of leaves and roots. *Hypopygus* are not fished by humans for food or other reasons, and members of the Tukano and Tuyuka tribes in the Brazilian Amazon have taboos against their consumption (Lima et al. 2005).

Iracema (36 cm TL)

Distinguished from other rhamphichthyids by:
one series of irregular to roundish dark blotches on the side of the body along the lateral line; the loss of scales on the anterior portion of the body; an intermediate number (240–257) of anal-fin rays (vs. 139–211 in *Gymnorhamphichthys* and 304–470 in *Rhamphichthys*); and 15–16 pectoral-fin rays (vs. 10–13 in *Gymnorhamphichthys* and 17–22 in *Rhamphichthys*). **SPECIES** One, *I. caiana*. Species information in Triques (1996) and Carvalho and Albert (2011). **DISTRIBUTION AND HABITAT** Known only from the type locality on the Jauaperi River, a tributary of Negro River in the Amazon basin. **BIOLOGY** No data available.

Rhamphichthys (40–100 cm TL)

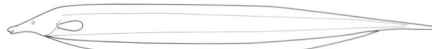

Distinguished from other members of
Rhamphichthyidae by: a long slender snout, more than one-third the head length; a high number of pectoral-fin rays (17–22), a skin fold inside the branchial opening; origin of anal fin anterior to vertical with branchial opening; >300 anal-fin rays; body entirely covered with scales. **SPECIES** Seven, six in the AOG region. Review of genus and key to the species in Carvalho (2013). **COMMON NAMES** *Itui-terçado*, *Peixe-tatu* (Brazil); *Macana cinturon* (Peru); Ossa knifefishes (English). **DISTRIBUTION AND HABITAT** Vegetated margins and benthos of large lowland rivers and wetlands throughout much of cis-Andean tropical South America, including the Orinoco, Essequibo, Amazon, Parnaiba, Paraná, and Paraguay basins, as well as coastal drainages of the Guianas. **BIOLOGY** Little is known of their reproductive or behavioral biology in the wild. In aquaria *Rhamphichthys* stay motionless for hours during the daytime, becoming active only at night (Kawasaki et al. 1996; JSA pers. obs.).

Steatogenys (21–41 cm TL)

Distinguished from other gymnotiform fishes
by: a short snout, less than one-third the head
length; a rounded snout in lateral profile; a
subterminal mouth; an accessory electric organ over each pectoral fin that extends posteriorly along the lateral line over the pectoral fin; paired filamentous accessory electric organs on the ventral surface of head; and an opaque body with multiple dark saddles and obliquely oriented bands with irregular margins on body and anal-fin membrane. **SPECIES** Three, all in the AOG region. **COMMON NAMES** *Sarapó pintado* (Brazil); *Macana sierra* (Peru); Leaf and Painted knifefishes (English). **DISTRIBUTION AND HABITAT** *S. duidae* (Leaf knifefish, 21 cm TL) inhabits small upland streams in the Amazon and Orinoco basins; *S. elegans* (Painted knifefish, 27 cm TL) inhabits large nutrient-rich and turbid whitewater rivers in the Amazon and Orinoco; and *S. ocellatus* (41 cm TL) inhabits nutrient-poor blackwater rivers, e.g., Negro and Tefé rivers. **BIOLOGY** *S. duidae* is a leaf mimic, with a deep, laterally compressed body, a color pattern strongly resembling a dead leaf, and a behavior of lying on its side in root tangles and leaf litter in the daytime (Sazima et al. 2006). *Steatogenys elegans* congregate into large schools in rivers. *Steatogenys duidae* forage along the substrate at night close to stream margins (Brejão et al. 2013), while *S. elegans* undergo daily migrations of several hundred meters from deeper portions of the river during the day onto flooded beaches at night (Steinbach 1970). The social role of electric signals is poorly understood. When presented with electric signals of the same frequency, *S. elegans* sometimes respond with "chirplet" responses, brief increases in EOD rate with decreases in amplitude (Field and Braun 2012).

FAMILY STERNOPYGIDAE—GLASS KNIFEFISHES, RATTAIL KNIFEFISHES

— *BRANDON T. WALTZ and JAMES S. ALBERT*

DIVERSITY 44 species in five genera (*Archolaemus, Distocyclus, Eigenmannia, Rhabdolichops,* and *Sternopygus*), with 30 species in the AOG region. In addition, †*Humboldtichthys kirschbaumi* from Bolivia is known only as a fossil (Gayet et al. 1994, Albert and Fink 2007). Sternopygids are highly conserved in overall external appearance, although they do differ visibly in aspects of head and snout shape, mouth position, eye size, squamation, and body coloration (Albert 2001, 2003). Some subterranean populations of *E. vicentespelaea* completely lack eyes (Bichuette and Trajano 2006).

GEOGRAPHIC DISTRIBUTION Freshwater systems throughout much of tropical and subtropical South America, from the pampas of Argentina and Uruguay, to rivers in the arid Caatinga thorn-forest of northeastern Brazil, to freshwater marshes and estuaries on the Pacific coast from northern Peru to Panama.

ADULT SIZES From 14 cm TL in *Eigenmannia muirapinima* from streams and caves in the upper Tocantins basin, to 70 cm TL in *Sternopygus macrurus* from the lowland Amazon basin (Viana and Lucena Frédou 2014), and 56 cm TL in *S. obtusirostris* from floodplain lakes in the eastern Amazon basin (Crampton et al. 2004a, Hulen et al. 2005).

DIAGNOSIS OF FAMILY Readily distinguished from other gymnotiforms by: expanded infraorbital laterosensory canal bones, and numerous villiform teeth on the premaxilla and dentary. Sternopygids can be further distinguished from other gymnotiform families by the following unique combination of characters: terminal or subterminal mouth (vs. superior mouth in *Gymnotus* and some sternarchelline apteronotids); relatively large eyes (reduced or absent in some populations of *E. vicentespelaea*); snout length moderate, preorbital distance from eye to tip of snout less than post-orbital distance from eye to back of head (preorbital distance longer than postorbital distance in most *Archolaemus*); absence of dark pigment saddles or narrow vertical bars (broad vertical bars present in *E. kirschbaum*); absence of accessory of electric organs; absence of a caudal fin (present in Apteronotidae); absence of fleshy strap of tissue (dorsal organ) on the dorsal midline of the posterior portion of the body (present in Apteronotidae).

SEXUAL DIMORPHISM Sexual differences in external morphology are not known in most sternopygid species. In *E. virescens* males grow to larger size than females, but this is not the case for *S. macrurus* (Hulen et al. 2005). Sexual dimorphism has been reported in *E. trilineata*, with females possessing greater body depth at maturity than males (Kirschbaum 1995). Mature males of *A. ferreirai* have darker pigmentation, and mature males of *A. luciae* develop a somewhat elongate snout. The electric organ discharge is sexually dimorphic in most, if not all, sternopygids (Crampton and Albert 2006). In some members of the *E. trilineata* group (synonymous with *E. virescens* of Kramer, 1987), mature males discharge at a slightly higher frequency than females or juveniles (Kramer 1987), and in *S. macrurus* at a lower frequency (Fleishman 1992). In *S. macrurus* the individual electric pulses of sexually mature males are also longer and "louder" (have higher amplitude) than those of females (Mills and Zakon 1991).

HABITATS Inhabit almost all freshwater habitats of lowland (below about 500 m or 1,640 ft elevation) tropical South America, ranging from small, meandering rainforest streams to floodplain lakes and the rapids and deep channels of large rivers. *Sternopygus macrurus* is the most ecologically tolerant of all gymnotiform species, living in almost all lowland freshwater habitats from the Pacific slope of Colombia to northern Argentina. Members of the *Eigenmannia trilineata* species group have an even larger geographic distribution from Panama throughout most of the lowland humid Neotropics. Several sternopygids are specialized to live in the channels of large lowland Neotropical rivers, including *Distocyclus conirostris* and *Rhabdolichops* spp. *Archolaemus* spp. inhabit rapids of large rivers draining the Brazilian and Guiana shields, and *E. vicentespelaea* inhabit swiftly flowing streams in caves and at the surface in the karst landscape of the upper Tocantins basin.

FEEDING ECOLOGY Most sternopygids are generalist carnivores, consuming zooplankton as larvae and small juveniles, and small crustaceans, insects, and fish larvae as they grow. Larger individuals can become piscivorous, consuming adult fishes. *Sternopygus macrurus* is a highly generalist feeder as an adult, consuming small fishes and crustaceans, fruits, and insect larvae (Olaya-Nieto et al. 2009). Adult *Eigenmannia* feed at night on microcrustaceans (water fleas, ostracods, etc.) and insects by capturing prey items individually from the substrate (Giora et al. 2005, Brejão et al. 2013). *Rhabdolichops* are highly specialized zooplanktivores that filter food items from the water with elongate gill rakers (Lundberg and Mago-Leccia 1986).

BEHAVIOR All sternopygids are nocturnal, and all produce a weak (<1 volt) monophasic wave-type electric organ discharge, used for foraging, communication, and navigating in murky, turbid waters. In *E. virescens*, electric organ discharges are used in many social and sexual behaviors, including aggression, courtship, and spawning (Hagedorn and Heiligenberg 1985). Spawning is preceded by several nights of courtship during which males modulate their electric organ discharge to produce "chirps" by briefly increasing the discharge cycling frequency (repetition rate). Gravid females can be stimulated to spawn by hearing a recording of male courtship chirps. As with many floodplain fish species, *E. virescens* use aquatic surface respiration when oxygen levels are hypoxic or anoxic, and reduce their general activity levels (Crampton 1998a).

ADDITIONAL NOTES Sternopygids have high abundance and biomass in many aquatic ecosystems, and are important components of the food web in the Amazon and Orinoco rivers (Lundberg and Lewis 1987, Crampton 1996, Lewis Jr. et al. 2001, Gimenes et al. 2013). Most sternopygids are not commercially exploited or used in aquaculture. Large-bodied specimens of *Sternopygus macrurus* (>30 cm TL) are often sold in fish markets throughout lowland Amazonia. The "glass electric fish" is exported in large numbers in the commercial aquarium industry (Henderson and Crampton 1997). This species, with 2–3 thin dark pigment lines along each side of the body, is widely but incorrectly referred to as "*E. virescens*" in the literature, and is in fact a member of the *E. trilineata* species group.

KEY TO THE GENERA AND SPECIES GROUPS

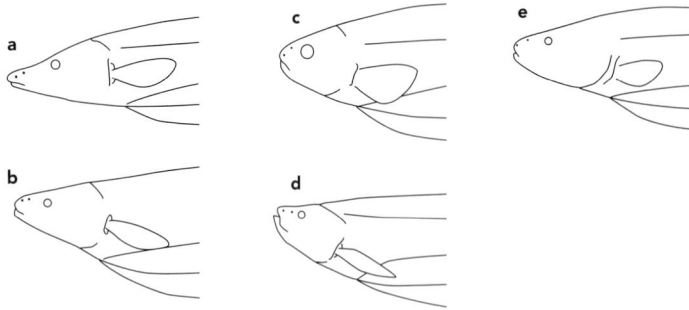

FIGURE 1 Head profiles of genera: (a) *Archolaemus blax*, (b) *Distocyclus conirostris*, (c) *Eigenmannia* sp., (d) *Rhabdolichops troscheli*, and (e) *Sternopygus macrurus*.

1a. Orbital margin free; eye not covered by skin and surrounded by groove; >12 precaudal vertebrae; brown or gray body color with fine, dark chromatophores 2
1b. Orbital margin covered by skin; <15 precaudal vertebrae; body color yellow, white, or translucent..................3

2a. Snout rounded, less than or equal to one-third head length (fig. 1e); nasal capsule not closer to snout tip than to eye; all anal-fin rays unbranched; thin white or yellow longitudinal stripe between axial and pterygiophore muscles; black humeral patch often present (not readily visible in very dark *S. macrurus*; absent in *S. branco*).. *Sternopygus*
2b. Snout pointed, more than one-third head length (fig. 1a); nasal capsule closer to snout tip than eye; posterior anal-fin rays branched; dark brown ground color over epaxial muscles and anal-fin pterygiophores; broad pale longitudinal band overlying hypaxial muscles; no humeral spot or white longitudinal stripe (may possess lighter-colored longitudinal band) .... *Archolaemus*

3a. Snout conical, approximately one-third head length (fig. 1b); small nasal capsules; nares closely spaced; anal fin with >200 rays; electric organ not visible over anal-fin pterygiophores .... *Distocyclus*
3b. Snout blunt, less than one-third head length (fig. 1c,d); moderate-sized nasal capsules, with nares not closely spaced; anal fin with <220 rays; electric organ visible as translucent brick-like array over caudal anal-fin pterygiophores 4

4a. Scales present on anterior middorsum; mouth terminal to subterminal; branchial opening small, less than length of snout; electrocytes square in profile; gill rakers not attached to gill arches...............................5
4b. Scales absent on anterior middorsum; mouth terminal to superior; branchial opening longer than length of snout; electrocytes large and rectangular, longer than deep; gill rakers attached to gill arches...............................6

5a. Body laterally compressed, semi translucent in life; dark, narrow longitudinal stripes; body shape slender, its depth at anal-fin origin 13–17% length to end of anal fin ....*Eigenmannia trilineata* group
5b. Body robust and opaque in life; longitudinal stripes absent; body shape deep, its depth ≥18% of length to end of anal fin, some individuals dark in color..*Eigenmannia humboldtii* group
5c. Body fairly laterally compressed, translucent white/yellow in life; longitudinal stripes absent; eye large (greater than or equal to snout length), long caudal filament (half of body length without head)..*Eigenmannia macrops* group

6a. Adults with black chromatophores on pectoral-fin rays; pectoral fins longer than head length .... *Rhabdolichops nigrimans* group
6b. Dark pigments absent on pectoral-fin rays; pectoral fins shorter than head length .... *Rhabdolichops troscheli* group

GENUS ACCOUNTS

Archolaemus (18–38 cm TL SL)

Distinguished from other sternopygid genera
by: pointed and long snout, more than one-
third of head length, with nasal capsule closer to snout tip than to eye (except *A. orientalis*);
adults with a free orbital rim (also present in *Sternopygus*); mobile teeth on premaxilla (only
anterior margin of first tooth row attaching firmly to premaxilla); pronounced gap between
posterior margin of upper lip and anterior margin of premaxilla; a sponge-like ventral margin of
upper lip with extruded papillae and fleshy elongate ridges; and broad, pale longitudinal band
overlying hypaxial muscles. *Archolaemus* can also be separated from *Eigenmannia*, *Distocyclus*, and
Rhabdolichops by a dark brown ground color, and from *Sternopygus* by branched posterior anal-fin
rays (vs. unbranched in the latter). **SPECIES** Six, including five species in the AOG region. Among
other traits, these species differ in the relative proportions of the preorbital and postorbital
regions of the head and the relative size of the eyes. Vari et al. (2012) provide an identification
key and diagnoses of all *Archolaemus* species. **COMMON NAMES** *Sarapó torrente* (Brazil); *Cuchillo
torrente* (Venezuela); Glass knifefishes (English). **DISTRIBUTION AND HABITAT** Swiftly flowing rivers
and rapids of the Brazilian and Guiana shields. **BIOLOGY** No data available.

Distocyclus (29–50 cm TL)

Distinguished from other sternopygid genera
by: snout conical (vs. rounded in lateral view),
approximately one-third head length; small nasal capsules, nares closely spaced; a single row of
teeth, with ≤6 teeth, limited to the anterior portion of dentary (vs. ≥11 teeth in *Eigenmannia*);
anal fin with >200 rays; and electric organ not visible over anal-fin pterygiophores (Dutra
et al. 2014, Meunier et al. 2014). **SPECIES** Three, including two species in the AOG region.
DISTRIBUTION AND HABITAT *D. conirostris* (50 cm TL) inhabits large rivers throughout the Amazon
and Orinoco basins in Brazil, Colombia, Peru, Ecuador, and Venezuela, and the Essequibo basin
in Guyana; and *D. guchereauae* (29 cm TL) in the Maroni basin in French Guiana. *Distocyclus*
species inhabit floodplain lakes, flooded forests, and river channels. **BIOLOGY** No data available.

Eigenmannia (12–49 cm TL)

Distinguished from other sternopygid genera
by: a short snout, less than one-third head
length, and relatively large eyes (except *E. vicentespelaea*). **SPECIES** 17 species in three species
groups, of which 9 species inhabit the AOG region (Peixoto et al. 2015). Species of the
E. humboldtii group grow to larger body sizes (>45 cm TL), possess a relatively deep body at
maturity (body depth >11% TL), and a darker body coloration in some specimens. Species of
the *E. trilineata* group grow to a smaller adult body size, possess a relatively slender body (its
depth <11% TL), and a semitransparent body with 2 or 3 thin dark pigment lines along each
side of the body. The *E. trilineata* group includes several genetically distinct but morphologically
cryptic species (Silva et al. 2009a, Moysés et al. 2010, Henning et al. 2011). *Japigny kirschbaum*
(Meunier et al. 2011) is here treated in the genus *Eigenmannia* according to results of a
combined molecular and morphological phylogenetic analysis by Tagliacollo et al. (2016).
COMMON NAMES *Sarapó transparente* (Brazil); *Cuchillo transparente* (Colombia and Venezuela);
Macana transparente (Ecuador and Peru); *Loga loga* (Guyana); Glass knifefishes (English).
DISTRIBUTION AND HABITAT Widespread throughout the lowlands of temperate and tropical South
America, from the pampas of Argentina to the Caatinga of northeastern Brazil, the Baudó and
San Juan basins on the Pacific coast of Colombia, and the Tuyra basin in Panama. There are
two species of the *E. humboldtii* group in Greater Amazonia: *E. nigra* from the upper Orinoco
and upper Negro rivers in Brazil and Venezuela (48 cm TL); and *E. limbata* from the lowlands

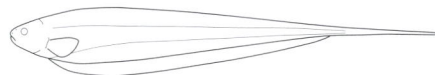

of the Amazon and Orinoco basins in Brazil, Colombia, Ecuador, and Venezuela (49 cm TL). There are six species of the *E. trilineata* group in Greater Amazonia. *Eigenmannia macrops*, belonging to a monotypic species group, is known from the Amazon and Orinoco lowlands of Bolivia, Brazil, Colombia, Ecuador, Peru, and Venezuela, and the Essequibo basin in Guyana (25 cm TL). *Eigenmannia kirschbaum* is known from the Mana and Maroni basins in French Guiana (22 cm TL). *Eigenmannia* species inhabit most freshwater systems throughout this broad geographic range, usually slowly moving water in streams and floodplains and at the margins of large rivers. *Eigenmannia* species are commonly found near floating vegetation and near rooted vegetation close to shore. **BIOLOGY** Adult *Eigenmannia* feed at night on microcrustaceans (water fleas, ostracods, etc.) and insects by capturing prey items individually from the substrate (Giora et al. 2005, Brejão et al. 2013). *Eigenmannia macrops* occurs in large numbers in deep channels of lowland rivers in Greater Amazonia, moving to shallower marginal habitats at night (JSA pers. obs.; J. Zuanon, pers. comm.).

Rhabdolichops (18–49 cm TL)

Distinguished from other sternopygid genera by: absence of scales on the anterior middorsum; moderate body cavity length (12–15 precaudal vertebrae) (Hulen et al. 2005); enlarged and barrel-shaped electrocytes that can be observed externally (with transmitted light) at the back of the body and tail; and an elongate premaxilla that extends laterally and posteriorly around the oral margin (Lundberg and Mago-Leccia 1986, Albert 2001). Some *Rhabdolichops* are semitransparent (similar to the *Eigenmannia macrops* and *virescens* groups). **SPECIES** 10 species in two species groups, all in the AOG region. The two species of the *R. nigrimans* group have long pectoral fins, exceeding total head length, and dark pigments on the pectoral fins as adults. The eight species of the *R. troscheli* group have shorter pectoral fins (less than total head length), and lack dark pigments on the pectoral fins. Lundberg and Mago-Leccia (1986) and Correa et al. (2006) provide identification keys and diagnoses of *Rhabdolichops* species. **DISTRIBUTION AND HABITAT** The two species of *R. nigrimans* group are both found in the AOG: *R. lundbergi* (24 cm TL) from the whitewater Amazon River in the central Amazon basin; and *R. nigrimans* (42 cm TL) from blackwater rivers in the central Amazon basin. Most members of the *R. troscheli* group (except *R. navalha* and *R. zareti*) have broad geographic distributions in large rivers of the Amazon and Orinoco basins; *R. navalha* is restricted to blackwater rivers in the central Amazon basin and *R. zareti* to whitewater rivers in the Orinoco basin. Species typically inhabit channels of large lowland rivers. **BIOLOGY** *Rhabdolichops* species feed on small aquatic animals and some feed mostly or entirely on zooplankton (Lundberg and Mago-Leccia 1986, Crampton 1998b).

Sternopygus (23–58 cm TL)

Distinguished from other sternopygid genera by: all anal-fin rays unbranched (shared with the apteronotid *Sternarchorhamphus*). Most species possess a thin white or yellow longitudinal stripe between the axial and pterygiophore muscles, and a black humeral patch (not readily visible in very dark specimens of *S. macrurus*); *S. branco* has very little pigmentation, and *S. astrabes* has dark pigment saddles across the dorsal midline (Hulen et al. 2005). Also separated from other sternopygid genera by: eyes with a free orbital rim (also present in *Archolaemus*) versus eye covered by skin in other sternopygids. *Sternopygus* is readily separated from *Archolaemus* by its shorter and rounder snout (with a concave forehead in mature *S. xingu*), and absence of a broad pale lateral stripe. All *Sternopygus* species except *S. astrabes* attain large body sizes of >40 cm TL total length. **SPECIES** Eight, including four in the AOG region. Hulen et al. (2005) provide an identification key and diagnoses of all *Sternopygus* species. **COMMON NAMES** *Sarapó limpo* (Brazil); *Macana búfalo* (Peru); *Pejeraton, Ratona* (Colombia, Venezuela); Rat-tail, Gold-line knifefish (English). **DISTRIBUTION AND HABITAT** The most geographically

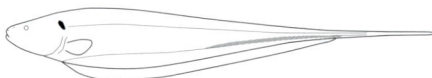

widespread genus of all Sternopygidae, with species present throughout the Amazon, Orinoco, and Guyana basins, as well as trans-Andean waters of the Maracaibo and Magdalena basins, and the Pacific slope of Colombia, Panama, Ecuador, and Peru. Four species are present in the AOG: *S. branco* from the Amazon River in Brazil (49 cm TL); *S. macrurus* from across most of tropical South America, including La Plata River in northern Argentina, the São Francisco and Parnaíba basins of northwestern Brazil, and the Rio Baudó on the Pacific coast of Colombia (70 cm TL); *S. obtusirostris* from the Amazon of Brazil (56 cm TL); and *S. xingu* from the clearwater Xingu and Tocantins basins of the Brazilian Shield (53 cm TL). Members of this genus can be found in upland streams and rivers, floodplains, and the margins and bottom of large rivers.

BIOLOGY *S. macrurus* is the most ecologically tolerant of all gymnotiform species, living in almost all lowland freshwater habitats from high-gradient mountain streams to the floodplains and benthos of large lowland rivers. *Sternopygus macrurus* ranges in color pattern from jet black in the upper Negro River of Brazil, where the characteristic humeral blotch is not easily visible, to whitish pink in the upper Juruá (Yurua) River of Peru, and with individuals ranging from the presence of a well-developed white lateral stripe to none at all. Some populations of *S. macrurus* from coastal rivers in the Brazilian state of Pará migrate at night into brackish-water mangroves where they forage on marine shrimps, presumably without the use of their electric sense that does not function well in salty water (J. Ready pers. comm.). *Sternopygus macrurus* is a highly generalist feeder as an adult, consuming small fishes and crustaceans, fruits, and insect larvae (Olaya-Nieto et al. 2009).

FAMILY ANABLEPIDAE—FOUR-EYED FISHES
— *PETER VAN DER SLEEN and JAMES S. ALBERT*

Family includes 18 species in three genera: *Anableps* (Four-eyed fishes, 3 species), *Jenynsia* (One-sided livebearers, 14 species) and *Oxyzygonectes* (White-eye, 1 species). Only the genus *Anableps* inhabits the AOG region.

Anableps (22–32 cm SL)

Recognized by their distinct eyes, with each eye divided lengthwise forming two pupils, one dorsal and one ventral. The eyes are prominently raised above the top of head, and individuals often swim with the center of the eye at the water's surface and are capable of simultaneous aerial and aquatic vision (Ghedotti 2003). Males have a urogenital papilla or gonopodium, which is naturally oriented to the left or right depending on the individual (Ghedotti 2003). **SPECIES** Three, including two in the AOG region: *A. anableps* and *A. microlepis*. **COMMON NAMES** *Quatro-olhos*, *Tariota* (Brazil); *Grosjé vaz*, *Kutali* (French Guiana), *Cipotero escamoso* (Venezuela). **DISTRIBUTION AND HABITAT** Coastal drainages of Central America and northern South America from Venezuela to the Amazon River delta. Inhabit muddy tidal flats, mangroves, full seawater and freshwaters. Some *Anableps* species undertake regular intertidal migrations (Brenner and Krumme 2007). **BIOLOGY** They are surface feeders that eat predominantly terrestrial insects. Occasionally capture insects in the air. Also consume diatoms and invertebrates on tidally exposed silt flats (Zahl et al. 1977, Miller 1979, Ghedotti 1998). They are viviparous (e.g., Burns and Flores 1981, Garman 1986, Oliveira et al. 2011b).

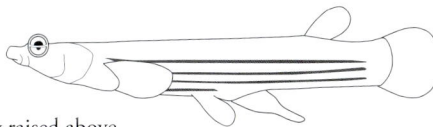

FAMILY CYPRINODONTIDAE—PUPFISHES
— *PETER VAN DER SLEEN and JAMES S. ALBERT*

Family includes 139 species in nine genera worldwide. Only the genus *Orestias* inhabits the AOG region, occurring in high-elevation lakes and streams in the Andean mountains.

Orestias (3.5–27 cm SL)

Characterized by: absence of pelvic fins; reduced body squamation; and a unique squamation and head pore pattern characterized by a dorsal median ridge of scales from the head to the dorsal fin and a lyre-shaped arrangement of neuromasts on the dorsal surface of the head (Parenti 1984). Females can be much larger than males. **SPECIES** 45, including at least 6 species in the headwaters of the Amazon basin. Review of genus and key to the species in Parenti (1984), except for *O. piacotensis* (Vila 2006) and *O. gloriae* (Vila et al. 2012). **COMMON NAMES** *Carachi, Challhua* (Peru). **DISTRIBUTION AND HABITAT** Endemic to high-altitude lakes and tributary streams of the Peruvian, Bolivian, and Chilean Andes. The greatest diversity of *Orestias* is concentrated in the Lake Titicaca basin, but the *Orestias* species in the lake do not represent a single evolutionary radiation (Lüssen et al. 2003). The "carachito" *Orestias* cf. *agassizii* (7.0 cm SL) exhibits most of the geographic and elevational range of the genus as a whole, including Andean lakes of southeastern Peru, eastern Bolivia, and northern Chile, including Lake Titicaca and the upper Amazon basins, and rivers from 2,800 m (c. 9,200 ft) at Urubamba, Peru, to 5,500 m (c. 18,000 ft) elevation at Cailloma on the Rio San Ignacio, Peru. **BIOLOGY** Planktivorous or omnivorous. Egg-laying, with external fertilization (males without a gonopodium).

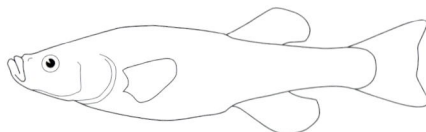

FAMILY POECILIIDAE—LIVEBEARERS
— *PAULO H. F. LUCINDA and PETER VAN DER SLEEN*

DIVERSITY Approximately 220 species in 28 genera and three subfamilies worldwide: Aplocheilichthyinae, Procatopodinae, and Poeciliinae (Lucinda and Reis 2005). The family includes well-known aquarium fish such as guppies, mollies, mosquito fishes, platies, and swordtails. Twenty species in 7 genera in the AOG region.
COMMON NAMES *Barrigudinho, Lebiste, Guaru* (Brazil); *Gupi* (Peru).
GEOGRAPHIC DISTRIBUTION North America, Middle America (including the Caribbean Islands), and South America to northern Argentina, Congo basin and the African rift lakes, Dar es Salaam and Madagascar (Parenti 1981). The subfamily Aplocheilichthyinae is restricted to Africa, the Poeciliinae inhabit the Americas, and Procatopodinae are mainly from Africa, but also include the South American genus *Fluviphylax*.
ADULT SIZES Small to very small; the species occurring in the AOG region range in size from 1.5 cm SL (*Fluviphylax palikur*) to 6.0 cm SL (female *Poecilia reticulata*).
DIAGNOSIS OF FAMILY Characterized by: (1) a highly inserted pectoral fin, (2) pelvic fins that migrate anteriorly during growth, (3) recessed supraorbital pores 2b through 4a (see figure, right; after Rodriguez 1997), and (4) pleural ribs on the first several hemal arches as well as a series of other internal characters (Parenti 1981, Costa 1998b, Ghedotti 2000). The subfamily Poeciliinae is additionally characterized by (1) the possession of a gonopodium (see figure below), formed by the modified male anal-fin rays 3, 4, and 5; (2) internal fertilization; (3) viviparity, with facultative viviparity in *Tomeurus gracilis*; and (4) ventral portion of proximal anal-fin radials 6–10 in adult males not laterally compressed and without anterior and posterior flanges (Lucinda 2003, Lucinda and Reis 2005).
SEXUAL DIMORPHISM Pronounced in some species, including brighter body and fin colors in males. Females are often larger than males. In the subfamily Poeciliinae, males possess a gonopodium, which is used for internal fertilization.
HABITATS Fresh and brackish waters, a few species have been reported to occur in salt waters in coastal areas (e.g., *Gambusia rhizophorae, Poecilia latipinna, Poecilia vandepolli, Poeciliopsis turrubarensis, Poeciliopsis elongata*).

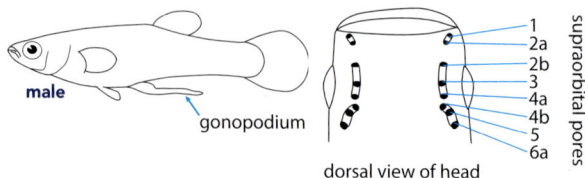

male

gonopodium

dorsal view of head

1
2a
2b
3
4a
4b
5
6a

supraorbital pores

FEEDING ECOLOGY Generally surface feeders and omnivorous, with diets including zooplankton, small insects, and detritus.

BEHAVIOR Although the family is known as Livebearers, species of the subfamily Procatopodinae are egg-scattering with external fertilization, and facultative viviparity is a trait in the most early-branching poeciliine, *Tomeurus gracilis*. Males of some poeciliines exhibit courtship displays in an attempt to persuade females to be cooperative during copulation. Other poeciliines possess a behavior called gonopodial thrusting, during which the male furtively introduces the gonopodium tip in the genital opening of the female without her cooperation. Species with courtship displays are usually sexually dichromatic and have short gonopodia, whereas species exhibiting gonopodial thrusting behavior are sexually monochromatic and have long gonopodia.

KEY TO THE GENERA

1a. Dorsal fin small, composed of just 4–6 rays, and set rather far back on the body, posterior to middle of the body (fig. 1a) **2**
1b. Dorsal fin with ≥7 rays, set on middle of body or just slightly anterior or posterior to middle (fig. 1b) **3**

2a. Eyes extremely large (about 50% of the head length; fig. 1a); no gonopodium (anal fin not modified) in males ..... *Fluviphylax*
2b. Eyes large (but <50% of head length); gonopodium present in males (and with elaborate modifications) ...... *Tomeurus gracilis*

3a. Adult males with elongate pelvic fins (see fig. 4); males with 6 pelvic-fin rays; membranous appendix at tip of R3 absent; fleshy palp on R3 present (fig. 3a) **4**
3b. Adult males without elongate pelvic fins; males with 4 or 5 pelvic-fin rays; membranous appendix at tip of R3 present (see fig. 6); fleshy palp on R3 absent **6**

4a. Second ray of pelvic fin in males with triangular, comb-like shape (fig. 4a; after Rodriguez 1997) .... *Pamphorichthys*
4b. Second ray of pelvic fin in males without a triangular, comb-like shape (fig. 4b; after Rodriguez 1997) **5**

5a. Males with 6th and 7th anal-fin rays and pelvic fins reaching the base of the gonopodial palp; males usually polychromic; males and females usually with shoulder spot **7**
5b. Males with 6th and 7th anal-fin rays and pelvic fins not reaching the base of the gonopodial palp; males and females more or less monomorphic .... *Poecilia*

6a. Membranous appendix at tip of R3 single (fig. 6a; after Lucinda 2005); males with 4 pelvic-fin rays; body usually with several dark bars along sides; males with a dark brown spot posterior to anal-fin base continuous ventrally side by side and meeting ventral median line of caudal peduncle .... *Cnesterodon septentrionalis*
6b. Membranous appendix at tip of R3 bifid (fig. 6b; after Lucinda 2008); males with 5 pelvic-fin rays; a single roundish to rounded and well-defined ocellated lateral spot on each side of the body; dark brown spot posterior or anal-fin base in males absent .... *Phalloceros leticiae*

7a. Gonopodium palp extremely long and slender, R3 with very long comb-like spines .... *Acanthophacelus reticulatus* (see genus description of *Poecilia*)
7b. Gonopodium palp short and not slender, R3 with thorn-like spines .... *Micropoecilia*

GENUS ACCOUNTS

Cnesterodon (3.0 cm SL)

Distinguished by the following unique characters: dark brown spot posterior to anal-fin base of males continuous ventrally side by side and meeting ventral median line of caudal peduncle; large bony basal process on first anal-fin proximal radial in adult males; unpaired appendix at tip of anal-fin ray 3 (R3); and distal segment at tip of anterior ramus of anal-fin ray 5 (R5a) transformed into a posteriorly oriented triangular spine (Lucinda 2005). Other characters that could aid identification: preopercular canal absent or opened in a shallow groove; width of first pelvic-fin ray in adult males decreasing abruptly at distal portion, distal slender portion short; second pelvic-fin ray unbranched in adult males; lateral process on base of fifth median anal-fin radial very large in adult males; 9 anal-fin rays in males; dorsal expansion present on R5p; more distal elements of branches of R6 totally fused; absence of orbital bones (Lucinda 2005). **SPECIES** Ten, including one species in the AOG region, *C. septentrionalis*. Key to species in Lucinda (2005), except for *C. pirai* (Aguilera et al. 2009). **DISTRIBUTION AND HABITAT** Southern South America, in the upper Araguaia basin (*Cnesterodon septentrionalis*), the Paraná-Paraguay system, the Uruguay basin, and along coastal drainages from São Paulo to Argentina, as well as in small drainage basins of western Argentina (Lucinda 2005). **BIOLOGY** Omnivorous. As with most other poeciliines, gonopodium of males is used for internal fertilization. Females give birth to live young.

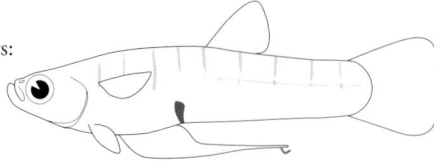

Fluviphylax (1.5–2.0 cm SL)

Distinguished by: very large eyes (~50% of the head length); miniaturization (maximum adult size 22 mm SL); reduced dorsal process of maxilla; fourth ceratobranchial teeth absent; distinct narrow process on the anterior portion of opercle; 17–20 caudal-fin rays; reduced cephalic sensory system; and melanophores concentrated on dorsal and ventral midlines of body (Costa 1996, Costa and Le Bail 1999). Genus also differs from other poeciliid genera occurring in the AOG by the absence of a gonopodium in males. **SPECIES** Five, all in the AOG region. See key to the species in Costa (1996), except for *F. palikur* (Costa and Le Bail 1999). **DISTRIBUTION AND HABITAT** Endemic to Amazon and Orinoco region, and coastal drainages of the Brazilian state of Amapá. Habitats include lakes, swamps, and floating meadows. **BIOLOGY** Food items include microalgae, detritus, and small-bodied invertebrates (Goulding et al. 1988).

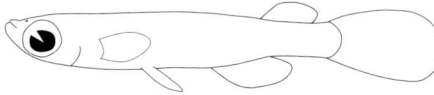

Micropoecilia (1.5–5.0 cm SL)

Micropoecilia is tentatively diagnosed by the following characters: males with more dorsal fin rays than females; males with anal-fin rays 6 and 7 and pelvic fins extending to the base of the gonopodium palp; ray 3 of the gonopodium with 10–16 rose thorn-like spines; anterior portion of first gonapophysis gently curved ventrally; and tip of anal-fin ray 5 without a posteriorly oriented claw (Meyer 1993, Costa and Sarraf 1997, Bragança and Costa 2011, Bragança et al. 2012b). The validity of this genus has long been debated, and researchers have failed to reach a consensus on the matter. Some color patterns (e.g., variegated color pattern in both males and females with a relatively well-developed dark humeral spot on the side of the body) have been suggested as diagnostic for the genus by Meyer (1993); however, these are not observed in all species of *Micropoecilia*, and the anal-fin ray characters used to diagnose the genus have also been observed in some other poeciliid species (Costa and Sarraf 1997). **SPECIES** Seven, including six species in the AOG region: *M. bifurca*, *M. branneri*, *M. minima*, *M. parae*, *M. picta*, and *M. waiapi*.

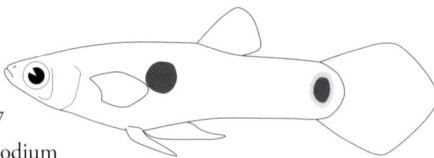

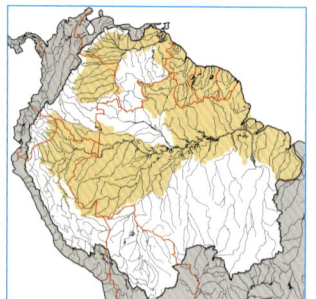

DISTRIBUTION AND HABITAT Amazonas and Orinoco basin and coastal drainages of Brazil (Amapá, Pará, Maranhã, Piauí, and Ceará states), French Guiana, Guyana, Suriname, Trinidad and Tobago, and Venezuela usually in small swamps and shallow slow-flowing creeks and coastal brackish waters. **BIOLOGY** Omnivorous. Gonopodium of males is used for internal fertilization. *Micropoecilia picta* embryos can make up 25% of the weight of the mother; 11–25 offspring are produced (Keith et al. 2000). *Micropoecilia parae* produces 5–15 young after about 24 days of gestation. Life histories of *Micropoecilia* were studied by Meredith et al. (2010) and Pires et al. (2010).

Pamphorichthys (1.7–2.5 cm SL)

Distinguished by: minute and double-spined processes on last distal segments of third gonopodial ray; lateral bone process comb-shaped in subdistal segments of second pelvic-fin ray of adult males (absent in *P. pertapeh*); distal half of second pelvic-fin ray of adult males covered by soft tissue and forming a fleshy appendix; second ray of pelvic fin of adult males separated from rays 3–5 by a broad deep notch in the membrane in rays 2 and 3; pelvic-fin rays 3, 4, and 5 of males joined by a membrane (Rosen and Bailey 1963, Costa 1991, Meyer 1993, Rodriguez 1997, Figueiredo 2008). Other characters that might aid identification: outer row teeth subcylindrical and pointed; serrae in ray 4p of the gonopodium (see location of ray 4p in fig. 3a); distal segments to the serrae taller than long (absent in *P. hasemani*); and a dark zigzag stripe along the flanks (Rosen and Bailey 1963, Rodriguez 1997). **SPECIES** Six, including three species in the AOG region: *P. araguaiensis*, *P. minor*, and *P. scalpridens*. **DISTRIBUTION AND HABITAT** Amazon, Tocantins, Tapajós, Xingu, Paraguay-Paraná, Parnaíba, and São Francisco basins. **BIOLOGY** Gonopodium of males is used for internal fertilization and females give birth to live young. Reproduction biology of *P. hollandi* was studied by Casatti et al. (2006). See Meredith et al. (2011) for life history of *Pamphorichthys* species.

Phalloceros (1.7 cm SL)

Distinguished by a paired appendix at the tip of R3 of the gonopodium (Lucinda 2008). Other characters that can aid identification: width of first pelvic-fin ray decreasing abruptly at distal portion, which is slender and long in adult males; large membranous tip anterior to R4 and R5; ≥8 subdistal posteriorly oriented spines on R4p of the gonopodium (see location of ray 4p in fig. 3a); elongate and dorsal protuberance along R4p (just behind the posteriorly oriented series of spines); distal portion of R6 not expanded (Lucinda 2008). **SPECIES** 22, including one (*P. leticiae*) in the AOG region, which is quite distinct in having a roundish to rounded and well-defined ocellated lateral spot. See review of genus and key to the species in Lucinda (2008). **DISTRIBUTION AND HABITAT** Mainly southern and southeastern basins of South America. Some species have been introduced elsewhere. *Phalloceros leticiae* inhabits the upper Araguaia basin. **BIOLOGY** Omnivorous. Gonopodium of males is used for internal fertilization. Females give birth to live young. Life histories of four *Phalloceros* species were described and discussed by Arias and Reznick (2000), Machado et al. (2002) and Almeida-Silva and Mazzoni (2014).

Poecilia (4.0–6.0 cm SL)

Distinguished from other poeciliid genera in the AOG region by: ≥7 dorsal-fin rays (vs. 4–6 rays in *Fluviphylax* and *Tomeurus*); the second ray of the pelvic fin in males without a triangular-shaped bony process (vs. present in *Pamphorichthys*); fleshy palp on gonopodial ray 3 (vs. fleshy palp absent in *Cnesterodon*, *Phalloceros*, *Fluviphylax*, and *Tomeurus*); sixth and seventh anal-fin rays and pelvic fins not reaching the base of the gonopodial palp (vs. reaching in *Acanthophacelus* and *Micropoecilia*). **SPECIES** Around 40, including at least two species in the AOG region: *P. reticulata* and *P. vivipara*. Overview of the genus in Figueiredo (2003) and Poeser (2003). **DISTRIBUTION AND HABITAT** Slowly moving or standing waters with dense vegetation throughout the humid Neotropics and adjacent subtropical areas, from southeastern USA to northern Argentina (Poeser 2003). **BIOLOGY** *Poecilia reticulata* (the guppy) is one of the most popular aquarium fishes in the world, and has been introduced to all continents (except Antarctica) and many oceanic islands. Natural diet includes zooplankton, small insects (e.g., mosquito larvae) and detritus. Gonopodium of males is used for internal fertilization. Females can store sperm for later fertilization. After a gestation period of several weeks, females give birth to live young. No parental care is exercised and parents may even prey on their own young.

Tomeurus (3.5 cm SL)

Distinguished from other poeciliids by: complex gonopodium tip; a keel of scales on the ventral margin of the caudal peduncle; and a dorsal fin with 6 rays (in both males and females), which is set rather far back on the body (Eigenmann 1909, Lucinda and Reis 2005). Other characters that might aid identification: 3 pelvic-fin rays; second pelvic-fin ray in adult males unbranched; width of first pelvic-fin ray in adult males decreasing abruptly at distal portion, distal slender portion short; 9 or 10 pectoral-fin rays; 11 anal-fin rays in females; and males with 8 anal-fin rays (Lucinda and Reis 2005). **SPECIES** One, *T. gracilis*. **DISTRIBUTION AND HABITAT** Most of the rivers draining the Guiana and Brazilian shields, including the Guamá and Tocantins Rivers in northeastern Brazilian Amazon, the Courantyne basin in Suriname, the Cuyuni, Mazaruni, and Essequibo rivers in Guyana, and other small coastal drainages of Venezuela and Brazil (Lucinda and Reis 2005). Usually found in muddy creeks or along the sandy-muddy edges of shallow estuarine zones. **BIOLOGY** Facultative viviparous and with internal fertilization, laying fertilized eggs or giving birth to free-swimming young (Breder and Rosen 1966). Lives in schools of about several dozen individuals. The eggs can be laid prematurely and attached to plants (Keith et al. 2000).

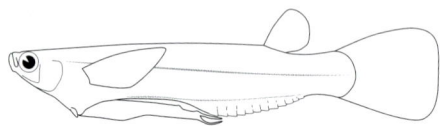

FAMILY RIVULIDAE—RIVULINE KILLIFISHES

— *PEDRO F. AMORIM and PEDRO H. N. BRAGANÇA*

DIVERSITY About 340 species in 32 genera and three subfamilies worldwide, including 160 species in 22 genera in the AOG region.
COMMON NAMES *Rivulos* (Brazil; Spanish).
GEOGRAPHIC DISTRIBUTION Nearly all basins of tropical South America, both insular and continental Central America, and southern North America (Costa 2008).
ADULT SIZES Generally small-bodied, with most species attaining total lengths between 5.0 and 8.0 cm TL, although some species reach 15 cm, and some miniature species never exceed 2.5 cm total length.

DIAGNOSIS OF FAMILY Easily recognized among cyprinodontiforms by continuous branchiostegal and opercular membranes, reduced laterosensory system on preopercle, and several characters related to the bony structures of head and fins (Costa 1998a). However, the elaborate and vivid color patterns in males are the most conspicuous features to identify rivulids among other Neotropical fishes. Species of the subfamily Rivulinae (containing all AOG genera, except *Cynolebias*, *Hypsolebias*, *Kryptolebias*, *Simpsonichthys*, and *Spectrolebias*) differ from other rivulids by a greater number of pelvic-fin rays (7 or 8), and in a circular pattern of frontal scales (Costa 1998a). The subfamily Cynolebiasinae (with *Cynolebias*, *Hypsolebias*, *Simpsonichthys*, and *Spectrolebias* in the AOG) is diagnosed by a cylindrical urogenital papilla in males and a prominent pocket-like urogenital papilla in females; reduced supraorbital squamation; reduced caudal-fin squamation; vertical bars on body sides in juveniles, and many other osteological characters related to head structure (Costa 1998a).

SEXUAL DIMORPHISM Marked sexual dimorphism in most species, including an elaborate and bright color pattern in adult males, differences in fin shapes, and in some species even differences in number of rays.

HABITATS Rivulids occupy a variety of freshwater and brackish water habitats, with greatest diversity in purely freshwater environments (Costa 2006b, 2014). Many rivulids present a seasonal life cycle, inhabiting temporary pools that completely dry out for part of the year, where only the eggs survive the dry seasons, lying in diapause underground for months until they hatch in the next rainy season (Myers 1942). This kind of seasonal development has evolved separately in three independent groups of rivulids. Although a seasonal life cycle is generally associated with semiarid environments with a well-marked rainy season (such as the Brazilian Caatinga and Cerrado biomes), some seasonal rivulids inhabit areas of Amazon rainforest characterized by higher rainfall.

FEEDING ECOLOGY Little or no specific information is available about feeding ecology for most rivulids. Most rivulids are generalists feeding on small arthropods, such as crustaceans, insects, and other aquatic animals such as nematodes and mollusks (Shibatta and Rocha 2001, Shibatta and Bennemann 2003, Keppeler et al. 2015).

BEHAVIOR Always oviparous. In the variety of habitats occupied by rivulids, particular adaptations have arisen in some groups, such as: internal fertilization in the genera *Campellolebias* and *Cynopoecilus* (not occurring in the Amazon; Costa 1995c, b), miniaturization in some species of *Laimosemion*, *Melanorivulus*, and *Simpsonichthys* (Costa 2006c, 2007a), production of sound in some species of *Cynolebias* (Costa et al. 2010), and simultaneous hermaphroditism with self-fertilization in the genus *Kryptolebias*, the only known vertebrate able to self-fertilize (Harrington 1961, Costa 2011a).

ADDITIONAL NOTES Exceptional among Neotropical fishes for the elaborate and vivid color patterns of adult males that make them a widely popular group in ornamental fish keeping around the world (Costa 2003). Several books with species-level information, photographs, and illustrations are available (e.g., Wildekamp 1995, 2004, Seegers 2000).

KEY TO THE GENERA

1a. Transverse dark bar over eye absent (fig. 1a).. **2**
1b. Transverse dark bar over eye present (fig. 1b).. **5**

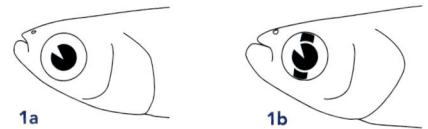
1a   1b

2a. Black margin on dorsal, anal, and caudal fins of females........ *Melanorivulus*
2b. Absence of black margin on dorsal, anal, and caudal fins of females **3**

3a. Scaled chin in both males and females; longitudinal lines of red dots generally present along flanks in males.... *Anablepsoides*
3b. Absence of scaled chin in both males and females; longitudinal lines of red dots absent along flanks in males.....................**4**

4a. Intense yellow or orange pigmentation of anal fin in adult females...*Laimosemion*
4b. Both males and females with orange pectoral fins; or with a dark supraorbital bar and dark longitudinal line on middle-ventral portion of flank; or with a dark brown marble pigmentation along flank and a tubular nostril in both males and hermaphrodites and a dark spot surrounded by yellow ring on side of caudal peduncle in hermaphrodites .... *Kryptolebias*

5a. Dorsal and anal fins anteriorly positioned and symmetrically disposed (fig. 5a) **6**
5b. Posteriorly positioned dorsal-fin origin (fig. 5b)................. **10**

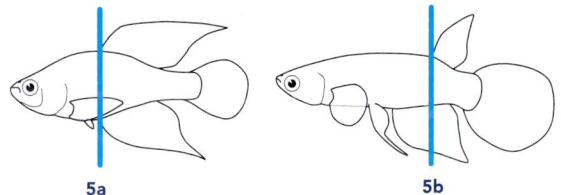
5a   5b

6a. Males with dorsal and anal fins extremely elongate and pointed, with membrane extensions between elongate rays.. **7**
6b. Anal fins not extremely elongate in males; when dorsal fin is elongate, there are no membrane extensions between rays..**8**

7a. Both males and females with an orange suborbital bar; males with a lozenge-shaped caudal fin, with two elongate filaments on the posterior tip; an expanded opercular membrane; 3 longitudinal rows of orange spots on the anterodorsal portion of body; and a metallic blue branchiostegal membrane .... *Maratecoara*
7b. Both males and females without an orange suborbital bar; males with a long pectoral fin; reduced caudal fin squamation; a slightly posteriorly directed black suborbital bar; and absence of alternate black and white spots on basal and posterior part of anal fin .... *Terranatos*

8a. Anterodistal portion of anal fin in females distinctively thickened; small blue iridescent spots often present on posterior portion of anal fin in females .... *Hypsolebias*
8b. Anterodistal portion of anal fin in females not distinctively thickened; small blue spots never present on posterior portion of anal fin in females **9**

9a. Pelvic fin and pelvic girdle well developed; red bars never present on opercle in males .... *Spectrolebias*
9b. Pelvic fin and pelvic girdle vestigial or absent; red bars present on opercle in males .... *Simpsonichthys*

10a. Yellowish-white or pale yellow stripe with a broad red upper margin on ventral portion of caudal fin in males **11**
10b. Absence of a yellowish-white or pale yellow stripe with a broad red upper margin on ventral portion of caudal fin in males **12**

11a. Elongate posterior extension on lower border of caudal fin in males and presence of 2 well-defined oblique reddish stripes on opercular region in males; an ocellus (black spot) absent on caudal fin in females .... *Micromoema*
11b. Absence of an elongate posterior extension on lower border of caudal fin in males; absence of 2 well-defined oblique reddish stripes on opercular region of males; presence of a caudal ocellus in females .... *Renova*

12a. Adipose ridge present in predorsal region of mature males .... *Rachovia*
12b. Absence of an adipose ridge on predorsal region of mature males **13**

13a. Tip of pelvic fin in males bearing a single filamentous ray; pelvic fins medially fused in both males and females **14**
13b. Tip of pelvic fin in males not bearing a single filamentous ray; pelvic fins not medially fused **15**

14a. Deep and compressed body; elongate filamentous rays on dorsal and anal fins in males, anal fin with 22–26 rays; males with a horizontal reddish-brown stripe close to dorsal border of caudal fin and a white and light green to black border on lower edge of pectoral fins .... *Gnatholebias*
14b. Robust and cylindrical body; very small fin extensions in males; anal-fin base short with 14–15 anal-fin rays; males without a horizontal reddish-brown stripe close to dorsal border of caudal fin and without a white and light green to black border on lower edge of pectoral fins .... *Llanolebias*

15a. Dark blue anal fin with a narrow white line along distal margin in males .... *Papiliolebias*
15b. Absence of dark blue anal fin with a narrow white line along distal margin in males **16**

16a. Males with a slightly anteriorly directed black suborbital bar followed by transversal dark bar over eye, and alternate black-and-white to yellowish-white spots on basal and posterior part of anal fin **17**
16b. Males without a slightly anteriorly directed black suborbital bar and without alternate black-and-white to yellowish-white spots on basal and posterior part of anal fin absent **18**

17a. Head with an angular ventral profile (fig. 17a); a transversal arrangement of frontal scales; anterior pointed extension in pupil (fig. 17a); males with rays of the anterior portion of anal fin longer than posterior portion .... *Plesiolebias*
17b. Head with a concave ventral profile (fig. 17b); a circular arrangement of frontal scales; males with an elongate and pointed anal fin; bright red pigmentation in anterior margin of dorsal fin (in larger males); a green and red stripe on distal margin of anal fin; black spots on pectoral fin; flank with small yellow spots arranged in tortuous oblique rows; and a large bright blue humeral blotch ...... *Pituna*

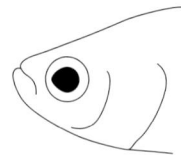 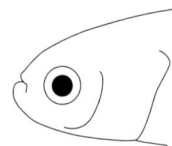

17a   **17b**

18a. Males with metallic orange humeral spot and black vertical bars on pectoral fin .... *Pterolebias*
18b. Males without metallic orange humeral spot and without black vertical bars on pectoral fin **19**

19a. Males and females with an elongate and pointed pectoral fin, distal portion narrow, forming a long pointed tip, in males surpassing anal-fin origin; presence of 2 bars on opercular region in males .... *Moema*
19b. Short, rounded or elliptical pectoral fin not surpassing anal-fin base; absence of 2 bars on opercular region; pectoral fin without a long pointed tip **20**

20a. Number of supraorbital neuromasts 21–30 (e.g., as in fig. 20a; after Costa 1995a) .... *Cynolebias*
20b. Number of supraorbital neuromasts 6–10 (e.g., as in fig. 20b) **21**

20a   20b

21a. Short snout in both males and females; dark anal and caudal fins, with a yellow to white border close to anal-fin base and a similar light border on lower margin of caudal fin in males; transversal bars on caudal fin in juveniles and females; males with a median line above caudal fin; black spots on pectoral fin in males; absence of longitudinal lines composed of red dots along flank of males................. *Neofundulus*
21b. Elongated snout; absence of dark anal and caudal fins with a yellow to white border close to anal-fin base and on the lower margin of caudal fin in males; absence of transversal bars on caudal fin in juveniles and females; pectoral fin without black spots in males; longitudinal lines composed of red dots along flank of males ....*Trigonectes*

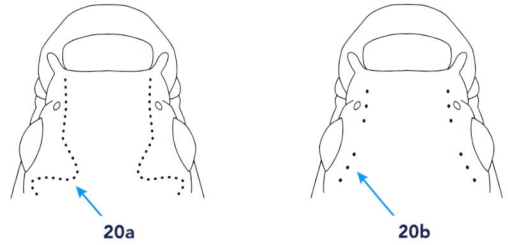

GENUS ACCOUNTS

NOTES ON ILLUSTRATIONS All drawings represent males. In some genera (e.g., *Spectrolebias*) the variance in color patterns between species is high, making it impossible to summarize colorations in a single, simplified drawing. In other cases, more details are given, as long as they are consistent among species within a genus. In the few monotypic genera, or in genera with only a single species in the AOG, as much detail as possible is provided. This variation across genera has resulted in large differences in the level of detail between drawings, which do not necessarily reflect actual differences in the abundance of coloration between genera.

Anablepsoides (3.0–12 cm SL)

Characterized by: a scaled chin, and the males of most of the species have red dots arranged in longitudinal lines along the flanks (Costa 2011b). The fins are short and rounded, except the caudal fin that in some species can be elliptical and long. **SPECIES** 51, including 48 species in the AOG region. No key to the species available. Three species groups are known (Costa et al. 2013): the *A. limoncochae* group from the western Amazon, the *A. urophthalmus* group from the central and eastern Amazon basin, the Guianas, and eastern Venezuela, and the *A. ornatus* group from the central Amazon and Guiana shield. The *A. urophthalmus* and *A. limoncochae* groups have red dots along the flank in males, short unpaired fins, an E-pattern in the frontal squamation, dorsal fin with 6–9 rays, and anal fin with 12–15 rays. Species from the *A. urophthalmus* and *A. limoncochae* species groups also share absence of contact organs on the scales, and absence of a red stripe on the middle of the dorsal fin. Members of the *A. limoncochae* group have a yellow pigmentation on the upper and lower margins of the caudal fin in males. The *A. ornatus* species group has only five species, characterized by their small sizes, depressed head, and long caudal fin. Species information in Costa (2006b, 2008, 2011b) and Costa et al. (2013). **DISTRIBUTION AND HABITAT** Throughout the AOG region, Lesser Antilles, and small coastal rivers of northeastern Brazil. Species of *Anablepsoides* inhabit shallow places near the banks of streams and swamps, in forest or open savanna areas. **BIOLOGY** All species are perennials, with a life cycle extending for more than one year.

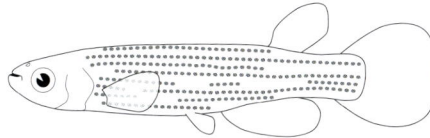

Cynolebias (5.0–12 cm SL)

Characterized by: a rounded supraorbital spot in both males and females; dark brown to black blotches scattered on the dorsoposterior portion of the head in both males and females; light blue lines between rays on the distal portion of anal fin of males; and number of supraorbital neuromasts between 21 and 30 (Costa 2001). **SPECIES** 14, with one species in the AOG region. Key to the species in Costa (2001), except *C. parnaibensis* (Costa et al. 2010). **DISTRIBUTION AND HABITAT** Most species in the São Francisco basin. Only one species, *C. griseus*, in the upper Tocantins and Amazon basins. Inhabits temporary pools. **BIOLOGY** All *Cynolebias* are seasonal. In some species of *Cynolebias* the males produce sounds during courtship behavior (Costa et al. 2010).

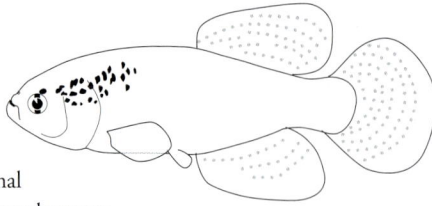

Gnatholebias (5.0–6.5 cm SL)

Characterized by: a deep and laterally compressed body; 22–26 anal-fin rays; elongate filamentous rays on the dorsal and anal fin in males; a horizontal reddish-brown stripe close to the dorsal border of caudal fin in males; a white and light green border followed by a narrow black margin on the lower edge of the pectoral fin in males. **SPECIES** Two, both in the AOG region: *G. hoignei* and *G. zonatus*. Species information in Thomerson (1974), Hrbek and Taphorn (2008), and Costa (2008, 2014). **DISTRIBUTION AND HABITAT** Restricted to the Llanos wetlands in the Orinoco basin. *Gnatholebias hoignei* inhabits deep seasonal pools in shaded areas, whereas *G. zonatus* occurs in shallow seasonal pools directly exposed to sunlight (Costa 2008). **BIOLOGY** Both species have a seasonal life cycle.

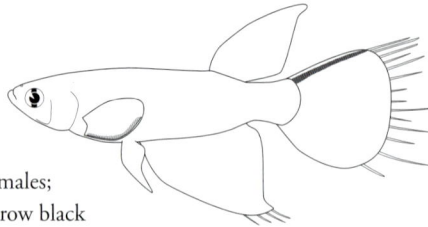

Hypsolebias (1.5–4.5 cm SL)

Characterized by two characters found in females: a thickened anterodistal portion of anal fin and a small blue spot often present on posterior portion of anal fin. **SPECIES** 47 species, including six in the AOG region. No key to the species available; species information in Costa (2006a, 2007a). **DISTRIBUTION AND HABITAT** Greatest diversity in the Caatinga of northeastern Brazil, with six species in the Tocantins basin (*H. brunoi, H. flammeus, H. marginatus, H. multiradiatus, H. notatus*, and *H. radiosus*). **BIOLOGY** All species have a seasonal life cycle.

Kryptolebias (2.8–5.0 cm TL)

Distinguished from all other rivulids by osteological characters (Costa 2004b). Externally, the Amazonian species are distinguished from other rivulids by the following coloration patterns: *K. sepia* has an orange pectoral fin in both males and females; *K. campelloi* has a dark supraorbital bar and a dark longitudinal line in the middle-ventral portion of the flank; and *K. marmoratus* has a dark brown marble pigmentation along flank, a dark spot surrounded by yellow ring is present on the side of the caudal peduncle, and the anterior nostril is tubular (Costa 2011a). **SPECIES** Eight, including three species in Amazon. No key to the species available; species information in Costa (2004a, b, 2011a). **DISTRIBUTION AND HABITAT** Inhabits both brackish and freshwater environments, and the only rivulid genus found in Amazon estuarine waters. *Kryptolebias sepia* inhabits the Tapanahony basin (Suriname), *K. campelloi* coastal basins of northern Brazil, and *K. marmoratus* coastal river basins between the mouth of

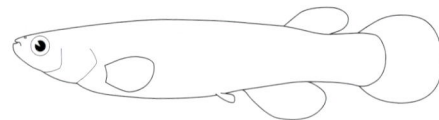

the Orinoco and northern Brazil. **BIOLOGY** Some *Kryptolebias* from brackish water are remarkable for having populations without females, and only males and hermaphrodites (Costa 2011a). The hermaphrodites are internal self-fertilizing, possessing both ovary and testis.

Laimosemion (1.5–4.0 cm SL)

Characterized by the intense red or orange coloration of the anal fin of females. *Laimosemion* is also characterized by the variable morphology of the caudal, anal, and dorsal fins. Some species have a rounded, subtruncate or lyre-shaped caudal fin, and short, rounded or elongate anal and dorsal fins. Miniature species are present in *Laimosemion*, with some species (*L. ubim, L. jauaperi, L. uatuman, L. romeri*, and *L. kirovsky*) not reaching 2.5 cm SL. Many *Laimosemion* species exhibit reductions, structural simplifications, and other morphological novelties (Costa and Lazzarotto 2014). **SPECIES** 26, all in the AOG region. No key to the species available; species information in Costa (2006b, 2008) and Costa and Lazzarotto (2014). **DISTRIBUTION AND HABITAT** Inhabit the Amazon and Orinoco basins and Guianas, generally inhabiting shallow streams or pools inside the forest. **BIOLOGY** Not seasonal.

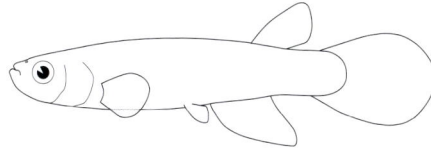

Llanolebias (4.0–5.0 cm SL)

Characterized by: very small fin extension in males; 14–15 anal-fin rays; a robust and cylindrical body (Hrbek and Taphorn 2008). Resembling *Rachovia* but distinguished by absence of a fatty predorsal ridge in older males (as is commonly present in *Rachovia* species), 7 pelvic-fin rays (8 pelvic-fin rays in *Rachovia*), contact organs in the pectoral fin (absent in *Rachovia*), and dark pigmentation pattern on upper portion of caudal-fin base in females (Costa 2005). **SPECIES** One, *L. stellifer*. Species information in Hrbek and Taphorn (2008). **DISTRIBUTION AND HABITAT** Restricted to the Llanos in the Orinoco basin. Found in the margin of seasonal pools shaded by dense vegetation (Hrbek and Taphorn 2008). **BIOLOGY** Seasonal lifestyle (Hrbek and Taphorn 2008, Costa 2014).

Maratecoara (2.0–3.0 cm SL)

Characterized by the following characters in males: elongate dorsal and anal fins, tips reaching beyond posterior margin of caudal fin; caudal fin lanceolate, tip with 2 or 3 filamentous rays; long opercular membranes with blue iridescence, extending to anterior portion of pectoral fins; and a metallic blue flank, with orange-golden spots (Costa 2007b). **SPECIES** Four, all in the AOG region. Species key and species information in Costa (2007b). **DISTRIBUTION AND HABITAT** Southern tributaries of the Amazon basin; *M. lacortei* restricted to the Araguaia basin, *M. formosa* and *M. splendida* the Tocantins basin, and *M. gasmonei* the Xingu basin. Inhabit seasonal pools, freshwater lagoons, and swamps. **BIOLOGY** All species have a seasonal life cycle (Costa 2007b).

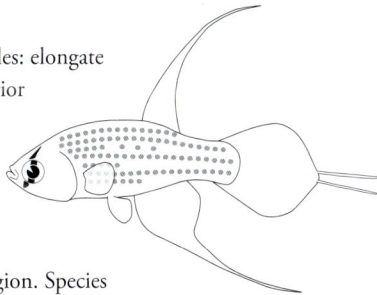

Melanorivulus (2.0–4.0 cm SL)

Characterized by: a distinct black margin on the dorsal, anal, and caudal fins of females; and a chevron pattern on the side of the body in males. All species except *M. schuncki* can also be recognized by: two dark gray to black oblique bars on the postorbital region in both males and females and pelvic fin with black margin in females. **SPECIES** 39 species, including 18 species in the Amazon basin. No key to the species available; species information in Costa (2006b, 2008, 2011b). **DISTRIBUTION AND HABITAT** Throughout Amazon, Paraná-Paraguay, São Francisco, and Parnaíba basins. Highest species diversity in southern tributaries of the Amazon basin (i.e., Tocantins, Araguaia, Xingu, and Tapajós). Only one species (*M. schuncki)* occurs both north and south of the Amazon River (Bragança et al. 2012a). Inhabit shallow clearwater swampy areas with orange clay as substrate and generally associated with the Buriti palm-tree, *Mauritia flexuosa* (Costa 2011b). **BIOLOGY** Not seasonal.

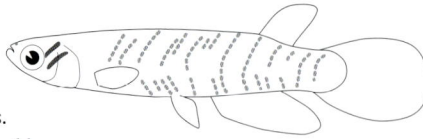

Micromoema (2.5 cm SL)

Distinguished by the elongate rays in the ventral lobe of the caudal fin in males (resembling that of the Middle American poeciliid *Xiphophorus*). **SPECIES** One, *M. xiphophora*. Species information in Thomerson and Taphorn (1992) and Costa (1998a). **DISTRIBUTION AND HABITAT** Temporary pools in the upper Orinoco basin, Venezuela. **BIOLOGY** Seasonal life cycle.

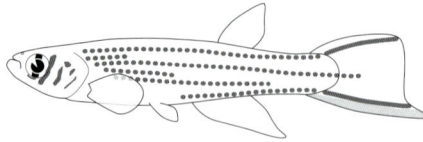

Moema (3.0–9.0 cm SL)

Distinguished by the opercle shape, which has its ventroposterior part expanded and the dorsal portion distinctively longer than the ventral portion (see right figure of left opercle; Costa 1998a), and a long and pointed pectoral fin, with a narrow distal portion (Costa 2014). In addition, males of some species have elongate fin rays in the dorsal and ventral lobes of the caudal fin (Costa 2014). **SPECIES** 17. Recently, the genus *Aphyolebias* was considered a synonym of *Moema* based on a combined morphological and molecular phylogenetic analysis (Costa 2014). No key to the species available; species information in Costa (2008, 2014). **DISTRIBUTION AND HABITAT** All species except one (*M. heterostigma*) occur in the Amazon basin, inhabiting temporary pools and swamps. **BIOLOGY** All species have a seasonal life cycle.

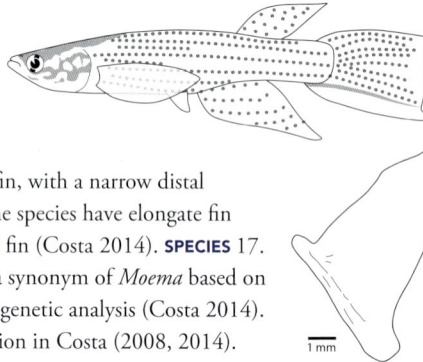

Neofundulus (2.0–6.0 cm SL)

Distinguished by: dark anal and caudal fins in males, with a yellow to white border close to the anal-fin base and a similar light border on the lower margin of the caudal fin; black spots on the pectoral fin of males; transversal bars on the caudal fin are present in juveniles and females (Costa 1998a). **SPECIES** Five. No key to the species available; species information in Costa (1992, 1998a). **DISTRIBUTION AND HABITAT** Four species in the Paraguay basin and one species (*N. guaporensis*) in the Amazon basin (Madeira River). No information available about its natural habitat. **BIOLOGY** All species have a seasonal life cycle.

Papiliolebias (2.5–3.5 cm SL)

Distinguished by: 9 pelvic-fin rays; and a dark blue anal fin in males, with a white line along its distal margin. **SPECIES** Four. No key to the species available; species information in Costa (1998a, c). **DISTRIBUTION AND HABITAT** *P. ashleyae* and *P. fracescae* from the Mamoré basin; *P. bitteri* and *P. hatinne* from the Paraguay basin. All species inhabit temporary pools (Costa 1998a). **BIOLOGY** All species have a seasonal life cycle.

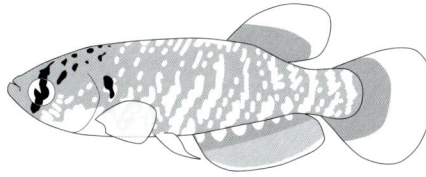

Pituna (1.5–4.0 cm SL)

Characterized by the following characters in males: dark brown flanks, with oblique rows of golden spots; a metallic blue humeral spot; dorsal and anal fins pointed; and distal portion of dorsal fin red (Costa 2007b). **SPECIES** Six. Key to the species and species information in Costa (2007b). **DISTRIBUTION AND HABITAT** Four species inhabit the Amazon basin: *P. poranga* and *P. obliquoseriata* are known from the middle Araguaia, *P. compacta* from middle Tocantins and *P. xinguensis* from middle Xingu. All species inhabit temporary pools (Costa 2007b). **BIOLOGY** All species have a seasonal life cycle.

Plesiolebias (1.5–3.0 cm SL)

Characterized by: longer rays on the anterior portion of the anal fin compared with the posterior portion; and G-shaped pattern of scales on the frontal bone region (Costa 2007b). All species of *Plesiolebias* are miniatures. **SPECIES** Eight, including seven in the AOG region. Key to the species in Costa (2007b). Species information also in Costa (1998a, c). **DISTRIBUTION AND HABITAT** Portions of the Brazilian Shield including the middle Xingu River (*P. altamira*), middle Araguaia River (*P. aruana, P. lacerdai,* and *P. fragilis*), middle Tocantins River (*P. xavantei, P. filamentosus,* and *P. canabravensis*), and upper Paraguay River (*P. glaucopterus*). Species inhabit temporary swamps (Costa 1998c). **BIOLOGY** All species have a seasonal life cycle.

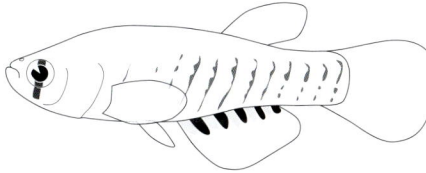

Pterolebias (3.5–5.0 cm SL)

Characterized by several characters in males, including: black vertical bars on the pectoral fin; a metallic-orange humeral spot; and elongate pelvic fins (Costa 2005). **SPECIES** Two, including one in the AOG region. Species information in Costa (2005, 2008). **DISTRIBUTION AND HABITAT** *P. phasianus* in the Paraná basin; *P. longipinnis* in the lower Amazon, Madeira, and Paraná basins. Inhabits seasonal pools in areas directly exposed to sunlight or close to the forest margin (Costa 2008). **BIOLOGY** All species have a seasonal life cycle.

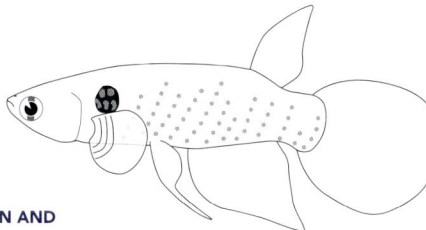

Rachovia (3.5–9.0 cm SL)

Characterized by: an adipose ridge on the predorsal region in older males (Hrbek and Taphorn 2008); and absence of a yellowish-white or pale yellow stripe with a broad red upper margin on the ventral portion of the caudal fin in males, as is present in similar-looking *Micromoema* and *Renova* species (both found in the Orinoco basin). SPECIES 11, including three in the AOG region. Recently, *Austrofundulus* was considered a synonym of *Rachovia* based on a combined morphological and molecular phylogenetic analysis (Costa 2014). No key to the species available; species information in Costa (2008, 2014) and Hrbek and Taphorn (2008). DISTRIBUTION AND HABITAT Distributed in the Magdalena and adjacent coastal basins, the plains adjacent to Lake Maracaibo, the Orinoco basin (*R. maculipinnis* and *R. transilis*), and the upper Branco basin (*R. rupununi*). *Rachovia* species inhabit temporary swamps in the Llanos and in the savannas of the upper Branco River (Costa 2008). BIOLOGY All species have a seasonal life cycle.

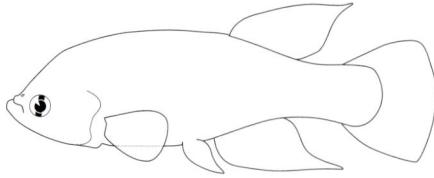

Renova (4.5 cm SL)

Characterized by: a rod-shaped tip of anal fin of males; yellowish-white or pale yellow stripe with a broad red upper margin on ventral portion of caudal fin in males; 3 longitudinal rows of dark spots from the anterior portion of body sides to caudal-fin base (dark reddish brown in males, dark brown in females), alternating with shorter rows of similar spots restricted to the anterior portion of the body; a narrow brown, short, anteriorly directed suborbital bar in both male and female; and absence of 2 well-defined oblique reddish-brown stripes on the preopercular region of males (Costa 1998a). SPECIES One, *R. oscari*. Species information in Costa (1998a). DISTRIBUTION AND HABITAT Seasonally flooded swamps in the upper Orinoco basin, Venezuela (Costa 2008). BIOLOGY Seasonal.

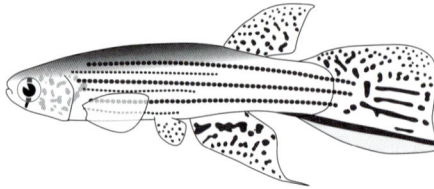

Simpsonichthys (2.5–3.0 cm SL)

Characterized by a highly reduced or absent pelvic fin and pelvic-fin girdle. Males with red bars on opercle (Costa 2007a) and a red flank with bright blue bars on the anterior portion and vertical rows of blue dots on the posterior portion (except in *S. cholopteryx*, which has bright blue bars on the entire flank) (Costa 2006a). Some species are among the smallest-bodied rivulids. SPECIES Eight, including one species *S. cholopteryx* in the AOG region. No key to the species available; species information in Costa (2006a, 2007a). DISTRIBUTION AND HABITAT Upper Araguaia, Paraná, and São Francisco basins. BIOLOGY All species have a seasonal life cycle.

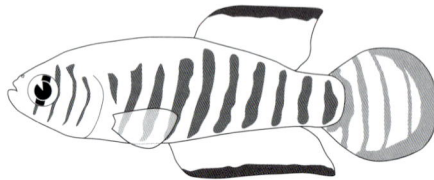

Spectrolebias (1.5–3.5 cm SL)

Similar to *Simpsonichthys* and *Hypsolebias* by having eyes placed dorsolaterally on head, anterior and posterior sections of supraorbital series of neuromasts continuous, and anal fin hyaline in females. Differs from the aforementioned genera by having the pelvic fin and pelvic girdle well developed in both males and females, and males always lacking

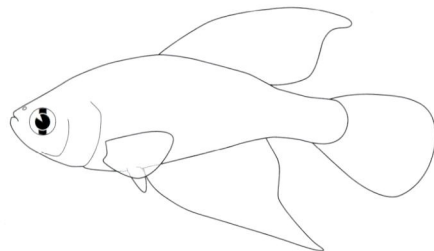

red bars on the opercle (Costa 2007a). Despite the low number of species, *Spectrolebias* show a wide variety of body shapes and color patterns. **SPECIES** 10, including 9 in the AOG region. Key to the species in Costa (2007a), species information also in Costa (2006a). **DISTRIBUTION AND HABITAT** Xingu (*S. reticulatus*), Mamoré (*S. brousseaui, S. filamentosus,* and *S. pilltie*), Tocantins (*S. inaequipinnatus, S. costae,* and *S. semiocellatus*), Araguaia (*S. costae* and *S. semiocellatus*), and Paraguay rivers (*S. chacoensis*). **BIOLOGY** All species have a seasonal life cycle.

Terranatos (3.0 cm SL)

Characterized by: extremely elongate and anteriorly positioned dorsal fin in males, symmetrically disposed to the anal fin; a posteriorly directed black suborbital bar in both male and female; reduced caudal-fin squamation in both male and females; and a long pectoral fin in males (Costa 1998a). **SPECIES** One, *T. dolichopterus* (Saberfin killifish). Species information in Costa (1998a, 2008). **DISTRIBUTION AND HABITAT** Seasonal flooded swamps in the Llanos and rainforests of the Orinoco basin (Costa 2008). **BIOLOGY** All species have a seasonal life cycle.

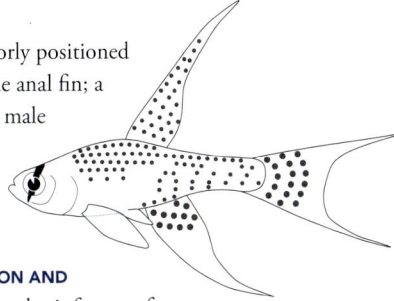

Trigonectes (4.0–8.0 cm SL)

Characterized by: rounded and elongate pectoral fins; filamentous pelvic fins in males; absence of two bars on the opercular region; longitudinal lines of red dots along flank of males; and relatively large adult body sizes, reaching ≤8 cm SL (Costa 1998a). **SPECIES** Six, including four species in the Amazon basin. No key to the species available; species information in Costa (1998a, 2014). **DISTRIBUTION AND HABITAT** Guaporé basin (*T. macrophthalmus*), Mamoré basin (*T. rogoague*), Araguaia basin (*T. rubromarginatus*), and Tocantins basin (*T. strigabundus*). All inhabit seasonal pools and swamps (Costa 2008). **BIOLOGY** All species have a seasonal life cycle.

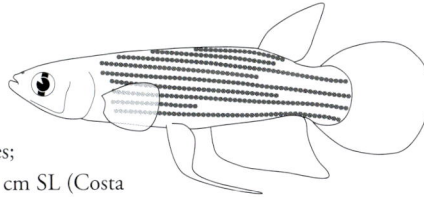

FAMILY CICHLIDAE—CICHLIDS
— *SVEN O. KULLANDER, HERNÁN LÓPEZ-FERNÁNDEZ, and PETER VAN DER SLEEN*

DIVERSITY Cichlidae is one of the most species-rich fish families in the world, with more than 1,650 described species and hundreds of additional undescribed species. The family is distributed in tropical and subtropical freshwaters of South and Central America, Africa, Madagascar, and portions of southern India and the Middle East. Cichlids are most diverse in South and Central America and Africa, with spectacular radiations in large tectonic lakes of East Africa (Fryer and Iles 1972, Greenwood 1974, Salzburger et al. 2002, Seehausen 2006) and crater lakes of Nicaragua (Barlow and Munsey 1976, Barluenga et al. 2006) and West Africa (Seehausen and Wagner 2014, Wagner et al. 2014).

Neotropical cichlids (subfamily Cichlinae) constitute the third-most species-rich group of South American freshwater fishes, following the Characidae and Loricariidae (Albert et al. 2011c). More than 415 South American cichlid species are currently recognized, allocated to 44 genera in seven tribes.

Many more species are already known that have not yet been formally described, and additional genera may be recognized following continuous systematic revision. Currently, the tribe Astronotini includes one genus and at least three valid species and perhaps five species in total (Colatreli et al. 2012); Chaetobranchini includes four species in two genera; Cichlasomatini includes about 80 described species in 12 genera; Cichlini includes 15 species in the genus *Cichla* (Willis et al. 2012b); Geophagini includes about 252 named species in 16 genera; Heroini includes about 176 described species in 26 genera, most of them in Central and North America, with only 57 species in 11 genera in South America; and Retroculini includes four species in the genus *Retroculus.*

COMMON NAMES Acará, Cará (Brazil); Bujurqui (Peru)

GEOGRAPHIC DISTRIBUTION Neotropical cichlids are distributed throughout tropical South America, the Greater Antilles, and Central America. The northern limit is reached in the Rio

Grande on the border between Mexico and the USA with *Herichthys cyanoguttatus*. Introduced cichlid species are reported mainly from southern South America, including the medium-sized predatory cichlid *Nandopsis managuensis* and several species of tilapias, mainly *Oreochromis niloticus*, *Tilapia buttikoferi*, and *Tilapia rendalli*. *Oreochromis niloticus* is perhaps the only regularly occurring non-native cichlid species in tropical South America and is common in estuarine regions of the Guianas. It is easily recognized by the cycloid scales, numerous vertical stripes on the caudal fin, a black spot at the upper corner of the gill cover and at the base of the soft dorsal fin, and by having teeth with two cusps. *Oreochromis mossambicus* is established in coastal areas of Suriname and can be distinguished from *O. niloticus* by absence of stripes from the caudal fin, and by a black and red body color.

ADULT SIZES Neotropical cichlids range in body size from the tiny species of *Taeniacara* and *Apistogramma*, many of which are adult at sizes of about 2.0–3.0 cm SL, to several species of peacock basses (*Cichla* spp.) that reach almost 100 cm SL.

DIAGNOSIS OF FAMILY Among Neotropical fishes, cichlids can be recognized externally by the possession of 7–24 (usually 13–16) spines in the dorsal fin, 2–12 (usually 3, rarely more than 5) spines in the anal fin; and a single nostril on each side of the head. The lateral line is usually (the exception are some species of *Cichla* and *Teleocichla*) divided into an anterior upper portion ending below the end of the dorsal-fin base, and a posterior lower portion running along the middle of the caudal peduncle (Kullander 2003a).

The most conspicuous external character separating Afro-Asian from Neotropical cichlids is that the former have a black spot on the upper posterior corner of the gill cover, covering a small area where scales are absent. In contrast, Neotropical cichlids have a fully scaled opercle typically without a distinct black spot at the upper posterior corner of the opercle. In most Neotropical cichlids, the lower lip fold typically covers the posterior part of the upper lip, so-called American-type lips, whereas in most African cichlid (and the Neotropical genera *Cichla*, *Retroculus*, and *Astronotus*) the upper and lower lips simply meet at the rictus, without any overlap, so-called African-type lips.

Neotropical cichlids range in body shape from the deep and highly laterally compressed bodies of heroines like *Mesonauta*, *Pterophyllum*, and *Symphysodon* that inhabit lakes and river margins, to the elongate and cylindrical bodies of some geophagines like *Teleocichla* and *Crenicichla* that inhabit rapids. Most cichlids attain moderate adult body sizes, and many species have a slightly laterally compressed, approximately ovate body shape in lateral view.

SEXUAL DIMORPHISM Many cichlid species are sexually dimorphic, with sexes differing in color and size. Sexual color differences tend to be subtle in Neotropical cichlids, with the exceptions of *Apistogramma* and *Crenicichla*. Usually it is the male that is the larger and more colorful sex. All cichlids exhibit some form of parental care including guarding both the eggs and the free-swimming offspring. In some groups, most notably *Apistogramma*, harem breeding is frequent. Eggs are typically deposited on a hard substrate and both parents share the guarding, which may last for several weeks. Oral incubation, or mouth brooding, has been recorded for many species of *Geophagus*, *Gymnogeophagus*, *Satanoperca*, and some *Bujurquina*, and also for one species each of *Apistogramma*, *Aequidens*, and *Heros*.

HABITATS Neotropical cichlids typically occupy lentic habitats within rivers and streams from the Andean piedmont to coastal regions, but several species of *Crenicichla*, *Cichla*, and the genera *Teleocichla* and *Retroculus* are rheophilic, specialized for living in river rapids.

FEEDING ECOLOGY Most Neotropical cichlids feed on a variety of invertebrates and some plant matter. *Cichla* and large species of *Crenicichla*, *Caquetaia*, *Astronotus*, and *Acaronia* feed on fishes and large invertebrates. *Chaetobranchopsis*, *Chaetobranchus*, and *Satanoperca acuticeps* are plankton feeders (Kullander 2003a). Many species in the subfamily Geophaginae, e.g., *Geophagus* and *Satanoperca* species, are sediment-sifters and have an "epibranchial lobe" (an anteroventral expansion of the first epibranchial bone), which could be an adaptation for this mode of feeding (or for mouth brooding; López-Fernández et al. 2012b, 2014). *Uaru* is the only Neotropical cichlid that is strongly herbivorous, ingesting algae and parts of higher plants.

ADDITIONAL NOTES Many Neotropical cichlids are important food fishes locally, or are valued game fish. Neotropical cichlids also include many familiar aquarium species, like the Freshwater angelfishes (*Pterophyllum*), the Discus fishes (*Symphysodon*), and the Oscars (*Astronotus*). Species of *Cichla* are avidly sought in sport fishing.

NOTES ON CHARACTERS USED TO IDENTIFY GENERA

Cichlid genera are commonly distinguished on the basis of anatomical characteristics, but most genera can easily be identified by external characters. Important external characters include counts of fin spines and scales, and color pattern.

Scale counts

Scales along the side are counted in the row above that containing the lower lateral line. This scale row is known for short as the E1 row, which means that it is the first row above (epaxial) the middle row (row 0) on the side. The middle row is the one containing the lower (posterior) lateral line.

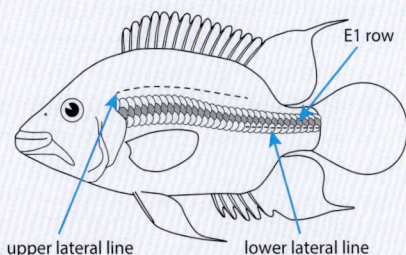

E1 row

upper lateral line   lower lateral line

Fin spines

With a single exception, cichlids have three or more spines in the anal fin, and the numbers can often be used for a first sorting. The most common count is three spines, commonly found in geophagines and cichlasomatines, although particular species in these groups may have four spines or more. Five or more spines are found in *Apistogrammoides*, *Chaetobranchopsis*, and the tribe Heroini. Principal caudal-fin rays are all branched rays plus one unbranched dorsally and ventrally and are invariably 16, except in *Nannacara* with 14.

Gill rakers

Gill raker counts discriminate particularly plankton-filtering genera with numerous gill rakers. Gill rakers are counted only on the lower part (the ceratobranchial) on the outside of the first gill arch.

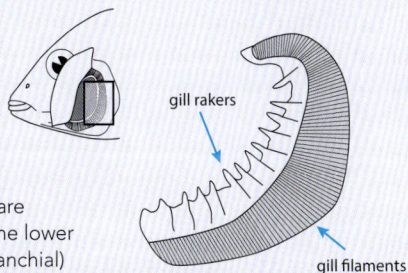

gill rakers

gill filaments

Lateral line

Cichlids typically have a divided lateral line whereby the anterior (upper) part is longer and runs high on the side, whereas the lower part is short and runs along the middle of the side. The lower lateral line is often continued on the caudal fin where it may be branched into dorsal, median and/or ventral branches that typically extend for only one or a few scales. In juveniles and small species, the lateral line

lateral line openings on head (*Apistogramma*)

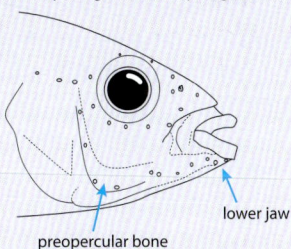

lower jaw

preopercular bone

is seen as small pits, one on each scale, whereas in adults these pits become short bony tubes that span most of the length of each scale. The only Neotropical cichlid species with continuous lateral lines are found in the genera *Cichla* and *Teleocichla*.

Lateral line openings on the head are frequently used markers of genus affiliation. They can be difficult to trace on live or fresh fish in field conditions but are conspicuous in preserved specimens. Afro-Asian and some Neotropical cichlids have six openings of the lateral line system running through the preopercular bone, whereas the majority of Neotropical cichlids have five openings. Most Neotropical cichlids have four openings on each side of the lower jaw, but reductions occur in smaller species. The small bones bordering the lower edge of the orbit (infraorbitals) are usually palpable in fresh specimens, but also best observed in preserved specimens. These are plate-like bones with openings on the middle or at the ends of each. The anteriormost is relatively large and more or less square and has typically four openings; it is followed by typically five ossicles, some of which may be absent or fused with each other forming a single, longer bone.

Predorsal scales

In some genera, scales in the predorsal region are arranged in a single row along the midline. In others, they are partially arranged in a single row and partially as two overlapping scale series, giving a so-called triserial predorsal scale pattern (see also fig. 28a in key to the genera). In yet others, scales in the predorsal region are irregularly arranged.

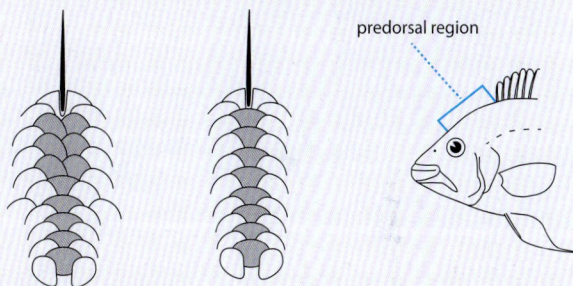

predorsal region

Color patterns

Color pattern offers the most important characters for distinguishing species but also provides characters for separating genera. Commonly, dark coloration formed by the presence of melanophores is the most useful because these are the only colors that remain visible in preserved fish and which are also usually visible in living and fresh specimens. An *ocellus* or ocellar marking is common in South American cichlids and consists of a black spot with a light, contrasting ring around it; ocelli are sometimes also called "eye-spots." A dark spot is commonly present on the middle of the side (a *midlateral spot*), and another one at the base of the caudal fin (*caudal spot*). The basic color pattern includes 7–9 dark vertical *bars* and a horizontal *lateral band* along the middle of the side. On the head there are usually dark stripes emanating from the orbit and running to the mouth (*preorbital stripe*), preopercle (*suborbital stripe*), opercular tip (*postorbital stripe*) and occiput (*supraorbital stripe*).

KEY TO THE GENERA

1a. Lower lip not folded over upper lip at rictus ("African lips";
fig. 1a): 7 preopercular foramina.. **2**
1b. Upper lip covered by lower lip at rictus ("American lips";
fig. 1b); 6 preopercular foramina **6**

2a. Scales very small, >60 in the horizontal series
above that including the lower lateral line:
dorsal-fin outline with distinct incision at junction
of spinous and soft dorsal-fin portions (fig. 2a)........ *Cichla*
2b. Scales small or large, ≤40 body scales in the
horizontal series above that including the lower
lateral line; spinous and soft dorsal fins without
marked indention at junction (fig. 2b)............................. **3**

3a. Scales cycloid; numerous small bicuspid and tricuspid teeth in jaws **4**
3b. Scales ctenoid on side; may be cycloid on head, chest, and abdomen; jaw teeth caniniform **5**

4a. Gill rakers 27–33 externally on first ceratobranchial; caudal fin with vertical stripes .... *Oreochromis*
(introduced and non-native; not included in genus accounts)
4b. Gill rakers 8–11 externally on first ceratobranchial; caudal fin without vertical stripes .... *Tilapia*
(introduced and non-native; not included in genus accounts)

5a. Anal fin with 3 spines and 6–7 soft rays, dorsal fin with 14–16 spines and 10–12 soft rays; unpaired fins
sparsely scaled, with a few rows of small scales never forming a thick pad; no ocellated markings anywhere
on the body or fins .... *Retroculus*
5b. Anal fin with 3 spines and 15–17 soft rays, dorsal fin with 12–14 spines and 19–22 soft rays; unpaired
fins with dense scale cover basally, forming a thick pad; ocellus on upper portion of caudal peduncle and
sometimes at the base of the dorsal fin.. *Astronotus*

6a. Gill rakers long and slender, ≥50 externally on first ceratobranchial... **7**
6b. Gill rakers short, <20 on first ceratobranchial.. **8**

7a. Scales absent from dorsal and anal fins; 3 anal-fin spines; caudal ocellus present or absent .... *Chaetobranchus*
7b. Unpaired fins with wide basal scale cover; 4–6 anal-fin spines; caudal ocellus absent .... *Chaetobranchopsis*

8a. Prominent flattened, lobe-like expansion present on the first gill arch
(fig. 8a)..**38** (geophagines)
8b. Upper limb of first gill arch slender, without a lobe..**9**

9a. Anal fin with 3 or 4 spines **10**
9b. Anal fin with >4 spines .... **19** (heroines + *Cichlasoma amazonarum*)

10a. Slender body; >45 body scales in the horizontal series above those including
the lower lateral line...**11**
10b. Moderately deep body; ≤30 body scales in the horizontal series above that
including the lower lateral line **12**

11a. Preopercular margin smooth; short and downturned snout; pelvic fin long, with the middle ray longest,
reaching to (or almost to) the genital papilla, thickened outer margin, and relatively stiff..................................... *Teleocichla*
11b. Preopercular margin smooth or serrated; long to short snout, not particularly downturned; pelvic fin not
stiff, short, not reaching to vent or genital papilla.. *Crenicichla*

12a. Preopercular and prootic margins strongly serrated; "checkerboard" color pattern on flanks **13**
12b. Preopercular margin smooth, occasionally with a few minute serrations; prootic margin smooth **14**

13a. Rounded caudal fin; absence of filamentous pelvic fins in males .... *Crenicara*
13b. Lanceolate or lyrate caudal fin; filamentous pelvic fins in males.. *Dicrossus*

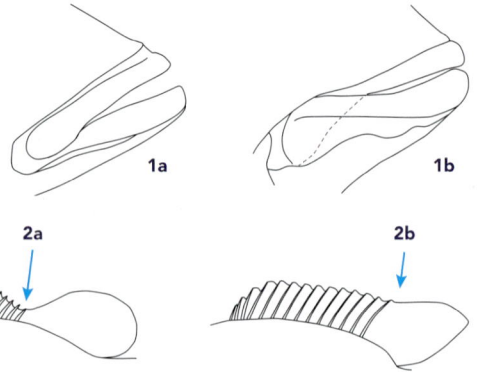

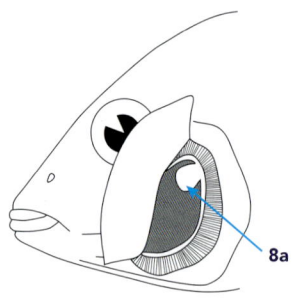

14a. Predorsal scales usually >10 and overlapping across midline **15**
14b. Predorsal scales 7–8 and either in a single series along midline, or anterior 3–4 scales on the midline,
followed by 4–5 scale pairs overlapping midline .... **28** (cichlasomatines)

15a. 22 body scales in the horizontal series above that including the lower lateral line............................ *Acaronia*
15b. 24–30 body scales in the horizontal series above that including the lower lateral line.......................... **16**

16a. Dorsal-fin spines 7–9; gill rakers absent from external face of first gill arch; body slender...................... *Biotoecus*
16b. Dorsal-fin spines 12–17; gill rakers present on external face of first gill arch; body deep **17**

17a. Dorsal-fin spines 12–13; 29–30 body scales in the horizontal series above that including the lower lateral line ... *Acarichthys*
17b. Dorsal-fin spines 15–17; 24–26 body scales in the horizontal series above those including the lower lateral line................ **18**

18a. Unique color pattern formed by two vertical bars, one from nape through preopercle to edge of isthmus
through the eye, and another on the flank at midlength (midlateral bar) ....*Guianacara*
18b. Color pattern not as above; if bar on head present, only in juveniles and not reaching isthmus (*M.
mazarunii*); midlateral bar never present. All species restricted to upper Mazaruni River, Guyana .... *Mazarunia*

Heroines

19a. Predorsal scales large, ~8 along midline, 3–4 anterior on midline, posterior overlapping pairs
.... *Cichlasoma amazonarum* (eastern Amazon populations; others with 3 anal spines)
19b. Predorsal scales small, >8 along midline, not arranged in single row along midline.......................... **20**

20a. Pelvic-fin first ray greatly elongate, much beyond the anal-fin base **21**
20b. Pelvic fin shorter, reaching at most to anal-fin base... **22**

21a. Body round; dorsal and anal soft rays much prolonged, beyond spines; pattern of contrasting well-defined
black vertical bars on silvery body .... *Pterophyllum*
21b. Body deep, not round; dorsal and anal soft rays not much longer than spines; color pattern of irregular
dark bars and a dark horizontal band or row of spots from eye to dorsal-fin base.............................. *Mesonauta*

22a. Jaw teeth few, ≤8 anteriorly in each jaw; dorsal-fin spines 8–10................................... *Symphysodon*
22b. Jaw teeth >10 anteriorly in each jaw; dorsal-fin spines >10.. **23**

23a. Premaxillary ascending processes reaching behind orbit; enlarged canines anteriorly in jaws...................*Caquetaia*
23b. Premaxillary ascending processes short, reaching about orbit; no enlarged canines anteriorly in jaws..................... **24**

24a. Scales small, >40 in E1 row.. **25**
24b. Scales large, <40 in E1 row (27–34)... **26**

25a. Anal-fin spines 5, scales in E1 row 44–47.. *Hoplarchus*
25b. Anal-fin spines 7–8, scales in E1 row 48–57... *Uaru*

26a. Jaw teeth with lingual cusp on anterior teeth; dark spot at base of soft dorsal fin; body sides with vertical
bars, but no black spot on middle of side .... *Heros*
26b. Jaw teeth simple unicuspids; no dark blotch at base of soft dorsal fin; large dark spot at middle or upper half of side...... **27**

27a. Coloration with distinguishable but often diffuse vertical, black bars from base of dorsal fin to ventral
edge of the body over light background, occasionally reaching base of anal fin; midlateral spot entirely
overlapping fourth bar counting from posterior edge of dorsal fin .... *Heroina*
27b. Coloration variable, generally marbled but never with clearly distinguishable bars, usually with dark to
very dark background; if midlateral spot visible, not reaching above upper lateral line and is generally
associated with a thin midlateral band (*H. temporalis*) or forming a vertically elongate spot bisected by the
upper lateral line (*H. coryphaenoides*) .... *Hypselecara*

Cichlasomatines

28a. Triserial predorsal scale pattern (fig. 28a) **29**
28b. Uniserial predorsal scale pattern (fig. 28b) **32**

29a. Preopercle with 2–4 scales; dark spot at base of
caudal fin absent or present at middle of base.............. **30**
29b. Preopercle naked (rarely 1 scale); dark spot
present on dorsal half of caudal-fin base **31**

30a. 3 or 4 scales on preopercle; unpaired soft
fins naked or only moderately scaly .... *Laetacara*
30b. 2 scales on horizontal limb of preopercle;
extensively scaled unpaired soft fins.............. *Cleithracara*

31a. Scales present on bases of soft dorsal and anal fins...*Cichlasoma*
31b. Dorsal and anal fins naked ....*Aequidens*

32a. Preopercle scaled; caudal ocellus absent **33**
32b. Preopercle without scales; caudal ocellus usually present.. **34**

33a. 14 principal caudal-fin rays..*Nannacara*
33b. 16 principal caudal-fin rays..*Ivanacara*

34a. Scales present at base of soft dorsal fin...*Krobia*
34b. Scales absent from dorsal and anal fins **35**

35a. E1 scales 21–22; short black stripes basally on dorsal fin interradial membranes...............................*Rondonacara*
35b. E1 scales 22–15, no short black stripes basally on dorsal fin interradial membranes.............................. **36**

36a. Lateral band running from head to dorsal part of caudal-fin base ....*Aequidens paloemeuensis*
36b. Lateral band or series of blotches running from head to or toward end of dorsal-fin base................................ **37**

37a. Dark band from orbit running caudodorsad (i.e.,
toward dorsal fin) to nape (fig. 37a) ....*Tahuantinsuyoa*
37b. Dark lateral band continuing rostrodorsad (i.e., away
from dorsal fin) across nape (fig. 37b).................................*Bujurquina*

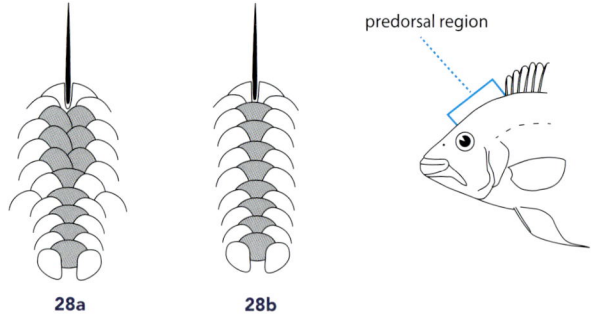

Geophagines

38a. Dorsal and anal fins extensively scaled ....*Geophagus*
38b. Scales absent from dorsal and anal fins **39**

39a. 13–17 gill rakers externally on first ceratobranchial; ocellar mark at caudal-fin base in adults.............................*Satanoperca*
39b. 0–6 gill rakers externally on first ceratobranchial; no ocellar mark at caudal-fin base **40**

40a. Anal-fin spines 6–9 ....*Apistogrammoides*
40b. Anal-fin spines 3–4 .... **41**

41a. 27–29 body scales in the horizontal series above that including the lower lateral line; body color plain, with
a black stripe through the eye from the nape to the preopercular corner and an ocellated black spot on the
posterior flanks...*Biotodoma*
41b. ≤26 body scales in the horizontal series above those including the lower lateral line; body color not as above.................. **42**

42a. Body elongate; a broad dark band along middle of side; circumpeduncular scales 12; principal caudal-fin
rays 14; dentary lateralis foramina 3 on each side; lateral line pores as single pits ....*Taeniacara*
42b. Elongate or deep-bodied; color pattern varies; circumpeduncular scales 16, rarely 12, principal caudal-fin
rays 16; dentary lateralis foramina 4–5 on each side; lateral line pores as short tubes...................................... **43**

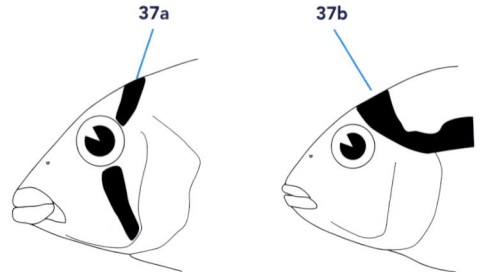

43a. First dorsal spine inserted behind vertical line from hind margin of gill cover; 12–14 dorsal-fin spines, ratio of dorsal spines to dorsal rays <2:1; terminal scale of the upper lateral line at 1–1.5 scales distance from the dorsal fin; 25–27 scales in a horizontal series above that containing the lower lateral line............ *Mikrogeophagus*

43b. First dorsal spine inserted before margin of gill cover; 14–18 dorsal-fin spines, ratio of dorsal spines to dorsal rays generally >2:1; terminal scale of the upper lateral line at 0.5 scales distance from the dorsal fin; 20–24 scales in a horizontal series above that containing the lower lateral line ....*Apistogramma*

GENUS ACCOUNTS

Acarichthys (13 cm SL)

Outwardly similar to *Geophagus* but without an epibranchial lobe, and swim bladder does not extend beyond the abdominal cavity. Distinguished from other cichlids by: a relatively deep body with a straight, steep, predorsal contour; deep lachrymal bone; and a small mouth. From *Geophagus* it can be identified externally by the few dorsal spines (12–14) and broad-tipped pelvic fins with the branches of the first ray equal in length or the inner slightly longer. E1 scales 29–30. Dorsal fin with 12–14 spines and 11–12 soft rays. Anal fin with 3 spines and 8 or 9 soft rays. Caudal fin emarginate. Serrations are absent from the preopercle, present on the supracleithrum (upper part of shoulder girdle). The preserved color pattern is simple, with a dark stripe from the occiput through the eye down to the preopercle, black anterior dorsal-fin rays and a black spot on the middle of the side. Living specimens are colorful, with orange anterior side and a general iridescent blue on the head, body, and fins. In large specimens the posterior 5–6 soft dorsal-fin rays are spectacularly produced and tipped with bright red. Young specimens with black spot at the beginning of the dorsal fin followed by a pink blotch. Sexual dimorphism includes differences in body shape and color (Leibel 1984). **SPECIES** One; *A. heckelii* (Threadfin acara). **DISTRIBUTION AND HABITAT** Mainstream Amazon River in Peru, Colombia, and Brazil, including lower parts of the Putumayo, Trombetas, Negro, and Xingu rivers; lower Tocantins and Capim rivers; Branco River in Brazil and Guyana, and the Essequibo River in Guyana. **BIOLOGY** Reproduction has been studied in the wild in a spring-fed pond in Guyana (Cichocki 1979). Females have territories and dig long tunnels into the bottom, leading to a breeding chamber. Several tunnels may reach the breeding chamber. Females select males and after an extended courtship lay around 2,000 eggs in the breeding chamber, presumably on the walls and roof. Fanning the eggs is primarily the responsibility of the female, while the male actively patrols the territory above the nest-tunnel complex. Male and female defend their fry using the nest as the focal point until the young are 8–12 mm SL.

Acaronia (8.0–16 cm SL)

Similar in shape to *Chaetobranchus flavescens* and the young resemble *Bujurquina* in coloration, having a dark stripe running from the head obliquely upward to the upper part of the caudal peduncle. Large adults are marked by a small caudal ocellus, an ocellated midlateral spot, and spots and stripes on the head margined with silvery lines in life (Kullander 1986). Important identification characters are the proportionally large head, eye, and gape, accompanied by a long ascending premaxillary process reaching beyond the middle of the orbit. *Acaronia* species have an extraordinary ability to extend their oral jaws. The scales are large (22, rarely 23 E1 scales, very low for a large cichlid species), vertebrae few (12+12); and the

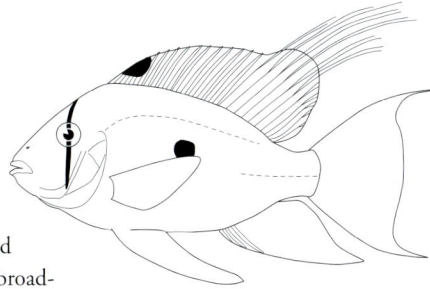

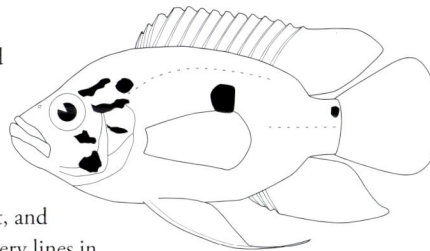

lateral line with prominent flank lateral line tubes and large cephalic lateralis foramina. *Acaronia* is distinguished from the otherwise superficially most similar genus, *Chaetobranchus*, especially by the comparatively strong and not as numerous gill rakers: 1–2 epibranchial and 9–13 cerato- and hypobranchial rakers, lower limb rakers contiguous to form a long tooth plate in large specimens. Dorsal fin with 12–14 spines and 9–10 soft rays. Anal fin with 3 spines and 7–9 soft rays. Caudal fin rounded. Males are bigger than females (Keith et al. 2000). **SPECIES** Two, both in the AOG region: *A. nassa* and *A. vultuosa*. Species information in Kullander (1986, 1989b). **COMMON NAMES** *Boca de Juquiá* (Brazil); Bigeye cichlid (English). **DISTRIBUTION AND HABITAT** *A. nassa* widespread in Amazon basin and Guianas; *A. vultuosa* in the Orinoco and upper Negro basins. Both species are encountered more commonly in lentic habitats, preferably with aquatic vegetation. Juveniles are found amid submerged leaf litter banks. **BIOLOGY** *A. nassa* is an ambush predator, with a diet including mainly shrimps, insects, insect larvae, and small fishes (Axelrod 1993, Keith et al. 2000) and a biparental substrate brooder (Stawikowski and Werner 1998); in French Guiana reproduction takes place during the rainy season (Keith et al. 2000).

Aequidens (10–20 cm SL)

Distinguished by a moderately large and deep body with a generalized body shape similar to *Cichlasoma*. Differs from *Cichlasoma* by the naked dorsal and anal fins (vs. scaled in *Cichlasoma*) and long caudal peduncle, including 2–3 vertebrae instead of none (Kullander and Nijssen 1989). Further distinguished by: triserial predorsal scale pattern; usually 13+13 (sometimes 12+13) vertebrae; usually 24 E1 scales, rarely 25; absence of preopercular scales; three anal-fin spines; comparatively small mouth with unicuspid teeth arranged in several rows; few gill rakers externally on the first ceratobranchial (4–6); rounded to subtruncate caudal fin; 1–2 tooth plates on 4th ceratobranchial; a cheek spot like *Cichlasoma*; a midlateral spot; and a dorsally positioned more or less ocellated blotch at the base of the caudal fin: dorsal fin with 14–16 spines and 10–12 soft rays; anal fin with 3 spines and 8–10 soft rays (Kullander 1986). No obvious external sexual dimorphism in most species; in some species males have the dorsal, anal, and pelvic fins prolonged (Kullander 1995) and a brighter coloration (Keith et al. 2000). **SPECIES** 16, with at least 12 species in AOG including *A. rivulatus* (Green terror) and *A. tetramerus* (Saddle cichlid). **DISTRIBUTION AND HABITAT** Amazon, Tocantins-Araguaia, Orinoco, and Paraguay basins and the Guianas. *Aequidens* species are found predominantly in small forest streams and lakes. **BIOLOGY** Limited studies suggest opportunistic feeding on small fishes, insects, insect larvae, shrimps, and some plant matter (Keith et al. 2000). Both parents guard the eggs and young; mouth brooding has been reported in at least one species (Stawikowski and Werner 1998).

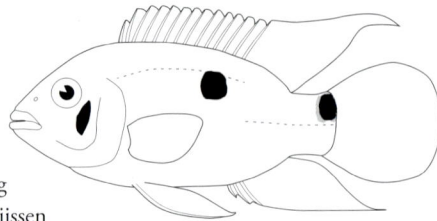

Apistogramma (2.0–8.0 cm SL)

Characterized by: small adult body sizes; moderately elongate body shape; and several reductions including a relatively small epibranchial lobe with a few marginal gill rakers, few or no gill rakers on the first ceratobranchial (≤5), and reduction of several skull bones (Kullander and Nijssen 1989). Scales large and few, 21–23 in the E1 row; lateral line represented mostly by pored instead of tubed scales; dorsal fin with 14–18, usually 15 spines, and 5–8 soft rays; anal fin typically with 3 spines, but a few species (*A. luelingi* and *A. commbrae*) with 4 spines; soft anal-fin rays 4–8, usually 6 or 7; preopercle and posttemporal margins smooth or serrate; caudal fin is rounded or subtruncate, commonly lanceolate or emarginate with marginal streamers in males; scales absent from fins except basal

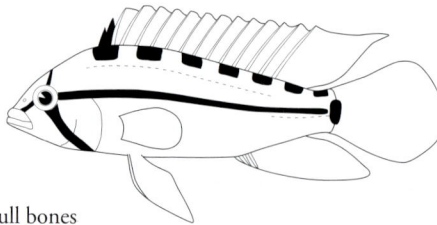

portion of caudal fin. Color pattern includes 7 vertical bars, the first above the opercle, last on the caudal peduncle, a midlateral spot in the third bar, and a dark spot at the base of the middle caudal-fin rays. Bars faint or absent and commonly there is a dark band or row of dark blotches along the middle of the side. The head is marked by dark stripes radiating from the orbit toward the occiput, opercular margin, mouth, and junction of sub-and interopercle, respectively. The anterior 2–3 dorsal-fin membranes are usually black. Ocellated spots are absent but iridescent blue or green stripes are commonly present on the sides of the head in living individuals. Sexual dimorphism is generally marked: males grow larger than females and often present relatively larger fins, commonly featuring prolonged rays in the caudal fin, produced dorsal-fin lappets, and filamentous extension of the first pelvic-fin ray, as well as brighter coloration; females have species-specific brooding color pattern consisting of contrasting yellow ground color and intense black or dark brown markings (Kullander 1980). **SPECIES** 86, but several recently described species of doubtful validity. The genus was revised by Kullander (1980). Regional species-level revisions include: Kullander (1986) for the Peruvian Amazon, Kullander and Nijssen (1989) for Suriname, and Kullander (1984) for the Paraguay basin. Several popular books with species-level information and photographs are available, e.g., Koslowski (2002) and Römer (2001, 2006). **COMMON NAMES** *Acará anão* (Brazil); *Bujurquita, Apistograma* (Peru); Dwarf cichlid (English). **DISTRIBUTION AND HABITAT** Tropical South America east of the Andes, including the Orinoco and Amazon basins and coastal drainages in the Guianas, as well as the Tocantins, Maranhão, Paraguay, Uruguay, and lower Paraná rivers. The genus is poorly represented in the Amazon tributaries draining the Brazilian and Guianas highlands (Kullander 1980). *Apistogramma* are found in small forest streams or in dense vegetation along lake and river margins. **BIOLOGY** In most *Apistogramma* species, spawning takes place in concealed spaces like leaf litter aggregations or crevices in woody debris, and eggs are deposited hanging from the roof of the space. Different breeding strategies exist. Some species form pairs at least for the breeding season; other species breed in polygamous harems. Usually, it is the female that guards the eggs, larvae, and free-swimming young, whereas the male defends the territory (Koslowski 2002). Mouth brooding reported from at least one species. Omnivorous, feeding on plant debris, algae, and small terrestrial and aquatic arthropods (Henderson and Walker 1990, Silva 1993, Sánchez et al. 2003).

Apistogrammoides (2.7 cm SL)

Similar to *Apistogramma*, differing by: a long anal fin with 6–9 spines (3–4 in *Apistogramma*); a wide interorbital space (11.9–14.1% of SL; generally <11.4% in *Apistogramma*); and a unique caudal-fin and caudal-peduncle marking consisting of a triangular spot on the caudal peduncle and three elongate contrasting spots on the caudal-fin base (Kullander 1986). Scales in E1 row 21–22. The lateral line is represented mostly by pored scales. Scales are absent from the fins except basally on the caudal fin. Dorsal fin with 16–18 spines and 4–5 soft rays, modally 16 spines and 5 soft rays. Anal fin with 6–9 spines and 5–6 rays, modally 7 spines and 6 soft rays. Caudal fin rounded. Serrations are absent from the preopercle and other head bones. The sexes (specimens >1.0 cm SL) differ in coloration and the longer pelvic, dorsal, and anal fins in males. **SPECIES** One; *A. pucallpaensis*. **DISTRIBUTION AND HABITAT** Amazon basin, from the middle Ucayali River (Peru) to Leticia (Colombia), and a recent record from the Tefé region in Mamirauá Reserve, Japurá basin in Brazil (J. Zuanon pers. comm.). Collected in pools with abundant floating vegetation and in floating meadows in rivers (Kullander 1986). **BIOLOGY** According to aquarium observations, eggs, numbering about 20–100, are laid in concealed spaces and attached to the underside of the ceiling substrate. Eggs and larvae are guarded by the female, whereas the male defends the territory and participates in guarding the free-swimming fry (Koslowski 2002).

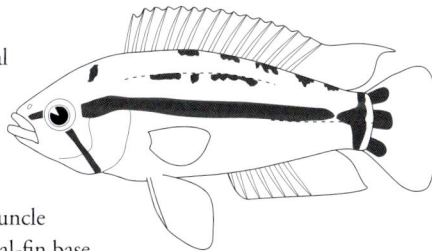

Astronotus (25–40 cm SL)

Readily identified by: a distinct color pattern composed of dark brown blotches and vertical bars on the sides, and usually an ocellus with bright red or orange center on the caudal-fin base, and long, extensively scale-sheathed dorsal and anal-fin bases (Kullander 1986). The following characters might also help identification (Kullander 1986): "African type" lips; large gill rakers that bear numerous spines on the exposed surface; lachrymal with only 3 foramina. E1 scales 34–40. Scales cycloid anteriorly on side and chest and above upper lateral line. Dorsal fin with 12–14 spines and 19–22 soft rays. Anal fin with 3 spines and 15–17 soft rays. Caudal fin rounded. No obvious external sexual dimorphism. Juveniles with very different color patterns from adults, including white and black blotches, wavy bands, and spotted heads. **SPECIES** Three, all in the AOG region: *A. crassipinnis*, *A. ocellatus*, and *A. rubroocellatus*. **COMMON NAMES** *Acará-açu*, *Corró-baiano* (Brazil); *Acarahuazú* (Peru); Oscar (English). **DISTRIBUTION AND HABITAT** Widespread in South America, including the Amazon and Orinoco basins and French Guiana. Introduced in many areas. Prefer quiet shallow waters, e.g., floodplain lakes. **BIOLOGY** Little is known about the ecology of *A. crassipinnis* or *A. rubroocellatus*. *Astronotus ocellatus* on the other hand is a popular food and game fish in South America and also an aquarium fish. The species has been reported to feed on small fish, crayfish, worms, and insect larvae (Keith et al. 2000, Galvis et al. 2006). *Astronotus ocellatus* is a substrate spawner that forms pairs and guards eggs and fry (Paiva and Nepomuceno 1989, Galvis et al. 2006). The bright ocelli, or eyespots, near the base of the caudal fin in *Astronotus ocellatus* have been suggested to mimic the head and thereby reduce fin-nipping by piranhas (Winemiller 1990).

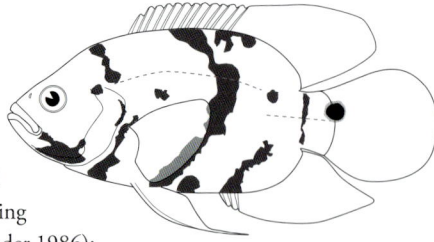

Biotodoma (10 cm SL)

Diagnostic color pattern being plain with a black stripe through the eye from the nape to the preopercular corner, an ocellated black spot on the posterior flanks, and thin blue stripes on the snout (Kullander 1986). Body shape like *Geophagus* but epibranchial lobe is somewhat smaller than *Geophagus*, and cheek squamation extends more anteriorly. The vertebral count is lower (14+14; versus 14–15+18–19 in *Geophagus*). Ribs are absent from the caudal vertebrae and there are no caudal swim bladder extensions. There are only 3 postlachrymal infraorbitals, the last associated with the sphenotic, the middle long and with a middle foramen (Kullander 1986). E1 scales 27–29. Dorsal fin with 15 spines and 9–11, usually 12 soft rays in *B. cupido*, 13–14 spines and 11–12 soft rays, modal count 14 spines and 11 soft rays in *B. wavrini*. Anal fin with 3 spines and 8–10, usually 9 soft rays. Caudal fin emarginate with pointed corners. The preopercle, pterotic, and supracleithrum have smooth margins. The sexes are isomorphic, except that in *B. cupido* at least males have more intense blue snout coloration (Cichocki 1977). **SPECIES** Two; *B. cupido* and *B. wavrini*. **COMMON NAMES** *Acará chibante* (Brazil); *Ciclido cupido* (Peru); Green-streaked eartheater (English). **DISTRIBUTION AND HABITAT** *B. cupido* in the Amazon basin of Bolivia, Brazil, and Peru, and the Essequibo basin in Guyana; *B. wavrini* in the Orinoco basin and the middle and upper Rio Negro basin in Brazil. Collected in streams, sandy beaches, and lakes, mainly in dark-colored waters (Kullander 1986). **BIOLOGY** Parental behavior of *B. cupido* has been studied in a small creek entering the lower Essequibo River in Guyana (Cichocki 1977). Water conditions is this area are affected by semidiurnal tides. *Biotodoma cupido* is a substrate brooder with strong parental care, including intense aggression toward brood predators and orally transferring the brood to a secondary nest depression in deeper water at very low tide.

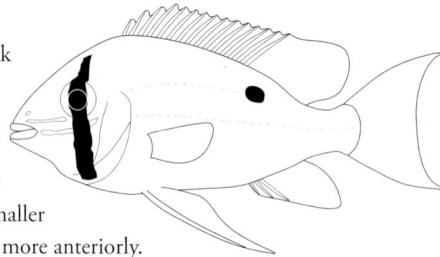

Biotoecus (3.0–4.0 cm SL)

Small adult body sizes; elongate body shape; semitransparent; and relatively modestly colored in life. The most obvious distinguishing character is the dorsal fin with 7–9 spines and 13–16 soft rays, contrasting with all other Neotropical cichlids, which have more spines than soft rays in the dorsal fin. Several distinguishing characters are reductions, such as reduced lateral line system involving 3 instead of >3 lachrymal foramina, loss of infraorbital ossicles, flank lateral lines with pored instead of tube-bearing scales except occasionally the first scale in the upper lateral line; and scales absent from most of the predorsal and prepelvic region (Kullander 1989a). E1 row scales 28–29. Scales are absent from the fins except from the base of the caudal fin. Anal fin with 2 spines and 7–8 soft rays (*B. dicentrarchus*) or 3 spines and 6–8 soft rays (*B. opercularis*). Caudal fin emarginate. Serrations are absent from the preopercle and other head bones. Gill rakers are absent externally on first ceratobranchial. The color pattern is simple, with a dark spot on the gill cover, middle of side, and caudal-fin base, but may also feature 6 faint vertical bars each with a dark spot at the midline of the side. Males are smaller and develop more colorful and extended pelvic and caudal fins than females. **SPECIES** Two, both in the AOG region: *B. dicentrarchus* and *B. opercularis*, described and reviewed by Kullander (1989a). **DISTRIBUTION AND HABITAT** *B. dicentrarchus* in the Orinoco basin (from the Inírida River, Colombia, to Maripa, Venezuela); *B. opercularis* in the Amazon basin (in the central Amazon, Branco, and middle and lower Negro Rivers). Both species occupy shallow waters in streams and riverside lagoons. **BIOLOGY** Aquarium studies show that *B. dicentrarchus* breeds in pairs, depositing eggs at the ceiling of a cave; the female tends eggs and yolk-sac larvae. Both parents care for the free-swimming young (Staeck 2002).

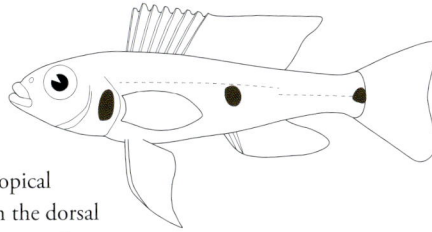

Bujurquina (5.0–11 cm SL)

Distinguished externally from *Aequidens* and similar genera by a uniserial predorsal scale pattern of 8 scales; accessory lateral line scales between caudal-fin rays D1 and D2; an oblique lateral band running from the opercular cleft to or toward either the end of the soft dorsal fin or the junction of the dorsal fin and the caudal peduncle; and the continuation of the lateral band on the head obliquely forward across the nape. Most species possess a distinct but small, often ocellated spot at the base of the caudal fin, and the caudal fin is usually emarginate, in some species with produced marginal rays. E1 scales 22–25, usually 24. Scales are absent from the unpaired fins, except basally on the caudal fin. Dorsal fin with 13–15 spines and 9–11 soft rays, modally 14 spines and 9 or 10 soft rays. Anal fin with 3 spines and 5–8, modally 7 soft rays. The distribution of the genus is complementary to that of *Krobia*, which also has a slanting lateral band, consideration of which facilitates field identification. In the sympatric *Tahuantinsuyoa* there is a dark stripe from the eye running obliquely caudad to the front, instead of a dark stripe running obliquely forward across it. The sexes are isomorphic. **SPECIES** 18, and several undescribed; key to the described species in Arbour et al. (2014). **DISTRIBUTION AND HABITAT** Most species inhabit the western Amazon basin in Peru and Ecuador. Other species in tributaries and main channel of the Amazon River, as well as western Orinoco, Paraguay, and lower Paraná basins. The typical biotopes are small forest streams and margins of small rivers. **BIOLOGY** Breed in pairs, each spawn producing ~400 eggs which are attached to loose leaves that may be pulled around, possibly in reaction to predators or water level changes. Both parents defend the eggs and take the yolk-sac larvae in their mouths (Kullander 1986, Stawikowski and Werner 1998).

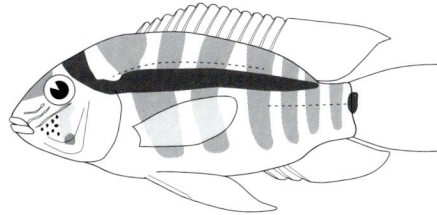

Caquetaia (16–26 cm SL)

Medium-sized predators with strongly compressed body and very large mouth with protractile jaws. Distinguished by a very long, ascending premaxillary process reaching behind the middle of the eye. The teeth in the outer row in both jaws are spaced, enlarged canines, the remaining teeth much shorter and forming a narrow band. E1 scales 28–32. Dorsal fin with 15 or 16 (rarely 17) spines and 10–13 soft rays; anal fin with 5 or 6 (rarely 7) spines and 9–10 soft rays. The color pattern is species specific, with two or three dark vertical bars and a dark vertical stripe on the head in *C. myersi*, and several vertical bars, a midlateral blotch, and a caudal ocellus in *C. kraussii* and *C. spectabilis*, the latter lacking large dark spots on the head present in *C. kraussii*. Sexes are isomorphic. **SPECIES** Four, but *C. umbrifera*, distributed west of the Andes, may not belong in this genus. **DISTRIBUTION AND HABITAT** Two species found in the Amazon basin: *C. myersi* in the Putumayo and Napo basins, and *C. spectabilis* in the Madeira, Uatumã, Branco, Trombetas, Araguari (Amapá), Tapajós, and Xingu basins. An additional species, *C. kraussii*, native to rivers draining to the Caribbean in Venezuela and Colombia, has been introduced in the Orinoco basin in Venezuela. Found in rivers, preferring areas abundant with aquatic plants (Galvis et al. 1997, Kullander 2003a). **BIOLOGY** More than 1,000 eggs are deposited on open substrate and guarded by both parents. Larvae are transferred to a depression in the sand or to various hiding places by the female (Stawikowski and Werner 1998). The food consists mainly of other fishes and benthic invertebrates (Galvis et al. 1997, Kullander 2003a)

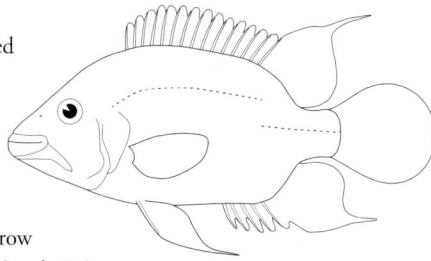

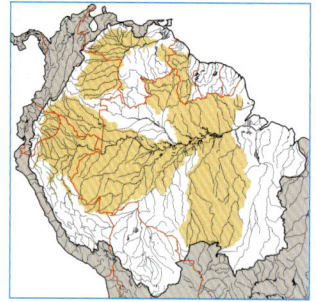

Chaetobranchopsis (12 cm SL)

Deep-bodied with broad head and numerous long gill rakers specialized for filter-feeding. Similar to *Chaetobranchus* (see below) but with extensively scaled unpaired fins; 5–6 (rarely 4) instead of 3 anal-fin spines; and lacking a basibranchial tooth plate. E1 scales 25–26. Dorsal fin with 13–16 spines and 12–14 soft rays. Anal fin with 4–6 spines and 15–18 soft rays. The color pattern includes two dark horizontal bands along the side and a more or less distinct black spot in the upper band. There is no ocellated mark at the base of the caudal fin. No obvious external sexual dimorphism. **SPECIES** Two, both in the AOG region: *C. orbicularis* (15 dorsal-fin spines and 6 anal-fin spines) and *C. australis* (14 dorsal- and 5 anal-fin spines; Eigenmann et al. 1907). **COMMON NAMES** *Acará tucuma* (Brazil); *Bujurqui vaso* (Peru). **DISTRIBUTION AND HABITAT** *C. orbicularis* in the Amazon River from the mouth of the Negro River to Marajó Island, and in Amapá, Brazil; *C. australis* in the upper Madeira and Paraguay basins. Occurs in lakes, swamps, and flooded areas with shallow, almost stagnant water, which may be either clear or loaded with sediments (Keith et al. 2000). **BIOLOGY** Hiding among aquatic plants and tree roots. The weak teeth and long and dense gill rakes suggest microphagous and planktivorous feeding, possibly including small crustaceans (Keith et al. 2000).

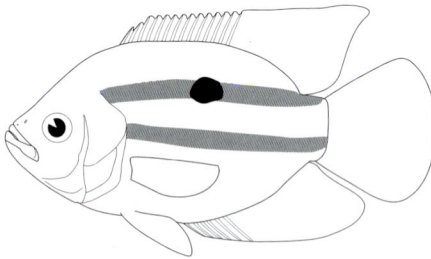

Chaetobranchus (21–25 cm SL)

Similar to *Chaetobranchopsis*, but more elongate, with naked unpaired fins, 3 instead of 4–6 anal-fin spines, and a basibranchial tooth plate (unique for cichlids). Both genera share many long and slender gill rakes (>50 on first gill arch; Kullander 1986). The position of the caudal-fin lateral lines in *Chaetobranchus* is unusual among cichlids: extending to or beyond the scale cover of fin (Kullander 1986). E1 scales 25–26. Scales absent from fins except basally on caudal fin. Dorsal fin with 13 spines and 13–14 soft rays. Anal fin with 3 spines and 10–12 soft rays. Caudal fin subtruncate, but individual rays may be projecting beyond fin membranes. Teeth conical, slightly recurved, in narrow bands in both jaws. Gill rakers externally on first ceratobranchial 50–60. Overall dark with a black spot on the middle of the side, or light with several vertical bars. Males are larger than females (Keith et al. 2000). **SPECIES** Two, both in the AOG region: *C. flavescens* and *C. semifasciatus*. **DISTRIBUTION AND HABITAT** *C. semifasciatus* limited to the central Amazon, between Óbidos and Tefé; *C. flavescens* widely distributed in the AOG region. Found in swamps, lakes, and flooded areas with either turbid or clear stagnant water (Kullander 1986, Keith et al. 2000). **BIOLOGY** Feed predominantly on plankton (Keith et al. 2000).

Cichla (26–100 cm SL)

Largest body size of South American cichlids. Muscular predatory fishes with a large gape specialized for preying on fish and large invertebrates. Easily distinguished from all other South American cichlids by the shape of the dorsal fin: the spines increase in length to about the fifth spine, after which there is a gradual decrease to a very short penultimate spine; the soft fin is again about as high as the anterior spinous part. Other distinguishing characteristics useful for field identification include: densely scaled caudal and anal fins; large mouth with lower jaw extending beyond the upper jaw; "African-type" lips; small scales; frequently continuous lateral line; and a large conspicuous ocellus at the base of the caudal fin (Kullander and Ferreira 2006). Kullander and Ferreira (2006) list nine diagnostic characters for the genus, including a prominent rostral fold on maxilla; unique color pattern with 3 dark vertical bars on side, developing from 3 dark blotches in juveniles; elongate depression behind nostril without opening; and a distinct vertical line across the cheek about four scales anterior to the preopercle, apparently representing a free neuromast (lateral line sensory organs) row. E1 scales 79–109. Lateral line divided or continuous. Dorsal fin with 14–16 spines and 15–18 soft rays, modal count 15 spines and 17 soft rays. Anal fin with 3 spines and 10–12, usually 11 soft rays. Caudal fin slightly emarginate to subtruncate. Soft dorsal, anal, and caudal fins with numerous small scales. Teeth small, pointed, slightly recurved, in wide bands in both jaws. Sexual dimorphism: males develop a frontal hump prior to spawning. **SPECIES** Eight, all native to the AOG region (Willis et al. 2012b). Key to the species, species descriptions, and distribution maps in Kullander and Ferreira (2006) and Willis et al. (2012b). **COMMON NAMES** *Tucunaré* (Brazil, Peru); *Pavón* (Colombia, Venezuela); *Lukanani* (Guyana); *Tukunari* (Suriname). **DISTRIBUTION AND HABITAT** Widely distributed in the Amazon, Tocantins, and Orinoco basins, as well as in the coastal rivers in the Guianas. Introduced in many places outside this native distribution range because of its attraction as a sport fish. Found in many habitats, including rivers, close to rapids, and in floodplain lakes. **BIOLOGY** Diurnal, feeding mainly on fish (Winemiller et al. 1997). For some species, reproduction is not markedly seasonal and spawning can take place several times per year (Keith et al. 2000, Chellappa et al. 2003). Parents guard the eggs and fry (Zaret 1977).

Cichlasoma (7.5–14 cm SL)

Small to moderate body sizes, recognized most easily by a layer of scales on the bases of the soft dorsal and anal fins in combination with large scales (E1 scales 22–25, usually 23–24), usually triserial, occasionally uniserial predorsal scale pattern, short caudal peduncle not including any complete vertebral centrum, rounded caudal fin (Kullander 1983, Kullander and Nijssen 1989), and characteristic color pattern including a dark spot posterodorsally on the cheek, a black midlateral spot, a dark stripe from the head to the midlateral spot and often continued to the caudal-fin base, a black light-ringed spot on the caudal-fin base, and richly spotted caudal fin. The anal fin has 3 spines in species in the Paraguay, Paraná, Uruguay, and Tocantins basins, and ≥4 spines in other areas. (Kullander 1986). Dorsal fin with 13–18 spines, usually 14–15, and 8–12 soft rays, usually 10 or 11. Anal fin with 3–6 spines and 7–9 soft rays. Males grow slightly larger than females (Keith et al. 2000). **SPECIES** 12. The genus was revised by Kullander (1983), reducing the nearly pan-Neotropical cichlid genus with >100 species to include only 12, possibly 13 species east of the Andes (Kullander 2003a). **DISTRIBUTION AND HABITAT** Much of tropical South America, including the Amazon and Orinoco basins (at least 5 species), coastal rivers in the Guianas (3 species), and Trinidad. Found mainly in lakes, inundated areas, and short-drainage streams close to floodplains; nearly absent from terra firme streams. **BIOLOGY** omnivores; reported food items include: mollusks, shrimp, fish, aquatic and terrestrial insects, vegetal debris, filamentous algae, and seeds (Lowe-McConnell 1969, Keith et al. 2000).

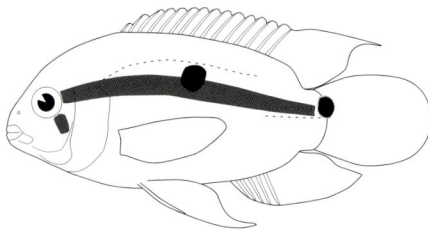

Cleithracara (7.0 cm SL)

Small adult body size with a distinctive color pattern consisting of a broad black, vertically oriented crescent extending over the nape, eye, and gill cover, and a large black, light-margined blotch (sometimes a pair of ocelli or an irregular vertical blotch) on the posterior part of the body, superimposed on a plain beige ground color. The body is strongly compressed and the caudal peduncle extremely short. The genus can be recognized also by two scales on the horizontal limb of the preopercle, triserial predorsal squamation, extensively scaled unpaired soft fins, rounded caudal fin, and absence of accessory caudal-fin lateral lines (Kullander and Nijssen 1989). *Cleithracara* is similar to *Laetacara* in having a deep notch in the dorsal margin of the anterior ceratohyal (Kullander and Nijssen 1989). E1 scales 22, rarely 23. Dorsal fin with 14–16 spines and 10–13 soft rays. Anal fin with 3 spines and 10–12, usually 10 soft rays. Sexes are isomorphic. **SPECIES** One, *C. maronii* (Keyhole cichlid). **DISTRIBUTION AND HABITAT** From the Orinoco delta in Venezuela east to the Ouanary in French Guiana (Kullander and Nijssen 1989); recorded from southern Trinidad, W.I., but possibly only intermittent there. Inhabits small creeks, with little current and rich in decaying wood (Keith et al. 2000). Occasionally found in brackish water. **BIOLOGY** Feeds on invertebrates, including insect larvae and crustaceans (Mol 2012b). If threatened, individuals can swim very close against a submerged log, changing their color pattern to resemble the substrate (Keith et al. 2000). Spawns on hard surfaces; produces 400–600 eggs; both parents guard eggs and larvae (Stawikowski and Werner 1998, Keith et al. 2000).

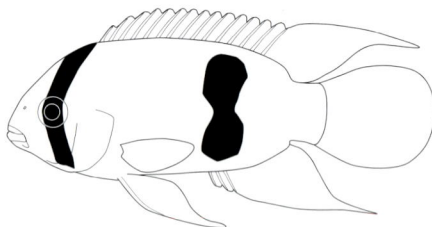

Crenicara (9.0–10 cm SL)

Crenicara, *Dicrossus*, and some *Mazarunia* species possess a strongly decurved snout with a subinferior mouth; a wide postlabial skin fold that sheaths the upper lip at least partially; and a laminar expansion of epibranchial 1 associated with the pharyngobranchial 2 arm (Kullander 1990a). A reduced everted fold of the lower lip and loss of the second dentary lateralis canal opening are unique to *Crenicara* and *Dicrossus* (Kullander 2011). Both genera have distinctly serrated preopercle and pterotic, and usually the supracleithrum, but the latter is smooth-margined in *Crenicara punctulatum*. Species of *Crenicara* and some species of *Dicrossus* also share an unusual color pattern consisting of several dark, rounded to almost square spots on the side of the body. Species of *Crenicara* reach larger sizes, 80–100 mm SL (vs. ~40–56 mm in *Dicrossus*), and can be differentiated from *Dicrossus* by an elongate rounded caudal fin (vs. commonly lanceolate or lyrate in *Dicrossus*); absence of a filamentous pelvic fin in males (vs. filamentous in *Dicrossus* except in *D. foirni*); and a longer snout, deeper body, and shorter caudal peduncle (Kullander 2011). E1 scales 26–28. Dorsal fin with 14–17 spines and 7–9 soft rays; usually 15 spines in *C. latruncularium*, 17 spines in *C. punctulatum*. Anal fin with 3 spines and 6–8 soft rays. The color pattern overall dark, with two rows of dark blotches, one along the dorsum and one along the middle of the side, the latter commonly replaced by a dark band with contrasting light stripe along its upper margin. Sexually mature females have red pelvic fins (Kullander 1986). **SPECIES** Two, both in the AOG region: *C. latruncularium* and *C. punctulatum*. **DISTRIBUTION AND HABITAT** *C. punctulatum* is widespread through the Amazon basin. Primarily a black water species found mostly in forest streams and on sandy river beaches, near fallen leaves or among plants (Kullander 1986). *Crenicara latruncularium* inhabits the Amazon River tributary Guaporé and Mamoré basins. **BIOLOGY** *Crenicara punctulatum* is well known for protogyny, living in groups with several females and one male. When the male is removed, one of the females develops male characters. At spawning about 200 eggs laid on wood or stone substrates. Brood care is exclusively female. All offspring are born females; at about 8 months of age one female changes into a dominant male (Carruth 2000).

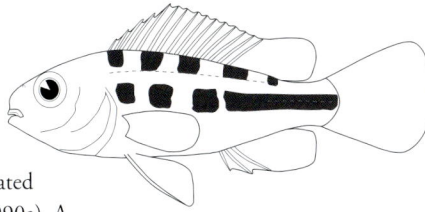

Crenicichla (5.0–30 cm SL)

Characterized by: an elongate body, usually with a large gape and strong dentition; a high number of vertebrae (lowest number 32); small and numerous E1 scales (33 to about 130); and high dorsal fin-ray count (rarely as few as 16, usually ~20–24 spines; usually ~15 rays). Almost all species have serrated preopercular margin, and some have serrated sphenotic. The lower jaw is prognathous in species feeding on fish or large invertebrates, but many species from fast-running water have downturned mouths with short lower jaw and feed on benthic invertebrates. The eye is conspicuously large in most species. The caniniform jaw teeth are either all depressible, folding backward, or those in the outer row are relatively stout and fixed. The lateral line is divided, and the lateral line scales are relatively longer than the adjacent scales and bear tubes. Some species have only cycloid scales, but in most species only the head scales, and the occipital and chest and abdominal scales are cycloid, the remaining scales ctenoid. Scales are absent from the fins, except on the base of the caudal fin. Anal fin with 3 spines and 8–13 soft rays. Gill rakers externally on first ceratobranchial 6–11, mostly 8–10, strongly denticulated. Gill rakers are present also on the sides of the lower pharyngeal jaw. The color pattern varies considerably, but juveniles of most species are distinctive in having a black horizontal stripe along the side. Adult coloration uniform grayish, striped, or barred. Most species possess an ocellated marking at the base of the caudal fin. *Crenicichla* are similar to *Teleocichla* but lack the

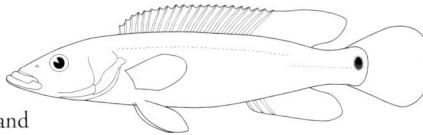

specializations for rheophily in that genus, and particularly the pelvic fin is shorter and mostly rounded. Sexual dimorphism is expressed in most species by the presence of one or more dark or ocellated spots in the dorsal fin and red or orange abdomen or abdominal sides in females. Males tend to grow larger, and when small spots are present in the dorsal fin, males retain them, while they disappear in breeding females. **SPECIES** 91, outdated key in Ploeg (1991). **COMMON NAMES** *Cacunda, Jacundá, Joaninha* (Brazil); *Añashua* (Peru); Pike cichlids (English). **DISTRIBUTION AND HABITAT** Most of tropical South America east of the Andes, including the Amazon and Orinoco basins, as well as all of the Guianas, the Paraguay, Paraná, and Uruguay basins, and coastal drainages between the Amazon and Paraná rivers. Found in many habitats. Benthopelagic; often close to sunken branches, rocks, or leaf litter. **BIOLOGY** Often ambush predators; reported food items include fish, fish scales, mollusks, shrimp, and aquatic insect larvae (e.g., Kullander 1990b, Kullander and De Lucena 2006, Montaña and Winemiller 2009). *Crenicichla* species breed in pairs or harems, usually with the female being the more active guarder of eggs and young.

Dicrossus (3.5–7.0 cm SL)

A reduced everted fold of the lower lip and loss of the second dentary lateralis canal opening are unique to *Crenicara* and *Dicrossus* (Kullander 2011). *Crenicara* and some *Dicrossus* species share an unusual color pattern of several dark, rounded to almost square spots on the side of the body. Both these genera have distinct serrations along the free margin of the preopercle and pterotic and in most species also the supracleithrum. *Dicrossus* can be differentiated from *Crenicara* by a lanceolate or lyrate caudal fin (vs. rounded in *Crenicara*); the filamentous pelvic fin of males in *Dicrossus* (except in *D. foirni*); and a shorter snout, more rounded body, and longer caudal peduncle (Kullander 2011). E1 row scales 24–26. Dorsal fin with 14–16 spines and 7–9 soft rays. Anal fin with 3 spines and 6–8, usually 7 soft rays. The color pattern, which is highly species specific, consists chiefly of 7 slanting vertical bars, each containing 2 or 3 blotches, a dark blotch on the gill cover, and a more or less distinct black spot at the caudal-fin base. Females have red pelvic fins when breeding. **SPECIES** Four or five, key to the species in Kullander (2011). **DISTRIBUTION AND HABITAT** Orinoco and Negro basins, central Amazon and lower Tapajós basin. Species distribution map in Kullander (2011). Found in small streams and in shallow water along riverbanks, mainly in clear- or blackwater streams with leaf litter (Schindler and Staeck 2008, Römer et al. 2010). **BIOLOGY** Aquarium observations of *D. filamentosus, D. foirni, D. maculatus,* and *D. warzeli* show them to be concealment spawners with exclusively maternal brood care (Warzel 1996).

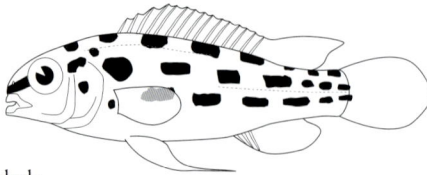

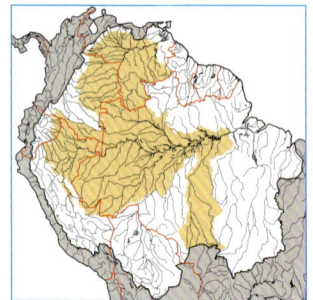

Geophagus (7.5–28 cm SL)

Geophagus is a catch-all group for three genera, of which one is restricted to northwestern trans-Andean South America, and another to Atlantic coast drainages in Brazil and Uruguay. True *Geophagus* found in the AOG region are characterized by: a prominent epibranchial lobe with several gill rakers along the margin; absence of scales on the anterior half of the cheek; a basal layer of scales on the dorsal fin; long projections of the swim bladder into the caudal region supported by epihemal ribs; deep lachrymal bone; and relatively steep frontal contour in adults. The color pattern includes a black spot on the middle of the side, and often a dark stripe below the eye or a black spot on the preopercle. Vertical bars are faint and a caudal spot is absent. Gill rakers are absent from the lower pharyngeal tooth

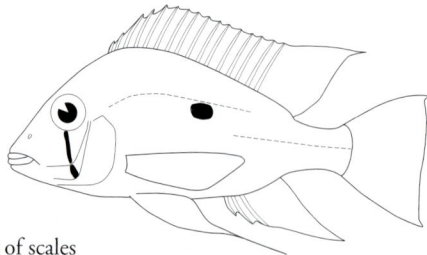

plate. E1 scales 29–34. Scales are present basally on the posterior part of the dorsal fin and basally on the caudal fin, present or absent on the base of the anal fin, and commonly present on the base of the pectoral fin; large specimens of some species present very thick and wide scale covers on dorsal and caudal fins. The development of fin squamation is correlated with size, and dorsal-fin scales may be absent in small specimens of large species and adults of small species. Dorsal fin with 15–19 spines and 9–13 soft rays. Anal fin with 3 spines and 7–9 soft rays. Caudal fin subtruncate or emarginate. Smooth opercular and pectoral girdle bones. Gill rakers externally on first ceratobranchial 10–15. Males are more intensely colored than females, often a little larger, and develop longer fin extensions. In addition, some dominant males develop a nuchal hump. **SPECIES** 27 species, genus is in need of taxonomic revision. Regional revisions are available for Peru (Kullander 1986), Suriname (Kullander and Nijssen 1989), Venezuela (López-Fernández and Taphorn 2004), and Guyana (Hauser and López-Fernández 2013). **DISTRIBUTION AND HABITAT** Widely distributed in the AOG region. Found in many habitats, but preferring areas with sandy bottoms (Keith et al. 2000, Hauser and López-Fernández 2013). **BIOLOGY** Most species search for food by sifting through the sand (López-Fernández et al. 2012b, 2014). Omnivorous or invertivorous generally specializing in benthic food items; food item include mollusks, crustaceans, insects, detritus, and plant matter (e.g., Mazzoni and Costa. 2007, Bastos et al. 2011). Parents form pairs and are either ovophile (brooding eggs, larvae, and small free-swimming fry) or larvophile (brooding hatchlings and small free-swimming fry) mouth brooders (Kullander and Nijssen 1989).

Guianacara (10 cm SL)

Characterized by: body strongly compressed laterally; relatively large and deep head and steep frontal contour; and a distinctive color pattern with plain sides except for a vertical bar traversing the middle of the fish that can be permanently limited to or reduced in adults to a midlateral spot, a complete black stripe extending from occiput to eye and from eye to preopercle, black anterior dorsal-fin membranes present but may disappear in adults of some species, and absence of a spot at the caudal-fin base. Species of the subgenus *Guianacara* have a black bar from the middle of the spinous dorsal-fin base down across the side, but *Guianacara (Oelemaria) oelemariensis* has a large black blotch on the middle of the side. In *Guianacara (Guianacara)* the midbody bar may be reduced to a blotch dorsally on the side in large specimens. The pelvic fin has a broad tip with the inner branch of the first ray longer than the outer branch. The major osteological diagnostic character is the wide dorsal shelf of the urohyal bone, but externally the coloration pattern is unique to the genus. E1 scales 24–26. Scales are absent from the fins except the caudal-fin base. Dorsal fin with 14–16 spines and 9–12 soft rays, usually 15 spines and 10 soft rays. Anal fin with 3 spines and 7–9, usually 8 soft rays. Caudal fin truncate or subtruncate. The preopercle is occasionally serrated, other head and shoulder girdle bones with smooth margin. **SPECIES** Five, key to the species in Arbour and López-Fernández (2011). **DISTRIBUTION AND HABITAT** Restricted to the Guiana Shield area (Guianas, Venezuela, and Brazil). Species are generally found in clearwater streams with moderate current, but also in seasonally flooded lagoons (Arbour and López-Fernández 2011). No known external sexual dimorphism. **BIOLOGY** Feed on small benthic invertebrates (Keith et al. 2000). The males have harems (Axelrod 1993). In *G. owroewefi*, breeding may occur 3–4 times per year. Spawning takes place in crevices of rocks; mainly females care for the eggs and young. Males defend the nest. Parental care in terms of food supply; protection lasts for 1–3 months (Keith et al. 2000).

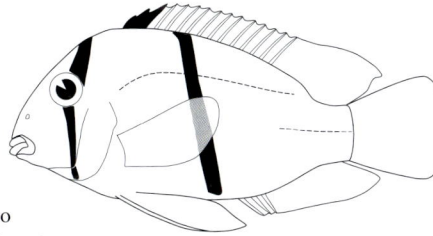

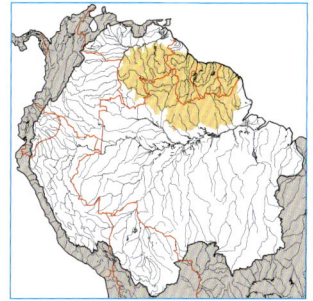

Heroina (10 cm SL)

Similar to *Caquetaia* and differing from all other cichlids with >4 anal-fin spines distributed east of the Andes in having the median pair of teeth in the upper jaw elongate and the corresponding teeth in the lower jaw shortened. Differs from *Caquetaia* in having much shorter anteriorly isognathous jaws, and a shorter premaxillary ascending process, which reaches only to the anterior third of the orbit instead of beyond the middle of the orbit. In other respects, superficially similar to *Heros* or *Hypselecara* in having a deep body, strongly compressed laterally, and short caudal peduncle, containing the last half-centrum, but no additional vertebral centra. The caudal prolongation of the swim bladder does not extend past the third hemal spine and is not supported by ribs (Kullander 1996b). E1 row scales 27–29, usually 28. The soft-rayed portions of the unpaired fins are densely scaled basally. Dorsal fin with 16–17 spines and 12–13 soft rays. Anal fin with 7 spines and 10–11 soft rays. The color pattern is distinctive in the complete absence of dark markings on the head, the subdivision of the third dark vertical bar into two, and the breeding color pattern of females, consisting in intensification of the middle and ventral portions of the dark vertical bars (Kullander 1996b). **SPECIES** One, *H. isonycterina*. **DISTRIBUTION AND HABITAT** Restricted to the western Amazon basin in Colombia, Peru, and Ecuador. Found in headwaters of major river drainages. **BIOLOGY** Biparental substrate brooder (Stawikowski and Werner 1998).

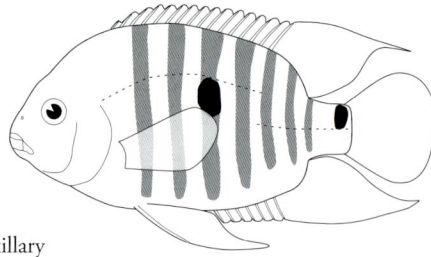

Heros (12–20 cm SL)

Characterized by: a deep body shape, compressed laterally; narrow in frontal aspect, with keeled nape and narrowly flattened chest; and small mouth, in some species with fleshy lips. Similar to *Symphysodon* from which it differs by having more teeth (20–30 in the outer series in the upper jaw, and 20–40 in the outer series of the lower jaw), and more spines (15–16 vs. 9–10) and fewer soft rays (12–14 vs. 29–31) in the dorsal fin. Scales in E1 row 27–29. Soft-rayed portions of dorsal and anal fins, and base of caudal fin with dense layer of scales basally. Anal fin with 7–8 spines and 11–13 soft rays. The outer teeth are mostly unicuspid, inner teeth often bicuspid, occasionally tricuspid. The color pattern is characteristic, with 6 or 7 dark vertical bars on the side, of which the one between the anterior part of the soft dorsal fin and the middle of the soft anal fin is particularly strongly pigmented and terminating in a dark blotch on each fin base, except in one species in which the bar does not reach onto the anal fin. **SPECIES** Five. **DISTRIBUTION AND HABITAT** Wide distribution in South America, including much of the Amazon and Orinoco basins, the Tocantins basin, the Essequibo basin, and coastal streams in French Guiana. **BIOLOGY** Most species are biparental substrate brooders; eggs are deposited on stones or roots and defended by both parents; parental care can last up to 6 weeks. At least one species is a mouth brooder (Stawikowski and Werner 1998). *Heros* species are omnivorous, feeding on shrimp, insects, plants, fruits and seeds (Keith et al. 2000).

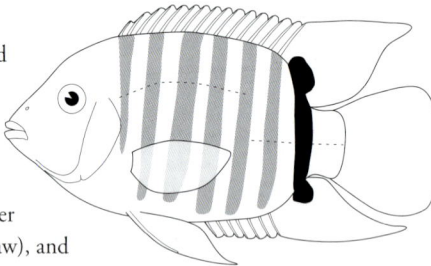

Hoplarchus (32 cm SL)

Similar to *Uaru* with a deep, laterally compressed body and steep frontal contour, but with a different color pattern including a row of dark spots along the middle of the side, a caudal ocellus, and several dark vertical bars. The scales are small, 38–47 in the E1 row, gradually smaller above the upper lateral line. The lower lip is continuous. The teeth fixed, simple, pointed, apically recurved, those of labiad series much stronger than those of inner band of teeth. Dorsal fin with 15 spines and 11–13 soft rays. Anal fin with 5 spines and 9–10 soft rays. Caudal fin subtruncate. The bases of the unpaired fins are scaled. Adults are iridescent green with bright red eyes and characterized by a dark stripe along the ventral margin of the cheek squamation; juveniles have a unique "camouflage" pattern of green splotches alternating with light-colored bands, which gradually changes over to a solid iridescent green. When frightened or when spawning, adults exhibit the characteristic juvenile pattern (Leibel 1994). There is no apparent sexual dimorphism, although males may be larger and have a slightly more rounded profile than females. **SPECIES** One, *H. psittacus*. **DISTRIBUTION AND HABITAT** Blackwater rivers in the Amazon basin (Brazil and Colombia), including the Negro River; also in tributaries of the upper and middle Orinoco basin. Found in rivers, swamps, and lakes. **BIOLOGY** Feeds on aquatic insects and crustaceans during high water, also on fishes during low-water seasons (Stawikowski and Werner 1998). Parents take care of the fry for some weeks (Leibel 1994).

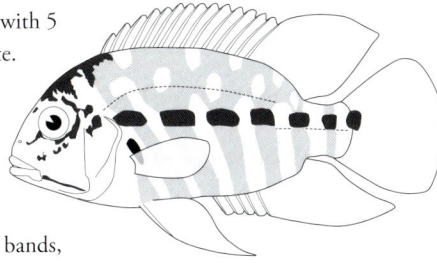

Hypselecara (15–30 cm SL)

Differ from *Mesonauta, Pterophyllum*, and *Symphysodon* in absence of prolongation of the swim bladder into the caudal region and absence of caudal ribs, and well-developed unicuspid dentition. From *Uaru* and *Hoplarchus* they can be distinguished by larger scales (25–34 vs. more than 37 E1 scales) and color pattern. *Hypselecara* agrees in most counts with *Heroina*, but differs in dentition, in not having the middle pair of teeth in the lower jaw shortened, and in color pattern. The basic color pattern includes a horizontal lateral band and a large blotch above the lateral band on the middle the side, whereas *Heroina* never shows a lateral band. Dorsal fin with 16–17 spines and 10–14 soft rays. Anal fin with 6–8 spines and 8–11 soft rays. Caudal fin subtruncate or rounded. Males are slightly larger and may develop a nuchal hump. **SPECIES** Two: *H. coryphaenoides* and *H. temporalis*. **DISTRIBUTION AND HABITAT** Amazon and Orinoco basins. *Hypselecara coryphaenoides* prefers blackwater rivers; *H. temporalis* inhabits both black and turbid (white) waters. **BIOLOGY** Eggs are deposited on vertical substrates and both parents guard eggs and larvae (Stawikowski and Werner 1998).

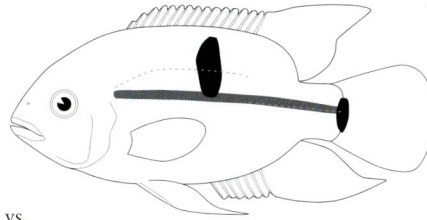

Ivanacara (4.0–5.0 cm SL)

Similar to *Nannacara* in the presence of large scales on the preopercle but have 16 instead of 14 principal caudal-fin rays, and the second anteriormost frontal lateralis canal opening is present. In contrast to *Nannacara* they also have a discontinuous instead of continuous lower lip fold, and the color pattern is marked by vertical bars on the side instead of a horizontal dark lateral band. E1 scales 22 or 24. Scales are present on the bases of the soft-rayed portions of the dorsal and anal fins in *I. adoketa* but absent in *I. bimaculata*. Dorsal fin with 16–17 spines and 7–9 soft rays. Anal fin with 3 spines and 8 soft rays. Caudal fin rounded. **SPECIES** Two: *I. bimaculata* and *I. adoketa*. **DISTRIBUTION AND HABITAT** Western Guianas; *I. bimaculata* in the Potaro and Essequibo basins of Guyana, and *I. adoketa* in the Negro basin of Brazil. Species distribution map in Kullander and Prada-Pedreros (1993). **BIOLOGY** Eggs in *I. adoketa* are deposited on roots; both parents guard the eggs and larvae (Stawikowski and Werner 1998).

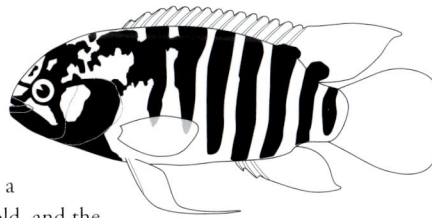

Krobia (8.5–13 cm SL)

Similar to *Bujurquina* and *Tahuantinsuyoa*, and
some species of *Aequidens* in general shape,
meristics, uniserial predorsal scale pattern,
a furrow for the hyoid artery on the dorsal
surface of the anterior ceratohyal, and in having
a slanting lateral band. In *Krobia* the lateral band
typically ends on the caudal-fin base, although this state
can be difficult to distinguish from the state in the other genera in which it ends dorsally on
the caudal peduncle. Scales are absent from the anal fin but a few scales may be present basally
on the soft dorsal fin. E1 scales 22–24, usually 23. Dorsal fin with 13–15, usually 14 spines
and 9–11, usually 9 or 10 soft rays. Anal fin with 3 spines and 7–9, modally 8 soft rays. Three
distinctive facial stripes between the eyes and a suborbital stripe are present at all sizes. Males
are a little bigger and more colorful than females; they also have elongate anal, dorsal, and
pelvic fins (Keith et al. 2000). **SPECIES** Four; notes on species differences in Steele et al. (2013).
DISTRIBUTION AND HABITAT Coastal drainages of the Guianas and adjacent northern tributaries
in the Amazon basin and in the Xingu basin (Steele et al. 2013). Found in small, slow-flowing,
shallow creeks (Keith et al. 2000). **BIOLOGY** Feeds on small crustaceans and insect larvae. Eggs
are laid on a flat stone and guarded by the parents (Keith et al. 2000).

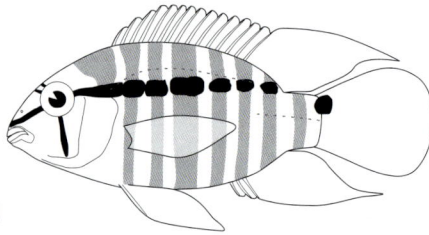

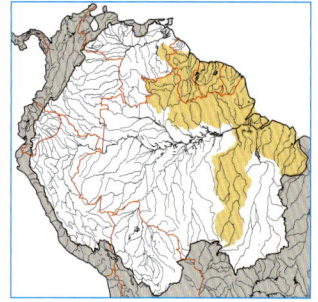

Laetacara (3.5–11 cm SL)

Characterized by: small to medium adult body
sizes; only two rows of scales on the cheek and
with one row of scales on the preopercle, thus
somewhat similar to *Cleithracara*, *Nannacara*,
and *Ivanacara*, sharing also with these genera
absence of an ocellus at the caudal-fin base (vs. ocellus
present in most cichlasomatines with three anal spines). Also
shares with *Cleithracara* a deep notch in the dorsal margin of the anterior ceratohyal. The generic
name, translating to smiling cichlids, relates to a dark stripe across the snout, bordered above
by a light stripe. The color pattern otherwise includes a midlateral spot and a dark band from
the eye to the midlateral spot. E1 scales 24 in *L. flavilabris*, 23 in *L. fulvipinnis*, 22, rarely 21 in
remaining species. Scales are absent from the dorsal fin, or a few scales are present basally. Dorsal
fin with 14–17 spines and 7–10 soft rays. Anal fin with 3 spines and 7–9 soft rays. Caudal fin
rounded. Males and females are similar in color, but the males are slightly larger. **SPECIES** Six or
seven. **DISTRIBUTION AND HABITAT** Widespread in the Amazon, Tocantins, Orinoco, Paraguay,
and Paraná basins and coastal rivers of Amapá state, Brazil. Habitats include small streams and
lakes, with dense leaf litter and/or abundant aquatic vegetation (Staeck and Schindler 2007,
Ottoni and Costa 2009, Ottoni et al. 2012) **BIOLOGY** Monogamous; eggs are deposited on a flat
surface, which both parents guard (Stawikowski and Werner 1998, Staeck and Schindler 2007).
Omnivorous; food items include aquatic insects, crustaceans, algae, and vegetal debris (Souza
Filho and Casatti 2009). *Laetacara dorsigera* is known to leap from the water to land on nearby
buoyant vegetation to escape aquatic predators (Sazima and Machado 1989).

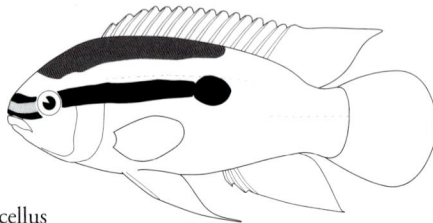

Mazarunia (5.0–8.5 cm SL)

Three species confined to the upper Mazaruni basin in Guyana. Each species has a distinctive
coloration and body shape, young specimens relatively deep-bodied, adults moderately elongate.
The genus is distinguished from all other cichlid genera by the unique loss of infraorbital 6
(López-Fernández et al. 2012a). The most closely related genus is *Guianacara*. All species of
Mazarunia can be distinguished from *Guianacara* by their distinct color patterns, including
crossbars (most obvious in juveniles) or a large dark spot on the side of the body (not in

M. charadrica), and absence of a distinct infraorbital bar in adults (vs. always present in *Guianacara*). E1 scales 24–26, usually 25. Scales absent from fins except base of caudal fin. Dorsal fin 15–17 spines and 7–8 soft rays. Anal fin with 3, rarely 4 spines, and 7–8 soft rays. Caudal fin subtruncate or slightly emarginate. No apparent external sexual dimorphism. **SPECIES** Three; species information in López-Fernández et al. (2012a). **DISTRIBUTION AND HABITAT** Upper reaches of Mazaruni River, Essequibo basin, Guyana. Occurring in habitats with both fast and slow current (López-Fernández et al. 2012a). **BIOLOGY** The three species may partition habitats, with *M. charadrica* more abundant in areas with faster currents, and *M. mazarunii* and *M. pala* preferring more lentic areas. All species are benthic invertivores (López-Fernández et al. 2012a).

Mesonauta (7.0–10 cm SL)

Characterized by: medium adult body sizes; strongly compressed and deep body; very long, thick first pelvic-fin ray shared only with *Pterophyllum* among cichlids (Kullander and Silfvergrip 1991). Distinguished from *Pterophyllum* by: a more slender shape (depth ≤61% SL vs. >65%); larger scales (E1 scales ≤27 vs. ≥27; usually 25–26); and shorter vertical fins (dorsal fin with 14–16 spines and 9–12 soft rays instead of 11–13 spines and 19–26 soft rays; anal fin with 6–9 spines and 9–14 rays instead of 5–6 spines and 19–28 soft rays) (Kullander and Silfvergrip 1991). Furthermore, *Pterophyllum* species have a pronounced pattern of dark vertical bars, whereas *Mesonauta* species have an oblique band from the mouth to the soft dorsal fin, less intensely pigmented and often irregular vertical bars below the band, and a prominent ocellated spot on the caudal-fin base (Kullander and Silfvergrip 1991). Unpaired fins are scaled basally. The outer row teeth are bi- or tricuspid. Scales are commonly present on the preopercle. Caudal fin rounded. No known sexual dimorphism. **SPECIES** Six, key to the species (except *M. guyanae*) in Kullander and Silfvergrip (1991). **DISTRIBUTION AND HABITAT** Amazon and Orinoco basins, middle and upper Paraguay basin, and the Essequibo basin in Guyana. Rough map of species distribution in Kullander and Silfvergrip (1991). Found in many types of habitats, including streams, lakes, and sandy shores. Often found close to tree litter or vegetation in slow water (Kullander and Silfvergrip 1991). **BIOLOGY** Juveniles show leaf-mimicking behavior, likely to reduce predation (Pires et al. 2015). *Mesonauta acora* has been reported to jump out of the water when disturbed (Kullander and Silfvergrip 1991). Eggs are deposited on plants near the surface and both parents care for the eggs and larvae (Stawikowski and Werner 1998, Pires et al. 2015)

Mikrogeophagus (3.0–6.0 cm SL)

Similar to small-sized *Geophagus* or *Biotodoma*, with a small epibranchial lobe, but a more rounded outline and an approximately square-shaped lachrymal bone (vs. lachrymal bone deeper than wide). Black stripe from the occiput to the eye and from the eye down to the preopercle and a black blotch on the middle of the side. The first few dorsal-fin membranes are black. There is no black blotch at the caudal-fin base, but a form known only from aquarium literature has a large black spot on the caudal peduncle. In the smaller *M. ramirezi* (to 40 mm SL) some anterior dorsal-fin

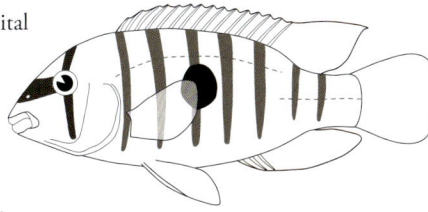

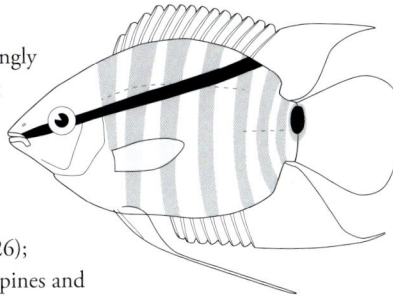

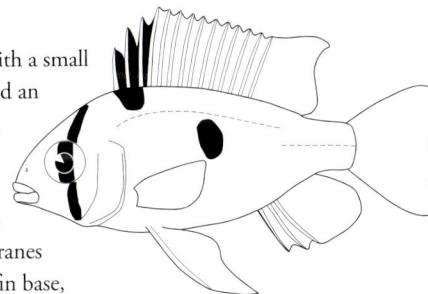

spines and lappets are distinctly prolonged beyond the rest, whereas in the larger *M. altispinosus* (to 60 mm SL) the first few spines are very long, slightly shorter from the fifth, but the lappets are not produced. E1 scales 25–26. Scales are absent from the fins except basally on the caudal fin. Dorsal fin with 12–14 spines and 9–10 soft rays in *M. ramirezi*; 14–16 spines and 8–10 soft rays in *M. altispinosus*. Anal fin with 3 spines and 7–9, usually 8 soft rays. Caudal fin subtruncate or slightly emarginate. The preopercle is usually finely serrated. Gill rakers externally on the first ceratobranchial 5–7. Only very slight sexual dimorphism: males with longer pelvic fins, females with irregular blue spotting in the lateral spot (in *M. ramirezi*). **SPECIES** Two, of which *M. ramirezi* can be readily identified by its bright color pattern. **DISTRIBUTION AND HABITAT** Disjunct distribution; *M. altispinosus* in the Mamoré and Guaporé basins of Bolivia and Brazil; *M. ramirezi* in the Llanos of the Orinoco basin in Colombia and Venezuela. **BIOLOGY** Monogamous, open breeders on horizontal surfaces or in pits and the brood care is biparental (Kullander 1977).

Nannacara (3.0–4.0 cm SL)

Characterized by: small adult body sizes and large scales on the preopercle, similar to *Ivanacara, Laetacara,* and *Cleithracara* (Kullander and Prada-Pedreros 1993). Unique in having only 14 principal caudal-fin rays (vs. 16 in all other cichlids), and absence of the second anteriormost frontal lateral line canal opening (present in all other cichlids). The lower lip fold is continuous, unlike in most cichlids in which it is interrupted anteriorly (Kullander and Prada-Pedreros 1993). E1 scales 21–22. Scales are absent from the fins except the base of the caudal fin. Dorsal fin with 15–18 spines and 7–8 soft rays. Anal fin with 3 spines and 6–9 soft rays, except *N. quadrispinae* with 4 anal-fin spines. Caudal fin rounded. Males are larger and with bright metallic green or blue sides, females smaller and pale brownish with intense dark markings, particularly when breeding. **SPECIES** Four. **DISTRIBUTION AND HABITAT** A narrow band along the northeastern coast of South America, from the lower Orinoco River south to the southern bank of the Amazon River near Belém. Prefer areas with abundant leaf litter (Staeck and Schindler 2004). **BIOLOGY** Polygynous substrate spawners, in which the male defends a territory containing several potential spawning sites; each of these sites may serve as the focus of a smaller territory occupied by a female (Staeck and Schindler 2004). In some species parental care is exclusively maternal, in others both parents take care of the eggs and fry (Stawikowski and Werner 1998, Staeck and Schindler 2004).

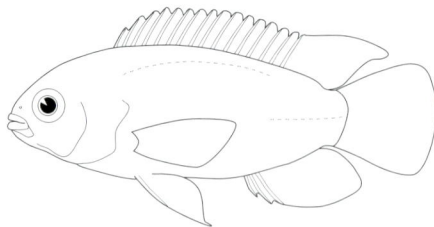

Pterophyllum (5.0–7.5 cm SL)

Readily distinguished by their characteristic body shape, which is strongly compressed laterally and with strongly arched dorsal and ventral contours. The rounded outline of the body is extended into a triangular appearance by the shape of the vertical fins, with spines gradually increasing in length from a minute first spine to a long ultimate, a few anterior soft rays very long; the height between the tips of the longest rays of the dorsal and anal fin exceeds the length of the body (Kullander 1986). The scales are small, about 27–48 in the E1 row. The preopercle is scaled on the horizontal limb. Dorsal fin with 11–13 spines and 19–26 soft rays. Anal fin with 5–6 spines and 19–28 soft rays. Caudal fin emarginate with long marginal rays. The pelvic fin is considerably elongate, a character shared with *Mesonauta* (Kullander

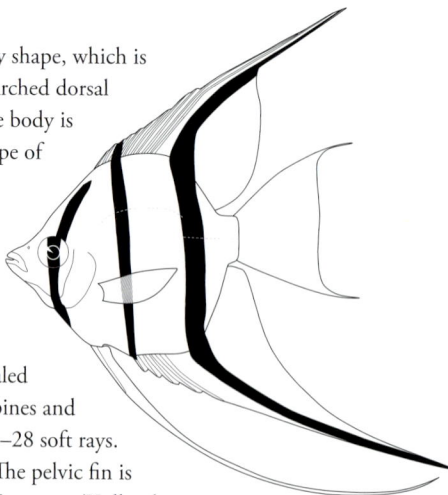

1986). The color is silvery or whitish with broad contrasting black vertical bars (Kullander 1986). No external sexual dimorphism. **SPECIES** Three: *P. altum* (Altum angelfish), *P. leopoldi* (Teardrop angelfish), and *P. scalare* (Freshwater angelfish). Review of species in Kullander (1986). Many popular publications on species and aquarium bred varieties, e.g., Göbel and Mayland (1998). **COMMON NAMES** *Acará bandeira* (Brazil); *Pez angel, Scalare* (Peru); *Kweyu* (French Guiana); Freshwater angelfishes (English). **DISTRIBUTION AND HABITAT** Lowland Amazon and Orinoco basins (to Puerto Ayacucho, Venezuela) and some coastal rivers in the Guianas. Found in rivers, swamps, and flooded areas, with little or no water movement and dense vegetation (Keith et al. 2000), with *P. altum* in the upper Orinoco basin (above Puerto Ayacucho) and the upper Negro River in southern Venezuela, southeastern Colombia, and extreme northern Brazil. *P. leopoldi* in the central and eastern Amazon basin and Essequibo basin, and *P. scalare* in the lowland Amazon basin and the Essequibo and Oyapock basins of the Guianas. **BIOLOGY** Among the most popular and well-known aquarium fish in the world. Body shape and color patterns serve as camouflage in dense vegetation. They form monogamous pairs during the brooding period. Eggs are attached to the surface of aquatic vegetation in a nest area and both parents guard the eggs and young (Yamamoto et al. 1999, Cacho et al. 2007).

Retroculus (14–20 cm SL)

Characterized by: 5 dentary lateral line pores; 7 preopercular lateral line pores; "African type" lips; and numerous epibranchial gill rakers (Kullander 1998). Superficially resembles *Geophagus* but without the anteroventral expansion on the first epibranchial, and without a midlateral blotch. Specializations such as the wide and short pelvic fin with middle ray longest, minute ventral scales, long snout, well-developed papillae on the gill arches and thick lips reflect rheophily and bottom feeding. E1 scales 34–41. Dorsal fin with 14–17 spines and 10–12 soft rays. Anal fin with 3 spines and 6–7 soft rays. Caudal fin slightly emarginate or subtruncate. Adult males (of *R. lapidifer*) are larger than females, with extended dorsal, anal, and pelvic fins (Weidner 1999). **SPECIES** Four; review of species in Gosse (1971) and Landim et al. (2015). **COMMON NAMES** *Acará bicudo* (Brazil). **DISTRIBUTION AND HABITAT** Restricted to rapids or running water in a few clearwater rivers on the Guiana and Brazilian shields (Gosse 1971, Keith et al. 2000). *Retroculus lapidifer* has been collected in the Tocantins and Capim drainages; *R. xinguensis* in the Xingu and Arapiuns drainages; *R. septentrionalis* in the Oyapock and Araguari drainages; and *R. acherontos* in the upper Tocantins drainage. **BIOLOGY** *R. lapidifer* and *R. xinguensis* are bottom feeders, with chironomid, trichopteran, and ephemeropteran larvae dominating the diet (Moreira and Zuanon 2002). The reproductive behavior is incompletely known. Aquarium and field observations show that at least *R. lapidifer* excavates bottom substrate to a pit in which eggs are deposited and then covered with sand or pebbles. Both parents guard the nest.

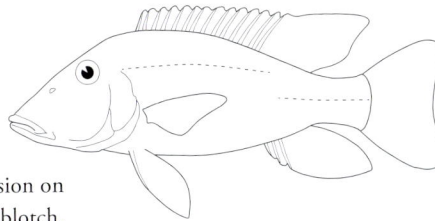

Rondonacara (7.0 cm SL)

Characterized by a small adult body size, and a unique color pattern composed of a lateral band of distinct dark spots, distinct vertical bars, a prominent suborbital stripe at all sizes, a dark vertical stripe at the base of the caudal fin, and black stripes at the base of each dorsal-fin spine and soft ray. E1 scales 21–22. Predorsal squamation uniserial. Scales absent from fins except basally on caudal fin. Dorsal fin with 13–14 spines and 9–10, occasionally 11soft rays. Anal fin with 3 spines and 8–9 soft rays. Caudal fin rounded.

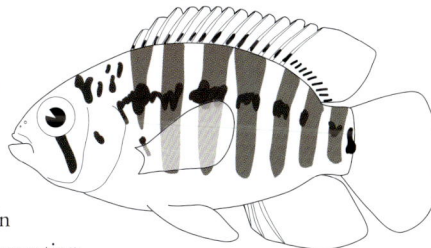

SPECIES One, *R. hoehnei*. Species information in Ottoni and Mattos (2015). **DISTRIBUTION AND HABITAT** Small streams and pools in the Rio das Mortes drainage, headwaters of the Araguaia River. **BIOLOGY** Biparental substrate brooder, depositing eggs on an exposed surface (Stawikowski and Werner 1998).

Satanoperca (14–26 cm SL)

Characterized by an ocellated blotch dorsally at the base of the caudal fin, a color marking common among South American cichlids, but absent in all other Geophagini (Kullander 2012). *Satanoperca* has a well-developed epibranchial lobe and is uniquely distinguished by having almost all of the 15–22 external first ceratobranchial gill rakers attached to the skin over the bases of the gill filaments (Kullander and Nijssen 1989). Other characters distinguishing the genus from one or more of the other genera of geophagines: jaw dentition principally uni- or biserial (vs. several series in *Geophagus*); naked dorsal and anal fins (vs. scaly dorsal and anal fins in *Geophagus*); upper or both lip folds continuous (vs. discontinuous lip folds in *Geophagus*); cheek naked only ventrally (vs. anterior half naked in *Geophagus*); gill rakers absent from the lower pharyngeal tooth plate; preopercle and supracleithrum frequently serrated (Kullander 1986, Kullander and Nijssen 1989). E1 scales 26–31. Scales are absent from the fins except basally on the caudal fin. Dorsal fin with 13–17 spines and 8–14 soft rays; modally 15 spines and 9–10 soft rays in the *S. jurupari* group, 13 spines and 12 soft rays in *S. acuticeps*, 14 spines and 13 soft rays in *S. lilith* and *S. daemon*. Anal fin with 3 spines and 6–9 soft rays. Caudal fin truncate or subtruncate. Sexes monomorphic. **SPECIES** Eight; species-level phylogeny in Willis et al. (2012a). **COMMON NAMES** *Papa terra* (Brazil); *Bujurqui vaso* (Peru); Demon eartheater (English). **DISTRIBUTION AND HABITAT** Most river basins east of the Andes, including all of the Orinoco, Amazon, and Tocantins basins, as well as the Paraguay basin, and probably introduced in the Paraná basin. The genus is absent from coastal Atlantic rivers south of the mouth of the Amazon, but present in the state of Amapá and coastal rivers in the Guianas (Kullander 2012). Most species inhabit quiet sections of rivers, in streams or lagoons with sandy or muddy bottoms (Keith et al. 2000). **BIOLOGY** With the exception of *S. acuticeps* which is a midwater feeder, species of *Satanoperca* forage by plunging their mouth into the sandy or muddy bottom, filling it with sediment, and sorting out food inside the mouth, but expelling the bulk of the mouthful through the opercular openings and so producing a cloud of sediment (Sazima 1986). Diet includes seeds, detritus, vegetal matter, crustaceans, and insects (Sazima 1986, Keith et al. 2000, Galvis et al. 2006, Cassemiro et al. 2008). Some species are reported to be (biparental) ovophile and/or larvophile mouth brooders, i.e., mouth brooding from the egg stage, or from the hatchling stage. Underwater observation of *S. pappaterra* found that it discourages piranha attacks by watching the predator, a tactic used when 2–5 fish feed in groups and the individuals take turns as the "watchman." Confronts an approaching piranha by charging toward the predator with an open mouth and erect dorsal fin (Sazima and Machado 1990).

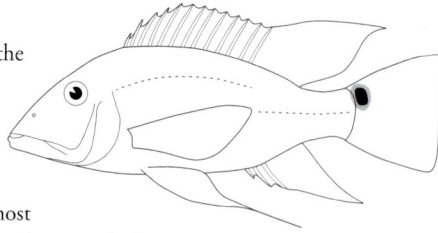

Symphysodon (12–14 cm SL)

Characterized by a unique body shape, strongly compressed laterally and rounded in overall body outline. The body scales are very small and finely ctenoid, E1 scales 48–62, and the scale cover extends far out on the dorsal and anal fins. The mouth is relatively smaller than in other cichlids, with thick lips and highly reduced dentition. Also the pharyngeal jaws are small and bear a reduced number of teeth (Kullander 1986). The dorsal and anal fins are long: dorsal fin with 8–10 spines and 28–33

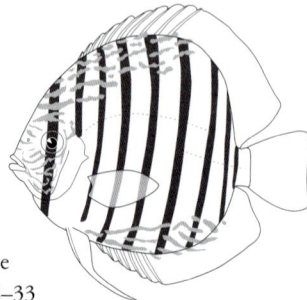

soft rays, anal fin with 6–9 spines and 27–31 soft rays (Kullander 1986). Unique traits include: the 3 extrascapulars (other cichlids have 2, rarely 1) and edentulous second pharyngobranchial (Kullander 1986). Caudal fin subtruncate. No obvious external sexual dimorphism. **SPECIES** Three; all in the AOG region: *S. aequifasciatus* (13.7 cm SL), the Brown discus from the Amazon River east of the confluences of the Solimões and Negro rivers; *S. discus* (12.3 cm SL), the Red or Heckel discus from central and eastern Amazon near its confluence with the Negro and Trombetas rivers; and *S. tarzoo* (13.2 cm SL), the Green discus from the central Amazon near its confluences with the Madeira and Purus rivers. See reviews in Kullander (1986, 1996a), Ready et al. (2006), and Bleher et al. (2007), with different views of species composition. **COMMON NAMES** *Acará disco* (Brazil); *Pez disco* (Peru). **DISTRIBUTION AND HABITAT** Restricted to floodplain habitats of the lowland Amazon basin (Bleher et al. 2007, Crampton 2008). **BIOLOGY** Diet includes algal periphyton, fine organic detritus, plant matter, and small aquatic invertebrates (Crampton 2008). During the high-water period, individuals forage alone or in small groups in the flooded forests. At low water levels large aggregations assemble in fallen tree crowns along lake margins (Crampton 2008). Breeding occurs at the beginning of the flood season (Crampton 2008). They form monogamous pairs with biparental care of their broods (Matthaeus 1992). *Symphysodon* is well known for contacting behavior, in which the fry feed on parental mucus for the first few weeks to obtain nutrients and immunoglobulins (Hildemann 1959, Buckley et al. 2010).

Taeniacara (3.0 cm SL)

Characterized by a small adult body size, similar to *Apistogramma* but more slender than most *Apistogramma* species and with further reductions to the lateral line system. Distinguished from *Apistogramma* by having only 3 dentary lateralis foramina and the loss of the anguloarticular canal, middle pterotic foramen, distal extrascapular, and the second frontal lateralis foramen (Kullander 1987). E1 scales 21–22. Most or all lateral line scales with a pore; tube-shaped scales, if present, only in the upper line. Scales are absent from the fins except basally on the caudal fin. Dorsal fin with 15–17 spines and 6–7 soft rays, modally 15 spines and 7 soft rays. Anal fin with 3 spines and 6 soft rays. Caudal fin lanceolate in males, rounded in females. Serrations are absent from the preopercle and other head bones. Gill rakers externally on first ceratobranchial absent or a single gill raker present. Vertical bars are absent. The color pattern is dominated by a broad black stripe from the eye to the base of the caudal fin. Adult males are slightly larger than females and have long lanceolate caudal fin and elongate pelvic fin. **SPECIES** One, *T. candidi*. **DISTRIBUTION AND HABITAT** Small forest streams with clearwater in the middle and lower Negro, lower reaches of the Tapajós, Madeira, Trombetas, and Branco river drainages, and lakes and smaller rivers along the Amazon River from the mouth of the Negro River to the region of the Tapajós and Trombetas rivers. **BIOLOGY** Aquarium observations suggest that *T. candidi* breeds in pairs, depositing eggs on the underside of a roofed space, where they are guarded by the female. The male may care in guarding the free-swimming young (Koslowski 2002).

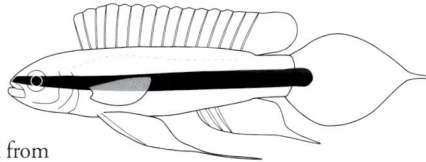

Tahuantinsuyoa (8.0–12 cm SL)

Similar to *Bujurquina* in general appearance, presence of uniserial predorsal squamation, a notch in the dorsal margin of the anterior ceratohyal bone, a small caudal ocellus and a lateral band from the head obliquely upward to the end of the dorsal-fin base. Differs from *Bujurquina* by having a dark band running from the eye backward, slanting to the occiput instead of a dark band from the eye running obliquely forward to or across the front. *Tahuantinsuyoa* are also slightly more elongate, and the dorsal accessory caudal-fin lateral line

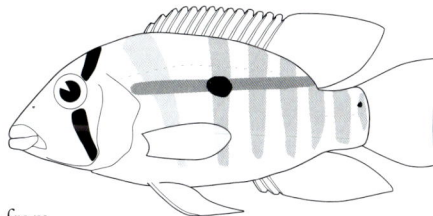

runs between rays D2 and D3, or frequently between D3 and D4 instead of between rays D1 and D2. E1 scales 24–26, usually 25. Scales are absent from the fins except at the base of the caudal fin. Dorsal fin with 13–15 spines and 10–12 soft rays, modally 14 spines and 11 soft rays. Anal fin with 3 spines and 7–9, usually 8 soft rays. Sexes are isomorphic. **SPECIES** Two; species information in Kullander (1991). **DISTRIBUTION AND HABITAT** Known only from the left bank Ucayali tributaries Aguaytía and Pachitea in the Peruvian Amazon drainage. Both species inhabit clearwater rivers with sand and gravel bottom (Kullander 1986, 1991). **BIOLOGY** *T. macantzatza* deposits 30–60 eggs on a transportable substrate, and both parents mouth-brood the larvae from hatching after about two days (Werner and Minde 1990).

Teleocichla (4.0–14 cm SL)

Similar to *Crenicichla* with a slender body, small scales, many dorsal-fin spines, and relatively large eye. Distinguished from *Crenicichla* by: a short and downturned snout and the shape of the pelvic fin, which is long, with the middle ray longest, reaching to (or almost to) the genital papilla, has a more or less obviously thickened outer margin, and is relatively stiff. In contrast, the pelvic fin of a *Crenicichla* can be spread with ease simply by articulating the spine, but in *Teleocichla* the fin retains its elongate shape even when the spine is raised maximally. In some species the lateral line is continuous from the head to the caudal fin, a character otherwise shared only by some species of *Cichla* among Neotropical cichlids. Sexual dimorphism includes differences in coloration. **SPECIES** Eight; species information in Kullander (1988), Zuanon and Sazima (2002), and Varella and Moreira (2013). Molecular analyses suggest that *Teleocichla* is one of the several lineages making up the genus *Crenicichla* (Kullander et al. 2010, Piálek et al. 2012). **DISTRIBUTION AND HABITAT** Xingu, Tapajós, Tocantins, Jari, and Araguari rivers in Brazil. Restricted to areas with fast current, e.g., in rapids. **BIOLOGY** Snout and mouth morphology suggest that *Teleocichla* species obtain their food from the upper layer of the substrate (Kullander 1988). *Teleocichla centisquama* has been reported to forage on rock surfaces and picks mostly benthic prey, including microcrustaceans, aquatic insects, and minute clams (Zuanon and Sazima 2002).

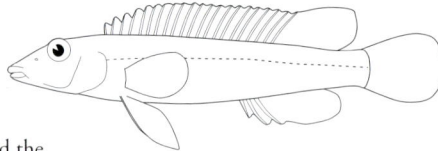

Uaru (19–25 cm SL)

Characterized by a strongly compressed, disk-shaped body and very short head, similar to *Symphysodon* or *Heros* but distinguished by color pattern and meristics. Juveniles with beige to brown sides with numerous white spots that fade with size; adults characterized by beige to pale brown body with strong black markings. Jaw dentition distinct: anterior teeth gradually and greatly enlarged and procumbent, distally compressed, but strengthened medially by a narrow ledge with 2 or 3 small projections. Posterior teeth simple and pointed. Scales small and ctenoid on sides of body; 47–48 in E1 row in *U. amphiacanthoides*, 56–57 in *U. fernandezyepezi*. Scales present on soft dorsal and anal fins, and base of caudal fin. Dorsal fin with 16 spines and 14–15 soft rays in *U. amphiacanthoides*, 15 spines and 15–16 soft rays in *U. fernandezyepezi*; anal fin with 8 spines and 13–14 soft rays in *U. amphicanthoides*, 7 spines and 14–15 soft rays in *U. fernandezyepezi*. No obvious external sexual dimorphism. **SPECIES** Two (Stawikowski 1989). Adult *U. amphiacanthoides* (illustrated) with a large black triangular blotch on the side, a black spot or vertical bar on the caudal-fin base, a small black spot on the base of the pectoral fin, and a dark blotch close to the eye. Adult *U. fernandezyepezi* with a black bar covering the caudal-fin base, a black band descending from the soft dorsal-fin base but not reaching to the anal fin, a round black spot immediately below the end of the spinous dorsal

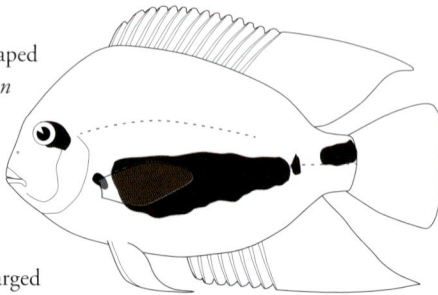

fin, a large black blotch on the pectoral-fin base, and a black blotch associated with the eye (Stawikowski 1989). **DISTRIBUTION AND HABITAT** *U. amphiacanthoides* from the Amazon basin, *U. fernandezyepezi* the Atabapo drainage in Venezuela. **BIOLOGY** Parents care for ≤200 eggs and larvae. Similar to *Symphysodon* and many other cichlids in contacting behavior, i.e., the fry feed from mucus on the body of the parents for the first few weeks (Hildemann 1959, Buckley et al. 2010). Adults feed on worms, crustaceans, insects, and plants (Stawikowski and Werner 1998).

FAMILY POLYCENTRIDAE—NEW WORLD LEAF-FISHES
— *PETER VAN DER SLEEN and JAMES S. ALBERT*

DIVERSITY Three species in two genera, all restricted to South America: *Monocirrhus polyacanthus, Polycentrus schomburgkii*, and *P. jundia*. Polycentridae is not closely related to the Asian and African leaf-fishes (Nandidae) and has been treated as a separate family in recent fish classifications (Near et al. 2013, Van Der Laan et al. 2014).

COMMON NAMES *Peixe folha* (Brazil); *Poisson feuille* (French Guiana); *Pez hoja* (Spanish).

GEOGRAPHIC DISTRIBUTION Amazon basin and Atlantic coastal rivers of Venezuela, the Guianas, and the state of Amapá in Brazil.

ADULT SIZES 3.0–8.0 cm SL.

DIAGNOSIS OF FAMILY All species characterized by a strongly compressed body; large mouth and head, with extremely protractile upper jaws in *Monocirrhus*; a color pattern mimicking dead leaves; dorsal fin with 16–18 spines and 7–13 rays; anal fin with 11–13 spines and 7–14 rays; and no lateral line on the body side (occasionally present in *Monocirrhus*) (Liem 1970, Britz and Kullander 2003).

KEY TO THE GENERA

1a. Short skin flap projecting from the lower jaw (fig. 1a); marbled color pattern without horizontal series of spots; caudal fin almost entirely covered with scales; ventral margin of preopercle smooth; posterior lateral line occasionally present...............*Monocirrhus*
1b. No skin flap projecting from the lower jaw; absence of scales on caudal-fin base; color pattern with longitudinal series of dark spots, surrounded by pale spots; ventral margin of preopercle serrated (fig. 1b; after Coutinho and Wosiacki 2014); posterior lateral line lacking..*Polycentrus*

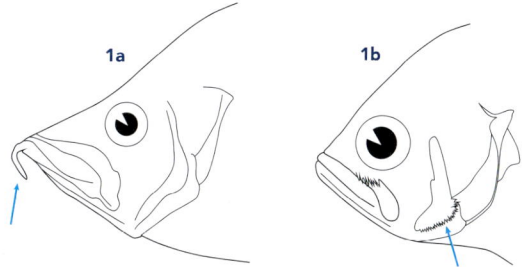

GENUS ACCOUNTS

Monocirrhus (8.0 cm SL)

Characterized by a strongly compressed body laterally, resembling a dead leaf both in color pattern and shape, including a short skin flap projecting from the lower jaw that resembles the stalk of the leaf (Britz and Kullander 2003). Three-quarters of caudal fin covered with scales; ventral margin of preopercle smooth; posterior lateral line occasionally present (Coutinho and Wosiacki 2014). **SPECIES** One, *M. polyacanthus* (Amazon leaf-fish). **DISTRIBUTION AND HABITAT** Amazon basin in Brazil, Bolivia, Colombia, Peru, and Venezuela. Found in streams and lake and river margins, in areas with little or no flow and with submerged vegetation or leaf litter. **BIOLOGY** Feed primarily on fish, but also invertebrates (Catarino and Zuanon 2010). May consume fish of their own length. They move slowly

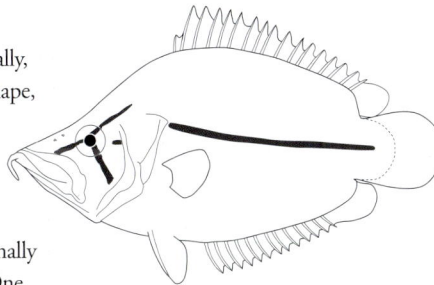

toward prey fish, apparently mimicking the movement pattern of a drifting leaf. Pectorals and soft parts of the dorsal and anal fins are highly transparent and the only fins that move when approaching prey, which are sucked in very quickly by protraction of the mouth (Liem 1970). An egg clutch is deposited under leaves of aquatic plants; males exhibit parental care of eggs and larvae (Barlow 1967).

Polycentrus (3.0–6.0 cm SL)

Characterized by: a distinct color pattern formed by vertical series of dark spots, surrounded by pale spots; 11–13 spines in anal fin; a single perforated scale on upper lateral line; small serrations on ventral margin of preopercle; absence of mandibular barbel; absence of scales on caudal-fin base; and lack of posterior lateral line. Males and females with different color patterns during the breeding season (Barlow 1967). In *Polycentrus jundia*, males also have a discrete adipose swelling between the occipital and coronal region, which is absent in females (Coutinho and Wosiacki 2014).

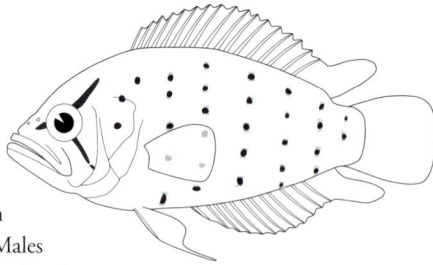

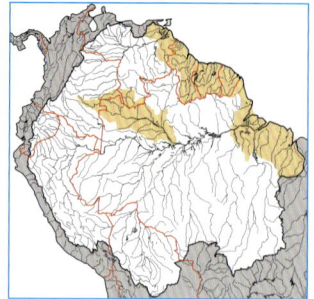

SPECIES Two, both in the AOG region: *P. jundia* and *P. schomburgkii* (Guyana leaf-fish). For species differences see Coutinho and Wosiacki (2014). **DISTRIBUTION AND HABITAT** *P. schomburgkii* from coastal drainages from the states of Amapá and Pará in Brazil, north to Trinidad and Tobago, and adjacent areas of Venezuela, in clear and turbid water, and often brackish water; *P. jundia* from floodplain streams of the lower and upper Rio Negro in Brazil. **BIOLOGY** Species mimic dead leaves, which they use to approach prey fishes. May consume fish of its own size; prey are sucked in very quickly by protraction of the mouth (Liem 1970). Territorial males of *P. schomburgkii* are nearly black with prominent white spots; the mature females are yellowish pink to white, with the abdomen distended by eggs. Egg clutch is deposited at the roof of small crevices or at the underside of woody debris or broad leaves; males exhibit parental care of eggs and larvae (Barlow 1967).

FAMILY SCIAENIDAE—DRUMS OR CROAKERS
— *PETER VAN DER SLEEN*

DIVERSITY 283 species in 66 genera, with 15 species in 4 genera in the AOG region: *Pachypops* (3 species), *Pachyurus* (6 species), *Petilipinnis* (one species, *P. grunniens*), and *Plagioscion* (5 species).

COMMON NAMES *Corvina* (Brazil, Peru, Venezuela); *Roncador* (Spanish).

GEOGRAPHIC DISTRIBUTION Distributed mainly in the warm temperate and tropical regions of the Atlantic, Indian, and Pacific oceans. Freshwater species occurring in South America are found throughout the Magdalena, Orinoco, Amazon, São Francisco, and Paraná basins and coastal rivers in the Guianas.

ADULT SIZES 5.6 cm SL in *Pachypops pigmaeus* from the Amazon basin to 200 cm in some marine species, with *Plagioscion squamosissimus* (max length 80 cm SL) as the largest species in the AOG region.

DIAGNOSIS OF FAMILY Head short to medium-sized, usually with bony ridges on top of the skull; sensory pores present at the tip of the snout (rostral pores, 3–7), and on the lower margin of the snout (marginal pores, 2 or 5); tip of lower jaw (chin) with 2–6 mental pores, some species with barbels; teeth usually small, villiform, set in bands on the jaws (but conical and

caniniform teeth in *Plagioscion*); roof of the mouth toothless (vomer and palatines toothless); dorsal fin long, continuous, with deep notch between anterior (spinous) and posterior (soft) portions; anal fin with 2 (rarely 1) spines; soft ray portion of the dorsal fin much longer than the soft ray portion of anal fin; caudal fin never forked, usually pointed in juveniles, becoming emarginate, truncate, rounded to rhomboidal, or S-shaped in adults; a single continuous lateral line extending to the hind margin of the caudal fin (Chao 2002).

SEXUAL DIMORPHISM Not pronounced.

HABITATS Found mainly in estuarine environments, but also on continental shelves, and around coral and rocky reefs. Relatively few species are restricted to freshwaters.

FEEDING ECOLOGY Generalist predators that eat fishes and benthic invertebrates.

BEHAVIOR Common names reflect their sound-producing behavior, especially evident during mating. The gas bladders of sciaenids are often complex, with secondary chambers and numerous extensions, enhancing their auditory abilities, and complex drumming muscles for sound production (Ramcharitar et al. 2006).

KEY TO THE GENERA

NOTE An important feature to differentiate between genera is the morphology of the swim bladder, but these characteristics are not included here (see details e.g., in Casatti 2002a).

1a. A single anal-fin spine .... *Petilipinnis*
1b. Two anal-fin spines **2**

2a. Chin, underside of lower jaw, with small barbel(s)..*Pachypops*
2b. Chin without barbels (barbels present in *Pachyurus adspersus* from drainages of southeastern Brazil)...............................**3**

3a. Mouth oblique; lateral line with a much thickened appearance, pored lateral line scales completely concealed by layers of smaller scales; dark axillary spot present...*Plagioscion*
3b. Mouth inferior; lateral line not appearing thickened, pored lateral line scales with intercalated scales but never concealed by small scales; dark axillary spot absent ....*Pachyurus*

GENUS ACCOUNTS

Pachypops (5.6–25 cm SL)

Distinguished from other South American sciaenids by: an inferior mouth (vs. oblique in *Plagioscion*); 3 mental barbels (vs. absent in most *Pachyurus* and usually 1 barbel in *Petilipinnis*); and 2 anal-fin spines (vs. 1 in *Petilipinnis*) (Casatti 2002b). **SPECIES** Three, all in the AOG region: *P. fourcroi*, *P. pigmaeus*, and *P. trifilis*. See review of genus and key to the species in Casatti (2002b). **DISTRIBUTION AND HABITAT** Throughout the Orinoco, Amazon, Essequibo, Corantijn, and Approuague drainages. Species are found mainly in rivers and lakes, but also in estuaries. **BIOLOGY** Benthivorous; prey on fish. Some species form small schools (Keith et al. 2000).

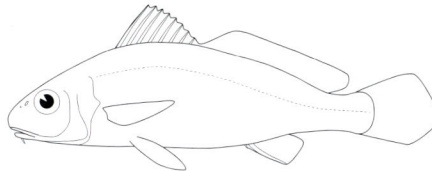

Pachyurus (8.0–30 cm SL)

Distinguished from other South American sciaenids by: an inferior mouth (vs. oblique in *Plagioscion*); absence of mental barbels (present in *P. adspersus*); and 2 anal-fin spines. **SPECIES** Ten, including six species in the AOG region. Review of genus in Casatti (2001), except for *P. stewarti* (Casatti and Chao 2002). **DISTRIBUTION AND HABITAT** Orinoco, Amazon, Tocantins, São Francisco, Paraíba do Sul, Doce, Mucuri, Uruguay, and lower Paraná basins. **BIOLOGY** Benthivorous; prey on fishes and insects (e.g., Pacheco et al. 2008).

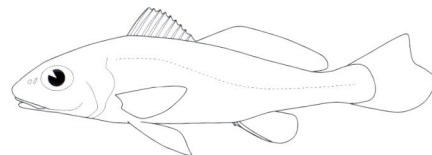

Petilipinnis (28 cm SL)

Distinguished from other South American sciaenids by: an inferior mouth (vs. oblique in *Plagioscion*); 1 anal-fin spine (vs. 2 in the other genera); 1–3 mental barbels (vs. absent in *Plagioscion* and most *Pachyurus*) (Casatti 2002a). **SPECIES** One, *P. grunniens*; see species information in Casatti (2002a). **DISTRIBUTION AND HABITAT** Amazon basin in Brazil and Essequibo basin in Guyana. Typically found in shallow sandy beaches of clearwater rivers. **BIOLOGY** Usually collected in small numbers, indicating that the species does not form large schools (J. Zuanon pers. comm.).

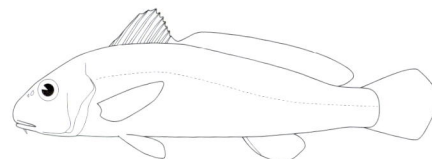

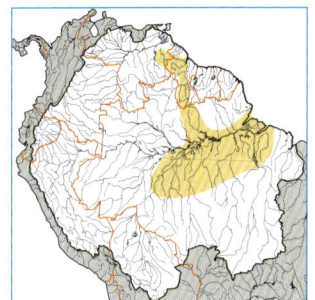

Plagioscion (28–80 cm SL)

Distinguished from other South American sciaenids by: an obliquely oriented terminal mouth in lateral view (vs. inferior mouth); large scales in lateral line covered by smaller scales, arranged in a rosette-like pattern (vs. lateral line covered by a single scale); conical teeth present on premaxilla and dentary (vs. absent); elongate gill rakers on first gill arch (vs. short gill rakers on first gill arch); a dark axillary spot (absent in other AOG genera); absence of mental barbels (vs. mental barbels present in *Pachypops*, *Petilipinnis*, and *Pachyurus adspersus*) (Casatti 2005). **SPECIES** Seven, including five in the AOG region. Review of genus and key to the species in Casatti (2005). **DISTRIBUTION AND HABITAT** Native to the Magdalena, Amazon, Orinoco, lower Paraná basins and rivers of the Guianas, and introduced into the upper Paraná and São Francisco basins (Casatti 2005). Found mainly in big rivers; also in brackish waters. AOG species include: *P. auratus*, the Black curbinata (35 cm SL) from the Orinoco, Essequibo, and Amazon basins; *P. casattii* (33 cm SL) from the Orinoco basin; *P. montei* (28 cm SL) from the Amazon basin in Brazil and Peru; *P. squamosissimus* (80 cm SL) from the AOG region and the Paraná-Paraguay basin; and *P. surinamensis*, the Pacora (70 cm SL) from the Amazon and Magdalena basins and coastal rivers of Suriname. **BIOLOGY** Juveniles feed on insect larvae and crustaceans; adults are mainly piscivores (Keith et al. 2000).

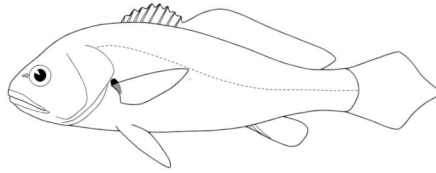

FAMILY ELEOTRIDAE—SLEEPERS

— *RODRIGO CAIRES and AURYCÉIA GUIMARÃES DA COSTA*

ILLUSTRATIONS BY RODRIGO CAIRES

DIVERSITY 150 species in 35 genera worldwide, with 22 species in six genera in Neotropical freshwaters. Here, Eleotridae is treated as a member of the Gobioidei following recent classifications and practice (Van Der Laan et al. 2014). Some researchers have included this group as subfamily Eleotrinae within Gobiidae (e.g., Thacker 2003, 2009), or as a family more closely related to Gobiidae than to other gobioids (e.g., Gill and Mooi 2012, Agorreta et al. 2013).

COMMON NAMES *Amorés pixuna*, *Peixe dorminhoco* (Brazil); *Dormeurs* (French), *Dormilónes*, *Guavinas* (Spanish); Sleepers, Gudgeons (English).

GEOGRAPHIC DISTRIBUTION Globally distributed in marine, brackish, and fresh waters. At least eight eleotrid species have been reported in the AOG region, including *Dormitator maculatus*, *Eleotris amblyopsis*, *E. pisonis*, four species of *Microphilypnus*, and two species of *Leptophilypnion*. In addition, several eleotrid species in the genera *Dormitator*, *Eleotris*, *Microphilypnus*, and *Gobiomorus* are known from streams and estuaries of the Atlantic coast (Camargo and Isaac 1998, Barletta et al. 2003, Andrade-Tubino et al. 2008). Here we provide information on ten eleotrid species in five genera that are found in the AOG region. Two additional marine genera that are (occasionally) found in estuaries near the Amazon mouth are also included in the identification key, but not further treated.

ADULT SIZES Small to medium body sizes (5–20 cm TL; *Gobiomorus* to 60 cm TL), although *Leptophilypnion* reaches a maximum length of <1 cm, and *Microphilypnus* reaches 2.7 cm TL.

DIAGNOSIS OF FAMILY Eleotrids retain the primitive condition of 6 branchiostegal rays, whereas Gobiidae have 5 branchiostegal rays. Eleotrids are recognized by: body stout, trunk covered with cycloid or ctenoid scales, without posterior lateral line; head short, snout blunt or conical, with or without head pores in preopercular and supraocular sensory canals (postocular canal generally absent, and preopercle canal short, restricted to posterior border), but with several rows of sensory papillae; gill openings attached to isthmus at a vertical below or behind eye; teeth small, conical, arranged in 2–3 rows; two dorsal fins, the first usually with 5–7 spines; second dorsal fin and anal fin with 1 flexible spine and 5–11 rays; caudal peduncle longer than base of dorsal or anal fins; pectoral fins with 11–25 rays; pelvic fins separate, with 1 spine and 5 rays (Murdy and Hoese 2003a, Thacker 2011a). Male urogenital papilla long, conical, and with a smooth tip; female urogenital papilla long and stout, covered with fleshy fringes on tips around urogenital aperture.

SEXUAL DIMORPHISM Most eleotrids from the Amazon basin do not exhibit sexually dimorphic coloration, with the exceptions of *Microphilypnus tapajosensis*, in which males have many irregular brown stripes on the trunk and the head is densely peppered with melanophores, whereas females are pale with few melanophores on the trunk and head (Caires 2013), and *Dormitator*, in which reproductive males have more brilliant colors and a bigger head than females.

HABITATS Amazonian eleotrids are benthic fishes usually found in slow-moving waters such as lakes, side channels, creeks, and estuaries, and frequently encountered on or between submerged leaves (Winemiller and Ponwith 1998).

FEEDING ECOLOGY Amazonian eleotrids are usually omnivorous, feeding on a variety of small invertebrates and, in some cases, plant material or mud (Teixeira 1994). Larger specimens may also ambush smaller fishes and they are usually voracious predators.

BEHAVIOR Information on the behavior of these taxa is scarce, but they often lie quietly on the substrate, as suggested by the English common name. Some species are mimetic and benefit from being confounded with other light-colored, translucent-bodied fishes and shrimps (Carvalho et al. 2006).

KEY TO THE GENERA

1a. Interorbital space and snout serrated, with a row of short spines (fig. 1a); large black spot on pectoral-fin base (fig. 1aa) .... *Butis*
(1 species, *B. koilomatodon*, native from Indo-Pacific Ocean and exotic in South Atlantic, not treated herein)
1b. Interorbital space and snout smooth; no large black spot on pectoral-fin base **2**

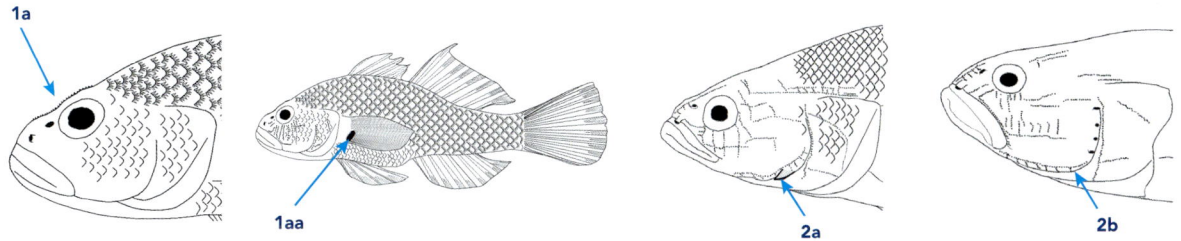

2a. Spine present on anteroventral portion of preopercle (fig. 2a) ....*Eleotris*
2b. Spine absent on preopercle (fig. 2b)...**3**

3a. Adult size large, reaching 36–60 cm total length; snout pointed, nearly 1.5 times larger than eye; head densely covered with scales below eye ....*Gobiomorus*
3b. Adult size <30 cm total length; snout moderately pointed equal to or slightly larger than eye; head barely or not covered with scales below eye...**4**

4a. Adult with minute size (<1 cm total length); pelvic fins long, innermost ray filamentous (fig. 4a); trunk naked anteriorly; 12–18 scales on lateral rows.................................... *Leptophilypnion*
4b. Adult size >1 cm; pelvic fins without filamentous rays (fig. 4b), innermost ray shorter than outermost rays; trunk fully scaled; 21–110 scales on lateral rows **5**

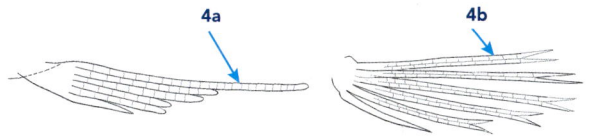

5a. First dorsal fin with 6 spines; body not deep; adult size <3 cm; ground color translucent or pale with irregular dark markings ....*Microphilypnus*
5b. First dorsal fin with 7 spines; body stout, adult size >3 cm; ground color dark brown or gray, at least in adults **6**

6a. Small cycloid scales on trunk, 80–110 in lateral row ....*Guavina* (estuarine taxon, not treated herein)
6b. Large ctenoid scales, 25–35 in lateral row ....*Dormitator*

GENUS ACCOUNTS

Dormitator (10–30 cm SL)

Characterized by: body stout; trunk covered with longitudinal series of 25–35 large, ctenoid scales; head short, upper profile convex, more so in larger males, with scales on nape, preopercle, opercle, and below eye; two series of longitudinal papillae below eyes; gill openings attached to isthmus at a vertical that passes behind eye; teeth small, conical, in 2–3 rows; 7 spines in first dorsal fin, 1 spine and 7–9 rays in second dorsal fin, and 1 spine and 9–10 rays in anal fin (Smith 1997). **SPECIES** Five, including one species in the Amazon basin: *D. maculatus* (Fat sleeper). **COMMON NAMES** *Cundundé* (Brazil); *Monengue* (Peru). **DISTRIBUTION AND HABITAT** *D. maculatus* in coastal waters of the western Atlantic from the USA to Uruguay, including lower reaches of rivers draining the Guiana Shield and the Amazon and Tocantins rivers. It inhabits brackish water estuaries and mangroves, as well as freshwater marshes, muddy ponds, creeks, and channels. **BIOLOGY** Euryhaline species (Nordlie and Haney 1993) and amphidromous. Omnivorous, feeding on plants, small invertebrates, and fine particulate organic matter (Teixeira 1994). May ascend upstream, but migrations are not common (Robins and Ray 1986). Reproduces one time per year, usually in the dry season, and matures at 5.1 cm SL (Teixeira 1994). During the reproductive season adults change color (Keith et al. 2000) and guard a nest with eggs on leaves; they do not burrow into sand (Carvalho-Filho 1999).

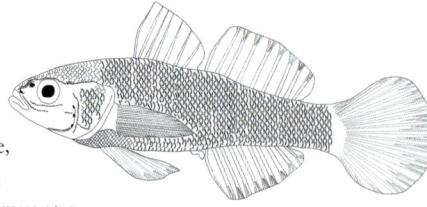

Eleotris (8.0–25 cm SL)

Characterized by: body stout; trunk covered with 40–86 small, ctenoid scales; strong spine on anteroventral border of preopercle; head moderately long, snout conical, scales on nape and opercle; one series of longitudinal papillae below eyes, connected by 5 series of transversal rows of papillae, 3 of them crossing longitudinal papillae; gill openings attached to isthmus at a vertical that passes behind eye; teeth small, conical, in 2–3 rows, those from anteriormost row enlarged in some species; 6 spines in first dorsal fin, 1 flexible spine and 8 rays in second dorsal fin and anal fin; 11–18 pectoral-fin rays (Pezold and Cage 2002). **SPECIES** Seven, including two species in the AOG region: *E. amblyopsis* (Large-scaled spinycheek sleeper) and *E. pisonis* (Spinycheek sleeper). See Pezold and Cage (2002) for species information. **COMMON NAMES** *Amoré-pixuna* (Brazil). **DISTRIBUTION AND HABITAT** *E. amblyopsis* occurs in western Atlantic from southern USA to Brazil (coast of Rio de Janeiro). It has been recorded from the Essequibo and possibly the lower Xingu basin. *Eleotris pisonis* is limited to South America, occurring from the Orinoco to Rio Grande do Sul, noted also in Brazil from lowlands and tributaries of the Essequibo, and lower portions of the Xingu and Tocantins rivers. Inhabit muddy or sandy bottoms in estuaries, creeks, and small channels in fresh or brackish waters. **BIOLOGY** Amphidromous. Carnivore, feeding on invertebrates, mainly Diptera and Crustacea, and small fish, such as small eleotrid and poeciliid fishes (Robins and Ray 1986, Perrone and Vieira 1991). Gonadal development takes place during the dry season. Average size at maturity is 4.3 cm for females and 5.7 cm for males (Teixeira 1994). Eggs are laid on leaves of aquatic plants; nests are guarded by males (Perrone and Vieira 1991, Carvalho-Filho 1999).

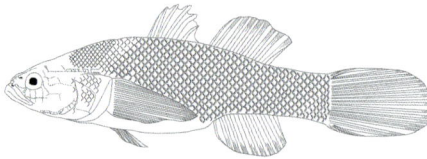

Gobiomorus (36–60 cm SL)

Characterized by: body large, compressed, fully scaled trunk with 40–65 ctenoid scales on lateral rows; head very large, with 2 low longitudinal crests on top, cycloid scales on opercle, below eye, nape, and snout; snout long, nearly 1.5 times length of eye; 3 longitudinal rows of papillae below eye, the first one wide, with more than one series of small papillae; gill openings attached to isthmus at a vertical that passes behind eye; teeth conical; 6 spines in first dorsal fin, 1 spine and 8 rays in second dorsal fin, 1 spine and 9 rays in anal fin, 15 rays in pectoral fin (Robins and Ray 1986, Smith 1997). **SPECIES** Three, including one in the Guianas: *G. dormitor*. **DISTRIBUTION AND HABITAT** *G. dormitor* is known from southern USA (Florida, Louisiana, Texas) to coastal rivers in Guyana and Venezuela, possibly also in Brazil (Smith 1997). Adults are found in clearwater streams and estuaries. **BIOLOGY** Amphidromous. Carnivore, feeding on relatively large invertebrates and fishes (Bussing 1998). Adults live in lotic environments, under logs and rocks, but migrate to estuaries in rainy season to breed; maturity at 15 cm total length (Bacheler et al. 2004). Males build nests 60–120 cm long in rock crevices, where females lay 4,000–6,000 eggs (McKaye et al. 1979).

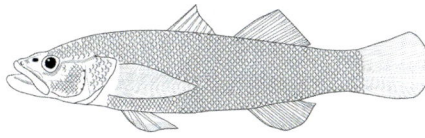

Leptophilypnion (0.8–1.0 cm SL)

Characterized by: body minuscule, slightly compressed; trunk anteriorly naked, covered posteriorly with 12–18 ctenoid scales on lateral rows; head short and robust, without scales, snout blunt; eye very large; some minute papillae below eye, not forming defined rows; gill openings attached to isthmus at a vertical that passes behind eye; teeth small, conical, in two rows; 6 spines in first dorsal fin, 7 rays in second dorsal fin, 1 spine and 5–6 rays in anal fin, 13 rays in pectoral fin (Roberts 2013). **SPECIES** Two, both in the Amazon basin: *L. fittkaui* and *L. pusillus*. Species information in Roberts (2013). **DISTRIBUTION AND HABITAT** *L. fittkaui* from the Negro basin and near the Negro-Purus confluence; *L. pusillus* was recorded only in a small stream near the mouth of the Tapajós River. Found in shallow margins of streams over sandy bottom with deposited leaf litter. **BIOLOGY** Usually collected in association with *Microphilypnus* (J. Zuanon pers. comm.).

Microphilypnus (1.5–3.0 cm SL)

Characterized by: body minute, slightly compressed; trunk covered with large, ctenoid scales, 21–32 on lateral rows; head moderately large and robust; eye large; snout blunt or moderately pointed; cycloid or ctenoid scales on predorsal region and opercle; 5–6 rows of vertical sensory papillae below eye, two horizontal rows below eye; gill openings attached to isthmus at a vertical that passes below eye or posteriorly; teeth small, conical, in two rows; 6 spines in first dorsal fin, 1 spine and 5–9 rays in second dorsal fin, 1 spine and 5–9 rays in anal fin, 11–15 rays in pectoral fin (Caires and De Figueiredo 2011). Color pale or translucent, with small dark spots; 3 dark longitudinal series of spots along anal-fin base and ventral margin of caudal peduncle. *Microphilypnus tapajosensis* presents sexual dimorphism in its coloration (Caires 2013). **SPECIES** Four, all in the Amazon and Orinoco basins. **DISTRIBUTION AND HABITAT** Restricted to freshwater, very widespread and found in all tributaries of the Amazon and Orinoco rivers, and in the Paraguay basin. *Microphilypnus tapajosensis* is endemic to middle Tapajós River; *M. acangaquara* inhabits the Tapajós, and perhaps also present in Aripuanã River, upper Madeira River; *M. macrostoma* was recorded from the Orinoco, Negro, and lower Amazon rivers; and *M. ternezti* occurs in all Amazon tributaries and the Orinoco and Paraguay basins. Generally found in clear- or darkwater small streams, channels, or lakes near sandy substrate, macrophytes, or fallen leaves. **BIOLOGY** Abundant locally where they occur. Carnivorous, feeding on microcrustaceans (Bergleiter 1999). Nonmigratory and territorial. Mature at 1.5 cm SL, reproduction otherwise unknown. Species are known to stay close to shrimp and other small translucent fishes, possibly obtaining a benefit against predation by being confounded with these animals (Carvalho et al. 2006).

FAMILY GOBIIDAE—GOBIES

— *RODRIGO CAIRES and AURYCÉIA GUIMARÃES DA COSTA*

ILLUSTRATIONS BY RODRIGO CAIRES

DIVERSITY Gobiidae (gobies) is the largest family of the sub-order Gobioidei (which also includes sleepers and allies) and one of the most diverse groups of fishes, with 260 genera and >2,000 species worldwide. Five species in three genera are found in the AOG region.

COMMON NAMES *Amborês, Babosas* (Brazil); *Gobios* (Spanish); Gobies (English, French).

TAXONOMIC NOTE The classification of Gobioidei has changed substantially in the last two decades (Thacker 2003, 2009, 2011b, Gill and Mooi 2012). Gobiidae is traditionally divided into the subfamilies Amblyopinae, Oxudercinae, Sicydiinae, Gobionellinae, and Gobiinae, but many authors now consider Kraemeriidae and Microdesmidae as subfamilies within Gobiidae based on DNA data (Thacker 2009, Thacker and Roje 2011, Tornabene et al. 2013), and others regard Gobionellinae as a separate family (Pezold 2011). Here we follow the most common classification of Gobiidae in recent practice (Van Der Laan et al. 2014).

GEOGRAPHIC DISTRIBUTION Gobies are found around the globe in tropical and subtropical coastal waters, with some freshwater species. Kullander (2003b, 2009) recorded the following freshwater species from cis-Andean South America, one of them a sicydiine, *Sicydium punctatum*, and the remaining ones gobionellines: *Awaous tajasica, Awaous flavus, Ctenogobius claytonii, Evorthodus lyricus, Gobioides broussonnetii,* and *Gobioides grahamae.* Of these species, *E. lyricus* and *S. punctatum* are not known from the AOG region. Further information on these taxa, including five species in three genera, is provided below. Although the richness of gobiid fish fauna in the Amazon basin is relatively low, many species are commonly found in coastal rivers and estuaries near the AOG region, such as the gobiines *Akko dionaea, Barbulifer enigmaticus, Bathygobius soporator, Coryphopterus glaucofraenum,* and *Microgobius meeki,* the microdesmid *Microdesmus longipinnis,* and the gobionellines *Ctenogobius boleosoma, C. smaragdus, C. stigmaticus, C. thoropsis, Evorthodus lyricus, Gnatholepis thompsoni, Gobionellus oceanicus,* and *G. stomatus* (see also Birdsong and Robins 1995, Camargo and Isaac 1998, 2001, Barletta et al. 2003, Andrade-Tubino et al. 2008). Most of these species are not treated in detail herein, but it is possible that some of these taxa may be occasionally encountered in the lower Amazon basin. These species are therefore included in the identification key.

ADULT SIZES Usually small (2–20 cm SL), but members of the genus *Gobioides* reach 60 cm SL.

DIAGNOSIS OF FAMILY Gobiidae are most easily identified by 5 branchiostegal rays (vs. 6 in most eleotrids and other gobioids) and a caudal peduncle shorter than the length of the dorsal- or anal-fin bases. Many gobiids (and all species in AOG) also have the pelvic fins completely united by a membrane along their entire length, forming a cup-like disk. Gobiids are also recognized by: body usually stout or slightly compressed (but eel-like in *Gobioides*); trunk covered with cycloid or ctenoid scales, without a lateral line; head short, snout blunt or conical; head pores in preopercular, supraocular (2 anterior interorbital pores in sicydiines and gobionellines and one in gobiines, absent in some species), and postocular sensory canals; rows of sensory papillae below eye; gill openings attached to isthmus at a vertical below or posterior to eye, teeth small, conical, in 2–3 rows, sometimes anterior row with larger teeth in males; 2 separate dorsal fins in most specimens, 6–7 spines in first dorsal fin and 10–17 rays on second dorsal fin and anal fins (Murdy and Hoese 2003b). Male urogenital papilla long, conical, and with a smooth tip, female urogenital papilla long and stout, covered with fleshy fringes on tips around urogenital aperture.

SEXUAL DIMORPHISM Sexual dimorphism in color observed in some species, otherwise males may have anterior spines in first dorsal fin prolonged as a filament.

HABITATS Gobiids from Amazon basin are benthic fishes found in slowly moving waters such as lakes, channels, creeks, and estuaries.

FEEDING ECOLOGY They are usually carnivores, feeding on a variety of small invertebrates, but some species may eat algae, particulate organic matter, periphyton, or other vegetal material. Information on behavior of these taxa is scarce.

BEHAVIOR Often found on the substrate, where males guard a territory and aggressively defend it against conspecifics; males also build a nest where females lay their eggs. Some species (e.g., those from genus *Coryphopterus*) are protogynic hermaphrodites (Mazzoldi et al. 2011). Also, South American goby taxa such as *Microgobius* possibly live together with alpheid crustaceans in a symbiotic relationship (Carvalho-Filho 1999).

KEY TO THE GENERA

1a. Body long, depth <10% of body length; dorsal fins connected by membrane, dorsal and anal fins united with caudal-fin base; eye very small, its length >10 times in head length..2

1b. Body short, depth >10% of body length; dorsal fins separate; last dorsal- and anal-fin rays separated from caudal fin base; eye diameter <10 times in head length 3

2a. First dorsal fin with 6 spines; 8–11 gill rakers on
lower arch; upper jaw slightly inclined (fig. 2a);
jaws with 1–3 tooth rows, with 9–46 usually
small, caniniform teeth ....**Gobioides**

2b. First dorsal fin with 7 spines; 6 gill rakers on
lower arch; upper jaw strongly inclined (fig. 2b);
jaws with 1–2 tooth rows, most exposed externally
with 8–16 very large, caniniform teeth........................ **Akko**
(1 species endemic to the mouth of Amazon River,
A. dionaea, not treated herein)

3a. Mouth inferior **4**
3b. Mouth subterminal **5**

4a. Teeth compressed with bilobed
tips (fig. 4a); scales absent below
eye; 2 dusky spots at base of caudal
fin..**Evorthodus**
(1 species found in coastal rivers near Amazon
mouth, *E. lyricus*, not treated herein)

4b. Teeth conical (fig. 4b); scales present
below eye; 1 dark stripe below eye,
without dark spots at base of caudal
fin..**Gnatholepis**
(1 species found in coastal rivers near Amazon
mouth, *G. thompsoni*, not treated herein)

5a. Shoulder girdle, under gill cover, with
fleshy lobes (fig. 5a); 6 dark brown
stripes on trunk ....**Awaous**

5b. Shoulder girdle, under gill cover,
without fleshy lobes; dark stripes on
trunk absent **6**

6a. Upper 3–5 pectoral rays filamentous,
distally separated from other rays (fig. 6a)..........**Bathygobius**
(1 species found in coastal rivers near Amazon mouth, *B.
soporator*, not treated herein)

6b. Rays not filamentous, all of them united by
a membrane **7**

7a. 6 spines in first dorsal fin.. **8**
7b. 7 spines in first dorsal fin....................................... **10**

8a. Anterior interorbital pore single (fig. 8a); 22–28
(usually ≤26) scales in lateral series on trunk........................ **Coryphopterus**
(1 species found in coastal rivers near Amazon mouth,
C. glaucofraenum, not treated herein)

8b. Paired anterior interorbital pores (fig. 8b); 28–90
(usually >30) scales in lateral series on trunk **9**

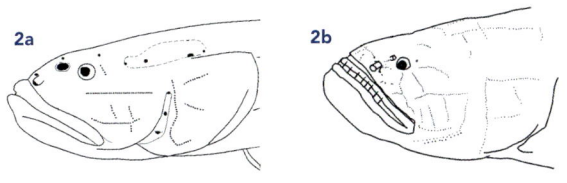

2a   2b

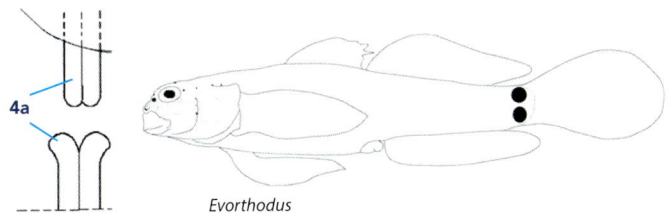

4a

Evorthodus

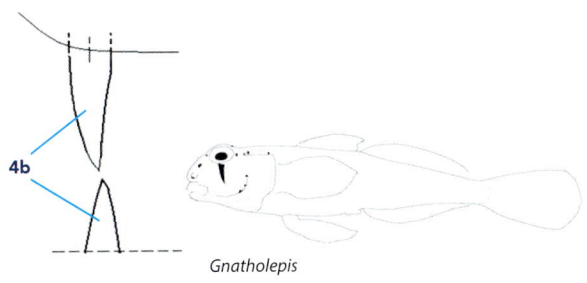

4b

Gnatholepis

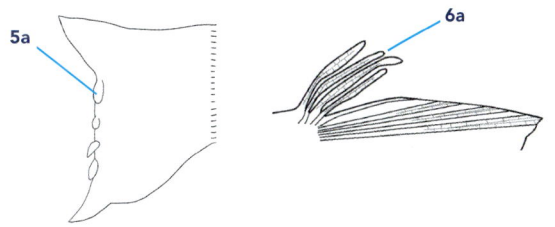

5a

6a

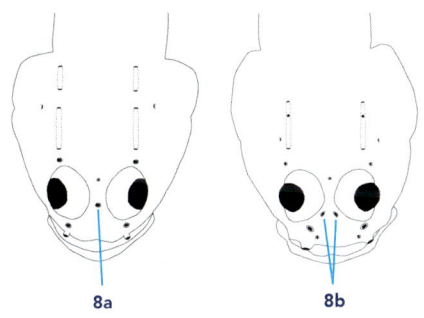

8a   8b

9a. Sensory canal behind the eye large, reaching to scapular region above posterior edge of gill cover, with 4 pores (fig. 9a); 14–15 elements in anal fin................ *Gobionellus*
(2 species found in coastal rivers near Amazon mouth, *G. oceanicus* and *G. stomatus*, not treated herein)

9b. Sensory canal behind the eye short, with 2 pores terminating above the preopercle (fig. 9b); 11–13 elements in anal fin..*Ctenogobius*

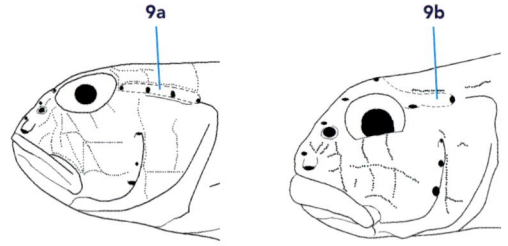

9a   **9b**

10a. Body compressed; mouth very large, inclined, with caniniform teeth in males; second dorsal fin with 1 flexible spine and 14–18 rays .... *Microgobius*
(1 species found in coastal rivers near Amazon mouth, currently identified as *M. meeki*, not treated herein)

10b. Body stout; mouth not inclined, without caniniform teeth; second dorsal fin with 1 flexible spine and 10–13 rays **11**

11a. Body naked; barbels on lower jaw and below eye; interorbital pore absent .... *Barbulifer*
(1 species found in coastal rivers near Amazon mouth, *B. enigmaticus*, not treated herein)

11b. Body with scales at least on caudal fin base; no barbels; a single interorbital pore .... *Gobiosoma*
(1 species found in coastal rivers near Amazon mouth, currently identified as *G. hemigymnum*, not treated herein)

GENUS ACCOUNTS

Awaous (10–15 cm SL)

Characterized by: 6 spines in first dorsal fin; 1 flexible spine and 9–10 rays in second dorsal and anal fins; 15–17 pectoral rays; pectoral girdle with fleshy lobes under gill cover; pelvic fins completely united by a membrane along entire length, forming a cup-like disk; body long, somewhat compressed; trunk covered with 52–60 small ctenoid scales; head long, upper profile slightly convex, with scales on nape, but naked on opercle and below eye; supraocular canal with 2 anterior pores in interorbital region; preopercular canal with 3 pores; postocular canal long from posterior border of eye to scapular region, with 5 pores; 2 postocular canals in *A. flavus*, the anterior with 3 pores and the posterior on scapular region and with 2 pores; 2 series of longitudinal and 5 series of transversal sensory papillae below eye; teeth small, conical, in 2–3 rows; 6–8 dark vertical bars on trunk (Watson 1996). **SPECIES** 20, including one species in the AOG region, *A. flavus*. Further information on *Awaous* species in Watson and Horsthemke (1995) and Watson (1996). **DISTRIBUTION AND HABITAT** Western South Atlantic from Colombia to the Amazon mouth, recorded from coastal rivers and from tributaries of the Orinoco, Essequibo, Atrato, and Tocantins rivers, possibly also in the lower Xingu River. Found in small creeks, beaches, flooded forest channels, and tidal mudflats (Watson and Horsthemke 1995). **BIOLOGY** Amphidromous. In the wild, this species usually ingests substrate; in captivity also consumes small insect larvae and crustaceans on substrate or water column. Spawns year-round; life span short (3–4 years); males aggressive and territorial, guarding a nest with eggs and defending them against predators and other males (Watson and Horsthemke 1995).

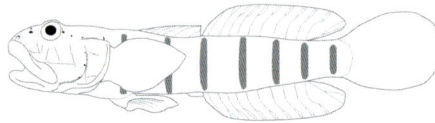

Ctenogobius (5.0–15 cm SL)

Characterized by: 6 spines in first dorsal fin; 11–13 rays in second dorsal and 12–14 rays in anal fin; 12–19 pectoral-fin rays; pectoral girdle without fleshy lobes under gill cover; pelvic fins completely united by a membrane along entire length, forming a cup-like disk; body moderately stout; trunk covered with 29–66 ctenoid scales (cycloid anteriorly in some species); head short, upper profile slightly convex, with or without scales on nape, but naked on opercle and below eye; supraocular canal with 2 anterior pores in interorbital region; preopercular canal with 3 pores; postocular canal short, with 2 pores; 2 longitudinal and several vertical rows of sensory papillae below eye; teeth small, conical, in 2–3 rows, anteriormost with larger teeth (Gilbert and Randall 1979). **SPECIES** 14, including at least two species in the AOG region: *C. fasciatus* and *C. thoropsis*. Further information on genus in Murdy and Hoese (2003b). Kullander (2009) reported *C. claytonii* from Delta Amacuro, but this species is now considered restricted to estuaries and rivers in the southwestern Gulf of Mexico (F. Pezold pers. comm.), thus such records by Kullander probably refer to *C. fasciatus*. **DISTRIBUTION AND HABITAT** *C. fasciatus* in Central America and South America from Dominica to Brazil; in the AOG region from the Delta Amacuro and coast of Suriname to the Marajó delta of the Amazon River, and mouth of Tocantins River. *Ctenogobius thoropsis* occurs from Suriname to the lower Amazon River. Fairly common in bays, estuaries, mangroves, lagoons, or coastal rivers. **BIOLOGY** Benthic fishes that stand on sandy bottoms and feed on small invertebrates, algae, diatoms, and detritus (Carvalho-Filho 1999). Reproduction unknown.

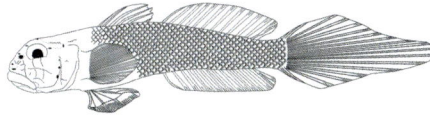

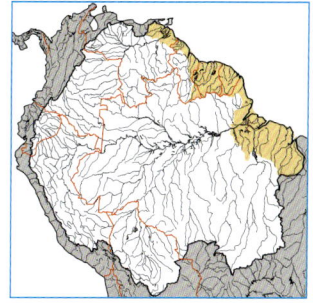

Gobioides (50–60 cm SL)

Characterized by: 6 spines in first dorsal fin, 15–17 rays in second dorsal fin, both fins connected by membrane; 15–17 rays in anal fin; the last rays of dorsal and anal fins united to caudal fin; 15–20 pectoral rays; pelvic fins completely united by a membrane along entire length, forming a cup-like disk; body very elongate; trunk covered with small, embedded cycloid scales; head long; mouth slightly inclined; 2–3 tooth series, the most external with 9–46 usually small, caniniform teeth; sensory pores small; pores present on supraocular (but absent in interorbital), preopercular, and postocular sensory canals; 2 longitudinal and 5 transversal rows of sensory papillae below eye (Murdy 1998). Color pallid, in *G. broussonnetii* often with a midlateral row of saddle-shaped purple or dark markings along trunk, accompanying intervertebral spaces. **SPECIES** Five, including two species in the AOG region: *G. broussonnetii* and *G. grahamae*. Further information on genus in Murdy (1998). **DISTRIBUTION AND HABITAT** *G. broussonnetii* occurs in western Atlantic from southern USA to Brazil (Santa Catarina) and *G. grahamae* off the Amazon mouth, Guyana and Suriname. Both species were also caught in coastal rivers in the Guianas. They live burrowed in low-salinity mudflats or coasts down to 3 m deep and with a sandy or muddy substrate. **BIOLOGY** Amphidromous. Feed primarily on diatoms, algae, and detritus (Birdsong and Robins 1995, Mata-Cortés et al. 2004). Species live burrowed into soft substrate and are more active at night (Carvalho-Filho 1999). Males mature as larger length than females; spawning is thought to occur in the sea (Mata-Cortés et al. 2004). *Gobioides* are intensively collected in the Amazon mouth and used as bait for fishing (e.g., on large catfishes).

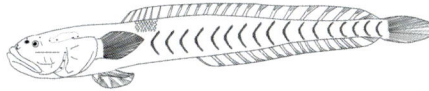

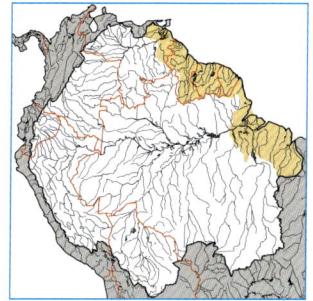

FAMILY BATRACHOIDIDAE—TOADFISHES

— *PETER VAN DER SLEEN and JAMES S. ALBERT*

DIVERSITY 83 species in 23 genera worldwide. Two freshwater species are found in the AOG, *Thalassophryne amazonica* and *Potamobatrachus trispinosus*, each in a different subfamily (Thalassophryininae and in Batrachoidinae, respectively), and each representing an independent freshwater lineage.

COMMON NAMES *Pacamão* (Brazil); *Pez sapo de agua dulce* (Peru).

GEOGRAPHIC DISTRIBUTION Toadfishes are found mostly in nearshore marine waters in the Atlantic, Indian, and Pacific oceans. Three freshwater species occur in South America.

ADULT SIZES Freshwater species in Amazon 5.0–10 cm SL.

DIAGNOSIS OF FAMILY Easily recognized by their characteristic body shape, with a broad and flat head, often with fleshy projections (in *Potamobatrachus*) and with eyes located on top of the head; 2 dorsal fins, the first consisting of 2 (*Thalassophryne*) or 3 (*Potamobatrachus*) strong and sharp spines; second dorsal fin with a large number of soft rays; pelvic fins jugular and inserted well in advance of pectoral fins, with 1 spine and 2 or 3 soft rays (Collette 2003a). In *Thalassophryne* the dorsal and preopercular spines are hollow and connected to venom glands (Haddad Jr. et al. 2003).

KEY TO THE GENERA

1a. Mouth superior; dorsal fin with 19–29 soft rays; anal fin with 18–19 soft rays; 2 dorsal spines ....*Thalassophryne*
1b. Mouth terminal; dorsal fin with 18–19 soft rays; anal fin with 15–16 soft rays; 3 dorsal spines ....*Potamobatrachus*

GENUS ACCOUNTS

Potamobatrachus (5 cm SL)

Characterized by: a dorsal fin with 18–19 soft rays; anal fin with 15–16 soft rays; and 3 dorsal-fin spines (vs. 2 in *Thalassophryne*). Body is light-colored and with 4 irregular brown saddles. **SPECIES** One, *P. trispinosus*. See description of species in Collette (1995). **DISTRIBUTION AND HABITAT** Araguaia and Tocantins basins in Brazil. Usually caught among rocks. **BIOLOGY** Poorly known. They are bottom dwellers and likely ambush predators (as most species of family), feeding on invertebrates and other fishes.

Thalassophryne (9–14 cm SL)

Characterized by: dorsal fin with 19–29 soft rays; anal fin with 18–19 soft rays; two dorsal-fin spines (vs. 3 in *Potamobatrachus*); body color brownish with many spots and several dark-colored saddles. **SPECIES** Five marine and one freshwater species. **DISTRIBUTION AND HABITAT** Distributed across the Western Atlantic with one species (*T. amazonica*) in the Amazon basin of Brazil, Colombia, Ecuador, and Peru. *Thalassophryne amazonica* prefers acidic waters with low conductivity and high temperature, and with muddy or sandy bottoms (Britz and Toledo-Piza 2012). A second species (*T. nattereri*) inhabits freshwater estuarine and brackish waters of the Brazilian coast. **BIOLOGY** Species lie motionless on sandy or muddy bottoms and are ambush predators, feeding mainly on invertebrates and other fishes. Relatively few eggs (~100) of ~5 mm are attached to the underside of objects with the aid of an adhesive disk on the ventral surface of a large yolk sac that persists after hatching (Britz and Toledo-Piza 2012). Hatched larvae remain attached until the yolk sac is resorbed and leave the nesting site as small juveniles with fully formed fins at ~18 mm SL. Eggs and attached larvae are guarded and fanned by the male. Both the dorsal spines and preopercular spine are hollow and connected to venom glands. Venom is injected in the flesh of the victim when pressure is applied to the gland. This can inflict extremely painful wounds to humans, causing dizziness, fever, and even necrosis of infected tissue (Haddad Jr. et al. 2003).

FAMILY BELONIDAE—NEEDLEFISHES

— *PETER VAN DER SLEEN and NATHAN R. LOVEJOY*

DIVERSITY 47 species in 10 genera.

COMMON NAMES *Peixe agulha* (Brazil); *Agoeja fiesie, Aiguillette* (French Guiana); *Pez aguja* (Peru).

GEOGRAPHIC DISTRIBUTION Most species are marine, but three freshwater genera (*Belonion, Potamorrhaphis,* and *Pseudotylosurus*) occur in freshwaters of South America. In addition, the marine species *Strongylura marina* is commonly found in estuaries in northern South America.

ADULT SIZES AOG species between 5 cm (*Belonion*) and 40 cm (*Pseudotylosurus*) SL.

DIAGNOSIS OF FAMILY Needlefishes are long and slender fishes that can be recognized by their elongate jaws that contain numerous sharp teeth. The anal and dorsal fins are relatively small and lie posteriorly; scales are small and cycloid; trunk lateral-line canal low on body; and nostrils located in a depression anterior to the eyes.

SEXUAL DIMORPHISM Not pronounced.

HABITATS Typically found close to the water surface.

FEEDING ECOLOGY *Potamorrhaphis* hunt close to the water surface for insects and other fishes during the day and remain close to the shore at night, usually aligning themselves with submerged roots, branches, or other vegetation. See notes on ecology and diet of Amazonian needlefishes in Goulding and Carvalho (1984).

KEY TO THE GENERA

1a. Upper jaw in adults elongate; adult size 10–40 cm SL...**2**
1b. Upper jaw in adults short; adult size 5 cm SL ....*Belonion*

2a. Elongate dorsal and anal fins; caudal fin rounded; caudal peduncle compressed...*Potamorrhaphis*
2b. Relatively short dorsal and anal fins; caudal fin forked; caudal peduncle lozenge-shaped in cross section ... *Pseudotylosurus*

GENUS ACCOUNTS

Belonion (5 cm SL)

Characterized by an elongate lower jaw and a short upper jaw; hence they resemble species of the closely related family Hemiramphidae (the halfbeaks). The halfbeak species in the region are larger, less elongate, and have an indented tail. However, the juveniles of most of the other species of needlefishes also have a similar "halfbeak" stage during their development. *Belonion* can be differentiated from juveniles of other Neotropical freshwater needlefish genera by: secondary tubes of the lateral line on body extending only ventrally or absent (vs. extend both dorsally and ventrally in other genera); absence of a fourth upper pharyngeal tooth plate (vs. present in the other genera); branchiostegal rays 6 or 7 (vs. >8 in *Potamorrhaphis*); pectoral-fin rays 5 or 6 (vs. ≥7 in other genera); elongate nasal papillae (vs. spatulate in *Pseudotylosurus*); principal caudal rays 7+8 or 6+7 (Collette 1966, Lovejoy and Collette 2001). **SPECIES** Two, *B. apodion* and *B. dibranchodon*; species descriptions in Collette (1966).

DISTRIBUTION AND HABITAT *B. apodion* from the Guaporé and Madeira basins, *B. dibranchodon* in the Atabapo and Negro basins, Brazil, Colombia, and Venezuela. Found close to the water surface.

BIOLOGY Feed mostly on small insect larvae and zooplankton (Goulding and Carvalho 1984).

Potamorrhaphis (10–30 cm SL)

Differs from *Pseudotylosurus* and *Belonion* by having longer dorsal and anal fins and a rounded caudal fin. Most species also have a dark band of spots or stripes running along the body. Both the upper and lower jaws are elongate in adults. Also distinguished from other Neotropical freshwater needlefish genera by: elongate nasal papillae (vs. spatulate in *Pseudotylosurus*); lateral line scales with both dorsally and ventrally directed short secondary tubes (vs. extending ventrally only or absent in *Belonion*); and the first scales of the lateral line displaced to the region between the opercle and the pectoral fin (Collette 1982, Lovejoy and Collette 2001, Sant'Anna et al. 2012). **SPECIES** Four, key to

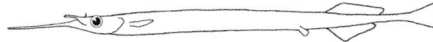

species in Sant'Anna et al. (2012). **DISTRIBUTION AND HABITAT** *P. eigenmanni* inhabits the upper Madeira basin and the Paraguay-Paraná basin; *P. guianensis* is found throughout the AOG region; *P. labiatus* in the Ucayali and Solimões drainages in the western Amazon; and *P. petersi* is restricted to the upper Negro and upper Orinoco. See distribution map of species in Lovejoy and De Araújo (2000) and Collette (1982). *Potamorrhaphis* species are commonly found in backwater lakes and streams (inhabiting the shore zones) and are usually absent in large rivers and open waters. **BIOLOGY** Known to form schools; insectivorous and piscivorous (Goulding and Carvalho 1984).

Pseudotylosurus (30–40 cm SL)

Characterized by relatively short dorsal and anal fins and a forked caudal fin. Both the upper and lower jaws are elongate in adults. In addition, *Pseudotylosurus* can be distinguished from other genera of freshwater Neotropical needlefishes by the spatulate shape of nasal papillae (vs. elongate) and a high density of pharyngeal teeth (vs. low) (Lovejoy and Collette 2001). **SPECIES** Two, *P. microps* and *P. angusticeps*; see species descriptions in Collette (1974). **DISTRIBUTION AND HABITAT** *P. microps* is widespread and found throughout South America (including most of the AOG); *P. angusticeps* from the upper Amazon basin (Peru and Ecuador) and the Paraná-Paraguay basin. Commonly found in muddy beaches of whitewater rivers in the dry season (J. Zuanon pers. comm.). **BIOLOGY** Oviparous. Eggs attached to objects in the water by tendrils from the egg's surface (Breder and Rosen 1966). Both species are insectivorous and piscivorous (Goulding and Carvalho 1984).

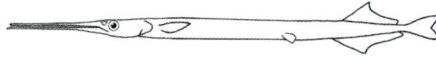

FAMILY HEMIRAMPHIDAE—HALFBEAKS

— *PETER VAN DER SLEEN and NATHAN R. LOVEJOY*

Family includes 62 species in 8 genera worldwide. Most species are marine, but some inhabit freshwaters. One species in the Amazon basin, *Hyporhamphus brederi*.

Hyporhamphus (10 cm SL)

Characterized by an elongate body, resembling needlefishes (family Belonidae), but differing by having only the lower jaw elongate and a short triangular upper jaw. Other diagnostic characters include: pectoral and pelvic fins short; pelvic fins in abdominal position, with 6 soft rays; no spines in fins; dorsal and anal fins posterior in position; nostrils in a pit anterior to the eyes; teeth minute or absent; premaxillae pointed anteriorly; scales relatively large and cycloid; trunk lateral line low on body (Collette 2003b). **SPECIES** 39, with one species in AOG, *H. brederi*. **COMMON NAMES** *Meio bico* (Brazil); *Agujeta*, *Pez pajarito* (Peru). **DISTRIBUTION AND HABITAT** Marine environments around the tropics and subtropics. One freshwater species, *H. brederi*, occurs in the lower Orinoco and lower Amazon. The marine species *H. roberti* (20 cm SL) is also found in estuaries and river mouths of northern South America. **BIOLOGY** Poorly known. Other members of this genus form large schools and are omnivorous, feeding on floating vegetation, as well as small crustaceans and fishes.

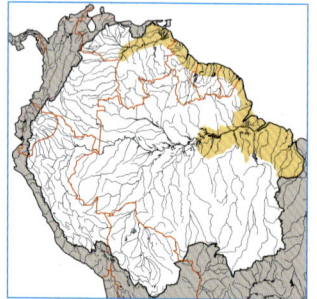

FAMILY SYNBRANCHIDAE—SWAMP EELS

— *PETER VAN DER SLEEN and JAMES S. ALBERT*

Family includes 23 species in four genera worldwide, and is found throughout the tropics and subtropics in fresh and occasionally brackish water. Three species in a single genus in the AOG region.

Synbranchus (100–150 cm TL)

Completely lacking pectoral, pelvic, and caudal fins as adults, and with rudimentary dorsal and anal fins (reduced to a rayless midsagittal ridge). However, larvae of ≤15 mm have large pectoral fins that they shed overnight after about two weeks. Eyes very small; gill membranes fused, leaving a single ventral slit-like opening to the gill chamber (Rosen and Greenwood 1976, Kullander 2003c). **SPECIES** Three, all in the AOG region. Species-level taxonomy is poorly understood and species richness will likely increase in the future. **COMMON NAMES** *Muçum* (Brazil); *Atinga* (Peru). **DISTRIBUTION AND HABITAT** *S. marmoratus* is found throughout Central and South America (from Mexico to northern Argentina); *S. madeira* (Rosen and Rumney 1972) is widespread in the Amazon basin, especially in river systems draining the lowlands; and *S. lampreia* (Favorito et al. 2005) on Marajó Island, Pará, Brazil. **BIOLOGY** Air breathers with highly vascularized linings of the mouth and pharynx, allowing them to thrive in deoxygenated waters (like swamps) and to make short travels over land. Swamp eels are burrowers and also hibernate in burrows when water levels drop (Boujard et al. 1997). The lack of paired fins is likely an adaptation for burrowing in soft substrates. They are mainly nocturnal and capture prey by sucking it in or by ripping off body pieces with strong corkscrew-like movements. At least *S. marmoratus* is a protogynous hermaphrodite; females transition into male fish (known as secondary males) at a length of between 45 and 60 cm (Lo Nostro and Guerreo 1996).

FAMILY SYNGNATHIDAE—PIPEFISHES

— *PETER VAN DER SLEEN and JAMES S. ALBERT*

DIVERSITY 298 species in 57 genera worldwide, with around 20 species restricted to freshwaters (mostly in the Indo-Pacific), and three species in two genera in the AOG region.

COMMON NAMES *Peixe cachimbo* (Brazil); *Pez pipa* (Peru).

ADULT SIZES AOG species 16–21 cm TL.

GEOGRAPHIC DISTRIBUTION Mostly marine, some in brackish and fresh water. Atlantic, Indian, and Pacific oceans.

DIAGNOSIS OF FAMILY Easily recognized by their distinct body shape and behavior. They have an elongate body encased in bony rings; elongate snouts; fused jaws; lack dorsal-fin spines; soft dorsal fin variable in size, with 15–60 rays; anal fin small to absent; pelvic fins absent; caudal fin present or absent; and restricted gill openings.

KEY TO THE GENERA

1a. Caudal fin with 9 rays; anal fin small (usually with 4 rays); trunk rings 17–22; tail rings 20–26; total rings 39–45........ *Microphis*

1b. Caudal fin with 10 rays; anal fin absent; trunk rings usually 13; tail rings 31–38; total rings 44–51 .... *Pseudophallus*

GENUS ACCOUNTS

Microphis (21 cm SL)

Characterized by: elongate body encased in
bony rings, trunk usually covered by 17–22 rings, tail with 20–26 rings, total rings 39–45;
pelvic fins absent; anal fin small, usually with 4 rays; caudal fin with 9 rays. Mature males with
a brood pouch. **SPECIES** 19, including one species in the AOG region: *M. brachyurus* (Short-
tailed pipefish). Key to the species in Dawson (1984), key to the subspecies of *M. brachyurus*
in Dawson (1979). **DISTRIBUTION AND HABITAT** *M. brachyurus* from the Indo-West Pacific and
Western Atlantic, in marine, brackish, and freshwaters. **BIOLOGY** Feed on worms, crustaceans, and
zooplankton, using their small mouths to pick animals from the water column or substrate. Eggs
are deposited in a brood pouch under the tail of males, where they are retained until hatching.

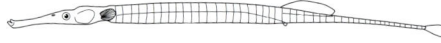

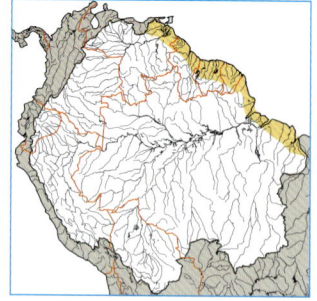

Pseudophallus (16–18 cm SL)

Characterized by: elongate body encased in
bony rings, trunk usually covered in 13 rings, tail with 31–38 rings, total rings 44–51; anal
and pelvic fins absent; caudal fin with 10 rays; a mottled brown color, females with a (faintly)
striped pigmentation pattern. Mature males with brood pouch, females with an enlarged anal
papilla. **SPECIES** Four, including two species in the AOG region: *P. mindii* and *P. brasiliensis*.
Key to the species in Dawson (1974). **DISTRIBUTION AND HABITAT** Atlantic, Caribbean, and
Pacific coastal rivers of Central and South America. Found mainly in freshwater habitats, but
sometimes in estuaries and mangroves; planktonic juveniles have been collected well offshore at
sea. **BIOLOGY** Feed on small crustaceans, using their small mouths to pick animals from the water
column or substrate. Eggs are deposited in a brood pouch under the tail of males, where they
are retained until hatching. In the Corantijn River (Suriname), the sex ratio and difference in
coloration suggested that female *P.* aff. *brasiliensis* live in a sex-role reversed harem structure with
territorial females competing for males that brood their eggs (Mol 2012a).

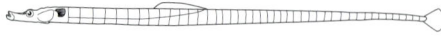

FAMILY ACHIRIDAE—AMERICAN SOLES
— *PETER VAN DER SLEEN and JAMES S. ALBERT*

DIVERSITY 35 species in nine genera, including 11 species in
three genera in the AOG region.
COMMON NAMES *Linguado lixa, Soia* (Brazil); *Lenguado* (Peru).
GEOGRAPHIC DISTRIBUTION Mainly shore fishes found on both
sides of the Americas in the Atlantic and eastern Pacific oceans.
Most species of the genus *Apionichthys* and *Hypoclinemus
mentalis* are restricted to freshwater and found in rivers of
northern South America. Some marine species of the genus
Achirus occasionally enter freshwater.
ADULT SIZES Medium sizes, most species between 15 and
30 cm SL.

DIAGNOSIS OF FAMILY American soles can be distinguished
from other flatfish families by: the eyes on the right side of the
body (shared with Soleidae); a conspicuous fringed fleshy rim
of the lower lip on the eyed side; posterior nares forming a
wide longitudinal slit above the posterior end of the upper lip,
hidden by the lower lip when the mouth is closed (the anterior
nares is tubular); dorsal and anal fins usually separate from the
caudal fin; and pectoral fins small or absent (Ramos 2003a).

KEY TO THE GENERA

1a. Supracranial area and corresponding portion of dorsal fin projecting over the anterior region of the cranium, extending beyond the mouth and concealing the anterior margin of the mouth (fig. 1a; after Ramos 2003b).................................... *Apionichthys*

1b. Supracranial area not projecting over the mouth (fig. 1b)..**2**

2a. Patch with villiform teeth present only on jaws of the blind side ....*Achirus*

2b. Patch with villiform teeth present on jaws of both eyed and blind sides .... *Hypoclinemus mentalis*

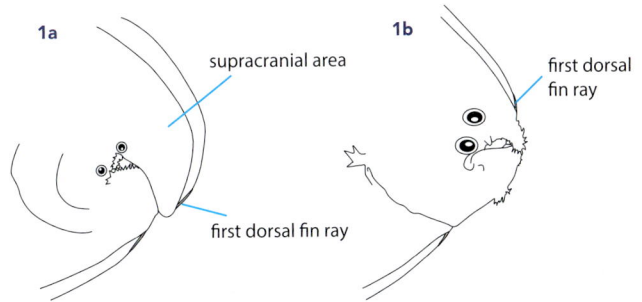

1a

supracranial area

first dorsal fin ray

1b

first dorsal fin ray

GENUS ACCOUNTS

Achirus (11–37 cm SL)

Characterized by: eyes relatively large and placed almost above each other; villiform teeth (arranged in patches) present only on jaws of the blind side. **SPECIES** Nine, including two species in the AOG region. **DISTRIBUTION AND HABITAT** Most species occur in coastal marine and estuarine waters. *Achirus novoae* inhabits freshwaters of the Orinoco River and its delta, and *A. achirus*, coastal marine waters under the influence of South American rivers, and freshwater areas of the Tocantins and Orinoco basins (Ramos et al. 2009). **BIOLOGY** Occur on sand-mud bottoms and feed on invertebrates (especially crustaceans) and small fishes. They are often completely covered with sediment with only the eyes above the substrate, presumably for protection from predators and to ambush prey (Keith et al. 2000).

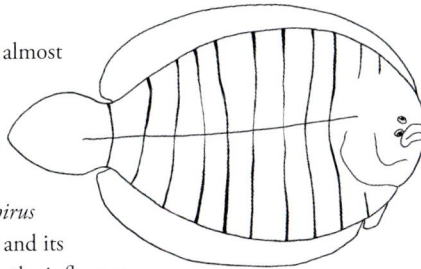

Apionichthys (4.0–24 cm SL)

Characterized by: supracranial area and corresponding portion of dorsal fin projecting over the anterior region of the cranium, extending beyond and concealing the anterior margin of the mouth (as a consequence the first dorsal-fin ray lies ventral to the mandibular symphysis); pelvic fin projects anteriorly, with the first pelvic-fin rays inserted just below the ventral end of the mandibular symphysis; branchiostegal membrane united to pelvic-thoracic region, partially or completely concealing the eye-sided gill opening; and a concavity behind the eye-sided anterior nostril, resulting from a depression of tissue just posterior to the nostril and elevation of tissue just anterior of the fixed eye (Ramos 2003b). In addition, in most species the dorsal and anal fins are united to the caudal fin by a narrow membrane (except in *A. finis*, where the fins are free); eyes are in general small; body coloration is variable, including a lighter body with many irregular, scattered dark blotches; a very dark body coloration or a uniform body coloration with several dark stripes (Ramos 2003b). **SPECIES** Eight, all species at least occasionally in freshwaters of the AOG region, including *A. dumerili* (Longtail sole). Species information and a key to the species in Ramos (2003b). The genus includes the species formerly known as *Soleonasus finis* (now *A. finis*), *Pnictes asphyxiatus* (now *A. asphyxiatus*), and *Achiropsis nattereri* (now *A. nattereri*).

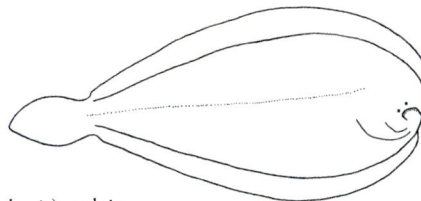

DISTRIBUTION AND HABITAT Northern South America. Most species are restricted to freshwater; some are also found in estuaries and marine areas under the influence of rivers. *Apionichthys* species have been caught with bottom trawl nets in the deep main channel of large Amazon rivers. Species distribution maps in Ramos (2003b). **BIOLOGY** Carnivorous, feeding on benthic invertebrates and fishes.

Hypoclinemus (21 cm SL)

Characterized by: eyes relatively large and placed almost above each other; body sand-colored with many dark spots; villiform teeth (arranged in patches) present on jaws of both eyed and blind sides. **SPECIES** One, *H. mentalis.* **DISTRIBUTION AND HABITAT** Throughout the Amazon basin (except upper portion of Tapajós and Xingu rivers), the Orinoco basin (upper tributaries: Guaviare, Casiquiare, Venturi), and throughout the Essequibo basin (Ramos 2003a). **BIOLOGY** Carnivorous, feeding on benthic invertebrates and fishes.

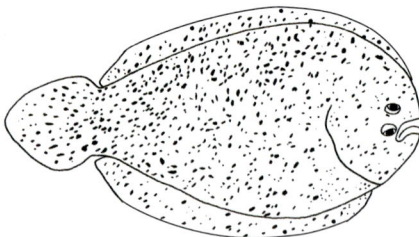

FAMILY TETRAODONTIDAE—PUFFERFISHES

— *PETER VAN DER SLEEN and JAMES S. ALBERT*

Family includes 199 species in 28 genera worldwide, mostly marine, but including 29 species in freshwater systems on three continents (e.g., Cooke et al. 2012), and three described species in one genus (*Colomesus*) in the AOG region. Some marine species can also be found in brackish waters in the Amazon and Orinoco deltas and coastal rivers of the Guianas (e.g., *Sphoeroides testudineus*; 20–30 cm).

Colomesus (13–35 cm SL)

Characterized by: rounded body; absence of pelvic fins; dorsal and anal fins located far posteriorly on body; large eyes; teeth fused into four ridges (two above and two below); and olfactory organs with prominent flaps. **SPECIES** Three, all in the AOG region.
COMMON NAMES *Baiacu-de-água-doce; Corrotucho* (Brazil); *Pez globito* (Peru). **DISTRIBUTION AND HABITAT** *C. asellus* (Amazon puffer; 13 cm SL) from throughout most of the AOG region; *C. psittacus* (Banded puffer; 29 cm SL) from marine, brackish, and freshwaters along the coast from Trinidad to Suriname and near the Amazon mouth; and *C. tocantinensis* (Tocantins puffer; 35 cm SL) from the middle Tocantins basin. **BIOLOGY** Feed mainly on small benthic aquatic animals, including crustaceans, gastropods, and bivalves, which they crush with their powerful teeth (Keith et al. 2000, Krumme et al. 2007). If threatened, pufferfishes can inflate themselves by swallowing water (or air) into a thin-walled chamber opening ventrally from the stomach. Inflation is helped by absence of ribs and pelvic fins. Species are also toxic, especially the liver and ovaries (de Freitas 2006, Oliveira et al. 2006). Often solitary or in groups of 2–3 individuals (Keith et al. 2000), but *C. asellus* may form groups of hundreds of individuals close to muddy beaches along the Amazon River during the dry season (J. Zuanon pers. obs.).

GLOSSARY OF TECHNICAL TERMS

abdomen: belly or lower surface of a fish; ventral area between pelvic fins and anus.

acute: pointed or sharp.

adipose eyelid: transparent tissue partially covering the eye in some fishes.

adipose fin: unpaired median fin, situated posterior to the main dorsal fin in some fishes (see fig. 1), usually rayless (but a rayed adipose fin present in *Colossoma*, *Phractocephalus*, and a few other genera).

adnate: joined together congenitally.

allopatric: occurring in different geographic areas (compare with sympatric).

amphidromous: fishes that regularly migrate between freshwater and the sea (in both directions), but not for the purpose of breeding, as in anadromous and catadromous species

anadromous: fishes that ascend rivers to spawn (opposite of catadromous: migrating from freshwater to the sea to spawn).

anal fin: unpaired median fin on the ventral body margin, usually just behind the anus (see fig. 1).

anguilliform: shaped like or resembling an eel (see fig. 2).

anterior: pertaining to the front portion, toward the nose (see fig. 3).

apical end: end furthest from the attachment.

apophysis: a protuberance or projection on the surface of certain bones.

axillary: referring to the base of the pectoral fins (compare with inguinal).

axillary process: a modified, usually elongate, scale or scales at the upper or anterior base of the pectoral fins.

band: oblique or curved, linear color marking on the body surface.

bar: vertical color marking (see fig. 4).

barbel: elongate, fleshy, tentacle-like sensory projection, found on the head, usually near the mouth.

basal: at the base; the portion of an appendage closest to the body.

base of fin: part of the fin attached to the body.

benthic: living, feeding, and resting on the bottom, often covering themselves in sediment (e.g., flatfishes and stingrays).

blotch: irregularly shaped color mark (see fig. 4).

branchial: relating to the gills.

branchial arch: (or gill arch) one of the bony or cartilaginous arches that form the support for the gills.

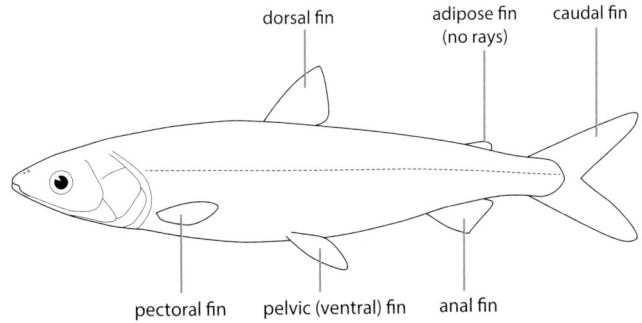

FIGURE 1 The fins of fishes are either paired or unpaired. Paired fins are located on both sides of the body: the pectoral and pelvic fins. Unpaired fins (also called median fins) are located along the middle of the body: the dorsal, adipose, caudal, and anal fins.

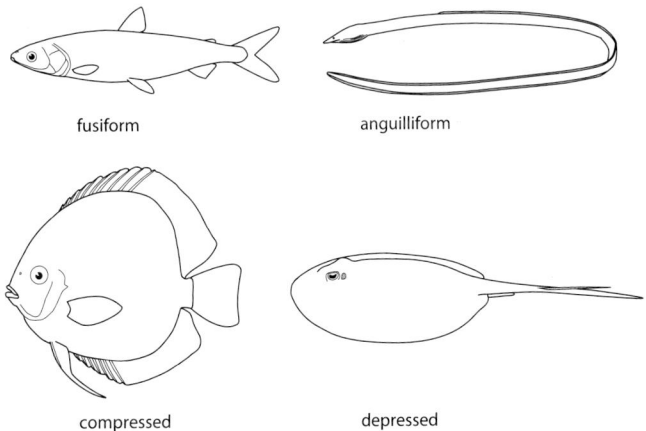

fusiform anguilliform

compressed depressed

FIGURE 2 AOG fishes come in a dazzling array of body shapes. Some of the primary examples are **fusiform** (cigar-shaped, often tapering both anteriorly and posteriorly), **anguilliform** (greatly elongate and usually tubular), **compressed** (flattened laterally and often deep-bodied), and **depressed** (flattened dorsoventrally).

branchiocranium: one of three main divisions of the fish skull, constituting the mandibular and hyoid arches, and the branchial or gill arches. The other skull divisions are the neurocranium (cartilage bones that form the braincase) and dermatocranium (dermal bones including skull roofing bones, opercular series, tooth bearing bones, and others).

branchiostegal membranes: membranes on the ventral interior surface of the gill cover supported by branchiostegal rays (see fig. 5).

branchiostegal rays: bony rays supporting the branchiostegal membranes inside the lower part of the gill cover (see fig. 6).

buccal: referring to the mouth.

caniniform teeth: large teeth that are conical and pointed (without one or more cusps). They are usually larger than the surrounding teeth (see fig. 7b).

carnivorous: feeding on animals.

catadromous: species usually found in freshwater as adults, but which spawn in the ocean; e.g., true eels (*Anguilla*).

caudal fin: tail fin (see fig. 1).

caudal peduncle: tail base; posterior part of the body between the rear parts of the dorsal and anal fins, and the caudal fin (see fig. 8).

caudal spot: spot at base (origin) of caudal fin (see fig. 4).

cephalic: pertaining to the head.

cephalic shield: the large bony dorsal covering on the head of fork-tailed sea catfishes (Ariidae).

ceratobranchial: bone or cartilage in the branchial arch below the epibranchial.

chromatophores: pigment cells that can be altered in shape and size to produce color change.

circumpeduncular scales: the transverse series of scales that completely encircle the tail base.

cis-Andean: east from the Andes (vs. trans-Andean: west from the Andes).

cleithral: pertaining to the cleithrum or area of the cleithrum, typically the largest bone of a series of bones that support the pectoral fin (pectoral-girdle bones) (see fig. 9).

compressed: laterally flattened body form (see fig. 2).

concave: bowed or curved inward; opposite of convex.

convex: bowed or curved outward; opposite of concave.

coracoid: paired bones of the pectoral girdle that usually contact one another at the ventral midline, and that also articulate with the scapula and cleithrum (see fig. 9).

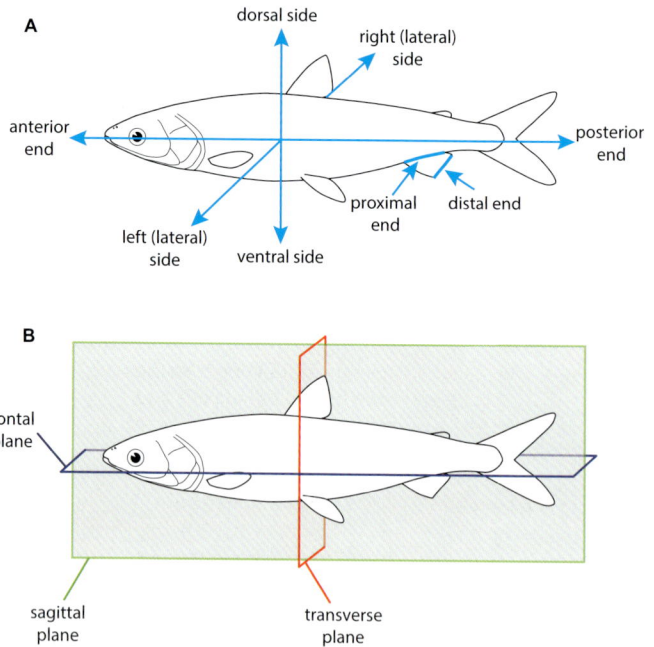

FIGURE 3 Terminology used to indicate directions and position of structures in animals, with (A) body axes and (B) body planes.

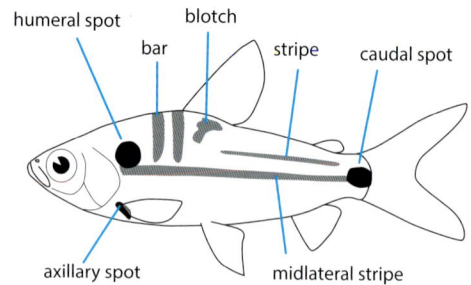

FIGURE 4 Some terminology used to describe color patterns: a spot is a circular or elongate color mark; bars are verticwal color markings; a blotch is an irregularly shaped color mark; stripes are elongate, straight-sided, horizontal color markings.

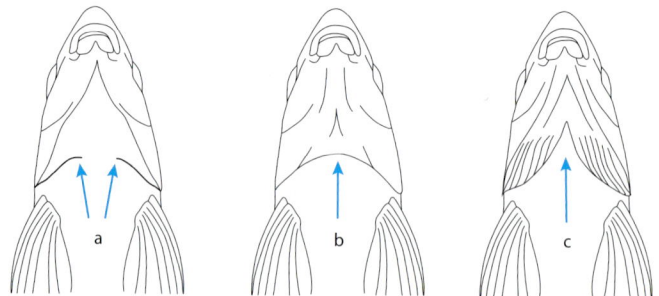

FIGURE 5 Gill (branchiostegal) membranes on isthmus: (a) right and left gill membranes bound to, but not free from isthmus; (b) gill membranes broadly joined across, but free from, isthmus; and (c) gill membranes separate and free from isthmus. Illustration based on Cross (1967).

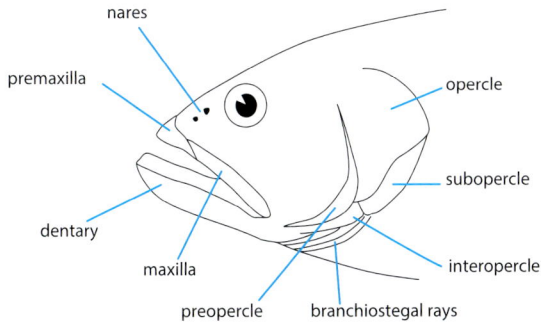

FIGURE 6 In bony fishes (Osteichthyes) the upper portion of the oral jaw is formed by two bones, the **maxilla** and the **premaxilla**. The lower jaw is formed by three bones, the largest and most anterior of which is the **dentary**. All these bones can bear teeth. In many specialized groups only the premaxilla bears teeth and the maxilla is toothless. The **operculum** is the hard bony flap covering and protecting the gills. In most fish, the rear edge of the operculum marks the division between the head and the body. The operculum is composed of four bones: the **opercle**, **preopercle**, **interopercle**, and **subopercle**. The **branchiostegal rays** are bony rays that support the branchiostegal membranes inside the lower part of the gill cover.

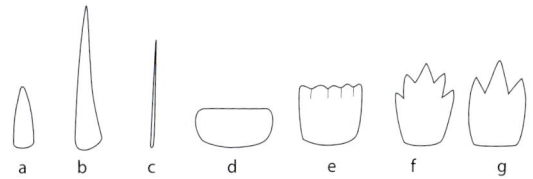

FIGURE 7 Some of the main teeth types: (a) **conical** teeth; (b) **caniniform** teeth are large elongate and conical teeth, frequently located near the anterior end of the mouth; (c) **villiform** teeth are small, fine teeth, often arranged as a brush; (d) **molariform** teeth (molars) are pavement-like crushing teeth, often located in the throat; (e) **incisiform** teeth (incisors) are large teeth with flattened cutting surfaces; flattened cutting teeth with (f) 5 points (or **pentacuspid**) and (g) 3 points (**tricuspid**).

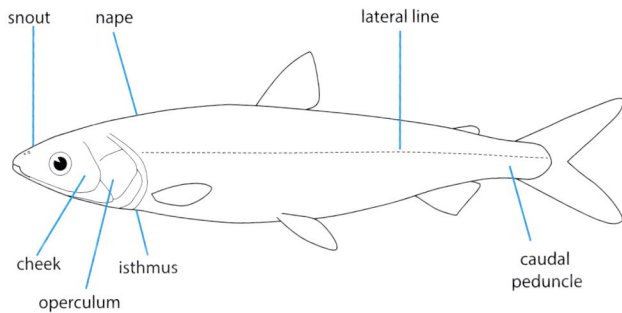

FIGURE 8 Some external regions of fishes have specific names: the **nape** is the part of the back that extends from the margin of the skull to the dorsal-fin origin (i.e., the region of the head above and behind the eyes); the **isthmus** is the area of the throat ventral to the gill openings; the **operculum** is a hard bony flap covering and protecting the gills; the **caudal peduncle** is the part of the body between the rear parts of the dorsal and anal fins, and the caudal fin. The **lateral line** is a sensory canal along the midside of the body with a series of pores through specialized lateral line scales.

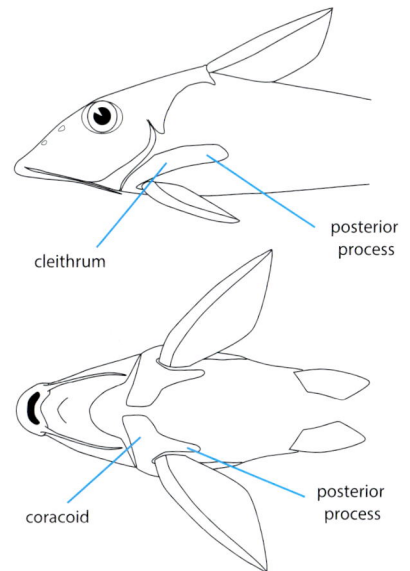

FIGURE 9 The pectoral girdle supports the pectoral fin and consists of the cleithral and coracoid bones. The posterior processes of the cleithrum and coracoid are visible under the skin in many catfishes; their shape can help to distinguish between genera and species.

cranium: the skull; with three major components: the neurocranium (cartilage bones that form the braincase), dermatocranium (dermal bones including skull roofing bones, opercular series, tooth bearing bones, and others) and the branchiocranium (mandibular and hyoid arches, and branchial or gill arches).

crenate: having an edge that is round-toothed or scalloped.

crenulate: having an edge that is highly scalloped or wavy.

cryptic coloration: coloring that conceals or disguises an animal's shape.

cryptic species: species that are morphologically indistinguishable, but that nevertheless represent different species based on genetic, chromosomal, or other biological differences.

ctenoid scale: scale in which the exposed part and/ or posterior margin have small, tooth-like spines or projections called *ctenii* (singular *cteni*) (see fig. 10). Giving a rough texture to the surface of the scales.

cusps or cuspids: principal projecting point of a tooth (see fig. 7f,g). Teeth may have a single point or cusp (unicuspid), or several cusps (multicuspid). Teeth with two points are called bicuspid, with three points tricuspid, etc.

cycloid scale: scale with smooth posterior margin, without spines on posterior margin (see fig. 10).

demersal: dwelling at or near the bottom.

dentary: the main tooth-bearing bone of the lower jaw (see figs. 6, 11).

denticle: small tooth-like structure.

dentigerous: bearing teeth.

depressed: flattened from top to bottom; body shape much wider than deep (see fig. 2).

dermal: pertaining to the skin.

detritivore: (or detritus feeder) an animal that feeds on detritus.

detritus: organic-rich debris formed from fragments of dead organic material, most typically of plant origin, but also of small animals.

dichromic: two color patterns, each with different coloration or ornamentation (usually referring to difference between male and female).

distal: near outer edge; far end from point of attachment or center of body. Opposite of proximal: the site where an appendage (e.g., a fin) joins the body (see fig. 3).

diverticulum: an out-pocketing or blind sac.

dorsal: toward the back or upper part of the body (see fig. 3).

dorsal fin: median fin along the back (see fig. 1).

dorsal-fin insertion: the position of the first (anterior) dorsal-fin ray.

dorsoventral: stretching from dorsal to ventral surface.

durophagous: consuming hard-shelled or exoskeleton-bearing organisms such as mollusks.

ectopterygoid: in teleost fishes, one of the bones on the roof of the mouth that connects the upper and lower oral jaws.

edentulous: without teeth.

elongate: extended or drawn out.

emarginate: margin slightly concave; used to describe the posterior margin of a caudal fin that is inwardly curved (see fig. 12).

embedded: covered by skin (usually refers to scales).

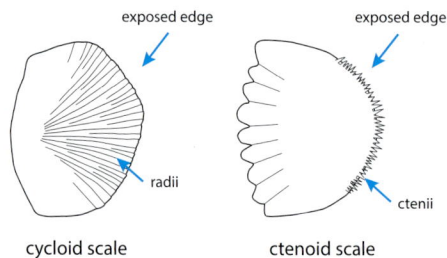

FIGURE 10 Cycloid or ctenoid scales are present in Teleostei, the vast majority of bony fishes. Ctenoid scales have ctenii (teeth) on the posterior (exposed) border; cycloid scales lack ctenii. However, including all scales with spines on their posterior margins under the term ctenoid is an oversimplification. Three different general types of spiny scales exist: (i) **crenate**, with simple marginal indentations and projections; (ii) **spinoid**, with spines continuous with the main body of the scale, and (iii) **stenoid**, with ctenii formed as separate ossifications distinct from the main body of scale (Roberts 1993).

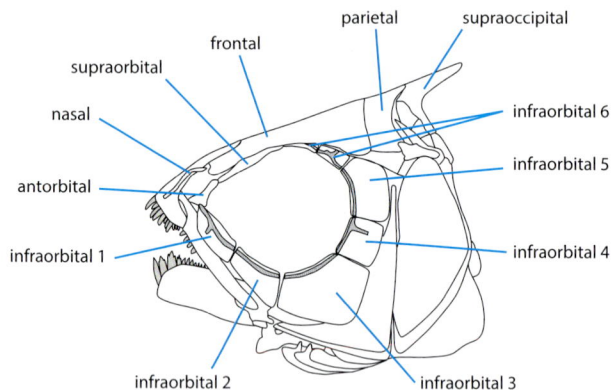

FIGURE 11 Head bones in a generalized characoid. The **infraorbital bones** surround the eye and have a laterosensory canal (indicated in gray on these bones). Most characiforms have 6 infraorbital bones; the first one is located just ventral to the antorbital and extends to a varying degree along the longitudinal length of this bone. The **supraorbital**, when present (absent in the Characidae), is situated dorsal or anterodorsal to the orbit. Illustration after Weitzman (1962).

endemic: in biogeography, native to, or with a geographic range restricted to, a certain region.

entire: smooth or straight margin.

estuary: part of the mouth or lower course of a river in which the river's current meets the sea's tide.

extant: a taxon whose members are living at the present time (opposite to extinct).

evertible: able to turn inside-out; e.g., evertible cheek odontodes of tribe Ancistrini, Loricariidae.

falcate (falciform): deeply indented or sickle-shaped, e.g., the edge of a fin.

filiform teeth: slender, thread-like teeth found in many loricariid catfishes.

fin formula (or fin count): of the three unpaired fins (dorsal, anal, and caudal), only the formula of the first two is usually given. The first two or three, sometimes more, unbranched (or simple) rays are numbered in small Roman numerals: ii, iii, iv, and so on; in certain cases there exists a strong, spinous ray, which is numbered in Roman capitals: I. The soft, branched rays are numbered in Arabic numbers. The formula of the two paired fins (pectoral and ventral/pelvic) usually shows one simple ray/spinous ray followed by a number of branched rays.

finlets: small separate dorsal and anal fins.

fish (pl. fishes): nontetrapod craniates; i.e., animals with a cranium, excluding terrestrial vertebrates.

fontanel: oval or triangular hole in the top of the skull covered only by skin.

foramen (pl. foramina): a small opening, orifice, or perforation. Used when something such as a nerve, blood vessel, or notochord passes through the opening.

forked: branched; in reference to the caudal fin, a fin shape with distinct upper and lower lobes, and the posterior margin of each lobe relatively straight or gently curved (see fig. 12).

frontal: a large pair of skull bones that articulate medially, generally located dorsal to the orbit (or eye) (see fig. 11).

hyaline: translucent, without marks.

hyoid: referring to the series of bones behind the gill cover that suspend the branchiostegal rays and connect to the gill arches.

furcate: forked.

fusiform: cigar-shaped or spindle-shaped, tapering at both ends (see fig. 2).

gape: margin, outline, or rim of the mouth opening.

gas bladder: see swim bladder.

genital papilla: a small, fleshy projection at the genital pore (located immediately behind the anus).

gill arch: (or branchial arch) the bony skeleton that supports the gill filaments and gill rakers.

gill cover: (or operculum) bony flap covering the outside of the gill chamber (see figs. 6, 8).

gill filaments: principal site of gas exchange in the gills (see fig. 13).

gill membrane: membranes along the posterior and ventral margin of the gill cover or operculum (see fig. 5).

gill rakers: bony tooth-like projections along the front edge of the gill arch (on the opposite side from the red gill filaments) that act as a sieve and help to prevent food from escaping through the gill opening (see fig. 13).

gonads: reproductive organs located inside the body cavity.

gonopodium: front anal-fin rays 3, 4, and 5 of male livebearers (Poeciliidae and Anablepidae), modified to serve as an intromittent organ for internal fertilization.

herbivorous: feeding on plant matter.

hermaphrodite: having both male and female sex organs (gonads) in one individual (not necessarily at the same time).

holotype: a single physical example (or illustration) of an organism designated as the official name-bearing specimen in the formal taxonomic description of a species.

humeral hiatus: see pseudotympanum.

humeral spot: a black or dark spot or blotch in the humeral region, at the upper edge of the pectoral-fin base (on the flank behind the opercle), occurring in many characids and in the gymnotiform *Sternopygus* (see fig. 4).

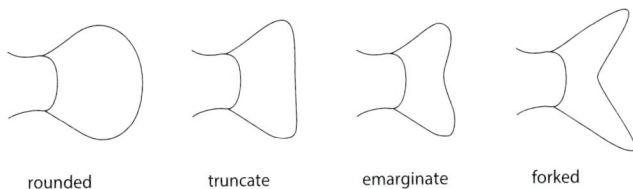

FIGURE 13 Gill rakers are present in some fishes. They are bony tooth-like projections along the front edge of the gill arch (on the opposite side from the red gill filaments) that help prevent food from escaping through the gill opening.

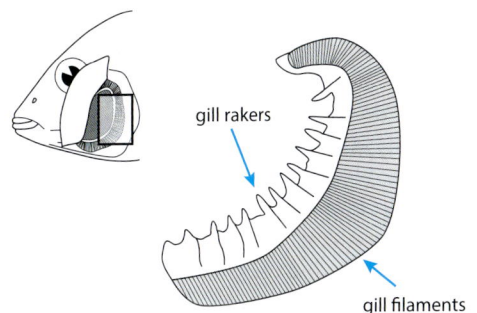

rounded   truncate   emarginate   forked

FIGURE 12 The shape of the caudal fin is highly variable and roughly related to swimming behavior, with fast-swimming fishes usually having deeply forked caudal fins with stiff upper and lower lobes, and slow-moving fishes often with rounded caudal fins.

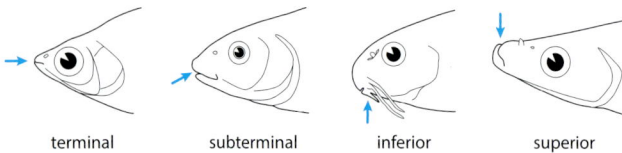

FIGURE 14 Mouth positions: a **terminal** mouth is located at the tip of the snout; a **subterminal** mouth is located below the tip of the snout; an **inferior** mouth opens ventrally and is positioned well posterior to the snout tip; a **superior** mouth opens dorsally.

terminal   subterminal   inferior   superior

hypertrophy (hypertrophied): increase in the volume of an organ or tissue due to the enlargement of its component cells.

hypobranchial: the ventralmost paired bones in each gill arch.

hypomaxilla: see supramaxilla.

hypural plate: bones that support the caudal-fin rays.

incisiform teeth: teeth that are compressed from front to back, and with the cutting edge straight or denticulate, similar to mammalian incisors (see fig. 7e).

inferior: usually referring to mouth position on underside of head (see fig. 14).

infraorbital bones: small, flat bones that border the orbit and contain the infraorbital laterosensory canal (see fig. 11).

infraorbital canal: the sensory canal of the lateral line system that runs around the eye within small flat bones called infraorbitals that border the orbit (see fig. 11).

inguinal: referring to the base of the pelvic fins (compare with axillary).

insertion: anterior or posterior point of attachment of a fin to the body.

integument: referring to the skin.

interdorsal: space on the back between the bases of the first and second dorsal fin, between the dorsal and adipose fins.

intermandibular: between the mandibles.

interopercle: lower anterior bone of the gill cover (see fig. 6).

Interorbital (distance): space on top of the head between the eyes.

intromittent organ: modified anal-fin rays used for depositing sperm into the female reproductive tract via internal fertilization (e.g., Auchenipteridae and Poeciliidae); see gonopodium.

isthmus: area of the throat ventral to the gill openings (see fig. 8).

jugular: pertaining to the throat region

juvenile: immature individual that looks externally like the adults of the species and different from larvae or fry.

keeled: a fleshy or bony ridge in the form of a keel.

labial: referring to structures that are part of the lips or mouth or to the position of the teeth inside the mouth (as opposed to lingual).

lacrimal: region between eye and nostril. The lacrimal bone of sarcopterygian (lobed-fin fishes), and the antorbital bone of actinopterygians (ray-fin fishes), is the anteriormost member of the infraorbital series.

lamella: a thin plate-like structure.

lanceolate: spear- or lance-shaped.

lateral: on or to the side of the body, either left or right (see fig. 3).

lateral band: a stripe of color along the side of the body (see fig. 4).

lateral line: a sensory canal along the midside of the body with a series of pores through specialized lateral line scales (see fig. 8). This canal is the posterior extension of the lateralis system and contains sense organs (neuromasts) that detect disturbances in the water, thereby helping fish to detect currents, capture prey, maintain position in a school, and avoid obstacles and predators.

lateral scale count: number of scales along lateral line, or along midline if lateral line is absent or incomplete.

lateral scales: row of scales along the lateral body surface from rear end of gill cover to base of caudal fin.

lateralis system: sensory system consisting of a series of pores and canals on the head, body, and sometimes caudal fin of a fish, used to detect vibrations in the water.

lentic: still or standing water, as in pools or lakes (compare with lotic).

lepidotrichia: bony, bilaterally paired, segmented fin rays found in bony fishes.

lingual: pertaining to the tongue; opposite of labial when referring to the position of teeth.

lotic: flowing or moving water, as in rivers and streams (compare with lentic).

lunate: crescent-shaped; caudal-fin shape that is deeply emarginate with narrow lobes.

mammiliform teeth: nipple-like teeth present outside mouth and on snout tip in scale-eating fishes.

mandible: the lower oral jaw, composed of three bones in boney fishes: dentary usually with teeth, articular that articulates with the quadrate bone of the suspensorium, and the angular.

mandibular: associated with the lower portion of the oral jaw.

maxilla (pl. maxillae): one of the two bones that make up each half of the upper jaw, displaced inward by the premaxilla in specialized Teleostei (see fig. 6). It may or may not bear teeth.

maxillary barbel: the tentacle-like protuberance attached to each end of the upper lip in catfishes.

medial: in the middle plane or axis of the body.

median: middle or toward the midline.

median fins: (or unpaired fins) those positioned on the midline (median plane): the dorsal, anal, and caudal fins (see fig. 1).

melanophore: cell containing melanin, a dark brown or black pigment.

membrane: a thin sheet of molecules (e.g., cell membrane) or cells; often refers to thin sheet of tissue between fin spines and rays.

mental: the mandibular symphysis at the tip of the chin.

mesethmoid: the most anterodorsal skull bone, at the front tip of the skull.

midlateral scales: referring to the longitudinal series of scales from the upper edge of the operculum or upper pectoral fin base to the base of the caudal fin. Often used in regard to fishes without a lateral line, e.g., gobies.

mimicry: morphological similarity of one species to another which protects one or both.

molar: a low, blunt, rounded tooth for crushing and grinding (see fig. 7d).

molariform teeth: see molar.

monophyletic group (or clade): a group formed by taxa descending from a single ancestor.

monotypic: a genus that includes only one species, or a family containing only one genus.

naked: without scales.

nape: the part of the back that extends from the margin of the skull to the dorsal-fin origin; also, the region of the head above and behind the eyes (see fig. 8).

nares: nasal opening or nostril (see fig. 6).

neotype: the specimen selected to serve as a replacement for the holotype if the original type or types have been lost or destroyed.

neurocranium: one of three main divisions of the fish skull, constituting the cartilage bones that form the braincase. The other skull divisions are the dermatocranium (dermal bones including skull roofing bones, opercular series, tooth-bearing bones, and others), and branchiocranium (including the mandibular and hyoid arches, and the branchial or gill arches).

niche: (1) The ecological role of a species in an ecosystem; (2) the set of resources it consumes and habitats it occupies; (3) the functional position of an organism in a community, including its interactions with all physical, chemical, and biological components of the environment.

nocturnal: active at night.

nostril: nasal opening or nares (see fig. 6).

nuchal: related to, or situated in the region of the nape (dorsal surface just behind the head).

nuptial: associated with, or occurring during the breeding season.

occipital process: extension of the (supra-)occipital bone at the rear edge of the skull.

occiput: back part of the head or skull where it contacts the first vertebra.

ocellus: a round eye-like spot or marking; usually dark, bordered by a light ring.

ocular side: in flatfishes, the side where the eyes are located; the opposite side is called the blind side.

omnivorous: feeding on both plant and animal matter.

opercle: the large posterior upper bone of the gill cover (see fig. 6).

operculum: the hard bony flap covering and protecting the gills (see figs. 6, 8). In most fish, the rear edge of the operculum roughly marks the division between the head and the body. The operculum is composed of four bones: the opercle, preopercle, interopercle, and subopercle (see fig. 6).

orbit: the cavity or socket of the head in which the eye is situated.

orbital: referring to the eye.

orifice: any opening or hole in the body, including the mouth, gills, anus, and urogenital pore.

ossified: turned into bone or bony tissue.

oviparity (oviparous): reproduction mode in which eggs are fertilized externally and developments occurs outside the female's body.

ovoviviparity (ovoviviparous): reproduction mode in which embryos develop inside eggs that remain in the mother's body until they are ready to hatch.

paired fin: fins found on both sides of the body; the pectoral and pelvic fins (see fig. 1).

palate: the roof of the mouth.

palatine: one of a pair of bones on the roof of the mouth, behind and lateral to the vomer, sometimes bearing teeth (see fig. 15).

papilla: small lumps of dermal tissue, containing taste buds. In fish, they are typically found on the floor of the mouth, or on the upper lip.

papillose: with papillae.

parietal: a bone of the upper posterior part of the skull.

pectoral fin: paired fins behind the gill cover, attached to the shoulder girdle (see fig. 1).

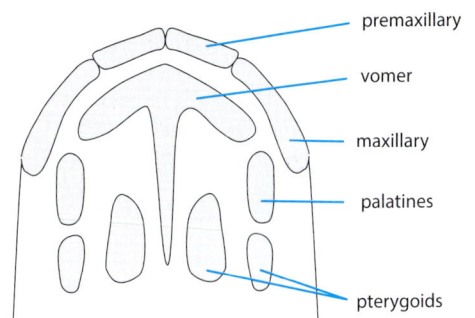

FIGURE 15 Tooth-bearing bones of the roof of the mouth.

pelagic fish: living near the surface or in the water column, but not on the bottom.

pelvic fins: (or ventral fins) paired fins on the lower part of the body in front of the anus, attached to the pelvis (see fig. 1).

perennial stream: (or perennial river) has continuous flow in parts of its streambed year-round during years of normal rainfall.

periphyton: (or aufwuchs) organisms attached to or clinging to the stems and leaves of plants or other objects projecting above bottom sediments.

pharyngeal: of, or near, the pharynx, at back of mouth and throat.

pharyngeal teeth: teeth in the throat (i.e., on the elements of the last gill arch or pharyngeal arch).

phylogeny: the history of the evolution of a species or group, especially in reference to lines of descent and relationships among broad groups of organisms (i.e., the relationships of groups of organisms as reflected in their evolutionary history).

piscivorous: fish eating.

pit lines: very tiny holes in the skin, usually situated on the top and sides of the head.

plankton: diverse group of small organisms that live suspended in the water column. When autotrophic (i.e., photosynthesizing) they are called phytoplankton, whereas small animals are called zooplankton.

polyandry: a form of polygamy in which females mate with two or more males during the same mating season.

polygamy: in animals, a pattern of mating in which an individual has more than one sexual partner.

polygyny: in animals, a pattern of mating in which a male has more than one female partner.

polyphyletic group: groups containing the descendants of different ancestors; often characterized by one or more characters of convergent evolution; e.g., elongate "eel-shaped" fishes; "basses."

pore: tiny opening in the skin, usually involved with sensory perception in fishes; e.g., narial, laterosensory.

pored scale: scale with a pore, e.g., the lateral line scales.

postcleithrum (plural postcleithra): dermal bone of the pectoral girdle posterior to the cleithrum.

posterior: pertaining to the rear body portion, toward the tail (see fig. 3).

postorbital: behind the eyes.

predorsal scales: the series of scales along the middorsal line between the head and the origin of the dorsal fin.

premaxilla: anterior bone in upper jaw; in teleost fishes it extends backward and bears most or all of the teeth of the upper jaw (see fig. 6).

preopercle: a boomerang-shaped, upper anterior bone of the gill cover, the edges of which form the posterior and lower margins of the cheek (see fig. 6). The upper vertical margin is sometimes called the upper limb, and the lower horizontal edge the lower limb.

preopercular spine(s): spine(s) along the rear or lower edges of the preopercle.

preorbital: region in front of the eye.

principal caudal-fin ray: large branched and unbranched caudal-fin rays that reach the rear margin of the fin.

procurrent caudal-fin rays: small, nonsegmented rays anterior to the large, segmented caudal-fin rays (see fig. 16).

protractile mouth: protrusible mouth (see fig. 17).

protrusible: capable of projection; in reference to mouth with upper lip not attached to the snout, and which may be extended far forward to catch prey (see fig. 17).

proximal: the site where an appendage (e.g., a fin) joins the body. Opposite of distal: the point furthest from the point of attachment to the body (see fig. 3).

pseudotympanum: a translucent, triangular area in the humeral region, behind the opercle, without muscle and where the anterior part of the gas bladder may be in direct contact with the skin. It is hypothesized to function as a hearing aid in many juvenile characoids and is present in adults of certain species, notably members of the Cheirodontinae and Gymnotiformes.

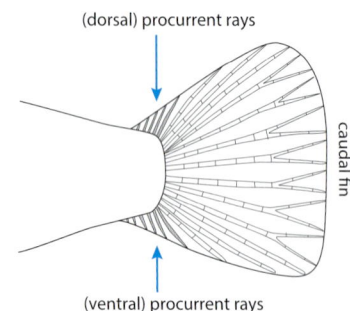

(dorsal) procurrent rays

caudal fin

(ventral) procurrent rays

FIGURE 16 Procurrent rays are accessory rays that are defined as the nonsegmented rays anterior to the large segmented caudal-fin rays. Their number and shape is sometimes used to differentiate between species and genera.

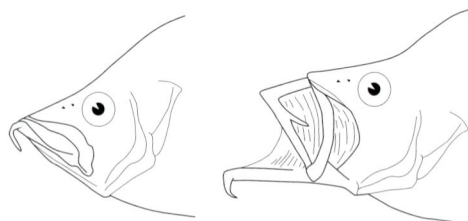

FIGURE 17 In most teleost fishes the upper jaw is highly mobile and the upper lip is not firmly attached to the snout. In some of these species the upper jaw may be extended far forward, to catch prey (a protractile or protrusible mouth). Illustrated is the extremely protrusible jaw in the Amazon leaf-fish (*Monocirrhus polyacanthus*); right figure after Liem (1970).

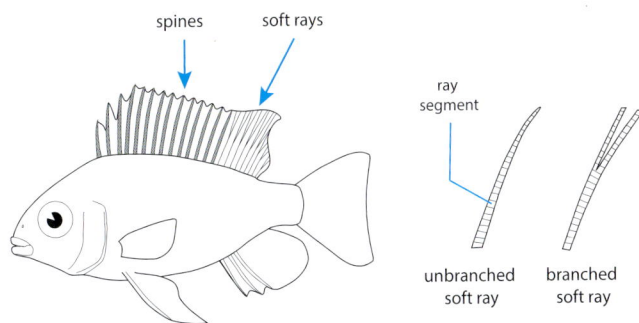

FIGURE 18 Fins can have spines and soft rays. Spines lack segments, are stiff and never branched. Soft rays have well-defined segments, can be branched or unbranched, and are typically flexible.

pterygiophores: bones that support the median dorsal and anal fins. There are two to three of them: proximal, middle, and distal. In spinous fins, the distal is often fused to the middle or not present at all.

pterygoid bones: paired sets of bones forming part of the palate (roof of the mouth); bearing teeth in some fishes (referred to a pterygoid teeth; see fig. 15).

radial: (1) A bony or cartilaginous support for paired fins. Usually three: proximal, middle, and distal radials. If the proximal radial is notably larger it is spoken of as a basal. (2) Direction away from the axis running anteroposteriorly through the middle of the organism or structure; peripheral.

radii (of scales): furrows in the scales, usually radiating from their center (see fig. 10).

ray: segmented supporting bony element of fins (see fig. 18).

reticulate: color markings forming a net-like or chain-like pattern.

rheophilic: organisms that inhabit swiftly flowing waters; riffles and rapids in English; *cascatas* and *cachoeiras* in Portuguese; *cascadas* and *rápidos* in Spanish.

rictus: the corner of the mouth where the upper and lower jaws meet.

riffle: fast-flowing, shallow segment of a stream or river where the surface of the water is broken over rocks or debris.

riparian (riverine): pertaining to a riverbank.

riverine: pertaining to a river.

rostral: toward the front (nose) end of the fish.

rostrum: the area of the snout; an extension of the snout.

rudimentary: (or vestigial) imperfectly or incompletely developed.

saddle: color mark, more or less ovoid or rectangular, across the dorsal midline (back).

sagittal plane: vertical plane that passes from anterior to posterior, dividing the body into right and left halves.

scalation: arrangement of scales; squamation.

scapula: a flat bone on the upper part of the pectoral girdle.

scapular: pertaining to the shoulder region.

scute: thickened, hardened scale, or external bony plate.

serrate: with saw-like teeth along a margin.

sexual dimorphism: external differences between the two sexes of a species (e.g., in size or color).

simple fin ray: single element, unbranched for all or most of its length.

SL: (standard length) length from tip of snout to the base of the caudal fin.

snout: the area of the head in front of the eye and above the mouth (see fig. 8).

soft ray: a fin support element that is segmented and usually flexible and branched (see fig. 18).

spatulate: flattened.

sphenotic: a bone of the skull between and behind the eyes.

spine: a sharp bony projection; hard, unbranched ray in a fin (see fig. 18).

spinules: small spines or thorns.

spiracle: opening, usually behind the eye, of a duct leading to the gill chamber in sharks and rays.

spot: small circular or irregular color mark (see fig. 4).

stripe: elongate, straight-sided, horizontal color marking (see fig. 4).

subopercle: lower rear bone in the gill cover (see fig. 6).

suborbital: the area below the eye.

suborbital bones: bones below the eye sockets (see lachrymal).

subterminal: usually referring to the mouth position, with the opening of the mouth positioned posterior and slightly below the tip of the snout (see fig. 14).

superior: usually referring to the mouth position, being directed upward, with the snout behind the anterior opening of the mouth (see fig. 14).

supra: above

supracleithrum: paired dorsal-most bone of the secondary pectoral girdle above the cleithrum.

supramaxilla: (also known as hypomaxilla d) 1 or 2 bones above the maxilla, between the posterior tip of the premaxilla and the expanded blade of the maxilla, found in primitive bony fishes and retained in Clupeiformes.

supraoccipital: a bone on the dorsal side of the great foramen of the skull, usually forming a part of a fused occipital region in adults, often presenting a distinct ossification in juveniles.

supraorbital: the region bordering the upper edge of the eye.

suprapreopercle: small, paired dermal tube bone carrying the lateral line canal across the gap between the preopercle and supratemporal, often absent.

supratemporal canal: portion of the lateral line system that connects the lateral canals of the two sides by crossing the top of the head at the occiput or behind that point.

suture: rigid joint between hard parts of animals.

swim bladder: (or air bladder, gas bladder) a gas-filled sac located in the abdominal cavity, used by different fishes in one or several functions, including: buoyancy, respiration, sound production, and sound detection.

sympatric: occurring in same geographic area (compare with allopatric).

symphysis: the articulation of paired bones at or along the midline (e.g., mental symphysis of the two dentaries at the chin).

synonym: an invalid scientific name for a species that goes by a different (previously accepted) scientific name.

terminal: pertaining to the end. Mouth terminal: opening of the mouth even with the tip of the snout, the upper and lower jaws being equally far forward (see fig. 14).

thoracic: position of the pelvic-fin insertion generally below or slightly posterior to the pectoral-fin insertion.

thorax (thoracic): referring to the anterior region of the body cavity.

TL: (total length) length from tip of snout to the tip of the caudal fin with the caudal-fin lobe pressed flat.

trans-Andean: west of the Andes (vs. cis-Andean: east of the Andes).

transverse scales: row of scales from anal-fin origin to dorsal fin (or middle of back).

truncate: terminating abruptly in a square or flattened end (see fig. 12).

trunk: portion of the body between the head and the anus.

tubercle: epidermal projection on head, body, or fins of males of some species that facilitates contact with females during spawning or is used during aggressive interactions with other fish.

type locality: the locality where the type specimen was collected.

type specimen: (or holotype) an individual plant or animal specimen chosen by taxonomists to serve as the basis for naming and describing a new species or variety.

unculiferous: bearing unculi (horny projections arising from a single cell).

uniserial: arranged in a single row.

unpaired fins: fins on the midline of the body: the dorsal, adipose, caudal, and anal fins (see fig. 1).

vent: anus.

ventral: below; on the lower half of the body, or abdominal part of the body (see fig. 3). Opposite of dorsal.

ventral fins: pelvic fins (see fig. 1).

vertical fins: (or median fins, or unpaired fins) the dorsal, caudal, and anal fins (see fig. 1).

vestige: (or rudimental) a degenerate or imperfectly developed organ or structure that has little or no apparent utility, but that in an earlier stage of the individual or in preceding evolutionary forms of the organism performed a useful function.

villiform teeth: slender teeth forming velvety or brush-like bands (see fig. 7c).

viviparity: reproduction mode in which development and growth of the embryo occur internally with nourishment from the mother.

vomer: an unpaired median bone in the front of the roof of the mouth (occasionally bearing teeth, called vomerine teeth; see fig. 15).

Weberian ossicles: chain of small bones connecting the swim bladder to the inner ear for sound transmission. The possession of Weberian ossicles characterizes ostariophysan fishes, including the following groups naturally occurring in the AOG region: Characiformes, Siluriformes, Gymnotiformes.

Abe, K. T., T. C. Mariguela, G. S. Avelino, R. M. C. Castro, and C. Oliveira. 2013. Multilocus molecular phylogeny of Gasteropelecidae (Ostariophysi: Characiformes) reveals the existence of an unsuspected diversity. Molecular Phylogenetics and Evolution **69**:1209–1214.

Abe, K. T., T. C. Mariguela, G. S. Avelino, F. Foresti, and C. Oliveira. 2014. Systematic and historical biogeography of the Bryconidae (Ostariophysi: Characiformes) suggesting a new rearrangement of its genera and an old origin of Mesoamerican ichthyofauna. BMC Evolutionary Biology **14**.

Abell, R., M. L. Thieme, C. Revenga, M. Bryer, M. Kottelat, N. Bogutskaya, B. Coad, N. Mandrak, S. C. Balderas, W. Bussing, M. L. J. Stiassny, P. Skelton, G. R. Allen, P. Unmack, A. Naseka, R. Ng, N. Sindorf, J. Robertson, E. Armijo, J. V. Higgins, T. J. Heibel, E. Wikramanayake, D. Olson, H. L. López, R. E. Reis, J. G. Lundberg, M. H. Sabaj Pérez, and P. Petry. 2008. Freshwater ecoregions of the world: a new map of biogeographic units for freshwater biodiversity conservation. BioScience **58**:403–414.

Adriaens, D., J. N. Baskin, and H. Coppens. 2010. Evolutionary morphology of trichomycterid catfishes: about hanging on and digging in. Pages 337–362 in J. S. Nelson, H. P. Schultze, and M. V. H. Wilson, editors. Origin and Phylogenetic Interrelationships of Teleosts. Verlag Dr. Friedrich Pfeil, München, Germany.

Adriaens, D., T. Geerinckx, J. Vlassenbroeck, L. Van Hoorebeke, and A. Herrel. 2009. Extensive jaw mobility in suckermouth armored catfishes (Loricariidae): A morphological and kinematic analysis of substrate scraping mode of feeding. Physiological and Biochemical Zoology **82**:51–62.

Agorreta, A., D. San Mauro, U. Schliewen, J. L. van Tassell, M. Kovacic, R. Zardoya, and L. Rueber. 2013. Molecular phylogenetics of Gobioidei and phylogenetic placement of European gobies. Molecular Phylogenetics and Evolution **69**:619–633.

Agostinho, A. A., L. C. Gomes, H. I. Suzuki, and H. F. Júlio Jr. 2003. Migratory fishes of the Upper Paraná River Basin, Brazil. Pages 19–98 in J. Carolsfeld, B. Harvey, C. Ross, and A. Baer, editors. Migratory fishes of South America: Biology, Fisheries and Conservation Status. IDRC and World Bank.

Agostinho, A. A., E. E. Marques, C. S. Agostinho, D. A. Almeida, R. J. Oliveira, and J. R. B. d. Melo. 2007. Fish ladder of Lajeado Dam: migrations on one-way routes? Neotropical Ichthyology **5**:121–130.

Agostinho, A. A., F. M. Pelicice, and L. C. Gomes. 2008. Dams and the fish fauna of the Neotropical region: impacts and management related to diversity and fisheries. Brazilian Journal of Biology **68**:1119–1132.

Agostinho, C. S., A. A. Agostinho, E. E. Marques, and L. M. Bini. 1997. Abiotic factors influencing piranha attacks on netted fish in the upper Parana River, Brazil. North American Journal of Fisheries Management **17**:712–718.

Aguilera, G., J. M. Mirande, and M. M. Azpelicueta. 2011. A new species of Heptapterus Bleeker 1858 (Siluriformes, Heptapteridae) from the río Salí basin, north-western Argentina. Journal of Fish Biology **78**:240–250.

Aguilera, G., J. M. Mirande, and M. M. de las Azpelicueta. 2009. A new species of Cnesterodon (Cyprinodontiformes: Poeciliidae) from a small tributary of arroyo Cuñá-Pirú, río Paraná basin, Misiones, Argentina. Zootaxa **2195**:34–42.

Akagi, H., O. Malm, Y. Kinjo, M. Harada, F. J. P. Branches, W. C. Pfeiffer, and H. Kato. 1995. Methylmercury pollution in the Amazon, Brazil. Science of the Total Environment **175**:85–95.

Akama, A., and C. J. Ferraris Jr. 2011. Spinipterus, a new genus of small, spiny catfish (Siluriformes: Auchenipteridae) from the Peruvian Amazon. Zootaxa **2992**:52–60.

Akama, A., and C. J. Ferraris Jr. 2003. Entomocorus malaphareus, a new species of auchenipterid catfish (Osteichthyes: Siluriformes) from the lower and middle reaches of the rio Amazonas. Neotropical Ichthyology **1**:77–82.

Albert, J. S. 2001. Species diversity and phylogenetic systematic of American knifefishes (Gymnotiformes, Teleostei). Miscellaneous Publication, Museum of Zoology, University of Michigan **190**:1–127.

Albert, J. S. 2003. Family Sternopygidae. Pages 493–497 in R. E. Reis, S. O. Kullander, and C. J. Ferraris Jr., editors. Check List of the Freshwater Fishes of South and Central America. EDIPUCRS, Porto Alegre, Brazil.

Albert, J. S., H. L. Bart Jr., and R. E. Reis. 2011a. Species richness and cladal diversity. Pages 89–104 in J. S. Albert and R. E. Reis, editors. Historical Biogeography of Neotropical Freshwater Fishes. University of California Press, Berkeley.

Albert, J. S., T. P. Carvalho, P. Petry, M. A. Holder, E. L. Maxime, J. Espino, I. Corahua, R. Quispe, B. Rengifo, H. Ortega, and R. E. Reis. 2011b. Aquatic biodiversity in the Amazon: habitat specialization and geographic isolation promote species richness. Animals **1**:205–241.

Albert, J. S., and W. Crampton. 2003. Family Hypopomidae. Pages 494–502 in R. E. Reis, S. O. Kullander, and C. J. Ferraris Jr., editors. Check List of the Freshwater Fishes of South and Central America. EDIPUCRS, Porto Alegre, Brazil.

Albert, J. S., and W. G. R. Crampton. 2001. Five new species of Gymnotus (Teleostei: Gymnotiformes) from an upper Amazon floodplain, with descriptions of electric organ discharges and ecology. Ichthyological Exploration of Freshwaters **12**:241–266.

Albert, J. S., and W. G. R. Crampton. 2005. Electroreception and electrogenesis. Pages 431–470 in D. H. Evans and J. B. Claiborne, editors. The Physiology of Fishes, 3rd edition. CRC Press, Boca Raton, FL.

Albert, J. S., and W. G. R. Crampton. 2006. Pariosternarchus amazonensis: A new genus and species of Neotropical electric fish (Gymnotiformes: Apteronotidae) from the Amazon River. Ichthyological Exploration of Freshwaters **17**:267–274.

Albert, J. S., and W. G. R. Crampton. 2009. A new species of electric knifefish, genus Compsaraia (Gymnotiformes: Apteronotidae) from the Amazon River, with extreme sexual dimorphism in snout and jaw length. Systematics and Biodiversity **7**:81–92.

Albert, J. S., W. G. R. Crampton, D. H. Thorsen, and N. R. Lovejoy. 2005. Phylogenetic systematics and historical biogeography of the Neotropical electric fish Gymnotus (Teleostei: Gymnotiformes). Systematic and Biodiversity **2**:375–417.

Albert, J. S., and W. L. Fink. 2007. Phylogenetic relationships of fossil Neotropical electric fishes (Osteichthyes: Gymnotiformes) from the Upper Miocene of Bolivia. Journal of Vertebrate Paleontology **27**:17–25.

Albert, J. S., D. M. Johnson, and J. H. Knouft. 2009. Fossils provide better estimates of ancestral body size than do extant taxa in fishes. Acta Zoologica **90**:357–384.

Albert, J. S., P. Petry, and R. E. Reis. 2011c. Major biogeographic and phylogenetic patterns. Pages 21–58 *in* J. S. Albert and R. E. Reis, editors. Historical Biogeography of Neotropical Freshwater Fishes. University of California Press, Berkeley, USA.

Albert, J. S., and R. E. Reis, editors. 2011. Historical Biogeography of Neotropical Freshwater Fishes. University of California Press, Berkeley, USA.

Alcaraz, H. S. V., W. J. Da Graça, and O. A. Shibatta. 2008. *Microglanis carlae*, a new species of bumblebee catfish (Siluriformes: Pseudopimelodidae) from the río Paraguay basin in Paraguay. Neotropical Ichthyology **6**:425–432.

Alexander, R. M. 1965. Structure and function in the catfish. Journal of Zoology **148**:88–152.

Alexandrou, M. A., C. Oliveira, M. Maillard, R. A. R. McGill, J. Newton, S. Creer, and M. I. Taylor. 2011. Competition and phylogeny determine community structure in Mullerian co-mimics. Nature **469**:84-88.

Alkins-Koo, M. 2000. Reproductive timing of fishes in a tropical intermittent stream. Environmental Biology of Fishes **57**:49–66.

Almeida, R. G. 1984. Biologia alimentar de três espécies de *Triportheus* (Pisces: Characoidei, Characidae) do lago Castanho, Amazonas. Acta Amazonica **14**:48–76.

Almeida, V. L. L. D., N. S. Hahn, and A. E. A. D. M. Vazzoler. 1997. Feeding patterns in five predatory fishes of the high Parana River floodplain (PR, Brazil). Ecology of Freshwater Fish **6**:123–133.

Almeida-Silva, P. H., and R. Mazzoni. 2014. Life history aspects of *Phalloceros anisophallos* Lucinda, 2008 (Osteichthyes, Poeciliidae) from Córrego Andorinha, Ilha Grande (RJ, Brazil). Studies on Neotropical Fauna and Environment **49**:191–198.

Almirón, A., J. Casciottai, J. Bechara, F. Ruíz Díaz, C. Bruno, S. D'Ambrosio, P. Solimano, and P. Soneira. 2007. *Imparfinis mishky* (Siluriformes, Heptapteridae) a new species from the rios Parana and Uruguay basins in Argentina. Revue Suisse de Zoologie **114**:817–824.

Alvarenga, É. R., N. Bazzoli, G. B. Santos, and E. Rizzo. 2006. Reproductive biology and feeding of *Curimatella lepidura* (Eigenmann & Eigenmann) (Pisces, Curimatidae) in Juramento reservoir, Minas Gerais, Brazil. Revista Brasileira de Zoologia **23**:314–322.

Alves, C. B. M., F. Vieira, A. L. B. Magalhaes, and M. F. G. Brito. 2007. Impacts of non-native fish species in Minas Gerais, Brazil: present situation and prospects. Pages 291–314 *in* T. M. Bert, editor. Ecological and Genetic Implications of Aquaculture Activities. Kluwer Scientific Publications, Dordrecht, The Netherlands.

Alves, G. H. Z., R. M. Tófoli, G. C. Novakowski, and N. S. Hahn. 2011. Food partitioning between sympatric species of *Serrapinnus* (Osteichthyes, Cheirodontinae) in a tropical stream. Acta Scientiarum–Biological Sciences **33**:153–159.

Anderson, J. T., J. S. Saldana-Rojas, and A. S. Flecker. 2009. High-quality seed dispersal by fruit-eating fishes in Amazonian floodplain habitats. Oecologia **161**:279–290.

Andrade, M. C. 2013. Revisão taxonômica de *Tometes* Valenciennes, 1850 (Characiformes: Serrasalmidae) das drenagens do Escudo das Guianas. Unpublished MSc thesis. Universidade Federal do Pará, Belém, Brazil.

Andrade, M. C., T. Giarrizzo, and M. Jégu. 2013. *Tometes camunani* (Characiformes: Serrasalmidae), a new species of phytophagous fish from the Guiana Shield, rio Trombetas basin, Brazil. Neotropical Ichthyology **11**:297–306.

Andrade, M. C., M. Jégu, and T. Giarrizzo. 2016a. A new large species of *Myloplus* (Characiformes, Serrasalmidae) from the Rio Madeira basin, Brazil. ZooKeys **571**:153–167.

Andrade, M. C., M. Jégu, and T. Giarrizzo. 2016b. *Tometes kranponhah* and *Tometes ancylorhynchus* (Characiformes: Serrasalmidae), two new phytophagous serrasalmids, and the first *Tometes* species described from the Brazilian Shield. Journal of Fish Biology **89**:467–494.

Andrade, M. C., A. J. S. Jesus, and T. Giarrizzoa. 2015. Length-weight relationships and condition factor of the eaglebeak pacu *Ossubtus xinguense* Jégu, 1992 (Characiformes, Serrasalmidae), an endangered species from Rio Xingu rapids, northern Brazil. Brazilian Journal of Biology **75**:102–105.

Aquino, A. E., and S. A. Schaefer. 2002. Revision of *Oxyropsis* Eigenmann and Eigenmann, 1889 (Siluriformes, Loricariidae). Copeia:374–390.

Andrade-López, J., and A. Machado-Allison. 2009. Aspectos morfológicos y ecológicos de las especies de Heptapteridae y Auchenipteridae presentes en el Morichal Nicolasito (Río Aguaro, Estado Guárico, Venezuela). Boletín de la Academia de Ciencias Físicas Matemáticas y Naturales de Venezuela **69**:35–52.

Andrade-Tubino, M. F., A. L. R. Ribeiro, and M. Vianna. 2008. Organização espaço-temporal das ictiocenoses demersais nos ecossistemas estuarinos brasileiros: uma síntese. Oecologia Brasiliensis **12**:640–661.

Aquino, A. E., and S. A. Schaefer. 2010. Systematics of the genus *Hypoptopoma* Günther, 1868 (Siluriformes, Loricariidae). Bulletin of the American Museum of Natural History **336**:1–110.

Aranha, J. M. R., J. H. C. Gomes, and F. N. O. Fogaça. 2000. Feeding of two sympatric species of *Characidium*, *C. lanei* and *C. pterostictum* (Characidiinae) in a coastal stream of Atlantic forest (southern Brazil). Brazilian Archives of Biology and Technology **43**:527–531.

Araujo-Lima, C., and M. Goulding. 1997. So Fruitful a Fish: Ecology, Conservation, and Aquaculture of the Amazon's Tambaqui. Columbia University Press, New York.

Arbour, J. H., and H. López-Fernández. 2011. *Guianacara dacrya*, a new species from the rio Branco and Essequibo river drainages of the Guiana Shield (Perciformes: Cichlidae). Neotropical Ichthyology **9**:87–96.

Arbour, J. H., R. E. B. Salazar, and H. López-Fernández. 2014. A new species of *Bujurquina* (Teleostei: Cichlidae) from the río Danta, Ecuador, with a key to the species in the genus. Copeia:79–86.

Arce, H. M., R. E. Reis, A. J. Geneva, and M. H. Sabaj Pérez. 2013. Molecular phylogeny of thorny catfishes (Siluriformes: Doradidae). Molecular Phylogenetics and Evolution **67**:560–577.

Arcila, D., R. P. Vari, and N. A. Menezes. 2013. Revision of the Neotropical genus *Acrobrycon* (Ostariophysi: Characiformes: Characidae) with description of two new species. Copeia:604–611.

Ardila-Rodríguez, C. A. 1978. Contribución al estudio de la ictiogeografia venezolana de la *Piabucina pleurotaenia* (Regan). Memoria de la Sociedad de Ciencias Naturales La Salle **38**:77–84.

Arias, A. L., and D. Reznick. 2000. Life history of *Phalloceros caudiomaculatus*: a novel variation on the theme of livebearing in the family Poeciliidae. Copeia:792–798.

Armbruster, J. W. 1997. Phylogenetic relationships of the sucker-mouth armored catfishes (Loricariidae) with particular emphasis on the Ancistrinae, Hypostominae, and Neoplecostominae. Unpublished PhD thesis. University of Illinois, Urbana-Champaign.

Armbruster, J. W. 1998a. Modifications of the digestive tract for holding air in loricariid and scoplacid catfishes. Copeia:663–675.

Armbruster, J. W. 1998b. Phylogenetic relationships of the suckermouth armored catfishes of the *Rhinelepis* group (Loricariidae: Hypostominae). Copeia:620–636.

Armbruster, J. W. 2002. *Hypancistrus inspector*: a new species of suckermouth armored catfish (Loricariidae: Ancistrinae). Copeia:86–92.

Armbruster, J. W. 2003. The species of the *Hypostomus cochliodon* group (Siluriformes: Loricariidae). Zootaxa **249**:1–60.

Armbruster, J. W. 2004a. Phylogenetic relationships of the suckermouth armoured catfishes (Loricariidae) with emphasis on the Hypostominae and the Ancistrinae. Zoological Journal of the Linnean Society **141**:1–80.

Armbruster, J. W. 2004b. *Pseudancistrus sidereus*, a new species from southern Venezuela (Siluriformes: Loricariidae) with a redescription of *Pseudancistrus*. Zootaxa **628**:1–15.

Armbruster, J. W. 2005. The loricariid catfish genus *Lasiancistrus* (Siluriformes) with descriptions of two new species. Neotropical Ichthyology **3**:549–569.

Armbruster, J. W. 2008. The genus *Peckoltia* with the description of two new species and a reanalysis of the phylogeny of the genera of the Hypostominae (Siluriformes: Loricariidae). Zootaxa **1822**:1–76.

Armbruster, J. W. 2011. Global catfish biodiversity. American Fisheries Society Symposium **77**:15–37.

Armbruster, J. W., and M. Hardman. 1999. Redescription of *Pseudorinelepis genibarbis* (Loricariidae: Hypostominae) with comments on behavior as it relates to air-holding. Ichthyological Exploration of Freshwaters **10**:53–61.

Armbruster, J. W., N. K. Lujan, and D. C. Taphorn. 2007. Four new *Hypancistrus* (Siluriformes: Loricariidae) from Amazonas, Venezuela. Copeia:62–79.

Armbruster, J. W., and L. M. Page. 1996. Redescription of *Aphanotorulus* (Teleostei: Loricariidae) with description of one new species, *A. ammophilus*, from the río Orinoco Basin. Copeia:379–389.

Armbruster, J. W., and L. M. Page. 2006. Redescription of *Pterygoplichthys punctatus* and description of a new species of *Pterygoplichthys* (Siluriformes: Loricariidae). Neotropical Ichthyology **4**:401–409.

Armbruster, J. W., and F. Provenzano. 2000. Four new species of the suckermouth armored catfish genus *Lasiancistrus* (Loricariidae: Ancistrinae). Ichthyological Exploration of Freshwaters **11**:241–254.

Armbruster, J. W., M. H. Sabaj, M. Hardman, L. M. Page, and J. H. Knouft. 2000. Catfish genus *Corymbophanes* (Loricariidae: Hypostominae) with description of one new species: *Corymbophanes kaiei*. Copeia:997–1006.

Armbruster, J. W., and D. C. Taphorn. 2011. A new genus and species of weakly armored catfish from the upper Mazaruni River, Guyana (Siluriformes: Loricariidae). Copeia:46–52.

Armbruster, J. W., and D. C. Taphorn. 2013. Description of *Neblinichthys peniculatus*, a new species of loricariid catfish from the río Paragua drainage of Venezuela. Neotropical Ichthyology **11**:65–72.

Armbruster, J. W., D. C. Werneke, and M. Tan. 2015. Three new species of saddled loricariid catfishes, and a review of *Hemiancistrus*, *Peckoltia*, and allied genera (Siluriformes). ZooKeys:97–123.

Arratia, G. A. 1990. The South American Trichomycterinae (Teleostei: Siluriformes), a problematic group. Pages 395–403 *in* G. Peters and R. Hutterer, editors. Vertebrates in the Tropics. Museum Alexander Koenig, Bonn.

Arratia, G. A., and L. Huaquin. 1995. Morphology of the lateral lines system and of the skin of diplomystid and certain primitive loricarioid catfishes and systematic and ecological considerations. Bonner Zoologische Monographien **36**:1–110.

Arrington, D. A., and K. O. Winemiller. 2003. Diel changeover in sandbank fish assemblages in a Neotropical floodplain river. Journal of Fish Biology **63**:442–459.

Assunção, M. I. S., and H. O. Schwassmann. 1995. Reproduction and larval development of *Electrophorus electricus* on Marajó Island (Pará, Brazil). Ichthyological Exploration of Freshwaters **6**:175–184.

Axelrod, H. R. 1993. The most complete colored lexicon of cichlids. T.F.H. Publications, Neptune City, NJ, USA.

Azevedo, P. G., R. M. C. Melo, and R. J. Young. 2011. Feeding and social behavior of the piabanha, *Brycon devillei* (Castelnau, 1855) (Characidae: Bryconinae) in the wild, with a note on following behavior. Neotropical Ichthyology **9**:807–814.

Azpelicueta, M. M., G. Aguilera, and J. M. Mirande. 2011. *Heptapterus mbya* (Siluriformes: Heptapteridae), a new species of catfish from the Parana river basin, in Argentina. Revue Suisse De Zoologie **118**:319–327.

Bacheler, N. M., J. W. Neal, and R. L. Noble. 2004. Reproduction of a landlocked diadromous fish population: bigmouth sleepers *Gobiomorus dormitor* in a reservoir in Puerto Rico. Caribbean Journal of Science **40**:223–231.

Baicere-Silva, C. M., R. C. Benine, and I. Quagio-Grassiotto. 2011a. *Markiana nigripinnis* (Perugia, 1891) as a putative member of the subfamily Stevardiinae (Characiformes: Characidae): spermatic evidence. Neotropical Ichthyology **9**:371–376.

Baicere-Silva, C. M., K. M. Ferreira, L. R. Malabarba, R. C. Benine, and I. Quagio-Grassiotto. 2011b. Spermatic characteristics and sperm evolution on the subfamily Stevardiinae (Ostariophysi: Characiformes: Characidae). Neotropical Ichthyology **9**:377–392.

Balassa, G. C., R. Fugi, N. S. Hahn, and A. B. Galina. 2004. Dieta de espécies de Anostomidae (Teleostei, Characiformes) na área de influência do reservatório de Manso, Mato Grosso, Brasil. Iheringia, Séries Zoologia, Porto Alegre **94**:77–82.

Ballen, G. A., and J. I. Mojica. 2014. A new trans-Andean stick catfish of the genus *Farlowella* Eigenmann & Eigenmann, 1889 (Siluriformes: Loricariidae) with the first record of the genus for the río Magdalena basin in Colombia. Zootaxa **3765**:134–142.

Ballen, G. A., and R. P. Vari. 2012. Review of the Andean armored catfishes of the genus *Dolichancistrus* Isbrücker (Siluriformes: Loricariidae). Neotropical Ichthyology **10**:499–518.

Bambaradeniya, C. N. B. 2002. The status and implications of alien invasive species in Sri Lanka. Zoos' Print Journal **17**:930–935.

Barbarino-Duque, A., and D. C. Taphorn. 1995. Especies de la Pesca Deportiva, Una Guía de Identificación y Reglamentación de los Peces de Agua Dulce en Venezuela. UNELLEZ-Fundación Polar, Caracas.

Barbarino Duque, A., and K. O. Winemiller. 2003. Dietary segregation among large catfishes of the Apure and Arauca rivers, Venezuela. Journal of Fish Biology **63**:410–427.

Barbieri, G., and J. Garavello. 1981. Sobre a dinâmica da reprodução e da nutrição de *Leporinus friderici* (Bloch, 1794) na represa do Lobo. Brotas, Oitirapina, SP (Pisces, Anostomidae). II Seminário Regional de Ecologia (UFSCar, São Carlos):347–387.

Barletta, M., A. Barletta-Bergan, U. Saint-Paul, and G. Hubold. 2003. Seasonal changes in density, biomass, and diversity of estuarine fishes in tidal mangrove creeks of the lower Caete Estuary (northern Brazilian coast, east Amazon). Marine Ecology Progress Series **256**:217–228.

Barletta, M., V. E. Cussac, A. A. Agostinho, C. Baigún, E. K. Okada, A. C. Catella, N. F. Fontoura, P. S. Pompeu, L. F. Jiménez-Segura, V. S. Batista, C. A. Lasso, D. Taphorn, and N. N. Fabré. 2016. Fishery ecology in South American river basins. Pages 311–348 *in* J. F. Craig, editor. Freshwater Fisheries Ecology. John Wiley and Sons.

Barlow, G. W. 1967. Social behavior of a South American leaf fish, *Polycentrus schomburgkii*, with an account of recurring pseudofemale behavior. American Midland Naturalist **78**:215–234.

Barlow, G. W., and J. W. Munsey. 1976. The red devil-Midas-arrow cichlid species complex in Nicaragua. Pages 359–369 *in* T. B. Thorson, editor. Investigations of the Ichthyology of Nicaraguan Lakes. University of Nebraska Press, Lincoln, USA.

Barluenga, M., K. N. Stölting, W. Salzburger, M. Muschick, and A. Meyer. 2006. Sympatric speciation in Nicaraguan crater lake cichlid fish. Nature **439**:719–723.

Barroca, T., G. Santos, N. Duarte, and E. Kalapothakis. 2012. Evaluation of genetic diversity and population structure in a commercially important freshwater fish *Prochilodus costatus* (Characiformes, Prochilodontidae) using complex hypervariable repeats. Genetics and Molecular Research **11**:4456–4467.

Barthem, R., and M. Goulding. 1997. The Catfish Connection: Ecology, Migration, and Conservation of Amazon Predators. Columbia University Press, New York.

Barthem, R., and M. Goulding. 2007. An unexpected ecosystem. The Amazon as revealed by fisheries. Amazon Conservation Association; Missouri Botanical Garden Press.

Baskin, J. N. 1973. Structure and relationships of the Trichomycteridae. Unpublished PhD thesis. City University of New York, NY.

Bastos, R. F., M. V. Condini, A. S. Varela Jr., and A. M. Garcia. 2011. Diet and food consumption of the pearl cichlid *Geophagus brasiliensis* (Teleostei: Cichlidae): relationships with gender and sexual maturity. Neotropical Ichthyology **9**:825–830.

Bastos, W. R., J. G. Dorea, J. V. E. Bernardi, L. C. Lauthartte, M. H. Mussy, L. D. Lacerda, and O. Malm. 2015. Mercury in fish of the Madeira River (temporal and spatial assessment), Brazilian Amazon. Environmental Research **140**:191–197.

Begossi, A., R. A. M. Silvano, B. D. do Amaral, and O. T. Oyakawa. 1999. Uses of fish and game by inhabitants of an extractive reserve (Upper Juruá, Acre, Brazil). Environment, Development and Sustainability **1**:73–93.

Beltrão, H., and J. Zuanon. 2012. *Hemiodus langeanii* (Characiformes: Hemiodontidae), a new species from rio Amana, rio Maues-Acu drainage, Amazon basin, Brazil. Neotropical Ichthyology **10**:255–262.

Bemis, W. E., and G. V. Lauder. 1986. Morphology and function of the feeding apparatus of the lungfish, *Lepidosiren paradoxa* (Dipnoi). Journal of Morphology **187**:81–108.

Benine, R. C., R. M. C. Castro, and J. Sabino. 2004. *Moenkhausia bonita*: A new small characin fish from the Rio Paraguay basin, southwestern Brazil (Characiformes: Characidae). Copeia:68–73.

Benine, R. C., G. A. M. Lopes, and E. Ron. 2010. A new species of *Ctenobrycon* Eigenmann, 1908 (Characiformes: Characidae) from the río Orinoco basin, Venezuela. Zootaxa **2715**:59–67.

Benine, R. C., T. C. Mariguela, and C. Oliveira. 2009. New species of *Moenkhausia* Eigenmann, 1903 (Characiformes: Characidae) with comments on the *Moenkhausia oligolepis* species complex. Neotropical Ichthyology **7**:161–168.

Benine, R. C., B. F. Melo, R. M. C. Castro, and C. Oliveira. 2015. Taxonomic revision and molecular phylogeny of *Gymnocorymbus* Eigenmann, 1908 (Teleostei, Characiformes, Characidae). Zootaxa **3956**:1–28.

Bergleiter, S. 1999. Zur ökologischen Struktur einer zentralamazonischen Fischzönose. Zoologica Stuttgart **149**:1–191.

Bergmann, L. A. C. 1988. *Schizodon jacuiensis* sp. n., um novo anostomídeo do sul do Brasil e redescrição de *Schizodon kneri* (Steindachner, 1875) e *S. platae* (Garman, 1890). (Pisces, Characiformes, Anostomidae). Comunicações do Museu de Ciências da PUCRS, Série Zoologia **1**:13–28.

Berry, R. J. 2011. Individual differences in electric fishes: an animal model of personality. Unpublished PhD thesis. City University of New York.

Bertaco, V. A., and F. R. Carvalho. 2010. New species of *Hasemania* (Characiformes: Characidae) from central Brazil, with comments on the endemism of upper rio Tocantins basin, Goias state. Neotropical Ichthyology **8**:27–32.

Bertaco, V. A., and L. R. Malabarba. 2007. A new species of *Hasemania* from the upper Rio Tapajós drainage, Brazil (Teleostei: Characiformes: Characidae). Copeia:350–354.

Bertaco, V. A., and L. R. Malabarba. 2010. A review of the cis-Andean species of *Hemibrycon* Günther (Teleostei: Characiformes: Characidae: Stevardiinae), with description of two new species. Neotropical Ichthyology **8**:737–770.

Bertmar, G., B. G. Kapoor, and R. V. Miller. 1969. Epibranchial organs in lower teleostean fishes: an example of structural adaptation. Pages 1–48 *in* W. J. L. Felts and R. J. Harrison, editors. International Review of General and Experimental Zoology.

Bessa, E., L. N. Carvalho, J. Sabino, and P. Tomazzelli. 2011. Juveniles of the piscivorous dourado *Salminus brasiliensis* mimic the piraputanga *Brycon hilarii* as an alternative predation tactic. Neotropical Ichthyology **9**:351–354.

Betancur-R, R., A. Acero P, E. Bermingham, and R. Cooke. 2007. Systematics and biogeography of New World sea catfishes (Siluriformes: Ariidae) as inferred from mitochondrial, nuclear, and morphological evidence. Molecular Phylogenetics and Evolution **45**:339–357.

Bichuette, M. E., and E. Trajano. 2006. Morphology and distribution of the cave knifefish *Eigenmannia vicentespelaea* Triques, 1996 (Gymnotiformes: Sternopygidae) from Central Brazil, with an expanded diagnosis and comments on subterranean evolution. Neotropical Ichthyology **4**:99–105.

Birdsong, R. S., and C. R. Robins. 1995. New genus and species of seven-spined goby (Gobiidae: Gobiosomini) from the offing of the Amazon River, Brazil. Copeia:676–683.

Birindelli, J. L. O. 2010. Relações filogenéticas da superfamília Doradoidea (Ostariophysi, Siluriformes). Unpublished PhD thesis. Universidade de São Paulo, São Paulo.

Birindelli, J. L. O. 2014. Phylogenetic relationships of the South American Doradoidea (Ostariophysi: Siluriformes). Neotropical Ichthyology 12:451–564.

Birindelli, J. L. O., A. Akama, and H. A. Britski. 2012a. Comparative morphology of the gas bladder in driftwood catfishes (Siluriformes: Auchenipteridae). Journal of Morphology 273:651–660.

Birindelli, J., and H. Britski. 2013. Two new species of Leporinus (Characiformes: Anostomidae) from the Brazilian Amazon, and redescription of Leporinus striatus Kner 1858. Journal of Fish Biology 83:1128–1160.

Birindelli, J. L. O., H. A. Britski, and F. C. T. Lima. 2013a. New species of Leporinus from the Rio Tapajós Basin, Brazil, and redescription of L. moralesi (Characiformes: Anostomidae). Copeia:238–247.

Birindelli, J. L. O., D. F. Fayal, and W. B. Wosiacki. 2011. Taxonomic revision of thorny catfish genus Hassar (Siluriformes: Doradidae). Neotropical Ichthyology 9:515–542.

Birindelli, J. L. O., F. C. T. Lima, and H. A. Britski. 2012b. New species of Pseudanos Winterbottom, 1980 (Characiformes: Anostomidae), with notes on the taxonomy of P. gracilis and P. trimaculatus. Zootaxa 3425:55–68.

Birindelli, J. L. O., L. A. W. Peixoto, W. B. Wosiacki, and H. A. Britski. 2013b. New species of Hypomasticus Borodin, 1929 (Characiformes: Anostomidae) from tributaries of the lower Rio Amazonas, Brazil. Copeia:464–469.

Birindelli, J. L. O., and M. H. S. Pérez. 2011. Ossancora, new genus of thorny catfish (Teleostei: Siluriformes: Doradidae) with description of one new species. Proceedings of the Academy of Natural Sciences of Philadelphia 161:117–152.

Birindelli, J. L. O., M. H. Sabaj, and D. C. Taphorn. 2007. New species of Rhynchodoras from the río Orinoco, Venezuela, with comments on the genus (Siluriformes: Doradidae). Copeia:672–684.

Birindelli, J. L. O., L. M. Sarmento-Soares, and F. C. T. Lima. 2015. A new species of Centromochlus (Siluriformes, Auchenipteridae, Centromochlinae) from the middle Rio Tocantins basin, Brazil. Journal of Fish Biology 87:860–875.

Birindelli, J. L. O., L. M. Sousa, and M. H. Sabaj Pérez. 2008. New species of thorny catfish, genus Leptodoras Boulenger (Siluriformes: Doradidae), from Tapajós and Xingu basins, Brazil. Neotropical Ichthyology 6:465–480.

Birindelli, J. L. O., A. M. Zanata, L. M. Sousa, and A. L. Netto-Ferreira. 2009. New species of Jupiaba Zanata (Characiformes: Characidae) from Serra do Cachimbo, with comments on the endemism of upper rio Curuá, rio Xingu basin, Brazil. Neotropical Ichthyology 7:11–18.

Birindelli, J. L. O., and J. Zuanon. 2012. Systematics of the Jaguar catfish genus Liosomadoras Fowler, 1940 (Auchenipteridae: Siluriformes). Neotropical Ichthyology 10:1–11.

Blanco, H. V., G. J. Solipá, C. W. Olaya-Nieto, F. F. Segura-Guevara, S. B. Brú-Cordero, and G. Tordecilla-Petro. 2005. Crecimiento y mortalidad de la Yalúa (Cyphocharax magdalenae Steindachner, 1878) en el río Sinú, Colombia. Revista MVZ Córdoba 10:555–563.

Bleher, H., K. N. Stölting, W. Salzburger, and A. Meyer. 2007. Revision of the genus Symphysodon Heckel, 1840 (Teleostei: Perciformes: Cichlidae) based on molecular and morphological

characters. Aqua, International Journal of Ichthyology 12:133–174.

Bloom, D. D., and N. R. Lovejoy. 2012. Molecular phylogenetics reveals a pattern of biome conservatism in New World anchovies (family Engraulidae). Journal of Evolutionary Biology 25:701–715.

Bloom, D. D., and N. R. Lovejoy. 2014. The evolutionary origins of diadromy inferred from a time-calibrated phylogeny for Clupeiformes (herring and allies). Proceedings of the Royal Society B-Biological Sciences 281.

Bockmann, F. A. 1994. Description of Mastiglanis asopos, a new pimelodid catfish from northern Brazil, with comments on phylogenetic relationships inside the subfamily Rhamdiinae (Siluriformes, Pimelodidae). Proceedings of the Biological Society of Washington 107:760–777.

Bockmann, F. A. 1998. Análise filogenética da família Heptapteridae (Teleostei, Ostariophysi, Siluriformes) e redefenição de seus gêneros. Unpublished PhD thesis. Universidade de São Paulo, São Paulo, Brazil.

Bockmann, F. A., and R. M. C. Castro. 2010. The blind catfish from the caves of Chapada Diamantina, Bahia, Brazil (Siluriformes: Heptapteridae): Description, anatomy, phylogenetic relationships, natural history, and biogeography. Neotropical Ichthyology 8:673–706.

Bockmann, F. A., and C. J. Ferraris Jr. 2005. Systematics of the Neotropical catfish genera Nemuroglanis Eigenmann and Eigenmann 1889, Imparales Schultz 1944, and Medemichthys Dahl 1961 (Siluriformes: Heptapteridae). Copeia:124–137.

Bockmann, F. A., and G. Guazzelli. 2003. Family Heptapteridae (Heptapterids). Pages 406–431 in R. E. Reis, S. O. Kullander, and C. J. Ferraris Jr., editors. Check List of the Freshwater Fishes of South and Central America. EDIPUCRS, Porto Alegre, Brazil.

Bockmann, F. A., and J. G. Lundberg. 2006. New species of the South American catfish genus Gladioglanis Ferraris and Mago-Leccia, 1989 (Siluriformes: Heptapteridae): adding pieces to a phylogenetic puzzle. Joint Meeting of Ichthyologists and Herpetologists. University of New Orleans/Tulane University.

Bockmann, F. A., and A. M. Miquelarena. 2008. Anatomy and phylogenetic relationships of a new catfish species from northeastern Argentina with comments on the phylogenetic relationships of the genus Rhamdella Eigenmann and Eigenmann 1888 (Siluriformes, Heptapteridae). Zootaxa 1780:1–54.

Bockmann, F. A., and R. E. Reis. 2011. Two new, beautifully-colored species of the Neotropical catfish Cetopsorhamdia Eigenmann and Fisher, 1916 (Siluriformes: Heptapteridae) from western Brazil, with a cladistic analysis of the genus. Joint Meeting of Ichthyologists and Herpetologists, Minneapolis, USA.

Bockmann, F. A., and V. Slobodian. 2013. Heptapteridae. Pages 12–71 in L. J. Queiroz, G. Torrente-Vilara, W. M. Ohara, T. H. da Silva Pires, J. Zuanon, and C. R. da Costa Doria, editors. Peixes do Rio Madeira. Dialeto, São Paulo, Brazil.

Boeseman, M. 1971. The "comb-toothed" Loricariinae of Surinam, with reflections on the phylogenetic tendencies within the family Loricariidae (Siluriformes, Siluroidei). Zoologische Verhandelingen 116:1–56.

Bogota-Gregory, J. D., and J. A. Maldonado-Ocampo. 2006. Primer registro de Lepidosiren paradoxa Fitzinger, 1837 en la cuenca del Orinoco (PNN El Tuparro, Vichada, Colombia). Biota Colombiana 7:301–304.

Böhlke, J. E., and J. E. McCosker. 1975. The status of the Ophichthid eel genera *Caecula* Vahl and *Sphagebranchus* Bloch, and the description of a new genus and species from fresh waters in Brazil. Proceedings of the Academy of Natural Sciences of Philadelphia **127**:1–11.

Bojsen, B. H. 2005. Diet and condition of three fish species (Characidae) of the Andean foothills in relation to deforestation. Environmental Biology of Fishes **73**:61–73.

Borodin, N. A. 1929. Notes on some species and subspecies of the genus *Leporinus* Spix. Memoirs of the Museum of Comparative Zoology **50**:269–290.

Boujard, T., M. Pascal, F. J. Meunier, and P. Y. Le Bail. 1997. Poissons de Guyane. Guide écologique de l'Approuague et de la réserve des Nouragues. Institut National de la Recherche Agronomique, Paris, France.

Boujard, T., D. Sabatier, R. Rojas-Beltran, M. F. Prevost, and J. F. Renno. 1990. The food habits of three allochthonous feeding characoids in French Guiana. Revue d'Ecologie (la Terre et la Vie) **45**:247–258.

Bowen, S. H. 1983. Detritivory in Neotropical fish communities. Environmental Biology of Fishes **9**:137–144.

Bowen, S. H., A. A. Bonetto, and M. O. Ahlgren. 1984. Microorganisms and detritus in the diet of a typical Neotropical riverine detritivore, *Prochilodus platensis* (Pisces: Prochilodontidae). Limnology and Oceanography **29**:1120–1122.

Braga, F. M. S. 1990. Aspectos da reprodução e alimentação de peixes comuns em um trecho do rio Tocantins entre Imperatriz e Estreito, Estados do Maranhão e Tocantins, Brasil. Revista Brasileira de Biologia **50**:547–558.

Braga, R. A. 1956. Carater sexual secundario em pirambeba, "*Serrasalmus rhombeus*" (L., 1766) Lacepede, 1803 (Ostariophisis, Characidae, Serrasalminae). Revista Brasileira de Biologia **16**:167–180.

Bragança, P. H. N., P. F. Amorim, and W. J. E. M. Costa. 2012a. Geographic distribution, habitat, colour pattern variability and synonymy of the Amazon killifish *Melanorivulus schuncki* (Cyprinodontiformes: Rivulidae). Ichthyological Exploration of Freshwaters **23**:51–55.

Bragança, P. H. N., and W. J. E. M. Costa. 2011. *Poecilia sarrafae*, a new poeciliid from the Paraíba and Mearim river basins, northeastern Brazil (Cyprinodontiformes: Cyprinodontoidei). Ichthyological Exploration of Freshwaters **21**:369–376.

Bragança, P. H. N., W. J. E. M. Costa, and C. S. Gama. 2012b. *Poecilia waiapi*, a new poeciliid from the Jari River drainage, northern Brazil (Cyprinodontiformes: Cyprinodontoidei). Ichthyological Exploration of Freshwaters **22**:337–343.

Brandão, H., A. P. Vidotto-Magnoni, I. P. Ramos, and E. D. Carvalho. 2013. Biological attributes of *Steindachnerina insculpta* (Fernandez-Yepez, 1948) (Curimatidae) in two reservoirs of middle Paranapanema River, Brazil. Revista de Ciências Agro-Ambientais, Alta Floresta-MT **11**:67–75.

Breder, C. M. 1927. The fishes of the Chucunaque drainage, eastern Panama. Bulletin of the American Museum of Natural History **57**:91–176.

Breder, C. M., and D. E. Rosen. 1966. Modes of Reproduction in Fishes. T.F.H. Publications, Neptune City, NJ.

Brejão, G. L., P. Gerhard, and J. Zuanon. 2013. Functional trophic composition of the ichthyofauna of forest streams in eastern Brazilian Amazon. Neotropical Ichthyology **11**:361–373.

Brenner, M., and U. Krumme. 2007. Tidal migration and patterns in feeding of the four-eyed fish *Anableps anableps* L. in a north Brazilian mangrove. Journal of Fish Biology **70**:406–427.

Britski, H. A. 1977. Sobre o gênero *Colossoma* (Pisces, Characidae). Suplemento Ciência e Cultura (São Paulo) **29**:810.

Britski, H. A. 1993. Uma nova espécie de *Phenacorhamdia* da bacia do Alto Paraná (Pisces, Siluriformes). Comunicações do Museu de Ciências da Pontifícia Universidade Católica do Rio Grande do Sul, série zoologia **6**:41–50.

Britski, H. A. 1997. Descrição de um novo gênero de Hypoptopomatinae, com duas espécies novas (Siluriformes, Loricariidae). Papéis Avulsos de Zoologia **40**:231–255.

Britski, H. A., and A. Akama. 2011. New species of *Trachycorystes* Bleeker, with comments on other species of the genus (Ostariophysi: Siluriformes: Auchenipteridae). Neotropical Ichthyology **9**:273–279.

Britski, H. A., and J. L. O. Birindelli. 2008. Description of a new species of the genus *Leporinus* Spix (Characiformes: Anostomidae) from the rio Araguaia, Brazil, with comments on the taxonomy and distribution of *L. parae* and *L. lacustris*. Neotropical Ichthyology **6**:45–51.

Britski, H. A., J. L. O. Birindelli, and J. C. Garavello. 2011. *Synaptolaemus latofasciatus*, a new combination for *Leporinus latofasciatus* Steindachner, 1910 and its junior synonym *Synaptolaemus cingulatus* Myers and Fernández-Yepez, 1950 (Characiformes: Anostomidae). Zootaxa **3018**:59–65.

Britski, H. A., J. L. O. Birindelli, and J. C. Garavello. 2012. A new species of *Leporinus* Agassiz, 1829 from the upper Rio Paraná basin (Characiformes, Anostomidae) with redescription of *L. elongatus* Valenciennes, 1850 and *L. obtusidens* (Valenciennes, 1837). Papéis Avulsos de Zoologia **52**:441–475.

Britski, H. A., and J. C. Garavello. 2007. Description of two new sympatric species of the genus *Hisonotus* Eigenmann and Eigenmann, 1889, from upper Rio Tapajós, Mato Grosso state, Brazil (Pisces: Ostariophysi: Loricariidae). Brazilian Journal of Biology **67**:413–420.

Britski, H. A., and F. C. T. Lima. 2008. A new species of *Hemigrammus* from the upper Rio Tapajos basin in Brazil (Teleostei: Characiformes: Characidae). Copeia:565–569.

Britski, H. A., Y. Sato, and A. B. S. Rosa. 1988. Manual de Identificação de Peixes da Região de Três Marias (Com Chaves de Identificação para os Peixes da Bacia do São Francisco). 3rd edition. Câmara Dos Deputados/CODEVASF, Brasília, Brazil.

Britski, H. A., K. Z. S. Silimon, and B. S. Lopes. 1999. Peixes do Pantanal: Manual de Identificação. Embrapa, Brasília, D.F., Brazil.

Britto, M. R., and C. R. Moreira. 2002. *Otocinclus tapirape*: a new hypoptopomatine catfish from central Brazil (Siluriformes: Loricariidae). Copeia:1063–1069.

Britz, R., and S. O. Kullander. 2003. Family Polycentridae (Leaffishes). Pages 603–604 *in* R. E. Reis, S. O. Kullander, and C. J. Ferraris Jr., editors. Check List of the Freshwater Fishes of South and Central America. EDIPUCRS, Porto Alegre, Brazil.

Britz, R., and M. Toledo-Piza. 2012. Egg surface structure of the freshwater toadfish *Thalassophryne amazonica* (Teleostei: Batrachoididae) with information on its distribution and natural habitat. Neotropical Ichthyology **10**:593–599.

Brown, G. E., J. G. J. Godin, and J. Pedersen. 1999. Fin-flicking behaviour: a visual antipredator alarm signal in a characin fish, *Hemigrammus erythrozonus*. Animal Behaviour **58**:469–475.

Brum, S. M., V. M. F. da Silva, F. Rossoni, and L. Castello. 2015. Use of dolphins and caimans as bait for *Calophysus macropterus* (Lichtenstein, 1819) (Siluriforme: Pimelodidae) in the Amazon. Journal of Applied Ichthyology **31**:675–680.

Buck, S., and I. Sazima. 1995. An assemblage of mailed catfishes (Loricariidae) in southeastern Brazil: Distribution, activity, and feeding. Ichthyological Exploration of Freshwaters **6**:325–332.

Buckley, J., R. J. Maunder, A. Foey, J. Pearce, A. L. Val, and K. A. Sloman. 2010. Biparental mucus feeding: a unique example of parental care in an Amazonian cichlid. Journal of Experimental Biology **213**:3787–3795.

Buckup, P. A. 1993a. The monophyly of the Characidiinae, a Neotropical group of characiform fishes (Teleostei: Ostariophysi). Zoological Journal of the Linnean Society **108**:225–245.

Buckup, P. A. 1993b. Phylogenetic interrelationships and reductive evolution in Neotropical characidiin fishes (Characiformes, Ostariophysi). Cladistics **9**:305–341.

Buckup, P. A. 1993c. Review of the characidiin fishes (Teleostei: Characiformes), with descriptions of four new genera and ten new species. Ichthyological Exploration of Freshwaters **4**:97–154.

Buckup, P. A. 1998. Relationships of the Characidiinae and phylogeny of characiform fishes (Teleostei: Ostariophysi). Pages 123–144 in L. R. Malabarba, R. E. Reis, R. P. Vari, Z. M. S. Lucena, and C. A. S. Lucena, editors. Phylogeny and Classification of Neotropical Fishes. EDIPUCRS, Porto Alegre, Brazil.

Buckup, P. A. 2003. Family Crenuchidae (South American darters). Pages 87–95 in R. E. Reis, S. O. Kullander, and C. J. Ferraris Jr., editors. Check List of the Freshwater Fishes of South and Central America. EDIPUCRS, Porto Alegre, Brazil.

Buckup, P. A., C. Zamprogno, F. Vieira, and R. L. Teixeira. 2000. Waterfall climbing in *Characidium* (Crenuchidae: Characidiinae) from eastern Brazil. Ichthyological Exploration of Freshwaters **11**:273–278.

Bührnheim, C. M., T. P. Carvalho, L. R. Malabarba, and S. H. Weitzman. 2008. A new genus and species of characid fish from the Amazon basin: the recognition of a relictual lineage of characid fishes (Ostariophysi: Cheirodontinae: Cheirodontini). Neotropical Ichthyology **6**:663–678.

Bührnheim, C. M., and L. R. Malabarba. 2006. Redescription of the type species of *Odontostilbe* Cope, 1870 (Teleostei: Characidae: Cheirodontinae), and description of three new species from the Amazon basin. Neotropical Ichthyology **4**:167–196.

Bührnheim, C. M., and L. R. Malabarba. 2007. Redescription of *Odontostilbe pulchra* (Gill, 1858) (Teleostei: Characidae: Cheirodontinae), and description of two new species from the río Orinoco basin. Neotropical Ichthyology **5**:1–20.

Buitrago-Suarez, U. A., and B. M. Burr. 2007. Taxonomy of the catfish genus *Pseudoplatystoma* Bleeker (Siluriformes: Pimelodidae) with recognition of eight species. Zootaxa **1512**:1–38.

Buitrago, U., and G. Galvis. 1997. Description of some accessory structures of the urogenital system in the Neotropical family Astroblepidae (Pisces, Siluroidei). Revista de la Academia Colombiana de Ciencias Exactas, Físicas y Naturales **21**:347–352.

Burgess, W. E. 1989. An atlas of freshwater and marine catfishes, a preliminary survey of the Siluriformes. T.F.H. Publications, Neptune City, NJ.

Burgess, W. E. 1994. *Scobinancistrus aureatus*, a new species of loricariid catfish from the Rio Xingu (Loricariidae: Ancistrinae). Tropical Fish Hobbyist **43**:236–242.

Burns, J. R., and J. A. Flores. 1981. Reproductive biology of the cuatro ojos, *Anableps dowi* (Pisces: Anablepidae), from El Salvador and its seasonal variations. Copeia:25–32.

Burns, J. R., and S. H. Weitzman. 2005. Insemination in ostariophysan fishes. Pages 107–134 in M. C. Uribe and H. J. Grier, editors. Viviparous Fishes. New Life Publications, Homestead, FL, USA.

Burns, J. R., and S. H. Weitzman. 2006. Intromittent organ in the genus *Monotocheirodon* (Characiformes: Characidae). Copeia:529–534.

Burns, J. R., S. H. Weitzman, H. J. Grier, and N. A. Menezes. 1995. Internal fertilization, testis and sperm morphology in glandulocaudine fishes (Teleostei: Characidae: Glandulocaudinae). Journal of Morphology **224**:131–145.

Burns, J. R., S. H. Weitzman, L. R. Malabarba, and A. D. Meisner. 2000. Sperm modifications in inseminating ostariophysan fishes, with new documentation of inseminating species. Page 255 *in* Proceedings of the 6th International Symposium on the Reproductive Physiology of Fish. Institute of Marine Resources and University of Bergen, Bergen, Norway.

Burns, M. D., B. W. Frable, and B. L. Sidlauskas. 2014. A new species of *Leporinus* (Characiformes: Anostomidae), from the Orinoco basin, Venezuela. Copeia:206–214.

Bushmann, P. J., J. R. Burns, and S. H. Weitzman. 2002. Gill-derived glands in glandulocaudine fishes (Teleostei: Characidae: Glandulocaudinae). Journal of Morphology **253**:187–195.

Bussing, W. A. 1998. Peces de las aguas continentales de Costa Rica 2nd edition. Editorial de la Universidad de Costa Rica, San José.

Cacho, M. S. R. F., M. E. Yamamoto, and S. Chellappa. 2007. Mating system of the Amazonian cichlid angel fish, *Pterophyllum scalare*. Brazilian Journal of Biology **67**:161–165.

Caires, R. A. 2013. *Microphilypnus tapajosensis*, a new species of eleotridid from the Tapajós basin, Brazil (Gobioidei: Eleotrididae). Ichthyological Exploration of Freshwaters **24**:155–160.

Caires, R. A., and J. L. De Figueiredo. 2011. Review of the genus *Microphilypnus* Myers, 1927 (Teleostei: Gobioidei: Eleotridae) from the lower Amazon basin, with description of one new species. Zootaxa **3036**:39–57.

Calcagnotto, D., S. A. Schaefer, and R. DeSalle. 2005. Relationships among characiform fishes inferred from analysis of nuclear and mitochondrial gene sequences. Molecular Phylogenetics and Evolution **36**:135–153.

Calegari, B. B., R. E. Reis, and R. P. Vari. 2014. Miniature catfishes of the genus *Gelanoglanis* (Siluriformes: Auchenipteridae): monophyly and the description of a new species from the upper rio Tapajós basin, Brazil. Neotropical Ichthyology **12**:699–706.

Camargo, M., and V. Isaac. 1998. Population structure in fish fauna in the estuarine area of Caeté River, Pará, Brazil. Acta Scientiarum **20**:171–177.

Camargo, M., and V. Isaac. 2001. Os peixes estuarinos da região norte do Brasil: lista de espécies e considerações sobre sua distribuição geográfica. Boletim do Museu Paraense Emilio Goeldi **17**:133–157.

Campos-da-Paz, R. 1999. New species of *Megadontognathus* from the Amazon basin, with phylogenetic and taxonomic discussions on the genus (Gymnotiformes: Apteronotidae). Copeia:1041–1049.

Canistri, V. 1970. Alimentacion frutivora en *Colossoma brachypomus* (Osteichthyes–Cipriniformes–Characidae). Memoria de la Fundación La Salle de Ciencias Naturales **30**:196–205.

Caramaschi, É. P. 1986. Distribuição da ictiofauna de riachos das bacias do Tietê e do Paranapanema, junto ao divisor de águas (Botucatu, SP). Unpublished PhD thesis. Universidade Federal de São Carlos, São Carlos, Brazil.

Cardoso, A. L., J. C. Pieczarka, E. Feldberg, S. S. R. Milhomem, T. Moreira-Almeida, D. dos Santos Silva, P. C. da Silva, and C. Y. Nagamachi. 2011. Chromosomal characterization of two species of genus *Steatogenys* (Gymnotiformes: Rhamphichthyoidea: Steatogenini) from the Amazon basin: sex chromosomes and correlations with Gymnotiformes phylogeny. Reviews in Fish Biology and Fisheries **21**:613–621.

Cardoso, A. R. 2010. *Bunocephalus erondinae*, a new species of banjo catfish from southern Brazil (Siluriformes: Aspredinidae). Neotropical Ichthyology **8**:607–613.

Cardoso, Y. P., and J. I. Montoya-Burgos. 2009. Unexpected diversity in the catfish *Pseudancistrus brevispinis* reveals dispersal routes in a Neotropical center of endemism: the Guyanas Region. Molecular Ecology **18**:947–964.

Carolsfeld, J., B. Harvey, C. Ross, and A. Baer, editors. 2003. Migratory Fishes of South America: Biology, Fisheries and Conservation Status. World Fisheries Trust, Victoria, BC, Canada.

Carruth, L. L. 2000. Freshwater cichlid *Crenicara punctulata* is a protogynous sequential hermaphrodite. Copeia:71–82.

Carvalho, F. M., and E. K. Resende. 1984. Aspectos da biologia de *Tocantinsia depressa* (Siluriformes, Auchenipteridae). Amazoniana **8**:327–337.

Carvalho, F. R., V. A. Bertaco, and F. C. Jerep. 2010. *Hemigrammus tocantinsi*: a new species from the upper rio Tocantins basin, central Brazil (Characiformes: Characidae). Neotropical Ichthyology **8**:247–254.

Carvalho, F. R., and F. Langeani. 2013. *Hyphessobrycon uaiso*: new characid fish from the rio Grande, upper rio Paraná basin, Minas Gerais state (Ostariophysi: Characidae), with a brief comment about some types of *Hyphessobrycon*. Neotropical Ichthyology **11**:525–536.

Carvalho, L. N., R. Arruda, J. Raizer, and K. Del-Claro. 2007. Feeding habits and habitat use of three sympatric piranha species in the Pantanal wetland of Brazil. Ichthyological Exploration of Freshwaters **18**:109–116.

Carvalho, L. N., R. Arruda, and J. Zuanon. 2003. Record of cleaning behavior by *Platydoras costatus* (Siluriformes: Doradidae) in the Amazon basin, Brazil. Neotropical Ichthyology **1**:137–139.

Carvalho, L. N., J. Zuanon, and I. Sazima. 2006. The almost invisible league: crypsis and association between minute fishes and shrimps as a possible defence against visually hunting predators. Neotropical Ichthyology **4**:219–224.

Carvalho, M., and A. Datovo. 2012. A new species of cascudinho of the genus *Hisonotus* (Siluriformes: Loricariidae: Hypoptopomatinae) from the upper Rio Tapajós basin, Brazil. Copeia:266–275.

Carvalho, M. S., J. Zuanon, and E. J. G. Ferreira. 2014. Diving in the sand: the natural history of *Pygidianops amphioxus* (Siluriformes: Trichomycteridae), a miniature catfish of central Amazonian streams in Brazil. Environmental Biology of Fishes **97**:59–68.

Carvalho, T. 2013. Systematics and evolution of the toothless knifefishes Rhamphichthyoidea Mago-Leccia (Actinopterygii : Gymnotiformes): diversification in South American freshwaters. Unpublished PhD thesis. University of Louisiana at Lafayette.

Carvalho, T. P. 2008. A new species of *Corumbataia* (Siluriformes: Loricariidae: Hypoptopomatinae) from upper Rio Tocantins basin, central Brazil. Copeia:552–557.

Carvalho, T. P., and J. S. Albert. 2011. Redescription and phylogenetic position of the enigmatic Neotropical electric fish *Iracema caiana* Triques (Gymnotiformes: Rhamphichthyidae) using X-ray computed tomography. Neotropical Ichthyology **9**:457–469.

Carvalho, T. P., and J. S. Albert. 2015. A new species of *Rhamphichthys* (Gymnotiformes: Rhamphichthyidae) from the Amazon basin. Copeia:34–41.

Carvalho, T. P., A. R. Cardoso, J. P. Friel, and R. E. Reis. 2015. Two new species of the banjo catfish *Bunocephalus* Kner (Siluriformes: Aspredinidae) from the upper and middle rio São Francisco basins, Brazil. Neotropical Ichthyology **13**:499–512.

Carvalho, T. P., P. Lehmann A, and R. E. Reis. 2008. *Gymnotocinclus anosteos*, a new uniquely-plated genus and species of loricariid catfish (Teleostei: Siluriformes) from the upper rio Tocantins basin, central Brazil. Neotropical Ichthyology **6**:329–338.

Carvalho-Filho, A. 1999. Peixes: costa brasileira. Melro, São Paulo, Brazil.

Casatti, L. 2001. Taxonomia dos peixes neotropicais do gênero *Pachyurus* Agassiz, 1831 (Teleostei, Perciformes, Sciaenidae) e descrição de duas novas espécies. Comunicações do Museu de Ciências e Tecnologia da PUCRS, série Zoologia **14**:133–178.

Casatti, L. 2002a. *Petilipinnis*, a new genus for *Corvina grunniens* Schomburgk, 1843 (Perciformes, Sciaenidae) from the Amazon and Essequibo river basins and redescription of *Petilipinnis grunniens*. Papéis Avulsos de Zoologia **42**:169–181.

Casatti, L. 2002b. Taxonomy of the South American genus *Pachypops* Gill 1861 (Teleostei: Perciformes: Sciaenidae), with the description of a new species. Zootaxa **26**:1–20.

Casatti, L. 2003. Biology of a catfish, *Trichomycterus* sp. (Pisces, Siluriformes), in a pristine stream in the Morro do Diabo State Park, southeastern Brazil. Studies on Neotropical Fauna and Environment **38**:105–110.

Casatti, L. 2005. Revision of the South American freshwater genus *Plagioscion* (Teleostei, Perciformes, Sciaenidae). Zootaxa **1080**:39–64.

Casatti, L., F. R. Carvalho, J. L. Veronezi Jr., and D. R. Lacerda. 2006. Reproductive biology of the Neotropical superfetaceous *Pamphorichthys hollandi* (Cyprinodontiformes: Poeciliidae). Ichthyological Exploration of Freshwaters **17**:59–64.

Casatti, L., and R. M. C. Castro. 1998. A fish community of the São Francisco River headwaters riffles, southeastern Brazil. Ichthyological Exploration of Freshwaters **9**:229–242.

Casatti, L., and R. M. C. Castro. 2006. Testing the ecomorphological hypothesis in a headwater riffles fish assemblage of the rio São Francisco, southeastern Brazil. Neotropical Ichthyology **4**:203–214.

Casatti, L., and N. L. Chao. 2002. A new species of *Pachyurus* Agassiz 1831 (Teleostei: Perciformes: Sciaenidae) from the río Napo basin, eastern Ecuador. Zootaxa **38**:1–7.

Casatti, L., R. M. Romero, F. B. Teresa, J. Sabino, and F. Langeani. 2010. Fish community structure along a conservation gradient in Bodoquena Plateau streams, central west of Brazil. Acta Limnologica Brasiliensia **22**:50–59.

Cassemiro, F. A. S., T. F. L. V. B. Rangel, F. M. Pelicice, and N. S. Hahn. 2008. Allometric and ontogenetic patterns related to feeding of a Neotropical fish, *Satanoperca pappaterra* (Perciformes, Cichlidae). Ecology of Freshwater Fish **17**:155–164.

Castello, L., and M. N. Macedo. 2016. Large-scale degradation of Amazonian freshwater ecosystems. Global Change Biology **22**:990–1007.

Castro, R. M. C. 1981. *Engraulisoma taeniatum*, um novo gênero e espécie de Characidae da bacia do Rio Paraguai (Pisces, Ostariophysi). Papéis Avulsos do Departamento de Zoologia, Secretaria da Agricultura, São Paulo **34**:135–139.

Castro, R. M. C., and R. P. Vari. 2003. Family Prochilodontidae. Pages 65–70 *in* R. E. Reis, S. O. Kullander, and C. J. Ferraris Jr., editors. Check List of the Freshwater Fishes of South and Central America. EDIPUCRS, Porto Alegre, Brazil.

Castro, R. M. C., and R. P. Vari. 2004. Detritivores of the South American fish family Prochilodontidae (Teleostei: Ostariophysi: Characiformes): a phylogenetic and revisionary study. Smithsonian Contributions to Zoology:1–189.

Catania, K. 2014. The shocking predatory strike of the electric eel. Science **346**:1231–1234.

Catarino, M. F., and J. Zuanon. 2010. Feeding ecology of the leaf fish *Monocirrhus polyacanthus* (Perciformes: Polycentridae) in a terrafirme stream in the Brazilian Amazon. Neotropical Ichthyology **8**:183–186.

Catella, A. C., and M. Petrere Jr. 1998. Body-shape and food habits of fish from Baía da Onça, a pantanal floodplain lake, Brazil. Verhandlungen Internationale Vereinigung für Theoretische und Angewandte Limnologie **26**:2203–2208.

Ceneviva-Bastos, M., and L. Casatti. 2007. Feeding opportunism of *Knodus moenkhausii* (Teleostei, Characidae): an abundant species in streams of the northwestern in the state of São Paulo, Brazil. Iheringia–Serie Zoologia **97**:7–15.

Cetra, M., and M. Petrere Jr. 2001. Small-scale fisheries in the middle River Tocantins, Imperatriz (MA), Brazil. Fisheries Management and Ecology **8**:153–162.

Chamon, C. C., and L. H. R. Py-Daniel. 2014. Taxonomic revision of *Spectracanthicus* Nijssen & Isbrücker (Loricariidae: Hypostominae: Ancistrini), with description of three new species. Neotropical Ichthyology **12**:1–25.

Chang, F., and E. Castro. 1999. *Crossoloricaria bahuaja*, a new loricariid fish from Madre de Dios, southeastern Peru. Ichthyological Exploration of Freshwaters **10**:81–88.

Chao, N. L. 2002. Sciaenidae. Pages 1583–1653 *in* K. E. Carpenter, editor. The Living Marine Resources of the Western Central Atlantic. Volume 3: Bony Fishes. FAO, Rome.

Chao, N. L., and S. Prada-Pedreros. 1995. Protection of Aquatic Biodiversity: Proceedings of the World Fisheries Congress (Theme 3):242–261.

Chará, J. D., D. J. Baird, T. C. Telfer, and E. A. Rubio. 2006. Feeding ecology and habitat preferences of the catfish genus *Trichomycterus* in low-order streams of the Colombian Andes. Journal of Fish Biology **68**:1026–1040.

Charvet-Almeida, P., M. L. Góes de Araújo, and M. P. de Almeida. 2005. Reproductive aspects of freshwater stingrays (Chondrichthyes: Patamotrygonidae) in the Brazilian Amazon basin. Journal of Northwest Atlantic Fishery Science **35**:165–171.

Chellappa, S., M. R. Câmara, N. T. Chellappa, M. C. M. Beveridge, and F. A. Huntingford. 2003. Reproductive ecology of a Neotropical cichlid fish, *Cichla monoculus* (Osteichthyes: Cichlidae). Brazilian Journal of Biology **63**:17–26.

Chernoff, B., and A. Machado-Allison. 1990. Characid fishes of the genus *Ceratobranchia*, with descriptions of new species from Venezuela and Peru. Proceedings of the Academy of Natural Sciences of Philadelphia **172**:261–290.

Chernoff, B., and A. Machado-Allison. 1999. *Bryconops colaroja* and *B. colanegra*, two new species from the Cuyuní and Caroní drainages of South America (Teleostei: Characiformes). Ichthyological Exploration of Freshwaters **10**:355–370.

Chernoff, B., and A. Machado-Allison. 2005. *Bryconops magoi* and *Bryconops collettei* (Characiformes: Characidae), two new freshwater fish species from Venezuela, with comments on *B. caudomaculatus* (Günther). Zootaxa **1094**:1–23.

Chernoff, B., A. Machado-Allison, P. Willink, J. Sarmiento, S. Barrera, N. Menezes, and H. Ortega. 2000. Fishes of three Bolivian rivers: diversity, distribution and conservation. Interciencia-Caracas **25**:273–283.

Chockley, B. R., and J. W. Armbruster. 2002. *Panaque changae*, a new species of catfish (Siluriformes: Loricariidae) from eastern Peru. Ichthyological Exploration of Freshwaters **13**:81–90.

Cichocki, F. P. 1977. Tidal cycling and parental behavior of the cichlid fish *Biotodoma cupido*. Environmental Biology of Fishes **1**:159–169.

Cichocki, F. P. 1979. Cladistic history of cichlid fishes and reproductive strategies of the American genera *Acarichthys*, *Biotodoma*, and *Geophagus*. Unpublished PhD dissertation, University of Michigan.

Coates, C. W., and R. T. Cox. 1945. A comparison of length and voltage in the electric eel, *Electrophorus electricus* (Linnaeus). Zoologica: Scientific Contributions of the New York Zoological Society **30**:89–93.

Colatreli, O. P., N. V. Meliciano, D. Toffoli, I. P. Farias, and T. Hrbek. 2012. Deep phylogenetic divergence and lack of taxonomic concordance in species of *Astronotus* (Cichlidae). International Journal of Evolutionary Biology. Article ID 915265.

Collette, B. B. 1966. *Belonion*, a new genus of fresh-water needlefishes from South America. American Museum Novitates **2274**:1–22.

Collette, B. B. 1974. South American freshwater needlefishes (Belonidae) of the genus *Pseudotylosurus*. Zoologische mededelingen **48**:169–186.

Collette, B. B. 1982. South American freshwater needlefishes of the genus *Potamorrhaphis* (Beloniformes: Belonidae). Proceedings of the Biological Society of Washington **95**:714–747.

Collette, B. B. 1995. *Potamobatrachus trispinosus*, a new freshwater toadfish (Batrachoididae) from the Rio Tocantins, Brazil. Ichthyological Exploration of Freshwaters **6**:333–336.

Collette, B. B. 2003a. Family Batrachoididae. Pages 509–510 *in* R. E. Reis, S. O. Kullander, and C. J. Ferraris Jr., editors. Check List of the Freshwater Fishes of South and Central America. EDIPUCRS, Porto Alegre, Brazil.

Collette, B. B. 2003b. Family Hemirhamphidae. Pages 589–590 *in* R. E. Reis, S. O. Kullander, and C. J. Ferraris Jr., editors. Check List of the Freshwater Fishes of South and Central America. EDIPUCRS, Porto Alegre, Brazil.

Cooke, G. M., N. L. Chao, and L. B. Beheregaray. 2012. Natural selection in the water: freshwater invasion and adaptation by water colour in the Amazonian pufferfish. Journal of Evolutionary Biology 25:1305–1320.

Corrêa, C. E., N. S. Hahn, and R. L. Delariva. 2009. Extreme trophic segregation between sympatric fish species: the case of small sized body Aphyocharax in the Brazilian Pantanal. Hydrobiologia 635:57–65.

Correa, E., and H. Ortega. 2010. Diversidad y variación estacional de peces en la cuenca baja del río Nanay, Perú. Revista Peruana de Biología 17:37–42.

Correa, S. B. 2012. Trophic ecology of frugivorous fishes in floodplain forests of the Colombian Amazon. PhD dissertation. Texas A&M University, College Station, USA.

Correa, S. B., J. K. Araujo, J. M. F. Penha, C. N. Cunha, P. R. Stevenson, and J. T. Anderson. 2015a. Overfishing disrupts an ancient mutualism between frugivorous fishes and plants in Neotropical wetlands. Biological Conservation 191:159–167.

Correa, S. B., R. Betancur-R., B. Mérona, and J. W. Armbruster. 2014. Diet shift of red belly pacu Piaractus brachypomus (Cuvier, 1818) (Characiformes: Serrasalmidae), a Neotropical fish, in the Sepik-Ramu River basin, Papua New Guinea. Neotropical Ichthyology 12.

Correa, S. B., R. Costa-Pereira, T. Fleming, M. Goulding, and J. T. Anderson. 2015b. Neotropical fish-fruit interactions: eco-evolutionary dynamics and conservation. Biological Reviews 90:1263–1278.

Correa, S. B., W. G. R. Crampton, and J. S. Albert. 2006. Three new species of the Neotropical electric fish Rhabdolichops (Gymnotiformes: Sternopygidae) from the central Amazon, with a new diagnosis of the genus. Copeia:27–42.

Correa, S. B., W. G. R. Crampton, L. J. Chapman, and J. S. Albert. 2008. A comparison of flooded forest and floating meadow fish assemblages in an upper Amazon floodplain. Journal of Fish Biology 72:629–644.

Costa, R. M. R., and L. A. F. Mateus. 2009. Reproductive biology of pacu Piaractus mesopotamicus (Holmberg, 1887) (Teleostei: Characidae) in the Cuiabá River basin, Mato Grosso, Brazil. Neotropical Ichthyology 7:447–458.

Costa, W. J. E. M. 1991. Description d'une nouvelle espèce du genre Pamphorichthys (Cyprinodontiformes: Poeciliidae) du bassin de l'Araguaia, Brésil. Revue Française d'Aquariologie Herpétologie 18:39–42.

Costa, W. J. E. M. 1992. Descrição de uma nova espécie do gênero Neofundulus (Cyprinodontiformes: Rivulidae), da bacia do rio São Francisco, Brasil. Revista Brasileira de Biologia 52:615–618.

Costa, W. J. E. M. 1994. A new genus and species of Sarcoglanidinae (Siluriformes: Trichomycteridae) from the Araguaia basin, central Brazil, with notes on subfamilial phylogeny. Ichthyological Exploration of Freshwaters 5:207–216.

Costa, W. J. E. M. 1995a. Pearl Killifishes: The Cynolebiatinae. TFH Publications, Neptune City, NJ, USA

Costa, W. J. E. M. 1995b. Revision of the Neotropical annual fish genus Cynopoecilus (Cyprinodontiformes: Rivulidae). Copeia 2:456–465.

Costa, W. J. E. M. 1995c. Revision of the Neotropical fish genus Campellolebias (Cyprinodontiformes: Rivulidae), with notes on phylogeny and biogeography of the Cynopoecilinae. Cybium 19:349–369.

Costa, W. J. E. M. 1996. Relationships, monophyly and three new species of the Neotropical miniature poeciliid genus Fluviphylax (Cyprinodontiformes: Cyprinodontoidei). Ichthyological Exploration of Freshwaters 7:111–130.

Costa, W. J. E. M. 1998a. Phylogeny and classification of Rivulidae revisited: origin and evolution of annualism and miniaturization in rivulid fishes (Cyprinodontiformes: Aplocheiloidei). Journal of Comparative Biology 31:33–92.

Costa, W. J. E. M. 1998b. Phylogeny and classification of the Cyprinodontiformes (Euteleostei: Atherinomorpha): a reappraisal. Pages 537–560 in L. R. Malabarba, R. E. Reis, R. P. Vari, Z. M. S. Lucena, and C. A. S. Lucena, editors. Phylogeny and Classification of Neotropical Fishes. EDIPUCRS, Porto Alegre, Brazil.

Costa, W. J. E. M. 1998c. Revision of the Neotropical annual fish genus Plesiolebias (Cyprinodontiformes: Rivulidae). Ichthyological Exploration of Freshwaters 8:131–134.

Costa, W. J. E. M. 2001. The Neotropical annual fish genus Cynolebias (Cyprinodontiformes: Rivulidae): phylogenetic relationships, taxonomic revision and biogeography. Ichthyological Exploration of Freshwaters 12:333–383.

Costa, W. J. E. M. 2003. Family Rivulidae (South American annual fishes). Pages 526–548 in R. E. Reis, S. O. Kullander, and C. J. Ferraris Jr., editors. Check List of the Freshwater Fishes of South and Central America. EDIPUCRS, Porto Alegre, Brazil.

Costa, W. J. E. M. 2004a. Kryptolebias, a substitute name for Cryptolebias Costa, 2004 and Kryptolebiatinae, a substitute name for Cryptolebiatinae Costa, 2004 (Cyprinodontiformes: Rivulidae). Neotropical Ichthyology 2:107–108.

Costa, W. J. E. M. 2004b. Relationships and redescription of Fundulus brasiliensis (Cyprinodontiformes: Rivulidae), with description of a new genus and notes on the classification of the Aplocheiloidei. Ichthyological Exploration of Freshwaters 15:105–120.

Costa, W. J. E. M. 2005. The Neotropical annual killifish genus Pterolebias Garman (Teleostei: Cyprinodontiformes: Rivulidae): Phylogenetic relationships, descriptive morphology, and taxonomic revision. Zootaxa 1067:1–36.

Costa, W. J. E. M. 2006a. Descriptive morphology and phylogenetic relationships among species of the Neotropical annual killifish genera Nematolebias and Simpsonichthys (Cyprinodontiformes: Aplocheiloidei: Rivulidae). Neotropical Ichthyology 4:1–26.

Costa, W. J. E. M. 2006b. Relationships and taxonomy of the killifish genus Rivulus (Cyprinodontiformes: Aplocheiloidei: Rivulidae) from the Brazilian Amazonas river basin, with notes on historical ecology. Aqua, Journal of Ichthyology and Aquatic Biology 11:133–175.

Costa, W. J. E. M. 2006c. The South American annual killifish genus Austrolebias (Teleostei: Cyprinodontiformes: Rivulidae): Phylogenetic relationships, descriptive morphology and taxonomic revision. Zootaxa 1213:1–162.

Costa, W. J. E. M. 2007a. Taxonomic revision of the seasonal South American killifish genus Simpsonichthys (Teleostei: Cyprinodontiformes: Aplocheiloidei: Rivulidae). Zootaxa 1669:1–134.

Costa, W. J. E. M. 2007b. Taxonomy of the plesiolebiasine killifish genera Pituna, Plesiolebias and Maratecoara (Teleostei: Cyprinodontiformes: Rivulidae), with descriptions of nine new species. Zootaxa 1410:1–41.

Costa, W. J. E. M. 2008. Catalog of aplocheiloid killifishes of the world. UFRJ Depto de Zoologia, Rio de Janeiro.

Costa, W. J. E. M. 2011a. Identity of Rivulus ocellatus and a new name for a hermaphroditic species of Kryptolebias from south-eastern Brazil (Cyprinodontiformes: Rivulidae). Ichthyological Exploration of Freshwaters **22**:185–192.

Costa, W. J. E. M. 2011b. Phylogenetic position and taxonomie status of *Anablepsoides, Atlantirivulus, Cynodonichthys, Laimosemion* and *Melanorivulus* (Cyprinodontiformes: Rivulidae). Ichthyological Exploration of Freshwaters **22**:233–249.

Costa, W. J. E. M. 2014. Phylogeny and evolutionary radiation in seasonal rachovine killifishes: biogeographical and taxonomical implications. Vertebrate Zoology **64**:177–192.

Costa, W. J. E. M., and F. A. Bockmann. 1993. Un nouveau genre neotropical de la famille des Trichomycteridae (Siluriformes, Loricarioidei). Revue Française d'Aquariologie Herpétologie **20**:43–46.

Costa, W. J. E. M., P. H. N. Bragança, and P. F. Amorim. 2013. Five new species of the killifish genus *Anablepsoides* from the Brazilian Amazon (Cyprinodontiformes: Rivulidae). Vertebrate Zoology **63**:283–293.

Costa, W. J. E. M., and H. Lazzarotto. 2014. *Laimosemion ubim*, a new miniature killifish from the Brazilian Amazon (Teleostei: Rivulidae). Ichthyological Exploration of Freshwaters **24**:371–378.

Costa, W. J. E. M., and P. Y. Le Bail. 1999. *Fluviphylax palikur*: a new poeciliid from the rio Oiapoque basin, northern Brazil (Cyprinodontiformes: Cyprinodontoidei), with comments on miniaturization in *Fluviphylax* and other Neotropical freshwater fishes. Copeia:1027–1034.

Costa, W. J. E. M., T. P. A. Ramos, L. C. Alexandre, and R. T. C. Ramos. 2010. *Cynolebias parnaibensis*, a new seasonal killifish from the Caatinga, Parnaíba River basin, northeastern Brazil, with notes on sound producing courtship behavior (Cyprinodontiformes: Rivulidae). Neotropical Ichthyology **8**:283–288.

Costa, W. J. E. M., and A. Sarraf. 1997. *Poecilia (Lebistes) minima*, a new species of Neotropical poeciliid fish from the Brazilian Amazon. Ichthyological Exploration of Freshwaters **8**:185–191.

Costa, W. J. E. M., and E. O. Vicente. 1994. Une nouvelle espèce du genre *Melanocharacidium* (Characiformes: Crenuchidae) du bassin du rio Araguaia, Brésil central. Revue Française d'Aquariologie Herpétologie **20**:67–70.

Costa-Pereira, R., and F. Severo-Neto. 2012. Dining out: *Bryconops caudomaculatus* jumps out of water to catch flies. Revista Chilena de Historia Natural **85**:241–244.

Coutinho, D. P., and W. B. Wosiacki. 2014. A new species of leaffish *Polycentrus* Müller & Troschel, 1849 (Percomorpha: Polycentridae) from the rio Negro, Brazil. Neotropical Ichthyology **12**:747–754.

Covain, R., and S. Fisch-Muller. 2007. The genera of the Neotropical armored catfish subfamily Loricariinae (Siluriformes: Loricariidae): a practical key and synopsis. Zootaxa **1462**:1–40.

Covain, R., and S. Fisch-Muller. 2012. Molecular evidence for the paraphyly of *Pseudancistrus sensu lato* (Siluriformes, Loricariidae), with revalidation of several genera. Cybium **36**:229–246.

Covain, R., S. Fisch-Muller, J. I. Montoya-Burgos, J. H. Mol, P. Y. L. E. Bail, and S. Dray. 2012. The Harttiini (Siluriformes, Loricariidae) from the Guianas: a multi-table approach to assess their diversity, evolution, and distribution. Cybium **36**:115–161.

Covain, R., S. Fisch-Muller, C. Oliveira, J. H. Mol, J. I. Montoya-Burgos, and S. Dray. 2016. Molecular phylogeny of the highly diversified catfish subfamily Loricariinae (Siluriformes: Loricariidae) reveals incongruences with morphological

classification. Molecular Phylogenetics and Evolution **94**:492–517.

Cox-Fernandes, C., A. Nogueira, and J. A. Alves-Gomes. 2014. *Procerusternarchus pixuna*, a new genus and species of electric knifefish (Gymnotiformes: Hypopomidae, Microsternarchini) from the Negro River, South America. Proceedings of the Academy of Natural Sciences of Philadelphia **163**:95–118.

Cramer, C. A. 2014. Redescription of *Panaqolus purusiensis* (LaMonte, 1935) (Siluriformes: Loricariidae) with identification key to the species of the genus. Neotropical Ichthyology **12**:61–70.

Crampton, W. G., C. D. de Santana, J. C. Waddell, and N. R. Lovejoy. 2016a. Phylogenetic systematics, biogeography, and ecology of the electric fish genus *Brachyhypopomus* (Ostariophysi: Gymnotiformes). PLoS ONE **11**:e0161680.

Crampton, W. G. R. 1996. Gymnotiform fish: an important component of Amazonian floodplain fish communities. Journal of Fish Biology **48**:298–301.

Crampton, W. G. R. 1998a. Effects of anoxia on the distribution, respiratory strategies and electric signal diversity of gymnotiform fishes. Journal of Fish Biology **53**:307–330.

Crampton, W. G. R. 1998b. Electric signal design and habitat preferences in a species rich assemblage of gymnotiform fishes from the upper Amazon basin. Anais da Academia Brasileira de Ciencias **70**:805–847.

Crampton, W. G. R. 2007. Diversity and adaptation in deep-channel Neotropical electric fishes. Pages 283–339 *in* P. Sebert, D. W. Onyango, and B. G. Kapoor, editors. Fish Life in Special Environments. Science Publishers, Enfield, NH, USA.

Crampton, W. G. R. 2008. Ecology and life history of an Amazon floodplain cichlid: The discus fish *Symphysodon* (Perciformes: Cichlidae). Neotropical Ichthyology **6**:599–612.

Crampton, W. G. R. 2011. An ecological perspective on diversity and distributions. Pages 165–192 *in* J. S. Albert and R. E. Reis, editors. Historical Biogeography of Neotropical Freshwater Fishes. University of California Press, Berkeley, USA.

Crampton, W. G. R., and J. S. Albert. 2003. A redescription of *Gymnotus coropinae*, an often misidentified species of gymnotid fish, with notes on ecology and electric organ discharge. Zootaxa **348**:1–20.

Crampton, W. G. R., and J. S. Albert. 2004. Redescription of *Gymnotus coatesi* (Gymnotiformes, Gymnotidae): a rare species of electric fish from the lowland Amazon basin, with descriptions of osteology, electric signals, and ecology. Copeia:525–533.

Crampton, W. G. R., and J. S. Albert. 2006. Evolution of electric signal diversity in gymnotiform fishes. Pages 641–725 *in* F. Ladich, S. P. Collin, P. Moller, and B. G. Kapoor, editors. Communication in Fishes. Science Publishers, Enfield, NH, USA..

Crampton, W. G. R., J. K. Davis, N. R. Lovejoy, and M. Pensky. 2008. Multivariate classification of animal communication signals: a simulation-based comparison of alternative signal processing procedures using electric fishes. Journal of Physiology Paris **102**:304–321.

Crampton, W. G. R., C. D. de Santana, J. C. Waddell, and N. R. Lovejoy. 2016b. A taxonomic revision of the Neotropical electric fish genus *Brachyhypopomus* (Ostariophysi: Gymnotiformes: Hypopomidae), with descriptions of 15 new species. Neotropical Ichthyology **14**:e150146.

Crampton, W. G. R., and C. D. Hopkins. 2005. Nesting and paternal care in the weakly electric fish *Gymnotus*

(Gymnotiformes: Gymnotidae) with descriptions of larval and adult electric organ discharges of two species. Copeia:48–60.

Crampton, W. G. R., K. G. Hulen, and J. S. Albert. 2004a. Redescription of *Sternopygus obtusirostris* Steindachner (Gymnotiformes, Sternopygidae) from the Amazon basin, with descriptions of ecology and electric organ discharges. Ichthyological Exploration of Freshwaters **15**:121–134.

Crampton, W. G. R., D. H. Thorsen, and J. S. Albert. 2004b. *Steatogenys ocellatus*: a new species of Neotropical electric fish (Gymnotiformes: Hypopomidae) from the lowland Amazon basin. Copeia:78–91.

Crampton, W. G. R., D. H. Thorsen, and J. S. Albert. 2005. Three new species from a diverse, sympatric assemblage of the electric fish *Gymnotus* (Gymnotiformes: Gymnotidae) in the lowland Amazon Basin, with notes on ecology. Copeia:82–99.

Cross, F. B. 1967. Handbook of fishes of Kansas. Museum of Natural History, University of Kansas, Lawrence, USA.

Curtis, C. C., and P. K. Stoddard. 2003. Mate preference in female electric fish, *Brachyhypopomus pinnicaudatus*. Animal Behaviour **66**:329–336.

D'Agosta, F. C. P. 2011. Taxonomia e relações filogenéticas do gênero *Astyanacinus* Eigenmann, 1907 (Characiformes: Characidae). Unpublished MSc. thesis. Universidade de São Paulo FFCLRP, Ribeirão Preto, SP, Brazil.

da Cruz, A. L., H. R. da Silva, L. M. Lundstedt, A. R. Schwantes, G. Moraes, W. Klein, and M. N. Fernandes. 2013. Air-breathing behavior and physiological responses to hypoxia and air exposure in the air-breathing loricariid fish, *Pterygoplichthys anisitsi*. Fish Physiology and Biochemistry **39**:243–256.

da Graça, W. J., C. S. Pavanelli, and P. A. Buckup. 2008. Two new species of *Characidium* (Characiformes: Crenuchidae) from Paraguay and Xingu river basins, State of Mato Grosso, Brazil. Copeia:326–332.

da Silva, C. C., E. J. G. Ferreira, and C. P. de Deus. 2008. Diet of five species of Hemiodontidae (Teleostei, Characiformes) in the area of influence of the Balbina reservoir, Uatuma River, State of Amazonas, Brazil. Iheringia Serie Zoologia **98**:464–468.

da Silva, P. C., C. Y. Nagamachi, D. dos Santos Silva, S. S. Rodrigues Milhomem, A. L. Cardoso, J. A. de Oliveira, and J. C. Pieczarka. 2013. Karyotypic similarities between two species of *Rhamphichthys* (Rhamphichthyidae, Gymnotiformes) from the Amazon basin. Comparative Cytogenetics **7**:279–291.

Dagosta, F. C. P., and A. Datovo. 2013. Monophyly of the Agoniatinae (Characiformes: Characidae). Zootaxa **3646**:265–276.

Dagosta, F. C. P., and M. N. De Lima Pastana. 2014. New species of *Creagrutus* Günther (Characiformes: Characidae) from rio Tapajós basin, Brazil, with comments on its phylogenetic position. Zootaxa **3765**:571–582.

Dahl, G. 1971. Los Peces del norte de Colombia. INDERENA, Bogota.

Dala-Corte, R. B., and C. B. Fialho. 2013. Reproductive tactics and development of sexually dimorphic structures in a stream-dwelling characid fish (*Deuterodon stigmaturus*) from Atlantic Forest. Environmental Biology of Fishes **97**:1119–1127.

Darwin, C. R. 1882. The Descent of Man and Selection in Relation to Sex. John Murray, London.

Datovo, A. 2014. A new species of *Ituglanis* from the Rio Xingu basin, Brazil, and the evolution of pelvic fin loss in trichomycterid catfishes (Teleostei: Siluriformes: Trichomycteridae). Zootaxa **3790**:466–476.

Datovo, A., and M. C. C. de Pinna. 2014. A new species of *Ituglanis* representing the southernmost record of the genus, with comments on phylogenetic relationships (Teleostei: Siluriformes: Trichomycteridae). Journal of Fish Biology **84**:314–327.

Datovo, A., and M. I. Landim. 2005. *Ituglanis macunaima*, a new catfish from the rio Araguaia basin, Brazil (Siluriformes: Trichomycteridae). Neotropical Ichthyology **3**:455–464.

Dawson, C. E. 1974. *Pseudophallus brasiliensis* (Pisces: Syngnathidae), a new freshwater pipefish from Brazil. Proceedings of the Biological Society of Washington **87**:405–410.

Dawson, C. E. 1979. Review of the polytypic doryrhamphine pipefish *Oostethus brachyurus* (Bleeker). Bulletin of Marine Science **29**:465–480.

Dawson, C. E. 1984. Revision of the genus *Microphis* Kaup (Pisces: Syngnathidae). Bulletin of Marine Science **35**:117–181.

de Almeida-val, V. M. F., A. R. C. Gomes, and N. P. Lopes. 2005. Metabolic and physiological adjustments to low oxygen and high temperature in fishes of the Amazon. Pages 443–500 *in* A. L. Val, V. M. F. de Almeida-val, and D. J. Randall, editors. The Physiology of Tropical Fishes. Elsevier.

de Carvalho, M. R., and N. R. Lovejoy. 2011. Morphology and phylogenetic relationships of a remarkable new genus and two new species of Neotropical freshwater stingrays from the Amazon basin (Chondrichthyes: Potamotrygonidae). Zootaxa **2776**:13–48.

de Carvalho, M. R., N. R. Lovejoy, and R. S. Rosa. 2003. Family Potamotrygonidae (River stingrays). Pages 22–28 *in* R. E. Reis, S. O. Kullander, and C. J. Ferraris Jr., editors. Check List of the Freshwater Fishes of South and Central America. EDIPUCRS, Porto Alegre, Brazil.

de Carvalho, M. R., and M. P. Ragno. 2011. An unusual, dwarf new species of Neotropical freshwater stingray, *Plesiotrygon nana* sp. nov., from the upper and mid Amazon basin: the second species of *Plesiotrygon* (Chondrichthyes: Potamotrygonidae). Papéis Avulsos de Zoologia **51**:101–138.

de Crop, W., E. Pauwels, L. Van Hoorebeke, and T. Geerinckx. 2013. Functional morphology of the Andean climbing catfishes (Astroblepidae, Siluriformes): alternative ways of respiration, adhesion, and locomotion using the mouth. Journal of Morphology **274**:1164–1179.

de Freitas, J. C. 2006. Eating habits: are we safe to consume freshwater puffer fish from the Amazon region in Brazil? Journal of Venomous Animals and Toxins Including Tropical Diseases **12**:153–155.

de Jesus, M. J., and C. C. Kohler. 2004. The commercial fishery of the Peruvian Amazon. Fisheries **29**:10–16.

de Lima Filho, J. A. J., J. Martins, R. Arruda, and L. N. Carvalho. 2012. Air-breathing behavior of the jeju fish *Hoplerythrinus unitaeniatus* in Amazonian streams. Biotropica **44**:512–520.

de Melo, C. E., J. D. Lima, and E. F. da Silva. 2009. Relationships between water transparency and abundance of Cynodontidae species in the Bananal floodplain, Mato Grosso, Brazil. Neotropical Ichthyology **7**:251–256.

de Mérona, B., R. Vigouroux, and V. Horeau. 2003. Changes in food resources and their utilization by fish assemblages in a large tropical reservoir in South America (Petit-Saut Dam, French Guiana). Acta Oecologica **24**:147–156.

de Morais, L. T., and J. Raffray. 1999. Movements of *Hoplias aimara* during the filling phase of the Petit-Saut dam, French Guyana. Journal of Fish Biology **54**:627–635.

de Oliveira, R. R., L. R. Py-Daniel, J. Zuanon, and M. S. Rocha. 2012. A new species of the ornamental catfish genus *Peckoltia* (Siluriformes: Loricariidae) from Rio Xingu basin, Brazilian Amazon. Copeia:547–553.

De Pinna, M. C. C. 1989. A new Sarcoglanidinae catfish, phylogeny of its subfamily, and an appraisal of the phyletic status of the Trichomycterinae (Teleostei, Trichomycteridae). American Museum Novitates **2950**:1–39.

De Pinna, M. C. C. 1998. Phylogenetic relationships of Neotropical Siluriformes (Teleostei: Ostariophysi): historical overview and synthesis of hypotheses. Phylogeny and Classification of Neotropical Fishes:279–330.

De Pinna, M. C. C., and H. A. Britski. 1991. *Megalocentor*, a new genus of parasitic catfish from the Amazon basin: the sister group of *Apomatoceros* (Trichomycteridae: Stegophilinae). Ichthyological Exploration of Freshwaters **2**:113–128.

De Pinna, M. C. C., C. J. Ferraris Jr., and R. P. Vari. 2007. A phylogenetic study of the Neotropical catfish family Cetopsidae (Osteichthyes, Ostariophysi, Siluriformes), with a new classification. Zoological Journal of the Linnean Society **150**:755–813.

De Pinna, M. C. C., and A. L. Kirovsky. 2011. A new species of sand-dwelling catfish, with a phylogenetic diagnosis of *Pygidianops* Myers (Siluriformes: Trichomycteridae: Glanapteryginae). Neotropical Ichthyology **9**:493–504.

De Pinna, M. C. C., and W. C. Starnes. 1990. A new genus and species of Sarcoglanidinae from the río Mamoré, Amazon basin, with comments on subfamilial phylogeny (Teleostei, Trichomycteridae). Journal of Zoology, London **222**:75–88.

De Pinna, M. C. C., and K. O. Winemiller. 2000. A new species of *Ammoglanis* (Siluriformes: Trichomycteridae) from Venezuela. Ichthyological Exploration of Freshwaters **11**:225–264.

De Pinna, M. C. C., and W. B. Wosiacki. 2003. Family Trichomycteridae (Pencil or parasitic catfishes). Pages 270–290 *in* R. E. Reis, S. O. Kullander, and C. J. Ferraris Jr., editors. Check List of the Freshwater Fishes of South and Central America. EDIPUCRS, Porto Alegre, Brazil.

De Pinna, M. C. C., and J. Zuanon. 2013. The genus *Typhlobelus*: monophyly and taxonomy, with description of a new species with a unique pseudotympanic structure (Teleostei: Trichomycteridae). Copeia:441–453.

De Santana, C. D., and W. G. R. Crampton. 2007. Revision of the deep-channel electric fish genus *Sternarchogiton* (Gymnotiformes: Apteronotidae). Copeia:387–402.

De Santana, C. D., and W. G. R. Crampton. 2010. A review of the South American electric fish genus *Porotergus* (Gymnotiformes: Apteronotidae) with the description of a new species. Copeia:165–175.

De Santana, C. D., and W. G. R. Crampton. 2011. Phylogenetic interrelationships, taxonomy, and reductive evolution in the Neotropical electric fish genus *Hypopygus* (Teleostei, Ostariophysi, Gymnotiformes). Zoological Journal of the Linnean Society **163**:1096–1156.

De Santana, C. D., and C. C. Fernandes. 2012. A new species of sexually dimorphic electric knifefish from the Amazon basin, Brazil (Gymnotiformes: Apteronotidae). Copeia:283–292.

De Santana, C. D., and R. P. Vari. 2009. The South American electric fish genus *Platyurosternarchus* (Gymnotiformes: Apteronotidae). Copeia:233–244.

De Santana, C. D., and R. P. Vari. 2010a. Electric fishes of the genus *Sternarchorhynchus* (Teleostei, Ostariophysi,

Gymnotiformes); phylogenetic and revisionary studies. Zoological Journal of the Linnean Society **159**:223–371.

De Santana, C. D., and R. P. Vari. 2010b. New rheophilic species of electric knifefish from the rapids and waterfalls of the lower Rio Xingu, Brazil (Gymnotiformes: Apteronotidae). Copeia:160–164.

De Souza, L. S., M. R. S. Melo, C. C. Chamon, and J. W. Armbruster. 2008. A new species of *Hemiancistrus* from the rio Araguaia basin, Goiás state, Brazil (Siluriformes: Loricariidae). Neotropical Ichthyology **6**:419–424.

Devincenzi, G. J., and G. W. Teague. 1942. Ictiofauna del rio Uruguai medio. Anales del Museo de Historia Natural de Montevideo (série 2) **5**:1–100.

Dias, J. H., S. G. C. Britto, N. C. Vianna, and J. C. Garavello. 2004. Biological and ecological aspects of *Pinirampus pirinampu* (Spix, 1829), Siluriformes, Pimelodidae, in Capivara reservoir, Paranapanema River, southern Brazil. Acta Limnologica Brasiliensia **16**:293–304.

Dias, T. S., and C. B. Fialho. 2009. Feeding biology of four sympatric Cheirodontinae species (Characiformes, Characidae) from Ceará Mirim River, Rio Grande do Norte, Brazil. Iheringia–Serie Zoologia **99**:242–248.

Diaz-Sarmineto, J. A., and R. Alvarez-León. 2003. Migratory fishes of the Colombian Amazon. Pages 303–344 *in* Migratory Fishes of South America. International Bank for Reconstruction and Development/World Bank, Washington, DC.

DoNascimiento, C. 2015. Morphological evidence for the monophyly of the subfamily of parasitic catfishes Stegophilinae (Siluriformes, Trichomycteridae) and phylogenetic diagnoses of its genera. Copeia:933–960.

DoNascimiento, C., and J. G. Lundberg. 2005. *Myoglanis aspredinoides* (Siluriformes: Heptapteridae), a new catfish from the río Ventuari, Venezuela. Zootaxa **1009**:37–49.

DoNascimiento, C., and N. Milani. 2008. The Venezuelan species of *Phenacorhamdia* (Siluriformes: Heptapteridae), with the description of two new species and a remarkable new tooth morphology for siluriforms. Proceedings of the Academy of Natural Sciences of Philadelphia **157**:163–180.

DoNascimiento, C., and F. Provenzano. 2006. The genus *Henonemus* (Siluriformes: Trichomycteridae) with a description of a new species from Venezuela. Copeia:198–205.

Dorn, E., and F. Schaller. 1972. Über den lautapparat von Amazonasfischen. 2. Die Naturwissenschaften **59**:169–170.

Dotzer, P., and T. Weiner. 2003. *Harttia guianensis*. Die Pflege und die Zucht sind swar schwierig, aber doch möglich. Das Aquarium **407**:20–26.

Duponchelle, F., F. Lino, N. Hubert, J. Panfili, J.-F. Renno, E. Baras, J. P. Torrico, R. Dugue, and J. Nunez. 2007. Environment-related life-history trait variations of the red-bellied piranha *Pygocentrus nattereri* in two river basins of the Bolivian Amazon. Journal of Fish Biology **71**:1113–1134.

Duque, A., D. Taphorn, and K. Winemiller. 1998. Ecology of the coporo, *Prochilodus mariae* (Characiformes, Prochilodontidae), and status of annual migrations in western Venezuela. Environmental Biology of Fishes **53**:33–46.

Dutra, G. M., C. D. De Santana, R. P. Vari, and W. B. Wosiacki. 2014. The South American electric glass knifefish genus *Distocyclus* (Gymnotiformes: Sternopygidae): redefinition and revision. Copeia:345–354.

Eigenmann, C. H. 1909. Reports on the expedition to British Guiana of the Indiana University and the Carnegie Museum; 1908. Annals of the Carnegie Museum **6**:4–54.

Eigenmann, C. H. 1912a. The freshwater fishes of British Guiana, including a study of the ecological grouping of species, and the relation of the fauna of the plateau to that of the lowlands. Memoirs of the Carnegie Museum **5**:1–578.

Eigenmann, C. H. 1912b. Some results from an ichthyological reconnaissance of Colombia, South America. Part I Indiana University Studies **8**:1–27.

Eigenmann, C. H. 1915a. The Cheirodontinae, a subfamily of minute characid fishes of South America. Memoirs of the Carnegie Museum 7:1–99.

Eigenmann, C. H. 1915b. The Serrasalminae and Mylinae. Annals of the Carnegie Museum **9**:226–262.

Eigenmann, C. H. 1917. The Pygidiidae. Proceedings of the Indiana Academy of Science **27**:59–66.

Eigenmann, C. H. 1918a. The American Characidae. Part 2. Memoirs of the Museum of Comparative Zoology **43**:103–208.

Eigenmann, C. H. 1918b. The Pygidiidae, a family of South American catfishes. Memoirs of the Carnegie Museum 7:259–398.

Eigenmann, C. H. 1921. The American Characidae. Part 3. Memoirs of the Museum of Comparative Zoology **43**:209–310.

Eigenmann, C. H. 1922. On a new genus and two new species of Pygidiidae, a family of South American nematognaths. Bijdragen tot de Dierkunde **22**:113–114.

Eigenmann, C. H. 1925. A review of the Doradidae, a family of South American Nematognathi, or catfishes. Transactions of the American Philosophical Society (N. S.) **22**:280–365.

Eigenmann, C. H. 1927. The American Characidae. Part 4. Memoirs of the Museum of Comparative Zoology **43**:311–428.

Eigenmann, C. H., and R. S. Eigenmann. 1888. American Nematognathi. American Naturalist **22**:647–649.

Eigenmann, C. H., and R. S. Eigenmann. 1890. Preliminary notes on South American Nematognathi or catfishes. Occasional Papers of the California Academy of Sciences **1**:1–508.

Eigenmann, C. H., W. L. McAtee, and D. P. Ward. 1907. On further collections of fishes from Paraguay. Annals of the Carnegie Museum **4**:110–157.

Eigenmann, C. H., and G. S. Myers. 1929. The American Characidae. Part 5. Memoirs of the Museum of Comparative Zoology **43**:429–558.

Ellis, M. D. 1911. On the species of *Hasemania*, *Hyphessobrycon*, and *Hemigrammus* collected by J. D. Haseman for the Carnegie Museum. Annals of the Carnegie Museum **8**:148–163.

Eschmeyer, W. N., R. Fricke, and R. van der Laan, editors. 2016. Catalogue of Fishes: Genera, Species, References. (http://researcharchive.calacademy.org/research/ichthyology/catalog/fishcatmain.asp). Electronic version accessed April 2016.

Esguicero, A. L., and M. S. Arcifa. 2010. Biology and population features of a rare species of Pseudopimelodidae from the upper Parana river basin. Biota Neotropica **10**:161–167.

Evers, H. G., and I. Seidel. 2005. Catfish Atlas Vol. 1: South American Catfishes of the Families Loricariidae, Cetopsidae, Nematogenyidae and Trichomycteridae. Mergus Verlag, Melle, Germany.

Farina, S. C., T. J. Near, and W. E. Bemis. 2015. Evolution of the branchiostegal membrane and restricted gill openings in actinopterygian fishes. Journal of Morphology **276**:681–694.

Farrell, A. P. 1978. Cardiovascular events associated with air breathing in two teleosts, *Hoplerythrinus unitaeniatus* and *Arapaima gigas*. Canadian Journal of Zoology **56**:953–958.

Favorito, S. E., A. M. Zanata, and M. I. Assumpção. 2005. A new *Synbranchus* (Teleostei: Synbranchiformes: Synbranchidae) from Ilha de Marajó, Pará, Brazil, with notes on its reproductive biology and larval development. Neotropical Ichthyology **3**:319–328.

Fernandes, C. C. 1997. Lateral migration of fishes in Amazon floodplains. Ecology of Freshwater Fish **6**:36–44.

Fernandez, J. M., and S. H. Weitzman. 1987. A new species of *Nannostomus* (Teleostei: Lebiasinidae) from near Puerto Ayacucho, río Orinoco drainage, Venezuela. Proceedings of the Biological Society of Washington **100**:164–172.

Fernández, L., and G. Miranda. 2007. A catfish of the genus *Trichomycterus* from a thermal stream in southern South America (Teleostei, Siluriformes, Trichomycteridae), with comments on relationships within the genus. Journal of Fish Biology **71**:1303–1316.

Fernandez, L., L. J. Saucedo, F. M. Carvajal-Vallejos, and S. A. Schaefer. 2007. A new phreatic catfish of the genus *Phreatobius* Goeldi 1905 from groundwaters of the Iténez River, Bolivia (Siluriformes: Heptapteridae). Zootaxa **1626**:51–58.

Fernández, L., and S. A. Schaefer. 2003. *Trichomycterus yuska*, a new species from high elevations of Argentina (Siluriformes: Trichomycteridae). Ichthyological Exploration of Freshwaters **14**:353–360.

Fernández, L., and S. A. Schaefer. 2009. Relationships among the Neotropical candirus (Trichomycteridae, Siluriformes) and the evolution of parasitism based on analysis of mitochondrial and nuclear gene sequences. Molecular Phylogenetics and Evolution **52**:416–423.

Fernández, L., and R. P. Vari. 2000. New species of *Trichomycterus* (Teleostei: Siluriformes: Trichomycteridae) lacking a pelvic fin and girdle from the Andes of Argentina. Copeia:990–996.

Fernández, L., and R. P. Vari. 2012. New species of *Trichomycterus* (Teleostei: Siluriformes) from the Andean cordillera of Argentina and the second record of the genus in thermal waters. Copeia:631–636.

Ferraris, C. J., and F. Mago-Leccia. 1989. A new genus and species of pimelodid catfish from the río Negro and río Orinoco drainages of Venezuela (Siluriformes: Pimelodidae). Copeia:166–171.

Ferraris Jr., C. J. 2000. The deep-water South American catfish genus *Pseudepapterus* (Ostariophysi: Auchenipteridae). Ichthyological Exploration of Freshwaters **11**:97–112.

Ferraris Jr., C. J. 2003a. Subfamily Loricariinae. Pages 330–350 *in* R. E. Reis, F. Kullander, and C. J. Ferraris Jr., editors. Check List of the Freshwater Fishes of South and Central America. EDIPUCRS, Porto Alegre, Brazil.

Ferraris Jr., C. J. 2003b. Family Auchenipteridae. Pages 470–482 *in* R. E. Reis, S. O. Kullander, and C. J. Ferraris Jr., editors. Check List of the Freshwater Fishes of South and Central America. EDIPUCRS, Porto Alegre, Brazil.

Ferraris Jr., C. J. 2007. Checklist of catfishes, recent and fossil (Osteichthyes: Siluriformes), and catalogue of siluriform primary types. Zootaxa **1418**:1–628.

Ferraris Jr., C. J., and B. A. Brown. 1991. A new species of *Pseudocetopsis* from the río Negro drainage of Venezuela (Siluriformes: Cetopsidae). Copeia:161–165.

Ferraris Jr., C. J., and R. P. Vari. 1999. The South American catfish genus *Auchenipterus* Valenciennes, 1840 (Ostariophysi:

Siluriformes: Auchenipteridae): monophyly and relationships, with a revisionary study. Zoological Journal of the Linnean Society **126**:387–450.

Ferraris Jr., C. J., R. P. Vari, and S. J. Raredon. 2005. Catfishes of the genus *Auchenipterichthys* (Osteichthyes: Siluriformes: Auchenipteridae); a revisionary study. Neotropical Ichthyology **3**:89–106.

Ferreira, E. J. G. 1984. A ictiofauna da represa hidrelétrica de Curuá-Una, Santarém, Pará. I: Lista e distribuição das especies. Amazoniana **8**:351–363.

Ferreira, E. J. G., J. A. S. Zuanon, B. Forsberg, M. Goulding, and R. Briglia-Ferreira. 2007. Rio Branco: peixes, ecologia e conservação de Roraima. Amazon Conservation Association (ACA)–Instituto Nacional de Pesquisas da Amazônia (INPA)– Sociedade Civil Mamirauá.

Ferreira, F. S., W. Vicentin, F. E. S. Costa, and Y. R. Súarez. 2014. Trophic ecology of two piranha species, *Pygocentrus nattereri* and *Serrasalmus marginatus* (Characiformes, Characidae), in the floodplain of the Negro River, Pantanal. Acta Limnologica Brasiliensia **26**.

Ferreira, G. L., V. G. Lopes, J. H. Cantarino Gomes, and C. W. Castelo Branco. 2013. Condition factor and diet of the catfish *Loricariichthys castaneus* (Castelnau, 1855): gender differences in three tropical reservoirs. Studies on Neotropical Fauna and Environment **48**:190–198.

Ferreira, K. M., and F. M. Carvajal. 2007. *Knodus shinahota* (Characiformes: Characidae) a new species from the río Shinahota, río Chapare basin (Mamoré system), Bolivia. Neotropical Ichthyology **5**:31–36.

Ferreira, K. M., and F. C. T. Lima. 2006. A new species of *Knodus* (Characiformes: Characidae) from the Rio Tiquié, upper Rio Negro system, Brazil. Copeia:630–639.

Ferreira, K. M., N. A. Menezes, and I. Quagio-Grassioto. 2011. A new genus and two new species of Stevardiinae (Characiformes: Characidae) with a hypothesis on their relationships based on morphological and histological data. Neotropical Ichthyology **9**:281–298.

Ferreira da Silva, G. d. S., H. Giusti, A. P. Sanchez, J. M. do Carmo, and M. L. Glass. 2008. Aestivation in the South American lungfish, *Lepidosiren paradoxa*: effects on cardiovascular function, blood gases, osmolality and leptin levels. Respiratory Physiology & Neurobiology **164**:380–385.

Ferriz, R. A., C. A. Villar, D. Colautti, and C. Bonetto. 2000. Alimentacion de *Pterodoras granulosus* (Valenciennes) (Pisces, Doradidae) en la baja cuenca del Plata. Revista del Museo Argentino de Ciencias Natureles **2**:151–156.

Fichberg, I., and C. C. Chamon. 2008. *Rineloricaria osvaldoi* (Siluriformes: Loricariidae): A new species of armored catfish from rio Vermelho, Araguaia basin, Brazil. Neotropical Ichthyology **6**:347–354.

Fichberg, I., O. T. Oyakawa, and M. C. C. de Pinna. 2014. The end of an almost 70-year wait: a new species of *Spatuloricaria* (Siluriformes: Loricariidae) from the Rio Xingu and Rio Tapajós basins. Copeia:317–324.

Field, C., and C. B. Braun. 2012. The role of a chirp-like signal in the weakly electric fish *Steatogenys* sp. Frontiers in Behavioral Neuroscience Conference Abstract: Tenth International Congress of Neuroethology.

Figueiredo, C. A. 2003. Análise cladística da subfamília Poeciliidae (Cyprinodontiformes: Poeciliidae) com ênfase nasinter-relações

dos gêneros *Poecilia*, *Limia* e *Pamphorichthys*. Unpublished PhD thesis. Universidade de São Paulo, São Paulo, Brazil.

Figueiredo, C. A. 2008. A new *Pamphorichthys* (Cyprinodontiformes: Poeciliidae: Poeciliini) from central Brazil. Zootaxa **1918**:59–69.

Figueiredo, C. A., and M. R. Britto. 2010. A new species of *Xyliphius*, a rarely sampled banjo catfish (Siluriformes: Aspredinidae) from the rio Tocantins-Araguaia system. Neotropical Ichthyology **8**:105–112.

Figueiredo-Silva, S. L., M. Camargo, and R. A. Estupiñán. 2012. Fishery management in a conservation area: the case of the Oiapoque River in northern Brazil. Cybium **36**:17–30.

Finger, S., and M. Piccolino. 2011. The Shocking History of Electric Fishes: From Ancient Epochs to the Birth of Modern Neurophysiology. Oxford University Press, New York.

Fink, W. L. 1993. Revision of the piranha genus *Pygocentrus* (Teleostei, Characiformes). Copeia:665–687.

Fink, W. L., and A. Machado-Allison. 1992. Three new species of piranhas from Brazil and Venezuela (Teleostei: Characiformes). Ichthyological Exploration of Freshwaters **3**:55–71.

Fisch-Muller, S., J. I. Montoya-Burgos, P. Y. Le Bail, and R. Covain. 2012. Diversity of the Ancistrini (Siluriformes: Loricariidae) from the Guianas: the *Panaque* group, a molecular appraisal with descriptions of new species. Cybium **36**:163–193.

Flecker, A. S. 1996. Ecosystem engineering by a dominant detritivore in a diverse tropical stream. Ecology **77**:1845–1854.

Fleishman, L. J. 1992. Communication in the weakly electric fish *Sternopygus macrurus*. Journal of Comparative Physiology A **170**:335–348.

Fonteles, S. B. A., C. E. Lopes, A. Akama, F. M. C. Fernandes, F. Porto-Foresti, J. A. Senhorini, M. d. F. Z. Daniel-Silva, F. Foresti, and L. F. de Almeida-Toledo. 2008. Cytogenetic characterization of the strongly electric Amazonian eel, *Electrophorus electricus* (Teleostei, Gymnotiformes), from the Brazilian rivers Amazon and Araguaia. Genetics and Molecular Biology **31**:227–230.

Franchina, C., and C. Hopkins. 1996. The dorsal filament of the weakly electric Apteronotidae (Gymnotiformes; Teleostei) is specialized for electroreception. Brain Behavior and Evolution **47**:165–178.

Freeman, B., L. G. Nico, M. Osentoski, H. L. Jelks, and T. M. Collins. 2007. Molecular systematics of Serrasalmidae: deciphering the identities of piranha species and unraveling their evolutionary histories. Zootaxa **1484**:1–38.

Freitas, C. E. C., F. K. Siqueira-Souza, R. Humston, and L. E. Hurd. 2012. An initial assessment of drought sensitivity in Amazonian fish communities. Hydrobiologia **705**:159–171.

Friel, J. P. 1994. A phylogenetic study of the Neotropical banjo catfishes (Teleostei: Siluriformes: Aspredinidae). Unpublished PhD thesis. Duke University, Durham, NC, USA.

Friel, J. P. 1995. *Acanthobunocephalus nicoi*, a new genus and species of miniature banjo-catfish from the upper Orinoco and Casiquiare rivers, Venezuela. (Siluriformes: Aspredinidae). Ichthyological Exploration of Freshwaters **6**:89–95.

Friel, J. P. 2003. Family Aspredinidae (Banjo catfishes). Pages 261–267 *in* R. E. Reis, S. O. Kullander, and C. J. Ferraris Jr., editors. Check List of the Freshwater Fishes of South and Central America. EDIPUCRS, Port Alegre, Brazil.

Friel, J. P. 2008. *Pseudobunocephalus*, a new genus of banjo catfish with the description of a new species from the Orinoco River system of Colombia and Venezuela (Siluriformes: Aspredinidae). Neotropical Ichthyology **6**:293–300.

Friel, J. P., and J. G. Lundberg. 1996. *Micromyzon akamai*, gen. et sp. nov., a small and eyeless banjo catfish (Siluriformes: Aspredinidae) from the river channels of the lower Amazon basin. Copeia:641–648.

Fryer, G., and T. D. Iles. 1972. The cichlid fishes of the Great Lakes of Africa. Their biology and evolution. Oliver & Boyd, Edinburgh.

Fugi, R., A. A. Agostinho, and N. S. Hahn. 2001. Trophic morphology of five benthic-feeding fish species of a tropical floodplain. Revista Brasileira de Biologia **61**:27–33.

Fugi, R., K. D. G. Luz-Agostinho, and A. A. Agostinho. 2008. Trophic interactions between an introduced (peacock bass) and a native (dogfish) piscivorous fish in a Neotropical impounded river. Hydrobiologia **607**:143–150.

Fuller, I. A. M., and H.-G. Evers. 2005. Identifying Corydoradinae Catfish. Verslag A.C.S. GmbH, Offenbach, Germany.

Fuller, P. L., L. G. Nico, and J. D. Williams. 1999. Nonindigenous Fishes Introduced to Inland Waters of the United States. American Fisheries Society, Bethesda, MD, USA.

Galetti, M., C. I. Donatti, M. A. Pizo, and H. C. Giacomini. 2008. Big fish are the best. Biotropica **40**:386–389.

Galvis, G., J. I. Mojica, and M. Camargo. 1997. Peces del Catatumbo. Asociación Cravo Norte, Santafé de Bogotá, D.C.

Galvis, G., J. I. Mojica, S. Duque, C. Castellanos, P. Sánchez, M. Arce, A. Guitierrez, L. F. Jiménez, M. Santos-Acevedo, S. Vejarano, F. Arbeláez, E. Prieto, and M. Leiva. 2006. Peces del medio Amazonas, Región de Leticia. Conservación Internacional, Bogotá, D.C. Colombia.

Gama, C. S., and E. P. Caramaschi. 2001. Alimentação de *Triportheus albus* (Cope 1971) (Osteichthyes, Characiformes) face à implantação do AHE Serra da Mesa no rio Tocantins. Revista Brasileira de Zoociências **3**:159–170.

Garavello, J. C. 1994. Descrição de uma nova espécie do gênero *Schizodon* Agassiz da bacia do Rio Uruguai, Brasil (Ostariophysi, Anostomidae). Comunicações do Museu de Ciências da PUCRS, Série Zoologia **7**:179–193.

Garavello, J. C. 2000. Two new species of *Leporinus* Spix with a review of the blotched species of the Rio Orinoco system and rescription of *Leporinus muyscorum* Steindachner (Characiformes: Anostomidae). Proceedings of the Academy of Natural Sciences of Philadelphia **150**:193–201.

Garavello, J. C., and H. A. Britski. 1990. Duas novas espécies do gênero *Schizodon* Agassiz da bacia do alto Paraná, Brasil, América do Sul (Ostariophysi, Anostomidae). Naturalia (São Paulo) **15**:153–170.

Garavello, J. C., and H. A. Britski. 2003. Family Anostomidae. Pages 71–84 *in* R. E. Reis, S. O. Kullander, and C. J. Ferraris Jr., editors. Check List of the Freshwater Fishes of South and Central America. EDIPUCRS, Porto Alegre, Brazil.

Garavello, J. C., J. P. Garavello, and A. K. Oliveira. 2010. Ichthyofauna, fish supply and fishermen activities on the mid-Tocantins River, Maranhão state, Brazil. Brazilian Journal of Biology **70**:575–585.

Garg, T. K., F. X. Valdez Domingos, V. M. F. Almeida-Val, and A. L. Val. 2010. Histochemistry and functional organization of the dorsal skin of *Ancistrus dolichopterus* (Siluriformes: Loricariidae). Neotropical Ichthyology **8**:877–884.

Garman, S. 1986. Cross fertilization and sexual rights and lefts. American Naturalist **30**:232.

Garrone Neto, D., and L. N. Carvalho. 2011. Nuclear-follower foraging associations among Characiformes fishes and Potamotrygonidae rays in clean waters environments of Teles Pires and Xingu rivers basins, midwest Brazil. Biota Neotropica **11**:359–362.

Gayet, M., F. J. Meunier, and F. Kirschbaum. 1994. *Ellisella kirschbaumi* Gayet & Meunier, 1991, gymnotiforme fossile de Bolivie et ses relations phylogénétiques au sein des formes actuelles. Cybium **18**:273–306.

Gee, J. H. 1976. Buoyancy and aerial respiration: factors influencing the evolution of reduced swim-bladder volume of some Central American catfishes (Trichomycteridae, Callichthyidae, Loricariidae, Astroblepidae). Canadian Journal of Zoology **54**:1030–1037.

Gee, J. H., and J. B. Graham. 1978. Respiratory and hydrostatic functions of the intestine of the catfishes *Hoplosternum thoracatum* and *Brochis splendens* (Callichthyidae). Journal of Experimental Biology **74**:1–16.

Geerinckx, T., and D. Adriaens. 2006. The erectile cheek-spine apparatus in the bristlenose catfish *Ancistrus* (Loricariidae, Siluriformes), and its relation to the formation of a secondary skull roof. Zoology **109**:287–299.

Geerinckx, T., A. Herrel, and D. Adriaens. 2011. Suckermouth armored catfish resolve the paradox of simultaneous respiration and suction attachment: a kinematic study of *Pterygoplichthys disjunctivus*. Journal of Experimental Zoology Part A: Ecological Genetics and Physiology **315 A**:121–131.

German, D. P. 2009. Inside the guts of wood-eating catfishes: can they digest wood? Journal of Comparative Physiology B: Biochemical, Systemic, and Environmental Physiology **179**:1011–1023.

Géry, J. 1960a. Contributions a l'étude des poissons characoïdes (No. 7). Validité de *Leporinus despaxi* Puyo et du sous-genre *Hypomasticus* Borodin. Bulletin du Muséum National d'Histoire Naturelle (Sér. 2) **32**:222–229.

Géry, J. 1960b. New Cheirodontinae from French Guiana. Senckenbergiana Biologica **41**:15–39.

Géry, J. 1964a. Poissons characoides de l'Amazonie péruvienne. Beitrage zur Neotropischen Fauna **4**:1–44.

Géry, J. 1964b. Poissons Characoides nouveaux ou non signalés de l'Ilha do Bananal, Brésil. Vie Milieu Suppl **16**:447–471.

Géry, J. 1965a. A new genus from Brazil–*Brittanichthys*–a new sexually-dimorphic characid genus with peculiar caudal ornament from Rio Negro, Brazil, with a discussion of certain cheirodontin genera and a description of two new species, *B. axelrodi* and *B. myersi*. Tropical Fish Hobbyist **13**:13–24, 61–69.

Géry, J. 1965b. Poissons characoïdes sud-américains du Senckenberg Muséum, II. Characidae et Crenuchidae de l'Igarapé Préto (Haute Amazonie). Senckenbergiana Biologica **46**:195–218.

Géry, J. 1966. *Hoplocharax goethei*, a new genus and species of South American characoid fishes, with a review of the sub-tribe Heterocharacini. Ichthyologica, the Aquarium Journal **38**:281–296.

Géry, J. 1972a. Poissons characoides des Guyanes. II. Famille des Serrasalmindae. Zoologische Verhandelingen, Leiden **122**:134–250.

Géry, J. 1972b. Poissons characoïdes des Guyanes: I. Généralités; II. Famille des Serrasalmidae. Zoologische Verhandelingen **122**:1–250.

Géry, J. 1973. New and little-known Aphyoditeina (Pisces, Characoidei) from the Amazon basin. Studies on the Neotropical Fauna **8**:81–137.

Géry, J. 1976. Les Genres de Serrasalmindae (Pisces, Characoidei). Bulletin Zoologisch Museum Universiteit van Amsterdam **5**:47–54.

Géry, J. 1977. Characoids of the World. T.F.H. Publications, Neptune City, NJ, USA.

Gery, J. 1980. Un nouveau poisson characoide occupant la niche des mangeurs d'ecailles dans le haut rio Tapajoz, Bresil: *Bryconexodon juruenae* n. g. sp. Revue Française d'Aquariologie Herpétologie **7**:1–8.

Géry, J. 1986. Notes de characologie néotropicale. I. Progrès dans la systématique des genres *Colossoma* et *Piaractus*. Revue Française d'Aquariologie Herpétologie **12 (1985)**:97–102.

Géry, J. 1987. Description d'une nouvelle espèce de poisson anostomidé (Ostariophysi, Characoidei) du Rio Mamoré, Bolivie: *Rhytiodus lauzannei* sp. n. Cybium **11**:365–373.

Géry, J. 1993. Description de trois espèces nouvelles du genre *Iguanodectes* (Pisces, Characiformes, Characidae), avec quelques données récentes sur les autres espèces. Revue Française d'Aquariologie Herpétologie **19**:97–105.

Géry, J., and H. Boutiere. 1964. *Petitella georgiae*, gen. et sp.nov. (Pisces, Cypriniformes, Characoidei). Vie et Milieu **17**:473–484.

Gery, J., and L. Lauzanne. 1990. Les types des espèces du genre *Salminus* Agassiz, 1829 (Ostariophysi, Characidae) du Muséum National d'Histoire Naturelle de Paris. Cybium **14**:113–124.

Géry, J., P. Le Bail, and P. Keith. 1999. *Cynodon meionactis* sp. n., un nouveau characidé endémique du bassin du Haut Maroni en Guyane, avec une note sur la validité du genre *Cynodon* (Teleostei: Ostariophysi: Characiformes). Revue Française d'Aquariologie Herpétologie **25**:69–77.

Géry, J., V. Mahnert, and C. Dlouhy. 1987. Poissons characoides non characidae du Paraguay (Pisces, Ostariophysi). Revue Suisse de Zoologie **94**:357–464.

Géry, J., P. Planquette, and P.-Y. Le Bail. 1991. Faune Characoide (Poissons Ostariophysaires) de l'Oyapoque, l'Approuague et la riviere de Kaw (Guyane Francaise). Cybium **15**:1–69.

Géry, J., and C. Poivre. 1979. Un curieux système d'accrochage par dents transitoires chez les jeunes *Rhaphiodon vulpinus* Agassiz (Pisces, Cypriniformes, Characidae). Revue Française d'Aquariologie et Herpétologie **6**:1–4.

Géry, J., and U. Römer. 1997. *Tucanoichthys tucano* gen. n. sp. n., a new miniature characid fish (Teleostei: Characiformes: Characidae) from the rio Uaupes basin in Brazil. Aqua, Journal of Ichthyology and Aquatic Biology **2**:65–72.

Géry, J., and A. Zarske. 2002. *Derhamia hoffmannorum* gen. et sp. n.: a new pencil fish (Teleostei, Characiformes, Lebiasinidae), endemic from the Mazaruni River in Guyana. Zoologische Abhandlungen **52**:35–47.

Gery, J. R. 1962. Notes on the ichthyology of Surinam and other Guianas. 10. The distribution pattern of the genus *Hemibrycon*, with a description of a new species from Surinam and an incursion into ecotaxonomy. Bulletin of Aquatic Biology **3**:65–80.

Gery, J. R. 1963. Paired frontal foramina in living teleosts: definition of a new family of characoid fishes, the crenuchidae. Nature **198**:502–503.

Ghedotti, M. J. 1998. Phylogeny and classification of the Anablepidae (Teleostei: Cyprinodontiformes). Pages 560–582 *in* L. R. Malabarba, R. E. Reis, R. P. Vari, Z. M. S. Lucena, and C. A. S. Lucena, editors. Phylogeny and Classification of Neotropical Fishes. EDIPUCRS, Porto Alegre, Brazil.

Ghedotti, M. J. 2000. Phylogenetic analysis and taxonomy of the poecilioid fishes (Teleostei: Cyprinodontiformes). Zoological Journal of the Linnean Society **130**:1–53.

Ghedotti, M. J. 2003. Family Anablepidae. Pages 582–585 *in* R. E. Reis, S. O. Kullander, and C. J. Ferraris Jr., editors. Check List of the Freshwater Fishes of South and Central America. EDIPUCRS, Porto Alegre, Brazil.

Gilbert, C. R., and J. E. Randall. 1979. Two new western Atlantic species of the gobiid fish genus *Gobionellus*, with remarks on characteristics of the genus. Northeast Gulf Science **3**:27–47.

Gill, A. C., and R. D. Mooi. 2012. Thalasseleotrididae, new family of marine gobioid fishes from New Zealand and temperate Australia, with a revised definition of its sister taxon, the Gobiidae (Teleostei: Acanthomorpha). Zootaxa **3266**:41–52.

Gill, T. 1872. Arrangement of the families of fishes, or classes Pisces, Marsipobranchii, and Leptocardii. Smithsonian Contribution Zoology **247**:1–49.

Gimenes, M. D. F., R. Fugi, A. Isaac, and M. R. D. Silva. 2013. Spatial, seasonal and ontogenetic changes in food resource use by a piscivore fish in two Pantanal lagoons, Brazil. Neotropical Ichthyology **11**:163–170.

Giora, J., C. B. Fialho, and A. P. S. Dufech. 2005. Feeding habit of *Eigenmannia trilineata* Lopez & Castello, 1966 (Teleostei: Sternopygidae) of Parque Estadual de Itapuã, RS, Brazil. Neotropical Ichthyology **3**:291–298.

Giora, J., L. R. Malabarba, and W. Crampton. 2008. *Brachyhypopomus draco*, a new sexually dimorphic species of Neotropical electric fish from southern South America (Gymnotiformes: Hypopomidae). Neotropical Ichthyology **6**:159–168.

Giora, J., H. M. Tarasconi, and C. B. Fialho. 2014. Reproduction and feeding of the electric fish *Brachyhypopomus gauderio* (Gymnotiformes: Hypopomidae) and the discussion of a life history pattern for gymnotiforms from high latitudes. PLoS ONE **9**:e106515.

Glaser, U., F. Schäfer, and W. Glaser. 1996. All *Corydoras*. Verslag A.C.S. GmbH, Mörfelden-Walldorf, Germany.

Göbel, M., and H. J. Mayland. 1998. South American Cichlids IV. A.C.S. (Aqualog), Mörfelden-Walldorf.

Godoy, M. P. 1975. Peixes do Brasil. Subordem Characoidei. Bacia de Rio Mogi Guassu. Vol II. Ed. Franciscana, Piracicaba.

Gomes, I. D., F. G. Araújo, R. J. Albieri, and W. Uehara. 2012. Opportunistic reproductive strategy of a non-native fish, the spotted metynnis *Metynnis maculatus* (Kner, 1858) (Characidae Serrasalminae) in a tropical reservoir in south-eastern Brazil. Tropical Zoology **25**:2–15.

Gomes, I. D., F. G. Araújo, W. Uehara, and A. Sales. 2011. Reproductive biology of the armoured catfish *Loricariichthys castaneus* (Castelnau, 1855) in Lajes reservoir, southeastern Brazil. Journal of Applied Ichthyology **27**:1322–1331.

Gonçalves, T. K., M. A. Azevedo, L. R. Malabarba, and C. B. Fialho. 2005. Reproductive biology and development of sexually dimorphic structures in *Aphyocharax anisitsi* (Ostariophysi: Characidae). Neotropical Ichthyology **3**:433–438.

Gosline, W. A. 1951. Notes on the characid fishes of the subfamily Serrasalminae. Proceedings of the California Academy of Sciences **27**:17–64.

Gosse, J. P. 1971. Révision du genre *Retroculus* (Castelnau, 1855), designation d'un neotype de *Retroculus lapidifer* (Castelnau, 1855) et description de deux espèces nouvelles. Bulletin, Institut royale des Sciences naturelles de Belgique **47**:1–13.

Gottsberger, G. 1978. Seed dispersal by fish in the inundated regions of Humaita, Amazonia. Biotropica **10**:170–183.

Goulding, M. 1980. The Fishes and the Forest: Explorations in Amazonian Natural History. University of California Press, Berkeley, USA.

Goulding, M. 1981. Man and Fisheries on an Amazon Frontier. W. Junk, The Hague.

Goulding, M. 1983a. Amazonian fisheries. Pages 189–210 *in* E. F. Moran, editor. The Dilemma of Amazonian Development. Westview Press, Boulder, CO, USA.

Goulding, M. 1983b. The role of fishes in seed dispersal and plant distribution in Amazonian floodplain ecosystems. Sonderbd. naturwiss. Ver. Hamburg **7**:271–283.

Goulding, M. 1988. Ecology and management of migratory food fishes of the Amazon basin. Pages 71–85 *in* F. Almeda and C. M. Pringle, editors. Tropical Rainforests: Diversity and Conservation. California Academy of Sciences, San Francisco, USA.

Goulding, M., R. Barthem, and E. J. G. Ferreira. 2003. The Smithsonian Atlas of the Amazon. Smithsonian Books, Washington, DC, USA.

Goulding, M., and M. L. Carvalho. 1984. Ecology of Amazonian needlefishes (Belonidae). Revista Brasileira de Zoologia **2**:99–111.

Goulding, M., M. L. Carvalho, and E. J. G. Ferreira. 1988. Río Negro, rich life in poor water. Amazonian diversity and foodchain ecology as seen through fish communities. SPB Academic Press, The Hague.

Goulding, M., and E. J. G. Ferreira. 1984. Shrimp-eating fishes and a case of prey-switching in Amazon rivers. Revista Brasileira de Zoologia **2**:85–97.

Gradwell, N. 1971. Observations on jet propulsion in banjo catfishes. Canadian Journal of Zoology **49**:1611–1612.

Graham, J. B. 1997. Air-breathing fishes. Evolution, diversity and adaptation. Academic Press, San Diego, USA.

Granado-Lorencio, C., J. Lobon Cervia, and C. R. M. Araujo Lima. 2007. Floodplain lake fish assemblages in the Amazon River: directions in conservation biology. Biodiversity and Conservation **16**:679–692.

Greenwood, P. H. 1974. The cichlid fishes of Lake Victoria, East Aftica: the biology and evolution of a species-flock. Bulletin of the British Museum (Natural History) Zoology **6**:1–134.

Gudger, E. W. 1930a. On the alleged penetration of the human urethra by an Amazonian catfish called candirú, with a review of the allied habits of the other members of the family Pygidiidae. Part I. The American Journal of Surgery **8**:170–188.

Gudger, E. W. 1930b. On the alleged penetration of the human urethra by an Amazonian catfish called candirú, with a review of the allied habits of the other members of the family Pygidiidae. Part II. The American Journal of Surgery **8**:443–457.

Guennec, B. L., and G. Loubens. 2004. Biologie de *Pellona castelnaeana* (Teleostei: Pristigasteridae) dans le bassin du Mamore (Amazonie bolivienne). Ichthyological Exploration of Freshwaters **15**:369–383.

Haddad Jr., V., P. P. O. Pardal, J. L. C. Cardoso, and I. A. Martins. 2003. The venomous toadfish *Thalassophryne nattereri* (niquim or miquim): report of 43 injuries provoked in fishermen of Salinópolis (Pará State) and Aracaju (Sergipe State), Brazil. Revista do Instituto de Medicina Tropical de São Paulo **45**:221–223.

Haddad Jr., V., and I. Sazima. 2003. Piranha attacks on humans in southeast Brazil: epidemiology, natural history, and clinical treatment, with description of a bite outbreak. Wilderness and Environmental Medicine **14**:249–254.

Hagedorn, M. 1988. Ecology and behavior of a pulse-type electric fish, *Hypopomus occidentalis* (Gymnotiformes, Hypopomidae), in a fresh-water stream in Panama. Copeia:324–335.

Hagedorn, M., and W. Heiligenberg. 1985. Court and spark: electric signals in the courtship and mating of gymnotoid fish. Animal Behaviour **33**:254–265.

Hahn, N. S., R. Fugi, and I. F. Adrian. 1991. Espectro e atividade alimentares do armadinho, *Trachydoras paraguayensis* (Doradidae, Siluriformes) em distintos ambientes do rio Paraná. Revista Unimar, Maringá **13**:177–194.

Hahn, N. S., R. Fugi, V. L. L. Almeida, M. R. Russo, and V. E. Loureiro. 1997. Dieta e atividade alimentar de peixes do reservatório de Segredo. Pages 141–162 *in* A. A. Agostinho and L. C. Gomes, editors. Reservatório de Segredo: bases ecológicas para o manejo. EDUEM, Maringá, Brazil.

Hahn, N. S., A. Monfredinho Jr., R. Fugi, and A. A. Agostinho. 1992. Aspectos da alimentação do armado, *Pterodoras granulosus* (Ostariophisi, Doradidae) em distintos ambientes do Alto Rio Paraná. Revista Unimar, Maringá **14**:163–176.

Hahn, N. S., C. S. Pavanelli, and E. K. Okada. 2000. Dental development and ontogenetic diet shifts of *Roeboides paranensis* Pignalberi (Osteichthyes, Characinae) in pools of the upper Rio Paraná floodplain (state of Paraná, Brasil). Revista Brasileira de Biologia **60**:93–99.

Hallwass, G., P. F. Lopes, A. A. Juras, and R. A. M. Silvano. 2011. Fishing effort and catch composition of urban market and rural villages in Brazilian Amazon. Environmental Management **47**:188–200.

Hardman, M., L. M. Page, M. H. Sabaj, J. W. Armbruster, and J. H. Knouft. 2002. Comparison of fish surveys in the Essequibo and other coastal drainages of Guyana in 1908 and 1998. Ichthyological Exploration of Freshwaters **13**:225–238.

Harrington, R. W. J. 1961. Oviparous hermaphroditic fish with internal self-fertilization. Science **134**:1749–1750.

Hashimoto, D. T., F. F. Mendonça, J. A. Senhorini, C. Oliveira, F. Foresti, and F. Porto-Foresti. 2011. Molecular diagnostic methods for identifying serrasalmid fish (Pacu, Pirapitinga, and Tambaqui) and their hybrids in the Brazilian aquaculture industry. Aquaculture **321**:49–53.

Hatanaka, T., F. Henrique-Silva, and P. Galetti Jr. 2006. Population substructuring in a migratory freshwater fish *Prochilodus argenteus* (Characiformes, Prochilodontidae) from the São Francisco River. Genetica **126**:153–159.

Hauser, F. E., and H. López-Fernández. 2013. *Geophagus crocatus*, a new species of geophagine cichlid from the Berbice River, Guyana, South America (Teleostei: Cichlidae). Zootaxa **3731**:279–286.

Hawlitschek, O., K. C. Yamamoto, and F. G. M. R. Carvalho-Neto. 2013. Diet composition of fish assemblage of Lake Tupe, Amazonas, Brazil. Revista Colombiana de Ciencia Animal **5**:313–326.

Henderson, P. A., and W. G. R. Crampton. 1997. A comparison of fish diversity and abundance between nutrient-rich and nutrient-poor lakes in the upper Amazon. Journal of Tropical Ecology **13**:175–198.

Henderson, P. A., and I. Walker. 1986. On the leaf litter community of the Amazonian blackwater stream Tarumazinho. Journal of Tropical Ecology **2**:1–16.

Henderson, P. A., and I. Walker. 1990. Spatial organization and population density of the fish community of the litter banks within a central Amazonian blackwater stream. Journal of Fish Biology **37**:401–411.

Henning, F., C. B. Moysés, D. Calcagnotto, A. Meyer, and L. F. De Almeida-Toledo. 2011. Independent fusions and recent origins of sex chromosomes in the evolution and diversification of glass knife fishes (*Eigenmannia*). Heredity **106**:391–400.

Herman, J. R. 1973. Candirú: urinophilic catfish. Its gift of urology. Urology **1**:265–267.

Heyd, A., and W. Pfeiffer. 2000. Über die lauterzeugung der welse (Siluroidei, Ostariophysi, Teleostei) und ihren zusammenhang mit der phylogenie und der schreckreaktion. Revue Suisse de Zoologie **107**:165–211.

Higuchi, H., J. L. O. Birindelli, L. M. Sousa, and H. A. Britski. 2007. *Merodoras nheco*, new genus and species from Rio Paraguay basin, Brazil (Siluriformes, Doradidae), and nomination of the new subfamily Astrodoradinae. Zootaxa **1446**:31–42.

Higuchi, H., H. A. Britski, and J. C. Garavello. 1990. *Kalyptodoras bahiensis*, a new genus and species of thorny catfish from northeastern Brazil (Siluriformes, Doradidae). Ichthyological Exploration of Freshwaters **1**:219–225.

Hildemann, W. H. 1959. A cichlid fish *Symphysodon discus* with unique nurture habits. American Naturalist **93**:27–34.

Hilton, E., and C. Cox Fernandes. 2006. Sexual dimorphism in 'Apteronotus' bonapartii (Castelnau, 1855) (Gymnotiformes: Apteronotidae). Copeia:826–833.

Hilton, E. J., C. Cox Fernandes, J. P. Sullivan, J. G. Lundberg, and R. Campos-da-Paz. 2007. Redescription of *Orthosternarchus tamandua* (Boulenger, 1898) (Gymnotiformes, Apteronotidae), with reviews of its ecology, electric organ discharges, external morphology, osteology, and phylogenetic affinities. Proceedings of the Academy of Natural Sciences of Philadelphia **156**:1–25.

Hoedeman, J. J. 1954. Notes on the ichthyology of Surinam (Dutch Guiana). 3. A new species and two new subspecies of Nannostomidi from the Surinam River. Beufortia **4**:81–89.

Hoese, G., A. Addison, T. Toulkeridis, and R. Toomey, III. 2015. Observation of the catfish *Chaetostoma microps* climbing in a cave in Tena, Ecuador. Subterranean Biology **15**:29–35.

Hood, J. M., M. J. Vanni, and A. S. Flecker. 2005. Nutrient recycling by two phosphorus-rich grazing catfish: the potential for phosphorus-limitation of fish growth. Oecologia **146**:247–257.

Hoorn, C., F. P. Wesselingh, H. Ter Steege, M. A. Bermudez, A. Mora, J. Sevink, I. Sanmartín, A. Sanchez-Meseguer, C. L. Anderson, J. P. Figueiredo, C. Jaramillo, D. Riff, F. R. Negri, H. Hooghiemstra, J. Lundberg, T. Stadler, T. Särkinen, and A. Antonelli. 2010. Amazonia through time: Andean uplift, climate change, landscape evolution, and biodiversity. Science **330**:927–931.

Hopkins, C. D. 1999. Design features for electric communication. Journal of Experimental Biology **202**:1217–1228.

Horeau, V., P. Cerdan, A. Chapeau, and S. Richard. 1998. Importance of aquatic invertebrates in the diet of rapids-dwelling fish in the Sinnamary River, French Guiana. Journal of Tropical Ecology **14**:851–864.

Horn, M. H. 1997. Evidence for dispersal of fig seeds by the fruiteating characid fish *Brycon guatemalensis* Regan in a Costa Rican tropical rain forest. Oecologia **109**:259–264.

Horn, M. H., S. B. Correa, P. Parolin, B. J. A. Pollux, J. T. Anderson, C. Lucas, P. Widmann, A. Tjiu, M. Galetti, and M. Goulding. 2011. Seed dispersal by fishes in tropical and temperate fresh waters: the growing evidence. Acta Oecologica **37**:561–577.

Hostache, G., and J. H. A. Mol. 1998. Reproductive biology of the Neotropical armoured catfish *Hoplosternum littorale*: a synthesis stressing the role of the floating bubble nest. Aquatic Living Resources **11**:173–185.

Howes, G. J. 1976. The cranial musculature and taxonomy of characoid fishes of the tribes Cynodontini and Characini. Bulletin of the British Museum of Natural History (Zoology) **29**:203–248.

Hrbek, T., and D. C. Taphorn. 2008. Description of a new annual rivulid killifish genus from Venezuela. Zootaxa **1734**:27–42.

Hubert, N., and J. F. Renno. 2010. Description of a new *Serrasalmus* species, *Serrasalmus odyssei* n. sp. Pages 52–59 *in* Evolution of the Neotropical Ichthyofauna–Molecular and Evolutionary Perspectives about the Origin of the Fish Communities in the Amazon. VDM Publishing House, Verlag Dr. Müller, Saarbücken, Germany.

Hulen, K. G., W. G. R. Crampton, and J. S. Albert. 2005. Phylogenetic systematics and historical biogeography of the Neotropical electric fish *Sternopygus* (Teleostei: Gymnotiformes). Systematics and Biodiversity **3**:407–432.

Hurtado-Sepulveda, N. A. 1984. *Mylesinus schomburgkii* (Teleostei, Characidae): primera cita para la ictiofauna de Venezuela. Memoria de la Fundación La Salle de Ciencias Naturales **44**:131–142.

Ibañez, C., P. A. Tedesco, R. Bigorne, B. Hugueny, M. Pouilly, C. Zepita, J. Zubieta, and T. Oberdorff. 2007. Dietary-morphological relationships in fish assemblages of small forested streams in the Bolivian Amazon. Aquatic Living Resources **20**:131–142.

Ibarra, M., and D. J. Stewart. 1989. Longitudinal zonation of sandy beach fishes in the Napo River basin, eastern Ecuador. Copeia:364–381.

Ikeziri, A. A. S. L., L. J. Queiroz, C. R. C. Doria, L. F. Fávaro, et al. 2008. Estrutura populacional e abundância do Apapá-Amarelo, *Pellona castelnaeana* (Valenciennes, 1847) (Clupeiformes, Pristigasteridae), na Reserva Extrativista do Rio Cautário, Rondônia. Revista Brasileira de Zoociências **10**.

Ingenito, L. F. S., and P. A. Buckup. 2005. A new species of *Parodon* from the Serra da Mantiqueira, Brazil (Teleostei: Characiformes: Parodontidae). Copeia:765–771.

Isbrücker, I. J. H. 1975. *Pseudohemiodon thorectes*, a new species of mailed catfish from the Rio Mamoré system. Bolivia (Pisces, Siluriformes, Loricariidae). Beaufortia **23**:85–92.

Isbrücker, I. J. H. 1979. Description préliminaire de nouveaux taxa de la famille des Loricariidae, poissons-chats cuirassés néotropicaux, avec un catalogue critique de la sous-famille nominale (Pisces, Siluriformes). Revue Française d'Aquariologie et Herpétologie **5**:86–116.

Isbrücker, I. J. H. 1980. Classification and catalogue of the mailed Loricariidae (Pisces, Siluriformes). Verslagen en Technische Gegevens **22**:1–181.

Isbrücker, I. J. H. 1981. Revision of *Loricaria* Linnaeus, 1758 (Pisces, Siluriformes. Loricariidae). Beaufortia **31**:5196.

Isbrücker, I. J. H., H. A. Britski, H. Nijssen, and H. Ortega. 1983. *Aposturisoma myriodon*, une espèce et un genre nouveaux de poisson-chat cuirassé, tribu Farlowellini Fowler, 1958 du bassin du Rio Ucayali, Pérou (Pisces, Siluriformes, Loricariidae). Revue Française d'Aquariologie et Herpétologie **10**:33–42.

Isbrücker, I. J. H., and H. Nijssen. 1976. The South American mailed catfishes of the genus *Pseudoloricaria* Bleeker, 1862 (Pisces, Siluriformes, Loricariidae). Beaufortia **25**:107–129.

Isbrücker, I. J. H., and H. Nijssen. 1978. The Neotropical mailed catfishes of the genera *Lamontichthys* P. de Miranda-Ribeiro, 1939 and *Pterosturisoma* n. gen., including the description of *Lamontichthys stibaros* n. sp. from Ecuador (Pisces, Siluriformes, Loricariidae). Bijdragen tot de Dierkunde **48**:57–80.

Isbrücker, I. J. H., and H. Nijssen. 1982. New data on *Metaloricaria paucidens* from French Guiana and Surinam (Pisces, Siluriformes, Loricariidae). Bijdragen tot de Dierkunde **52**:155–168.

Isbrücker, I. J. H., and H. Nijssen. 1988. Review of the South American characiform fish genus *Chilodus*, with description of a new species, *C. gracilis* (Pisces, Characiformes, Chilodontidae). Beaufortia **38**:47–56.

Isbrücker, I. J. H., and H. Nijssen. 1991. *Hypancistrus zebra*, a new genus and species of uniquely pigmented ancistrine loricariid fish from the Rio Xingu, Brazil (Pisces: Siluriformes: Loricariidae). Ichthyological Exploration of Freshwaters **1**:345–350.

Isbrücker, I. J. H., H. Nijssen, and P. Cala. 1988. *Lithoxancistrus orinoco*, noveau genre et espèce de Poisson-chat cuirasse de Rio Orinoco en Colombie (Pisces, Siluriformes, Loricariidae). Revue Française d'Aquariologie et Herpétologie **15**:13–16.

Isbrücker, I. J. H., I. Seidel, J. P. Michels, E. Schraml, and A. Werner. 2001. Diagnose vierzehn neuer Gattungen der Familie Loricariidae Rafinesque, 1815 (Teleostei, Ostariophysi). Harnischwelse **2**:17–24.

Ivanyisky, S. J., and J. S. Albert. 2014. Systematics and biogeography of Sternarchellini (Gymnotiformes: Apteronotidae): diversification of electric fishes in large Amazonian rivers. Neotropical Ichthyology **12**:565–584.

Janovetz, J. 2005. Functional morphology of feeding in the scale-eating specialist *Catoprion mento*. Journal of Experimental Biology **208**:4757–4768.

Jarduli, L. R., and O. A. Shibatta. 2013. Description of a new species of *Microglanis* (Siluriformes: Pseudopimelodidae) from the Amazon basin, Amazonas state, Brazil. Neotropical Ichthyology **11**:507–512.

Javonillo, R., L. R. Malabarba, S. H. Weitzman, and J. R. Burns. 2010. Relationships among major lineages of characid fishes (Teleostei: Ostariophysi: Characiformes), based on molecular sequence data. Molecular Phylogenetics and Evolution **54**:498–511.

Jégu, M. 1992a. *Ossubtus xinguense*, nouveaux genre et espece du Rio Xingu, Amazonie, Brésil (Teleostei: Serrasalmidae). Icthyological Exploration of Freshwaters **3**:235–252.

Jégu, M. 1992b. Variations du niveau marin et distribution des poisson d'eau douce en Amazonie orientale. Pages 281–297 *in* M. Prost, editor. Évolution des Littoraux de Guyane et de la Zone Caraibe Méridionale Pendant le Quaternaire, Symposium PICG 274/ORSTOM, 9–14 Nov 1990, Cayenne, Guyana.

Jégu, M. 2003. Subfamily Serrasalminae (pacus and piranhas). Pages 182–196 *in* R. R. Reis, S. O. Kullander, and C. J. Ferraris Jr., editors. Check List of Freshwater Fishes of South and Central America. EDIPUCRS, Porto Alegre, Brazil.

Jégu, M. 2004. Taxinomie des Serrasalminae phytophages et phylogénie des Serrasalminae (Teleostei: Characiformes: Characidae). Unpublished PhD thesis. Museum National d'Histoire Naturelle, Paris.

Jégu, M., G. M. dos Santos, and E. Ferreira. 1991a. Une nouvelle espèce de *Bryconexodon* (Pisces, Characidae) décrite du bassin du Trombetas (Para, Brésil). Journal of Natural History 773–782.

Jégu, M., N. Hubert, and E. Belmont-Jégu. 2004. Réhabilitation de *Myloplus asterias* (Müller & Troschel, 1844), espèce-type de *Myloplus* Gill, 1896 et validation du genre *Myloplus* Gill (Characidae: Serrasalminae). Cybium **28**:119–157.

Jégu, M., and P. Keith. 1999. Lower Oyapock River as northern limit for the Western Amazon fish fauna or only a stage in its northward progression. Comptes Rendus de l'Academie des Sciences Series III Sciences de la Vie **12.322**:1133–1143.

Jégu, M., and P. Keith. 2005. Threatened fishes of the world: *Tometes lebaili* (Jégu, Keith & Belmount-Jégu 2002) (Characidae; Serrasalminae). Environmental Biology of Fishes **72**:378.

Jégu, M., P. Keith, and P.-Y. L. Bail. 2003. *Myloplus planquettei* sp. n. (Teleostei, Characidae), une nouvelle espèce de grand Serrasalminae phytophage du bouclier guyanais. Revue Suisse de Zoologie **110**:833–853.

Jégu, M., P. Keith, and E. Belmont-Jégu. 2002a. Une nouvelle espèce de *Tometes* (Teleostei: Characidae: Serrasalminae) du bouclier guyanais, *Tometes lebaili* n. sp. Bulletin Français de la Pêche et de la Pisciculture **364**:23–48.

Jégu, M., E. L. M. Leão, and G. M. Santos. 1991b. *Serrasalmus compressus*, une espèce nouvelle du Rio Madeira, Amazonie (Pisces: Serralmidae). Ichthyological Exploration of Freshwaters **2**:97–108.

Jégu, M., and G. M. Santos. 1988. Une nouvelle espèce du genre *Mylesinus* (Pisces, Serrasalmidae), *M. paucisquamatus*, decrite du bassin du Rio Tocantins (Amazonie, Brésil). Cybium **12**:331–341.

Jégu, M., and G. M. Santos. 1990. Description d'*Acnodon senai* n. sp. du Rio Jari (Brésil, Amapà) et redescription d'*A. normani* (Teleostei, Serrasalmidae). Cybium **14**:187–206.

Jégu, M., and G. M. Santos. 2001. Mise au point à propos de *Serrasalmus spilopleura* Kner, 1858 et réhabilitation de *S. maculatus* Kner, 1858. Cybium **25**:119–143.

Jégu, M., and G. M. Santos. 2002. Révision du statut de *Myleus setiger* Müller & Troschel, 1844 et de *Myleus knerii* (Steindachner, 1881) (Teleostei: Characidae: Serrasalminae) avec une description complémentaire des deux espèces. Cybium **26**.

Jégu, M., G. M. Santos, and E. Belmont-Jégu. 2002b. *Tometes makue* n. sp. (Characidae: Serrasalminae), une nouvelle espèce du bouclier guyanais décrite des bassins du rio Negro (Brésil) et de l'Orénoque (Venezuela). Cybium **26**:253–274.

Jégu, M., G. M. Santos, and E. Ferreira. 1989. Une nouvelle espèce du genre *Mylesinus* (Pisces, Serrasalmidae), *M. paraschombugkii*, décrite des bassins du Trombetas et du Uatumã (Brésil, Amazonie). Revue d'Hydrobiologie Tropicale **22**:49–62.

Jégu, M., L. Tito-Morias, and G. Mendes-Santos. 1992. Redescription des types d'*Utiaritichthys sennaebragai* Miranda Ribeiro, 1937 et description d'une nouvelle espèce du Bassin Amazoniaen, *U. longidorsalis* (Charachiformes, Serrasalmidae). Cybium **16**:105–120.

Jégu, M., and J. Zuanon. 2005. Threatened fishes of the world: *Ossubtus xinguense* (Jégu 1992) (Characidae; Serrasalminae). Environmental Biology of Fishes **73**:414.

Jepsen, D. B. 1997. Fish species diversity in sand bank habitats of a Neotropical river. Environmental Biology of Fishes **49**:449–460.

Jepsen, D. B., and K. O. Winemiller. 2002. Structure of tropical river food webs revealed by stable isotope ratios. Oikos **96**:46–55.

Jerep, F. C., and L. R. Malabarba. 2011. Revision of the genus *Macropsobrycon* Eigenmann, 1915 (Characidae: Cheirodontinae: Compsurini). Neotropical Ichthyology **9**:299–312.

Johansen, K., and C. Lenfant. 1967. Respiratory function in the South American lungfish. Journal of Experimental Biology **46**:205–218.

Johansen, K., C. Lenfant, K. Schmidt-Nielsen, and J. A. Petersen. 1968. Gas exchange and control of breathing in the electric eel, *Electrophorus electricus*. Zeitschrift für vergleichende Physiologie **61**:137–163.

Junk, W. J. 1997. The Central Amazon Floodplain: Ecology of a Pulsing System. Springer, Berlin, Germany.

Junk, W. J., M. G. M. Soares, and P. B. Bayley. 2007. Freshwater fishes of the Amazon River basin: their biodiversity, fisheries, and habitats. Aquatic Ecosystem Health & Management **10**:153–173.

Kaatz, I. M., and P. S. Lobel. 1999. Acoustic behavior and reproduction in five species of *Corydoras* catfishes. Biology Bulletin **197**:241–242.

Kalous, L., A. T. Bui, M. Petrtýl, J. Bohlen, and P. Chaloupková. 2012. The South American freshwater fish *Prochilodus lineatus* (Actinopterygii: Characiformes: Prochilodontidae): new species in Vietnamese aquaculture. Aquaculture Research **43**:955–958.

Kawasaki, M., J. Prather, and Y. X. Guo. 1996. Sensory cues for the gradual frequency fall responses of the gymnotiform electric fish, *Rhamphichthys rostratus*. Journal of Comparative Physiology A **178**:453–462.

Keith, P., P. Y. Le Bail, and P. Planquette. 2000. Atlas des poissons d'eau douce de Guyane. Tome 2(I). Publications Scientifiques du Muséum National d'Histoire Naturelle, Paris.

Keppeler, F. W., L. E. K. Lanés, A. S. Rolon, C. Stenert, P. Lehmann, M. Reichard, and L. Maltchik. 2015. The morphology-diet relationship and its role in the coexistence of two species of annual fishes. Ecology of Freshwater Fish **24**:77–90.

Kirschbaum, F. 1995. Vergleichende daten zur Fortpflanzungsbiologie von drei Messerfisch-Arten (Gymnotiformes). Pages 91–114 *in* H. Greven and R. Riehl, editors. Fortpflanzungsbiologie der Aquarienfische. Birgit Schmettkamp Verlag, Bornheim, Germany.

Knöppel, H. A. 1970. Food of Central Amazonian fishes. Contribution to the nutrient-ecology of Amazonian rainforest-streams. Amazonia **2**:257–352.

Knöppel, H. A. 1972. Zur nahrung tropischer süsswasserfische aus Südamerika- einige aus-gewählte arten der Anostomidae, Curimatidae, Hemiodidae und Characidae (Pisces, Characoidei). Amazoniana **3**:231–246.

Koch, W. R. 2002. Revisão taxonômica do gênero *Homodiaetus* (Teleostei, Siluriformes, Trichomycteridae). Iheringia, Série Zoologia **92**:33–46.

Koslowski, I. 2002. Die Buntbarsche Amerikas. Band 2: Apistogramma & Co. Verlag Eugen Ulmer, Stuttgart, Germany.

Kottelat, M., and J. Freyhof. 2007. Handbook of European Freshwater Fishes. Publications Kottelat, Cornol and Freyhof, Berlin.

Kramer, B. 1987. The sexually dimorphic jamming avoidance response in the electric fish *Eigenmannia* (Teleostei, Gymnotiformes). Journal of Experimental Biology **130**:39–62.

Kramer, D. L., and M. J. Bryant. 1995a. Intestine length in the fishes of a tropical stream. I. Ontogenic allometry. Environmental Biology of Fishes **42**:115–127.

Kramer, D. L., and M. J. Bryant. 1995b. Intestine length in the fishes of a tropical stream. II. Relationships to diet: the long and short of a convoluted issue. Environmental Biology of Fishes **42**:129–141.

Krumme, U., H. Keuthen, M. Barlettta, W. Villwock, and U. Saint-Paul. 2005. Contribution to the feeding ecology of the predatory wingfin anchovy *Pterengraulis atherinoides* (L.) in north Brazilian mangrove creeks. Journal of Applied Ichthyology **21**:469–477.

Krumme, U., H. Keuthen, U. Saint-Paul, and W. Villwock. 2007. Contribution to the feeding ecology of the banded puffer fish *Colomesus psittacus* (Tetraodontidae) in north Brazilian mangrove creeks. Brazilian Journal of Biology **67**:383–392.

Kubitzki, K., and A. Ziburski. 1994. Seed dispersal in flood plain forests of Amazonia. Biotropica **26**:30–43.

Kullander, S. O. 1977. *Papiliochromis* gen.n., a new genus of South American cichlid fish (Teleostei, Perciformes). Zoologica Scripta **6**:253–254.

Kullander, S. O. 1980. A taxonomic study of the genus *Apistogramma* Regan, with a revision of Brazilian and Peruvian species (Teleostei: Percoidei: Cichlidae). Bonner Zoologische Monographien **14**:1–152.

Kullander, S. O. 1983. A revision of the South American cichlid genus *Cichlasoma*. Swedish Museum of Natural History, Stockholm.

Kullander, S. O. 1984. Cichlid fishes from the La Plata basin. Part V. Description of *Aequidens plagiozonatus* sp.n. (Teleostei, Cichlidae) from the Paraguay river system. Zoologica Scripta **13**:155–159.

Kullander, S. O. 1986. Cichlid fishes of the Amazon River drainage of Peru. Swedish Museum of Natural History, Stockholm.

Kullander, S. O. 1987. A new *Apistogramma* species (Teleostei, Cichlidae) from the Rio Negro in Brazil and Venezuela. Zoologica Scripta **16**:259–270.

Kullander, S. O. 1988. *Teleocichla*, a new genus of South American rheophilic cichlid fishes with six new species (Teleostei: Cichlidae). Copeia:196–230.

Kullander, S. O. 1989a. *Biotoecus* Eigenmann and Kennedy (Teleostei: Cichlidae): description of a new species from the Orinoco basin and revised generic diagnosis. Journal of Natural History **23**:225–260.

Kullander, S. O. 1989b. Description of a new *Acaronia* species from the Rio Orinoco and Rio Negro drainages. Zoologica Scripta **18**:447–452.

Kullander, S. O. 1990a. *Mazarunia mazarunii* (Teleostei: Cichlidae), a new genus and species from Guyana, South America. Ichthyological Exploration of Freshwaters **1**:3–14.

Kullander, S. O. 1990b. A new species of *Crenicichla* from the Rio Tapajós, Brazil, with comments on interrelationships of the small crenicichline cichlids. Ichthyological Exploration of Freshwaters **1**:85–93.

Kullander, S. O. 1991. *Tahuantinsuyoa chipi*, a new species of cichlid fish from the Rio Pachitea drainage in Peru. Cybium **15**:3–13.

Kullander, S. O. 1995. Three new cichlid species from southern Amazonia: *Aequidens gerciliae*, *A. epae* and *A. michaeli*. Ichthyological Exploration of Freshwaters **6**:149–170.

Kullander, S. O. 1996a. Eine weitere Übersicht der Diskusfische, Gattung *Symphysodon* Heckel. DATZ Sonderheft Diskus:10–16.

Kullander, S. O. 1996b. *Heroina isonycterina*, a new genus and species of cichlid fish from western Amazonia, with comments on cichlasomine systematics. Ichthyological Exploration of Freshwaters 7:149–172.

Kullander, S. O. 1998. A phylogeny and classification of the South American Cichlidae (Teleostei: Perciformes). Pages 461–498 in L. R. Malabarba, R. E. Reis, R. P. Vari, Z. M. S. Lucena, and C. A. S. Lucena, editors. Phylogeny and Classification of Neotropical fishes. EDIPUCRS, Porto Alegre, Brazil.

Kullander, S. O. 2003a. Cichlidae (cichlids). Pages 605–654 in R. E. Reis, S. O. Kullander, and C. J. Ferraris Jr., editors. Check List of the Freshwater Fishes of South and Central America. EDIPUCRS, Porto Alegre, Brazil.

Kullander, S. O. 2003b. Family Gobiidae (gobies). Pages 657–665 in R. E. Reis, S. O. Kullander, and C. J. Ferraris Jr., editors. Check List of the Freshwater Fishes of South and Central America. EDIPUCRS, Porto Alegre, Brazil.

Kullander, S. O. 2003c. Family Synbranchidae. Pages 594–595 in R. E. Reis, S. O. Kullander, and C. J. Ferraris Jr., editors. Check List of the Freshwater Fishes of South and Central America. EDIPUCRS, Porto Alegre, Brazil.

Kullander, S. O. 2009. Family: Gobiidae. Pages 90–91 in R. P. Vari, C. J. Ferraris Jr., A. Radosavljevic, and V. A. Funk, editors. Checklist of Freshwater Fishes of the Guiana Shield. Bulletin of the Biological Society of Washington 17.

Kullander, S. O. 2011. A review of Dicrossus foirni and Dicrossus warzeli, two species of cichlid fishes from the Amazon River basin in Brazil (Teleostei: Cichlidae). Aqua 17:73–94.

Kullander, S. O. 2012. A taxonomie review of Satanoperca (Teleostei: Cichlidae) from French Guiana, South America, with description of a new species. Cybium 36:247–262.

Kullander, S. O., and C. A. S. De Lucena. 2006. A review of the species of Crenicichla (Teleostei: Cichlidae) from the Atlantic coastal rivers of southeastern Brazil from Bahia to Rio Grande do Sul States, with descriptions of three new species. Neotropical Ichthyology 4:127–146.

Kullander, S. O., and C. J. Ferraris Jr. 2003. Family Engraulididae (anchovies). Pages 39–42 in R. E. Reis, S. O. Kullander, and C. J. Ferraris Jr., editors. Check List of the Freshwater Fishes of South and Central America. EDIPUCRS, Porto Alegre, Brazil.

Kullander, S. O., and E. J. G. Ferreira. 2006. A review of the South American cichlid genus Cichla, with descriptions of nine new species (Teleostei: Cichlidae). Ichthyological Exploration of Freshwaters 17:289–398.

Kullander, S. O., and H. Nijssen. 1989. The Cichlids of Surinam. E. J. Brill, Leiden, Netherlands.

Kullander, S. O., M. Norén, G. B. Fridriksson, and C. A. Santos de Lucena. 2010. Phylogenetic relationships of species of Crenicichla (Teleostei: Cichlidae) from southern South America based on the mitochondrial cytochrome b gene. Journal of Zoological Systematics and Evolutionary Research 48:248–258.

Kullander, S. O., and S. Prada-Pedreros. 1993. Nannacara adoketa, a new species of cichlid fish from the Rio Negro in Brazil. Ichthyological Exploration of Freshwaters 4:357–366.

Kullander, S. O., and A. M. C. Silfvergrip. 1991. Review of the South American cichlid genus Mesonauta Günther with descriptions of two new species. Revue Suisse de Zoologie 98:407–448.

Kutaygil, N. 1959. Insemination, sexual differentiation and secondary sex characters in Stevardia albipinnis Gill. Hidrobiologie, Istanbul Universitat Fen Fakultesi Mecumuasi, ser. B 24: 93–128.

Ladich, F. 2001. Sound-generating and -detecting motor system in catfish: design of swimbladder muscles in doradids and pimelodids. The Anatomical Records 263:297–306.

Landim, M. I., C. R. Moreira, and C. A. Figueiredo. 2015. Retroculus acherontos, a new species of cichlid fish (Teleostei) from the Rio Tocantins basin. Zootaxa 3973:369–380.

Langeani, F. 1998. Phylogenetic study of the Hemiodotidae (Ostariophysi, Characiformes). Pages 145–160 in L. R. Malabarba, R. E. Reis, R. P. Vari, Z. M. Lucena, and C. A. S. Lucena, editors. Phylogeny and Classification of Neotropical Fishes. EDIPURCS, Porto Alegre, Brazil.

Langeani, F. 1999a. Argonectes robertsi sp.n., um novo Bivibranchiinae (Pisces, Characiformes, Hemiodontidae) dos Rios Tapajós, Xingu, Tocantins e Capim, Drenagem do Rio Amazonas. Naturalia (Rio Claro):171–183.

Langeani, F. 1999b. New species of Hemiodus (Ostariophysi, Characiformes, Hemiodontidae) from the rRio Tocantins, Brazil, with comments on color patterns and tooth shapes within the species and genus. Copeia:718–722.

Langeani, F. 2003. Family Hemiodontidae (Hemiodontids). Pages 96–100 in R. E. Reis, S. O. Kullander, and C. J. Ferraris Jr., editors. Check List of the Freshwater Fishes of South and Central America. EDIPUCRS, Porto Alegre, Brazil.

Langeani, F. 2004. Hemiodus jatuarana, a new species of Hemiodontidae from the rio Trombetas, Amazon Basin, Brazil (Teleostei, Characiformes). Zootaxa 546:1–6.

Langeani, F. 2009. Phylogenetic relationships within the South American fish family Hemiodontidae (Teleostei, Ostariophysi, Characiformes). Tese de Livre Docência. UNESP, São José do Rio Preto, São Paulo, Brazil.

Langeani, F. 2013. Hemiodontidae. Pages 192–204 in L. J. Queiroz, G. Torrente-Vilara, W. M. Ohara, T. S. Pires, J. Zuanon, and C. R. C. Doria, editors. Peixes do rio Madeira. Dialeto, São Paulo, Brazil.

Langeani, F., and C. R. Moreira. 2013. Hemiodus iratapuru, a new species of Hemiodontidae from the Rio Jari, Amazon Basin, Brazil (Teleostei, Characiformes). Journal of Fish Biology 82:1259–1268.

Langeani, F., O. T. Oyakawa, and J. I. Montoya-Burgos. 2001. New species of Harttia (Loricariidae, Loricariinae) from the Rio São Francisco basin. Copeia:136–142.

Langeani-Neto, F. 1996. Estudo filogenético e revisão taxonômica da família Hemiodontidae Boulenger, 1904 (sensu Roberts, 1974) (Ostariophysi, Characiformes). Unpublished PhD thesis, São Paulo, Brazil.

Lasso, C. A., J. I. Mojica, J. S. Usma, J. A. Maldonado-O., C. DoNascimiento, D. C. Taphorn, F. Provenzano, O. Lasso-Alcala, G. Galvis, L. Vasquez, M. Lugo, A. Machado-Allison, R. Royero, C. Suarez, and A. Ortega-Lara. 2004. Peces de la cuenca del rio Orinoco. Parte I: Lista de especies y distribucion por subcuencas (Fishes of the Orinoco River basin. Part 1: list of fishes and distribution by sub-basin). Biota Colombiana 5:95–158.

Lasso, C. A., A. Rial B., and O. Lasso-Alcalá. 1997. Notes on the biology of the freshwater stingrays Paratrygon aiereba (Müller & Henle, 1841) and Potamotrygon orbignyi (Castelnau, 1855) (Chondrichthyes: Potamotrygonidae) in the Venezuelan Llanos. Aqua, International Journal of Ichthyology 2:39–52.

Latini, A. O., and M. Petrere Jr. 2004. Reduction of a native fish fauna by alien species: an example from Brazilian freshwater tropical lakes. Fisheries Management and Ecology 11:71–79.

Leão, E. L. M. 1996. Reproductive biology of piranhas (Teleostei, Characiformes). Pages 31–41 in A. L. Val, V. M. F. Almeida-Val, and D. J. Randall, editors. Physiology and Biochemistry of the Fishes of the Amazon. INPA, Manaus, Brazil.

Le Bail, P. Y., P. Keith, and P. Planquette. 2000. Atlas des Poissons d'eau douce de Guyane. Tome 2, fascicule II: Siluriformes. Patrimoines Naturels **43**.

Lehmann, P. A. 2006. *Otocinclus batmani*, a new species of hypoptopomatine catfish (Siluriformes: Loricariidae) from Colombia and Peru. Neotropical Ichthyology **4**:379–383.

Lehmann, P. A., H. Lazzarotto, and R. E. Reis. 2014. *Parotocinclus halbothi*, a new species of small armored catfish (Loricariidae: Hypoptopomatinae), from the Trombetas and Marowijne river basins, in Brazil and Suriname. Neotropical Ichthyology **12**:27–33.

Lehmann, P. A., F. Mayer, and R. E. Reis. 2010a. A new species of *Otocinclus* (Siluriformes: Loricariidae) from the Rio Madeira drainage, Brazil. Copeia:635–639.

Lehmann, P. A., F. Mayer, and R. E. Reis. 2010b. Re-validation of *Otocinclus arnoldi* Regan and reappraisal of *Otocinclus* phylogeny (Siluriformes: Loricariidae). Neotropical Ichthyology **8**:57–68.

Lehmann, P. A., and R. E. Reis. 2004. *Callichthys serralabium*: a new species of Neotropical catfish from the upper Orinoco and Negro rivers (Siluriformes: Callichthyidae). Copeia:336–343.

Lehmann, P. A., L. J. Schvambach, and R. E. Reis. 2015. A new species of the armored catfish *Parotocinclus* (Loricariidae: Hypoptopomatinae), from the Amazon basin in Colombia. Neotropical Ichthyology **13**:47–52.

Leibel, W. 1984. Heckel's thread-finned acara *Acarichthys heckelii* (Mueller and Troschel 1848). Freshwater and Marine Aquarium Magazine:15–19, 78–86.

Leibel, W. 1994. The real parrot cichlid, *Hoplarchus psittacus* Kaup 1860. Cichlid News Magazine **3**:19–23.

Leitão, R. P., É. P. Caramaschi, and J. Zuanon. 2007. Following food clouds: feeding association between a minute loricariid and a characidiin species in an Atlantic Forest stream, southeastern Brazil. Neotropical Ichthyology **5**:307–310.

Leite, R. G., and M. Jégu. 1990. Régime alimentaire de deux espèces d'*Acnodon* (Characiformes, Serrasalmidae) et habitudes Lépidophages de *A. normani*. Cybium **14**:353–359.

Lewis Jr., W. M., S. K. Hamilton, M. A. Rodríguez, J. F. Saunders III, and M. A. Lasi. 2001. Foodweb analysis of the Orinoco floodplain based on production estimates and stable isotope data. Journal of the North American Benthological Society **20**:241–254.

Liem, K. F. 1970. Comparative functional anatomy of the Nandidae (Pisces: Teleostei). Fieldiana Zoology **56**:1–166.

Liem, K. F., B. Eclancher, and W. L. Fink. 1984. Aerial respiration in the banded knife fish *Gymnotus carapo* (Teleostei: Gymnotoidei). Physiological Zoology **57**:185–195.

Lima, F. C. T. 2003a. Subfamily Bryconinae (characins, tetras). Pages 174–181 in R. E. Reis, S. O. Kullander, and C. Ferraris Jr., editors. Check List of the Freshwater Fishes of South and Central America. EDIPUCRS, Porto Alegre, Brazil.

Lima, F. C. T. 2003b. Subfamily Clupeacharacinae. Page 171 in R. E. Reis, S. O. Kullander, and C. Ferraris Jr., editors. Check List of the Freshwater Fishes of South and Central America. EDIPUCRS, Porto Alegre, Brazil.

Lima, R. S. 2003c. Subfamily Aphyocharacinae (characins). Page 729 in R. E. Reis, S. O. Kullander, and J. Ferraris Jr., editors.

Check List of the Freshwater Fishes of South and Central America. EDIPUCRS, Porto Alegre, Brazil.

Lima, F. C. T. 2004. *Brycon gouldingi*, a new species from the rio Tocantins drainage, Brazil (Ostariophysi: Characiformes: Characidae), with a key to the species in the basin. Ichthyological Exploration of Freshwaters **15**:279–287.

Lima, F. C. T. 2006. Revisão taxonômica e relações filogenéticas do gênero *Salminus* (Teleostei: Ostariophysi: Characiformes: Characidae). Unpublished PhD thesis. Universidade de São Paulo, Brazil.

Lima, F. C. T., and H. A. Britski. 2007. *Salminus franciscanus*, a new species from the rio São Francisco basin, Brazil (Ostariophysi: Characiformes: Characidae). Neotropical Ichthyology **5**:237–244.

Lima, F. C. T., H. A. Britski, and F. A. Machado. 2004. New *Knodus* (Ostariophysi: Characiformes: Characidae) from the upper Rio Paraguay Basin, Brazil. Copeia:577–582.

Lima, F. C. T., and R. M. C. Castro. 2000. *Brycon vermelha*, a new species of characid fish from the Rio Mucuri, a coastal river of eastern Brazil (Ostariophysi: Characiformes). Ichthyological Exploration of Freshwaters **11**:155–162.

Lima, F. C. T., L. R. Malabarba, P. A. Buckup, J. F. Pezzi da Silva, R. P. Vari, A. Harold, R. Benine, O. T. Oyakawa, C. S. Pavanelli, N. A. Menezes, C. A. S. Lucinda, M. C. S. L. Malabarba, Z. M. S. Lucena, R. E. Reis, F. Langeani, L. Cassati, V. A. Bertaco, C. Moreira, and P. H. F. Lucinda. 2003. Genera *incertae sedis* in Characidae. Pages 107–171 in R. E. Reis, S. O. Kullander, and C. J. Ferraris Jr., editors. Check List of the Freshwater Fishes of South and Central America. EDIPUCRS, Porto Alegre.

Lima, F. C. T., and C. R. Moreira. 2003. Three new species of *Hyphessobrycon* (Characiformes: Characidae) from the upper rio Araguaia basin in Brazil. Neotropical Ichthyology **1**:21–33.

Lima, F. C. T., T. H. S. Pires, W. M. Ohara, F. C. Jerep, F. R. Carvalho, M. M. F. Marinho, and J. Zuanon. 2013. Characidae. Pages 213–395 in L. J. Queiroz, G. Torrente-Vilara, W. M. Ohara, T. H. S. Pires, J. Zuanon, and R. C. R. Doria, editors. Peixes do rio Madeira. Dialeto, São Paulo, Brazil.

Lima, F. C. T., L. Ramos, T. Barreto, A. Cabalzar, G. Tenório, A. Barbosa, F. Tenório, and A. S. Resende. 2005. Peixes do alto Tiquié: Ictiologia e Conhecimentos dos Tuyuka e Tukano. Pages 111–282 in A. Cabalzar, editor. Peixe e gente no alto Rio Tiquié. Instituto Socioambiental, São Paulo, Brazil.

Lima, F. C. T., and A. C. Ribeiro. 2011. Continental-scale tectonic controls of biogeography and ecology. Pages 145–164 in J. S. Albert and R. E. Reis, editors. Historical Biogeography of Neotropical Freshwater Fishes. University of California Press, Berkeley, USA.

Lima, F. C. T., and L. M. Sousa. 2009. A new species of *Hemigrammus* from the upper rio Negro basin, Brazil, with comments on the presence and arrangement of fin hooks in the genus (Ostariophysi: Characiformes: Characidae). Aqua, International Journal of Ichthyology **15**:153–168.

Lima, F. C. T., and A. Zanata. 2003. Subfamily Agoniatinae. Page 170 in R. E. Reis, S. O. Kullander, and C. Ferraris Jr., editors. Check List of the Freshwater Fishes of South and Central America. EDIPUCRS, Porto Alegre, Brazil.

Lima, M. R. L., E. Bessa, D. Krinski, and L. N. Carvalho. 2012. Mutilating predation in the Cheirodontinae *Odontostilbe pequira* (Characiformes: Characidae). Neotropical Ichthyology **10**:361–368.

Lin, D. S. C., and E. P. Caramaschi. 2005. Responses of the fish community to the flood pulse and siltation in a floodplains lake of the Trombetas River, Brazil. Hydrobiologia **545**:75–91.

Littmann, M. W. 2007. Systematic review of the Neotropical shovelnose catfish genus *Sorubim* Cuvier (Siluriformes: Pimelodidae). Zootaxa **1422**:1–29.

Lo Nostro, F. L., and G. A. Guerreo. 1996. Presence of primary and secondary males in a population of the protogynous *Synbranchus marmoratus*. Journal of Fish Biology **49**:788–800.

Loeb, M. V. 2012. A new species of *Anchoviella* Fowler, 1911 (Clupeiformes: Engraulidae) from the Amazon Basin, Brazil. Neotropical Ichthyology **10**:13–18.

Loeb, M. V., and A. V. Alcântara. 2013. A new species of *Lycengraulis* Günther, 1868 (Clupeiformes: Engraulinae) from the Amazon basin, Brazil, with comments on *Lycengraulis batesii* (Günther, 1868). Zootaxa **3693**:200–206.

Loeb, M. V., and J. L. Figueiredo. 2014. Redescription of the freshwater anchovy *Anchoviella vaillanti* (Steindachner, 1908) (Clupeiformes: Engraulidae) with notes on the distribution of estuarine congeners in the Rio São Francisco basin, Brazil. Arquivos de Zoologia (Sao Paulo) **45**:33–40.

Loir, M., C. Cauty, P. Planquette, and P. Y. le Bail. 1989. Comparative study of the male reproductive tract in seven families of South American catfishes. Aquatic Living Resources **2**:45–56.

Londoño-Burbano, A., S. L. Lefebvre, and N. K. Lujan. 2014. New species of *Limatulichthys* Isbrücker & Nijssen (Loricariidae, Loricariinae) from the western Guiana shield. Zootaxa **3884**:360–370.

Londoño-Burbano, A., C. Román-Valencia, and D. C. Taphorn. 2011. Taxonomic review of Colombian *Parodon* (Characiformes: Parodontidae), with descriptions of three new species. Neotropical Ichthyology **9**:709–730.

López, H., and P. Nass. 1989. Etapas del desarrollo de *Mylossoma duriventris* (Characiformes, Characidae) de los Llanos de Venezuela. Acta Biologica Venezuelica **12**:121–126.

López-Fernández, H., and J. S. Albert. 2011. Paleogene radiations. Pages 105–118 *in* J. S. Albert and R. E. Reis, editors. Historical Biogeography of Neotropical Freshwater Fishes. University of California Press, Berkeley, USA.

López-Fernández, H., J. Arbour, S. Willis, C. Watkins, R. L. Honeycutt, and K. O. Winemiller. 2014. Morphology and efficiency of a specialized foraging behavior, sediment sifting, in Neotropical cichlid fishes. PLoS ONE **9**:e89832.

López-Fernández, H., and D. C. Taphorn. 2004. *Geophagus abalios*, *G. dicrozoster* and *G. winemilleri* (Perciformes: Cichlidae), three new species from Venezuela. Zootaxa **439**:1–27.

López-Fernández, H., D. C. Taphorn, and E. A. Liverpool. 2012a. Phylogenetic diagnosis and expanded description of the genus *Mazarunia* Kullander, 1990 (Teleostei: Cichlidae) from the upper Mazaruni River, Guyana, with description of two new species. Neotropical Ichthyology **10**:465–486.

López-Fernández, H., and K. O. Winemiller. 2003. Morphological variation in *Acestrorhynchus microlepis* and *A. falcatus* (Characiformes: Acestrorhynchidae), reassessment of *A. apurensis* and distribution of *Acestrorhynchus* in Venezuela. Ichthyological Exploration of Freshwaters **14**:193–208.

López-Fernández, H., K. O. Winemiller, C. Montaña, and R. L. Honeycutt. 2012b. Diet-morphology correlations in the radiation of South American geophagine cichlids (Perciformes: Cichlidae: Cichlinae). PLoS ONE **7**:e33997.

Loubens, G., and J. Panfili. 1997. Biologie de *Colossoma macropomum* (Teleostei: Serrasalmidae) dans le bassin du Mamore (Amazonie bolivienne). Ichthyological Exploration of Freshwaters **8**:1–22.

Loubens, G., and J. Panfili. 2001. Biology of *Piarctus brachypomus* (Teleostei: Serrasalmidae) in the Mamore basin (Bolivian Amazonia). Ichthyological Exploration of Freshwaters **12**:51–64.

Lovejoy, N. R., J. S. Albert, and W. G. R. Crampton. 2006. Miocene marine incursions and marine/freshwater transitions: evidence from Neotropical fishes. Journal of South American Earth Sciences **21**:5–13.

Lovejoy, N. R., and B. B. Collette. 2001. Phylogenetic relationships of New World needlefishes (Teleostei: Belonidae) and the biogeography of transitions between marine and freshwater habitats. Copeia:324–338.

Lovejoy, N. R., and M. L. G. De Araújo. 2000. Molecular systematics, biogeography and population structure of Neotropical freshwater needlefishes of the genus *Potamorrhaphis*. Molecular Ecology **9**:259–268.

Lovejoy, N. R., S. C. Willis, and J. S. Albert. 2011. Molecular signatures of Neogene biogeographical events in the Amazon fish fauna. Pages 405–417 *in* C. Hoorn and F. P. Wesselingh, editors. Amazonia: Landscape and Species Evolution: A Look into the Past. Wiley-Blackwell, Oxford.

Lowe-McConnell, R. H. 1964. The fishes of the Rupununi savanna district of British Guiana, South America. Part 1. Ecological groupings of fish species and effects of the seasonal cycle on the fish. Journal of the Linnean Society (Zoology) **45**:103–144.

Lowe-McConnell, R. H. 1969. The cichlid fishes of Guyana, South America, with notes on their ecology and breeding behaviour. Zoological Journal of the Linnean Society **48**:255–302.

Lowe-McConnell, R. H. 1975. Fish communities in tropical freshwaters. Longman Publishing, New York, USA.

Lowe-McConnell, R. H. 1987. Ecological studies in tropical fish communities. Cambridge University Press.

Lowe-McConnell, R. H. 1991. Natural history of fishes in Araguaia and Xingu Amazonian tributaries, Serra do Roncador, Mato Grosso, Brazil. Icththyological Exploration of Freshwaters **2**:63–82.

Lucena, C. A. S. 1987. Revisão e redefinição do gênero neotropical *Charax* Scopoli, 1777 com a descrição de quatro espécies novas (Pisces: Characiformes: Characidae). Comunicações do Museu de Ciências e Tecnologia da PUCRS:5–124.

Lucena, C. A. S. 1998. Relações filogenéticas e definição do gênero *Roeboides* Günther (Ostariophysi: Characiformes: Characidae). Comunicações do Museu de Ciências e Tecnologia da PUCRS, série Zoologia **11**:19–59.

Lucena, C. A. S. 2007. Revisão taxonômica das espécies do gênero *Roeboides* grupo-*affinis* (Ostariophysi, Characiformes, Characidae). Iheringia, Série Zoologia **97**:117–136.

Lucena, C. A. S., and P. H. F. Lucinda. 2004. Variação geográfica de *Roeboexodon geryi* (Myers) (Ostariophysi: Characiformes: Characidae). Lundiana **5**:73–78.

Lucena, C. A. S., and N. A. Menezes. 1998. A phylogenetic analysis of *Roestes* Günther and *Gilbertolus* Eigenmann, with a hypothesis on the relationships of the Cynodontidae and Acestrorhynchidae (Teleostei: Ostariophysi: Characiformes). Pages 261–278 *in* L. R. Malabarba, R. E. Reis, R. P. Vari, Z. M. S. Lucena, and C. A. S. Lucena, editors. Phylogeny and Classification of Neotropical Fishes. EDIPUCRS, Porto Alegre, Brazil.

Lucena, C. A. S., and N. A. Menezes. 2003. Subfamily Characinae. Pages 200–208 in R. E. Reis, S. O. Kullander, and C. J. Ferraris Jr., editors. Check List of the Freshwater Fishes of South and Central America. EDIPUCRS, Porto Alegre, Brazil.

Lucena, Z. M. S., and L. R. Malabarba. 2010. Descriçõe nove espécies novas de *Phenacogaster* (Ostariophysi:Characiformes:Characidae) e comentários sobre as demais espécies do género. Zoologia 27:263–304.

Lucinda, P. H. F. 2003. Family Poeciliidae. Pages 555–581 in R. E. Reis, S. O. Kullander, and C. J. Ferraris Jr., editors. Check List of the Freshwater Fishes of South and Central America. EDIPUCRS, Porto Alegre, Brazil.

Lucinda, P. H. F. 2005. Systematics of the genus *Cnesterodon* Garman, 1895 (Cyprinodontiformes: Poeciliidae: Poeciliinae). Neotropical Ichthyology 3:259–270.

Lucinda, P. H. F. 2008. Systematics and biogeography of the genus *Phalloceros* Eigenmann, 1907 (Cyprinodontiformes: Poeciliidae: Poeciliinae), with the description of twenty-one new species. Neotropical Ichthyology 6:113–158.

Lucinda, P. H. F., I. S. Freitas, A. B. Soares, E. E. Marques, C. S. Agostinho, and R. J. Oliveira. 2007. Fish, Lajeado Reservoir, rio Tocantins drainage, State of Tocantins, Brazil. Check List 3:70–83.

Lucinda, P. H. F., and R. E. Reis. 2005. Systematics of the subfamily Poeciliinae Bonaparte (Cyprinodontiformes: Poeciliidae), with an emphasis on the tribe Cnesterodontini Hubbs. Neotropical Ichthyology 3:1–60.

Lucinda, P. H. F., and R. P. Vari. 2009. New *Steindachnerina* species (Teleostei: Characiformes: Curimatidae) from the Rio Tocantins drainage. Copeia:142–147.

Lujan, N. K., M. Arce, and J. W. Armbruster. 2009. A new black *Baryancistrus* with blue sheen from the upper Orinoco (Siluriformes: Loricariidae). Copeia:50–56.

Lujan, N. K., and J. W. Armbruster. 2011. Two new genera and species of ancistrini (Siluriformes: Loricariidae) from the western Guiana Shield. Copeia:216–225.

Lujan, N. K., and J. W. Armbruster. 2012. Morphological and functional diversity of the mandible in suckermouth armored catfishes (Siluriformes: Loricariidae). Journal of Morphology 273:24–39.

Lujan, N. K., J. W. Armbruster, N. R. Lovejoy, and H. López-Fernández. 2015a. Multilocus molecular phylogeny of the suckermouth armored catfishes (Siluriformes: Loricariidae) with a focus on subfamily Hypostominae. Molecular Phylogenetics and Evolution 82:269–288.

Lujan, N. K., J. W. Armbruster, and M. H. Sabaj. 2007. Two new species of *Pseudancistrus* from southern Venezuela (Siluriformes: Loricariidae). Ichthyological Exploration of Freshwaters 18:163–174.

Lujan, N. K., and J. L. O. Birindelli. 2011. A new distinctively banded species of *Pseudolithoxus* (Siluriformes: Loricariidae) from the upper Orinoco River. Zootaxa 2941:38–46.

Lujan, N. K., and C. C. Chamon. 2008. Two new species of Loricariidae (Teleostei: Siluriformes) from main channels of the upper and middle Amazon Basin, with discussion of deep water specialization in loricariids. Ichthyological Exploration of Freshwaters 19:271–282.

Lujan, N. K., D. P. German, and K. O. Winemiller. 2011. Do wood-grazing fishes partition their niche?: Morphological and isotopic evidence for trophic segregation in Neotropical Loricariidae. Functional Ecology 25:1327–1338.

Lujan, N. K., M. Hidalgo, and D. J. Stewart. 2010. Revision of *Panaque* (*Panaque*), with descriptions of three new species from the Amazon Basin (Siluriformes, Loricariidae). Copeia:676–704.

Lujan, N. K., V. Meza-Vargas, V. Astudillo-Clavijo, R. Barriga-Salazar, and H. Lopez-Fernandez. 2015b. A multilocus molecular phylogeny for *Chaetostoma* clade genera and species with a review of *Chaetostoma* (Siluriformes: Loricariidae) from the Central Andes. Copeia:664–701.

Lujan, N. K., S. Steele, and M. Velasquez. 2013. A new distinctively banded species of *Panaqolus* (Siluriformes: Loricariidae) from the western Amazon Basin in Peru. Zootaxa 3691:192–198.

Lujan, N. K., K. O. Winemiller, and J. W. Armbruster. 2012. Trophic diversity in the evolution and community assembly of loricariid catfishes. BMC Evolutionary Biology 12.

Lundberg, J., and W. Lewis. 1987. A major food web component in the Orinoco River channel: evidence from planktivorous electric fishes. Science 237:81–83.

Lundberg, J., F. Mago-Leccia, and P. Nass. 1991a. *Exallodontus aguanai*, a new genus and species of Pimelodidae (Pisces: Siluriformes) from deep river channels of South America, and delimitation of the subfamily Pimelodinae. Proceedings of the Biological Society of Washington 104:840–869.

Lundberg, J., P. Nass, and F. Mago-Leccia. 1989. *Pteroglanis manni* Eigenmann and Pearson, a juvenile of *Sorubimichthys planiceps* (Agassiz), with a review of the nominal species of *Sorubimichthys* (Pisces: Pimelodidae). Copeia:332–344.

Lundberg, J. G. 2005. *Gymnorhamphichthys bogardusi*, a new species of sand knifefish (Gymnotiformes: Rhamphichthyidae) from the Rio Orinoco, South America. Notula Naturae 479:1–4.

Lundberg, J. G., and A. Akama. 2005. *Brachyplatystoma capapretum*: a new species of goliath catfish from the Amazon Basin, with a reclassification of allied catfishes (Siluriformes: Pimelodidae). Copeia:492–516.

Lundberg, J. G., A. H. Bornbusch, and F. Mago-Leccia. 1991b. *Gladioglanis conquistador* n. sp., from Ecuador with diagnoses of the subfamilies Rhamdiinae Bleeker and Pseudopimelodinae n. subf. (Siluriformes, Pimelodidae). Copeia:190–209.

Lundberg, J. G., and C. Cox Fernandes. 2007. A new species of South American ghost knifefish (Apteronotidae: *Adontosternarchus*) from the Amazon Basin. Proceedings of the Academy of Natural Sciences of Philadelphia 156:27–37.

Lundberg, J. G., and W. M. Dahdul. 2008. Two new cis-Andean species of the South American catfish genus *Megalonema* allied to trans-Andean *Megalonema xanthum*, with description of a new subgenus (Siluriformes: Pimelodidae). Neotropical Ichthyology 6:439–454.

Lundberg, J. G., C. C. Fernandes, J. S. Albert, and M. Garcia. 1996. *Magosternarchus*, a new genus with two new species of electric fishes (Gymnotiformes: Apteronotidae) from the Amazon River Basin, South America. Copeia:657–670.

Lundberg, J. G., C. C. Fernandes, R. Campos-Da-Paz, and J. P. Sullivan. 2013. *Sternarchella calhamazon* n. sp., the Amazon's most abundant species of apteronotid electric fish, with a note on the taxonomic status of *Sternarchus capanemae* steindachner, 1868 (gymnotiformes, apteronotidae). Proceedings of the Academy of Natural Sciences of Philadelphia 162:157–173.

Lundberg, J. G., and M. Littmann. 2003. Family Pimelodidae. Pages 432–446 in R. E. Reis, S. O. Kullander, and C. J. Ferraris Jr., editors. Check List of the Freshwater Fishes of South and Central America. EDIPUCRS, Porto Alegre, Brazil.

Lundberg, J. G., A. Machado-Allison, and R. F. Kay. 1986. Miocene characid fishes from Colombia: Evolutionary stasis and extirpation. Science **234**:208–209.

Lundberg, J. G., and F. Mago-Leccia. 1986. A review of *Rhabdolichops* (Gymnotiformes, Sternopygidae), a genus of South American freshwater fishes, with descriptions of four new species. Proceedings of the Academy of Natural Sciences of Philadelphia **138**:53–85.

Lundberg, J. G., L. G. Marshall, J. Guerrero, B. Horton, M. C. S. L. Malabarba, and F. Wesselingh. 1998. The stage for Neotropical fish diversification: a history of tropical South American rivers. Pages 13–48 in L. R. Malabarba, R. E. Reis, R. P. Vari, C. A. S. Lucena, and Z. M. S. Lucena, editors. Phylogeny and Classification of Neotropical Fishes. PUCRS, Porto Alegre, Brazil.

Lundberg, J. G., and L. A. McDade. 1986. On the South American catfish *Brachyrhamdia imitator* Myers (Siluriformes, Pimelodidae), with phylogenetic evidence for a large intrafamilial lineage. Notulae Naturae of the Academy of Natural Sciences of Philadelphia **463**:1–24.

Lundberg, J. G., and B. M. Parisi. 2002. *Propimelodus*, new genus, and redescription of *Pimelodus eigenmanni* Van der Stigchel 1946, a long-recognized yet poorly-known South American catfish (Pimelodidae: Siluriformes). Proceedings of the Academy of Natural Sciences of Philadelphia **152**:75–88.

Lundberg, J. G., M. H. S. Pérez, W. M. Dahdul, and O. A. Aguilera. 2010. The Amazonian Neogene fish fauna. Pages 281–301 in C. Hoorn and F. P. Wesselingh, editors. Amazonia, Landscape and Species Evolution: A Look into the Past. Wiley-Blackwell, Oxford.

Lundberg, J. G., J. P. Sullivan, and M. Hardman. 2011. Phylogenetics of the South American catfish family Pimelodidae (Teleostei: Siluriformes) using nuclear and mitochondrial gene sequences. Proceedings of the Academy of Natural Sciences of Philadelphia **161**:153–189.

Lundberg, J. G., J. P. Sullivan, R. Rodiles-Hernández, and D. A. Hendrickson. 2007. Discovery of African roots for the Mesoamerican Chiapas catfish, Lacantunia enigmatica, requires an ancient intercontinental passage. Proceedings of the Academy of Natural Sciences of Philadelphia **156**:39–53.

Lüssen, A., T. M. Falk, and W. Villwock. 2003. Phylogenetic patterns in populations of Chilean species of the genus *Orestias* (Teleostei: Cyprinodontidae): results of mitochondrial DNA analysis. Molecular Phylogenetics and Evolution **29**:151–160.

Luz-Agostinho, K. D. G., L. M. Bini, R. Fugi, A. A. Agostinho, and H. F. Júlio Jr. 2006. Food spectrum and trophic structure of the ichthyofauna of Corumbá reservoir, Paraná River basin, Brazil. Neotropical Ichthyology **4**:61–68.

Machado, A. B. M., G. M. Drummond, and A. P. Paglia, editors. 2008. Livro vermelho da fauna brasileira ameaçada de extinção. MMA/Fundação Biodiversitas, Brazil.

Machado, G., A. A. Giaretta, and K. G. Facure. 2002. Reproductive cycle of a population of the Guaru, *Phallocerus caudimaculatus* (Poeciliidae), in southeastern Brazil. Studies on Neotropical Fauna and Environment **37**:15–18.

Machado-Allison, A. 1982a. Estudios sobre la sistematica de la Subfamilia Serrasalminae (Teleostei, Characidae). Parte 1. Estudio comparado de los juveniles de las "cachamas" de Venezuela (generos *Colossoma* y *Piaractus*). Acta Biologica Venezuelica **11**:1–101.

Machado-Allison, A. 1982b. Studies on the systematics of the subfamily Serrasalminae (Pisces-Characidae). PhD dissertation. Unpublished PhD thesis. George Washington University, Washington, DC, USA

Machado-Allison, A. 1983. Estudios sobre la sistematica de la subfamilía Serrasalminae (Teleostei, Characidae). Parte 2. Discusión sobre la condición monofilética de la subfamilía. Acta Biologica Venezuelica **11**:145–195.

Machado-Allison, A. 1985. Estudios sobre la sistematica de la Subfamilia Serrasalminae (Teleostei, Characidae). Parte III. Sobre el estatus generico y relaciones filogeneticas de los generos *Pygopristis, Pygocentrus, Pristobrycon* y *Serrasalmus* (Teleostei–Characidae–Serrasalminae). Acta Biologica Venezuelica **12**:19–42.

Machado-Allison, A. 1986. Osteologia comparada del neurocraneo y branquicraneo en los generos de la subfamilia Serrasalminae (Teleostei–Characidae). Suplemento Acta Biologica Venezuelica **12**:1–75.

Machado-Allison, A. 1987. Los Peces de Los Llanos de Venezuela: Un Ensayo Sobre Su Historia Natural. Universidad Central de Venezuela, Caracas.

Machado-Allison, A. 2002. Los peces caribes de Venezuela: una aproximación a su estudio taxonómico. Boletin Academia de Ciencias Fisicas, Matematicas y Naturales de Venezuela **62**:35–88.

Machado-Allison, A., and O. Castillo. 1992. Studies on the systematics of the subfamily Serrasalminae. IV. The genus *Mylossoma*: basis for the revision of the group in South America. Acta Biologica Venezuelica **13**:1–34.

Machado-Allison, A., W. J. Fink, and M. E. Antonio. 1989. Revision del genero *Serrasalmus* Lacepede, 1803 y géneros relacionados en Venezuela: I. Notas sobre la morfología y sistemática de *Pristobrycon striolatus* (Steindachner, 1908). Acta Biologica Venezuelica **12**:140–171.

Machado-Allison, A., and W. L. Fink. 1995. Sinopsis de las Especies de La Subfamilia Serrasalminae Presentes en La Cuenca del Orinoco: Claves, Diagnosis e Ilustraciones. Universidad Central de Venezuela, Caracas, Venezuela.

Machado-Allison, A., and W. L. Fink. 1996. Los Peces Caribes de Venezuela: Diagnosis, Claves, Aspectos Ecológicos y Evolutivos. Universidad Central de Venezuela, Caracas.

Machado-Allison, A., and C. Garcia. 1986. Food habits and morphological changes during ontogeny in three serrasalmin fish species of the Venezuelan floodplains. Copeia:193–196.

Machado-Allison, A., H. López, W. L. Fink, and R. Rodenas. 1993. *Serrasalmus neveriensis* a new species of piranha of Venezuela and redescription of *Serrasalmus medinai* Ramirez 1965. Acta Biologica Venezuelica **14**:45–60.

Magalhães, A. L. B., I. B. Amaral, T. F. Ratton, and M. F. G. Brito. 2002. Ornamental exotic fishes in the Gloria Reservoir and Boa Vista Stream, Paraiba do Sul River Basin, state of Minas Gerais, southeastern Brazil. Comunicações do Museu de Ciências e Tecnologia da PUCRS, Série Zoologia, Porto Alegre **15**:265–278.

Mago-Leccia, F. 1978. Los peces de agua dulce de Venezuela. Cuadernos Lagoven, Caracas, Venezuela.

Mago-Leccia, F. 1994. Electric fishes of the continental waters of America. Biblioteca de la Academia de Ciencias Fisicas Matematicas y Naturales **29**:1–225.

Mago-Leccia, F., J. G. Lundberg, and J. N. Baskin. 1985. Systematics of the South American freshwater fish genus *Adontosternarchus* (Gymnotiformes, Apteronotidae). Contributions in Science, Natural History Museum of Los Angeles County **358**:1–19.

Mai, A. C. G., M. V. Condini, C. Q. Albuquerque, D. Loebmann, T. D. Saint'Pierre, N. Miekeley, and J. P. Vieira. 2014. High plasticity in habitat use of *Lycengraulis grossidens* (Clupeiformes, Engraulididae). Estuarine, Coastal and Shelf Science **141**:17–25.

Mai, A. C. G., and J. P. Vieira. 2013. Review and consideration on habitat use, distribution and life history of *Lycengraulis grossidens* (Agassiz, 1829) (Actinopterygii, Clupeiformes, Engraulididae). Biota Neotropica **13**:121–130.

Makrakis, M. C., L. E. Miranda, S. Makrakis, A. M. M. Xavier, H. M. Fontes, and W. G. Morlis. 2007a. Migratory movements of pacu, *Piaractus mesopotamicus*, in the highly impounded Paraná River. Journal of Applied Ichthyology **23**:700–704.

Makrakis, S., M. C. Makrakis, R. L. Wagner, J. H. P. Dias, and L. C. Gomes. 2007b. Utilization of the fish ladder at the Engenheiro Sergio Motta Dam, Brazil, by long distance migrating potamodromous species. Neotropical Ichthyology **5**:197–204.

Malabarba, L. R. 1998. Monophyly of the Cheirodontinae, characters and major clades (Ostariophysi: Characidae). Pages 193–233 in L. R. Malabarba, R. E. Reis, R. P. Vari, Z. M. S. Lucena, and C. A. S. Lucena, editors. Phylogeny and Classification of Neotropical Fishes. EDIPURUS, Porto Alegre, Brazil.

Malabarba, L. R. 2003. Subfamily Cheirodontinae (characins, tetras). Pages 215–221 in R. E. Reis, S. O. Kullander, and C. J. Ferraris Jr., editors. Check List of the Freshwater Fishes of South and Central America. EDIPUCRS, Porto Alegre, Brazil.

Malabarba, L. R., and F. C. Jerep. 2012. A new genus and species of cheirodontine fish from South America (Teleostei: Characidae). Copeia:243–250.

Malabarba, L. R., and Z. M. S. Lucena. 1995. *Phenacogaster jancupa*, new species, with comments on the relationships and a new diagnosis of the genus (Ostariophysi: Characidae). Ichthyological Exploration of Freshwaters **6**:337–344.

Malabarba, L. R., and R. P. Vari. 2000. *Caiapobrycon tucurui*, a new genus and species of characid from the rio Tocantins basin, Brazil (Characiformes: Characidae). Ichthyological Exploration of Freshwaters **11**:315–326.

Malabarba, L. R., and S. H. Weitzman. 1999. A new genus and species of South American fishes (Teleostei: Characidae: Cheirodontinae) with a derived caudal fin, including comments about inseminating cheirodontines. Proceedings of the Biological Society of Washington **112**:410–432.

Malabarba, L. R., and S. H. Weitzman. 2003. Description of a new genus with six new species from southern Brazil, Uruguay and Argentina, with a discussion of a putative characid clade (Teleostei: Characiformes: Characidae). Comunicações do Museu de Ciências e Tecnologia da PUCRS, Série Zoologia **16**:67–151.

Malabarba, M. C. S. L. 2004. Revision of the Neotropical genus *Triportheus* Cope, 1872 (Characiformes: Characidae). Neotropical Ichthyology **2**:167–204.

Maldonado-Ocampo, J. A., C. D. de Santana, and W. G. R. Crampton. 2011. On *Apteronotus magdalenensis* (Miles, 1945) (Gymnotiformes: Apteronotidae): A poorly known species endemic to the río Magdalena basin, Colombia. Neotropical Ichthyology **9**:505–514.

Maldonado-Ocampo, J. A., H. López-Fernández, D. C. Taphorn, C. R. Bernard, W. G. R. Crampton, and N. R. Lovejoy. 2014. *Akawaio penak*, a new genus and species of Neotropical electric fish (Gymnotiformes, Hypopomidae) endemic to the upper Mazaruni River in the Guiana Shield. Zoologica Scripta **43**:24–33.

Maldonado-Ocampo, J. A., and S. Prada-Pedreros. 1999. Habitos alimentarios en los peces *Catoprion mento* y *Papiliochromis ramirezi* de un estero del Munlelpio de Puerto López, en la Orinoquia Colombiana. Dahlia (Revista de la Asociación Colombiana de Ictiólogos) **3**:41–46.

Maldonado-Ocampo, J. A., R. P. Vari, and J. S. Usma. 2008. Checklist of the freshwater fishes of Colombia. Biota Colombiana **9**:143–237.

Mannheimer, S., G. Bevilacqua, É. P. Caramaschi, and F. R. Scarano. 2003. Evidence for seed dispersal by the catfish *Auchenipterichthys longimanus* in an Amazonian lake. Journal of Tropical Ecology **19**:215–218.

Marceniuk, A. P., R. Betancur-R, A. Acero-P, and J. Muriel-Cunha. 2012a. Review of the genus *Cathorops* (Siluriformes: Ariidae) from the Caribbean and Atlantic South America, with description of a new species. Copeia:77–97.

Marceniuk, A. P., R. Betancur-R, and A. Acero P. 2009. A new species of *Cathorops* (Siluriformes; Ariidae) from Mesoamerica, with redescription of four species from the eastern Pacific. Bulletin of Marine Science **85**:245–280.

Marceniuk, A. P., and N. A. Menezes. 2007. Systematics of the family Ariidae (Ostariophysi, Siluriformes), with a redefinition of the genera. Zootaxa **1416**:1–126.

Marceniuk, A. P., N. A. Menezes, and M. R. Britto. 2012b. Phylogenetic analysis of the family Ariidae (Ostariophysi: Siluriformes), with a hypothesis on the monophyly and relationships of the genera. Zoological Journal of the Linnean Society **165**:534–669.

Mariguela, T. C., M. A. Alexandrou, F. Foresti, and C. Oliveira. 2013a. Historical biogeography and cryptic diversity in the Callichthyinae (Siluriformes, Callichthyidae). Journal of Zoological Systematics and Evolutionary Research **51**:308–315.

Mariguela, T. C., R. C. Benine, K. T. Abe, G. S. Avelino, and C. Oliveira. 2013b. Molecular phylogeny of *Moenkhausia* (Characidae) inferred from mitochondrial and nuclear DNA evidence. Journal of Zoological Systematics and Evolutionary Research **51**:327–332.

Mariguela, T. C., F. F. Roxo, F. Foresti, and C. Oliveira. 2016. Phylogeny and biogeography of Triportheidae (Teleostei: Characiformes) based on molecular data. Molecular Phylogenetics and Evolution **96**:130–139.

Marinho, M. M. F. 2010. A new species of *Moenkhausia* Eigenmann (Characiformes: Characidae) from the rio Xingu basin, Brazil. Neotropical Ichthyology **8**:655–659.

Marinho, M. M. F., D. A. Bastos, and N. A. Menezes. 2013. New species of miniature fish from Marajó Island, Pará, Brazil, with comments on its relationships (Characiformes: Characidae). Neotropical Ichthyology **11**:739–746.

Marrero, C., and K. O. Winemiller. 1993. Tube-snouted gymnotiform and mormyriform fishes: convergence of a specialized foraging mode in teleosts. Environmental Biology of Fishes **38**:299–309.

Martín Salazar, F. J., I. J. H. Isbrücker, and H. Nijssen. 1982. *Dentectus barbarmatus*, a new genus and species of mailed catfish from the Orinoco Basin of Venezuela (Pisces, Siluriformes, Loricariidae). Beaufortia **32**:125–137.

Martins-Queiroz, M. F., L. A. D. F. Mateus, V. Garutti, and P. C. Venere. 2008. Reproductive biology of *Triportheus trifurcatus* (Castelnau, 1855) (Characiformes: Characidae) in the middle rio Araguaia, MT, Brazil. Neotropical Ichthyology **6**:231–236.

Masson, V. L. 2007. Taxonomia do gênero *Myoglanis* Eigenmann, 1912 (Siluriformes: Heptapteridae), com um estudo comparativo de sua musculatura cefálica superficial. Unpublished MSc thesis. Universidade de São Paulo, Ribeirão Preto, Brazil.

Mata-Cortés, S., J. A. Martínez-Pérez, and M. S. Peterson. 2004. Feeding habits and sexual dimorphism of the violet goby, *Gobioides broussonnetii* (Pisces, Gobiidae) in the estuarine system of Tecolutla, Veracruz, Mexico. Gulf and Caribbean Research **16**:89–93.

Matamoros, W. A., C. D. McMahan, P. Chakrabarty, J. S. Albert, and J. F. Schaefer. 2015. Derivation of the freshwater fish fauna of Central America revisited: Myers's hypothesis in the twenty-first century. Cladistics **31**:177–188.

Mateus, L. A. F., and J. M. F. Penha. 2007. Avaliação dos estoques pesqueiros de quatro espécies de grandes bagres (Siluriformes, Pimelodidae) na bacia do rio Cuiabá, Pantanal norte, Brasil, utilizando alguns Pontos de Referência Biológicos. Revista Brasileira de Zoologia **24**:144–150.

Matthaeus, W. 1992. Observations on the behaviour of *Symphysodon aequifasciatus*. Freshwater and Marine Aquarium Magazine **15**:12–16.

Mattos, J. L. O., F. P. Ottoni, and M. A. Barbosa. 2013. *Microglanis pleriqueater*, a new species of catfish from the São João river basin, eastern Brazil (Teleostei: Pseudopimelodidae). Ichthyological Exploration of Freshwaters **24**:147–154.

Mattox, G. M. T., A. G. Bifi, and O. T. Oyakawa. 2014a. Taxonomic study of *Hoplias microlepis* (Gunther, 1864), a trans-Andean species of trahiras (Ostariophysi: Characiformes: Erythrinidae). Neotropical Ichthyology **12**:343–352.

Mattox, G. M. T., R. Britz, and M. Toledo-Piza. 2016. Osteology of *Priocharax* and remarkable developmental truncation in a miniature Amazonian fish (Teleostei: Characiformes: Characidae). Journal of Morphology **277**:65–85.

Mattox, G. M. T., R. Britz, M. Toledo-Piza, and M. M. F. Marinho. 2013. *Cyanogaster noctivaga*, a remarkable new genus and species of miniature fish from the Rio Negro, Amazon basin (Ostariophysi: Characidae). Ichthyological Exploration of Freshwaters **23**:297–318.

Mattox, G. M. T., M. Hoffmann, and P. Hoffmann. 2014b. Ontogenetic development of *Heterocharax macrolepis* Eigenmann (Ostariophysi: Characiformes: Characidae) with comments on the form of the yolk sac in the Heterocharacinae. Neotropical Ichthyology **12**:353–363.

Mattox, G. M. T., and M. Toledo-Piza. 2012. Phylogenetic study of the Characinae (Teleostei: Characiformes: Characidae). Zoological Journal of the Linnean Society **165**:809–915.

Mattox, G. M. T., M. Toledo-Piza, and O. T. Oyakawa. 2006. Taxonomic study of *Hophas aimara* (Valenciennes, 1846) and *Hoplias macrophthalmus* (Pellegrin, 1907) (Ostariophysi, Characiformes, Erythrinidae). Copeia:516–528.

Mautari, K. C., and N. A. Menezes. 2006. Revision of the South American freshwater genus *Laemolyta* Cope, 1872 (Ostariophysi: Characiformes: Anostomidae). Neotropical Ichthyology **4**:27–44.

Maxime, E. L., and J. S. Albert. 2009. A new species of *Gymnotus* (Gymnotiformes: Gymnotidae) from the Fitzcarrald Arch of southeastern Peru. Neotropical Ichthyology **7**:579–585.

Maxime, E. L., F. C. T. Lima, and J. S. Albert. 2011. A new species of *Gymnotus* (Gymnotiformes: Gymnotidae) from Rio Tiquié in northern Brazil. Copeia:77–81.

Mazzoldi, C., R. A. Patzner, and M. B. Rasotto. 2011. Morphological organization and variability of the reproductive apparatus in gobies. Pages 367–402 *in* R. A. Patzner, J. L. van Tassell, M. Kovačić, and B. G. Kapoor, editors. Biology of Gobies. Science Publishers and CRC Press, New Jersey, USA.

Mazzoni, R., and L. D. S. Costa. 2007. Feeding ecology of streamdwelling fishes from a coastal stream in the southeast of Brazil. Brazilian Archives of Biology and Technology **50**:627–635.

McKaye, K. R., D. J. Weiland, and T. M. Lim. 1979. Comments on the breeding biology of *Gobiomorus dormitor* (Osteichthyes: Eleotridae) and the advantage of schooling behavior to its fry. Copeia:542–544.

Mees, G. F. 1967. Freshwater fishes of Suriname: the genus *Heptapterus* (Pimelodidae). Zoologische Mededelingen **42**:215–229.

Mees, G. F. 1987. The members of the subfamily Aspredininae, family Asprenidae in Suriname (Pisces, Nematognathi). Proceedings of the Koninklijke Nederlandse Akademie van Wetenschappen (Series C) **90**:173–192.

Mees, G. F., and P. Cala. 1989. Two new species of *Imparfinis* from northern South America (Pisces, Nematognathi, Pimelodidae). Proceedings of the Koninklijke Nederlandse Akademie van Wetenschappen (Series C) **92**:379–394.

Meinken, H. 1936. Über einige in letzter Zeit eingeführte Fische. Blätter für Aquärien–und Terrarkunde **47**:49–51.

Meinken, H. 1975. *Microschemobrycon meyburgi* n.sp. aus dem Rio Xeriuini. Senckenbergiana Biologica **56**:217–220.

Melo, B. F., R. C. Benine, T. C. Mariguela, and C. Oliveira. 2011. A new species of *Tetragonopterus* Cuvier, 1816 (Characiformes: Characidae: Tetragonopterinae) from the rio Jari, Amapá, northern Brazil. Neotropical Ichthyology **9**:49–56.

Melo, B. F., R. C. Benine, G. S. C. Silva, G. S. Avelino, and C. Oliveira. 2016. Molecular phylogeny of the Neotropical fish genus *Tetragonopterus* (Teleostei: Characiformes: Characidae). Molecular Phylogenetics and Evolution **94**:709–717.

Melo, B. F., Y. Sato, F. Foresti, and C. Oliveira. 2013. The roles of marginal lagoons in the maintenance of genetic diversity in the Brazilian migratory fishes *Prochilodus argenteus* and *P. costatus*. Neotropical Ichthyology **11**:625–636.

Melo, B. F., B. L. Sidlauskas, K. Hoekzema, R. P. Vari, and C. Oliveira. 2014. The first molecular phylogeny of Chilodontidae (Teleostei: Ostariophysi: Characiformes) reveals cryptic biodiversity and taxonomic uncertainty. Molecular Phylogenetics and Evolution **70**:286–295.

Melo, C. E., F. A. Machado, and V. Pinto-Silva. 2004. Feeding habits of fish from a stream in the savanna of central Brazil, Araguaia Basin. Neotropical Ichthyology **2**:37–44.

Melo, C. E., and C. P. Röpke. 2004. Feeding and distribution of piaus (Pisces, Anostomidae) in the Paníce do Bananal, Mato Grosso, Brazil. Revista Brasileira de Zoologica **21**:51–56.

Menezes, N. A. 1969. The food of *Brycon* and three closely related genera of the tribe Acestrorhynchini. Papéis Avulsos de Zoologia **22**:217–223.

Menezes, N. A. 1976. On the Cynopotaminae, a new subfamily of Characidae (Osteichthyes, Ostariophysi, Characoidei). Arquivos de Zoologia **28**:1–91.

Menezes, N. A. 1987. Three new species of the characid genus *Cynopotamus* Valenciennes, 1849, with remarks on the remaining species (Pisces, Characiformes). Beaufortia **37**:1–9.

Menezes, N. A. 1992. Redefinição taxonômica das espécies de *Acestrorhynchus* do grupo lacustris com a descrição de uma nova espécie (Osteichthyes, Characiformes, Characidae).

Comunicações do Museu de Ciências e Tecnologia da PUCRS, Série Zoologia **5**:39–54.

Menezes, N. A. 2003. Family Acestrorhynchidae. Pages 231–233 *in* R. E. Reis, S. O. Kullander, and C. J. Ferraris Jr., editors. Check List of the Freshwater Fishes of South and Central America. EDIPUCRS, Porto Alegre, Brazil.

Menezes, N. A. 2006. Description of five new species of *Acestrocephalus* Eigenmann and redescription of *A. sardina* and *A. boehlkei* (Characiformes: Characidae). Neotropical Ichthyology **4**:385–400.

Menezes, N. A. 2007. A new species of *Cynopotamus* Valenciennes, 1849 (Characiformes, Characidae) with a key to the species of the genus. Zootaxa **1635**:55–61.

Menezes, N. A., and C. A. S. de Lucena. 1998. Revision of the subfamily Roestinae (Ostariophysi: Characiformes: Cynodontidae). Ichthyological Exploration of Freshwaters **9**:279–291.

Menezes, N. A., and M. C. C. De Pinna. 2000. A new species of *Pristigaster*, with comments on the genus and redescription of *P. cayana* (Teleostei: Clupeomorpha: Pristigasteridae). Proceedings of the Biological Society of Washington **113**:238–248.

Menezes, N. A., K. M. Ferreira, and A. L. Netto-Ferreira. 2009a. A new genus and species of inseminating characid fish from the rio Xingu basin (Characiformes: Characidae). Zootaxa **2167**:47–58.

Menezes, N. A., and J. Géry. 1983. Seven new acestrorhynchin characid species (Osteichthyes, Ostariophysi, Characiformes) with comments on the systematics of the group. Revue Suisse de Zoologie **90**:563–592.

Menezes, N. A., and C. A. S. Lucena. 2014. A taxonomic review of the species of *Charax* Scopoli, 1777 (Teleostei, Characidae, Characinae) with description of a new species from the rio Negro bearing superficial neuromasts on body scales, Amazon basin, Brasil. Neotropical Ichthyology **12**:193–228.

Menezes, N. A., A. L. Netto-Ferreira, and K. M. Ferreira. 2009b. A new species of *Bryconadenos* (Characiformes: Characidae) from the rio Curuá, rio Xingu drainage, Brazil. Neotropical Ichthyology **7**:147–152.

Menezes, N. A., S. H. Weitzman, and I. Quagio-Grassiotto. 2013. Two new species and a review of the inseminating freshwater fish genus *Monotocheirodon* (Characiformes: Characidae) from Peru and Bolivia. Papéis Avulsos de Zoologia **53**:129–144.

Menezes, R. S. 1949. Alimentação de mandí bicudo, "*Hassar affinis*" (Steindachner), da bacia do Rio Parnaíba, Piauí (Actinopterygii, Doradidae, Doradinae). Revista Brasileira de Biologia **9**:93–96.

Menezes, R. S., and M. F. Menezes. 1948. Alimentação de "graviola", "*Platydoras costatus*" (Linnaeus) da Lagoa de Nazaré, Piauí (Actinopterygii, Doradidae). Revista Brasileira de Biologia **8**:255–260.

Menezes, R. S., and S. L. Oliveira e Silva. 1949. Alimentação de Voador, *Hemiodus parnaguae* Eigenmann & Henn, da bacia do rio Parnaíba, Piauí (Actinopterygii, Characidae, Hemiodontinae). Revista Brasileira de Biologia **9**:241–245.

Meredith, R. W., M. N. Pires, D. N. Reznick, and M. S. Springer. 2010. Molecular phylogenetic relationships and the evolution of the placenta in *Poecilia* (*Micropoecilia*) (Poeciliidae: Cyprinodontiformes). Molecular Phylogenetics and Evolution **55**:631–639.

Meredith, R. W., M. N. Pires, D. N. Reznick, and M. S. Springer. 2011. Molecular phylogenetic relationships and the coevolution of placentotrophy and superfetation in *Poecilia* (Poeciliidae:

Cyprinodontiformes). Molecular Phylogenetics and Evolution **59**:148–157.

Mérigoux, S., and D. Ponton. 1998. Body shape, diet and ontogenetic diet shifts in young fish of the Sinnamary River, French Guiana, South America. Journal of Fish Biology **52**:556–569.

Mérona, B. d., and J. Rankin-de-Mérona. 2004. Food resource partitioning in a fish community of the central Amazon floodplain. Neotropical Ichthyology **2**:75–84.

Meschiatti, A. J., M. S. Arcifa, and N. Fenerich-Verani. 2000. Fish communities associated with macrophytes in Brazilian floodplain lakes. Environmental Biology of Fishes **58**:133–143.

Meunier, F. J., M. Jégu, and P. Keith. 2011. A new genus and species of Neotropical electric fish, *Japigny kirschbaum* (Gymnotiformes: Sternopygidae), from French Guiana. Cybium **35**:47–53.

Meunier, F. J., M. Jégu, and P. Keith. 2014. *Distocyclus guchereauae* a new species of Neotropical electric fish, (Gymnotiformes: Sternopygidae), from French Guiana. Cybium **38**:223–230.

Meyer, M. K. 1993. Reinstatement of *Micropoecilia* Hubbs, 1926, with a redescription of *M. bifurca* (Eigenmann, 1909) from northeast South America (Teleostei, Cyprinodontiformes: Poeciliidae). Zoologische Abhandlungen (Dresden) **47**:121–130.

Miles, C. 1945. Some new recorded fishes from Magdalena River System. Caldasia **111**:453–463.

Milhomem, S. S. R., J. C. Pieczarka, W. G. R. Crampton, A. C. P. De Souza, J. R. Carvalho Jr., and C. Y. Nagamachi. 2007. Differences in karyotype between two sympatric species of *Gymnotus* (Gymnotiformes: Gymnotidae) from the eastern Amazon of Brazil. Zootaxa **1397**:55–62.

Milhomem, S. S. R., J. C. Pieczarka, W. G. R. Crampton, D. S. Silva, A. C. P. De Souza, J. R. Carvalho Jr., and C. Y. Nagamachi. 2008. Chromosomal evidence for a putative cryptic species in the *Gymnotus carapo* species-complex (Gymnotiformes, Gymnotidae). BMC Genetics **9**.

Miller, R. R. 1979. Ecology, habits and relationships of the Middle American cuatro ojos, *Anableps dowi* (Pisces: Anablepidae). Copeia:82–91.

Mills, A., and H. H. Zakon. 1991. Chronic androgen treatment increases action potential duration in the electric organ of *Sternopygus*. Journal of Neuroscience **11**:2349–2361.

Mills, D., and G. Vevers. 1989. The Tetra Encyclopedia of Freshwater Tropical Aquarium Fishes. Tetra Press, New Jersey, USA.

Miranda, J. C., and R. Mazzoni. 2003. Composição da Ictiofauna de três riachos do alto rio Tocantins, GO. Biota Neotropica **3**:1–11.

Miranda Ribeiro, A. 1911. Fauna brasiliense. Peixes. Tomo IV (A) [Eleutherobranchios aspirophoros]. Arq. Mus. Nac. Rio de Janeiro **16**:1–504.

Miranda Ribeiro, P. 1951. Notas para o estudo dos Pygidiidae brasileiros (Pisces–Pygidiidae–Stegophilinae) IV. Boletim do Museu Nacional, Zoología **106**:1–23.

Mirande, J. M. 2010. Phylogeny of the family Characidae (Teleostei: Characiformes): from characters to taxonomy. Neotropical Ichthyology **8**:385–568.

Mirande, J. M., F. C. Jerep, and J. A. Vanegas-Ríos. 2013. Phylogenetic relationships of the enigmatic *Carlastyanax aurocaudatus* (Eigenmann) with remarks on the phylogeny of the Stevardiinae (Teleostei: Characidae). Neotropical Ichthyology **11**:747–766.

Mojica, J. I., C. Castellanos, S. Usma, and R. Alvarez, editors. 2002. Libro Rojo de Peces Dulceacuicolas de Colombia. Instituto de Ciencias Naturales, Universidad Nacional de Colombia, Bogota.

Mol, J. H. 2006. Attacks on humans by the piranha *Serrasalmus rhombeus* in Suriname. Studies on Neotropical Fauna and Environment **41**:189–195.

Mol, J. H. 2012a. Occurrence of a freshwater pipefish *Pseudophallus cf. brasiliensis* (Syngnathidae) in Corantijn River, Suriname, with notes on its distribution, habitat, and reproduction. Cybium **36**:45–53.

Mol, J. H., B. Mérona, P. E. Ouboter, and S. Sahdew. 2007a. The fish fauna of Brokopondo Reservoir, Suriname, during 40 years of impoundment. Neotropical Ichthyology **5**:351–368.

Mol, J. H., K. Wan Tong You, I. Vrede, A. Flynn, P. Ouboter, and F. Van der Lugt. 2007b. RAP Bulletin of Biological Assessment 43. A Rapid Biological Assessment of the Lely and Nassau Plateaus, Suriname (with additional information on the Brownsberg Plateau). Conservation International, Arlington, VA, USA.

Mol, J. H. A. 1995. Interspecific competition, predation, and the coexistence of three closely related Neotropical armoured catfishes (Siluriformes-Callichthyidae). Unpublished PhD thesis. University of Wageningen, Wageningen, the Netherlands.

Mol, J. H. A. 1996. Reproductive seasonality and nest-site differentiation in three closely related armoured catfishes. Environmental Biology of Fishes **45**:363–381.

Mol, J. H. A. 2012b. The freshwater fishes of Suriname. Koninklijke Brill NV, Leiden, the Netherlands.

Montaña, C. G., C. A. Layman, and K. O. Winemiller. 2011. Gape size influences seasonal patterns of piscivore diets in three Neotropical rivers. Neotropical Ichthyology **9**:647–655.

Montaña, C. G., and K. O. Winemiller. 2009. Comparative feeding ecology and habitats use of *Crenicichla* species (Perciformes: Cichlidae) in a Venezuelan floodplain river. Neotropical Ichthyology **7**:267–274.

Montenegro, A. K. A., J. E. R. Torelli, M. C. Crispim, and A. M. A. Medeiros. 2011. Population and feeding structure of *Steindachnerina notonota* Miranda-Ribeiro, 1937 (Actinopterygii, Characiformes, Curimatidae) in Taperoá II dam, semi-arid region of Paraíba, Brazil. Acta Limnologica Brasiliensia **23**:233–244.

Montoya-Burgos, J. I. 2003. Historical biogeography of the catfish genus *Hypostomus* (Siluriformes: Loricariidae), with implications on the diversification of Neotropical ichthyofauna. Molecular Ecology **12**:1855–1867.

Moraes, M. B., and F. M. S. Braga. 2011. Biologia populacional de *Imparfinis minutus* (Siluriformes, Heptapteridae) na microbacia do Ribeirão Grande, serra da Mantiqueira oriental, Estado de São Paulo. Acta Scientiarum, Biological Sciences **33**:301–310.

Moreau, M. A., and O. T. Coomes. 2007. Aquarium fish exploitation in western Amazonia: Conservation issues in Peru. Environmental Conservation **34**:12–22.

Moreira, C. R. 2003. Subfamily Iguanodectinae. Pages 172–173 *in* R. E. Reis, S. O. Kullander, and C. J. Ferraris Jr., editors. Check List of the Freshwater Fishes of South and Central America. EDIPUCRS, Porto Alegre, Brazil.

Moreira, C. R. 2005. *Xenurobrycon coracoralinae*, a new glandulocaudine fish (Ostariophysi: Characiformes: Characidae) from central Brazil. Proceedings of the Biological Society of Washington **118**:855–862.

Moreira, C. R., and F. C. T. Lima. 2011. On the name of the lepidophagous characid fish *Roeboexodon guyanensis* (Puyo) (Teleostei: Characiformes: Characidae). Neotropical Ichthyology **9**:313–316.

Moreira, S. S., and J. Zuanon. 2002. Dieta de *Retroculus lapidifer* (Perciformes: Cichlidae), peixe reofílico do rio Araguaia, Estado do Tocantins, Brasil. Acta Amazonica **32**:691–705.

Moreira-Hara, S. S., J. A. S. Zuanon, and S. A. Amadio. 2009. Feeding of *Pellona flavipinnis* (Clupeiformes, Pristigasteridae) in a Central Amazonian floodplain. Iheringia–Serie Zoologia **99**:153–157.

Moysés, C. B., M. F. de Zambelli Daniel-Silva, C. E. Lopes, and L. F. de Almeida-Toledo. 2010. Cytotype-specific ISSR profiles and karyotypes in the Neotropical genus *Eigenmannia* (Teleostei: Gymnotiformes). Genetica **138**:179–189.

Murdy, E. O. 1998. A review of the gobioid fish genus *Gobioides*. Ichthyological Research **45**:121–133.

Murdy, E. O., and D. F. Hoese. 2003a. Eleotridae. Pages 1778–1780 *in* K. E. Carpenter, editor. The Living Marine Resources of the Western Central Atlantic. Volume 3: Bony fishes part 2 (Opistognathidae to Molidae). FAO Species Identification Guide for Fishery Purposes and American Society of Ichthyologist and Herpetologists Special Publication No. 5, Rome, Italy.

Murdy, E. O., and D. F. Hoese. 2003b. Gobiidae. Pages 1781–1796 *in* K. E. Carpenter, editor. The Living Marine Resources of the Western Central Atlantic. Volume 3: Bony fishes part 2 (Opistognathidae to Molidae). FAO Species Identification Guide for Fishery Purposes and American Society of Ichthyologist and Herpetologists Special Publication No. 5, Rome, Italy.

Muriel-Cunha, J., and M. C. C. de Pinna. 2005. New data on cistern catfish, *Phreatobius cisternarum*, from subterranean waters at the mouth of the Amazon River (Siluriformes, Incertae Sedis). Papéis Avulsos Zoologia **45**:327–339.

Myers, G. S. 1927. Descriptions of new South American fresh-water fishes collected by Dr. Carl Ternetz. Bulletin of the Museum of Comparative Zoology **68**:107–135.

Myers, G. S. 1942. Studies on South American freshwater fishes I. Stanford Ichthyological Bulletin **2**:89–114.

Myers, G. S. 1944. Two extraordinary new blind nematognath fishes from the Rio Negro, representing a new subfamily of Pygidiidae, with a rearrangement of the genera of the family, and illustrations of some previously described genera and species from Venezuela and Brazil. Proceedings California Academy of Sciences **23**:591–602.

Myers, G. S. 1950a. Studies on South American fresh-water fishes. II. The genera of anostomine characids. Stanford Ichthyological Bulletin **3**:184–198.

Myers, G. S. 1950b. Supplementary notes on the flying characid fishes, especially *Carnegiella*. Stanford Ichthyological Bulletin **3**:182–183.

Myers, G. S. 1966. Derivation of the freshwater fish fauna of Central America. Copeia:766–773.

Myers, G. S., and A. L. d. Carvalho. 1959. A remarkable new genus of anostomin characid fishes from the upper rio Xingú in central Brazil. Copeia:148–152.

Myers, G. S., and S. H. Weitzman. 1966. Two remarkable new trichomycterid catfishes from the Amazon basin in Brazil and Colombia. Journal of Zoology **149**:277–287.

Near, T. J., A. Dornburg, R. I. Eytan, B. P. Keck, W. L. Smith, K. L. Kuhn, J. A. Moore, S. A. Price, F. T. Burbrink, M. Friedman, and P. C. Wainwright. 2013. Phylogeny and tempo of diversification in the superradiation of spiny-rayed fishes. Proceedings of the

National Academy of Sciences of the United States of America **110**:12738–12743.

Nelson, G. J. 1984. Notes on the rostral organ of anchovies (family Engraulidae). Japanese Journal of Ichthyology **31**:86–87.

Nelson, J. A., D. A. Wubah, M. E. Whitmer, E. A. Johnson, and D. J. Stewart. 1999. Wood-eating catfishes of the genus *Panaque*: Gut microflora and cellulolytic enzyme activities. Journal of Fish Biology **54**:1069–1082.

Nelson, K. 1964. Behavior and morphology in the glandulocaudine fishes (Ostariophysi, Characidae). University of California Publications in Zoology **75**:59–152.

Netto-Ferreira, A. L. 2010. Revisão taxonômica e relações interespecíficas de Lebiasininae (Ostariophysi: Characiformes: Lebiasinidae). Unpublished PhD thesis. Universidade de São Paulo, São Paulo, Brazil.

Netto-Ferreira, A. L. 2012. Three new species of *Lebiasina* (Characiformes: Lebiasinidae) from the Brazilian Shield border at Serra do Cachimbo, Para, Brazil. Neotropical Ichthyology **10**:487–498.

Netto-Ferreira, A. L., M. P. Albrecht, J. L. Nessimian, and E. P. Caramaschi. 2007. Feeding habits of *Thoracocharax stellatus* (Characiformes: Gasteropelecidae) in the upper rio Tocantins, Brazil. Neotropical Ichthyology **5**:69–74.

Netto-Ferreira, A. L., J. L. O. Birindelli, L. M. Sousa, and N. A. Menezes. 2014. A new species of *Rhinopetitia* Géry 1964 (Ostariophysi: Characiformes: Characidae) from the rio Teles Pires, rio Tapajós basin, Brazil. Journal of Fish Biology **84**:1539–1550.

Netto-Ferreira, A. L., H. Lopez-Fernandez, D. C. Taphorn, and E. A. Liverpool. 2013. New species of *Lebiasina* (Ostariophysi: Characiformes: Lebiasinidae) from the upper Mazaruni River drainage, Guyana. Zootaxa **3652**:562–568.

Netto-Ferreira, A. L., and M. M. F. Marinho. 2013. New species of *Pyrrhulina* (Ostariophysi: Characiformes: Lebiasinidae) from the Brazilian Shield, with comments on a putative monophyletic group of species in the genus. Zootaxa **3664**:369–376.

Netto-Ferreira, A. L., O. T. Oyakawa, J. Zuanon, and J. C. Nolasco. 2011. *Lebiasina yepezi*, a new Lebiasininae (Characiformes: Lebiasinidae) from the Serra Parima-Tapirapeco mountains. Neotropical Ichthyology **9**:767–775.

Netto-Ferreira, A. L., A. M. Zanata, J. L. O. Birindelli, and L. M. Sousa. 2009. Two new species of *Jupiaba* (Characiformes: Characidae) from the rio Tapajós and rio Madeira drainages, Brazil, with an identification key to the species of the genus. Zootaxa **2262**:53–68.

Neuberger, A. L., E. E. Marques, C. S. Agostinho, and R. J. de Oliveira. 2007. Reproductive biology of *Rhaphiodon vulpinus* (Ostariophysi: Cynodontidae) in the Tocantins River basin, Brazil. Neotropical Ichthyology **5**:479–484.

Nico, L. G. 1990. Feeding chronology of juvenile piranhas, *Pygocentrus notatus*, in the Venezuelan llanos. Environmental Biology of Fishes **29**:51–57.

Nico, L. G. 1991. Trophic ecology of piranhas (Characidae: Serrasalminae) from savanna and forest regions in the Orinoco River basin of Venezuela. Unpublished PhD thesis. University of Florida, Gainesville.

Nico, L. G., and M. C. C. de Pinna. 1996. Confirmation of *Glanapteryx anguilla* (Siluriformes, Trichomycteridae) in the Orinoco River basin, with notes on the distribution and habitats of the Glanapteryginae. Ichthyological Exploration of Freshwaters **7**:27–32.

Nico, L. G., H. L. Jelks, and T. Tuten. 2009. Non-native suckermouth armored catfishes in Florida: description of nest burrows and burrow colonies with assessment of shoreline conditions. Aquatic Nuisance Species Research Program Bulletin **9**:1–30.

Nico, L. G., and D. C. Taphorn. 1986. Those bitin' fish from South America. Tropical Fish Hobbyist **34**:24–27,30–34,36,40–41,56–57.

Nico, L. G., and D. C. Taphorn. 1988. Food habits of piranhas in the low llanos of Venezuela. Biotropica **20**:311–321.

Nijssen, H., and I. J. H. Isbrücker. 1976. The South American plated catfish genus *Aspidoras* R. Von Ihering, 1907, with descriptions of nine new species from Brazil (Pisces, Siluriformes, Callichthyidae). Bijdragen tot de Dierkunde **46**:107–131.

Nijssen, H., and I. J. H. Isbrücker. 1988. Trois nouvelles espèces du genre *Apistoloricaria* de Colombie et du Pérou, avec illustration du dimorphisme sexuel secondaire des lèvres de A. condei (Pisces, Siluriformes, Loricariidae). Revue Française d'Aquariologie et Herpétologie **15**:33–38.

Nomura, H., and C. Hayashi. 1980. Caracteres meristicos e biologia do Saguiru, *Curimatus gilberti* (Quoy and Gaimard, 1824), do rio Morgado (Matao, Sao Paulo) (Osteichthys, Curimatidae). Revista Brasileira de Biologia **40**:165–176.

Nomura, H., and A. C. Taveira. 1978. Biologia do Saguiru, *Curimatus elegans* Steindachner, 1874 do Mogi Guacu, Sao Paulo (Osteichthys, Curimatidae). Revista Brasileira de Biologia **39**:331–339.

Nordlie, F. G., and D. C. Haney. 1993. Euryhaline adaptations in the fat sleeper, *Dormitator maculatus*. Journal of Fish Biology **43**:433–439.

Norman, J. A. 1929. The South American characid fishes of the subfamily Serrasalmoninae, with a revision of the genus *Serrasalmus* Lacepede. Proceedings of the Zoological Society of London **52 (1928)**:781–830.

Novoa, D. F. 1989. The multispecies fisheries of the Orinoco River: Development, present status, and management strategies. Pages 422–428 *in* D. P. Dodge, editor. Proceedings of the International Large River Symposium. Canadian Special Publication of Fisheries and Aquatic Sciences 106.

Ohara, W. M. 2012. *Engraulisoma taeniatum* Castro, 1981 (Characiformes: Characidae): Range extension with new records in the rio Madeira basin, Rondônia and Amazonas states, Brazil. Check List **8**:1313–1314.

Ohara, W. M., and J. Zuanon. 2013. Aspredinidae. Pages 108–141 *in* Q. L. J., T.-V. G., W. M. Ohara, T. H. S. Pires, J. Zuanon, and C. R. C. Doria, editors. Peixes do rio Madeira. Santo Antonio Energia, São Paulo.

Olaya-Nieto, C., P. Soto-Fernández, and J. Barrera-Chica. 2009. Feeding habits of mayupa (*Sternopygus macrurus* Bloch & Schneider, 1801) in the Sinu River, Colombia. Revista MVZ Cordoba **14**:1787–1795.

Oliveira, C., G. S. Avelino, K. T. Abe, T. C. Mariguela, R. C. Benine, G. Ortí, R. P. Vari, and R. M. C. Castro. 2011a. Phylogenetic relationships within the speciose family Characidae (Teleostei: Ostariophysi: Characiformes) based on multilocus analysis and extensive ingroup sampling. BMC Evolutionary Biology **11**.

Oliveira, C. L. C., L. R. Malabarba, and J. R. Burns. 2012. Comparative morphology of gill glands in externally fertilizing and inseminating species of cheirodontine fishes, with implications on the phylogeny of the family Characidae

(Actinopterygii: Characiformes). Neotropical Ichthyology 10:349–360.

Oliveira, J. S., S. C. R. Fernandes, C. A. Schwartz, C. Bloch Jr., J. A. Taquita Melo, O. Rodrigues Pires Jr., and J. Carlos de Freitas. 2006. Toxicity and toxin identification in *Colomesus asellus*, an Amazonian (Brazil) freshwater puffer fish. Toxicon 48:55–63.

Oliveira, R. D., J. M. Lopes, J. R. Sanches, A. L. Kalinin, M. L. Glass, and F. T. Rantin. 2004. Cardiorespiratory responses of the facultative air-breathing fish jeju, *Hoplerythrinus unitaeniatus* (Teleostei, Erythrinidae) exposed to graded ambient hypoxia. Comparative Biochemistry and Physiology A, Molecular & Integrative Physiology 139:479–485.

Oliveira, V. A., N. F. Fontoura, and L. F. A. Montag. 2011b. Reproductive characteristics and the weight-length relationship in *Anableps anableps* (Linnaeus, 1758) (Cyprinodontiformes: Anablepidae) from the Amazon estuary. Neotropical Ichthyology 9:757–766.

Orsi, M. L., and O. A. Shibbata. 1999. Crescimento de *Schizodon intermedius* Garavello & Britski (Osteichthyes, Anostomidae) de rio Taibagi (Sertanópolis, Paraná). Revista Brasileira de Zoologia 16:701–710.

Ortaz, M. 1992. Feeding habits of fish in a Neotropical mountain river. Biotropica 24:550–559.

Ortega, H., H. Guerra, and R. Ramírez. 2007. The introduction of nonnative fishes into freshwater systems of Peru. Pages 247–248 *in* T. M. Bert, editor. Ecological and Genetic Implications of Aquaculture Activities. Springer, Netherlands.

Ortega-Lara, A., N. Milani, C. DoNascimiento, F. Villa-Navarro, and J. A. Maldonado-Ocampo. 2011. Two new trans-Andean species of *Imparfinis* Eigenmann & Norris, 1900 (Siluriformes: Heptapteridae) from Colombia. Neotropical Ichthyology 9:777–793.

Ortí, G., A. Sivasundar, K. Dietz, and M. Jégu. 2008. Phylogeny of the Serrasalmidae (Characiformes) based on mitochondrial DNA sequences. Genetics and Molecular Biology 31:343–351.

Ottoni, F. P., P. H. N. Bragança, P. F. Amorim, and C. S. Gama. 2012. A new species of *Laetacara* from the northern Brazil coastal floodplains (Teleostei: Cichlidae). Vertebrate Zoology 62:181–188.

Ottoni, F. P., and W. J. E. M. Costa. 2009. Description of a new species of *Laetacara* Kullander, 1986 from central Brazil and re-description of *Laetacara dorsigera* (Heckel, 1840) (Labroidei: Cichlidae: Cichlasomatinae). Vertebrate Zoology 59:41–48.

Ottoni, F. R., and J. L. O. Mattos. 2015. Phylogenetic position and re-description of the endangered cichlid *Nannacara hoehnei*, and description of a new genus from Brazilian Cerrado (Teleostei, Cichlidae, Cichlasomatini). Vertebrate Zoology 65:65–79.

Oyakawa, O. T. 1993. Cinco espécies novas de *Harttia* Steindachner, 1876 da região sudeste do Brasil, e comentários sobre o gênero (Teleostei, Siluriformes, Loricariidae). Comunicaçoes do Museu de Ciências da PUCRS, Série Zoologia, Porto Alegre 6:3–27.

Oyakawa, O. T. 2003. Family Erythrinidae. Pages 238–240 *in* R. E. Reis, S. O. Kullander, and C. J. Ferraris Jr., editors. Check List of the Freshwater Fishes of South and Central America. EDIPUCRS, Porto Alegre, Brazil.

Oyakawa, O. T., I. Fichberg, and F. Langeani. 2013a. *Harttia absaberi*, a new species of loricariid catfish (siluriformes: Loricariidae: Loricariinae) from the upper rio Paraná basin, Brazil. Neotropical Ichthyology 11:779–786.

Oyakawa, O. T., and G. M. T. Mattox. 2009. Revision of the Neotropical trahiras of the *Hoplias lacerdae* species-group (Ostariophysi: Characiformes: Erythrinidae) with descriptions of two new species. Neotropical Ichthyology 7:117–140.

Oyakawa, O. T., M. Toledo-Piza, and G. M. T. Mattox. 2013b. Erythrinidae. Pages 70–76 *in* L. J. Queiroz, G. Torrente-Vilara, W. M. Ohara, T. H. Silva Pires, J. Zuanon, and C. R. C. Doria, editors. Peixes do rio Madeira, volume 2. Dialeto, Sao Paulo, Brazil.

Pacheco, A. C. G., M. P. Albrecht, and É. P. Caramaschi. 2008. Ecologia de duas espécies de *Pachyurus* (Perciformes, Sciaenidae) do rio Tocantins, na região represada pela UHE Serra da Mesa, Goiás. Iheringia, Série Zoologia 98:270–277.

Page, L. M., and B. M. Burr. 2011. Peterson Field Guide to Freshwater Fishes. Houghton Mifflin Harcourt.

Page, L. M., G. B. Mottesi, M. E. Retzer, P. A. Ceas, and D. C. Taphorn. 1993. Spawning habitat and larval development of *Chaetostoma stannii* (Loricariidae) from Rio Crucito, Venezuela. Ichthyological Exploration of Freshwaters 4:93–95.

Pagezy, H., and M. Jégu. 2002. Patrimonial value of herbivorous Serrasalminae fish of the upper Maroni stream (French Guyana): biological and socio-cultural approaches in the Wayana region. Bulletin Français de la Pêche et de la Pisciculture 364:49–69.

Paiva, M. P. 1958. Sôbre o contrôle da pirambeba, "*Serrasalmus rhombeus*" (L., 1766) Lacépède, 1803, no Açude Lima Campos (Icó, Ceará), através da pesca seletiva. Revista Brasileira de Biologia 18:251–266.

Paiva, M. P., and F. H. Nepomuceno. 1989. On the reproduction in captivity of the oscar *Astronotus ocellatus* (Cuvier) according to the mating methods (Pisces–Cichlidae). Amazoniana 10:361–378.

Paixão, A. C., and M. Toledo-Piza. 2009. Systematics of *Lamontichthys* Miranda-Ribeiro (Siluriformes: Loricariidae), with the description of two new species. Neotropical Ichthyology 7:519–568.

Palmeira, C. A. M., L. F. da Silva Rodrigues-Filho, J. B. de Luna Sales, M. Vallinoto, H. Schneider, and I. Sampaio. 2013. Commercialization of a critically endangered species (largetooth sawfish, *Pristis perotteti*) in fish markets of northern Brazil: Authenticity by DNA analysis. Food Control 34:249–252.

Parenti, L. R. 1981. A phylogenetic and biogeographic analysis of cyprinodontiform fishes (Teleostei, Atherinomorpha). Bulletin of the American Museum of Natural History 168:335–557.

Parenti, L. R. 1984. A taxonomic revision of the Andean killifish genus *Orestias* (Cyprinodontiformes, Cyprinodontidae). Bulletin of the American Museum of Natural History 178:107–214.

Parisi, B. M., J. G. Lundberg, and C. Donascimiento. 2006. *Propimelodus caesius* a new species of long-finned pimelodid catfish (Teleostei: Siluriformes) from the Amazon basin, South America. Proceedings of the Academy of Natural Sciences of Philadelphia 155:67–78.

Pavanelli, C. S. 2003. Family Parodontidae. Pages 46–50 *in* R. E. Reis, S. O. Kullander, and C. J. Ferraris Jr., editors. Check List of the Freshwater Fishes of South and Central America. EDIPUCRS, Porto Alegre, Brazil.

Pavanelli, C. S. 2006. New species of *Apareiodon* (Teleostei: Characiformes: Parodontidae) from the rio Piquiri, upper rio Paraná basin, Brazil. Copeia:89–95.

Pavanelli, C. S., and A. G. Bifi. 2009. A new *Tatia* (Ostariophysi: Siluriformes: Auchenipteridae) from the rio Iguaçu basin, Paraná State, Brazil. Neotropical Ichthyology 7:199–204.

Pavanelli, C. S., and H. A. Britski. 2003. *Apareiodon* Eigenmann, 1916 (Teleostei, Characiformes), from the Tocantins-Araguaia basin, with description of three new species. Copeia:337–348.

Pavanelli, C. S., R. P. Ota, and P. Petry. 2009. New species of *Metynnis* Cope, 1878 (Characiformes: Characidae) from the rio Paraguay basin, Mato Grosso State, Brazil. Neotropical Ichthyology 7:141–146.

Peixer, J., L. A. F. Mateus, and E. K. Resende. 2006. First gonadal maturation of *Pinirampus pirinampu* (Siluriformes: Pimelodidae) in the Pantanal, Mato Grosso do Sul State, Brazil. Brazilian Journal of Biology 66:317–323.

Peixoto, L. A. W., G. M. Dutra, and W. B. Wosiacki. 2015. The electric glass knifefishes of the *Eigenmannia trilineata* species-group (Gymnotiformes: Sternopygidae): monophyly and description of seven new species. Zoological Journal of the Linnean Society 175:384–414.

Peixoto, L. A. W., and W. B. Wosiacki. 2010. Description of a new species of *Tetranematichthys* (Siluriformes: Auchenipteridae) from the lower Amazon basin, Brazil. Neotropical Ichthyology 8:69–75.

Pereira, J. d. O., M. T. da Silva, L. J. Soares Vieira, and R. Fugi. 2011. Effects of flood regime on the diet of *Triportheus curtus* (Garman, 1890) in an Amazonian floodplain lake. Neotropical Ichthyology 9:623–628.

Pereira, P. R., C. S. Agostinho, R. J. de Oliveira, and E. E. Marques. 2007a. Trophic guilds of fishes in sandbank habitats of a Neotropical river. Neotropical Ichthyology 5:399–404.

Pereira, P. R., C. S. Agostinho, R. J. Oliveira, and E. E. Marques. 2007b. Trophic guilds of fishes in sandbank habitats of a Neotropical river. Neotropical Ichthyology 5:399–404.

Pereira, T. N. A., and R. M. C. Castro. 2014. A new species of *Utiaritichthys* Miranda Ribeiro (Characiformes: Serrasalmidae) from the Serra dos Parecis, Tapajós drainage. Neotropical Ichthyology 12.

Perrone, E. C., and F. Vieira. 1991. Hábito alimentar de *Eleotris pisonis* (Teleostei: Eleotrididae) na região estuarina do Rio Jucú, Espírito Santo, Brasil. Revista Brasileira de Biologia 51:867–872.

Peterson, C. C., and P. McIntyre. 1998. Ontogenetic diet shifts in *Roeboides affinis* with morphological comparisons Environmental Biology of Fishes 53:105–110.

Peterson, C. C., and K. O. Winemiller. 1997. Ontogenetic diet shift and scale-eating in *Roeboides dayi*, a Neotropical characid. Environmental Biology of Fishes 49:111–118.

Petrere Jr., M. 1985. A pesca comercial no Rio Solimdes-Amazonas e seus afluentes: analise dos informes do pescado desembarcado no Mercado Municipal de Manaus (1976–1978). Ciencia e Cultura 37:1987–1999.

Pezold, F. L. 2011. Systematics of Gobionellidae. Pages 87–98 *in* R. A. Patzner, J. L. van Tassel, M. Kovačić, and B. G. Kapoor, editors. The Biology of Gobies. CRC Press, New Hampshire, USA.

Pezold, F., and B. Cage. 2002. A review of the spinycheek sleepers, genus *Eleotris* (Teleostei: Eleotridae), of the western Hemisphere, with comparison to the west African species. Tulane Studies in Zoology and Botany 31:19–63.

Pfeiffer, W. 1977. The distribution of fright reaction and alarm substance cells in fishes. Copeia:653–665.

Piálek, L., O. Říčan, J. Casciotta, A. Almirón, and J. Zrzavý. 2012. Multilocus phylogeny of *Crenicichla* (Teleostei: Cichlidae), with biogeography of the *C. lacustris* group: Species flocks as a model for sympatric speciation in rivers. Molecular Phylogenetics and Evolution 62:46–61.

Piggott, M. P., N. L. Chao, and L. B. Beheregaray. 2011. Three fishes in one: cryptic species in an Amazonian floodplain forest specialist. Biological Journal of the Linnean Society 102:391–403.

Pimentel-Souza, F., and N. Fernandes-Souza. 1987. Electric organ discharge rhythms and social interactions in a weakly electric fish, *Rhamphichthys rostratus* (Rhamphichthyidae, Gymnotiformes) in an aquarium. Experimental Biology 46:169–176.

Piorski, N. M., J. C. Garavello, M. Arce, and M. H. Sabaj-Pérez. 2008. *Platydoras brachylecis*, a new species of thorny catfish (Siluriformes: Doradidae) from northeast Brazil. Neotropical Ichthyology 6:481–494.

Pires, M. N., J. Arendt, and D. N. Reznick. 2010. The evolution of placentas and superfetation in the fish genus *Poecilia* (Cyprinodontiformes: Poeciliidae: Subgenera *Micropoecilia* and *Acanthophacelus*). Biological Journal of the Linnean Society 99:784–796.

Pires, T. H. S., D. F. Campos, C. P. Röpke, J. Sodré, S. Amadio, and J. Zuanon. 2015. Ecology and life-history of *Mesonauta festivus*: biological traits of a broad ranged and abundant Neotropical cichlid. Environmental Biology of Fishes 98:789–799.

Pires, T. H. S., T. B. Farago, D. F. Campos, G. M. Cardoso, and J. Zuanon. 2016. Traits of a lineage with extraordinary geographical range: ecology, behavior and life-history of the sailfin tetra *Crenuchus spilurus*. Environmental Biology of Fishes 99:925–937.

Planquette, P., P. Keith, and P. Y. Le Bail. 1996. Atlas des poissons d'eau douce de Guyane. Tome 1. Publications scientifiques du Muséum national d'Histoire naturelle, Paris.

Ploeg, A. 1991. Revision of the South American cichlid genus *Crenicichla* Heckel, 1840, with descriptions of fifteen new species and consideration on species groups, phylogeny and biogeography (Pisces, Perciformes, Cichlidae). Unpublished PhD thesis. Universiteit van Amsterdam.

Poeser, F. N. 2003. From the Amazon River to the Amazon molly and back again. Unpublished PhD thesis. University of Amsterdam, Amsterdam, the Netherlands.

Pouilly, M., F. Lino, J.-G. Bretenoux, and C. Rosales. 2003. Dietary-morphological relationships in a fish assemblage of the Bolivian Amazonian floodplain. Journal of Fish Biology 62:1137–1158.

Pouilly, M., T. Yunoki, C. Rosales, and L. Torres. 2004. Trophic structure of fish assemblages from Mamoré River floodplain lakes (Bolivia). Ecology of Freshwater Fish 13:245–257.

Pound, K. L., W. H. Nowlin, D. G. Huffman, and T. H. Bonner. 2011. Trophic ecology of a nonnative population of suckermouth catfish (*Hypostomus plecostomus*) in a central Texas spring-fed stream. Environmental Biology of Fishes 90:277–285.

Power, M. E. 1984. Depth distributions of armored catfish: predator-induced resource avoidance? Ecology 65:523–528.

Prang, G. 2007. An industry analysis of the freshwater ornamental fishery with particular reference to the supply of Brazilian freshwater ornamentals to the UK market. Uakari 3:7–51.

Presswell, B., S. H. Weitzman, and T. Bergquist. 2000. *Skiotocharax meixon*, a new genus and species of fish from Guyana with discussion of its relationships (Characiformes: Crenuchidae). Ichthyological Exploration of Freshwaters 11:175–192.

Pretti, V. Q., D. Calcagnotto, M. Toledo-Piza, and L. F. de Almeida-Toledo. 2009. Phylogeny of the Neotropical genus *Acestrorhynchus* (Ostariophysi: Characiformes) based on nuclear and mitochondrial gene sequences and morphology: a total evidence approach. Molecular Phylogenetics and Evolution **52**:312–320.

Provenzano, F., and N. Milani. 2006. *Cordylancistrus nephelion* (Siluriformes, Loricariidae), a new and endangered species of suckermouth armored catfish from the Tuy River, north-central Venezuela. Zootaxa **1116**:29–41.

Provenzano, R. F., S. A. Schaefer, J. N. Baskin, and R. Royero-Leon. 2003. New, possibly extinct lithogenine loricariid (Siluriformes, Loricariidae) from northern Venezuela. Copeia:562–575.

Queiroz, L. J., G. Torrente-Vilara, W. M. Ohara, T. H. S. Pires, J. Zuanon, and C. R. D. Doria. 2013. Peixes do rio Madeira, Volume 2. Dialeto, São Paulo, Brazil.

Ramcharitar, J., D. P. Gannon, and A. N. Popper. 2006. Bioacoustics of fishes of the family Sciaenidae (croakers and drums). Transactions of the American Fisheries Society **135**:1409–1431.

Ramos, R. T. C. 2003a. Family Achiridae (American soles). Pages 666–669 *in* R. E. Reis, S. O. Kullander, and C. J. Ferraris Jr., editors. Check List of the Freshwater Fishes of South and Central America. EDIPUCRS, Porto Alegre, Brazil.

Ramos, R. T. C. 2003b. Systematic review of *Apionichthys* (Pleuronectiformes: Achiridae), with description of four new species. Ichthyological Exploration of Freshwaters **14**:97–126.

Ramos, R. T. C., T. P. A. Ramos, and P. R. D. Lopes. 2009. New species of *Achirus* (Pleuronectiformes: Achiridae) from Northeastern Brazil. Zootaxa **2113**:55–62.

Rapp Py-Daniel, L. H. 1981. *Furcodontichthys novaesi* n. gen., n. sp. (Osteichthyes, Siluriformes; Loricariidae) na bacia Amazônia, Brasil. Boletim do Museu Paraense Emilio Goeldi, Nova Serie, Zoologia **105**:1–17.

Rapp Py-Daniel, L. H., and C. Cox Fernandes. 2005. Sexual dimorphism in Amazonian Siluriformes and Gymnotiformes (Ostariophysi). Acta Amazonica **35**:97–110.

Rapp Py-Daniel, L. H., P. M. M. Ito, R. P. Ota, I. M. Soares, D. A. Bastos, and S. Hashimoto. 2015. 30 anos da maior coleção de peixes amazônicos: a coleção de peixes do INPA. Boletim Sociedade Brasileira de Ictiologia **116**:4–16.

Rapp Py-Daniel, L. H., and E. C. Oliveira. 2001. Seven new species of *Harttia* from the Amazonian-Guyana region (Siluriformes: Loricariidae). Ichthyological Exploration of Freshwaters **12**:79–96.

Rapp Py-Daniel, L. H., and V. Py-Daniel. 1984. Observações sobre *Spatuloricaria evansii* (Boulenger, 1892) (Osteichthyes; Loricariidae) e a sua predação em Simuliidae (Díptera; Culicomorpha). Boletim do Museu Paraense Emilio Goeldi, Zoologia **1**:207–218.

Rapp Py-Daniel, L. H., and J. Zuanon. 2005. Description of a new species of *Parancistrus* (Siluriformes: Loricariidae) from the rio Xingu, Brazil. Neotropical Ichthyology **3**:571–577.

Rapp Py-Daniel, L. R., J. Zuanon, and R. R. de Oliveira. 2011. Two new ornamental loricariid catfishes of *Baryancistrus* from rio Xingu drainage (Siluriformes: Hypostominae). Neotropical Ichthyology **9**:241–252.

Ray, C. K., and J. W. Armbruster. 2016. The genera *Isorineloricaria* and *Aphanotorulus* (Siluriformes: Loricariidae) with description of a new species. Zootaxa **4072**:501–539.

Rayner, J. 1986. Pleuston: animals which move in water and air. Endeavour **10**:58–64.

Ready, J. S., E. J. G. Ferreira, and S. O. Kullander. 2006. Discus fishes: mitochondrial DNA evidence for a phylogeographic barrier in the Amazonian genus *Symphysodon* (Teleostei: Cichlidae). Journal of Fish Biology **69**:200–211.

Regan, C. T. 1904. A monograph of the fishes of the family Loricariidae. Transactions of the Zoological Society of London **17**:191–350.

Reinert, T. R., and K. A. Winter. 2002. Sustainability of harvested pacú (*Colossoma macropomum*) populations in the northeastern Bolivian Amazon. Conservation Biology **16**:1344–1351.

Reis, R. E. 1989. Systematic revision of the Neotropical characid subfamily Stethaprioninae (Pisces, Characiformes). Comunicacoes do Museu de Ciencias da PUCRS Serie Zoologia **2**:3–86.

Reis, R. E. 1997. Revision of the Neotropical catfish genus *Hoplosternum* (Ostariophysi: Siluriformes: Callichthyidae), with the description of two new genera and three new species. Ichthyological Exploration of Freshwaters **7**:299–326.

Reis, R. E. 1998. Anatomy and phylogenetic analysis of the Neotropical callichthyid catfishes (Ostariophysi, Siluriformes). Zoological Journal of the Linnean Society **124**:105–168.

Reis, R. E. 2004. *Otocinclus cocama*, a new uniquely colored loricariid catfish from Peru (Teleostei: Siluriformes), with comments on the impact of taxonomic revisions to the discovery of new taxa. Neotropical Ichthyology **2**:109–115.

Reis, R. E., and T. A. K. Borges. 2006. The South American catfish genus *Entomocorus* (Ostariophysi: Siluriformes: Auchenipteridae), with the description of a new species from the Paraguay River basin. Copeia:412–422.

Reis, R. E., and C. C. Kaefer. 2005. Two new species of the Neotropical catfish genus *Lepthoplosternum* (Ostariophysi: Siluriformes: Callichthyidae). Copeia:724–731.

Reis, R. E., P. Y. Le Bail, and J. H. A. Mol. 2005. New arrangement in the synonymy of *Megalechis* Reis, 1997 (Siluriformes: Callichthyidae). Copeia:678–682.

Reis, R. E., and P. A. Lehmann. 2009. Two new species of *Acestridium* Haseman, 1911 (loricariidae: Hypoptopomatinae) from the Rio Madeira basin, Brazil. Copeia:446–452.

Reis, R. E., L. R. Malabarba, and C. A. S. Lucena. 2014. A new species of *Rhamdella* Eigenmann and Eigenmann, 1888 (Siluriformes: Heptapteridae) from the coastal basins of southern Brazil. Arquivos de Zoologia **45**:41–50.

Reis, R. E., and E. H. L. Pereira. 2000. Three new species of the loricariid catfish genus *Loricariichthys* (Teleostei: Siluriformes) from southern South America. Copeia:1029–1047.

Reis, R. E., E. H. L. Pereira, and J. W. Armbruster. 2006. Delturinae, a new loricariid catfish subfamily (Teleostei, Siluriformes), with revisions of *Delturus* and *Hemipsilichthys*. Zoological Journal of the Linnean Society **147**:277–299.

Rengifo, B., N. K. Lujan, D. Taphorn, and P. Petry. 2008. A new species of *Gelanoglanis* (Siluriformes: Auchenipteridae) from the Marañon River (Amazon Basin), northeastern Perú. Proceedings of the Academy of Natural Sciences of Philadelphia **157**:181–188.

Reno, P. L., M. A. McCollum, C. O. Lovejoy, and R. S. Meindl. 2000. Morphology and histology of the male reproductive system in two species of internally inseminating South American catfishes, *Trachelyopterus lucenai* and *T. galeatus* (Teleostei: Auchenipteridae). Journal of Morphology **246**:131–141.

Retzer, M. E. 2006. A new species of *Farlowella* Eigenmann and Eigenmann (Siluriformes: Loricariidae), a stickcatfish from Bolivia. Zootaxa **1282**:59–68.

Retzer, M. E., and L. M. Page. 1997. Systematics of the stick catfishes, *Farlowella* Eigenmann & Eigenmann (Pisces, Loricariidae). Proceedings of the Academy of Natural Sciences of Philadelphia 147:33–88.

Ribeiro, A. C., F. C. T. Lima, and E. H. L. Pereira. 2012. A new genus and species of a minute suckermouth armored catfish (Siluriformes: Loricariidae) from the Rio Tocantins drainage, central Brazil: the smallest known loricariid catfish. Copeia:637–647.

Ribeiro, F. R. V., W. S. Pedroza, and L. H. Rapp Py-Daniel. 2011. A new species of *Nemuroglanis* (Siluriformes: Heptapteridae) from the rio Guariba, rio Madeira basin, Brazil. Zootaxa **2799**:41–48.

Ribeiro, M. C. L. B., and M. Petrere Jr. 1990. Fisheries ecology and management of the jaraqui (*Semaprochilodus taeniurus*, *S. insignis*) in central Amazonia. Regulated Rivers: Research & Management 5:195–215.

Roberts, C. D. 1993. Comparative morphology of spined scales and their phylogenetic significance in the Teleostei. Bulletin of Marine Science 52:60–113.

Roberts, T. R. 1970. Scale-eating American characoid fishes with special reference to *Probolodus heterostomus*. Proceedings of the California Academy of Sciences **38**:383–390.

Roberts, T. R. 1971. *Micromischodus sugillatus*, a new hemiodontid characin fish from Brazil, and its relationship to the Chilodontidae. Breviora:1–25.

Roberts, T. R. 1972. Ecology of fishes in the Amazon and Congo basins. Bulletin of the Museum of Comparative Zoology 143:117–147.

Roberts, T. R. 1974. Osteology and classification of the Neotropical characoid fishes of the families Hemiodontidae (including Anodontinae) and Parodontidae. Bulletin of the Museum of Comparative Zoology at Harvard College 146:411–472.

Roberts, T. R. 1984. *Amazonsprattus scintilla*, new genus and species from the Rio Negro, Brazil, the smallest known clupeomorph fish. Proceedings of the California Academy of Sciences **43**:317–321.

Roberts, T. R. 2013. *Leptophilypnion*, a new genus with two new species of tiny Central Amazonian gobioid fishes (Teleostei: Gobioidei, Eleotridae). Aqua 19:85–98.

Roberts, T. R. 2015. Mimicry of a bean seed by the Amazonian aspredinid catfish *Amaralia hypsiura* (Kner 1855), with notes on vegetative camouflage by fishes. Aqua, International Journal of Ichthyology 21:3–15.

Robins, C. R., and G. C. Ray. 1986. A Field Guide to Atlantic Coast Fishes of North America. Houghton Mifflin, Boston, USA.

Rocha, M., H. Lazzarotto, and L. R. Py-Daniel. 2012. A new species of *Scoloplax* with a remarkable new tooth morphology within loricarioidea (Siluriformes: Scoloplacidae). Copeia:670–677.

Rocha, M. S. 2012. Sistemática da família Pimelodidae Swainson, 1838 (Teleostei: Siluriformes). Unpublished PhD thesis. Instituto Nacional de Pesquisas da Amazônia, Manaus, Brazil.

Rocha, M. S., R. R. de Oliveira, and L. H. R. Py-Daniel. 2007. A new species of *Propimelodus* Lundberg & Parisi, 2002 (Siluriformes: Pimelodidae) from rio Araguaia, Mato Grosso, Brazil. Neotropical Ichthyology 5:279–284.

Rocha, M. S., R. R. De Oliveira, and L. H. Rapp Py-Daniel. 2008a. *Scoloplax baskini*: A new spiny dwarf catfish from rio Aripuanã, Amazonas, Brazil (Loricarioidei: Scoloplacidae). Neotropical Ichthyology 6:323–328.

Rocha, M. S., R. O. Oliveira, and L. Rapp Py-Daniel. 2008b. A new species of *Gladioglanis* Ferraris and Mago-Leccia from rio Aripuanã, Amazonas, Brazil (Siluriformes: Heptapteridae). Neotropical Ichthyology 6:433–438.

Rodriguez, C. M. 1997. Phylogenetic analysis of the tribe Poeciliini (Cyprinodontiformes: Poeciliidae). Copeia:663–679.

Rodriguez, M. S., C. A. Cramer, S. L. Bonatto, and R. E. Reis. 2008. Taxonomy of *Ixinandria* Isbrücker & Nijssen (Loricariidae: Loricariinae) based on morphological and molecular data. Neotropical Ichthyology 6:367–378.

Rodriguez, M. S., M. L. S. Delapieve, and R. E. Reis. 2015. Phylogenetic relationships of the species of *Acestridium* Haseman (1911) (Siluriformes: Loricariidae). Neotropical Ichthyology **13**:325–340.

Rodriguez, M. S., H. Ortega, and R. Covain. 2011. Intergeneric phylogenetic relationships in catfishes of the Loricariinae (Siluriformes: Loricariidae), with the description of *Fonchiiloricaria nanodon*: a new genus and species from Peru. Journal of Fish Biology 79:875–895.

Rodríguez-Cattáneo, A., and A. A. Caputi. 2009. Waveform diversity of electric organ discharges: the role of electric organ auto-excitability in *Gymnotus* spp. Journal of Experimental Biology **212**:3478–3489.

Rodríguez-Cattáneo, A., P. Aguilera, E. Cilleruelo, W. G. R. Crampton, and A. A. Caputi. 2013. Electric organ discharge diversity in the genus *Gymnotus*: anatomo-functional groups and electrogenic mechanisms. Journal of Experimental Biology **216**:1501–1515.

Rodríguez Fernández, C. A. 1991. Bagres, malleros y cuerderos en el Bajo río Caquetá. Pages 152–152 Estudios en la Amazonia Colombiana II. TROPENBOS, Bogota.

Rodríguez-Olarte, D., and D. C. Taphorn. 2006. Abundance, feeding and reproduction of *Salminus* sp. (Pisces: Characidae) from mountain streams of the Andean piedmont in Venezuela. Neotropical Ichthyology 4:73–79.

Rodríguez-Olarte, D., D. C. Taphorn, and C. J. Marrero. 2001. Aspectos de la distribución y ecología reproductiva de *Psectrogaster ciliata* (Pisces: Curimatidae) en la Orinoquia Venezolana. Bioagro **13**:85–89.

Roepke, C. P., E. J. G. Ferreira, and J. Zuanon. 2014. Seasonal changes in the use of feeding resources by fish in stands of aquatic macrophytes in an Amazonian floodplain, Brazil. Environmental Biology of Fishes 97:401–414.

Román-Valencia, C. 1996. Historia natural del rollizo, *Piabucina sp.* (Pisces, Lebiasinidae) en la cuenca del río La Vieja, alto Cauca. Dahlia 1:89–96.

Román-Valencia, C. 1997. Dieta de una especie nueva de *Piabucina* (Pisces, Lebiasinidae) en alto Cauca, Colombia. Revista de Biologia Tropical **45**:1255–1256.

Román-Valencia, C. 2004. Sobre la bioecologia de *Lebiasina panamensis* (Pisces: Lebiasinidae) en la cuenca del río León, Caribe colombiano. Dahlia 7:33–35.

Román-Valencia, C., and A. Botero. 2006. Trophic and reproductive ecology of a species of *Hemibrycon* (Pisces: Characidae) in Tinajas creek, Quindío River drainage, upper Cauca basin, Colombia. Revista del Museo Argentino de Ciencias Naturales 8:1–8.

Román-Valencia, C., C. A. García-Alzate, R. I. Ruiz-C, and D. C. Taphorn B. 2010a. A new species of *Creagrutus* from the Güejar River, Orinoco basin, Colombia (Characiformes: Characidae). Ichthyological Exploration of Freshwaters 21:87–95.

Román-Valencia, C., C. A. García-Alzate, R. I. Ruiz-C, and D. C. Taphorn B. 2012. A new species of *Tyttocharax* (Characiformes: Characidae: Stevardiinae) from the Güejar River, Orinoco River basin, Colombia. Neotropical Ichthyology **10**:519–525.

Román-Valencia, C., C. A. García-Alzate, R. I. Ruiz-C, and D. C. Taphorn. 2010b. New species of *Hemibrycon* (Characiformes, Characidae) from the Roble River, Alto Cauca, Colombia, with a key of species known from the Magdalena-Cauca basin. Vertebrate Zoology **60**:99–105.

Román-Valencia, C., R. I. Ruiz-C, and A. Giraldo. 2008a. Diet and reproduction of two syntopic species, *Hemibrycon boquiae* and *Bryconamericus caucanus* (Pisces, Characidae) the Boquia gault, Quindio River, Alto Cauca, Colombia. Revista del Museo Argentino de Ciencias Naturales, Nueva Serie:55–62.

Román-Valencia, C., R. I. Ruiz-C, D. C. Taphorn B, and C. García-A. 2013. Three new species of *Bryconamericus* (Characiformes, Characidae), with keys for species from Ecuador and a discussion on the validity of the genus *Knodus*. Animal Biodiversity and Conservation **36**:123–139.

Román-Valencia, C., J. A. Vanegas-Ríos, and R. I. Ruiz-C. 2008b. A new species of fish of the genus *Bryconamericus* (Ostariophysi: Characidae) from the Magdalena River, with a key to Colombian species. Revista de Biologia Tropical **56**:1749–1763.

Römer, U. 2001. Baensch/Mergus Cichlid Atlas, Vol. 1. Mergus Verlag, Melle, Germany

Römer, U. 2006. Baensch/Mergus Cichlid Atlas, Vol. 2. Mergus Verlag, Melle, Germany.

Römer, U., I. J. Hahn, and P. M. Vergara. 2010. Description of *Dicrossus foirni* sp. n. and *Dicrossus warzeli* sp. n. (Teleostei: Perciformes: Cichlidae), two new cichlid species from the Rio Negro and the Rio Tapajós, Amazon drainage, Brazil. Vertebrate Zoology **60**:123–138.

Rosa, R. S., H. P. Castello, and T. B. Thorson. 1987. *Plesiotrygon iwamae*, a new genus and species of Neotropical freshwater stingray (Chondrichthyes: Potamotrygonidae). Copeia:447–458.

Rosa, R. S., P. Charvet-Almeida, and C. C. D. Quijada. 2010. Biology of the South American Potamotrygonid stingrays. Pages 241–281 *in* J. C. Carrier, J. A. Musick, and M. R. Heithaus, editors. Sharks and Their Relatives II: Biodiversity, Adaptive Physiology, and Conservation. CRC Press, Boca Raton, FL, USA.

Rosen, D. E., and R. M. Bailey. 1963. The poeciliid fishes (Cyprinodontiformes), their structure, zoogeography, and systematics. Bulletin of the American Museum of Natural History **126**:1–176.

Rosen, D. E., and P. H. Greenwood. 1976. A fourth Neotropical species of synbranchid eel and the phylogeny and systematics of synbranchiform fishes. Bulletin of the American Museum of Natural History **157**:1–69.

Rosen, D. E., and A. Rumney. 1972. Evidence of second species of *Synbranchus* (Pisces, Teleostei) in South America. American Museum Novitates:1–45.

Ross, R. A., and F. Schäfer. 2000. Freshwater Rays. Verslag A.C.S. GmbH, Mörfelden-Walldorf, Germany.

Roubach, R., E. S. Correia, S. Zaiden, Ricardo, C. Martino, and R. O. Cavalli. 2003. Aquaculture in Brazil. World Aquaculture **34**:28–35.

Roxo, F. F., G. S. C. Silva, L. E. Ochoa, and C. Oliveira. 2015. Description of a new genus and three new species of Otothyrinae (Siluriformes, Loricariidae). ZooKeys:103–134.

Ruiz, W. B. G., and O. A. Shibatta. 2010. A new species of *Microglanis* (Siluriformes, Pseudopimelodidae) from lower Rio Tocantins basin, Pará, Brazil, with description of superficial neuromasts and pores of lateral line system. Zootaxa **2632**:53–66.

Ruiz, W. B. G., and O. A. Shibatta. 2011. Two new species of *Microglanis* (Siluriformes: Pseudopimelodidae) from the upper-middle rio Araguaia basin, Central Brazil. Neotropical Ichthyology **9**:697–707.

Sá-Oliveira, J. C., R. Angelini, and V. J. Isaac-Nahum. 2014. Diet and niche breadth and overlap in fish communities within the area affected by an Amazonian reservoir (Amapa, Brazil). Anais da Academia Brasileira de Ciências **86**:383–405.

Sabaj, M. H. 2002. Taxonomy of the Neotropical Thorny Catfishes (Siluriformes: Doradidae) and revision of genus *Leptodoras*. Unpublished PhD thesis. University of Illinois.

Sabaj, M. H. 2005. Taxonomic assessment of *Leptodoras* (Siluriformes: Doradidae) with description of three new species. Neotropical Ichthyology **3**:637–678.

Sabaj, M. H., J. W. Armbruster, and L. M. Page. 1999. Spawning in *Ancistrus* (Siluriformes: Loricariidae) with comments on the evolution of snout tentacles as a novel reproductive strategy: larval mimicry. Ichthyological Exploration of Freshwaters **10**:217–229.

Sabaj, M. H., and C. J. Ferraris. 2003. Doradidae (thorny catfishes). Pages 456–469 *in* R. E. Reis, S. O. Kullander, and C. J. Ferraris Jr., editors. Check List of the Freshwater Fishes of South and Central America. EDIPUCRS, Porto Alegre.

Sabaj, M. H., D. C. Taphorn, and O. E. Castillo G. 2008. Two new species of thicklip thornycats, genus *Rhinodoras* (Teleostei: Siluriformes: Doradidae). Copeia:209–226.

Sabaj-Pérez, M. 2015. Where the Xingu bends and will soon break. American Scientist **103**:395–403.

Sabaj-Pérez, M. H., and J. L. O. Birindelli. 2008. Taxonomic revision of extant *Doras* Lacepède, 1803 (Siluriformes: Doradidae) with descriptions of three new species. Proceedings of the Academy of Natural Sciences of Philadelphia **157**:189–234.

Sabaj Pérez, M. H., H. Mariangeles Arce, L. M. Sousa, and J. L. O. Birindelli. 2014. *Nemadoras cristinae*, new species of thorny catfish (Siluriformes: Doradidae) with redescriptions of its congeners. Proceedings of the Academy of Natural Sciences of Philadelphia **163**:133–178.

Sabino, J., and I. Sazima. 1999. Association between fruit-eating fish and foraging monkeys in western Brazil. Ichthyological Exploration of Freshwaters **10**:309–312.

Saint-Paul, U. 1992. Status of aquaculture in Latin America. Journal of Applied Aquaculture **8**:21–39.

Salazar, V. L., and P. K. Stoddard. 2008. Sex differences in energetic costs explain sexual dimorphism in the circadian rhythm modulation of the electrocommunication signal of the gymnotiform fish *Brachyhypopomus pinnicaudatus*. Journal of Experimental Biology **211**:1012–1020.

Salcedo, N. J., and H. Ortega. 2015. A new species of *Chaetostoma*, an armored catfish (Siluriformes: Loricariidae), from the rio Maranon drainage, Amazon basin, Peru. Neotropical Ichthyology **13**:151–156.

Salzburger, W., A. Meyer, S. Baric, E. Verheyen, and C. Sturmbauer. 2002. Phylogeny of the Lake Tanganyika cichlid species flock and its relationship to the Central and East African haplochromine cichlid fish faunas. Systematic Biology **51**:113–135.

Sanchez, A. P., H. Giusti, M. Bassi, and M. L. Glass. 2005. Acid-base regulation in the South American lungfish *Lepidosiren paradoxa*: effects of prolonged hypercarbia on blood gases and pulmonary ventilation. Physiological and Biochemical Zoology **78**:908–915.

Sánchez, M. R., G. Galvis, and F. P. Victoriano. 2003. Relación entre caracteristicas del tracto digestivo y los habitos alimentarios de peces del río Yucao, sistema del río Meta (Colombia). Gayana **67**:75–86.

Sanderson, S. L., A. Y. Cheer, J. S. Goodrich, J. D. Graziano, and W. T. Callan. 2001. Crossflow filtration in suspension-feeding fishes. Nature **412**:439–441.

Sanderson, S. L., and R. Wassersug. 1990. Suspension-feeding vertebrates. Scientific American **262**:96–101.

Sanderson, S. L., and R. Wassersug. 1993. Convergent and alternative designs for vertebrate suspension feeding. Pages 37–112 in J. Hanken and B. K. Hall, editors. The Skull. Volume 3. Functional and Evolutionary Mechanisms. University of Chicago Press, Chicago, USA.

Sant'Anna, I. R. A., C. R. C. Doria, and C. E. C. Freitas. 2014. Pre-impoundment stock assessment of two Pimelodidae species caught by small-scale fisheries in the Madeira River (Amazon Basin–Brazil). Fisheries Management and Ecology **21**:322–329.

Sant'Anna, V. B., M. L. S. Delapieve, and R. E. Reis. 2012. A new species of Potamorrhaphis (Beloniformes: Belonidae) from the Amazon Basin. Copeia:663–669.

Santos, C. A. B., and R. R. Nóbrega-Alves. 2016. Ethnoichthyology of the indigenous Truká people, northeast Brazil. Journal of Ethnobiology and Ethnomedicine **12**:DOI 10.1186/s13002-13015-10076-13005.

Santos, C. N. 2014. Sistemática do genero Ochmacanthus: um grupo de bagres neotropicais lepidófagos (Teleostei: Siluriformes: Trichomycteridae). Unpublished MSc thesis. Museu de Zoologia da Universidade de Sao Paulo, Sao Paulo, Brazil.

Santos, G. B., and G. Barbieri. 1993. Idade e crescimento do "piau gordura", Leporinus piau Fowler, 1941, na represa de Três Marias (Estado de Minas Gerais) (Pisces, Ostariophysi, Anostomidae). Revista Brasileira de Biologia **53**:649–658.

Santos, G. M., E. J. G. Ferreira, and J. A. S. Zuanon. 2006. Peixes comerciais de Manaus. Ibama, PróVárzea, Manaus, Brazil.

Santos, G. M., M. Jégu, and B. Mérona. 1984. Catálogo de peixes comercia do Baixo Rio Tocantins–Projecto Tucuruí. Electronorte/CNPQ/Instituto Nacional de Pesquisas da Amazonia (INPA), Manaus, Brazil.

Santos, G. M., S. S. Pinto, and M. Jégu. 1997. Alimentacao do Pacu-Cana, Mylesinus paraschomburgkii (Teleostei, Serrasalmidae) em rios da Amazonia brasileira. Revista Brasileira de Biologia **57**:311–315.

Santos, G. M. d. 1980. Aspectos de sistemática e morfologia de Schizodon fasciatus Agassiz 1829, Rhytiodus microlepis Kner 1859 e Rhytiodus argenteofuscus Kner, 1829 (Osteichthyes, Characoidei, Anostomidae) do Lago Janauacá-Amazonas. Acta Amazonica **10**:635–649.

Santos, G. M. d. 1981. Estudo da alimentaçao e hábitos alimentares de Schizodon fasciatus Agassiz, 1829, Rhytiodus microlepis Kner, 1859, e R. argenteofuscus Kner, 1859 do Lago Janauacá-AM (Osteichthyes, Characoidei, Anostomidae). Acta Amazonica **11**:267–283.

Santos, G. M. d. 1982. Caracterização, hábitos alimentares e reprodutivos de quatro espécies de "aracus" e considerações ecologicas sobre o grupo no lago Janauacá-AM (Osteichthyes, Characoidei, Anostomidae). Acta Amazonica **12**:713–739.

Santos, G. M. d., and M. Jégu. 1987. Novas ocorrências de Gnathodolus bidens, Synaptolaemus cingulatus e descrição de duas espécies novas de Sartor (Characiformes, Anostomidae). Amazoniana **10**:181–196.

Santos, G. M. d., and M. Jégu. 1989. Inventário taxonômico e redescrição das espécies da anostomideos (Characiformes, Anostomidae) do baixo rio Tocantins, PA, Brasil. Acta Amazonica **19**:159–213.

Santos, G. M. d., and M. Jégu. 1996. Inventário taxonômico dos anostomídeos (Pisces, Anostomidae) da bacia do rio Uatamã–AM, Brasil, com descrição de duas espécies novas. Acta Amazonica **26**:151–184.

Santos, G. M. d., B. d. Mérona, A. A. Juras, and M. Jégu. 2004. Peixes do Baixo Rio Tocantins: 20 anos depois da Usina Hidrelétrica Tucuruí. Eletronorte, Brasília.

Santos, G. M. d., and P. S. Rosa. 1998. Alimentação de Anostomus ternetzi e Synaptolaemus cingulatus, duas espécies de peixes Amazônicos com boca superior. Revista Brasileira Biologica **58**:255–262.

Santos, G. M. d., and J. Zuanon. 2006. Anostomoides passionis, a new fish species from Rio Xingu, Brasil (Characiformes: Anostomidae). Zootaxa **1168**:59–68.

Sarmento-Soares, L. M., and J. L. O. Birindelli. 2015. A new species of the catfish genus Centromochlus (Siluriformes: Auchenipteridae: Centromochlinae) from the upper rio Paraná basin, Brazil. Neotropical Ichthyology **13**:77–86.

Sarmento-Soares, L. M., F. G. Cabeceira, L. N. Carvalho, J. Zuanon, and A. Akama. 2013. Centromochlus meridionalis, a new catfish species from the southern Amazonian limits, Mato Grosso state, Brazil (Siluriformes: Auchenipteridae). Neotropical Ichthyology **11**:797–808.

Sarmento-Soares, L. M., and R. F. Martins-Pinheiro. 2008. A systematic revision of Tatia (Siluriformes: Auchenipteridae: Centromochlinae). Neotropical Ichthyology **6**:495–542.

Sato, Y., N. Fenerich-Verani, A. P. O. Nuñer, H. P. Godinho, and J. R. Verani. 2003. Padrões reprodutivos de peixes da bacia do São Francisco. Pages 229–274 in H. P. Godinho and A. L. Godinho, editors. Águas, peixes e pescadores do São Francisco das Minas Gerais. PUC Minas, Belo Horizonte, Brazil.

Saul, W. G. 1975. An ecological study of fishes at a site in upper Amazonian Ecuador. Proceedings of the Academy of Natural Sciences of Philadelphia **127**:93–134.

Sazima, I. 1980. Behavior of two Brazilian species of parodontid fishes, Apareiodon piracicabae and A. ibitiensis. Copeia:166–169.

Sazima, I. 1983. Scale-eating in characoids and other fishes. Environmental Biology of Fishes **9**:87–101.

Sazima, I. 1986. Similarities in feeding behaviour between some marine and freshwater fishes in two tropical communities. Journal of Fish Biology **29**:53–65.

Sazima, I. 1988. Territorial behaviour in a scale-eating and a herbivorous Neotropical characiform fish. Revista Brasileira de Biologia **48**:189–194.

Sazima, I., and S. Andrade-Guimarães. 1987. Scavenging on human corpses as a source for stories about man-eating piranhas. Environmental Biology of Fishes **20**:75–77.

Sazima, I., L. N. Carvalho, F. P. Mendonça, and J. Zuanon. 2006. Fallen leaves on the water-bed: diurnal camouflage of three night active fish species in an Amazonian streamlet. Neotropical Ichthyology **4**:119–122.

Sazima, I., and F. A. Machado. 1982. Hábitos e comportamento de Roeboides prognathus, um peixe lepidófago (Osteichthyes, Characoidei). Boletim de Zoologia da Universidade de São Paulo **7**:37–56.

Sazima, I., and F. A. Machado. 1989. Better on land than underwater: a defensive tactic of the cichlid fish, *Laetacara dorsigera*. Ciência e Cultura **41**:1014–1016.

Sazima, I., and F. A. Machado. 1990. Underwater observations of piranhas in western Brazil. Environmental Biology of Fishes **28**:17–31.

Sazima, I., F. A. Machado, and J. Zuanon. 2000. Natural history of *Scoloplax empousa* (Scoloplacidae), a minute spiny catfish from the Pantanal wetlands in western Brazil. Ichthyological Exploration of Freshwaters **11**:89–95.

Sazima, I., and J. P. Pombal Jr. 1986. Um albino de *Rhamdella minuta*, com notas sobre comportamento (Osteichthyes, Pimelodidae). Revista Brasileira de Biologia **46**:377–381.

Scarabotti, P. A., J. A. López, R. Ghirardi, and M. J. Parma. 2011. Morphological plasticity associated with environmental hypoxia in characiform fishes from Neotropical floodplain lakes. Environmental Biology of Fishes **92**:391–402.

Schaan, A. B., J. Giora, and C. B. Fialho. 2009. Reproductive biology of the Neotropical electric fish *Brachyhypopomus draco* (Teleostei: Hypopomidae) from southern Brazil. Neotropical Ichthyology **7**:737–744.

Schaefer, S., S. H. Weitzman, and H. A. Britski. 1989. Review of the Neotropical catfish genus *Scoloplax* (Pisces: Loricarioidea: Scoloplacidae) with comments on reductive characters in phylogenetic analysis. Proceedings of the Academy of Natural Sciences of Philadelphia **141**:181–211.

Schaefer, S. A. 1997. The Neotropical cascudinhos: systematics and biogeography of the *Otocinclus* catfishes (Siluriformes: Loricariidae). Proceedings of the Academy of Natural Sciences of Philadelphia **148**:1–120.

Schaefer, S. A. 2003a. Family Astroblepidae. Pages 312–317 *in* R. E. Reis, S. O. Kullander, and C. J. Ferraris Jr., editors. Check List of the Freshwater Fishes of South and Central America. EDIPUCRS, Porto Alegre, Brazil.

Schaefer, S. A. 2003b. Relationships of *Lithogenes villosus* Eigenmann, 1909 (Siluriformes, Loricariidae): Evidence from high-resolution computed microtomography. American Museum Novitates **3401**:1–55.

Schaefer, S. A., P. Chakrabarty, A. J. Geneva, and M. H. Sabaj Pérez. 2011. Nucleotide sequence data confirm diagnosis and local endemism of variable morphospecies of Andean astroblepid catfishes (Siluriformes: Astroblepidae). Zoological Journal of the Linnean Society **162**:90–102.

Schaefer, S. A., and L. Fernández. 2009. Redescription of the pez graso, *Rhizosomichthys totae* (trichomycteridae), of lago de Tota, Colombia, and aspects of cranial osteology revealed by microtomography. Copeia:510–522.

Schaefer, S. A., and G. V. Lauder. 1996. Testing historical hypotheses of morphological change: biomechanical decoupling in loricarioid catfishes. Evolution **50**:1661–1675.

Schaefer, S. A., and F. Provenzano. 1993. The Guyana Shield *Parotocinclus*: systematics, biogeography, and description of a new Venezuelan species (Siluriformes: Loricariidae). Ichthyological Exploration of Freshwaters **4**:39–56.

Schaefer, S. A., and F. Provenzano. 1998. *Niobichthys ferrarisi*, a new genus and species of armored catfish from southern Venezuela (Siluriformes: Loricariidae). Ichthyological Exploration of Freshwaters **8**:221–230.

Schaefer, S. A., and F. Provenzano. 2008. The Lithogeninae (Siluriformes, Loricariidae): anatomy, interrelationships, and description of a new species. American Museum Novitates **3637**:1–49.

Schaefer, S. A., F. Provenzano, M. C. C. de Pinna, and J. N. Baskin. 2005. New and noteworthy venezuelan glanapterygine catfishes (siluriformes, trichomycteridae), with discussion of their biogeography and psammophily. Pages 1–27 American Museum Novitates.

Schaefer, S. A., and D. J. Stewart. 1993. Systematics of the *Panaque dentex* species group (Siluriformes: Loricariidae), wood-eating armored catfishes from tropical South America. Ichthyological Exploration of Freshwaters **4**:309–342.

Schaller, F. 1971. Über den Lautapparat von Amazonas-Fischen. Die Naturwissenschaften **58**:573–574.

Schaller, F. 1974. Wie trommeln Amazonasfische? Umschau in Wissenschaften und Technik **74**:249.

Schindler, I. 1994. Neue fundorte und anmerkungen zur lebendfarbung von *Gymnorhamphichthys rondoni* (Gymnotiformes, Rhamphichthyidae). Zeitschrift fuer Fischkunde **2**:193–202.

Schindler, I., and w. Staeck. 2008. *Dicrossus gladicauda* sp. n.–a new species of crenicarine dwarf cichlids (Teleostei: Perciformes: Cichlidae) from Colombia, South America. Vertebrate Zoology **58**:67–73.

Schmidt, R. E. 1993. Relationships and notes on the biology of *Paracanthopoma parva* (Pisces: Trichomycteridae). Ichthyological Exploration of Freshwaters **4**:185–191.

Schneider, C. H., M. C. Gross, M. L. Terencio, and J. I. R. Porto. 2012. Cryptic diversity in the mtDNA of the ornamental fish *Carnegiella strigata*. Journal of Fish Biology **81**:1210–1224.

Schubart, O., and A. L. Gomes. 1959. Descrição de "*Cetopsorhamdia iheringi*" sp. n. (Pisces, Nematognathi, Pimelodidae, Luciopimelodinae). Revista Brasileira de Biologia **19**:1–7.

Schultz, L. P. 1944a. The catfishes of Venezuela, with descriptions of thirty-eight new forms. Proceedings of United States National Museum **94**:173–338.

Schultz, L. P. 1944b. The fishes of the family Characinidae from Venezuela, with descriptions of seventeen new forms. Proceedings of United States National Museum **95**:235–367.

Schwassmann, H. O. 1971. Circadian activity patterns in gymnotid electric fish. Pages 186–199 *in* Biochronometry. National Academy of Sciences, Friday Harbor, Washington.

Schwassmann, H. O. 1984. Species of *Steatogenys* Boulenger (Pisces, Gymnotiformes, Hypopomidae). Boletim do Museu Paraense Emilio Goeldi, Nova Serie, Zoologia **1**:97–114.

Schwassmann, H. O. 1989. *Gymnorhamphichthys rosamariae*, a new species of knifefish (Rhamphichthyidae, Gymnotiformes) from the upper Rio Negro, Brazil. Studies on Neotropical Fauna and Environment **24**:157–167.

Seegers, L. 2000. Killifishes of the World: New World Killis. Verlag A.C.S., Mörfelden-Walldorf, Germany.

Seehausen, O. 2006. African cichlid fish: a model system in adaptive radiation research. Proceedings of the Royal Society B: Biological Sciences **273**:1987–1998.

Seehausen, O., and C. E. Wagner. 2014. Speciation in freshwater fishes. Annual Review of Ecology, Evolution, and Systematics **45**:621–651.

Shelden, F. F. 1937. Osteology, myology and probable evolution of the nematognath pelvic girdle. Annals of the New York Academy of Sciences **37**:1–96.

Shibatta, O. A. 2003a. Family Pseudopimelodidae. Pages 401–405 *in* R. E. Reis, S. O. Kullander, and C. J. Ferraris Jr., editors. Check List of the Freshwater Fishes of South and Central America. EDIPUCRS, Porto Alegre, Brazil.

Shibatta, O. A. 2003b. Phylogeny and classification of 'Pimelodidae'. Pages 385–400 *in* G. Arratia, B. G. Kapoor, M. Chardon, and R. Diogo, editors. Catfishes. Science Publishers, Enfield, NH, USA.

Shibatta, O. A. 2014. A new species of *Microglanis* (Siluriformes: Pseudopimelodidae) from the upper rio Tocantins basin, Goiás State, central Brazil. Neotropical Ichthyology **12**:81–87.

Shibatta, O. A., and S. T. Bennemann. 2003. Plasticidade alimentar em *Rivulus pictus* Costa (Osteichthyes, Cyprinodontiformes, Rivulidae) de uma pequena lagoa em Brasília, Distrito Federal, Brasil. Revista Brasileira de Zoologia **20**:615–618.

Shibatta, O. A., J. Muriel-Cunha, and M. C. C. De Pinna. 2007. A new subterranean species of *Phreatobius* Goeldi, 1905 (Siluriformes, *Incertae sedis*) from the southwestern Amazon basin. Papéis Avulsos de Zoologia **47**:191–201.

Shibatta, O. A., and C. S. Pavanelli. 2005. Description of a new *Batrochoglanis* species (Siluriformes, Pseudopimelodidae) from the rio Paraguai basin, state of Mato Grosso, Brazil. Zootaxa **1092**:21–30.

Shibatta, O. A., and A. J. Rocha. 2001. Alimentação em machos e fêmeas do pirá-brasília, *Simpsonichthys boitonei* Carvalho (Cyprinodontiformes, Rivulidae). Revista Brasileira de Zoologia **18**:381–385.

Sidlauskas, B. L. 2007. Testing for unequal rates of morphological diversification in absence of a detailed phylogeny: a case study from characiform fishes. Evolution **61**:299–316.

Sidlauskas, B., J. C. Garavello, and J. Jellen. 2007. A new *Schizodon* (Characiformes: Anostomidae) from the Río Orinoco system, with a redescription of *S. isognathus* from the Rio Paraguay system. Copeia:711–725.

Sidlauskas, B., J. Mol, and R. P. Vari. 2011. Dealing with allometry in linear and geometric morphometrics: a taxonomic case study in the *Leporinus cylindriformis* group (Characiformes: Anostomidae) with description of a new species from Suriname. Zoological Journal of the Linnean Society:103–130.

Sidlauskas, B., and G. M. d. Santos. 2005. *Pseudanos winterbottomi*, a new anostomine species (Teleostei: Characiformes: Anostomidae) from Venezuela and Brazil, and comments on its phylogenetic relationships. Copeia:109–123.

Sidlauskas, B., and R. P. Vari. 2008. Phylogenetic relationships within the South American fish family Anostomidae (Teleostei, Ostariophysi, Characiformes). Zoological Journal of the Linnean Society **154**:70–210.

Sidlauskas, B. L., and R. P. Vari. 2012. Diversity and distribution of anostomoid fishes (Teleostei: Characiformes) throughout the Guianas. Cybium **36**:71–103.

Silfvergrip, A. M. C. 1996. A systematic revision of the Neotropical catfish genus *Rhamdia* (Teleostei, Pimelodidae). Unpublished PhD thesis. Swedish Museum of Natural History, Stockholm, Sweden.

Silva, A., L. Quintana, M. Galeano, and P. Errandonea. 2003. Biogeography and breeding in Gymnotiformes from Uruguay. Environmental Biology of Fishes **66**:329–338.

Silva, C. P. D. 1993. Alimentação e distribuição espacial de algumas espécies de peixes do igarapé do Candirú, Amazonas. Acta Amazonica **23**:271–285.

Silva, D. S., S. S. R. Milhomem, J. C. Pieczarka, and C. Y. Nagamachi. 2009a. Cytogenetic studies in *Eigenmannia virescens* (Sternopygidae, Gymnotiformes) and new inferences on the origin of sex chromosomes in the *Eigenmannia* genus. BMC Genetics **10**:1–8.

Silva, G. S. C., B. F. Melo, C. Oliveira, and R. C. Benine. 2013. Morphological and molecular evidence for two new species of *Tetragonopterus* (Characiformes: Characidae) from central Brazil. Journal of Fish Biology **82**:1613–1631.

Silva, G. S. C., F. F. Roxo, and C. Oliveira. 2014. *Hisonotus acuen*, a new and phenotypically variable cascudinho (Siluriformes, Loricariidae, Hypoptopomatinae) from the upper rio Xingu basin, Brazil. ZooKeys **442**:105–125.

Silva, G. S. C., F. F. Roxo, and C. Oliveira. 2015. Two new species of *Pseudancistrus* (Siluriformes, Loricariidae) from the Amazon basin, northern Brazil. ZooKeys:21–34.

Silva, S. E., W. R. C. Assunção, C. Duca, and J. Penha. 2009b. Cost of territorial maintenance by *Parodon nasus* (Osteichthyes: Parodontidae) in a Neotropical stream. Neotropical Ichthyology **7**:677–682.

Simpson, G. G. 1983. Splendid Isolation: The curious history of South American mammals. Yale University Press, New Haven, USA.

Siqueira-Souza, F. K., and C. E. C. Freitas. 2004. Fish diversity of floodplain lakes on the lower stretch of the Solimões River. Brazilian Journal of Biology **64**:501–510.

Sivasundar, A., E. Bermingham, and G. Ortí. 2001. Population structure and biogeography of migratory freshwater fishes (*Prochilodus*: Characiformes) in major South American rivers. Molecular Ecology **10**:407–417.

Slobodian, V. 2013. Taxonomia, sistemática e biogeografia de *Brachyrhamdia* Myers, 1927 (Siluriformes: Heptapteridae), com uma investigação sobre seu mimetismo com outros siluriformes. Unpublished MSc thesis. Universidade de São Paulo, Ribeirão Preto, Brazil.

Slobodian, V., and F. A. Bockmann. 2013. A new *Brachyrhamdia* (Siluriformes: Heptapteridae) from Rio Japurá basin, Brazil, with comments on its phylogenetic affinities, biogeography and mimicry in the genus. Zootaxa **3717**:1–22.

Smith, C. L. 1997. National Audubon Society Field Guide to Tropical Marine Fishes of the Caribbean, the Gulf of Mexico, Florida, the Bahamas, and Bermuda. Alfred A. Knopf, New York, USA.

Smith, N. J. H. 1981. Man, Fishes, and the Amazon. Columbia University Press, New York, USA.

Soto, J. M. R., and N. Castro-Neto. 1998. Revisão dos registros de tubarão-touro, *Carcharhinus leucas* (Valenciennes, 1839) (Chondrichthyes, Carcharhinidae), em rios e lagunas brasileiras. Resumos Expandidos da 11a Semana Nacional de Oceanografia. Rio Grande, RS, Brasil.

Sousa, L. M. 2010. Revisão taxonômica e filogenia de Astrodoradinae (Siluriformes, Doradidae). Unpublished PhD thesis. University of São Paulo, São Paulo.

Sousa, L. M., and J. L. O. Birindelli. 2011. Taxonomic revision of the genus *Scorpiodoras* (Siluriformes: Doradidae) with resurrection of *Scorpiodoras calderonensis* and description of a new species. Copeia:121–140.

Sousa, L. M., and L. H. Rapp Py-Daniel. 2005. Description of two new species of *Physopyxis* and redescription of *P. lyra* (siluriformes: Doradidae). Neotropical Ichthyology **3**:625–636.

Souza Filho, P. S., and L. Casatti. 2009. Life history of *Laetacara* aff. *araguaiae* Ottoni & Costa, 2009 (Perciformes, Cichlidae) in two streams in northwestern São Paulo State, Brazil. Biota Neotropica **10**:153–158.

Souza-Shibatta, L., L. F. Pezenti, D. G. Ferreira, F. S. de Almeida, S. H. Sofia, and O. A. Shibatta. 2013. Cryptic species of the genus *Pimelodella* (Siluriformes: Heptapteridae) from the Miranda River, Paraguay River basin, Pantanal of Mato Grosso do Sul, central Brazil. Neotropical Ichthyology **11**:101–109.

Spadella, M. A., C. Oliveira, and I. Quagio-Grassiotto. 2008. Morphology and histology of male and female reproductive systems in the inseminating species *Scoloplax distolothrix* (Ostariophysi: Siluriformes: Scoloplacidae). Journal of Morphology **269**:1114–1121.

Spotte, S. 2002. Candiru: Life and Legend of the Bloodsucking Catfishes. Creative Art Book, Berkeley, USA

Spotte, S., P. Petry, and J. A. S. Zuanon. 2001. Experiments on the feeding behavior of the hematophagous candiru, *Vandellia* cf. *plazaii*. Environmental Biology of Fishes **60**:459–464.

Staeck, W. 2002. *Biotoecus dicentrarchus* Eingeführt und nachgezüchtet. Datz **55**:30–31.

Staeck, W., and I. Schindler. 2004. *Nannacara quadrispinae* sp. n.: a new dwarf cichlid fish (Teleostei: Perciformes: Cichlidae) from the drainage of the Orinoco delta in Venezuela. Zoologische Abhandlungen **54**:155–161.

Staeck, W., and I. Schindler. 2007. Description of *Laetacara fulvipinnis* sp. n. (Teleostei: Perciformes: Cichlidae) from the upper drainages of the rio Orinoco and rio Negro in Venezuela. Vertebrate Zoology **57**:63–71.

Starks, E. C. 1913. The fishes of the Stanford expedition to Brazil. Leland Stanford Jr. University Publications, University Series:1–77.

Starnes, W. C., and I. Schindler. 1993. Comments on the genus *Apareiodon* Eigenmann (Characiformes: Parodontidae) with the description of a new species from the Gran Sabana region of eastern Venezuela. Copeia:754–762.

Stawikowski, R. 1989. Ein neuer Cichlide aus dem oberen Orinoco-Einzug: *Uaru fernandezyepezi* n. sp. (Pisces: Perciformes: Cichlidae). Bonner Zoologische Beitraege **40**:19–26.

Stawikowski, R., and U. Werner. 1998. Die Buntbarsche Amerikas, Band 1. Verlag Eugen Ulmer, Stuttgart, Germany.

Steele, S. E., E. Liverpool, and H. López-Fernández. 2013. *Krobia petitella*, a new species of cichlid fish from the Berbice River in Guyana (Teleostei: Cichlidae). Zootaxa **3693**:152–162.

Steinbach, A. B. 1970. Diurnal movements and discharge characteristics of electric gymnotid fishes in the Rio Negro, Brazil. The Biological Bulletin **138**:200–210.

Stewart, D., and M. Pavlik. 1985. Revision of *Cheirocerus* (Pisces: Pimelodidae) from tropical freshwater of South America. Copeia:356–367.

Stewart, D. J. 1985. A review of the South American catfish tribe Hoplomyzontini (Pisces, Aspredinidae) with descriptions of new species from Ecuador. Fieldiana: Zoology, New Series **25**:1–19.

Stewart, D. J. 1986a. A new pimelodid catfish from the deep-river channel of the Río Napo, eastern Ecuador (Pisces: Pimelodidae). Proceedings of the Academy of Natural Sciences of Philadelphia **138**:46–52.

Stewart, D. J. 1986b. Revision of *Pimelodina* and description of a new genus and species from the Peruvian Amazon (Pisces: Pimelodidae). Copeia:653–672.

Stewart, D. J. 2013a. A new species of *Arapaima* (Osteoglossomorpha: Osteoglossidae) from the Solimões River, Amazonas State, Brazil. Copeia:470–476.

Stewart, D. J. 2013b. Re-description of *Arapaima agassizii* (Valenciennes), a rare fish from Brazil (Osteoglossomorpha: Osteoglossidae). Copeia:38–51.

Stropp, J., P. van der Sleen, C. A. Quesada, and H. ter Steege. 2014. Herbivory and habitat association of tree seedlings in lowland evergreen rainforest on white-sand and terra-firme in the upper Rio Negro. Plant Ecology & Diversity **7**:255–265.

Sullivan, J. P., J. G. Lundberg, and M. Hardman. 2006. A phylogenetic analysis of the major groups of catfishes (Teleostei: Siluriformes) using rag1 and rag2 nuclear gene sequences. Molecular Phylogenetics and Evolution **41**:636–662.

Sullivan, J. P., J. Muriel-Cunha, and J. G. Lundberg. 2013a. Phylogenetic relationships and molecular dating of the major groups of catfishes of the Neotropical superfamily Pimelodoidea (Teleostei, Siluriformes). Proceedings of the Academy of Natural Sciences of Philadelphia **162**:89–110.

Sullivan, J. P., J. Zuanon, and C. C. Fernandes. 2013b. Two new species and a new subgenus of toothed *Brachyhypopomus* electric knifefishes (Gymnotiformes, Hypopomidae) from the central Amazon and considerations pertaining to the evolution of a monophasic electric organ discharge. ZooKeys **327**:1–34.

Tagliacollo, V. A., M. J. Bernt, J. M. Craig, C. Oliveira, and J. S. Albert. 2016. Model-based total evidence phylogeny of Neotropical electric knifefishes (Teleostei, Gymnotiformes). Molecular Phylogenetics and Evolution **95**:20–33.

Tagliacollo, V. A., R. Souza-Lima, R. C. Benine, and C. Oliveira. 2012. Molecular phylogeny of Aphyocharacinae (Characiformes, Characidae) with morphological diagnoses for the subfamily and recognized genera. Molecular Phylogenetics and Evolution **64**:297–307.

Tan, M., and J. W. Armbruster. 2016. Two new species of spotted *Hypancistrus* from the Rio Negro drainage (Loricariidae, Hypostominae). Zookeys:123–135.

Taphorn, D. C. 1992. The characiform fishes of the Apure River drainage, Venezuela. BioLlania Edición Especial. No. 4 Monografias Cientificas del Museo de Ciencias Naturales, UNELLEZ, Guanara, Venezuela.

Taphorn, D. C., J. W. Armbruster, H. López-Fernández, and C. R. Bernard. 2010. Description of *Neblinichthys brevibracchium* and *N. echinasus* from the upper Mazaruni River, Guyana (Siluriformes: Loricariidae), and recognition of *N. roraima* and *N. yaravi* as distinct species. Neotropical Ichthyology **8**:615–624.

Taphorn, D. C., and C. G. Lilyestrom. 1980. *Piabucina pleurotaenia* Regan, a synonym of *P. erythrinoides* Valenciennes (Pisces: Lebiasinidae); its distribution, diet and habitat in Lake Maracaibo basin, Venezuela. Copeia:335–340.

Taphorn, D. C., and C. G. Lilyestrom. 1984a. Claves para los peces de agua dulce de Venezuela. Revista UNELLEZ Ciencia y Tecnología **2**:5–30.

Taphorn, D. C., and C. G. Lilyestrom. 1984b. *Lamontichthys maracaibero* y *L. llanero* dos especies nuevas para Venezuela (Pisces, Loricariidae). Revista UNELLEZ de Ciencia y Tecnologia **2**:93–100.

Taphorn, D. C., and C. Marrero. 1990. *Hoplomyzon sexpapilostoma*, a new species of Venezuelan catfish (Pisces: Aspredinidae), with comments on the Hoplomyzontini. Fieldiana: Zoology (New Series) **61**:1–9.

Taphorn, D. C. B., H. López-Fernández, and C. R. Bernard. 2008. *Apareiodon agmatos*, a new species from the upper Mazaruni River, Guyana (Teleostei: Characiformes: Parodontidae). Zootaxa **1925**:31–38.

Taylor, B. W., A. S. Flecker, and R. O. Hall. 2006. Loss of a harvested fish species disrupts carbon flow in a diverse tropical river. Science **313**:833–836.

Tchernavin, V. 1944. A revision of some Trichomycterinae based on material preserved in the British Museum (Natural History). Proceedings of the Zoological Society of London **114**:234–275.

Teixeira, I. 2014. Portaria n° 445, de 17 de dezembro de 2014. Diária Oficial da União–Seção 1, n° 245, p.127.

Teixeira, R. L. 1994. Abundance, reproductive period, and feeding habits of eleotrid fishes in estuarine habitats of northeast Brazil. Journal of Fish Biology **45**:749–761.

Teixeira, T. F., F. C. T. Lima, and J. Zuanon. 2013. A new *Hyphessobrycon* Durbin from the rio Teles Pires, rio Tapajós basin, Mato Grosso state, Brazil (Characiformes: Characidae). Copeia:612–621.

ter Steege, H., N. C. A. Pitman, O. L. Phillips, J. Chave, D. Sabatier, A. Duque, J. F. Molino, M. F. Prevost, R. Spichiger, H. Castellanos, P. von Hildebrand, and R. Vasquez. 2006. Continental-scale patterns of canopy tree composition and function across Amazonia. Nature **443**:444–447.

Tesk, A., L. S. de Matos, D. C. Parisotto, F. G. Cabeceira, and L. N. Carvalho. 2014. Diet of the knifefish *Gymnorhamphichthys petiti* Géry & Vu-Tân-Tuê, 1964 (Rhamphichthyidae) in streams of Teles Pires River basin, southern Amazon. Bioscience Journal **30**:1573–1577.

Thacker, C. E. 2003. Molecular phylogeny of the gobioid fishes (Teleostei: Perciformes: Gobioidei). Molecular Phylogenetics and Evolution **26**:354–368.

Thacker, C. E. 2009. Phylogeny of Gobioidea and its placement within Acanthomorpha, with a new classification and investigation of diversification and character evolution. Copeia:93–104.

Thacker, C. E. 2011a. Systematics of Butidae and Eleotridae. Pages 79–86 in R. A. Patzner, J. L. van Tassell, M. Kovačić, and B. G. Kapoor, editors. Biology of Gobies. Science Publishers and CRC Press, NJ, USA.

Thacker, C. E. 2011b. Systematics of Gobiidae. Pages 129–138 in R. A. Patzner, J. L. van Tassell, M. Kovačić, and B. G. Kapoor, editors. Biology of Gobies. Science Publishers and CRC Press, NJ, USA.

Thacker, C. E., and D. M. Roje. 2011. Phylogeny of Gobiidae and identification of gobiid lineages. Systematics and Biodiversity **9**:329–347.

Thatcher, V. E. 1995. *Anphira xinguensis* sp. nov. (Isopoda, Cymothoidae) a gill chamber parasite of an Amazonian serrasalmid fish, *Ossubtus xinguense* Jégu, 1992. Amazoniana **12**:293–303.

Thomas, M. R., and M. H. S. Pérez. 2010. A new species of whiptail catfish, genus *Loricaria* (Siluriformes: Loricariidae), from the rio Curu (Xingu basin), Brazil. Copeia:274–283.

Thomas, M. R., and L. H. Rapp Py-Daniel. 2008. Three new species of the armored catfish genus *Loricaria* (Siluriformes: Loricariidae) from river channels of the Amazon basin. Neotropical Ichthyology **6**:379–394.

Thomaz, A. T., D. Arcila, G. Ortí, and L. R. Malabarba. 2015. Molecular phylogeny of the subfamily Stevardiinae Gill, 1858 (Characiformes: Characidae): classification and the evolution of reproductive traits. BMC Evolutionary Biology **15**.

Thomerson, J. E. 1974. *Pterolebias hoignei*, a new annual cyprinodontid fish from Venezuela, with a redescription of *Pterolebias zonatus*. Copeia:30–38.

Thomerson, J. E., and D. C. Taphorn. 1992. Two new annual killifishes from Amazonas Territory, Venezuela (Cyprinodontiformes: Rivulidae). Ichthyological Exploration of Freshwaters **3**:377–384.

Thompson, A. W., R. Betancur-R., H. López-Fernández, and G. Ortí. 2014. A time-calibrated, multi-locus phylogeny of piranhas and pacus (Characiformes: Serrasalmidae) and a comparison of species tree methods. Molecular Phylogenetics and Evolution **81**:242–257.

Thorson, T. B. 1972. The status of the bull shark, *Carcharhinus leucas*, in the Amazon River. Copeia:601–605.

Thorson, T. B. 1974. Occurrence of the sawfish, *Pristis perotteti*, in the Amazon River, with notes on *P. pectinatus*. Copeia:560–564.

Thorson, T. B., J. K. Langhammer, and M. I. Oetinger. 1983. Reproduction and development of the South American freshwater stingrays, *Potamotrygon circularis* and *P. motoro* Environmental Biology of Fishes **9**:3–24.

Toledo-Piza, M. 2000a. The Neotropical fish subfamily Cynodontinae (Teleostei: Ostariophysi: Characiformes): A phylogenetic study and a revision of *Cynodon* and *Rhaphiodon*. American Museum Novitates **3286**:1–88.

Toledo-Piza, M. 2000b. Two new *Heterocharax* species (Teleostei: Ostariophysi: Characidae), with a redescription of *H. macrolepis*. Ichthyological Exploration of Freshwaters **11**:289–304.

Toledo-Piza, M. 2007. Phylogenetic relationships among *Acestrorhynchus* species (Ostariophysi: Characiformes: Acestrorhynchidae). Zoological Journal of the Linnean Society **151**:691–757.

Toledo-Piza, M., G. M. T. Mattox, and R. Britz. 2014. *Priocharax nanus*, a new miniature characid from the rio Negro, Amazon basin (Ostariophysi: Characiformes), with an updated list of miniature Neotropical freshwater fishes. Neotropical Ichthyology **12**:229–246.

Toledo-Piza, M., and N. A. Menezes. 1996. Taxonomic redefinition of the species of *Acestrorhynchus* of the microlepis group, with the description of *Acestrorhynchus apurensis*, a new species from Venezuela (Ostariophysi: Characiformes: Characidae). American Museum Novitates **3160**:1–23.

Toledo-Piza, M., N. A. Menezes, and G. M. Santos. 1999. Revision of the Neotropical fish genus *Hydrolycus* (Ostariophysi: Cynodontinae) with the description of two new species. Ichthyological Exploration of Freshwaters **10**:255–280.

Tondato, K. K., C. B. Fialho, and Y. R. Súarez. 2014. Reproductive ecology of *Odontostilbe pequira* (Steindachner, 1882) (Characidae, Cheirodontinae) in the Paraguay River, southern Pantanal, Brazil. Environmental Biology of Fishes **97**:13–25.

Tornabene, L., Y. Chen, and F. Pezold. 2013. Gobies are deeply divided: phylogenetic evidence from nuclear DNA (Teleostei: Gobioidei: Gobiidae). Systematics and Biodiversity **11**:345–361.

Torrente-Vilara, G., J. Zuanon, S. A. Amadio, and C. R. C. Doria. 2008. Biological and ecological characteristics of *Roestes molossus* (Teleostei: Cynodontidae), a night hunting characiform fish from upper Madeira River, Brazil. Ichthyological Exploration of Freshwaters **19**:103–110.

Torres-Mejia, M., E. Hernndez, and V. Senechal. 2012. A new species of *Astyanacinus* (Characiformes: Characidae) from the río Magdalena system, Colombia. Copeia:501–506.

Trindade, M. E. J., and R. Jucá-Chagas. 2008. Diet of two serrasalmin species, *Pygocentrus piraya* and *Serrasalmus brandtii* (Teleostei: Characidae), along a stretch of the rio de Contas, Bahia, Brazil. Neotropical Ichthyology **6**:645–650.

Triques, M. L. 1996. *Iracema caiana*, a genus and species of electrogenic Neotropical freshwater fish (Rhamphichthyidae: Gymnotiformes: Ostariophysi: Actionpterygii). Revue Française d'Aquariologie Herpétologie **23**:91–92.

Triques, M. L. 2005. Novas sinapomorfias para *Rhamphichthys* Muller & Troschel, 1848 (Teleostei: Rhamphichthyidae). Lundiana **6**:35–39.

Triques, M. L. 2007. *Hypopomus* Gill: nova apomorfia e notas sobre suas relações filogenéticas (Teleostei: Gymnotiformes: Rhamphichthyoidea). Revista Brasileira de Zoologia **24**:717–720.

Val, A. L., and V. M. F. de Almeida-Val. 1995. Fishes of the Amazon and Their Environment. Springer-Verlag, Berlin.

Van Der Laan, R., W. N. Eschmeyer, and R. Fricke. 2014. Family-group names of recent fishes. Zootaxa **3882**:1–230.

Vanegas-Rios, J. A. 2016. Taxonomic review of the Neotropical genus *Gephyrocharax* Eigenmann, 1912 (Characiformes, Characidae, Stevardiinae). Zootaxa **4100**:1–92.

Vanegas-Ríos, J. A., M. M. Azpelicueta, J. M. Mirande, and M. D. G. Gonzales. 2013a. *Gephyrocharax torresi* (Characiformes: Characidae: Stevardiinae), a new species from the río Cascajales basin, río Magdalena system, Colombia. Neotropical Ichthyology **11**:275–284.

Vanegas-Ríos, J. A., M. M. Azpelicueta, and H. Ortega. 2011. *Chrysobrycon eliasi*, new species of stevardiine fish (Characiformes: Characidae) from the río Madre de Dios and upper río Manuripe basins, Peru. Neotropical Ichthyology **9**:731–740.

Vanegas-Ríos, J. A., V. Meza-Vargas, and M. D. L. M. Azpelicueta. 2013b. Extension of geographic distribution of *Chrysobrycon hesperus* and *C. myersi* (Characiformes, Characidae, Stevardiinae) for several drainages flowing into the Amazon River basin in Peru and Colombia. Revista Mexicana de Biodiversidad **84**:384–387.

Varella, H. R., and C. R. Moreira. 2013. *Teleocichla wajapi*, a new species of cichlid from the rio Jari, Brazil, with comments on *T. centrarchus* Kullander, 1988 (Teleostei: Cichlidae). Zootaxa **3641**:177–187.

Vari, R. P. 1977. Notes on the characoid subfamily Iguanodectinae, with a description of a new species. American Museum Novitates **2612**:1–6.

Vari, R. P. 1978. The genus *Leptagoniates* (Pisces: Characoidei) with a description of a new species from Bolivia. Proceedings of the Biological Society of Washington **91**:184–190.

Vari, R. P. 1982a. *Curimatopsis myersi*, a new curimatid characiform fish (Pisces: Characiformes) from Paraguay. Proceedings of the Biological Society of Washington **95**:788–792.

Vari, R. P. 1982b. Systematics of the Neotropical characoid genus *Curimatopsis* (Pisces: Characoidei). Smithsonian Contributions to Zoology **373**:1–28.

Vari, R. P. 1983. Phylogenetic relationships of the families Curimatidae, Prochilodontidae, Anostomidae, and Chilodontidae (Pisces, Characiformes). Smithsonian Contributions to Zoology **378**:1–60.

Vari, R. P. 1984. Systematics of the Neotropical characiform genus *Potamorhina* (Pisces, Characiformes). Smithsonian Contributions to Zoology **400**:1–36.

Vari, R. P. 1985. A new species of *Bivibranchia* (Pisces: Characiformes) from Surinam, with comments on the genus. Proceedings of the Biological Society of Washington **98**:511–522.

Vari, R. P. 1986. *Serrabrycon mago*i, a new genus and species of scale-eating characid (Pisces: Characiformes) from the upper río Negro. Proceedings of the Biological Society of Washington **99**:328–334.

Vari, R. P. 1989a. A phylogenetic study of the Neotropical characiform family Curimatidae (Pisces: Ostariophysi). Smithsonian Contributions to Zoology **471**:1–71.

Vari, R. P. 1989b. Systematics of the Neotropical characiform genus *Curimata* Bosc (Pisces: Characiformes). Smithsonian Contributions to Zoology **474**:1–63.

Vari, R. P. 1989c. Systematics of the Neotropical characiform genus *Psectrogaster* Eigenmann and Eigenmann (Pisces: Characiformes). Smithsonian Contributions to Zoology **481**:1–43.

Vari, R. P. 1991. Systematics of the Neotropical characiform genus *Steindachnerina* Fowler (Pisces: Ostariophysi). Smithsonian Contributions to Zoology **507**:1–118.

Vari, R. P. 1992a. Systematics of the Neotropical characiform genus *Curimatella* Eigenmann and Eigenmann (Pisces: Ostariophysi), with summary comments on the Curimatidae. Smithsonian Contributions to Zoology **533**:1–48.

Vari, R. P. 1992b. Systematics of the Neotropical characiform genus *Cyphocharax* Fowler (Pisces: Ostariophysi). Smithsonian Contributions to Zoology **529**:1–137.

Vari, R. P. 1995. The Neotropical fish family Ctenoluciidae (Teleostei: Ostariophysi: Characiformes): supra and intrafamilial phylogenetic relationships, with a revisionary study. Smithsonian Contributions to Zoology **564**:1–97.

Vari, R. P. 2003. Family Curimatidae. Pages 51–64 *in* R. E. Reis, S. O. Kullander, and C. J. Ferraris Jr., editors. Check List of the Freshwater Fishes of South and Central America. EDIPUCRS, Porto Alegre, Brazil.

Vari, R. P., and B. B. Calegari. 2014. New species of the catfish genus *Tatia* (Siluriformes: Auchenipteridae) from the rio Teles Pires, upper rio Tapajós basin, Brazil. Neotropical Ichthyology **12**:667–674.

Vari, R. P., R. M. C. Castro, and S. J. Raredon. 1995. The Neotropical fish family Chilodontidae (Teleostei: Characiformes): a phylogenetic study and a revision of *Caenotropus* Günther. Smithsonian Contributions to Zoology **577**:1–32.

Vari, R. P., C. D. De Santana, and W. B. Wosiacki. 2012. South American electric knifefishes of the genus *Archolaemus* (Ostariophysi, Gymnotiformes): undetected diversity in a clade of rheophiles. Zoological Journal of the Linnean Society **165**:670–699.

Vari, R. P., and C. J. Ferraris Jr. 1999. The Neotropical catfish genus *Epapterus* Cope (Siluriformes: Auchenipteridae): a reappraisal. Proceedings of the Biological Society of Washington **111**:992–1007.

Vari, R. P., and C. J. Ferraris Jr. 2003. Family Cetopsidae. Pages 257–260 *in* R. E. Reis, S. O. Kullander, and C. J. Ferraris Jr., editors. Check List of the Freshwater Fishes of South and Central America. EDIPUCRS, Porto Alegre, Brazil.

Vari, R. P., and C. J. Ferraris Jr. 2006. The catfish genus *Tetranematichthys* (Auchenipteridae). Copeia:168–180.

Vari, R. P., and C. J. Ferraris Jr. 2009. New species of *Cetopsidium* (Siluriformes: Cetopsidae: Cetopsinae) from the upper Rio Branco system in Guyana. Neotropical Ichthyology **7**:289–293.

Vari, R. P., and C. J. Ferraris Jr. 2013. Two new species of the catfish genus *Tatia* (Siluriformes: Auchenipteridae) from the Guiana shield and a reevaluation of the limits of the genus. Copeia:396–402.

Vari, R. P., C. J. Ferraris Jr., and M. C. C. De Pinna. 2005. The Neotropical whale catfishes (Siluriformes: Cetopsidae: Cetopsinae), a revisionary study. Neotropical Ichthyology **3**:127–238.

Vari, R. P., C. J. Ferraris Jr., A. Radosavjevic, and V. A. Funk. 2009. Checklist of the freshwater fishes of Guiana Shield. Bulletin of the Biological Society of Washington **17**:1–94.

Vari, R. P., and M. Goulding. 1985. A new species of *Bivibranchia* (Pisces: Characiformes) from the Amazon River basin. Proceedings of the Biological Society of Washington **98**:1054–1061.

Vari, R. P., and A. S. Harold. 2001. Phylogenetic study of the Neotropical fish genera *Creagrutus* Günther and *Piabina* Reinhardt (Teleostei: Ostariophysi: Characiformes), with a revision of the cis-Andean species. Smithsonian Contributions to Zoology **613**:1–239.

Vari, R. P., and F. C. T. Lima. 2003. New species of *Creagrutus* (Teleostei: Characiformes: Characidae) from the Rio Uaupés basin, Brazil. Copeia:583–587.

Vari, R. P., and H. Ortega. 1986. The catfishes of the Neotropical family Helogenidae (Ostariophysi: Siluroidei). Smithsonian Contributions to Zoology **442**:1–20.

Vari, R. P., and H. Ortega. 1997. A new *Chilodus* species from southeastern Peru (Ostariophysi: Characiformes: Chilodontidae): description, phylogenetic discussion, and comments on the distribution of other chilodontids. Ichthyological Exploration of Freshwaters **8**:71–80.

Vari, R. P., and H. Ortega. 2000. *Attonitus*, a new genus of sexually dimorphic characiforms (Ostariophysi: Characidae) from western Amazonia; a phylogenetic definition and description of three new species. Ichthyological Exploration of Freshwaters **11**:113–140.

Vari, R. P., and S. J. Raredon. 1991. The genus *Schizodon* (Teleostei: Ostariophysi: Anostomidae) in Venezuela, a reappraisal. Proceedings of the Biological Society of Washington **104**:12–22.

Vari, R. P., and S. J. Raredon. 2003. Family Chilodontidae. Pages 85–86 *in* R. E. Reis, S. O. Kullander, and C. J. Ferraris Jr., editors. Check List of the Freshwater Fishes of South and Central America. EDIPUCRS, Porto Alegre, Brazil.

Vari, R. P., and R. E. Reis. 1995. *Curimata acutirostris*, a new fish (Teleostei: Characiformes: Curimatidae) from the rio Araguaia, Brazil: description and phylogenetic relationships. Ichthyological Exploration of Freshwaters **6**:297–304.

Vari, R. P., and D. J. Siebert. 1990. A new unusually sexually dimorphic species of *Bryconamericus* (Pisces: Ostariophysi: Characidae) from the Peruvian Amazon. Proceedings of the Biological Society of Washington **103**:516–524.

Vari, R. P., and A. M. Williams. 1987. Headstanders of the Neotropical anostomid genus *Abramites* (Pisces: Characiformes: Anostomidae). Proceedings of the Biological Society of Washington **100**:89–103.

Vazzoler, A. E. A. d. M., and N. A. Menezes. 1992. Sintese de conhecimento sobre o comportamento reprodutivo dos Characiformes da America do Sul (Teleostei, Ostariophysi). Revista Brasileira de Biologia **52**:627–640.

Veitenheimer, I. L., and M. C. D. Mansur. 1975. Primeiras observações de bivalves dulciaquícolas como alimento do "Armado-amarillo" *Rhinodoras dorbignyi* (Kroyer, 1855) Bleeker, 1862. Iheringia **46**:25–31.

Viana, A. P., and F. Lucena Frédou. 2014. Ichthyofauna as bioindicator of environmental quality in an industrial district in the Amazon estuary, Brazil. Brazilian Journal of Biology **74**:315–324.

Vieira, I., and J. Gery. 1979. Differential growth and nutrition in *Catoprion mento* (Characoidei): Scale-eating fish of Amazonia. Acta Amazonica **9**:143–146.

Vila, I. 2006. A new species of killifish in the genus *Orestias* (Teleostei: Cyprinodontidae) from the southern High Andes, Chile. Copeia:472–477.

Vila, I., S. Scott, M. A. Mendez, F. Valenzuela, P. Iturra, and E. Poulin. 2012. *Orestias gloriae*, a new species of cyprinodontid fish from saltpan spring of the southern High Andes (Teleostei: Cyprinodontidae). Ichthyological Exploration of Freshwaters **22**:345–353.

Villacorta-Correa, M. A., and U. Saint-Paul. 1999. Structural indexes and sexual maturity of tambaqui *Colossoma macropomum* (Cuvier, 1818) (Characiformes: Characidae) in central Amazon, Brazil. Brazilian Journal of Biology **59**:637–652.

von Ihering, R., and P. Azevedo. 1934. A curimatá dos açudes nordestinos (*Prochilodus argenteus*). Archivos do Instituto Biológico de São Paulo **5**:143–183.

Vono, V., and J. L. O. Birindelli. 2007. Natural history of *Wertheimeria maculata*, a basal doradid catfish endemic to eastern Brazil (Siluriformes: Doradidae). Ichthyological Exploration of Freshwaters **18**:183–191.

Wagner, C. E., L. J. Harmon, and O. Seehausen. 2014. Cichlid species-area relationships are shaped by adaptive radiations that scale with area. Ecology Letters **17**:583–592.

Walsh, S. J., F. R. V. Ribeiro, and L. H. R. Py-Daniel. 2015. Revision of *Tympanopleura* Eigenmann (Siluriformes: Auchenipteridae) with description of two new species. Neotropical Ichthyology **13**:1–46.

Wantzen, K. M., F. De Arruda Machado, M. Voss, H. Boriss, and W. J. Junk. 2002. Seasonal isotopic shifts in fish of the Pantanal wetland, Brazil. Aquatic Sciences **64**:239–251.

Warzel, F. 1996. A new checkerboard cichlid from the Rio Tapajós. Cichlids Yearbook **6**:80–82.

Watson, R. E. 1996. Revision of the subgenus *Awaous* (*Chonophorus*) (Teleostei: Gobiidae). Ichthyological Exploration of Freshwaters **7**:1–18.

Watson, R. E., and H. Horsthemke. 1995. Review of *Euctenogobius*, a monotypic subgenus of *Awaous*, with discussion of its natural history (Teleostei: Gobiidae). Ichthyological Exploration of Freshwaters **22**:83–92.

Weber, C., R. Covain, and S. Fisch-Muller. 2012. Identity of *Hypostomus plecostomus* (Linnaeus, 1758), with an overview of *Hypostomus* species from the Guianas (Teleostei: Siluriformes: Loricariidae). Cybium **36**:195–227.

Weidner, T. 1999. *Retroculus lapidifer*–Steinträger und Grubenlaicher. DATZ **52**:8–12.

Weitzman, M., and S. H. Weitzman. 2003. Family Lebiasinidae. Pages 241–250 *in* R. E. Reis, S. O. Kullander, and C. J. Ferraris Jr., editors. Check List of Freshwater Fishes of South and Central America EDIPUCRS, Porto Alegre, Brazil.

Weitzman, S. H. 1960. Further notes on the relationships and classification of the South American characid fishes of the subfamily Gasteropelecine. Stanford Ichthyological Bulletin **7**:217–239.

Weitzman, S. H. 1962. The osteology of *Brycon meeki*, a generalized characid fish, with an osteological definition of the family. Stanford Ichthyogical Bulletin **8**:3–77.

Weitzman, S. H. 1978. Three new species of fishes of the genus *Nannostomus* from the Brazilian states of Pará and Amazonas (Teleostei: Lebiasinidae). Smithsonian Contributions to Zoology **263**:1–14.

Weitzman, S. H. 1987. A new species of *Xenurobrycon* (Teleostei: Characidae) from the río Mamoré basin of Bolivia. Proceedings of the Biological Society of Washington **100**:112–120.

Weitzman, S. H., and J. S. Cobb. 1975. A revision of the South American fishes of the genus *Nannostomus* Günther (Family Lebiasinidae). Smithsonian Contributions to Zoology **186**:1–36.

Weitzman, S. H., and S. V. Fink. 1985. Xenurobryconin phylogeny and putative pheromone pumps in glandulocaudine fishes (Teleostei: Characidae). Smithsonian Contributions to Zoology **421**:1–121.

Weitzman, S. H., S. V. Fink, A. Machado-Allison, and R. Royero L. 1994. A new genus and species of Glandulocaudinae (Teleostei: Characidae) from southern Venezuela. Ichthyological Exploration of Freshwaters **5**:45–64.

Weitzman, S. H., and W. L. Fink. 1983. Relationships of the neon tetras, a group of South American freshwater fishes (Teleostei, Characidae), with comments on the phylogeny of New World characiforms. Bulletin of the Museum of Comparative Zoology **150**:339–395.

Weitzman, S. H., and R. H. Kanazawa. 1977. A new species of pygmy characoid fish from the Rio Negro and Rio Amazonas, South America (Teleostei: Characidae). Proceedings of the Biological Society of Washington **90**:149–160.

Weitzman, S. H., and N. A. Menezes. 1998. Relationships of the tribes and genera of the Glandulocaudinae (Ostariophysi: Characiformes: Characidae) with a description of a new genus, *Chrysobrycon*. Pages 171–192 *in* L. R. Malabarba, R. E. Reis, R. P. Vari, Z. M. Lucena, and C. A. S. Lucena, editors. Phylogeny and Classification of Neotropical Fishes. EDIPUCRS, Porto Alegre, Brazil.

Weitzman, S. H., N. A. Menezes, H. G. Evers, and J. R. Burns. 2005. Putative relationships among inseminating and externally fertilizing characids, with a description of a new genus and species of Brazilian inseminating fish bearing an anal-fin gland in males (Characiformes: Characidae). Neotropical Ichthyology **3**:329–360.

Weitzman, S. H., and H. Ortega. 1995. A new species of *Tyttocharax* (Teleostei: Characidae: Glandulocaudinae: Xenurobryconini) from the rio Madre de Dios basin of Peru. Ichthyological Exploration of Freshwaters **6**:129–148.

Weitzman, S. H., and L. Palmer. 1966. Do freshwater hatchetfishes really fly. Tropical Fish Hobbyist **45**:195–206.

Weitzman, S. H., and L. Palmer. 1997. A new species of *Hyphessobrycon* (Teleostei: Characidae) from the Neblina region of Venezuela and Brazil, with comments on the putative 'rosy tetra clade'. Ichthyological Exploration of Freshwaters **7**:209–242.

Weitzman, S. H., and L. Palmer. 2003. Family Gasteropelecidae. Pages 101–103 *in* R. E. Reis, S. O. Kullander, and C. J. Ferraris Jr., editors. Check List of the Freshwater Fishes of South and Central America. EDIPUCRS, Porto Alegre, Brazil.

Weitzman, S. H., and R. P. Vari. 1987. Two new species and a new genus of miniature characid fishes (Teleostei: Characiformes) from northern South America. Proceedings of the Biological Society of Washington **100**:640–652.

Weitzman, S. H., and R. P. Vari. 1988. Miniaturization in South American freshwater fishes; an overview and discussion. Proceedings of the Biological Society of Washington **101**:444–465.

Werneke, D. C., J. W. Armbruster, N. K. Lujan, and D. C. Taphorn. 2005a. *Hemiancistrus guahiborum*, a new suckermouth armored catfish from southern Venezuela (Siluriformes: Loricariidae). Neotropical Ichthyology **3**:543–548.

Werneke, D. C., M. H. Sabaj, N. K. Lujan, and J. W. Armbruster. 2005b. *Baryancistrus demantoides* and *Hemiancistrus subviridis*, two new uniquely colored species of catfishes from Venezuela (Siluriformes: Loricariidae). Neotropical Ichthyology **3**:533–542.

Werner, U., and U. Minde. 1990. Der Inka-Steinfisch, *Tahuantinsuyoa macantzatza*. Die Aquarien und Terrarien-Zeitschrift **43**:78–80.

Wetzel, J., J. P. Wourms, and J. Friel. 1997. Comparative morphology of cotylephores in *Platystacus* and *Solenostomus*: Modifications of the integument for egg attachment in skin-brooding fishes. Environmental Biology of Fishes **50**:13–25.

Whitehead, P. J. 1985. Clupeoid fishes of the world (suborder Clupeoidei). Part 1: Chirocentridae, Clupeidae and Pristigasteridae. FAO, Rome.

Whitehead, P. J., G. J. Nelson, and T. Wongratana. 1988. Clupeoid fishes of the world (suborder Clupeoidei). Part 2. Engraulididae. FAO, Rome.

Wiest, F. C. 1995. The specialized locomotory apparatus of the freshwater hatchetfish family Gasteropelecidae. Journal of Zoology **236**:571–592.

Wildekamp, R. H. 1995. A World of Killies. Atlas of the Oviparous Cyprinodontiform Fishes of the World. Volume II. The American Killifish Association, Mishawaka, Indiana.

Wildekamp, R. H. 2004. A World of Killies. Atlas of the Oviparous Cyprinodontiform Fishes of the World. Volume IV. The American Killifish Association, Mishawaka, Indiana.

Wiley, M. L., and B. B. Collette. 1970. Breeding tubercles and contact organs in fishes: their occurrence, structure and significance. Bulletin of the American Museum of Natural History **143**:143–216.

Willink, P. W., B. Chernoff, A. Machado-Allison, F. Provenzano, and P. Petry. 2003. *Aphyocharax yekwanae*, a new species of bloodfin tetra (Teleostei: Characiformes: Characidae) from the Guyana Shield of Venezuela. Ichthyological Exploration of Freshwaters **14**:1–8.

Willis, S. C., H. López-Fernández, C. G. Montaña, I. P. Farias, and G. Ortí. 2012a. Species-level phylogeny of 'Satan's perches' based on discordant gene trees (Teleostei: Cichlidae: Satanoperca Günther 1862). Molecular Phylogenetics and Evolution **63**:798–808.

Willis, S. C., J. Macrander, I. P. Farias, and G. Orti. 2012b. Simultaneous delimitation of species and quantification of interspecific hybridization in Amazonian peacock cichlids (genus *Cichla*) using multi-locus data. BMC Evolutionary Biology **12**:1–24.

Winemiller, K. O. 1989a. Development of dermal lip protuberances for aquatic surface respiration in South American characid fishes. Copeia:382–390.

Winemiller, K. O. 1989b. Patterns of variation in life history among South American fishes in seasonal environments. Oecologia **81**:225–241.

Winemiller, K. O. 1990. Caudal eyespots as deterrents against fin predation in the Neotropical cichlid *Astronotus ocellatus*. Copeia:665–673.

Winemiller, K. O., and A. Adite. 1997. Convergent evolution of weakly electric fishes from floodplain habitats in Africa and South America. Environmental Biology of Fishes **49**:175–186.

Winemiller, K. O., P. B. McIntyre, L. Castello, E. Fluet-Chouinard, T. Giarrizzo, S. Nam, I. G. Baird, W. Darwall, N. K. Lujan, I. Harrison, M. L. J. Stiassny, R. A. M. Silvano, D. B. Fitzgerald, F. M. Pelicice, A. A. Agostinho, L. C. Gomes, J. S. Albert, E. Baran, M. Petrere, C. Zarfl, M. Mulligan, J. P. Sullivan, C. C. Arantes, L. M. Sousa, A. A. Koning, D. J. Hoeinghaus, M. Sabaj, J. G. Lundberg, J. Armbruster, M. L. Thieme, P. Petry, J. Zuanon, G. T. Vilara, J. Snoeks, C. Ou, W. Rainboth, C. S. Pavanelli, A. Akama, A. van Soesbergen, and L. Saenz. 2016. Balancing hydropower and biodiversity in the Amazon, Congo, and Mekong. Science **351**:128–129.

Winemiller, K. O., and B. J. Ponwith. 1998. Comparative ecology of eleotrid fishes in Central American coastal streams. Environmental Biology of Fishes **53**:373–384.

Winemiller, K. O., D. C. Taphorn, and A. Barbarino-Duque. 1997. Ecology of *Cichla* (Cichlidae) in two blackwater rivers of Southern Venezuela. Copeia:690–696.

Winemiller, K. O., and H. Y. Yan. 1989. Obligate mucus-feeding in a South American trichomycterid catfish (Pisces: Ostariophysi). Copeia:511–514.

Winterbottom, R. 1980. Systematics, osteology and phylogenetic relationships of fishes of the ostariophysan subfamily Anostominae (Characoidei, Anostomidae). Life Sciences Contributions: Royal Ontario Museum **123**:1–112.

Wosiacki, W. B., D. P. Coutinho, and L. F. De Assis Montag. 2011. Description of a new species of sand-dwelling catfish of the genus *Stenolicmus*. Zootaxa **2752**:62–68.

Wosiacki, W. B., and D. P. SilvaMiranda. 2013. Description of a new small species of the genus *Cyphocharax* (Characiformes: Curimatidae) from the lower Amazon Basin. Copeia:627–633.

Xu, H., S. Qiang, P. Genovesi, H. Ding, J. Wu, L. Meng, and e. al. 2012. An inventory of invasive alien species in China. NeoBiota **15**:1–26.

Yamamoto, M. E., S. Chellappa, M. S. R. F. Cacho, and F. A. Huntingford. 1999. Mate guarding in an Amazonian cichlid, *Pterophyllum scalare*. Journal of Fish Biology **55**:888–891.

Yan, H. Y. 2009. A histochemical study on the snout tentacles and snout skin of bristlenose catfish *Ancistrus triradiatus*. Journal of Fish Biology **75**:845–861.

Zahl, P. A., J. J. A. McLaughlin, and R. J. Gomprecht. 1977. Visual versatility and feeding of the four-eyed fishes, *Anableps*. Copeia:791–793.

Zanata, A. M. 1997. *Jupiaba*, um novo gênero de Tetragonopterinae com osso pélvico em forma de espinho (Characidae, Characiformes). Iheringia, série Zoologia:99–136.

Zanata, A. M. 2000. Estudo das relações filogenéticas do gênero *Brycon* Müller and Troschel, 1844 (Characidae; Characiformes). Unpublished PhD thesis. Universidade de São Paulo, Instituto de Biociências, São Paulo.

Zanata, A. M., and F. C. T. Lima. 2005. New species of *Jupiaba* (Characiformes: Characidae) from Rio Tiquié, upper Rio Negro Basin, Brazil. Copeia:272–278.

Zanata, A. M., and J. P. Serra. 2010. *Hasemania piatan*, a new characid species (Characiformes: Characidae) from headwaters of rio de Contas, Bahia, Brazil. Neotropical Ichthyology **8**:21–26.

Zanata, A. M., and M. Toledo-Piza. 2004. Taxonomic revision of the South American fish genus *Chalceus* Cuvier (Teleostei: Ostariophysi: Characiformes) with the description of three new species. Zoological Journal of the Linnean Society **140**:103–135.

Zanata, A. M., and R. P. Vari. 2005. The family Alestidae (Ostariophysi, Characiformes): A phylogenetic analysis of a trans-Atlantic clade. Zoological Journal of the Linnean Society **145**:1–144.

Zaniboni Filho, E., S. Meurer, S. O. A., and A. P. de Oliverira Nuñer. 2004. Catálogo ilustrado de peixes do alto Rio Uruguai. Editora da UFSC: Tractebel Energia, Florianopolis, Brazil.

Zaret, T. M. 1977. Inhibition of cannibalism in *Cichla ocellaris* and hypothesis of predator mimicry among south american fishes. Evolution **31**:421–437.

Zaret, T. M., and R. S. Rand. 1971. Competition in tropical stream fishes: support for the competitive exclusion principle. Ecology **52**:336–342.

Zarke, A. 1997. *Geryichthys sterbai* gen. et spec. nov. and *Microcharacidium geryi* spec. nov.: Beschreibung einer neuen Gattung und zweier neuer Arten von Bodensalmern aus dem Einzugsgebiet des Rio Ucayali in Peru (Teleostei: Ostariophysi: Characiformes: Characidiidae). Zoologische Abhandlungen (Dresden) **49**:157–172.

Zarke, A. 2010. Der Kolibrisalmer- *Trochilocharax ornatus* gen. et spec. nov.- ein neuer Salmler aus Peru (Teleostei: Characiformes: Characidae). Vertebrate Zoology **60**:75–98.

Zarke, A., and J. Géry. 2006. Beschreibung einer neuen Salmler-Gattung und zweier neuer Arten (Teleostei: Characiformes: Characidac) aus Peru und Brasilien. Zoologische Abhandlungen (Dresden) **55**:31–49.

Zarske, A., and J. Géry. 1997. Rediscovery of *Agoniates halecinus* Müller & Troschel, 1845, with a supplementary description of *Agoniates anchovia* Eigenmann, 1914, and a definition of the genus (Teleostei: Ostariophysi: Characiformes: Characidae). Zoologische Abhandlungen (Dresden) **49**:173–184.

Zarske, A., and J. Géry. 2006. Zur Identität von *Copella nattereri* (Steindachner, 1876) einschließlich der Beschreibung einer neuen Art (Teleostei: Characiformes: Lebiasinidae). Zoologische Abhandlungen (Dresden) **56**:15–46.

Zarske, A., and J. Géry. 2008. Revision der neotypischen Gattung *Metynnis* COPE, 1878. II. Beschreibung zweier neuer Arten und zum Status von *Metynnis goeldii* Eigenmann, 1903 (Teleostei: Characiformes: Serrasalmidae). Vertebrate Zoology **58**:173–196.

Zebedin, A., and F. Ladich. 2013. Does the hearing sensitivity in thorny catfishes depend on swim bladder morphology? PLoS ONE **8**:1–8.

Zenaid, A. K., and R. Almeida Prado. 2012. Peixes fluviais do Brasil: espécies esportivas. Pescaventura, Campinas, Brazil.

Zuanon, J. 1999. História natural da ictiofauna de corredeiras do Rio Xingu, na região de Altamira, Pará. Unpublished PhD thesis. Universidade Estadual de Campinas, Campinas.

Zuanon, J., F. A. Bockmann, and I. Sazima. 2006a. A remarkable sand-dwelling fish assemblage from central Amazonia, with comments on the evolution of psammophily in South American freshwater fishes. Neotropical Ichthyology **4**:107–118.

Zuanon, J., L. N. Carvalho, and I. Sazima. 2006b. A chamaeleon characin: The plant-clinging and colour-changing *Ammocryptocharax elegans* (Characidiinae: Crenuchidae). Ichthyological Exploration of Freshwaters **17**:225–232.

Zuanon, J., and E. Ferreira. 2008. Feeding Ecology of Fishes in the Brazilian Amazon–A Naturalistic Approach. Pages 1–35 *in* J. E.

P. Cyprino, D. P. Bureau, and B. G. Kapoor, editors. Feeding and Digestive Functions of Fishes. CRC Press, Boca Raton, FL, USA.

Zuanon, J., L. Rapp Py-Daniel, and M. Jégu. 1993. Two new species of *Aguarunichthys* from the Amazon basin (Siluroidei: Pimelodidae). Ichthyological Exploration of Freshwaters **4**:251–260.

Zuanon, J., and I. Sazima. 2004a. Natural history of *Stauroglanis gouldingi* (Siluriformes: Trichomycteridae), a miniature sand-dwelling candiru from central Amazonian streamlets. Ichthyological Exploration of Freshwaters **15**:201–208.

Zuanon, J., and I. Sazima. 2004b. Vampire catfishes seek the aorta not the jugular: candirus of the genus *Vandellia*

(Trichomycteridae) feed on major gill arteries of host fishes. Journal of Ichthyology and Aquatic Biology **8**:31–36.

Zuanon, J., and I. Sazima. 2005a. Free meals on long-distance cruisers: the vampire fish rides giant catfishes in the Amazon. Biota Neotropica **5**:109–114.

Zuanon, J., and I. Sazima. 2005b. The ogre catfish: prey scooping by the auchenipterid *Asterophysus batrachus*. Aqua Journal of Ichthyology and Aquatic Biology **10**:15–22.

Zuanon, J. A. S., and I. Sazima. 2002. *Teleocichla centisquama*, a new species of rapids-dwelling cichlid from Xingu River, Amazonia (Perciformes: Cichlidae). Ichthyological Exploration of Freshwaters **13**:373–378.

PHOTO CREDITS

James Albert *Gymnotus carapo, Lepidosiren paradoxa, Rhamphichthys heleios, Steatogenys elegans*

Pierre-Yves Le Bail-INRA *Anchovia surinamensis, Piabucus dentatus*

Max Bernt *Magosternarchus raptor*

José Birindelli *Acanthodoras cataphractus, Ageneiosus ucaylensis, Anadoras weddellii, Auchenipterus nuchalis, Laemolyta taeniata, Leporellus vittatus, Leporinus fasciatus, Pseudanos irinae, Tatia intermedia, Pterodoras granulosus, Trachelyopterus coriaceus, Tenellus ternetzi,* figure of mouth positions in Anostomidae chapter.

Pedro Braganca *Laimosemion geayi, Melanorivulus zygonectes, Pituna xinguensis, Trigonectes rubromarginatus*

Ralf Britz *Thalassophryne amazonica*

Fernando M. Carvajal-Vallejos *Orestias agassizii*

Tiago Carvalho *Acestrorhynchus falcatus, Aphyocharax* sp., *Apionichthys finis, Apteronotus albifrons, Aspredo aspredo* (lateral and dorsal views), *Astyanax bimaculatus, Brachychalcinus copei, Brachyhypopomus beebei, Brachyplatystoma juruense, Bunocephalus knerii* (dorsal view), *Bunocephalus knerii* (lateral view), *Cetopsis coecutiens, Clupeacharax anchoveoides, Copeina guttata, Creagrutus ungulus, Crossoloricaria bahuaja, Curimatella meyeri, Dianema longibarbis, Eigenmannia virescens, Farlowella nattereri, Galeocharax gulo, Gymnotus chaviro, Hemisorubim platyrhynchos, Henonemus punctatus, Hoplias malabaricus, Hypoclinemus mentalis, Hypoptopoma incognitum, Hypostomus emarginatus, Lamontichthys filamentosus, Moenkhausia oligolepis, Otocinclus vittatus, Pachyurus* cf. *stewarti, Panaque schaeferi, Paragoniates alburnus, Pimelodus blochii, Plagioscion squamosissimus, Planiloricaria cryptodon, Prochilodus nigricans, Psectrogaster rutiloides, Pseudorinelepis genibarbis, Rhamdia quelen, Rhynchodoras boehlkei, Roeboides myersi, Salminus* sp., *Steindachnerina hypostoma, Sternopygus macrurus, Tetragonopterus argenteus, Thoracocharax stellatus, Triportheus albus, Xyliphius melanopterus*

Kenneth Catania *Electrophorus electricus*

Gregoire Germeau *Awaous flavus*

Jaime Hernandez *Pterengraulis atherinoides*

Peter and Martin Hoffmann *Corydoras eques, Hyphessobrycon jackrobertsi, Nannostomus nigrotaeniatus, Poecilocharax weitzmani*

Erling Holm *Cephalosilurus albomarginatus, Eleotris amblyopsis, Microphis brachyurus*

Michel Jégu *Chilodus punctatus*

Ben Lee *Gobioides broussonnetii*

Hernán López-Fernández *Acarichthys heckelii, Aequidens michaeli, Akawaio penak, Apistogramma steindachneri, Crenicichla lugubris, Crenicichla percna, Geophagus argyrostictus*

Paulo Lucinda *Phalloceros leticiae*

Alexandre Marceniuk *Cathorops agassizii,* figures of skulls and tooth plates in Ariidae chapter

Ivan Mikolji *Colossoma macropomum, Metynnis hypsauchen, Myloplus asterias, Pristobrycon maculipinnis, Pygocentrus cariba, Serrasalmus irritans, Serrasalmus rhombeus*

Willian M. Ohara *Batrochoglanis villosus, Phreatobius dracunculus*

Mark Sabaj Pérez *Anableps anableps, Ancistrus* sp., *Apareiodon orinocensis, Archolaemus janeae, Astroblepus* sp., *Bivibranchia bimaculata, Brittanichthys myersi, Bryconops melanurus, Caenotropus maculosus, Callichthys callichthys, Cetopsis oliveirai, Cetopsis* sp., *Cetopsorhamdia insidiosa, Chalceus epakros, Characidium zebra, Cichla melaniae, Copella compta, Crossoloricaria bahuaja* (mouth), *Cynodon meionactis, Engraulisoma taeniatum, Gymnorhamphichthys hypostomus, Helogenes marmoratus, Hydrolycus armatus, Leporacanthicus triactis, Leptocharacidium omospilus, Leptorhamdia* sp., *Microphilypnus ternetzi, Monocirrhus polyacanthus, Parodon guyanensis, Phenacorhamdia anisura, Pimelodella geryi, Plectrochilus* cf. *diabolicus, Polycentrus schomburgkii, Potamorrhaphis guianensis, Potamotrygon orbignyi, Pseudolithoxus dumas, Pseudopimelodus bufonius, Rhinosardinia amazonica, Sciades parkeri, Semaprochilodus kneri, Sternarchorhynchus marreroil* (male head), *Synbranchus marmoratus, Tomeurus gracilis, Trichomycterus quechuorus*

Clinton and Charles Robertson *Anodus elongatus, Brycon amazonicus, Carnegiella strigata, Fluviphylax* sp., *Hemiodus gracilis*

Peter van der Sleen *Adontosternarchus balaenops, Bagre marinus, Boulengerella cuvieri, Boulengerella lateristriga, Calophysus macropterus, Colomesus asellus, Hoplias malabaricus* (head), *Megalodoras uranoscopus, Osteoglossum bicirrhosum, Pellona flavipinnis, Platyurosternarchus macrostomus* (head), *Pristigaster cayana, Pseudostegophilus nemurus, Pseudotylosurus microps* (head), *Scoloplax distolothrix, Sternarchorhamphus muelleri* (head), *Sternarchorhynchus* sp. (head), *Stictorhinus potamius, Stictorhinus potamius* (head), *Synbranchus marmoratus* (head), *Vandellia sanguinea,* mouth of an unidentified loricariid, figure 8 (except bottom left photo) and figure 9.

Donald J. Stewart *Arapaima* sp.

Tessa de Vries Figure 8, bottom left photo

INDEX